MATHEMATICAL THEORY OF NON-LINEAR ELASTICITY

ELLIS HORWOOD SERIES IN MATHEMATICS AND ITS APPLICATIONS

Series Editor: Professor G.M. BELL, Chelsea College, University of London
Statistics and Operational Research
Edtior: B. W. CONOLLY, Chelsea College, University of London

Baldock, G. R. & Bridgeman, T.	Mathematical Theory of Wave Motion
de Barra, G.	Measure Theory and Integration
Berry, J. S., Burghes, D. N., Huntley, I. D., James, D. J. G. & Moscardini, A. O.	Teaching and Applying Mathematical Modelling
Burghes, D. N. & Borrie, M.	Modelling with Differential Equations
Burghes, D. N. & Downs, A. M.	Modern Introduction to Classical Mechanics and Control
Burghes, D. N. & Graham, A.	Introduction to Control Theory, including Optimal Control
Burghes, D. N., Huntley, I. & Mc Donald, J.	Applying Mathematics
Burghes, D. N. & Wood, A. D.	Mathematical Models in the Social, Management and Life Sciences
Butkovskiy, A. G.	Green's Functions and Transfer Functions Handbook
Butkovskiy, A. G.	Structure of Distributed Systems
Chorlton, F.	Textbook of Dynamics, 2nd Edition
Chorlton, F.	Vector and Tensor Methods
Duning-Davies, J.	Mathematical Methods for Mathematicians, Physical Scientists and Engineers
Eason, G., Coles, C. W., Gettinby, G.	Mathematics and Statistics for the Bio-sciences
Exton, H.	Handbook of Hypergeometric Integrals
Exton, H.	Multiple Hypergeometric Functions and Applications
Exton, H.	*q*-Hypergeometric Functions
Faux, I. D. & Pratt, M. J.	Computational Geometry for Design and Manufacture
Firby, P. A. & Gardiner, C. F.	Surface Topology
Gardiner, C. F.	Modern Algebra
Gasson:	Geometry of Spatial Forms
Goodbody, A. M.	Cartesian Tensors
Goult, R. J.	Applied Linear Algebra
Graham, A.	Kronecker Products and Matrix Calculus: with Applications
Graham, A.	Matrix Theory and Applications for Engineers and Mathematicians
Griffel, D. H.	Applied Functional Analysis
Hoskins, R. F.	Generalised Functions
Hunter, S. C.	Mechanics of Continuous Media, 2nd (Revised) Edition
Huntley, I. & Johnson, R. M.	Linear and Nonlinear Differential Equations
Jaswon, M. A. & Rose, M. A.	Crystal Symmetry: The Theory of Colour Crystallography
Kim, K. H. and Roush, F. W.	Applied Abstract Algebra
Kosinski, W.	Field Singularities and Wave Analysis in Continuum Mechanics
Marichev, O. I.	Integral Transforms of Higher Transcendental Functions
Meek, B. L. & Fairthorne, S.	Using Computers
Muller-Pfeiffer, E.	Spectral Theory of Ordinary Differential Operators
Nonweiler, T. R. F.	Computational Mathematics: An Introduction to Numerical Analysis
Oldknow, A. & Smith, D.	Learning Mathematics with Micros
Ogden, R. W.	Non-linear Elastic Deformations
Ratschek, H. & Rokne, Jon	Computer Methods for the Range of Functions
Scorer, R. S.	Environmental Aerodynamics
Smith, D. K.	Network Optimisation Practice: A Computational Guide
Srivastawa, H. M. and Manocha, H. L.	A Treatise on Generating Functions
Temperley, H. N. V. & Trevena, D. H.	Liquids and Their Properties
Temperley, H. N. V.	Graph Theory and Applications
Thom, R.	Mathematical Models of Morphogenesis
Thomas, L. C.	Games Theory and Applications
Townend, M. Stewart	Mathematics in Sport
Twizell, E. H.	Computational Methods for Partial Differential Equations
Wheeler, R. F.	Rethinking Mathematical Concepts
Willmore, T. J.	Total Curvature in Riemannian Geometry
Willmore, T. J. Hitchin, N.	Global Riemannian Geometry

MATHEMATICAL THEORY OF NON-LINEAR ELASTICITY

A. HANYGA
Institute of Geophysics
Polish Academy of Sciences
Warsaw

Translation Editor:
R. W. OGDEN
Professor of Mathematics, Brunel University
Uxbridge, Middlesex

ELLIS HORWOOD LIMITED
Publishers · Chichester

Halsted Press: a division of
JOHN WILEY & SONS
New York · Chichester · Brisbane · Toronto

PWN — Polish Scientific Publishers
Warsaw

English Edition first published in 1985 in coedition between
ELLIS HORWOOD LIMITED
Market Cross House, Cooper Street, Chichester, West Sussex, PO19 1EB, England
and
PWN—POLISH SCIENTIFIC PUBLISHERS
Warsaw, Poland.

Distributors:

Australia, New Zealand, South-east Asia:
Jacaranda-Wiley Ltd., Jacaranda Press,
JOHN WILEY & SONS INC.
GPO Box 859, Brisbane, Queensland 40001, Australia.

Canada:
JOHN WILEY & SONS CANADA LIMITED
22 Worcester Road, Rexdale, Ontario, Canada.

Europe, Africa:
JOHN WILEY & SONS LIMITED
Baffins Lane, Chichester, West Sussex, England.

Albania, Bulgaria, Cuba, Czechoslovakia, German Democratic Republic, Hungary, Korean People's Democratic Republic, Mongolia, People's Republic of China, Poland, Rumania, the USSR, Vietnam, Yugoslavia:
ARS POLONA—Foreign Trade Enterprise
Krakowskie Przedmieście 7, 00-068 Warszawa, Poland.

North and South America and the rest of the world:
Halsted Press: a division of
JOHN WILEY & SONS
605 Third Avenue, New York, NY 10016, USA

British Library Cataloguing in Publication Data

Hanyga, A.
Mathematical theory of nonlinear elasticity. —
(Ellis Horwood series in mathematics and its applications)
1. Elasticity 2. Differential equations 3. Nonlinear mechanics
531'.3823 QA931

Library of Congress Card No. 83-9140

ISBN 0-85312-227-X (Ellis Horwood Limited)
ISBN 0-470-27493-X (Halsted Press)

Printed in Poland

Table of Contents

Author's Preface

Current literature on continuum mechanics lacks a synthetic presentation of non-linear elasticity which would also meet the requirement of utility for engineers interested in solving their non-academic problems. Apart from elementary textbooks on continuum mechanics which approach this topic at best superficially, there are two groups of advanced monographs in this field. The axiomatically biased treatises differ from physical and mathematical reality for three main reasons. An axiomatically biased treatise serves the philosophy of a school and hence strongly depends on a number of questionable axioms. Many such treatises try to achieve maximal generality without regard for the mathematical implementability and physical applicability of the theory. Last but not least, a typical representative of this kind has a natural propensity for mathematical oversimplifications which manifests itself, in particular, in a predilection for smooth solutions and isolated discontinuities. No place is left for the analysis of the true structure of the solutions, which in fact can be much more complicated. The last shortcoming is in part an academic heritage and in part a consequence of the contemporary interest in general rheology.

In the other group we find specialized monographs and surveys motivated by practical needs of modern engineering. Although they are constrained to face mathematical and physical realities they fail to give a synthetic and comprehensive review of the theory adequate for their own methods.

Our aim is to fill the gap left by these two groups of monographs by developing a realistic and yet consistent mathematical framework for the theory of elasticity. Right at the beginning we must define the class of virtual configurations which accommodates the solutions of boundary-value and initial-value problems and hence depends on the constitutive assumptions. A specification of constitutive assumptions in the first chapter must anticipate the ensuing existence, uniqueness and stability theorems in the other chapters. Consequently one can hardly imagine a general theory of continuous media which would not be superficial.

In view of these remarks we have chosen elasticity in three dimensions as a testing ground for some general ideas that might be useful in other continuum theories after suitable mathematical implementation. The basic notions (e.g. virtual configurations) and the constitutive assumptions have been chosen in such a way that at least some "fundamental" boundary-value and initial-value problems can be reasonably posed and hopefully solved. This is our criterion of the acceptability of the hypotheses. Physical criteria of acceptability are beyond the scope of this experiment.

Although we do not attempt to solve (or show how to solve) any specific engineering problems, the general treatment of elasticity in this book comes close to the needs of modern applied mechanics (especially in its modern numerical aspects). These needs find their satisfaction in several specialized monographs and papers written in a similar spirit and hence provide a practical extension of this book.

The research on non-linear elasticity is far from being complete and we hope that many of the unsolved problems will be found challenging by applied mathematicians.

Lack of space has constrained us to concentrate on a few selected topics discussed in four chapters. In Chapter 1 (Section 1.2) we have attempted a correct formulation of the equations governing the behaviour of elastic bodies. At least a part of it can also be applied to simple materials.

Elastodynamics (Chapter 3) is conceptually the cleanest part of the theory. It is essentially based on the constitutive assumptions of hyperelasticity and strong ellipticity. Some additional assumptions ensuring global existence can be inferred from the blow-up and non-existence theorems of Section 3.5.

Geometrical analysis of elementary waves (shock and simple waves) is based on the author's papers. It is fairly exhaustive in the case of isotropic elasticity. Similar analysis for transversely isotropic and crystalline (anisotropic) elasticity would require more specific assumptions or consideration of lots of cases.

The discussion of more general solutions requires some knowledge about functions of essentially bounded variation. Section 3.2 contains a quite comprehensive exposition of this subject. Theorems relating directly to elastodynamics are preceded by an introduction to quasilinear hyperbolic systems of conservation laws. This background should provide a better insight into the problems of elastodynamics proper. Existence theorems in elastodynamics are still scarce. Existence of continuous solutions in the small has so far been proved for unbounded bodies only. Global existence of discontinuous solutions in 3 dimensions is still unsolved. In one dimension some difficulties result from the fact that the equations of elastodynamics are not strictly hyperbolic. Uniqueness of discontinuous solutions is deeply related to dissipation (Section 3.4). The condition defining the unique correct solution may depend on the "shock structure". This fact indicates the necessity of embedding elasticity in a richer continuum model which specifies the dissipative phenomena in the body.

Non-linear elastostatics (Chapter 2) defies a uniform approach. Its assumptions are more complicated than those of elastodynamics and largely remain controversial. Every line of attack on the boundary-value problems of non-linear elastostatics has its serious drawbacks. For this reason we have decided on a pluralistic exposition of this topic, including monotone potential operators and direct variational methods developed by John Ball. The assumption of monotonicity of the boundary-value problem has a limited range of validity, whereas direct variational methods are sufficiently general and directly related to numerical procedures. It is not certain whether they imply an equation of equilibrium in differential form.

The complementary energy principle (Section 2.6) has been attacked in a new way. By a recourse to convex analysis the difficulties associated with non-invertibility of stress-strain relations in non-linear elasticity have been avoided.

For lack of space only bilateral mixed boundary-value problems have been considered in Chapter 2.

Chapter 4 combines the group-theoretic and differential-geometric aspects of the theory of elasticity with its variational aspects. The former were treated in more detail by Truesdell, Truesdell and Noll, Wang and Truesdell—except for isolated Volterra dislocations. Our definition of reference configurations in Chapter 1 allows for a satisfactory representation of isolated Volterra dislocations. As regards continuous distributions of dislocations we have tried to clear up some common misunderstandings in this field. The combination of group-theoretical and variational aspects of the theory of elasticity is a positive step toward understanding of the dislocations, inhomogeneities and "forces" acting upon them.

The book is self-contained. Nearly all the mathematical theorems and lemmas quoted in the main text are recapitulated in Section 1.2. Those theorems that cannot be found in easily available monographs are accompanied by proofs. There is no reference monograph for the theory of functions of essentially bounded variation and therefore a substantial part of Section 3.2. has been devoted to a systematic presentation of this topic. Similar difficulties appear in Chapter 2 in connection with the non-linear superposition operator which provides a functional-analytic background for the stress-strain relations. In consequence the reader need not be familiar with specific theorems of modern mathematics, although a preparatory mathematical training is desirable.

The author is indebted to Professor Ray W. Ogden for reviewing the manuscript and correcting it as well as to Dr Małgorzata Seredyńska for her help in preparing it.

Chapter 1

Basic concepts of elasticity

1.1 MATHEMATICAL PRELIMINARIES

1.1.1 Manifolds and Fiber Bundles

Assuming that the reader is familiar with the basic notions and facts of differential geometry (manifolds, fiber bundles, calculus of k-forms, etc.) we shall recall some of them in order to avoid ambiguities of notation and terminology. For details reference may be made to Boothby (1975), Sternberg (1964).

Definition 1.1.1

The subset $\mathcal{N}$ of a differentiable manifold $\mathcal{M}$ is a **submanifold of** $\mathcal{M}$ if every point $P \in \mathcal{N}$ has a nbhd (neighbourhood) $\mathcal{U}$ in $\mathcal{M}$ endowed with a coordinate system $\varphi: \mathcal{U} \to \boldsymbol{R}^n$; $P' \mapsto (\xi^\alpha)$, such that $\varphi(\mathcal{U}) = \{(\xi^\alpha) \in \boldsymbol{R}^n |\ 0 < \xi^\alpha < 1\}$ and $\varphi(\mathcal{U} \cap \mathcal{N}) = \{(\xi^\alpha) \in \boldsymbol{R}^n |\ \xi^1 = a^1, \ldots, \xi^p = a^p, 0 < \xi^\alpha < 1 \text{ for } p < \alpha \leqslant n\}$, $a^k \in]0, 1[$ for $k = 1, \ldots, p$, $p = \dim \mathcal{N} \leqslant n$.

Remark

Let $i: \mathcal{N} \to \mathcal{M}$ be a smooth immersion of a differential manifold $\mathcal{N}$ in a differential manifold $\mathcal{M}$, i.e. $n = \dim \mathcal{N} \leqslant \dim \mathcal{M} = m$ with rank $D(\varphi \circ i \circ \overset{-1}{\psi})$ $\equiv n$ for any local maps $\psi: \mathcal{V} \to \boldsymbol{R}^n$, $\varphi: \mathcal{U} \to \boldsymbol{R}^m$ on $\mathcal{N}$, $\mathcal{M}$ respectively. In this case $i(\mathcal{N})$ is a submanifold of $\mathcal{M}$ iff the topology of $i(\mathcal{N})$ as a subset of $\mathcal{M}$ coincides with the topology induced by the mapping $i: \mathcal{N} \to i(\mathcal{N})$.

Theorem 1.1.2

Suppose that $\mathcal{Q}$ is a submanifold of $\mathcal{M}$ and $s: \mathcal{N} \to \mathcal{M}$ is a submersion at every point $P \in s^{-1}(\mathcal{Q})$ (i.e. $\dim \mathcal{N} \geqslant \dim \mathcal{M}$ and rank $D\big(\varphi \circ s \circ \overset{-1}{\psi}\big)(\psi(P))$ $= m\ \forall P \in s^{-1}(\mathcal{Q})\big)$. Then $s^{-1}(\mathcal{Q})$ is a submanifold of $\mathcal{N}$.

In particular $\mathcal{Q}$ may be a singleton $\{q\}$.

Definition 1.1.3

A **vector bundle with the standard fiber** $\mathscr{V}$ (a vector space) over a manifold $\mathscr{M}$ is a pair $\mathscr{E}$, $\mathscr{M}$ of differentiable manifolds endowed with a surjective submersion $\pi: \mathscr{E} \to \mathscr{M}$ and satisfying the following conditions:

(V) $\forall P \in \mathscr{M}$, $\mathscr{E}_P := \pi^{-1}(\{P\})$ is a vector space;

(LT) $\forall P \in \mathscr{M}$ there is an open neighbourhood $\mathscr{U}$ of P and a diffeomorphism $\varphi: \pi^{-1}(\mathscr{U}) \to \mathscr{U} \times \mathscr{V}$ (onto) such that $\varphi|\pi^{-1}(\{P\})$ is a vector space isomorphism. If pr_1, pr_2 denote the projections $\mathscr{U} \times \mathscr{V} \to \mathscr{U}$, $\mathscr{U} \times \mathscr{V} \to \mathscr{V}$, then $\mathrm{pr}_1 \circ \varphi = \pi$.

In particular a Cartesian product $\mathscr{M} \times \mathscr{V}$ is a vector bundle over $\mathscr{M}$ with the standard fiber $\mathscr{V}$ and $\pi: \mathscr{M} \times \mathscr{V} \to \mathscr{M}$; $(P, v) \mapsto P$.

A section s of $(\mathscr{E}, \mathscr{M}, \pi)$ is a smooth mapping $s: \mathscr{M} \to \mathscr{E}$ such that $\pi \circ s = \mathrm{id}_M$.

Let $\mathscr{V}$ be a finite-dimensional vector space. Let ψ be a coordinate system $\mathscr{U} \to \boldsymbol{R}^m$, defined on an open subset $\mathscr{U}$ of $\mathscr{M}$, and suppose $\varrho: \mathscr{V} \to \boldsymbol{R}^n$ is a Cartesian coordinate system on $\mathscr{V}$. Then $\Psi = (\psi \circ \pi, \varrho \circ \mathrm{pr}_2 \circ \varphi)$ is a coordinate system on $\pi^{-1}(\mathscr{U})$. Such coordinate systems, displaying the property (LT), will be said to **trivialize the vector bundle**. If $\psi: \mathscr{U} \to \boldsymbol{R}^m$ is a coordinate system on $\mathscr{M}$ and $\Psi = (\psi \circ \pi, \tilde{\varrho})$ maps $\pi^{-1}(\mathscr{U})$ onto $\mathscr{U} \times \boldsymbol{R}^n$ in such a way that $\tilde{\varrho}|\mathscr{E}_P$: $\mathscr{E}_P \to \boldsymbol{R}^n$ is an isomorphism of vector spaces then $(\pi, \tilde{\varrho})$ has the properties of φ in (LT), with $\boldsymbol{R}^n$ replacing $\mathscr{V}$.

For every manifold $\mathscr{M}$ there are some natural vector bundles over $\mathscr{M}$. For every natural vector bundle $(\mathscr{E}, \mathscr{M}, \pi)$ with the standard fiber $\boldsymbol{R}^n$ each coordinate system $\psi: \mathscr{U} \to \boldsymbol{R}^m$ on $\mathscr{M}$ defines a unique coordinate system $\Psi_\psi: \pi^{-1}(\mathscr{U}) \to \boldsymbol{R}^m \times \boldsymbol{R}^n$ and *ipso facto* a differentiable structure on $\mathscr{E}$.

Example 1

The tangent bundle $(T\mathscr{M}, \mathscr{M}, \pi)$ of a manifold $\mathscr{M}$.

Suppose $T_P\mathscr{M} = \pi^{-1}(\{P\})$ is the set of all the tangent vectors at $P \in \mathscr{M}$.

A curve c at P is a $\mathscr{C}^1$ immersion $c: \mathscr{I} \to \mathscr{M}$ such that $0 \in \mathscr{I} \subset \boldsymbol{R}$, $c(0) = P$. Two curves c_1, c_2 at $P \in \mathscr{M}$ have the same tangent vector at P: $c_1 \sim_P c_2$, iff in some coordinate system $\psi: \mathscr{U} \to \boldsymbol{R}^m$ on $\mathscr{M}$ with $P \in \mathscr{U}$

$$\frac{d\psi \circ c_1}{dt}(0) = \frac{d\psi \circ c_2}{dt}(0).$$

Clearly $\sim_P$ is an equivalence relation. By definition $T_P\mathscr{M}$ is the set of all the equivalence classes of curves at P.

Let ψ be an arbitrary local coordinate system at P, $[c_1]_{\sim P}$, $[c_2]_{\sim P} \in T_P\mathscr{M}$, $\alpha \in \boldsymbol{R}$. Then $[c_1]_{\sim P} + \alpha[c_2]_{\sim P}$ is the equivalence class of curves $t \mapsto \overset{-1}{\psi}(\psi \circ c_1(t) + \alpha\psi \circ c_2(t))$. It is easy to see that for any two coordinate systems ψ_1, ψ_2 at P the curves $t \mapsto \overset{-1}{\psi_i}(\psi_i \circ c_1(t) + \alpha\psi_i \circ c_2(t))$, $i = 1, 2$, are equivalent at P. Hence $T_P\mathcal{M}$ is a linear m-dimensional space.

Every equivalence class $v = [c]_{\sim P}$ of curves at P defines a homogeneous differential operator of first order at P. Indeed, if $f\colon \mathcal{U} \to \boldsymbol{R}$, then $v(f) :\equiv \dfrac{df \circ c_1}{dt}(0) = \dfrac{df \circ c_2}{dt}(0)$ for any two curves c_1, c_2 at P with the same tangent at P. Let $\psi\colon \mathcal{U} \to \boldsymbol{R}^m$; $P \mapsto (\xi^\alpha)$ be a coordinate system at P, $\psi(P) = 0$. Then $v(f) = v^\alpha \dfrac{\partial}{\partial \xi^\alpha} f \circ \overset{-1}{\psi}(0)$ if $\dfrac{d\psi \circ c}{dt}(0) = \{v^1, \ldots, v^m\}\ \forall c \in v$. The differential operators $\left.\dfrac{\partial}{\partial \xi^\alpha}\right|_P$, $\alpha = 1, \ldots, m$, form a basis of $T_P\mathcal{M}$.

Let $T\mathcal{M} = \bigcup_P T_P\mathcal{M}$ (disjoint union), $\pi(v) = P\ \forall v \in T_P\mathcal{M}$. For every coordinate system $\psi\colon \mathcal{U} \to \mathcal{V} \subset \boldsymbol{R}^m$ on $\mathcal{M}$ we can define a coordinate system $\hat{\psi}\colon \pi^{-1}(\mathcal{U}) \to \mathcal{V} \times \boldsymbol{R}^m$ on $T\mathcal{M}$ by assigning the coordinates $\{\xi^1, \ldots, \xi^m, v^1, \ldots, v^m\}$ to the vector $v = \bigcup_{\alpha=1}^{m} v^\alpha \left.\dfrac{\partial}{\partial \xi^\alpha}\right|_P \in T_P\mathcal{M}$, $\psi^\alpha(P) = \xi^\alpha$. A trivializing diffeomorphism φ_ψ is defined by $\varphi_\psi\colon \pi^{-1}(\mathcal{U}) \to \mathcal{U} \times \boldsymbol{R}^m$; $v \mapsto (\pi(v), \{v^1, \ldots, v^m\})$. For an arbitrary differentiable mapping of manifolds $\phi\colon \mathcal{M} \to \mathcal{N}$, $\phi_{*P}([c]_{\sim P}) = [\phi \circ c]_{\sim \phi(P)}$. A section of $T\mathcal{M}$ is a **vector field** X on $\mathcal{M}$. A vector field X on $\mathcal{M}$ can be considered as a differential operator of first order on $\mathcal{M}$. Thus $(Xf)(P) = X^\alpha(\xi^\beta)\dfrac{\partial}{\partial \xi^\alpha}(f \circ \psi^{-1})(\xi^\beta)$.

Example 2

The cotangent bundle $T_*\mathcal{M} = \bigcup_P T_{P*}\mathcal{M}$, where $T_{P*}\mathcal{M} = \pi^{-1}(\{P\})$ is the set of all linear homogeneous functionals $T_P\mathcal{M} \to \boldsymbol{R}$. The standard fiber is $(\boldsymbol{R}^m)^* \sim \boldsymbol{R}^m$. The functionals $d\xi^\alpha{}_P$, defined by $\left\langle d\xi^\alpha{}_P, \left.\dfrac{\partial}{\partial \xi^\beta}\right|_P \right\rangle = \delta^\alpha{}_\beta$ form a basis of $T_{P*}\mathcal{M}$. For every coordinate system $\psi\colon \mathcal{U} \to \boldsymbol{R}^m$ on $\mathcal{M}$ the corresponding coordinate system $\tilde{\psi}$ on $T_*\mathcal{M}$ assigns the coordinates $\{\xi^1, \ldots, \xi^m, \omega_1, \ldots, \omega_m\}$ to $\omega = \sum_{\alpha=1}^{m} \omega_\alpha d\xi^\alpha{}_P \in T_{P*}\mathcal{M}$, $\psi(P) = (\xi^\alpha)$. A section of $T_*\mathcal{M}$ is a **one-form,** $\omega = \sum_{\alpha=1}^{m} \omega_\alpha(\xi^\beta)\, d\xi^\alpha_p$ in local coordinates $\tilde{\psi}$.

We recall that the tensor product $\mathcal{V} \otimes \mathcal{W}$ of two vector spaces $\mathcal{V}$, $\mathcal{W}$ can be obtained from the set of all the formal linear combinations of elements of the Cartesian product $\mathcal{V} \times \mathcal{W}$ by means of the identifications $(\alpha v, w)$

$= (v, \alpha w) = \alpha(v, w)$, $(v_1+v_2, w) \equiv (v_1, w)+(v_2, w)$ and $(v, w_1+w_2) = (v, w_1)+(v, w_2)$, $\alpha \in \boldsymbol{R}$.

Definition 1.1.4

Let $\pi\colon \mathscr{E} \to \mathscr{M}$, $\varrho\colon \mathscr{F} \to \mathscr{M}$ be two vector bundles over $\mathscr{M}$ with standard fibers $\mathscr{V}$, $\mathscr{W}$ resp., $\mathscr{E}\otimes\mathscr{F} := \bigcup_{P\in\mathscr{M}} \mathscr{E}_P\otimes\mathscr{F}_P$. σ projects $\mathscr{E}_P\otimes\mathscr{F}_P$ onto $\{P\}$. Let $\varphi\colon \pi^{-1}(\mathscr{U}) \to \mathscr{U}\times\mathscr{V}$, $\psi\colon \varrho^{-1}(\mathscr{U}) \to \mathscr{U}\times\mathscr{W}$, $\varphi(v) = (\pi(v), \tilde{\varphi}(v))$, $\psi(w) = (\varrho(w), \tilde{\psi}(w))$. A **trivializing diffeomorphism** on $\mathscr{E}\otimes\mathscr{F}$ is $\chi\colon \sigma^{-1}(\mathscr{U}) \to \mathscr{U}\times\times(\mathscr{V}\otimes\mathscr{W})$; $v\otimes w \mapsto (\sigma(v\otimes w), \varphi(v)\otimes\tilde{\psi}(w))$. For details see Sternberg (1964).

1.1.2 Principal Bundles and Connections

Let M be an oriented manifold. Two coordinate systems $\varphi\colon \mathscr{U} \to \boldsymbol{R}^m$, $\psi\colon \mathscr{V} \to \boldsymbol{R}^m$, have a **compatible orientation** if $\mathscr{U}\cap\mathscr{V} = \emptyset$ or $\det D(\varphi \circ \overset{-1}{\psi}) > 0$ on $\mathscr{U}\cap\mathscr{V}$. A manifold is **orientable** if it can be covered by the domains of a family of compatible coordinate systems. The orientation is fixed by choosing such a family.

If $\psi\colon P \mapsto (\xi^\alpha)$ is a coordinate system compatible with the orientation of $\mathscr{M}$, then the orientation of the vector space $T_P\mathscr{M}$ is fixed by the ordered basis $\left\{\dfrac{\partial}{\partial\xi^1}, \ldots, \dfrac{\partial}{\partial\xi^m}\right\}$.

The **principal frame bundle** $\mathscr{F}\mathscr{M} = \bigcup_P \mathscr{F}_P\mathscr{M}$, $\sigma\colon \mathscr{F}\mathscr{M} \to \mathscr{M}$, where $\mathscr{F}_P\mathscr{M} = \sigma^{-1}(\{P\})$, is the set of invertible linear orientation preserving mappings $\mathbf{K}\colon T_P\mathscr{M} \to \boldsymbol{R}^m$.

We shall choose a fixed ordered basis $\mathbf{e}_1 = (1, 0, 0, \ldots)$, $\mathbf{e}_2 = (0, 1, 0, 0, \ldots)$, ..., etc., for $\boldsymbol{R}^m$. For every $\mathbf{K} \in \mathscr{F}_P\mathscr{M}$, $\{\overset{-1}{\mathbf{K}}\mathbf{e}_j|\ j = 1, \ldots, m\}$ is an ordered basis of $T_P\mathscr{M}$, called a **frame at** P. For any coordinate system $\psi\colon \mathscr{U} \to \boldsymbol{R}^m$ on $\mathscr{M}$ compatible with the orientation of $\mathscr{M}$ the corresponding coordinate system $\check{\psi}$ on $\mathscr{F}\mathscr{M}$ is $\mathbf{K} \mapsto (\psi \circ \sigma(\mathbf{K}), \mathbf{L})$, $\mathbf{L} := \mathbf{K}(\nabla\psi)^{-1} \in \mathrm{SL}(m)$. Here $\nabla\psi$ denotes the mapping $v \mapsto \nabla\psi$, v assigning to the tangent $v = [c]_{\sim P}$ the vector $\dfrac{d\psi \circ c}{dt}(0)$, *i.e.* $\nabla\psi(\xi) = \psi_{*P}$. $\mathscr{F}\mathscr{M}$ is thus a natural fiber bundle over $\mathscr{M}$ with the standard fiber $\mathrm{SL}(m)$.

The group $\mathrm{SL}(m)$ acts on the left on $\mathscr{F}\mathscr{M}$: $\mathbf{K} \to \mathbf{LK} \in \mathscr{F}_P\mathscr{M}$ and its action is free and transitive on fibers: $\forall \mathbf{K}, \tilde{\mathbf{K}} \in \mathscr{F}_P\mathscr{M}\ \exists!\ \mathbf{L} \in \mathrm{SL}(m)$, $\tilde{\mathbf{K}} = \mathbf{LK}$. Saying that $\mathrm{SL}(m)$ acts on the left we mean that the composition of mappings $\mathbf{K} \mapsto \mathbf{L}_1\mathbf{K}$, $\mathbf{L}_1\mathbf{K} \mapsto \mathbf{L}_2(\mathbf{L}_1\mathbf{K})$ coincides with the mapping $\mathbf{K} \mapsto (\mathbf{L}_2\mathbf{L}_1)\mathbf{K}$. We shall also note that the coordinate system $\check{\psi}$ maps $\mathbf{L}_1\mathbf{K}$ to $(\psi \circ \sigma(\mathbf{K}), \mathbf{L}_1\mathbf{L})$.

Let $\mathscr{G}$ be a Lie group. A fiber bundle $(\mathscr{F}, \mathscr{M}, \sigma)$ with the standard fiber $\mathscr{G}$, endowed with an action of $\mathscr{G}$ on the left which is free and transitive on each fiber $\sigma^{-1}(\{P\})$ and with trivializing maps φ: $\mathbf{K} \mapsto (\sigma(\mathbf{K}), \tilde{\varphi}(\mathbf{K}))$ such that $\varphi(g\mathbf{K}) = (\sigma(\mathbf{K}), g\tilde{\varphi}(\mathbf{K}))$, is said to be a **principal bundle with the structure group** $\mathscr{G}$. $\mathscr{F}\mathscr{M}$ is a principal bundle over $\mathscr{M}$ with the structure group SL(m).

Let $\mathscr{G}_o$ be a Lie subgroup of $\mathscr{G}$. Suppose that $\mathscr{F}$, $\mathscr{F}_o$ are principal bundles over $\mathscr{M}$ with the structure groups $\mathscr{G}$, $\mathscr{G}_o$, resp., and there is a differentiable mapping f: $\mathscr{F}_o \to \mathscr{F}$ such that $\sigma \circ f = \mathrm{id}_M$, $f(g\mathbf{K}_o) = gf(\mathbf{K}_o)$, $\forall \mathbf{K}_o \in \mathscr{F}_o$, $g \in \mathscr{G}_o$. We shall say that $(\mathscr{F}_o, f, \mathscr{F})$ is a **reduction of the principal bundle** $\mathscr{F}$ **to** $\mathscr{G}_o$.

$\mathscr{F}$ can be covered by a collection of trivializing maps φ_i: $\sigma^{-1}(\mathscr{U}_i) \to \mathscr{U}_i \times \mathscr{G}$; $\mathbf{K} \mapsto (\sigma(\mathbf{K}), \tilde{\varphi}_i(\mathbf{K}))$, $i \in \mathscr{I}$. If it is possible to cover $\mathscr{F}$ with a subcollection $\{(\mathscr{U}_i, \varphi_i)\}$, $i \in \mathscr{I}_o \subset \mathscr{I}$, of maps with the property that $\tilde{\varphi}_i(\mathbf{K})\varphi_j(\mathbf{K})^{-1} \in \mathscr{G}_o$ $\forall \mathbf{K} \in \sigma^{-1}(\mathscr{U}_i \cap \mathscr{U}_j)$, then $\mathscr{F}$ admits a reduction to $\mathscr{G}_o$. For a proof, see Nomizu (1956).

All the maps trivializing a principal bundle $\mathscr{F}$ can be generated by local sections $P \mapsto \tilde{\mathbf{K}}(P)$, $P \in \mathscr{U} \subset \mathscr{M}$, by setting $\varphi(\mathbf{K}) := g$ if $\mathbf{K} = g\tilde{\mathbf{K}}(P)$. In particular the trivializing maps of a principal frame bundle $\mathscr{F}\mathscr{M}$ over $\mathscr{M}$ can be generated by the local sections of the form $\tilde{\mathbf{K}}(P) = g(P) \nabla \psi(P)$, $P \in \mathscr{U}$, where ψ is a coordinate system $\mathscr{U} \to \boldsymbol{R}^m$ on $\mathscr{M}$ and $g(\cdot)$ is a smooth function $\mathscr{U} \to \mathscr{G}$.

Let us now turn to the connections. For more detailed and yet elementary expositions, see Nomizu (1956), Sternberg (1964), Lichnerowicz (1976), Wang and Truesdell (1973). The latter reference is particularly adapted to the applications in Chapter 4.

Two deeply interconnected notions are parallel transport and covariant derivative. The derivative of a field of geometric objects (= section of a vector bundle) is defined in terms of difference quotients. The latter are meaningless unless both terms of the difference belong to the same fiber of the vector bundle. Hence the covariant derivative of a section v defined on a curve $t \mapsto c(t) \in \mathscr{M}$ is given by the formula:

$$\frac{dv}{dt} = \lim_{\Delta \to 0} \frac{1}{\Delta} [P_{t,t+\Delta} v(t+\Delta) - v(t)],$$

where $P_{t,t+\Delta}$ is a linear mapping $\mathscr{E}_{c(t+\Delta)} \to \mathscr{E}_{c(t)}$ (actually a vector space isomorphism). It defines the parallel transport of $v \in \mathscr{E}$ along $c(\cdot)$. In order to define the parallel transport of $v_o \in \mathscr{E}_P$ along a curve $t \mapsto c(t)$, $c(0) = P$, one lifts the curve $c(t) \in \mathscr{M}$ obtaining a curve $v(t) \in \mathscr{E}$, $\pi(v(t)) = c(t)$, $v(0) = v_0$. The vector space isomorphism $v_o \mapsto v(t)$ of $\mathscr{E}_P$ onto $\mathscr{E}_{c(t)}$ is the map $P_{t,0}$. We now describe the recipes for lifting curves.

Every vector space $T_v\mathscr{E}$ contains a subspace $V_v\mathscr{E}$ tangent to the fiber $\mathscr{E}_{\pi(v)}$ through v. A subspace H_v of $T_v\mathscr{E}$ complementary to $V_v\mathscr{E}$ is isomorphic to $T_{\pi(v)}\mathscr{M}$ and the isomorphism is π_{*v}. This fact follows easily from the surjectivity of $\pi_{*v}: T_v\mathscr{E} \to T_{\pi(v)}\mathscr{M}$ and from the fact that $\ker \pi_{*v} = V_v\mathscr{E}$. Suppose that we have assigned such a horizontal space H_v to every $v \in \mathscr{E}$ and the assignment is smooth. The above-mentioned isomorphism determines a unique vector $\dot{v}(t) \in H_{v(t)}$ corresponding to the tangent $\dot{c}(t)$. Hence we shall derive a differential equation for $v(\cdot)$, for given $c(\cdot)$.

Let $\varphi: \pi^{-1}(\mathscr{U}) \to \mathscr{U} \times \mathscr{V}$; $v \mapsto (\pi(v), \tilde{\varphi}(v))$ be a trivializing map on $\mathscr{E}$ and let $\psi: \mathscr{U} \to \boldsymbol{R}^m$, $\chi: \mathscr{V} \to \boldsymbol{R}^n$ be coordinate systems on $\mathscr{M}$, $\mathscr{V}$ (the latter is Cartesian). In the coordinate system (ξ^α, v^a), $\alpha = 1, \ldots, m$, $a = 1, \ldots, n$, defined by $v \mapsto (\psi \circ \pi(v), \chi \circ \tilde{\varphi}(v))$, the vertical space $V_v\mathscr{E}$ is spanned by the vectors $\partial/\partial v^a$. Hence the complementary space H_v is spanned by a set $\{\partial/\partial\xi^\alpha - -\Gamma^a_\alpha(\xi^\beta, v^b)\partial/\partial v^a\}$ of linearly independent vectors at v. The lift of a vector $w \in T_v\mathscr{M}$ is by definition the unique vector $z \in H_v$ such that $\pi_{*v}z = w$.

Let $c(\cdot)$ be given by $t \mapsto \xi^\alpha(t)$ in the local coordinate system ψ. Since the tangent vector of $c(\cdot)$ is $\dot{\xi}^\alpha(t)\left.\dfrac{\partial}{\partial\xi^\alpha}\right|_{c(t)}$, the corresponding lifted vector $\dot{v}$ is

$$\dot{\xi}^\alpha(t)\left[\frac{\partial}{\partial\xi^\alpha} - \Gamma^a_\alpha(\xi^\beta, v^b)\frac{\partial}{\partial v^a}\right]$$

and

$$\dot{v}^a(t) = -\Gamma^a_\alpha\big(\xi^\beta(t), v^b(t)\big)\dot{\xi}^\alpha(t), \quad v^a(0) = v_o{}^a. \tag{1.1.1}$$

Equations (1.1.1) determine the lift $v(\cdot)$ of $c(\cdot)$. If the horizontal spaces are assigned smoothly on $\mathscr{E}$, i.e. the connection symbols $\Gamma_\alpha{}^a(\xi^\beta, v^b)$ are smooth functions of the coordinates, then (1.1.1) has a unique solution $v^a(t)$ for every smooth $\xi^\alpha(t)$ and $v_o{}^a$. The lifted curve $v(t)$ is given by $(\xi^\alpha(t), v^a(t))$ in coordinate form.

The assignment $v \mapsto H_v$ is called a **connection on** $\mathscr{E}$. The connection $v \mapsto H_v$ is said to be **integrable** if every $v_o \in \mathscr{E}$ has a nbhd $\pi^{-1}(\mathscr{U})$ such that H_v are tangent to the graph of a local section $P \mapsto s(P)$ defined on $\mathscr{U}$. In this case the lift of a curve $c(t) \in \mathscr{U}$ passing through v_o lies on the graph of s. If, moreover, $c(0) = c(1)$ then $v(0) = v(1)$. For a non-integrable connection it is, however, possible that $c(0) = c(1)$ but $v(0) \neq v(1)$. Hence the parallel transport $v(0) \mapsto v(1)$ from $\mathscr{E}_P$ to $\mathscr{E}_{P'}$ depends in general on the path $c(\cdot)$ from P to P'.

The connection $v \mapsto H_v$ is said to be **linear** if $\Gamma^a_\alpha(\xi^\beta, v^b) \equiv \Gamma^a_{\alpha b}(\xi^\beta)v^b$. In this case the mappings $P_{t,t'}: v(t) \mapsto v(t')$ are vector space isomorphisms.

In the cases encountered in practical applications the connections are linear and related to the group action on $\mathscr{E}$. Principal bundles and connections on principal bundles are powerful tools of the theory of linear connections.

In the case of a principal bundle $(\mathscr{F}, \mathscr{M}, \sigma)$ we shall define the connection $\mathbf{K} \mapsto H_{\mathbf{K}}$ in such a way that

(1) $T_{\mathbf{K}}\mathscr{F} = H_{\mathbf{K}} \oplus V_{\mathbf{K}}\mathscr{F}$;

(2) $L_{g*\mathbf{K}} H_{\mathbf{K}} = H_{g\mathbf{K}}$.

L_g denotes here the left action of $\mathscr{G}$ on $\mathscr{F}$, L_g: $\mathbf{K} \mapsto g\mathbf{K}$.

Let φ: $\sigma^{-1}(\mathscr{U}) \to \mathscr{U} \times \mathscr{G}$; $\mathbf{K} \mapsto (\sigma(\mathbf{K}), \tilde{\varphi}(\mathbf{K}))$, $\mathbf{K} = \tilde{\varphi}(\mathbf{K})\tilde{\mathbf{K}}(P)$ in terms of a local section $\tilde{\mathbf{K}}$: $\mathscr{U} \to \mathscr{E}$ of $\mathscr{E}$. The tangent mapping $\tilde{\varphi}_{*\mathbf{K}}$ of $\tilde{\varphi}$ maps $V_{\mathbf{K}}\mathscr{F}$ bijectively onto $T_{\tilde{\varphi}(\mathbf{K})}\mathscr{G}$ while $\sigma_{*\mathbf{K}}$ maps $H_{\mathbf{K}}$ bijectively onto $T_{\sigma(\mathbf{K})}\mathscr{M}$.

For $g_1 \in \mathscr{G}$ let R_{g_1}: $\mathscr{G} \to \mathscr{G}$; $g \mapsto gg_1$. The R_{g*e} maps $T_e\mathscr{G}$ bijectively onto $T_g\mathscr{G}$ and we can identify $T_g\mathscr{G}$ with $T_e\mathscr{G}$ by means of R_{g*e}. $T_e\mathscr{G}$ can be regarded as the Lie algebra $\mathfrak{g}$ of $\mathscr{G}$*.

Let $\pi_{\mathbf{K}}^V$ be the projection $T_{\mathbf{K}}\mathscr{F} = H_{\mathbf{K}} \oplus V_{\mathbf{K}}\mathscr{F} \to V_{\mathbf{K}}\mathscr{F}$. Since $\mathscr{G}$ acts freely and transitively on the fibers, each fiber $\mathscr{F}_{\sigma(\mathbf{K})}$ is isomorphic to the Lie group $\mathscr{G}$. If $\Lambda_{\mathbf{K}} g = g\mathbf{K}$ denotes the left action of $\mathscr{G}$ on $\mathscr{F}$, then $\Lambda_{\mathbf{K}*g}$ is an isomorphism of $T_g\mathscr{G}$ onto $V_{g\mathbf{K}}\mathscr{F}$. For the sake of convenience we shall define the elements γ of the Lie algebra $\mathfrak{g}$ of $\mathscr{G}$ as the right-invariant vector fields γ: $\mathscr{G} \to T\mathscr{G}$ on $\mathscr{G}$, so that $\gamma(g) = R_{g*e}\gamma(e)$. The mapping $\Lambda_{\mathbf{K}*e}$: $T_e\mathscr{G} \to V_{\mathbf{K}}\mathscr{F}$ induces a mapping $\mathfrak{g} \to V_{\mathbf{K}}\mathscr{F}$; $\gamma \mapsto X_\gamma(\mathbf{K}) \in V_{\mathbf{K}}\mathscr{F}$. The **fundamental fields** $X_\gamma(\mathbf{K})$ span $V_{\mathbf{K}}\mathscr{F}$ for every $\mathbf{K} \in \mathscr{F}$.

We are now ready to define the **connection form** $\omega_\varphi(\mathbf{K}) = \tilde{\varphi}_{*\mathbf{K}}\pi_{\mathbf{K}}^V$. It is a one-form on $\mathscr{F}$ with values in $\mathfrak{g}$. More precisely, for $w \in T_{\mathbf{K}}\mathscr{F}$ $\omega_\varphi(\mathbf{K})(w) \in T_{\tilde{\varphi}(\mathbf{K})}\mathscr{G}$ and hence $\omega_\varphi(\mathbf{K})(w) = \gamma(\tilde{\varphi}(\mathbf{K}))$ can be identified with $\gamma \in \mathfrak{g}$. Let φ_1 be another trivializing map on $\mathscr{F}$, $\varphi_1(\mathbf{K}) = (\sigma(\mathbf{K}), \tilde{\varphi}_1(\mathbf{K}))$, $\mathbf{K} = \tilde{\varphi}_1(\mathbf{K})\tilde{\mathbf{K}}_1(P)$, $\tilde{\mathbf{K}}_1(P) \equiv g(P)\tilde{\mathbf{K}}(P)$. Hence $\tilde{\varphi}_1(\mathbf{K}) = \tilde{\varphi}(\mathbf{K})g(P)^{-1}$, $P = \sigma(\mathbf{K})$. Since $\omega_{\varphi_1}(\mathbf{K})(w) = R_{g(P)^{-1}*} \circ \omega_\varphi(\mathbf{K})(w)$, $\omega_{\varphi_1}(\mathbf{K})(w)$ and $\omega_\varphi(\mathbf{K})(w)$ define the same element of $\mathfrak{g}$. With this in mind we shall drop the subscript φ.

Clearly $\omega(\mathbf{K})(w) = 0$ $\forall w \in H_{\mathbf{K}}$. If $\pi_{\mathbf{K}}^V w \neq 0$ then $\omega(\mathbf{K})(w) \neq 0$. Hence $H_{\mathbf{K}} = \ker\omega(\mathbf{K})$. It is easy to see that $\omega(\mathbf{K})(X_\gamma(\mathbf{K})) = \gamma$.

Since the decomposition of $T_{\mathbf{K}}\mathscr{F}$ into $H_{\mathbf{K}}$ and $V_{\mathbf{K}}\mathscr{F}$ is invariant with respect to the left action of $\mathscr{G}$ on $\mathscr{F}$, $\pi_{g\mathbf{K}}^V L_{g*\mathbf{K}} w = L_{g*\mathbf{K}}\pi_{\mathbf{K}}^V w$ $\forall w \in T_{\mathbf{K}}\mathscr{F}$. Also $\tilde{\varphi}_{*g\mathbf{K}} L_{g*\mathbf{K}} = L_{g*\tilde{\varphi}(\mathbf{K})}\tilde{\varphi}_{*\mathbf{K}}$ since $\tilde{\varphi}(g\mathbf{K}) = \varphi(L_g\mathbf{K}) = L_g\tilde{\varphi}(\mathbf{K})$ (the last expression involves the left action of $\mathscr{G}$ on itself: $L_g g' = gg'$). Hence $\omega(g\mathbf{K})(L_{g*\mathbf{K}} w) = L_{g*\tilde{\varphi}(\mathbf{K})}\tilde{\varphi}_{*\mathbf{K}}\pi_{\mathbf{K}}^V w = L_{g*}(\omega(\mathbf{K})(w)) = (\operatorname{ad} g) \circ \omega(\mathbf{K})(w)$.

We recall that the action $h \mapsto ghg^{-1} = L_g R_{g^{-1}} h$ on $\mathscr{G}$ induces the action of $\mathscr{G}$ on $\mathfrak{g}$: $\gamma \mapsto L_{g*} \circ R_{g^{-1}*} \circ \gamma = L_{g*} \circ \gamma =: (\operatorname{ad} g)(\gamma)$.

Let

$$\tilde{\omega}_\varphi(P)(v) :\equiv \omega(\tilde{\mathbf{K}}(P))(\tilde{\mathbf{K}}_{*P} v) \ \forall v \in T_P\mathscr{M}, \ P = \sigma(\mathbf{K}).$$

* In our case the Lie algebra consists of **right** invariant vector fields on $\mathscr{G}$. In differential geometry it is common to define Lie algebras as sets of **left**-invariant vector fields. On the other hand $\mathscr{G}$ is usually assumed to act on the **right** on $\mathscr{F}$. We have adopted the alternative convention adapted to the notations of Chapter 4.

For every $w \in T_{\mathbf{K}}\mathscr{F}$ we can find such a unique $g \in \mathscr{G}$ that $\mathbf{K} = g\tilde{\mathbf{K}}(P)$, a unique $v \in T_P\mathscr{M}$ and a unique $\gamma \in \mathfrak{g}$ such that $w = L_{g*\tilde{\mathbf{K}}(P)}\tilde{\mathbf{K}}_{*P}v + X_\gamma(\mathbf{K})$. This follows from the fact that $T_{\mathbf{K}}\mathscr{F} = V_{\mathbf{K}}\mathscr{F} \oplus L_{g*\tilde{\mathbf{K}}(P)}\tilde{\mathbf{K}}_{*P}(T_P\mathscr{M})$. Hence $\omega(\mathbf{K})(w) = \gamma + \omega(L_g\tilde{\mathbf{K}}(P))(L_{g*\tilde{\mathbf{K}}(P)}\tilde{\mathbf{K}}_{*P}v) = \gamma + (\operatorname{ad} g)(\omega_\varphi(P)(v))$. It follows that $\tilde{\omega}_\varphi$ determines ω completely on $\sigma^{-1}(\mathscr{U})$.

Introducing local coordinates $(\psi \circ \sigma, \chi \circ \varphi)$: $\mathbf{K} \mapsto (\xi^\alpha, \zeta^A)$, we have $\tilde{\omega}_\varphi(P) = \Gamma_\alpha{}^A(\xi^\beta)\, d\xi^\alpha \otimes e_A$, in terms of a basis e_A of the vector space $\mathfrak{g}$. Let $Y_A = X_{e_A}$, $v \in T_P\mathscr{M}$. We shall determine the lift $w \in H_{\mathbf{K}}$ of v from the equations

$$\omega(\mathbf{K})(w) = 0, \quad \sigma_{*\mathbf{K}}(w) = v, \tag{1.1.2}$$

the second of which implies that

$$w = v^\alpha \frac{\partial}{\partial \xi^\alpha}\bigg|_{\mathbf{K}} + w^A Y_A(\mathbf{K})$$

for some $w^A \in \boldsymbol{R}$. Note that $\left(\frac{\partial}{\partial \xi^\alpha}\bigg|_{\mathbf{K}}, Y_A(\mathbf{K})\right)$ is a basis of $T_{\mathbf{K}}\mathscr{F}$, $\sigma_{*\mathbf{K}}\frac{\partial}{\partial \xi^\alpha}\bigg|_{\mathbf{K}} = \frac{\partial}{\partial \xi^\alpha}\bigg|_P$. Let $\mathbf{K} = L_g\tilde{\mathbf{K}}(P)$. Noting that $L_g\overset{-1}{\varphi}(P, g') = \overset{-1}{\varphi}(P, gg')$ and hence $L_{g*\mathbf{K}}\frac{\partial}{\partial \xi^\alpha}\bigg|_{\mathbf{K}} = \frac{\partial}{\partial \xi^\alpha}\bigg|_{g\mathbf{K}}$, we have for $w \in T_{\mathbf{K}}\mathscr{F}$

$$\begin{aligned}
\omega(\mathbf{K})(w) &= \omega(\mathbf{K})\left(v^\alpha \frac{\partial}{\partial \xi^\alpha}\bigg|_{\mathbf{K}}\right) + w^A e_A \\
&= \omega\big(L_g\tilde{\mathbf{K}}(P)\big)\left(L_{g*\tilde{\mathbf{K}}(P)} v^\alpha \frac{\partial}{\partial \xi^\alpha}\bigg|_{\tilde{\mathbf{K}}(P)}\right) + w^A e_A \\
&= L_{g*} \circ \tilde{\omega}_\varphi(P)\left(v^\alpha \frac{\partial}{\partial \xi^\alpha}\bigg|_{\tilde{\mathbf{K}}(P)}\right) + w^A e_A.
\end{aligned}$$

The section $P \mapsto \tilde{\mathbf{K}}(P)$ in the coordinates (ξ^α, ζ^A) is represented by $P \mapsto (\psi(P), \chi(e))$, e = unit of $\mathscr{G}$. Hence $\tilde{\mathbf{K}}_{*P}v = v^\alpha \frac{\partial}{\partial \xi^\alpha}\bigg|_{\tilde{\mathbf{K}}(P)}$ for $v = v^\alpha \frac{\partial}{\partial \xi^\alpha}\bigg|_P$, and $0 = \omega(\mathbf{K})(w) = L_{g*} \circ \tilde{\omega}_\varphi(P)(v) + w^A e_A = \Gamma_\alpha{}^A v^\alpha L_{g*} e_A + w^A e_A$.

Let $\mathbf{K}(t) \in \mathscr{F}$ be a lift of the curve $P = c(t) \in \mathscr{M}$, $\xi^\alpha = \bar{\xi}^\alpha(t)$ in coordinate form. Then

$$\frac{d\mathbf{K}(t)}{dt} = \frac{d\bar{\xi}^\alpha}{dt} \frac{\partial}{\partial \xi^\alpha}\bigg|_{\mathbf{K}} + w^A Y_A(\mathbf{K}) \in T_{\mathbf{K}}\mathscr{F}$$

and, for $\mathbf{K} = L_g\tilde{\mathbf{K}}(P)$, $g = \tilde{\varphi}(\mathbf{K})$, $v = \dot{c}(t)$ (tangent to $c(t)$),

$$\frac{d\tilde{\varphi}(\mathbf{K}(t))}{dt} = \tilde{\varphi}_{*\mathbf{K}}\frac{d\mathbf{K}}{dt} = \tilde{\varphi}_{*\mathbf{K}}L_{g*\tilde{\mathbf{K}}(P)}\tilde{\mathbf{K}}_{*P}v + w^A e_A$$
$$= \tilde{\varphi}_{*\mathbf{K}}L_{g*\tilde{\mathbf{K}}(P)}\tilde{\mathbf{K}}_{*P}v - \Gamma_\alpha{}^A v^\alpha L_{g*} \circ e_A.$$

Since $\tilde{\varphi}(\tilde{\mathbf{K}}(P)) \equiv e$, $\tilde{\varphi}_{*\tilde{\mathbf{K}}(P)}\tilde{\mathbf{K}}_{*P} = 0$ and

$$\tilde{\varphi}(\mathbf{K}(t))^{-1}\frac{d\tilde{\varphi}(\mathbf{K}(t))}{dt} = -\Gamma_\alpha{}^A v^\alpha e_A. \tag{1.1.3}$$

In the case where $(\mathscr{F}, f, \mathscr{F}\mathscr{M})$ is a reduction of the frame bundle to a subgroup $\mathscr{G}$ of SL(m)

$$K^a{}_\alpha = Q^a{}_b\tilde{K}^b{}_\alpha(P), \quad \mathbf{Q} \in \mathscr{G} \subset \mathrm{SL}(m), \ \mathfrak{g} \subset M(m),$$

$$\overset{-1}{\mathbf{Q}}\frac{d\mathbf{Q}}{dt} = -\Gamma_\alpha^A e_A \frac{d\xi^\alpha}{dt} \tag{1.1.4}$$

(the l.h.s. (left-hand side) of (1.1.4) is a matrix product). We can write $\Gamma_\alpha{}^A e_A \in \mathfrak{g}$ in the indicial form

$$\Gamma_\alpha^A e_A = [\Gamma_\alpha{}^a{}_b] = \mathbf{\Gamma}_\alpha$$

so that

$$\overset{-1}{Q}{}^a{}_c\frac{dQ^c{}_b}{dt} = -\Gamma_\alpha{}^a{}_b(\xi^\beta)\frac{d\xi^\alpha}{dt}. \tag{1.1.5}$$

Let $\mathbf{K}(t) \in \mathscr{F}$ be a lift of a curve $P = c(t)$, $0 \leqslant t \leqslant 1$, in a subbundle $\mathscr{F}$ of $\mathscr{F}\mathscr{M}$, obtained by reduction, and let $v \in T_{c(0)}\mathscr{M}$. Then the parallel transport of v along $c(\cdot)$ is defined by

$$v(t) = \mathbf{K}(t)^{-1}\mathbf{K}(0)v. \tag{1.1.6}$$

Equation (1.1.6) is a very special case of transferring parallel transport from a principal bundle to associated vector bundles, cf. Nomizu (1956).

Let $v = v^\alpha \left.\frac{\partial}{\partial \xi^\alpha}\right|_{c(t)} = w^a \overset{-1}{\tilde{K}(\xi)}{}^\alpha{}_a \left.\frac{\partial}{\partial \xi^\alpha}\right|_{c(t)}$, and let $\mathbf{V}$, $\mathbf{W}$ be the column matrices $[v^\alpha]$, $[w^a]$, respectively. In matrix notation

$$\frac{d\mathbf{W}}{dt} = \mathbf{\Gamma}_\alpha\frac{d\xi^\alpha}{dt}\mathbf{W}, \quad \frac{d\mathbf{V}}{dt} = \tilde{\mathbf{\Gamma}}_\alpha\frac{d\xi^\alpha}{dt}\mathbf{V},$$

$$\tilde{\mathbf{\Gamma}}_\alpha(\xi^\beta) = \tilde{\mathbf{K}}(\xi)^{-1}\mathbf{\Gamma}_\alpha(\xi)\tilde{\mathbf{K}}(\xi) + (\tilde{\mathbf{K}}(\xi)^{-1})_{,\xi^\alpha}\tilde{\mathbf{K}}(\xi).$$

The transformation $\mathbf{\Gamma}_\alpha \mapsto \tilde{\mathbf{\Gamma}}_\alpha$ by means of a local section $\tilde{\mathbf{K}}$ is called a **local gauge**.

For any two connections Γ, Γ' on $\mathscr{F} \subset \mathscr{F}\mathscr{M}$

$$\tilde{\mathbf{\Gamma}}_\alpha(\xi^\beta) - \tilde{\mathbf{\Gamma}}'_\alpha(\xi^\beta) = \tilde{\mathbf{K}}(\xi)^{-1}[\mathbf{\Gamma}_\alpha(\xi) - \mathbf{\Gamma}'_\alpha(\xi)]\tilde{\mathbf{K}}(\xi) \in \tilde{\mathbf{K}}(\xi)^{-1}\mathfrak{G}\tilde{\mathbf{K}}(\xi),$$
$$\alpha = 1, \ldots, m.$$

If $c(1) = c(0)$, then

$$v(1) = \mathbf{K}(1)^{-1}\mathbf{K}(0)v(0) = \tilde{\mathbf{K}}\big(c(0)\big)^{-1}\mathbf{Q}(1)^{-1}\mathbf{Q}(0)\tilde{\mathbf{K}}\big(c(0)\big)v(0),$$
$$\mathbf{Q}(1)^{-1}\mathbf{Q}(0) \in \mathscr{G}.$$

The set of all the matrices $\mathbf{Q}(1)^{-1}\mathbf{Q}(0)$ associated with closed curves $c(\cdot)$ forms a group and is called the **holonomy group** of the connection. It is a subgroup of the structure group $\mathscr{G}$ of $\mathscr{F}$.

There is a theorem to the effect that there exist connections on every principal bundle $\mathscr{F}$. We shall sketch the proof in the next subsection. Obviously, there are many connections on a principal bundle.

1.1.3 Curvature and Torsion

For later reference we shall note the following facts. In our applications $\mathscr{G}$ appears merely as a transformation group of $\boldsymbol{R}^m$. Hence we can assume that $\mathscr{G} \subset \mathrm{SL}(m)$ is a matrix group without loss of generality (otherwise replace $\mathscr{G}$ by its representation on $\boldsymbol{R}^m$). In *matrix notation* $R_g h = hg$ and

$$\frac{\partial\big(h(\zeta)g\big)^k{}_l}{\partial h(\zeta)^r{}_s}\,\frac{\partial h(\zeta)^r{}_s}{\partial \zeta^A} = e_A(\zeta)^k{}_r g^r{}_l,$$

where $e_A(\zeta) = \partial h/\partial\zeta^A$ are the coordinate vectors on $\mathscr{G}$. Thus $R_{g*}w = wg$ *in matrix notation*. If the action of $\mathscr{G}$ on $\boldsymbol{R}^m$ is $f \mapsto gf$ in matrix notation, then the action of $\gamma \in \mathfrak{g}$ on $\boldsymbol{R}^m$ is given by $f \mapsto \gamma(e)f$ in matrix notation. Note that $\gamma(g) = \gamma(e)g$ in matrix notation.

A connection $\mathbf{K} \mapsto H_\mathbf{K}$ on a principal bundle $\mathscr{F}$ over $\mathscr{M}$ is said to be **integrable** if for every $\mathbf{K} \in \mathscr{F}$ there is a nbhd $\mathscr{U}$ of $\sigma(\mathbf{K})$ and a local section $\hat{\mathbf{K}}\colon \mathscr{U} \to \mathscr{F}$ of $\mathscr{F}$ such that $\mathbf{K} = \hat{\mathbf{K}}(P)$, $P = \sigma(\mathbf{K})$ and $H_\mathbf{K}$ is the tangent plane to the graph of $\hat{\mathbf{K}}$ at $\mathbf{K}$: $H_\mathbf{K} = \{\hat{\mathbf{K}}_{*P}v \mid v \in T_P\mathscr{M}\}$.

Since $L_{g*\mathbf{K}}H_\mathbf{K} = H_{g\mathbf{K}}$, it is clear that H is integrable iff there is a covering $\{\mathscr{U}_i\}$ of $\mathscr{M}$ by open domains of local sections $\hat{\mathbf{K}}_i\colon \mathscr{U}_i \to \mathscr{F}$ such that $H_{\hat{\mathbf{K}}_i(P)} = \hat{\mathbf{K}}_{i*P}(T_P\mathscr{M})$. It follows then that $H_\mathbf{K} = L_{g*\hat{\mathbf{K}}_i(P)}\hat{\mathbf{K}}_{i*P}(T_P\mathscr{M})$ for $\mathbf{K} = L_g\hat{\mathbf{K}}_i(P)$.

Let $\pi^H_\mathbf{K}$ denote the projection $T_\mathbf{K}\mathscr{F} \to H_\mathbf{K}$. The 2-form on $\mathscr{F}$ with values in $\mathfrak{g}$

$$\Omega(\mathbf{K})(v_1, v_2) := d\omega(\mathbf{K})(\pi^H_\mathbf{K}v_1, \pi^H_\mathbf{K}v_2) \tag{1.1.7}$$

is called the **curvature form** of the connection H.

Let $X_i\colon \mathscr{U} \to T\mathscr{F}$, $i = 1, 2$, be two locally defined vector fields on $\mathscr{F}$

(= sections of the bundle $T\mathscr{F}$ over $\mathscr{F}$), with $\mathscr{U} \subset \mathscr{F}$ open in $\mathscr{F}$, $X_i(\mathbf{K}) \in H_{\mathbf{K}}$ $\forall \mathbf{K} \in \mathscr{U}$, $i = 1, 2$. Then, by the standard calculus of differential forms

$$\begin{aligned} 2\Omega(X_1, X_2) &= 2d\omega(X_1, X_2) \\ &= X_1\big(\omega(X_2)\big) - X_2\big(\omega(X_1)\big) - \omega\big([X_1, X_2]\big) \\ &= -\omega\big([X_1, X_2]\big). \end{aligned}$$

Hence the Lie bracket $[X_1, X_2](\mathbf{K}) \in H_{\mathbf{K}}$ $\forall \mathbf{K} \in \mathscr{U}$ iff $\Omega(X_1, X_2) = 0$. By the Frobenius theorem $\mathbf{K} \mapsto H_{\mathbf{K}}$ is integrable iff $\Omega\,(X_1, X_2) = 0$ for any horizontal* vector fields X_i, $i = 1, 2$, i.e. iff $\Omega = 0$.

Note that $L_{g*\mathbf{K}}\pi^H_{\mathbf{K}} = \pi^H_{g\mathbf{K}} L_{g*\mathbf{K}}$ implies that Ω is invariant under L_g. Indeed

$$\begin{aligned} \Omega(L_g\mathbf{K})(L_{g*\mathbf{K}}v_1, L_{g*\mathbf{K}}v_2) &= d\omega(L_g\mathbf{K})(L_{g*\mathbf{K}}\pi^H_{\mathbf{K}}v_1, L_{g*\mathbf{K}}\pi^H_{\mathbf{K}}v_2) \\ &= L_{g*}\big(d\omega(\mathbf{K})(\pi^H_{\mathbf{K}}v_1, \pi^H_{\mathbf{K}}v_2)\big) \\ &= (\operatorname{ad} g) \circ \Omega(\mathbf{K})(v_1, v_2). \end{aligned}$$

Remark

We have used the fact that $\operatorname{ad} g$ is linear on $\mathfrak{g}$ and hence it commutes with d.

It is convenient to replace horizontal** L_g-invariant k-forms on $\mathscr{F}$ by equivalent k-forms on $\mathscr{M}$.

Let $\mathbf{K} = L_g\tilde{\mathbf{K}}(P)$, φ: $g\tilde{\mathbf{K}}(P) \mapsto (P, g) \in \mathscr{U} \times \mathscr{G}$ locally. For any $v_j \in T_{\mathbf{K}}\mathscr{F}$ we can find such $w_j \in T_P\mathscr{M}$, $P = \sigma(\mathbf{K})$ that

$$\pi^H_{\mathbf{K}}v_j = L_{g*\tilde{\mathbf{K}}(P)}\tilde{\mathbf{K}}_{*P}w_j, \quad j = 1, 2.$$

Hence

$$\begin{aligned} \Omega(\mathbf{K})(v_1, v_2) &= \Omega\big(L_g\tilde{\mathbf{K}}(P)\big)(\pi^H_{\mathbf{K}}v_1, \pi^H_{\mathbf{K}}v_2) \\ &= (\operatorname{ad} g) \circ \Omega\big(\tilde{\mathbf{K}}(P)\big)(\tilde{\mathbf{K}}_{*P}w_1, \tilde{\mathbf{K}}_{*P}w_2) \\ &= (\operatorname{ad} g) \circ \tilde{\Omega}_\varphi(P)(w_1, w_2) \\ &= L_{g*} \circ R_{g^{-1}*} \circ \big(\tilde{\Omega}_\varphi(P)(w_1, w_2)\big). \end{aligned}$$

$\tilde{\Omega}_\varphi(P)(w_1, w_2) := \Omega\big(\tilde{\mathbf{K}}(P)\big)(\tilde{\mathbf{K}}_{*P}w_1, \tilde{\mathbf{K}}_{*P}w_2)$ is a $\mathfrak{g}$-valued 2-form on $\mathscr{M}$.

Note that $\tilde{\Omega}_\varphi$ depends on the local trivializing map φ. It can however be replaced by a (3, 1) tensor field on $\mathscr{M}$ which does not depend on φ. Indeed, let Ω be any horizontal $\mathfrak{g}$-valued 2-form on $\mathscr{F}$ with the property $\Omega(L_g\mathbf{K})(L_{g*\mathbf{K}}v_1, L_{g*\mathbf{K}}v_2) = L_{g*} \circ R_{g^{-1}*}\Omega(\mathbf{K})(v_1, v_2)$. We shall set

$$C(P)(w_1, w_2, w_3) = \overset{-1}{\mathbf{K}} \circ \Omega(\mathbf{K})(v_1, v_2) \circ \mathbf{K}w_3 \in T_P\mathscr{M}$$

* A vector field on $\mathscr{F}$ is horizontal if $X(\mathbf{K}) \in H_{\mathbf{K}}$ $\forall \mathbf{K} \in \operatorname{dom} X$.

** A k-form Ω is said to be **horizontal** if $\Omega(\mathbf{K})(v_1, \ldots, v_k) = 0$ whenever some $v_j \in V_{\mathbf{K}}\mathscr{F}$ $\forall K \in \operatorname{dom} \Omega$.

for $w_1, w_2, w_3 \in T_P\mathscr{M}$, $\sigma_{*\mathbf{K}}v_i = w_i$, $i = 1, 2$. Note that the r.h.s. (right-hand side) of the definition of C is a composite mapping $T_P\mathscr{M} \to R^m \to R^m \to T_P\mathscr{M}$ and $\gamma = \Omega(\mathbf{K})(v_1, v_2)$ is identified with the matrix $\gamma(e) \in M(m)$.

Let $\sigma(\mathbf{K}') = \sigma(\mathbf{K}) = P$. Then we have $\mathbf{K}' = L_g\mathbf{K}$, $v_i' = L_{g*\mathbf{K}}v_i$, $i = 1, 2$, $\overset{-1}{\mathbf{K}'} \circ \Omega(\mathbf{K}')(v_1', v_2') \circ \mathbf{K}'w_3 = \overset{-1}{(g\mathbf{K})} \circ L_{g*} \circ R_{g^{-1}*} \circ \Omega(\mathbf{K})(v_1, v_2) \circ (g\mathbf{K})w_3 = \overset{-1}{\mathbf{K}} \circ \Omega(\mathbf{K})(v_1, v_2) \circ \mathbf{K}w_3$ since $(\mathrm{ad}\,g)(\gamma) = L_{g*} \circ R_{g^{-1}*}(\gamma) = g\gamma g^{-1}$ in matrix notation, identifying $\gamma \in \mathfrak{g}$ with $\gamma(e) \in M(m)$.

Let $\sigma_{*\mathbf{K}}v_i' = w_i$, $i = 1, 2$. Then $\sigma_{*\mathbf{K}}(v_i' - v_i) = 0$ and $v_i' - v_i \in V_{\mathbf{K}}\mathscr{F}$ for $i = 1, 2$. Hence $\Omega(\mathbf{K})(v_1', v_2') = \Omega(\mathbf{K})(v_1, v_2)$. Hence the right-hand side of the equation defining C does not depend on the choice of $\mathbf{K} \in \mathscr{F}_P$, v_i provided only $\sigma_{*\mathbf{K}}v_i = w_i$. Let $\tilde{\mathbf{K}}$: $\mathscr{U} \to \mathscr{F}$ be a local section of $\mathscr{F}$ defining the trivializing map φ. Since $\tilde{\mathbf{K}}_{*P}w_i$ is a lift of v_i,

$$C(P)(w_1, w_2, w_3) = \overset{-1}{\mathbf{K}} \circ \mathrm{ad}\,g\big(\tilde{\Omega}_\varphi(P)(w_1, w_2)\big) \circ \mathbf{K}w_3 \tag{1.1.8}$$

is independent of φ.

We can write

$$\tilde{\Omega}_\varphi(P) = R^A_{(\varphi)\alpha\beta}(\xi)d\xi^\alpha \wedge d\xi^\beta \otimes e_A, \tag{1.1.9}$$

or for $\mathscr{F} \subset \mathscr{F}\mathscr{M}$,

$$\Omega(P)^\gamma{}_\delta = R^\gamma{}_{\delta\alpha\beta}(\xi)d\xi^\alpha \wedge d\xi^\beta, \tag{1.1.10}$$

$$C(P)(w_1, w_2, w_3)^\gamma = R^\gamma{}_{\delta\alpha\beta}(\xi)w_3{}^\delta w_1{}^\alpha w_2{}^\beta. \tag{1.1.11}$$

In order to calculate the curvature $\tilde{\Omega}_\varphi$ for the connection $\tilde{\omega}_\varphi(v) = (\Gamma^A_\alpha v^\alpha + w^A)e_A$ on $\sigma^{-1}(\mathscr{U})$ it is necessary to replace the anholonomic coordinates w^A by the holonomic ones $d\zeta^A(w)$, associated with the coordinate system $(\xi^\alpha, \zeta^A) = \big(\psi \circ \sigma(\mathbf{K}), \chi \circ \tilde{\varphi}(\mathbf{K})\big)$ on $\sigma^{-1}(\mathscr{U})$. We shall set $\chi(e) = 0$. On $\varphi(\sigma^{-1}(\mathscr{U})) = \mathscr{U} \times \mathscr{G}$ the left action of $g \in \mathscr{G}$ is given by $(P, g_o) \mapsto (P, R_{g_o}g)$, hence the fundamental fields assume the form $(P, g_o) \mapsto (0, R_{g_o*}w)$, $w \in T_e\mathscr{G}$. The coordinates w^A refer to the vector decomposition $w = v^\alpha \frac{\partial}{\partial \xi^\alpha}\Big|_P + w^A R_{g*}e_A = v^\alpha \frac{\partial}{\partial \xi^\alpha}\Big|_P + w^A e_A g$, $e_A \in T_e\mathscr{G}$, using matrix notation on the right-hand side. In terms of the holonomic coordinates $\tilde{w}^A := d\zeta^A(w)$ on $T\mathscr{G}$ $w^A e_A g = \tilde{w}^A \frac{dg(\zeta)}{d\zeta^A}$, $w^A e_A = \tilde{w}^A \frac{dg(\zeta)}{d\zeta^A} g^{-1}$. Setting $dg := \frac{dg}{d\zeta^A} d\zeta^A$ (a matrix-valued one-form on $\mathscr{G}$ and on $\sigma^{-1}(\mathscr{U})$), $\tilde{\omega}_\varphi = \Gamma^A_\alpha(\xi)d\xi^\alpha \otimes e_A + (dg)g^{-1}$. On horizontal vectors the 2-form $d(\tilde{\omega}_\varphi)^k{}_l = d\Gamma_\alpha^k{}_l \wedge d\xi^\alpha - dg^k{}_p \wedge d(g^{-1})^p{}_l = d\Gamma_\alpha^k{}_l \wedge d\xi^\alpha + g^{-1p}{}_r g^{-1s}{}_l dg^k{}_p \wedge dg^r{}_s$ can be expressed in terms of the one-forms $d\xi^\alpha$. Thus

$$\tilde{\Omega}_\varphi{}^k{}_l = R_{(\varphi)}{}^k{}_{l\alpha\beta}(\xi) d\xi^\alpha \wedge d\xi^\beta = d\Gamma_\beta{}^k{}_l \wedge d\xi^\beta + \Gamma_\alpha{}^k{}_r \Gamma_\beta{}^r{}_l d\xi^\alpha \wedge d\xi^\beta$$

and hence

$$2R^k_{(\varphi)l\alpha\beta}(\xi) = \Gamma_\beta{}^k{}_{l,\xi^\alpha} - \Gamma_\alpha{}^k{}_{l,\xi^\beta} + 2\Gamma_{[\alpha}{}^k{}_{|r|}\Gamma_{\beta]}{}^r{}_l. \tag{1.1.12}$$

For $\mathscr{F} \subset \mathscr{F}\mathscr{M}$ this becomes

$$2R^\gamma{}_{\delta\alpha\beta} = \Gamma_\beta{}^\gamma{}_{\delta,\xi^\alpha} - \Gamma_\alpha{}^\gamma{}_{\delta,\xi^\beta} + 2\Gamma_{[\alpha|\rho|}^{\gamma}\Gamma^\rho{}_{\beta]\delta} \tag{1.1.13}$$

if the basis of $\boldsymbol{R}^m$ is taken to be $\{e_\alpha | \ \alpha = 1, \ldots, m\}$, while

$$\mathbf{K}\frac{\partial}{\partial \xi^\alpha}\bigg|_{\sigma(\mathbf{K})} = e_\alpha, \quad \alpha = 1, \ldots, m.$$

Let the principal bundle $\mathscr{F}$ be a subbundle of $\mathscr{F}\mathscr{M}$, $\sigma\colon \mathscr{F} \to \mathscr{M}$, $m = \dim\mathscr{M}$. For any vector $w \in T_\mathbf{K}\mathscr{F}$ we have $\sigma_{*\mathbf{K}} w \in T_P\mathscr{M}$ and we can define a natural one-form

$$\theta(\mathbf{K})(w) :\equiv \mathbf{K} \circ \sigma_{*\mathbf{K}} w \quad \forall w \in T_\mathbf{K}\mathscr{F} \tag{1.1.14}$$

with values in $\boldsymbol{R}^m$. Note that $\sigma(L_g\mathbf{K}) \equiv \sigma(\mathbf{K})$ implies that

$$\theta(L_g\mathbf{K})(L_{g*\mathbf{K}} w) = gK \circ \sigma_{*g\mathbf{K}} \circ L_{g*\mathbf{K}} w = g \circ \mathbf{K} \circ \sigma_{*\mathbf{K}} w = g\theta(\mathbf{K})(w).$$

Also $\theta(\mathbf{K})(w) = 0$ for $w \in V_\mathbf{K}\mathscr{F}$.

The $\boldsymbol{R}^m$-valued 2-form

$$\Theta(\mathbf{K})(w_1, w_2) := d\theta(\mathbf{K})(\pi^H_\mathbf{K} w_1, \pi^H_\mathbf{K} w_2) \tag{1.1.15}$$

is called the **torsion** of the connection H.

Again, for $\mathbf{K} = L_g\tilde{\mathbf{K}}(P)$, $\pi^H_\mathbf{K} w_i = L_{g*\tilde{\mathbf{K}}(P)}\tilde{K}_{*P} v_i$, $i = 1, 2$, $\Theta(\mathbf{K})(w_1, w_2) = g\tilde{\theta}(P)(v_1, v_2) = T^k{}_{\alpha\beta} d\xi^\alpha \wedge d\xi^\beta \otimes ge_k$.

For two horizontal vector fields $X_i\colon \mathscr{U} \to T\mathscr{F}$, $\mathscr{U} \subset \mathscr{M}$, with values in $H_\mathbf{K}$, $i = 1, 2$, $2\Theta(X_1, X_2) = 2d\theta(X_1, X_2) = X_1(\theta(X_2)) - X_2(\theta(X_1)) - \theta([X_1, X_2])$. Let $X_i(\mathbf{K})$ be the unique lift of $\overset{-1}{\mathbf{K}} w_i$, $w_i = \text{const} \in \boldsymbol{R}^m$, $i = 1, 2$, to $H_\mathbf{K}$. Then $\theta(X_i)(\mathbf{K}) \equiv w_i$. Hence for any two vector fields X_1, X_2 of the above form and such that $\pi^H_\mathbf{K}[X_1, X_2](\mathbf{K}) =$ lift of $\overset{-1}{\mathbf{K}} w_0$ we have $2\Theta(X_1, X_2)(\mathbf{K}) = -w_0$ (w_0 may depend on $\mathbf{K}$).

Suppose that H is locally integrable, $H_\mathbf{K} = L_{g*\hat{\mathbf{K}}(P)}\hat{\mathbf{K}}_{*P}(T_P\mathscr{M})$ for $\mathbf{K} = L_g\hat{\mathbf{K}}(P)$, $P = \sigma(\mathbf{K})$. In this case $X_i(\mathbf{K}) = L_{g*\hat{\mathbf{K}}(P)}\hat{\mathbf{K}}_{*P}\overset{-1}{\mathbf{K}} w_i = L_{g*\hat{\mathbf{K}}(P)}\hat{\mathbf{K}}_{*P}\hat{\mathbf{K}}(P)^{-1}g^{-1}w_i$, $i = 0, 1, 2$. By hypothesis X_1, X_2, and $[X_1, X_2]$ are horizontal with respect to the decomposition $\varphi\colon \sigma^{-1}(\mathscr{U}) \to \mathscr{U} \times \mathscr{G}$;

$\mathbf{K} \mapsto (\sigma(\mathbf{K}), g)$, $g = \mathbf{K}\hat{\mathbf{K}}(P)^{-1}$. In the coordinates (ξ^α, ζ^A) their components $X_1{}^A, X_2{}^A$, $[X_1, X_2]^A$ vanish. Hence $\pi^H_{\mathbf{K}}[X_1, X_2](\mathbf{K}) = [X_1, X_2](\mathbf{K})$ reduces to $X_1{}^\beta \frac{\partial}{\partial \xi^\beta} X_2{}^\alpha - X_2{}^\beta \frac{\partial}{\partial \xi^\beta} X_1{}^\alpha$, which is equal to $L_{g*\hat{\mathbf{K}}(P)}\hat{\mathbf{K}}_{*P}[\hat{\mathbf{K}}(\cdot)^{-1}g^{-1}w_1,$ $\hat{\mathbf{K}}(\cdot)^{-1}g^{-1}w_2](P)$. Hence $\hat{\mathbf{K}}(P)^{-1}w_0 = [\hat{\mathbf{K}}(\cdot)^{-1}w_1, \hat{\mathbf{K}}(\cdot)^{-1}w_2](P)$ for $g = e$. If $w_0 = 0$ for every $w_1, w_2 \in \boldsymbol{R}^m$, $P \in \mathscr{U}$, then it follows that $\hat{\mathbf{K}}^a{}_\alpha = \phi^a{}_{,\xi^\alpha}$ by a straightforward calculation.

Vanishing of curvature implies that $H_{\mathbf{K}}$ is tangent to (locally defined) sections $P \mapsto L_g\hat{\mathbf{K}}(P)$. If both curvature and torsion vanish then the local sections mentioned above are integrable (locally), $\hat{\mathbf{K}}(\xi)^a{}_\alpha = \dfrac{\partial \zeta^a}{\partial \xi^\alpha}$. These conditions are necessary and sufficient for the connection to be flat (the connection symbols Γ vanish in the coordinates ζ^a).

Torsion can be represented by a (2, 1) tensor independent of $\tilde{\mathbf{K}}$, φ. We shall set

$$T(P)(v_1, v_2) := \overset{-1}{\mathbf{K}} \circ \Theta(\mathbf{K})(w_1, w_2), \tag{1.1.16}$$

where w_1, w_2 are two arbitrary lifts of $v_1, v_2 \in T_P M$ (i.e. $\sigma_{*\mathbf{K}} w_i = v_i$), $\sigma(\mathbf{K}) = P$. It is easy to see that the right-hand side of (1.1.16) is independent of the choice of $\mathbf{K} \in \mathscr{F}_P, w_1, w_2$.

We shall now prove existence of connections on a principal bundle.

Let $\sigma\colon \mathscr{F} \to \mathscr{M}$ be a principal bundle with trivializing maps $\varphi_i\colon \sigma^{-1}(\mathscr{U}_i) \to \mathscr{U}_i \times \mathscr{G}$; $\mathbf{K} \mapsto (\sigma(\mathbf{K}), \tilde{\varphi}_i(\mathbf{K}))$, $\mathbf{K} = L_{\tilde{\varphi}_i(\mathbf{K})}\hat{\mathbf{K}}_i(P)$. Let $\mathscr{H}_i$ be the horizontal distribution on $\sigma^{-1}(\mathscr{U}_i)$ tangent to the sections $P \mapsto L_g\hat{\mathbf{K}}_i(P)$. It satisfies the requirements of transversality to the fibers $\mathscr{F}_P$ and L_g-invariance ((1) and (2) in Section 1.1.2).

Let ω_i be the connection one-form of $\mathscr{H}_i$ on $\sigma^{-1}(\mathscr{U}_i)$. Let $\{f_i\}$ be a partition of unity on $\mathscr{M}$ subordinate to $\{\mathscr{U}_i\}$. We shall set

$$\omega(\mathbf{K})(w) = \sum_i f_i\big(\sigma(\mathbf{K})\big)\omega_i(\mathbf{K})(w).$$

Clearly $\omega(L_g\mathbf{K})(L_{g*\mathbf{K}}w) = \sum_i f_i\big(\sigma(L_g\mathbf{K})\big)\ \mathrm{ad}\,g \circ \omega_i(\mathbf{K})(w) = \mathrm{ad}\,g \circ \omega(\mathbf{K})(w)$.

For $w \in V_{\mathbf{K}}$ $\omega_i(\mathbf{K})(w) = 0$ and hence $\omega(\mathbf{K})(w) = 0$. Hence ω is a connection one-form. This proves existence of connections on $\mathscr{F}$.

Let $\Gamma_{i\alpha}{}^A$ be the connection symbols for ω_i, expressed in terms of a basis of the Lie algebra $\mathfrak{g}$ of $\mathscr{G}$. It is easy to see that the connection symbols $\Gamma_\alpha{}^A$ of ω are given by $\sum_i f_i \Gamma_{i\alpha}{}^A$.

For future reference we shall note the following fact. Suppose that the subbundle $\mathscr{F}$ of the principal frame bundle $\mathscr{F}\mathscr{M}$ can be generated by a covering $\{\mathscr{U}_i\}$ of $\mathscr{M}$ with local sections $\mathbf{K}_i$: $\mathscr{U}_i \to \mathscr{F}$ of the form $K_i{}^a{}_\alpha = \partial\phi_i{}^a(\xi)/\partial\xi^\alpha$. On each $\sigma^{-1}(\mathscr{U}_i)$ we can construct a torsionless connection H_i. Since the correspondence between connection symbols and torsion is $\mathscr{F}$-linear, it follows that the patched-up connection H is torsionless too.

Let $P \in \mathscr{M}$ be fixed and let c: $[0, 1] \to \mathscr{M}$ be a closed curve with $c(1) = c(0) = P$. Parallel transport from $c(0) = P$ to $c(1) = P$ along c defines a vector space automorphism P_c: $\mathscr{F}_P \to \mathscr{F}_P$. It is easy to see that $P_c(g\mathbf{K}) = gP_c(\mathbf{K})$.

Let c_i: $[0, 1] \to \mathscr{M}$, $c_i(0) = c_i(1) = P$, $i = 1, 2$. We shall define the curve $c_2 * c_1$: $P = c_1(t)$ for $t \in [0, 1/2]$, $P = c_2(2t-1)$ for $t \in [1/2, 1]$. It is clear that $P_{c_2 * c_1} = P_{c_2}P_{c_1}$. The set of all automorphisms P_c: $\mathscr{F}_P \to \mathscr{F}_P$ induced by closed curves at P forms a group $\mathscr{G}_P$, called the **holonomy group of H at P.**

Let K_0 be fixed, $\sigma(\mathbf{K}_0) = P$, $P_c \in \mathscr{G}_P$. We can always find such a $g \in \mathscr{G}$ that $P_c(\mathbf{K}_0) = \overset{-1}{g}\mathbf{K}_0$. Let $P_{c_i}(\mathbf{K}_0) = g_i^{-1}\mathbf{K}_0$, $i = 1, 2$. Then $P_{c_2 * c_1}(\mathbf{K}_0) = (g_2 g_1)^{-1}\mathbf{K}_0$ and $\Phi_{\mathbf{K}_0}$: $P_c \to g$ is a homomorphism of $\mathscr{G}_P$ into $\mathscr{G}$.

If $\mathscr{M}$ is connected, then for any $Q \in \mathscr{M}$ there is a curve d: $[0, 1] \to \mathscr{M}$, $d(0) = P$, $d(1) = Q$, joining Q to P, and for every closed curve c' at Q, $c'(0) = c'(1) = Q$, there is a closed curve c at P, $c(0) = c(1) = P$, given by the formulae:

$$c(t) :\equiv \begin{cases} d(3t) & \text{for } 0 \leqslant t \leqslant 1/3, \\ c'(3t-1) & \text{for } 1/3 \leqslant t \leqslant 2/3, \\ d(3-3t) & \text{for } 2/3 \leqslant t \leqslant 1. \end{cases}$$

The mapping H_d: $P_{c'} \mapsto P_c$ is a group homomorphism. Obviously there are also homomorphisms $\mathscr{G}_P \to \mathscr{G}_Q$. Hence $\mathscr{G}_P$, $\mathscr{G}_Q$ are isomorphic.

Let $\mathbf{K}_1$ be the result of parallel transport of $\mathbf{K}_0$ along d from P to Q, $\mathbf{K}_2 = h\mathbf{K}_1$. Parallel transport of $\mathbf{K}_2$ along the reversed d (i.e. $t \mapsto d(1-t)$), c' at P, and finally along d back to Q yields successively $h\mathbf{K}_0$ at P, $P_{c'}(h\mathbf{K}_0) = hP_{c'}(\mathbf{K}_0) = h\Phi_{\mathbf{K}_0}(P_{c'})(\mathbf{K}_0)$ at P, $h\Phi_{\mathbf{K}_0}(P_{c'})(\mathbf{K}_1) = h\Phi_{\mathbf{K}_0}(P_{c'})(h^{-1}\mathbf{K}_2)$ at Q. Hence, setting c equal to the composite path at Q, we have $\Phi_{\mathbf{K}_2}(P_c) = h\Phi_{\mathbf{K}_0}(P_{c'})h^{-1}$. Hence the homomorphism $\Phi_{\mathbf{K}_2}$ differs from $\Phi_{\mathbf{K}_0}$ by an inner automorphism of $\mathscr{G}$.

The maps $\Phi_{\mathbf{K}_0}$ permit an identification of $\mathscr{G}_P$ with a Lie subgroup $\mathscr{G}_0$ of the structure group (Nomizu, 1956; Lichnerowicz, 1976).

There is a principal bundle $\mathscr{F}_0$ with the structure group $\mathscr{G}_0$, a mapping f: $\mathscr{F}_0 \to \mathscr{F}$ such that $(\mathscr{F}_0, f, \mathscr{F})$ is a reduction of $\mathscr{F}$ and a connection H^0 on $\mathscr{F}_0$ such that $f_{*\mathbf{K}} H^0_{\mathbf{K}} = H_{\mathbf{K}}$ $\forall \mathbf{K} \in \mathscr{F}_0$ (Sternberg, 1964; Nomizu, 1956).

Finally, the theorem of Ambrose and Singer (Sternberg, 1964) states that the Lie algebra of $\mathscr{G}_P$ is spanned by elements of the form $\Omega(\mathbf{K})(w_1, w_2)$, $\forall w_1, w_2 \in T_{\mathbf{K}}\mathscr{F}$, $\sigma(\mathbf{K}) = P$.

1.1.4 Elements of Measure Theory and Integration

We shall only recall some basic facts from measure theory for easy reference.

Let $\bar{\boldsymbol{R}} = \boldsymbol{R} \cup \{\infty\}$.

There are several approaches to the measure and integration theory. Ours is adapted to the applications in Chapter 3.

Let $\mathfrak{M}$ be a σ-algebra of subsets $\mathscr{A}, \mathscr{B}, \ldots$ of a set $\mathscr{X}$, i.e. $\forall \mathscr{A}, \mathscr{B} \in \mathfrak{M}$ $\mathscr{A} \cap \mathscr{B} \in \mathfrak{M}$, for every countable family of $\mathscr{A}_i \in \mathfrak{M}$ also $\bigcup_{i=1}^{\infty} \mathscr{A}_i \in \mathfrak{M}$; and, for every $\mathscr{A} \in \mathfrak{M}$ also $\mathscr{A}^c := \mathscr{X} \setminus \mathscr{A} \in \mathfrak{M}$. Since we assume that $\mathfrak{M}$ is non-empty, for some $\mathscr{A} \in \mathfrak{M}$, $\varnothing = \mathscr{A} \cap \mathscr{A}^c \in \mathfrak{M}$, $\mathscr{X} = \mathscr{A} \cup \mathscr{A}^c \in \mathfrak{M}$.

The **Boolean algebra** of a topological space X is the smallest σ-algebra of subsets of $\mathscr{X}$ containing all the open (or all the closed) subsets of $\mathscr{X}$.

A **measure** μ on a σ-algebra $\mathfrak{M}$ of sets is a σ-additive function on $\mathfrak{M}$ with values in $\bar{\bar{\boldsymbol{R}}}_+ := \bar{\boldsymbol{R}}_+ \cup \{\infty\}$, such that $\mu(\mathscr{A}) < \infty$ for some $\mathscr{A} \in \mathfrak{M}$. By **$\sigma$-additivity** of μ we mean the property

$$\mu\Big(\bigcup_{i=1}^{N} \mathscr{A}_i\Big) = \sum_{i=1}^{N} \mu(\mathscr{A}_i) \tag{1.1.17}$$

if $\mathscr{A}_i \cap \mathscr{A}_j = \varnothing$ for $i \neq j \leqslant N$ and $N \leqslant \infty$.

μ is said to be **additive** if (1.1.17) holds merely for $N < \infty$. An additive function $\mu\colon \mathfrak{M} \to \bar{\bar{\boldsymbol{R}}}_+$ is **σ-additive** iff $\lim_{i\to\infty} \mu(\mathscr{A}_i) = \mu\Big(\bigcup_{i=1}^{\infty} \mathscr{A}_i\Big)$ for every monotone sequence of sets $\mathscr{A}_i$:

$$\mathscr{A}_i \subset \mathscr{A}_{i+1}.$$

A vector-valued measure $\mathbf{f}$ on a σ-algebra $\mathfrak{M}$ of subsets of $\mathscr{X}$ is a function $\mathbf{f}$ on $\mathfrak{M}$ with values in a vector space $\mathscr{V}$ satisfying (1.1.18) or (1.1.17). Let $|\cdot|$ be the norm on $\mathscr{V}$.

Let $\mathbf{f}$ be a vector-valued measure on a σ-algebra $\mathfrak{M}$, $\mathscr{A} \subset \mathscr{X}$,

$$v(\mathbf{f}, \mathscr{A}) := \sup\Big\{\sum_{=1}^{n} |\mathbf{f}(\mathscr{A}_i)|\ \Big|\ n \in \boldsymbol{Z}_+, \mathscr{A}_i \subset \mathscr{A} \text{ for } i = 1, \ldots, n,$$
$$\mathscr{A}_i \cap \mathscr{A}_j = \varnothing \text{ for } i \neq j, i, j \leqslant n\Big\}.$$

The set function $\mathscr{A} \mapsto v(\mathbf{f}, \mathscr{A})$ is called the **variation of the measure f**. We shall occasionally write $|\mathbf{f}| = v(\mathbf{f}, \mathscr{X})$, $|\mathbf{f}|(\mathscr{A}) = v(\mathbf{f}, \mathscr{A})$.

It can be proved that $v(\mathbf{f}, \cdot)$ is a measure on the σ-algebra $\mathfrak{M}$. Indeed, let $\mathscr{A} = \bigcup_{i=1}^{\infty} \mathscr{A}_i$, $\mathscr{A}_i \cap \mathscr{A}_j = \emptyset$ for $i \neq j$, $\mathscr{A}_i \in \mathfrak{M}$. Suppose $\mathscr{B}_1, \ldots, \mathscr{B}_n$ are disjoint subsets of $\mathscr{A}$ and let $\mathscr{A}_{ji} := \mathscr{B}_j \cap \mathscr{A}_i$. Since

$$\mathbf{f}(\mathscr{B}_j) = \sum_{i=1}^{\infty} \mathbf{f}(\mathscr{A}_{ji}),$$

we have

$$\sum_{j=1}^{n} |\mathbf{f}(\mathscr{B}_j)| \leqslant \sum_{j=1}^{n} \sum_{i=1}^{\infty} |\mathbf{f}(\mathscr{A}_{ji})| \leqslant \sum_{i=1}^{\infty} v(\mathbf{f}, \mathscr{A}_i)$$

and

$$v(\mathbf{f}, \mathscr{A}) \leqslant \sum_{i=1}^{\infty} v(\mathbf{f}, \mathscr{A}_i).$$

On the other hand $\forall i, \varepsilon > 0\ \exists j, n_i < \infty$, $\{\mathscr{A}_{ij} \in \mathfrak{M} |\ \mathscr{A}_{ij} \subset \mathscr{A}, \mathscr{A}_{ij} \cap \mathscr{A}_{ik} = \emptyset$ for $j \neq k, j, k \leqslant n_i\}$ such that

$$\sum_{j=1}^{n_j} |\mathbf{f}(\mathscr{A}_{ij})| \leqslant v(\mathbf{f}, \mathscr{A}_i) \leqslant \sum_{j=1}^{n_i} |\mathbf{f}(\mathscr{A}_{ij})| + \frac{\varepsilon}{2^i}.$$

Hence

$$\sum_{i=1}^{N} v(\mathbf{f}, \mathscr{A}_i) \leqslant \sum_{i=1}^{N} \sum_{j=1}^{n_i} |\mathbf{f}(\mathscr{A}_{ij})| + \varepsilon \sum_{i=1}^{N} \frac{1}{2^i} \leqslant v(\mathbf{f}, \mathscr{A}) + \varepsilon.$$

Hence

$$\sum_{i=1}^{\infty} v(\mathbf{f}, \mathscr{A}_i) \leqslant v(\mathbf{f}, \mathscr{A}) + \varepsilon$$

and

$$\sum_{i=1}^{\infty} v(\mathbf{f}, \mathscr{A}_i) \leqslant v(\mathbf{f}, \mathscr{A}).$$

$v(\mathbf{f}, \cdot)$ is a non-negative measure on $\mathfrak{M}$.

Definition 1.1.5

Every set $\mathscr{A} \in \mathfrak{M}$ is said to be **measurable** (or **f-measurable**).

Definition 1.1.6

A vector-valued measure **f** on $\mathfrak{M}$ is said to be **Borel regular** if

(i) every open subset of $\mathscr{X}$ belongs to $\mathfrak{M}$,*

(ii) $\forall \mathscr{A} \in \mathfrak{M}\ \forall \varepsilon > 0\ \exists$ open $\mathscr{G}$, closed $\mathscr{F} \subset \mathscr{X}$ such that $\mathscr{F} \subset \mathscr{A} \subset \mathscr{G}$, $v(\mathbf{f}, \mathscr{G} \setminus \mathscr{F}) < \varepsilon$.

Definition 1.1.7

Let f be a Borel regular vector-valued measure. A subset $\mathscr{A}$ of $\mathscr{X}$ is said to be **of measure zero** if $\forall \varepsilon > 0\ \exists$ open $\mathscr{G} \subset \mathscr{X}$ such that $\mathscr{A} \subset \mathscr{G}$, $v(\mathbf{f}, \mathscr{G}) < \varepsilon$.

We shall extend $\mathfrak{M}$ including all the sets of measure zero in it.

Theorem 1.1.8

Let **f** be a Borelregular measure defined on $\mathfrak{M}$ (including sets of measure zero).

A set $\mathscr{A} \in \mathfrak{M}$ iff there is a Borel set $\mathscr{B}$** and a set $\mathscr{N}$ of measure zero such that $\mathscr{A} = \mathscr{B} \cup \mathscr{N}$.

Proof

Since **f** is Borel-regular, $\mathfrak{M}$ contains the σ-algebra of Borel sets, on account of (i). Hence every set $\mathscr{A} = \mathscr{B} \cup \mathscr{N}$, where $\mathscr{B}$ is Borel and $\mathscr{N}$ has measure zero, belongs to $\mathfrak{M}$.

Conversely, let $\mathscr{A} \in \mathfrak{M}$. Suppose $\mathscr{F}_n$ is a closed set and let $\mathscr{G}_n$ be an open set such that $v(\mathbf{f}, \mathscr{G}_n \setminus \mathscr{F}_n) < 1/n$, $\mathscr{F}_n \subset \mathscr{A} \subset \mathscr{G}_n$. Let $\mathscr{F}'_n = \bigcup_{k=1}^{n} \mathscr{F}_k$, $\mathscr{G}'_n = \bigcap_{k=1}^{n} \mathscr{G}_k$. Clearly $\mathscr{F}'_n \nearrow \mathscr{F}$, $\mathscr{G}'_n \searrow \mathscr{G}$, $v(\mathbf{f}, \mathscr{G} \setminus \mathscr{F}) = \lim_{n \to \infty} v(\mathbf{f}, \mathscr{G}'_n \setminus \mathscr{F}'_n) = 0$, since $\mathscr{G}'_n \subset \mathscr{G}_n$, $\mathscr{F}'_n \supset \mathscr{F}_n$. Hence $\mathscr{A} = \mathscr{F} \cup (\mathscr{A} \setminus \mathscr{F})$, $\mathscr{F}$ is Borel, $\mathscr{A} \setminus \mathscr{F}$ has measure zero. □

Theorem 1.1.9

A regular Borel measure is defined by its values on all open sets (or on all closed sets).

Proof

If **f** is known on open sets, it is known on all Borel sets. Indeed, for any Borel

* Hence every closed subset of $\mathscr{X}$ belongs to $\mathfrak{M}$.

** This means that $\mathscr{B}$ belongs to the σ-algebra of Borel subsets of $\mathscr{X}$.

set $\mathscr{A}$ and a sequence of closed sets $\mathscr{F}_n \to \mathscr{F}$*, $\mathscr{A}\setminus\mathscr{F}$ of measure zero,

$$\mathbf{f}(\mathscr{A}) = \mathbf{f}(\mathscr{F}) + \mathbf{f}(\mathscr{A}\setminus\mathscr{F}) = \lim_{n\to\infty} \mathbf{f}(\mathscr{F}_n).$$

If $\mathscr{A} \in \mathfrak{M}$, then $\mathscr{A} = \mathscr{B} \cup \mathscr{N}$, where $\mathscr{B}$ is Borel and $\mathscr{N}$ has measure zero. Hence $\mathbf{f}(\mathscr{A}) = \mathbf{f}(\mathscr{B})$. □

Let us consider all the rational intervals $[a, b]$ in $\boldsymbol{R}^n$, $[a, b] := \{x \in \boldsymbol{R}^n \backslash a^i < x^i < b^i, i = 1, \ldots, n\}$, with a^i, b^i rational for $i = 1, \ldots, n$. The set of all rational intervals in $\boldsymbol{R}^n$ is countable. Let $\mathscr{G}$ be open in $\boldsymbol{R}^n$. Since the rational intervals constitute a basis of the topology of $\boldsymbol{R}^n$, every $x \in \mathscr{G}$ is contained in an interval $\overline{[a, b]} \subset \mathscr{G}$. Hence G is the join of a countable family $\mathscr{J}$ of closed rational intervals. Choose the first two intervals $\mathscr{I}_1, \mathscr{I}_2 \in \mathscr{J}$. If $\mathscr{I}_1 \cap \mathscr{I}_2 \neq \emptyset$, then we can decompose $\overline{\mathscr{I}_1 \cup \mathscr{I}_2}$ into a finite number of non-intersecting intervals $\mathscr{J}_1, \ldots, \mathscr{J}_k$ such that $\overline{\mathscr{I}_1 \cup \mathscr{I}_2} = \overline{\mathscr{J}_1 \cup \ldots \cup \mathscr{J}_k}$. Proceeding recursively we find that $\mathscr{G}$ is the limit of a monotone sequence of sets $\overline{\bigcup_{i=1}^{k} \mathscr{J}_i} =: \mathscr{A}_k$ with $\mathscr{J}_i$ disjoint.

Let $L^n(\mathscr{J}_i) := \prod_{r=1}^{n} (b_i^r - a_i^r)$ if $\mathscr{J}_i = [a_i, b_i]$. Assuming that L^n is a measure, we have that $L^n(\mathscr{A}_k) := \sum_{r=1}^{k} L^n(\mathscr{J}_i) \to L^n(\mathscr{G})$. Assuming that L^n is Borel regular we can calculate it on the σ-algebra of sets $\mathscr{A} = \mathscr{B} \cup \mathscr{N}$, with $\mathscr{B}$ Borel and $\mathscr{N}$ of L^n measure zero. The elements of this algebra are known as **Lebesgue measurable sets.**

We shall also need the k-dimensional **Hausdorff measure** in $\boldsymbol{R}^n$, $k \leqslant n$. Let ω_k denote the Lebesgue measure of a unit sphere in $\boldsymbol{R}^k$. By integration it is easy to prove that

$$\omega_k = \frac{2\pi}{k}\omega_{k-2}, \qquad \omega_1 = 2,\ \omega_2 = \pi.$$

Let $\mathscr{A} \subset \boldsymbol{R}^n$ be covered by a sequence of balls $\mathscr{B}_1, \mathscr{B}_2, \ldots, \mathscr{B}_\nu$, of radii $r_1, r_2, \ldots, r_\nu$, resp.

$$H_\varepsilon^k(\mathscr{A}) := \inf\Big\{\omega_k \sum_{i=1}^{\nu} r_i^k \Big|\ \text{all finite coverings of } \mathscr{A} \text{ by balls of radii } r_i < \varepsilon\Big\}.$$

* For its existence, see the proof of Theorem 1.1.8.

The function $\varepsilon \mapsto H_\varepsilon^k$ is non-decreasing, hence the **exterior Hausdorff measure**

$$H^k(\mathscr{A}) = \lim_{\varepsilon \to 0} H_\varepsilon^k(\mathscr{A})$$

exists for every $\mathscr{A} \subset \boldsymbol{R}^n$.

A set $\mathscr{A} \subset \boldsymbol{R}^n$ is H^k**-measurable** if $\mathscr{A} = \mathscr{B} \cup \mathscr{N}$, $H^k(\mathscr{N}) = 0$ and $\mathscr{B}$ is a Borel subset of $\boldsymbol{R}^n$.

It can be shown that H^2 generalizes the area in $\boldsymbol{R}^3$ and H^1 generalizes the length, both defined in terms of the Riemann metric δ_{ij}. For details on Hausdorff measures, exterior measures and Carathéodory's definition of measurable sets, see Federer (1969)*.

1.1.5 Measurable Functions and Integration

Definition 1.1.10

Let **f** be a vector-valued measure defined on a σ-algebra $\mathfrak{M}$ of sets (including sets of zero **f** measure). A function $g\colon \Omega \to \boldsymbol{R}^m$, $\Omega \subset \boldsymbol{R}^n$, $\Omega \in \mathfrak{M}$, is said to be **f-measurable** if for every closed $\mathscr{F} \subset \boldsymbol{R}^m$ the set $g^{-1}(\mathscr{F}) \in \mathfrak{M}$.

Alternatively, g is **f**-measurable if there is a sequence of simple functions $g_n = \sum_{i=1}^{m_n} c_n^i \chi_{\mathscr{A}_i^n}$,** $\mathscr{A}_i^n \in \mathfrak{M}$, $m_n < \infty\ \forall n$, such that $g_n(x) \to g(x)$ **f**-a.e. on Ω.

Definition 1.1.11

Let $P(x)$ be a family of propositions parametrized by $x \in \Omega$. We shall say that $P(x)$ **f**-a.e. (almost everywhere) if $P(x)$ is true for $x \in \mathscr{A} \subset \Omega$ and $v(\mathbf{f}, \Omega \setminus \mathscr{A}) = 0$.

Theorem 1.1.12 (Lusin)

Let $\Omega \subset \boldsymbol{R}^m$. If $g\colon \Omega \to \boldsymbol{R}^n$ is L^m-measurable then $\forall \varepsilon > 0$ there is an L^m-measurable set $\mathscr{A} \subset \Omega$ such that $L^m(\mathscr{A}) < \varepsilon$ and g is continuous on $\Omega \setminus \mathscr{A}$ (i.e. $g(x_n) \to g(x)$ for every sequence $\{x_n\} \subset \Omega \setminus \mathscr{A}$ such that $x_n \mapsto x \in \Omega \setminus \mathscr{A}$).

* Federer calls a measure any subadditive set function which may be an exterior measure in the usual terminology.

** $\chi_{\mathscr{A}}(x) := \begin{cases} 1 & \text{for } x \in \mathscr{A}, \\ 0 & \text{for } x \notin \mathscr{A}. \end{cases}$

Theorem 1.1.13 (Egoroff)

Let μ be a (positive) measure on $\Omega \subset \boldsymbol{R}^m$, $\mu(\Omega) < \infty$. Let $\{g_n\}$ be a sequence of μ-measurable functions $\Omega \to \boldsymbol{R}^n$ such that $g_n(x) \to g(x)$ μ-a.e. in Ω. Then $\forall \varepsilon > 0$ there is a μ-measurable set $\mathscr{A} \subset \Omega$ such that $\mu(\mathscr{A}) < \varepsilon$ and $g_n(x) \to g(x)$ uniformly on $\Omega \setminus \mathscr{A}$.

Definition 1.1.14
Let $g\colon \Omega \to \boldsymbol{R}$ be $\mathbf{f}$-measurable with respect to a vector-valued measure $\mathbf{f}$. Let g_n be a sequence of simple functions converging to g $\mathbf{f}$-a.e. on Ω. If

$$\lim_{m,n\to\infty} \int_\Omega dv(\mathbf{f})\,|g_n - g_m| = 0 \tag{1.1.19}$$

we shall say that $\mathbf{g}$ is **f-summable** on Ω and set

$$\int_\Omega g d\mathbf{f} := \lim_{n\to\infty} \int_\Omega g_n d\mathbf{f} := \lim_{n\to\infty} \sum_{i=1}^{m_n} c_n^i \mathbf{f}(\mathscr{A}_n^i \cap \Omega) \tag{1.1.20}$$

for $g_n = \sum_{i=1}^{m_n} c_n^i \chi_{\mathscr{A}_n^i}$.

Note that the integrand in (1.1.19) is a simple function for every m, n and the integral is taken with respect to the positive measure $\mathscr{A} \mapsto v(\mathbf{f}, \mathscr{A})$, $\mathscr{A} \in \mathfrak{M}$.

Let $g\colon \Omega \to \boldsymbol{R}^m$ be measurable with respect to $\mathbf{f}$. We can define

$$\mathscr{A}_n^i := \{x \in \Omega |\ c_n^{i-1} < g(x) \leqslant c_n^i\}, \quad g_n = \sum_{i=1}^{m_n} c_n^i \chi_{\mathscr{A}_n^i}.$$

This way of partitioning the domain of integration goes back to Lebesgue.

The following facts are important.

$$\left|\int_{\mathscr{A}} g(x) d\mathbf{f}\right| \leqslant \int_{\mathscr{A}} |g(x)| dv(\mathbf{f}).$$

For a summable function g on Ω, $\mathbf{F}_g\colon \mathscr{A} \mapsto \int_{\mathscr{A}} g d\mathbf{f}$, $\mathscr{A} \subset \Omega$, $\mathscr{A} \in \mathfrak{M}$, is a set function on $\mathfrak{M}$. It is absolutely continuous with respect to $\mathbf{f}$:

$$v(\mathbf{f}, \mathscr{A}) = 0 \quad \text{implies that} \quad \mathbf{F}_g(\mathscr{A}) = \mathbf{0}$$

and $\mathbf{F}_g$ is a vector-valued measure whose variation is given by $\int_{\mathscr{A}} |g| dv(\mathbf{f})$.

Definition 1.1.15

Let $\mathfrak{M}_i$ be a σ-algebra of subsets of $\mathscr{X}_i$, and let μ_i be a measure on $\mathfrak{M}_i$, $i = 1, 2$. We shall set $\mathfrak{M}_1 \times \mathfrak{M}_2 := \{\mathscr{A}_1 \times \mathscr{A}_2 | \ \mathscr{A}_i \in \mathfrak{M}_i\}$, $(\mu_1 \otimes \mu_2)(\mathscr{A}_1 \times \times \mathscr{A}_2) := \mu_1(\mathscr{A}_1)\mu_2(\mathscr{A}_2)$. It can be proved that there is a unique measure defined on the smallest σ-algebra of subsets of $\mathscr{X}_1 \times \mathscr{X}_2$ which coincides with $\mu_1 \otimes \mu_2$ on $\mathfrak{M}_1 \times \mathfrak{M}_2$. We shall denote this measure by $\mu_1 \otimes \mu_2$ too (cf. Dunford and Schwartz, 1958).

Example 1.1.16

$L^3(\mathscr{A} \times \mathscr{I}) = L^2(\mathscr{A})L^1(\mathscr{I})$, for every L^2-measurable $\mathscr{A} \subset \boldsymbol{R}^2$ and L^1 measurable $\mathscr{I} \subset \boldsymbol{R}^1$.

Theorem 1.1.17 (Fubini)

Suppose that μ_i is a measure on the σ-algebra $\mathfrak{M}_i$ of subsets of $\mathscr{X}_i$, $i = 1, 2$, and $f\colon \mathscr{X}_1 \times \mathscr{X}_2 \to \boldsymbol{R}$ is $\mu_1 \otimes \mu_2$-summable. Then $x_1 \mapsto \int_{\mathscr{X}_2} d\mu_2 f(x_1, x_2)$ is defined μ_1-a.e. on $\mathscr{X}_1$ and μ_1-summable on $\mathscr{X}_1$. Moreover the double integral $\int_{\mathscr{X}_1} d\mu_1 \int_{\mathscr{X}_2} d\mu_2 f(x_1, x_2)$ exists and equals $\int_{\mathscr{X}_1 \times \mathscr{X}_2} d(\mu_1 \otimes \mu_2) f(x_1, x_2)$.

Theorem 1.1.18 (Radon–Nikodym)

If $\mathbf{f}$ is a vector measure defined on a σ-algebra $\mathfrak{M}$ of subsets of $\mathscr{X}$, μ is a measure on $\mathfrak{M}$ vanishing outside a subset $\mathscr{X}_o$ of $\mathscr{X}$ of semifinite measure* and $\mathbf{f}$ is absolutely continuous with respect to μ ($v(\mathbf{f}, \mathscr{A}) = 0$ is implied by $\mu(\mathscr{A}) = 0$), then there is a vector function $\mathbf{g}\colon \mathscr{X} \to \boldsymbol{R}^m$ such that

$$\mathbf{f}(\mathscr{A}) = \int_{\mathscr{A}} \mathbf{g}(x) d\mu.$$

Remarks 1.1.19

(1) Under the hypotheses of Theorem 1.1.18

$$\int_{\mathscr{A}} h d\mathbf{f} = \int_{\mathscr{A}} h \mathbf{g} d\mu$$

for every $\mathbf{f}$-integrable function $h\colon \mathscr{A} \to \boldsymbol{R}$. That h is $\mathbf{f}$-integrable iff $h\mathbf{g}$ is μ-integrable follows from a result of Dunford and Schwartz (1958) (Corollary 3.10.6).

(2) Let $\mathscr{X}$ be a separable metric space with a finite Borel measure $\mu \geqslant 0$. For $n = 1, 2, 3, \ldots$ let $\mathfrak{M}_n$ be a countable (or finite) family of disjoint Borel sets such that $\mathfrak{M}_n$ covers $\mathscr{X}$. Suppose that each $\mathscr{A} \in \mathfrak{M}_{n+1}$ is contained in

* This means that $\mathscr{X}_o$ is the join of a sequence of measurable sets $\mathscr{A}_n$ of finite measure μ.

some $\mathscr{B} \in \mathfrak{M}_n$ and $\sup\{\text{diameter}(\mathscr{A}) |\ \mathscr{A} \in \mathfrak{M}_n\} \to 0$ as $n \to \infty$. We shall note that such a sequence $\{\mathfrak{M}_n\}$ exists provided $\mathscr{X}$ is separable.

Let Φ be a σ-additive function of Borel subsets of $\mathscr{X}$, $|\Phi(\mathscr{X})| < \infty$. We shall define

$$d_n(x) := \begin{cases} \Phi(\mathscr{A})/\mu(\mathscr{A}) & \text{if } x \in \mathscr{A} \in \mathfrak{M}_n,\ \mu(\mathscr{A}) \neq 0, \\ +\infty & \text{if } x \in \mathscr{A} \in \mathfrak{M}_n,\ \mu(\mathscr{A}) = 0,\ \Phi(\mathscr{A}) \geqslant 0, \\ -\infty & \text{if } x \in \mathscr{A} \in \mathfrak{M}_n,\ \mu(\mathscr{A}) = 0,\ \Phi(\mathscr{A}) < 0. \end{cases}$$

The derivative $(\mu, \mathfrak{M})D\Phi(x) := \limsup_{n\to\infty} d_n(x)$ provided

$$\limsup_{n\to\infty} d_n(x) = \liminf_{n\to\infty} d_n(x) \in \boldsymbol{R} \cup \{+\infty, -\infty\}.$$

Let $\mathscr{E}_+ := \{x \in \mathscr{X} |\ (\mu, \mathfrak{M})D\Phi(x) = \infty\}$, $\mathscr{E}_- := \{x \in \mathscr{X} |\ (\mu, \mathfrak{M})D\Phi(x) = -\infty\}$. Then

(i) $(\mu, \mathfrak{M})D\Phi(x)$ exists μ-a.e.;

(ii) $\Phi(\mathscr{A}) = \Phi(\mathscr{A}\cap\mathscr{E}_+) + \Phi(\mathscr{A}\cap\mathscr{E}_-) + \int_{\mathscr{A}} d\mu(x)(\mu, \mathfrak{M})D\Phi(x)$;

(iii) $|\Phi|(\mathscr{A}) = \Phi(\mathscr{A}\cap\mathscr{E}_+) + |\Phi(\mathscr{A}\cap\mathscr{E}_-)| + \int_{\mathscr{A}} d\mu(x)|(\mu, \mathfrak{M})D\Phi(x)|$;

(iv) $\mu(\mathscr{E}_+) = 0$, $\mu(\mathscr{E}_-) = 0$ and the first two terms on the right-hand side of (ii) are equal to the singular part of Φ with respect to μ;

(v) if $\mathscr{E}$ is the set of $x \in \mathscr{X}$ such that $(\mu, \mathfrak{M})D\Phi(x)$ does not exist, then $\Phi(\mathscr{E}) = 0$;

(vi) if $\{\mathfrak{N}_n\}$ is another sequence having the properties of $\{\mathfrak{M}_n\}$ listed above, then $(\mu, \mathfrak{N}_n)D\Phi(x) = (\mu, \mathfrak{M})D\Phi(x)$ μ-a.e. in $\mathscr{X}$ and Φ vanishes on the set $\mathscr{F}$ of points $x \in X$ where this equation breaks down.

At least the assertions (i)–(iii) are associated with the name of de la Vallée-Poussin, who proved them for L^n instead of μ. For the proof of this theorem, see Saks (1937).

(3) The assumption that $\mathfrak{M}_n$ is a disjoint family is inconvenient in some applications. For example it is often desirable to define the derivative of Φ with respect to μ in terms of concentric balls shrinking to a point.

From 2.8.16, 2.9.1, 2.9.5–7, 2.9.10, 2.2.3 and 2.2.2 of Federer (1969) one readily deduces the following theorem.

Let $\mathscr{X}$ be a metric space with two Borel regular non-negative measures Φ and μ, such that $\Phi(\mathscr{A}) < \infty$, $\mu(\mathscr{A}) < \infty$ for every bounded Borel set $\mathscr{A}$. Let $\mathfrak{M}$ be a family of Borel subsets of $\mathscr{X}$ such that $\mathfrak{M}$ covers $\mathscr{X}$ and for every $x \in \mathscr{X}$

$$\inf\{\text{diameter}(\mathscr{A}) |\ x \in \mathscr{A} \in \mathfrak{M}\} = 0. \tag{*}$$

Suppose that every subfamily $\mathfrak{N}$ of $\mathfrak{M}$ covering a subset $\mathscr{U}$ of $\mathscr{X}$ and satisfying (*) for every $x \in \mathscr{U}$ contains a countable disjoint subfamily covering μ-a.a. points of $\mathscr{U}$.*

We define

$$(\mu, \mathfrak{M})_* D\Phi(x) = \lim_{\mathscr{A} \searrow x} \frac{\Phi(\mathscr{A})}{\mu(\mathscr{A})},$$

the limit being taken for all $\mathscr{A} \in \mathfrak{M}$ such that $x \in \mathscr{A}$ and diameter $(\mathscr{A}) \to 0$.

The theorem asserts that

(i′), (iv′) $(\mu, \mathfrak{M})_* D\Phi(x) \geqslant 0$ exists and is finite μ-a.e. Moreover, it is measurable and

(ii′) $\int_{\mathscr{A}} d\mu(x)(\mu, \mathfrak{M})_* D\Phi(x) = \inf\{\Phi(\mathscr{B}) | \ \mathscr{B}$ is Borel and $\mu(\mathscr{A} \setminus \mathscr{B}) = 0\}$,

(vi′) for any two families $\mathfrak{M}, \mathfrak{N}$ with the properties listed above $(\mu, \mathfrak{M})_* D\Phi(x) = (\mu, \mathfrak{N})_* D\Phi(x)$ for μ-a.a. $x \in \mathscr{X}$.

If Φ is absolutely continuous with respect to μ, then $\Phi(\mathscr{A} \setminus \mathscr{E}_+) = 0$ in (ii′) and the right-hand side of (ii′) equals $\Phi(\mathscr{A})$.

For a signed measure Φ the Jordan decomposition yields (i′), (iv′) with $(\mu, \mathfrak{M})_* D\Phi(x)$ of arbitrary sign as well as (ii′) for absolutely continuous Φ.

For $\mathscr{X} = \boldsymbol{R}$ the family of all open balls has the properties required of $\mathfrak{M}$ (Federer, 1969, 2.8.18).

(4) Let V be a (separable normed) vector space. For $\mathbf{f} \in \mathscr{L}^1_{loc}(\mathscr{X}, V; \mu)$ the complement of the set $\mathscr{G}$ of points $x \in \mathscr{X}$ such that

$$\lim_{\mathscr{A} \searrow x} \mu(\mathscr{A})^{-1} \int_{\mathscr{A}} d\mu(z) \, |\mathbf{f}(z) - \mathbf{f}(x)| = 0$$

has measure $\mu = 0$. The points of $\mathscr{G}$ are called **Lebesgue points** of $(\mathbf{f}, \mu)$ (Federer, 1969, 2.9.9). If $g \in \mathscr{L}^1_{loc}(\mathscr{X}, \bar{\boldsymbol{R}}; \mu)$, then $\lim_{\mathscr{A} \searrow x} \mu(\mathscr{A})^{-1} \int_{\mathscr{A}} d\mu(z) \, g(z) = g(x)$ for μ-a.a. $x \in \mathscr{X}$ (Federer, 1969, 2.9.8).

From this equation and (ii), (ii′) in the preceding remarks it follows that $(\mu, \mathfrak{M})_* D\Phi(x) = (\mu, \mathfrak{M}) D\Phi(x)$ for μ-a.a. $x \in \mathscr{X}$.

(5) We shall also recall the Lebesgue derivation theorem for $\mu = L^n$.

(a) If Φ is a Borel measure on an open subset Ω of $\boldsymbol{R}^n$ such that $\Phi(\mathscr{A})$ is finite for every Borel $\mathscr{A} \subset \Omega$, then the limit $\Phi'(x) = \lim_{\mathscr{A} \searrow x} L^n(A)^{-1} \Phi(\mathscr{A})$ exists for L^n-a.a. $x \in \Omega$ (Dunford and Schwartz, 1958, Theorem 3.12.6); $\mathscr{A}$ denotes here cubes shrinking to x.

* According to a celebrated theorem of Vitali the family of all cubes (or all balls) covering a bounded $\mathscr{X} \subset \boldsymbol{R}^n$ has these properties.

(b) If $\Phi(\mathscr{A}) := \int_{\mathscr{A}} dL^n f$, $f \in \mathscr{L}^1(\Omega; L^n)$, then $\Phi' = f$ L^n-a.e. (Dunford and Schwartz, 1958, Theorem 3.12.8).

(a) follows with some extra hypotheses from Remarks (2) and (3) (in the latter case use the Vitali theorem). It is also true for functions **of intervals** in $\boldsymbol{R}^n$ having **bounded variation** (for the definition of the variation of a function Φ of intervals it is enough to apply the natural extension of Φ to finite joins of intervals).

(b) follows from Remark 1.1.19 (4). Note that Φ is absolutely continuous with respect to L^n in this case* and hence, in particular, has bounded variation.

A set function Φ of bounded variation satisfies the equation $\Phi(\mathscr{A}) \equiv \int_{\mathscr{A}} dL^n \Phi'(x)$ iff it is absolutely continuous with respect to L^n.

Theorem 1.1.20 (Fatou's lemma)
Let μ be a non-negative measure on a set $\mathscr{X}$ and suppose that $\{g_n\}$ is a sequence of non-negative real-valued measurable functions $\mathscr{X} \to \boldsymbol{R}$. Then

$$\liminf_{n\to\infty} \int_{\mathscr{X}} d\mu\, g_n \leqslant \int_{\mathscr{X}} d\mu \liminf_{n\to\infty} g_n .$$

For increasing sequences $\{g_n\}$ the right-hand side equals the left-hand side.

Theorem 1.1.21 (the Lebesgue bounded convergence theorem)
Let $\mathbf{f}$ be a vector-valued measure on $\mathscr{X}$. If a sequence of **f-measurable functions** $g_n \to g$ **f**-a.e. on $\mathscr{X}$ and there is an **f**-summable function h: $\mathscr{X} \to \bar{\boldsymbol{R}}_+$ such that $|g_n(x)| \leqslant h(x)$ **f**-a.e., then g is **f**-summable and

$$\lim_{n\to\infty} \int_{\mathscr{X}} g_n\, d\mathbf{f} = \int_{\mathscr{X}} g\, d\mathbf{f}.$$

1.1.6 Elements of General Functional Analysis

We shall state some theorems of functional analysis which will be used in this book, mostly without proof.

Theorem 1.1.22 (Hahn–Banach; cf. Yosida, 1965, Section 4.1.)
Let $\mathscr{X}$ be a real Banach space** and let $\mathscr{M}$ be a linear subspace of $\mathscr{X}$. Suppose that f_0: $\mathscr{M} \to \boldsymbol{R}$ is a linear functional continuous with respect to the norm

* Cf. Lemma 1.1.63 below.

** We recall that normed vector space is a vector space endowed with a norm $x \mapsto \|x\| \geqslant 0$. with the properties: $\|x\| = 0$ implies $x = 0$, $\|x+y\| \leqslant \|x\|+\|y\|$, $\|\lambda x\| = |\lambda|\, \|x\|$ $\forall \lambda \in \boldsymbol{R}$. A sequence $\{x_n\}$ in a normed vector space is fundamental if $\|x_n - x_m\| \to 0$ when $n, m \to \infty$.

of $\mathscr{X}$. Then there is a continuous linear functional $f \in \mathscr{X}^*$ defined on $\mathscr{X}$ such that $f|\mathscr{M} = f_0$ and $||f||_{\mathscr{X}^*} = \sup\{|f_0(x)| \; ||x|| = 1, \; x \in \mathscr{M}\}$. Since $\mathscr{X}^*$ with the strong topology is a Banach space we can define $\mathscr{X}^{**} = (\mathscr{X}^*)^*$. $\mathscr{X}$ is said to be **reflexive** if $\mathscr{X}^{**} = \mathscr{X}$ (note that in general $\mathscr{X} \subset \mathscr{X}^{**}$).

Theorem 1.1.23 (Eberlein–Shmul'yan; cf. Yosida, 1965, Sections 5.2, 5.1)
Every bounded subset $\mathscr{A}$ of a reflexive Banach space is relatively sequentially weakly compact (i.e. if $\{x_n\} \subset \mathscr{A}$ then a subsequence $\{x_{n_k}\}$ converges weakly to some $x_0 \in \mathscr{X}$, $f(x_{n_k} - x_0) \to 0$ for $k \to \infty$ for all $f \in \mathscr{X}^*$).*

Every weakly convergent sequence $\{x_n\}$ in a Banach space is bounded.

Theorem 1.1.24 (cf. Dunford and Schwartz, 1958, Theorem 5.3.13; Yosida, 1965, Section 5.1)
Every closed convex subset $\mathscr{A}$ of a reflexive Banach space $\mathscr{X}$ is sequentially weakly closed (i.e. if $\{x_n\} \subset \mathscr{A}$ has the property that $f(x_n - x_m) \to 0$ $\forall f \in \mathscr{X}^*$ then $\exists\, x_0 \in \mathscr{A}$ such that $f(x_n - x_0) \to 0$ $\forall f \in \mathscr{X}^*$). In particular a reflexive Banach space is sequentially weakly closed.

Theorem 1.1.25 (Mazur; cf. Yosida, 1965, Section 5.1)
If $x_n \rightharpoonup x_0$ in a Banach space $\mathscr{X}$ then $\forall\, \varepsilon > 0$ $\exists\, n \in \boldsymbol{Z}_+$, $\{\vartheta_j^{(n)}\}$ such that $\sum_{j=1}^{n} \vartheta_j^{(n)} = 1$ and $\left\|\sum_{j=1}^{n} \vartheta_j^{(n)} x_j - x_0\right\| < \varepsilon$.

Theorem 1.1.26 (Yosida, 1965, Section 5.1)
If $||x_n|| < M$ $\forall\, n \in \boldsymbol{Z}_+$ and $f(x_n) \to f(x_0)$ for all f belonging to a dense subset of $\mathscr{X}^*$ (in the strong topology of $\mathscr{X}^*$), then $x_n \rightharpoonup x_0$.

Theorem 1.1.27 (the Banach–Steinhaus or the resonance theorem; Yosida, 1965, Section 2.1)
Let $\{T_\lambda | \lambda \in \Lambda\}$ be a family of bounded linear operators mapping a Banach space $\mathscr{X}$ into a normed space $\mathscr{Y}$. If $\{||T_\lambda x|| \; \lambda \in \Lambda\}$ is bounded for every $x \in \mathscr{X}$ then $\{||T_\lambda|| \; \lambda \in \Lambda\}$ is bounded.**

A normed vector space is a Banach space if every fundamental sequence $\{x_n\}$ converges to a limit x_0 in this space: $||x_n - x_0|| \to 0$ for $n \to \infty$. $\mathscr{X}^*$ is the set of all linear continuous functionals on the Banach space $\mathscr{X}$, endowed with the strong topology unless otherwise stated. The strong topology on $\mathscr{X}^*$ is defined by the norm $||f||_{\mathscr{X}^*} := \sup\{|f(x)| \; ||x|| = 1\}$,

* We shall use the notation $x_n \rightharpoonup x_0$ for weak convergence.

** The operator norm is defined by the formula

$$||T|| = \sup\{||Tx|| \; ||x|| = 1\}.$$

Theorem 1.1.28 (Dunford and Schwartz, 1958, Theorem 5.3.15)
Let $\mathscr{X}$, $\mathscr{Y}$ be Banach spaces and $T: \mathscr{X} \to \mathscr{Y}$ linear. Then T is weakly continuous iff it is strongly continuous.

Theorem 1.1.29 (Banach; cf. Yosida, 1965, Section 2.5)
Let $\mathscr{X}$, $\mathscr{Y}$ be Banach spaces. If $T: \mathscr{X} \to \mathscr{Y}$ is surjective, linear and continuous then it is open (i.e. it maps open sets onto open sets).

Theorem 1.1.30 (lemma of Ehrling)
Let $\mathscr{X}_1, \mathscr{X}_2, \mathscr{X}_3$ be real or complex Banach spaces with the norms $||u||_1$, $||u||_2$, $||u||_3$, resp. Suppose that $\mathscr{X}_1 \subset \mathscr{X}_2 \subset \mathscr{X}_3$, the imbedding (identity map) $\iota_1: \mathscr{X}_1 \to \mathscr{X}_2$ is completely continuous* and the imbedding $\iota_2: \mathscr{X}_2 \to \mathscr{X}_3$ is continuous. Then for every $\varepsilon > 0$ there is a $c(\varepsilon) > 0$ such that for every $u \in \mathscr{X}_1$

$$||u||_2 \leqslant \varepsilon ||u||_1 + c(\varepsilon)||u||_3. \tag{1.1.21}$$

Proof
Suppose that there is an $\varepsilon > 0$ such that for every $c > 0$ the inequality (1.1.21) breaks down for some $u \in \mathscr{X}_1$. Hence $\forall\, n \in \boldsymbol{Z}_+\ \exists\, u_n \in \mathscr{X}_1$ such that $||u_n||_2 > \varepsilon ||u_n||_1 + n||u_n||_3$.

Let $v_n := ||u_n||_1{}^{-1}\, u_n$. Obviously

$$||v_n||_2 > \varepsilon + n||v_n||_3. \tag{1.1.22}$$

Since $||v_n||_1 \leqslant 1$ it follows that $\exists\, L > 0$ $||v_n||_2 < L$ and $L > \varepsilon + n||v_n||_3$. For $n \to \infty$ it follows that $||v_n||_3 \to 0$.

Since $\{||v_n||_1\}$ is bounded, $\{v_n\}$ is compact in $\mathscr{X}_2$ and a subsequence $\{v_k'\}$ of $\{v_n\}$ converges in $\mathscr{X}_2$ and $\mathscr{X}_3$: $v_k' \to v$. But we know that $v_k' \to 0$ in $\mathscr{X}_3$, hence $v = 0$. But $||v_k'||_2 > \varepsilon > 0$, which is impossible. □

Theorem 1.1.31
Let $\mathscr{V}$, $\mathscr{Y}$ be two Banach spaces such that $\mathscr{V}$ is a dense subset of $\mathscr{Y}$ and $\exists\, K > 0$

$$||v||_{\mathscr{Y}} \leqslant K||v||_{\mathscr{V}} \ \forall v \in \mathscr{V}$$

($||\cdot||_{\mathscr{Y}}$, $||\cdot||_{\mathscr{V}}$ denote the norms on $\mathscr{Y}$, $\mathscr{V}$, resp.).

If $\mathscr{V}$ is reflexive then $\mathscr{Y}^*$ is a dense subset of $\mathscr{V}^*$ and the embedding $\mathscr{Y}^* \to \mathscr{V}^*$ is continuous.

* A mapping $f: \mathscr{X} \to \mathscr{Y}$ is said to be **compact** if it maps bounded sets onto relatively compact sets. A mapping $f: \mathscr{X} \to \mathscr{Y}$ is said to be **completely continuous** if it is continuous and compact.

Proof
Let $f \in \mathscr{Y}^*$. The corresponding element of $\mathscr{V}^*$ is $g := f|\mathscr{V}$. Indeed, g is continuous on $\mathscr{V}$:

$$|\langle g|v\rangle| = |\langle f|v\rangle| \leqslant ||f||_{\mathscr{Y}^*}||v||_{\mathscr{Y}} \leqslant K||f||_{\mathscr{Y}^*}||v||_{\mathscr{V}}. \tag{1.1.23}$$

Conversely, f is the unique extension of g by continuity. Hence the mapping $f \mapsto g$ is injective. We shall write $g = f$ henceforth, so that $\mathscr{Y}^* \subset \mathscr{V}^*$ is an embedding. This embedding is continuous since $||f||_{\mathscr{V}^*} \leqslant K||f||_{\mathscr{Y}^*}$, by (1.1.23).

Suppose now that for some $v \in \mathscr{V} \subset \mathscr{Y}$

$$\langle f|v\rangle = 0 \quad \forall f \in \mathscr{Y}^*.$$

It follows that $v = 0$ in $\mathscr{Y}$, and hence in $\mathscr{V}$, by the Hahn-Banach theorem. Hence every functional $v \in \mathscr{V} = \mathscr{V}^{**}$ on $\mathscr{V}^*$ that vanishes on $\mathscr{Y}^* \subset \mathscr{V}^*$, vanishes on $\mathscr{V}^*$ as well. By the Hahn–Banach theorem $\mathscr{Y}^*$ is dense in $\mathscr{V}^*$.⋆ □

Suppose that $\mathscr{Y} = \mathscr{H}$ is a complex Hilbert space. We can identify $\mathscr{H}$ with $\mathscr{H}^*$ by the Riesz' theorem (Yosida, 1965): every continuous linear functional f on $\mathscr{H}$ has the form

$$f(v) = \langle w|v\rangle \quad \forall v \in \mathscr{H},$$

where the rhs denotes the scalar product of $v, w \in \mathscr{H}$. Hence the duality $\langle \cdot | \cdot \rangle$ between $\mathscr{V}$, $\mathscr{V}^*$ when restricted to $\mathscr{V} \times \mathscr{H}$ is identical with the scalar product on $\mathscr{H}$. For example, let $\mathscr{V} = W_o^{1,p}(\Omega)$, $p \geqslant 2$, $\mathscr{H} = \mathscr{L}^2(\Omega)$. Then the duality between $\mathscr{V} = W_o^{1,p}(\Omega)$ and $\mathscr{V}^* = W^{-1,p}(\Omega)$ is given by the integral

$$\langle f|v\rangle = \int_\Omega dL^n\, vf.$$

For $v \in \mathscr{C}_o^1(\Omega)$ this is the action of a distribution f on a function v. For v, $f \in \mathscr{H}$ it is the scalar product in $\mathscr{L}^2(\Omega)$. For general $f \in \mathscr{V}^*$, $v \in \mathscr{V}$ the right-hand side is not to be taken too literally.

Theorem 1.1.32 (the Lax–Milgram lemma)
Let X be a Hilbert space with the scalar product $\langle \cdot | \cdot \rangle$.

If $B(\cdot, \cdot)$ is a bilinear (or sesquilinear⋆⋆ in the case of a complex Hilbert space) form, which is bounded:

$$|B(x, y)| \leqslant M||x||\,||y||$$

and coercive:

⋆ Suppose that $\overline{\mathscr{Y}^*} \neq \mathscr{V}^*$ and $g \in \mathscr{V}^* \setminus \overline{\mathscr{Y}^*}$, $g \neq 0$. Then there is a functional w on $\mathscr{V}^*$ such that $w|\mathscr{Y}^* = 0$ and $\langle w, g\rangle \neq 0$, by the Hahn–Banach theorem.

⋆⋆ A form $(x, y) \to B(x, y)$ is said to be bilinear if $B(tx, y) = B(x, ty) = tB(x, y)$ and sesquilinear if $B(tx, y) = tB(x, y)$, $B(x, ty) = \bar{t}B(x, y)$, $t \in \mathbf{C}$.

$$B(x, x) \geqslant \gamma \|x\|^2, \quad \gamma > 0,$$

then there is a bounded linear operator $S\colon \mathscr{X} \to \mathscr{X}$ with a bounded inverse $\overset{-1}{S}$, such that $\langle x|y\rangle \equiv B(x, Sy)$, $\|S\| \leqslant \gamma^{-1}$, $\|\overset{-1}{S}\| \leqslant M$.

Proof

Let us consider the equation $\langle x|y\rangle = B(x, y^*)$ $\forall\, x \in \mathscr{X}$, for a given $y \in \mathscr{X}$ and unknown $y^* \in \mathscr{X}$. Let $\mathscr{D}$ be the set of $y \in \mathscr{X}$ for which a solution y^* exists. Clearly for $y = 0$ we have a solution $y^* = 0$. Hence $\mathscr{D} \neq \emptyset$.

Suppose that y_1, y_2 are two different solutions for a given $y \in \mathscr{D}$. Then $B(x, y_1 - y_2) = 0$ for all $x \in \mathscr{X}$, in particular for $x = y_1 - y_2$. By coercivity $y_1 = y_2$. Hence there is a linear operator S, defined on $\mathscr{D}$, with $y^* = Sy$, so that

$$\gamma \|Sy\|^2 \leqslant B(Sy, Sy) = \langle Sy|y\rangle \leqslant \|Sy\|\,\|y\|$$

and $\|Sy\| \leqslant \gamma^{-1}\|y\|$.

We shall prove that $\mathscr{D} = \mathscr{X}$.

Firstly, $\mathscr{D}$ is a closed linear manifold in $\mathscr{X}$. Indeed, if $y_n \in \mathscr{D}$, $y_n \to y_0$, then Sy_n converge to some $w \in \mathscr{X}$, $B(x, Sy_n) = \langle x|y_n\rangle \to B(x, w) = \langle x|y_o\rangle$, so that $Sy_o = w$.

Suppose that $\mathscr{D} \neq .\mathscr{X}$ In this case $\exists\, w \in \mathscr{D}^\perp$, $w \neq 0$. Let $F(z) :\equiv B(z, w)$. F is continuous and linear on $\mathscr{X}$, hence by Riesz' theorem $F(z) = B(z, w) = \langle z|r\rangle$ for some $r \in \mathscr{X}$ and all $z \in \mathscr{X}$. Therefore $r \in \mathscr{D}$, $w \in \mathscr{D}^\perp$, $Sr = w$, $\gamma\|w\|^2 \leqslant B(w, w) = \langle w|r\rangle = 0$ and $w = 0$ in contradiction with the hypothesis. Thus $\mathscr{D} = \mathscr{X}$.

We proved that $Sy = 0$ implies $y = 0$. Hence S is invertible and by Riesz' theorem $\mathscr{D}(\overset{-1}{S}) = X$. Also $|\langle z|\overset{-1}{S}x\rangle| = |B(z, x)| \leqslant M\|z\|\,\|x\|$, $\|\overset{-1}{S}\| \leqslant M$. □

The Lax–Milgram lemma can be significantly sharpened by weakening the hypotheses (Babuška and Aziz, 1972, Theorem 5.2.1).

The Lax–Milgram lemma is an important tool in proving existence and uniqueness of solutions to elliptic systems. In this connection we shall consider the abstract equation

$$B(u, v) = f(u) \quad \forall u \in \mathscr{X}$$

for a given $f \in \mathscr{X}^*$. By Riesz' theorem $\exists !\, r \in \mathscr{X}$ such that $f(u) = \langle r|u\rangle$ $\forall\, u \in X$. By the Lax–Milgram lemma $v = Sr$,

$$\|v\|_{\mathscr{X}} \leqslant \gamma\|r\|_{\mathscr{X}} = \gamma\|f\|_{\mathscr{X}^*}.$$

Theorem 1.1.33 (Dunford and Schwartz, 1958, Theorems 6.5.6 and 6.5.2) If the linear operator $T\colon \mathscr{X} \to \mathscr{Y}$ is completely continuous then $x_n \rightharpoonup x_o$ implies $Tx_n \to Tx_o$.

Theorem 1.1.34 (Schauder; Dunford and Schwartz, 1958, Theorem 6.5.2) A continuous linear operator $T\colon \mathscr{X} \to \mathscr{Y}$ is completely continuous iff its adjoint* $T^\dagger\colon \mathscr{Y}^* \to \mathscr{X}^*$ is completely continuous.

Definition 1.1.35

Let $T\colon \mathscr{X} \to \mathscr{Y}$ be a linear and continuous operator mapping a complex Banach space $\mathscr{X}$ into a complex Banach space $\mathscr{Y}$. A complex number λ belongs to sp$T$ (the **spectrum** of T) if one of the following conditions is satisfied:

(i) $(T-\lambda E)^{-1}$ does not exist;

(ii) the domain of definition of $(T-\lambda E)^{-1}$ is not dense in $\mathscr{X}$;

(iii) the operator $(T-\lambda E)^{-1}$ is unbounded.

For precise definitions of inverse operators see Yosida (1965); for resolvents, see Dunford and Schwartz (1958).

The spectrum of a linear continuous operator turns out to be identical with the spectrum of its adjoint (Dunford and Schwartz, 1958).

The points λ of sp$T$ satisfying the condition (i) above are said to belong to the **point spectrum** of T.

Theorem 1.1.36 (Riesz–Schauder theory; Dunford and Schwartz, 1958) Let $\mathscr{X}, \mathscr{Y}$ be two real or complex Banach spaces and $T\colon \mathscr{X} \to \mathscr{Y}$ linear and continuous completely.

(i) The spectrum of $T^\dagger$ is identical with the spectrum of T;

(ii) the spectrum of T is at most countable and has no accumulation points except perhaps at 0 (in this case $0 \in \mathrm{sp}T$);

(iii) for each $\lambda \in \mathrm{sp}T$ there is a positive integer k such that $(T-\lambda E)^k x = 0$, $(T-\lambda E)^{k-1} x \neq 0$ for x belonging to a finite dimensional subspace of $\mathscr{X}$.

The Riesz–Schauder theory for complex Hilbert spaces is somewhat different. In this case the adjoint operator is defined by the formula $\langle T^\dagger x, y\rangle = \langle x, Ty\rangle$, where $\langle\cdot,\cdot\rangle$ denotes the scalar product in the complex Hilbert space $\mathscr{X}$ and we assume that $\mathscr{Y} = \mathscr{X}$. The correspondence $T \mapsto T^\dagger$ is antilinear ($(\lambda T)^\dagger = \bar{\lambda} T^\dagger$) in this case. Hence sp$T^\dagger$ = complex conjugate of sp$T$.

Definition 1.1.37

A continuous linear operator $T\colon \mathscr{X} \to \mathscr{X}$ on a complex Hilbert space $\mathscr{X}$ is said to be **normal** if $T^\dagger T = TT^\dagger$.

* The adjoint operator $T^\dagger$ of a continuous linear operator $T\colon \mathscr{X} \to \mathscr{Y}$ is defined by the formula $\langle T^\dagger x|y\rangle = \langle x|Ty\rangle$ for all $y \in \mathscr{X}$, $x \in \mathscr{Y}^*$.

Definition 1.1.38

A complex number λ is an **eigenvalue** of a linear operator $T\colon \mathscr{X} \to \mathscr{X}$ if there is a non-zero vector $x \in \mathscr{X}$ such that $Tx = \lambda x$. The vectors satisfying the above equation for a fixed eigenvalue λ form a closed linear manifold in $\mathscr{X}$, called the **eigenspace** of T corresponding to λ.

Theorem 1.1.39 (Dunford and Schwartz, 1958, Theorem 10.4.3)

Let T be a completely continuous normal linear operator on a complex Hilbert space $\mathscr{X}$. Then

(i) $\mathrm{sp}T^\dagger$ is complex conjugate of $\mathrm{sp}T$;

(ii) $\mathrm{sp}T$ is a countable subset of $\boldsymbol{C}$ without accumulation points except perhaps at 0;

(iii) every non-zero $\lambda \in \mathrm{sp}T$ is an eigenvalue with a finite-dimensional eigenspace;

(iv) all the eigenvectors of T form an orthogonal basis of $\mathscr{X}$;

(v) the equation $Tu - \lambda u = f$ has a unique solution if $\lambda \overline{\in} \mathrm{sp}T$; if $\lambda \in \mathrm{sp}T$ then this equation has a solution (defined modulo the eigenspace of λ) provided $\langle f, x\rangle = 0 \;\forall\, x \in \ker(T^\dagger - \bar{\lambda}E)$.

1.1.7 Sobolev Spaces. The Gårding Lemma

Let $\mathbf{u}\colon \Omega \to \boldsymbol{R}^m$, $\Omega \subset \boldsymbol{R}^n$. We shall define the norms

$$||\mathbf{u}||_p := \Big(\int_\Omega dL^n |\mathbf{u}|^p\Big)^{1/p}, \qquad ||\mathbf{u}||_\infty := \operatorname*{ess\,sup}_\Omega |\mathbf{u}|,$$

where $1 \leqslant p < \infty$ and

$$\operatorname*{ess\,sup}_\Omega f := \inf\{a \in \boldsymbol{R} |\; L^n(\{x \in \Omega |\; f(x) > a\}) > 0\}$$

for any real-valued and measurable $f\colon \Omega \to \boldsymbol{R}$. The set $\mathscr{L}^p(\Omega)$, $\mathscr{L}^\infty(\Omega)$ of measurable functions $\mathbf{u}\colon \Omega \to \boldsymbol{R}^m$ with $||\mathbf{u}||_p < \infty$, $||\mathbf{u}||_\infty < \infty$ resp. is a Banach space with the norm $||\cdot||_p$, $||\cdot||_\infty$. $\mathscr{L}^2(\Omega)$ is additionally a real Hilbert space (or a complex Hilbert space) with the scalar product $\int_\Omega dL^n\, \mathbf{u}\cdot\mathbf{v}$ (or $\int_\Omega dL^n\, \mathbf{u}\cdot\bar{\mathbf{v}}$). $\mathscr{C}_0^\infty(\Omega)$ is dense in $\mathscr{L}^p(\Omega)$.

The **Minkowski inequality** is the triangle inequality for $||\cdot||_p$:

$$||\mathbf{u}+\mathbf{v}||_p \leqslant ||\mathbf{u}||_p + ||\mathbf{v}||_p \;\; \forall \mathbf{u}, \mathbf{v} \in \mathscr{L}^p(\Omega),\; 1 \leqslant p \leqslant \infty.$$

The **Hölder inequality** asserts that

$$\Big|\int_\Omega dL^n \mathbf{u}\cdot\mathbf{v}\Big| \leqslant ||\mathbf{u}||_p ||\mathbf{v}||_q$$

$$\forall \mathbf{u} \in \mathscr{L}^p(\Omega),\; \mathbf{v} \in \mathscr{L}^q(\Omega),\; 1 < p < \infty,\; q^{-1}+p^{-1} = 1.$$

Consequently $\mathscr{L}^q(\Omega)$ is the conjugate space for $\mathscr{L}^p(\Omega)$, with $\langle \mathbf{u}, \mathbf{v} \rangle = \int_\Omega dL^n\, \mathbf{u} \cdot \mathbf{v}$. $\mathscr{L}^\infty$ is conjugate for $\mathscr{L}^1$, but $(\mathscr{L}^\infty)^* \neq \mathscr{L}^1$.

For a bounded Ω we have continuous embeddings $\mathscr{L}^\infty \subset \mathscr{L}^p \subset \mathscr{L}^{p'} \subset \mathscr{L}^1$ for $p \geqslant p' \geqslant 1$.

Definition 1.1.40

Let Ω be a bounded domain in $\boldsymbol{R}^n$. Then $W^{m,p}(\Omega)$ is the set of locally integrable functions $\mathbf{u}\colon\ \Omega \to \boldsymbol{R}^r$ such that $\mathbf{u}$ and its partial derivatives $D^\alpha \mathbf{u}$ in the sense of distributions belong to $\mathscr{L}^p(\Omega)$ for all multiindices α such that $|\alpha| :\equiv \sum_{i=1}^n \alpha_i \leqslant m$.

According to an alternative definition $W^{m,p}(\Omega)$ is the completion of $\mathscr{C}^m(\Omega)$ with respect to the norm

$$||\mathbf{u}||_{m,p} := \Big\{ \sum_{0 \leqslant |\alpha| \leqslant m} ||D^\alpha \mathbf{u}||_p{}^p \Big\}^{1/p}.$$

Here $||\cdot||_p$ denotes the norm on $\mathscr{L}^p(\Omega)$,

$$||\mathbf{u}||_p{}^p := \int_\Omega dL^n |\mathbf{u}|^p.$$

For the equivalence of these definitions of $W^{m,p}(\Omega)$ see Adams (1975), Theorem 3.16.

$W_o^{m,p}(\Omega)$ is defined as the completion of $\mathscr{C}_o^\infty(\Omega)$ with respect to the norm $||\cdot||_{m,p}$. The identity mapping $W_o^{m,p}(\Omega) \to W^{m,p}(\Omega)$ is an isometrical embedding.

Let $f\colon \bar{\mathscr{U}} \to \boldsymbol{R}^s$ be a Lipschitz continuous function on a closed domain $\bar{\mathscr{U}}$, $|f(x)-f(y)| \leqslant K|x-y|$ for $|x-y|$ sufficiently small. By $\operatorname{Lip} f$ we shall denote the greatest lower bound of real numbers K satisfying the above inequality.

Definition 1.1.41

A domain Ω in $\boldsymbol{R}^n$ has the **cone property** if there is a finite cone $C := \mathscr{K}(0,\ R) \cap \cap \{tx|\ x \in \mathscr{K}(x_o, r),\ t > 0\}$, $r < |x_o|$, such that every point $x \in \Omega$ is the vertex of a cone $C(x)$ contained in Ω and congruent to C.

Theorem 1.1.42 (Adams, 1975, Theorem 4.22)

For any domain Ω and a bounded domain $\mathscr{U}, \bar{\mathscr{U}} \subset \Omega$, there is a domain $\mathscr{V}$ with the cone property and such that $\mathscr{U} \subset \mathscr{V}, \bar{\mathscr{V}} \subset \Omega$.

Proof

Since $\overline{\mathscr{U}}$ is compact and $\partial\Omega$ is closed, there is a $\delta > 0$ such that the distance $\operatorname{dist}(\overline{\mathscr{U}}, \partial\Omega) > \delta$. Set $\mathscr{V} := \{y \in \boldsymbol{R}^n |\ \exists\, x \in \mathscr{U}\ |y-x| < \delta\}$. □

Definition 1.1.43

The domain $\Omega \subset \boldsymbol{R}^n$ has the **strong local Lipschitz property** provided there exist positive numbers δ, M, a positive integer ν, a locally finite open cover $\{\mathscr{U}_i |\ i \in \boldsymbol{Z}_+\}$ and an associated sequence $\{f_i |\ i \in \boldsymbol{Z}_+\}$ of Lipschitz continuous functions of $n-1$ real variables such that the following conditions are satisfied:

(i) the intersection of $\nu+1$ different sets $\mathscr{U}_j$ is empty;

(ii) for every pair of points $x, y \in \Omega_\delta := \{z \in \Omega |\ \operatorname{dist}(z, \partial\Omega) < \delta\}$ such that $|x-y| < \delta$ there is a $j \in \boldsymbol{Z}_+$ such that $x, y \in \{z \in \mathscr{U}_j |\ \operatorname{dist}(z, \partial\mathscr{U}_j) > \delta\}$;

(iii) $\operatorname{Lip} f_j \leqslant M\ \forall j$;

(iv) there is a Cartesian coordinate system on $\mathscr{U}_j$, $(\zeta_{j1}, \ldots, \zeta_{jn})$ say, such that $\Omega \cap \mathscr{U}_j$ is given by the inequality $\zeta_{jn} > f_j(\zeta_{j1}, \ldots, \zeta_{j,n-1})$.

Definition 1.1.44

A domain Ω has the **uniform $\mathscr{C}^m_-$ ($\mathscr{C}^{m,\mu}_-$) regularity property** if there exists a locally finite open cover $\{\mathscr{U}_j |\ j \in \boldsymbol{Z}_+\}$ of $\partial\Omega$ and a corresponding sequence of $\mathscr{C}^m$- ($\mathscr{C}^{m,\mu}$-) coordinate transformations $\varphi_j\colon \mathscr{U}_j \to \mathscr{K}(0, 1)$ (onto $\mathscr{K}(0, 1)$) such that

(i) $\varphi_j, \overset{-1}{\varphi_j}$ are $\mathscr{C}^m$-smooth on $\overline{\mathscr{U}}_j, \overline{\mathscr{K}(0, 1)}$ resp. and the derivatives of the component functions $\varphi_{j,k}, \overset{-1}{\varphi}_{j,k}$ up to order m inclusively are pointwise bounded by a constant $M > 0$ (or $\varphi_{j,k}, \overset{-1}{\varphi}_{j,k}$ are $\mathscr{C}^{m,\mu}$-smooth on $\overline{\mathscr{U}}_j, \overline{\mathscr{K}(0, 1)}$ resp., their derivatives up to order m are bounded by a constant M and the derivatives of order m satisfy the inequalities

$$\frac{|D^\alpha \varphi_{j,k}(x) - D^\alpha \varphi_{j,k}(y)|}{|x-y|^\mu} \leqslant M, \qquad \frac{|D^\alpha \overset{-1}{\varphi}_{j,k}(x) - D^\alpha \overset{-1}{\varphi}_{j,k}(y)|}{|x-y|^\mu} \leqslant M \tag{1.1.24}$$

for $x, y \in \overline{\mathscr{U}}_j$, $x, y \in \overline{\mathscr{K}(0, 1)}$ resp.);

(ii) $\exists\ \delta > 0\ \bigcup_{j=1}^{\infty} \overset{-1}{\varphi_j}\left(\mathscr{K}\left(0, \frac{1}{2}\right)\right) \supset \Omega_\delta$;

(iii) $\exists\ \nu \in \boldsymbol{Z}_+$ such that the intersection of $\nu+1$ different sets $\mathscr{U}_j$ is empty;

(iv) $\forall j$, $\varphi_j(\mathscr{U}_j \cap \Omega) = \{y \in \mathscr{K}(0, 1) | y_n > 0\}$.

In the above definitions

$$\Omega_\delta := \{y \in \Omega |\ \operatorname{dist}(y, \partial\Omega) < \delta\}.$$

The transformation $\varphi_j\colon \mathscr{U}_j \to \mathscr{K}(0, 1)$ is said to be $\mathscr{C}^{m,\mu}$-smooth if $\varphi_{j,k}$, $\overset{-1}{\varphi}_{j,k}$ are $\mathscr{C}^m$-smooth and their derivatives of order m satisfy equations (1.1.24) (for M depending on j, k).

Uniform $\mathscr{C}^m$-regularity property for $m \geqslant 1$ entails strong local Lipschitz property and the latter entails the cone property. A bounded domain satisfying the cone property is the join of a finite collection of domains satisfying the strong local Lipschitz property (Adams, 1975).

Theorem 1.1.45 (Rellich–Kondrachov; Adams, 1975, Theorem 6.2; Nečas, 1966, Theorem 2.6.1)
Let Ω be a domain in $\boldsymbol{R}^n$.

(1) If Ω is bounded and has the cone property, $j \geqslant 0$, m, $p \geqslant 1$, $mp < n$, $1 \leqslant s < np(n-mp)^{-1}$★ or $mp \geqslant n$, $1 \leqslant s$ then the embedding $W^{j+m,p}(\Omega) \to W^{j,s}(\Omega)$ is compact.

(2) If Ω is bounded and has the strong local Lipschitz property and $j \geqslant 0$, m, $p \geqslant 1$, $mp > n$ then the embedding $W^{j+m,p}(\Omega) \to \mathscr{C}^j(\Omega)$ is compact; if moreover $(m-1)p \leqslant n$, then the embedding $W^{j+m,p}(\Omega) \to \mathscr{C}^{j,\mu}(\Omega)$ is compact for $0 < \mu < m-n/p$.

(3) If Ω has the cone property, $j \geqslant 0$, m, $p \geqslant 1$, $mp > n$ or $m = n$, $p = 1$, then the embeddings $W^{j+m,p}(\Omega) \to \mathscr{C}^j_b(\Omega)$, $W^{j+m,p}(\Omega) \to W^{j,s}(\Omega)$ are compact for $1 \leqslant s \leqslant \infty$.

For an arbitrary domain $\Omega \subset \boldsymbol{R}^n$ the above embeddings are compact with W replaced by W_o.

□

In the above theorem $\mathscr{C}^j_b(\Omega)$ denotes the space of $\mathscr{C}^j$-smooth bounded functions on Ω with bounded derivatives up to order j with the norm

$$\|u\|_{\mathscr{C}^j_b} = \sup\{|D^\alpha \mathbf{u}(x)| \mid x \in \Omega,\ 0 \leqslant |\alpha| \leqslant j\}.$$

For bounded Ω the subscript "b" on $\mathscr{C}^j_b$ can be dropped. $\mathscr{C}^{j,\mu}_b$ is the Banach space of $\mathscr{C}^j$-smooth bounded functions with bounded derivatives up to order j and Hölder-continuous derivatives of order j with the Hölder exponent μ, endowed with the norm

$$\|u\|_{\mathscr{C}^{j,\mu}} := \sup\{|D^\alpha u(x)| \mid x \in \Omega, 0 \leqslant |\alpha| \leqslant j\} + \sup\left\{\frac{|D^\alpha u(x) - D^\alpha u(y)|}{|x-y|^\mu} \,\middle|\, x, y \in \Omega, x \neq y, |\alpha| = j\right\}.$$

★ If $1 \leqslant s \leqslant np(n-mp)^{-1}$ then the embedding is continuous.

Definition 1.1.46
$W^{m,p}(\Omega)$ is the conjugate Banach space of $W_o^{-m,q}(\Omega)$ for $m < 0$, $q^{-1} = 1-p^{-1}$.

Spaces $W^{m,p}(\Omega)$ with non-integer m can be defined by interpolation in such a way that the embedding theorem 1.1.45 remains valid. (Adams (1975) for $p \neq 2$, Lions and Magenes (1968) for $p = 2$.)

Theorem 1.1.47 (Trace theorem)
If the domain $\Omega \subset \boldsymbol{R}^n$ has the uniform $\mathscr{C}^m$-regularity property and $\partial\Omega$ is bounded then the mapping

$$\gamma_0 \colon \mathscr{C}^m(\bar{\Omega}) \to \prod_{k=0}^{m-1} W^{m-1/p-k,\,p}(\partial\Omega);^{\star}$$

$$u \mapsto \left(u|\partial\Omega, \frac{\partial u}{\partial \nu}, \ldots, \frac{\partial^{m-1} u}{\partial \nu^{m-1}}\right)^{\star\star}$$

extends by continuity to $\gamma \colon W^{m,p}(\Omega) \to \prod_{k=0}^{m-1} W^{m-1/p-k,p}(\partial\Omega)$. Moreover $\ker\gamma$ $= W_0^{m,p}(\Omega)$, $W^{m,p}(\Omega)/W_0^{m,p}(\Omega)^{\star\star\star}$ is homeomorphic to $\prod_{k=0}^{m-1} W^{m-1/p-k,p}(\partial\Omega)$, and $\|\gamma u\| \leqslant K\|u\|_{m,p}$ for some fixed $K > 0$.

Remarks
$\partial\Omega$ is a manifold with maps $\varphi_j \colon \mathscr{U}_j \to \mathscr{K}(0, 1)$. The $W^{m,p}$-norm on $\partial\Omega$ has to be defined in terms of these maps and a partition of unity subordinate to the cover $\{\mathscr{U}_j\}$. Although the norm of $W^{m,p}(\partial\Omega)$ depends on the choice of these maps the topology of $W^{m,p}(\partial\Omega)$ does not (cf. analogous considerations in Section 1.2).

In particular $\gamma_0 \colon u \mapsto u|\partial\Omega$ extends to a surjective mapping $W^{m,p}(\Omega) \to W^{m-1/p,p}(\partial\Omega)$.

The theorem remains valid if $\mathscr{C}^m$-regularity is replaced by $\mathscr{C}^{m-1,1}$-regularity (Nečas, 1966).

$\star$ Π denotes the Cartesian product (space of m-tuples).

$\star\star$ $\partial/\partial\nu$ denotes the normal derivative, $\partial^k/\partial\nu^k = \sum_{|\alpha|=k} \binom{k}{\alpha} \nu^\alpha D^\alpha u$ with $\nu^\alpha := \nu_1^{\alpha_1} \ldots \nu_n^{\alpha_n}$, $\binom{k}{\alpha} := \dfrac{k!}{\alpha_1! \ldots \alpha_n!}$, $\nu =$ the exterior unit normal on $\partial\Omega$.

$\star\star\star$ Denotes elements of $W^{m,p}$ taken modulo $W_o^{m,p}$, with the norm

$$\inf\{\|\mathbf{u}+\mathbf{v}\|_{m,p} |\ v \in W_o^{m,p}(\Omega)\}.$$

For domains Ω with the strong local Lipschitz property, $1 \leqslant p < n$, $s_o^{-1} := p^{-1} - (n-1)^{-1}q^{-1}$, $q := (p-1)/p$ there is a continuous trace γ_1: $W^{1,p}(\Omega) \to \mathscr{L}^{s_o}(\partial\Omega)$ (not surjective). For $s > s_o$, $1 < p < n$ the trace operator $W^{1,p}(\Omega) \to \mathscr{L}^s(\partial\Omega)$ is completely continuous (Nečas, 1966, Theorem 2.6.2).

The following continuous embeddings are also worth noting. For a domain $\Omega \subset R^n$ satisfying the cone property, $s > 0$, $1 < p \leqslant n$, and either $n > sp$, $p \leqslant r \leqslant np/(n-sp)$ or $n = sp$, $p \leqslant r < \infty$, $W^{s,p}(\Omega) \subset \mathscr{L}^r(\Omega)$. For $n < (s-j)p, j \geqslant 0, s > 0, 1 < p \leqslant n$, $W^{s,p}(\Omega) \subset \mathscr{C}_b^j(\Omega)$ (Adams, 1975, Theorem 7.57).

Traces can be calculated by pointwise limits. More precisely, we have

Theorem 1.1.48 (Mizohata, 1969)
For $\mathbf{u} \in W^{1,2}(\boldsymbol{R}^n_+)$ $\gamma\mathbf{u}(x_1, \ldots, x_{n-1}) = \lim_{x\to 0} \mathbf{u}(x_1, \ldots, x_{n-1}, x)$ and the limit exists for L^{n-1}-almost all $(x_1, \ldots, x_{n-1})$.

For $\mathbf{u} \in W^{1,2}(\Omega)$, with Ω bounded and $\mathscr{C}^2$-regular, the limit $\lim_{y\to x} \mathbf{u}(y)$, $x \in \partial\Omega$, along the normal to $\partial\Omega$, exists for H^{n-1}-a.e. $x \in \partial\Omega$, and coincides with $\gamma\mathbf{u}$.

Theorem 1.1.49 (Friedrichs' lemma; Morrey, 1966, Theorem 3.6.4)
If Ω has the uniform $\mathscr{C}^{0,1}$-property, $\Gamma_1 \subset \partial\Omega$, $H^{n-1}(\Gamma_1) > 0$ then

$$||\mathbf{u}||_{1,p}{}^p \leqslant K\Big[||\nabla\mathbf{u}||_p{}^p + \Big(\int_{\Gamma_1} (dH^{n-1}|\gamma\mathbf{u}|\Big)^p\Big] \quad \forall \mathbf{u} \in W^{1,p}(\Omega).$$

Theorem 1.1.50 (Poincaré-type lemmas; Morrey, 1966, Theorems 3.6.4–5)
For Ω having the uniform $\mathscr{C}^{0,1}$-regularity property and $\mathscr{U} \subset \Omega$ such that $L^n(\mathscr{U}) > 0$ $\exists$ $K > 0$ such that

(1) $||\mathbf{u}||_{1,p}{}^p \leqslant K\{||\nabla\mathbf{u}||_p{}^p + (\int_{\mathscr{U}} dL^n|\mathbf{u}|)^p\}$ $\forall$ $\mathbf{u} \in W^{1,p}(\Omega)$, $\mathscr{U} \subset \Omega$ such that $L^n(\mathscr{U}) > 0$;

(2) $||\mathbf{u}||_{1,p} \leqslant K||\nabla\mathbf{u}||_p$ if $\int_{\Omega} dL^n\, \mathbf{u} = \mathbf{0}$ or if $\mathbf{u}(x) = \mathbf{0}$ for $x \in \mathscr{U}$, $\mathscr{U} \subset \Omega$, $L^n(\mathscr{U}) > 0$.

Definition 1.1.51
Let Ω, $\mathscr{U}$ be two open subsets of $\boldsymbol{R}^n$. A bijective mapping h: $\Omega \mapsto \mathscr{U}$ is said to be **bi-Lipschitz** if h and $\overset{-1}{h}$ are (coordinatewise) Lipschitz continuous on $\bar{\Omega}$, $\bar{\mathscr{U}}$, resp.

Theorem 1.1.52 (Nečas, 1966, Lemmas 2.3.1–2)
Let Ω, $\mathscr{U}$ be bounded open subsets of $\boldsymbol{R}^n$, $p \geqslant 1$ and suppose that $h\colon \Omega \to \mathscr{U}$ is bi-Lipschitz. For every $\mathbf{u} \in W^{1,p}(\mathscr{U})$ (or $\mathbf{u} \in \mathscr{L}^p(\mathscr{U})$) $\mathbf{v} := \mathbf{u} \circ h \in W^{1,p}(\Omega)$ ($\mathbf{v} \in \mathscr{L}^p(\Omega)$, resp.) and there is a constant $K > 0$ independent of $\mathbf{u}$ such that $\|\mathbf{v}\|_{1,p,\Omega} < K\|\mathbf{u}\|_{1,p,\mathscr{U}}$ (or $\|\mathbf{v}\|_{p,\Omega} < K\|\mathbf{u}\|_{p,\mathscr{U}}$).

An analogous theorem for $h \in \mathscr{C}_b^m(\bar{\Omega})$, $\overset{-1}{h} \in \mathscr{C}_b^m(\bar{\mathscr{U}})$, $\mathbf{u}$, $\mathbf{v} \in W^{m,p}$ is proved in Chapter 3 of Adams (1975). It allows a definition of Sobolev spaces on manifolds (Section 1.2). Since all the notions appearing in Theorem 1.1.52 are invariant with respect to smooth coordinate transformations the theorem remains valid for arbitrary n-dimensional manifolds Ω, $\mathscr{U}$. For $h, \overset{-1}{h} \in \mathscr{C}_b^{m-1}(\bar{\Omega}) \cap \cap \mathscr{C}^{m-1,1}(\Omega)$ see Morrey (1966), Theorem 3.1.7.

Theorem 1.1.53 (Gårding's lemma)*
Let Ω be a bounded domain in $\boldsymbol{R}^3$, $u^k{}_{,\mu} := \dfrac{\partial u^k}{\partial x^\mu}$,

$$B(\mathbf{u}, \mathbf{v}) := \int_\Omega dL^3 A_k{}^\mu{}_l{}^\nu(\mathbf{x}) u^k{}_{,\mu} \overline{v^l}{}_{,\nu} \quad \text{for } \mathbf{u}, \mathbf{v} \in W^{1,2}(\Omega) \tag{1.1.25}$$

with $A_k{}^\mu{}_l{}^\nu(\cdot)$ continous in $\bar{\Omega}$, satisfying the inequality SE:

$$\operatorname{Re} A_k{}^\mu{}_l{}^\nu(x) p_\mu p_\nu a^k \overline{a^l} \geqslant M \mathbf{p}^2 |\mathbf{a}|^2$$

$\forall\, \mathbf{a} \in \boldsymbol{C}^3$, $\mathbf{p} \in \boldsymbol{R}^3$, $\mathbf{x} \in \Omega$ for some $M > 0$. Here $|\mathbf{a}|^2 = \sum_k |a^k|^2$. Then

$$\operatorname{Re} B(\mathbf{u}, \mathbf{u}) \geqslant K\|\mathbf{u}\|_{1,2}^2 - M'\|\mathbf{u}\|_2{}^2 \quad \forall \mathbf{u} \in W_o^{1,p}(\Omega)$$

with $K > 0$, $M' \geqslant 0$ independent of $\mathbf{u}$.

Proof
Let $\mathbf{u} \in W_o^{1,p}(\Omega)$.

Case 1. $A_k{}^\mu{}_l{}^\nu = \text{const.}$

Since Ω is bounded it is contained in a cube $\mathscr{Q}$: $|x^i| < L$. Expanding $\mathbf{u}\colon \mathscr{Q} \to \boldsymbol{R}^3$ (equal to 0 outside Ω) in a Fourier series

$$\mathbf{u}(\mathbf{x}) = (2\pi)^{-3/2} \sum_{p \in Z^3} \exp(i\mathbf{p} \cdot \mathbf{x}) \mathbf{c}(\mathbf{p}),$$

substituting into (1.1.25) and applying the Parseval formula we have

$$\begin{aligned} \operatorname{Re} B(\mathbf{u}, \mathbf{u}) &= \operatorname{Re} \sum_{p \in Z^3} A_k{}^\mu{}_l{}^\nu p_\mu p_\nu c^k(p) \overline{c^l(p)} \geqslant M \sum_{p \in Z^3} \mathbf{p}^2 |\mathbf{c}(\mathbf{p})|^2 \\ &= M\|\nabla \mathbf{u}\|_2{}^2. \end{aligned}$$

By Theorem 1.1.49 $\operatorname{Re} B(\mathbf{u}, \mathbf{u}) \geqslant K\|\mathbf{u}\|_{1,2}^2$.

* This is but a special case of Gårding's lemma. For more general versions, see Morrey (1966), Theorem 6.5.1, Fichera (1972), Nečas (1966), Theorem 1.5.2.

Case 2. $A_k{}^{\mu}{}_l{}^{\nu}$ is a continuous function of $\mathbf{x} \in \bar{\Omega}$, the support of $\mathbf{u}$, $\operatorname{supp}\mathbf{u} \subset \mathscr{K}(\mathbf{x}_o, r)$, r is sufficiently small.

Let M_o be the value of the constant K obtained above for $A_k{}^{\mu}{}_l{}^{\nu} := A_k{}^{\mu}{}_l{}^{\nu}(x_o)$. We have then

$$M_o\|\nabla\mathbf{u}\|_2{}^2 \leqslant \operatorname{Re}\int_{\Omega} dL^3 A_k{}^{\mu}{}_l{}^{\nu}(\mathbf{x}_o)u^k{}_{,\mu}\overline{u^l}_{,\nu} = \operatorname{Re}B(\mathbf{u},\mathbf{u})$$
$$+\operatorname{Re}\int_{\Omega} dL^3[A_k{}^{\mu}{}_l{}^{\nu}(\mathbf{x}_o)-A_k{}^{\mu}{}_l{}^{\nu}(\mathbf{x})]u^k{}_{,\mu}\overline{u^l}_{,\nu}.$$

Since $A_k{}^{\mu}{}_l{}^{\nu}(\mathbf{x})$ is continuous on a compact set $\bar{\Omega}$, we can choose r in such a way that $|A_k{}^{\mu}{}_l{}^{\nu}(\mathbf{x})-A_k{}^{\mu}{}_l{}^{\nu}(\mathbf{x}_o)| < \delta$ for $|\mathbf{x}-\mathbf{x}_o| < r$, with δ arbitrarily small and r independent of x_o. It follows that

$$\operatorname{Re}B(\mathbf{u},\mathbf{u}) \geqslant (M_o-9\delta)\|\nabla\mathbf{u}\|_2{}^2 = M\|\nabla\mathbf{u}\|_2{}^2 \geqslant K\|\mathbf{u}\|_{1,2}{}^2, \quad K>0,$$

provided $M_o > 9\delta$. We have used the inequality $2\operatorname{Re}(a\bar{b}) \leqslant |a|^2+|b|^2$, which implies that $\sum_{k,l,\mu,\nu} \operatorname{Re} u^k{}_{,\mu}\overline{u^l}_{,\nu} \leqslant 9\sum_{k,\mu} |u^k{}_{,\mu}|^2$

Case 3. $A_k{}^{\mu}{}_l{}^{\nu}$ continuous on $\bar{\Omega}$, $\mathbf{u} \in W_o^{1,2}(\Omega)$.

Let us fix $\delta > 0$ and choose r as before. Let $\{\varphi_s{}^2|\ s \leqslant q\}$ be a partition of unity subordinate to a covering $\{\mathscr{K}(\mathbf{x}_s, r)|\ s = 1, \ldots, q\}$ of $\bar{\Omega}$, $\sum_{s=1}^{q} \varphi_s{}^2 = 1$. We shall set $\mathbf{u}_s := \varphi_s\mathbf{u}$.

$$\operatorname{Re}B(\mathbf{u},\mathbf{u}) = \sum_s\int_{\Omega} dL^3\varphi_s{}^2A_k{}^{\mu}{}_l{}^{\nu}(\mathbf{x})u^k{}_{,\mu}\overline{u^l}_{,\nu}$$
$$\geqslant \sum_s\int_{\Omega} dL^3A_k{}^{\mu}{}_l{}^{\nu}(\mathbf{x})u^k_{s,\mu}\overline{u^l_{s,\nu}}$$
$$-\sum_s\int_{\Omega} dL^3A_k{}^{\mu}{}_l{}^{\nu}(\mathbf{x})[\varphi_{s,\mu}u^k\overline{u^l_{s,\nu}}+\varphi_{s,\nu}u^k_{s,\mu}\overline{u^l}]$$

since $A_k{}^{\mu}{}_l{}^{\nu}\varphi_{s,\mu}\varphi_{s,\nu}\operatorname{Re}(u^k\overline{u^l}) \geqslant 0$. Hence

$$\operatorname{Re}B(\mathbf{u},\mathbf{u}) \geqslant M\sum_s \|\nabla\mathbf{u}_s\|_2{}^2-2N\sum_s \|\nabla\mathbf{u}_s\|_2\|\mathbf{u}\|_2.$$

But

$$2N\|\nabla\mathbf{u}_s\|_2\|\mathbf{u}\|_2 = 2N\varepsilon\|\nabla\mathbf{u}_s\|_2\frac{1}{\varepsilon}\|\mathbf{u}\|_2$$
$$\leqslant N\left[\varepsilon^2\|\nabla\mathbf{u}_s\|_2{}^2+\frac{1}{\varepsilon^2}\|\mathbf{u}\|_2{}^2\right].$$

Noting that $\sum_s |\mathbf{u}_{s,\mu}|^2 = \sum_s \varphi_s{}^2|\mathbf{u}_{,\mu}|^2+2\operatorname{Re}\sum_s \varphi_s\varphi_{s,\mu}\mathbf{u}\cdot\bar{\mathbf{u}}_{,\mu}+\sum_s \varphi_{s,\mu}{}^2|\mathbf{u}|^2 \geqslant |\mathbf{u}_{,\mu}|^2$

(the middle term drops out) we obtain the estimate $\operatorname{Re}B(\mathbf{u},\mathbf{u}) \geqslant M'\|\nabla\mathbf{u}\|_2{}^2-$

$R\|\mathbf{u}\|_2^2$ with $M' = M - N\varepsilon^2 > 0$ for ε sufficiently small. Hence $\mathrm{Re}B(\mathbf{u},\mathbf{u}) \geqslant M'\|\mathbf{u}\|_{1,2} - (M'+R)\|\mathbf{u}\|_2^2$. □

In the case of elliptic operators of order $2m > 2$, $B(\mathbf{u},\mathbf{v}) = \sum_{k,l,\alpha,\beta} a_{kl}^{\alpha\beta}(\mathbf{x}) D^\alpha u^k \overline{D^\beta v^l}$ (sum over multiindices α, β with $|\alpha|, |\beta| \leqslant m$). Introducing the partition of unity we obtain the expression $\sum_{k,l} \sum_{|\alpha|=|\beta|=m} a_{kl}^{\alpha\beta}(\mathbf{x}) D^\alpha u_s^k \overline{D^\beta u_s^l}$ + analogous terms with either $|\alpha| < m$ or $|\beta| < m$. Assuming that the highest order coefficients $a_{kl}^{\alpha\beta}(\mathbf{x})$ are continuous and satisfy the condition SE:

$$\mathrm{Re} \sum_{\substack{k,l,\alpha,\beta \\ |\alpha|=|\beta|=m}} a_{kl}^{\alpha\beta} p^\alpha p^\beta a^k \overline{a^l} \geqslant M\mathbf{p}^{2m}|\mathbf{a}|^2, \quad p^\alpha := p_1^{\alpha_1} \dots p_n^{\alpha_n},$$

we conclude that $\mathrm{Re}B(\mathbf{u},\mathbf{u}) \geqslant M\|\mathbf{u}\|_{m,2}^2 - R\|\mathbf{u}\|_2^2$, $M > 0$. The terms with $|\alpha|+|\beta| < 2m$ can be estimated with the help of the Ehrling lemma (Theorem 1.1.30) as follows:

$$\begin{aligned} \geqslant -K\|\mathbf{u}\|_{m-1}\|\mathbf{u}\|_m &\geqslant -K\|\mathbf{u}\|_m[\delta\|\mathbf{u}\|_m + c(\delta)\|\mathbf{u}\|_o] \\ &\geqslant -K\delta\|\mathbf{u}\|_m^2 - Kc(\delta)\|\mathbf{u}\|_m\|\mathbf{u}\|_o \\ &\geqslant -K\delta\|\mathbf{u}\|_m^2 - Kc(\delta)\tfrac{1}{2}\{\varepsilon^{-2}\|\mathbf{u}\|_o^2 + \varepsilon^2\|\mathbf{u}\|_m^2\} \\ &= -(K\delta + \tfrac{1}{2}Kc(\delta)\varepsilon^2)\|\mathbf{u}\|_m^2 - \tfrac{1}{2}K\varepsilon^{-2}c(\delta)\|\mathbf{u}\|_o^2, \end{aligned}$$

with δ and ε arbitrarily small. To simplify the notations we have written $\|\mathbf{u}\|_m$ for $\|\mathbf{u}\|_{m,2}$ and $\|\mathbf{u}\|_o$ for $\|\mathbf{u}\|_2$ in the last estimate. □

Definition 1.1.54

$\mathscr{L}^p_{loc}(\Omega)$, $p \geqslant 1$, is the space of measurable $\mathbf{u}: \Omega \to \boldsymbol{R}$ such that for every compact $\mathscr{K}$ $\|u\|_{\mathscr{K},p} := \int_{\mathscr{K}} dL^n |u|^p < \infty$. The topology of $\mathscr{L}^p_{loc}(\Omega)$ is fixed by the seminorms* $\|\cdot\|_{\mathscr{K},p}$. We note that $\mathbf{u} = 0$ if $\|\mathbf{u}\|_{\mathscr{K},p} = 0$ for all compact $\mathscr{K} \subset \Omega$. A basis of nbhds of 0 is given by $\{\mathbf{u} \in \mathscr{L}^p_{loc} \mid \|\mathbf{u}\|_{\mathscr{K},p} < \varepsilon\}$, for all compact $\mathscr{K} \subset \Omega$ and $\varepsilon > 0$. Equivalently, $\mathbf{u}_n \to \mathbf{u}$ in $\mathscr{L}^p_{loc}(\Omega)$ if $\|\mathbf{u}_n - \mathbf{u}\|_{\mathscr{K},p} \to 0$ for all compact $\mathscr{K} \subset \Omega$.

$W^{m,p}_{loc}(\Omega)$ is the set of all measurable $\mathbf{u}: \Omega \to \boldsymbol{R}^m$ such that $\mathbf{u}$, $D^\alpha \mathbf{u} \in \mathscr{L}^p_{loc}(\Omega)$ for $|\alpha| \leqslant m$. The topology of $W^{m,p}_{loc}(\Omega)$ is fixed by the seminorms $\|\mathbf{u}\|_{\mathscr{K},m,p} := \left(\sum_{|\alpha|\leqslant m} \|D^\alpha \mathbf{u}\|_{\mathscr{K},p}{}^p\right)^{1/p}$.

* A **seminorm** p is a function $p: \mathscr{V} \to \overline{\boldsymbol{R}}_+$ on a vector space $\mathscr{V}$ with the following properties: $p(tx) = |t|p(x)$, $p(x+y) \leqslant p(x)+p(y)$. A **norm** p satisfies in addition the condition $p(x) = 0 \Rightarrow x = 0$.

Every $f \in \mathscr{L}^1_{loc}(\Omega)$ determines a unique distribution, which we shall also denote by f, according to the formula $\langle f|g\rangle := \int dL^n f(x) g(x)\ \forall\, g \in \mathscr{C}^\infty_o(\Omega)$. Distributional derivatives $D^\alpha f$ of f are defined by the formula $\langle D^\alpha f|g\rangle = (-)^{|\alpha|} \int dL^n f(x) D^\alpha g(x)\ \forall\, g \in \mathscr{C}^\infty_o(\Omega)$.

We shall need some absolute continuity properties of Sobolev functions. First of all we shall recall some basic facts concerning absolutely continuous functions of a real variable.

A function $f\colon I \to \boldsymbol{R}^n$, $I \subset \boldsymbol{R}$, is said to be **absolutely continuous** if $\forall\, a, b \in I$, $b > a$, $\forall\, \varepsilon > 0\ \exists\, \delta > 0$ such that for every finite sequence a_j, b_j, $j = 1, \ldots, m$, satisfying the inequalities $a \leqslant a_1 \ldots < b_{j-1} \leqslant a_j < b_j \leqslant a_{j+1} < \ldots \leqslant b$ and $\sum_{j=1}^{m} |b_j - a_j| < \delta$ we have $\sum_{j=1}^{m} |f(b_j) - f(a_j)| < \varepsilon$. It is clear that every Lipschitz continuous function is absolutely continuous.

If f is absolutely continuous on $I \subset \boldsymbol{R}$, then the derivative f' (defined in terms of difference quotients) exists a.e. on I, is a measurable function of $x \in I$, integrable on every compact $I' \subset I$, and $f(x) - f(a) = \int_a^x dy f'(y)$.

If $g\colon I \to \boldsymbol{R}^n$ is measurable and integrable on compact $I' \subset I$, then $f(x) := \int_a^x dy\, g(y)$ is absolutely continuous and $f'(x) = g(x)$ a.e. in I.

If f, g are absolutely continuous then the pointwise product fg is absolutely continuous. From these facts and the Leibniz rule it follows immediately that for absolutely continuous f, g integration by parts is allowed:

$$\int_a^b dx\, f(x) g'(x) = fg|_a^b - \int_a^b dx\, f'(x) g(x).$$

In the following paragraphs we shall denote the distributional and "classical" derivatives (limits of difference quotients) of a function f by f'_d, f'_c resp.

Let $f \in W^{1,p}_{loc}(I) \subset W^{1,1}_{loc}(I)$, $a \in I$, I denotes a bounded open interval in $\boldsymbol{R}$. The function $g(x) := \int_a^x dy f'_d(y) + c$ is absolutely continuous and $g'_c = f'_d$ a.e. on I. Let $\varphi\colon I \to \boldsymbol{R}$ be a smooth test function vanishing at the endpoints of I. We can integrate by parts in the formulae below, hence

$$\int_I dx\, \varphi g'_d = -\int_I dx\, \varphi' g = -\int_I dx\, \varphi'(x) \int_a^x dy\, f'_d(y) = \int_I dx\, \varphi(x) f'_d(x).$$

Hence $g'_d = f'_d$ and $h := f - g \in \mathscr{L}^1_{loc}(I)$ satisfies $h'_d = 0$. For any $\varphi \in \mathscr{C}^\infty_o(I)$ there is an $\varepsilon > 0$ such that $\operatorname{dist}(\operatorname{supp}\varphi, \partial I) = \varepsilon$. For $|z| < \varepsilon$ the function

$\phi(z) := \int_I dx\, \varphi(x+z)h(x) = \int_I dx\, \varphi(x)h(x-z)$ is differentiable and $\phi'(z) = \int_I dx\, \varphi'(x+z)h(x) = -\int_I dx\, \varphi(x+z)h'_d(x) = 0$. Hence $\phi = \text{const}$

$$\int_I dx\, \varphi(x)h(x-z) = \int_I dx\, \varphi(x)h(x) \quad \text{for } |z| < \varepsilon \tag{1.1.26}$$

for every $\varphi \in \mathscr{C}_o^\infty(I)$.

Let χ be the characteristic function of an L^1-measurable set $\mathscr{A} \subset I$. By Lusin's theorem (Theorem 1.1.11) for every $m \in \boldsymbol{Z}_+$ there is a continuous function φ_m such that $0 \leqslant \varphi_m \leqslant 1$ and $\chi(x) = \varphi_m(x)$ on $I \diagdown \mathscr{B}_m$, $L^1(\mathscr{B}_m) \leqslant 1/\text{m}$. We have the inequalities

$$\left|\int_I dx\, \chi(x)[h(x)-h(x+z)]\right| = \left|\int_I dx\, [\chi(x)-\varphi_m(x)][h(x)-h(x+z)]\right|$$
$$\leqslant \int_{\mathscr{B}_m} dx\, |h(x)-h(x+z)| \tag{1.1.27}$$

for sufficiently small $|z|$. Note that we can choose φ_m in such a way that they are contained in a closed subset $\mathscr{D} \subset I$ containing a nbhd of $\mathscr{A}$. By Theorem 1.1.63 below for every $\varepsilon > 0$ there is an $m_o \in \boldsymbol{Z}_+$ such that for $m > m_o$ the right-hand side of (1.1.27) is $< \varepsilon$. Hence $\int_{\mathscr{A}} dx[h(x)-h(x+z)] = 0$ for every measurable subset $\mathscr{A}$ of I different from I and sufficiently small $|z|$.

At every Lebesgue point x_o of h, $h(x_o) = \lim_{\Delta\to 0} \frac{1}{2\Delta} \int_{-\Delta}^{\Delta} dy\, h(x_o+y)$. If x_o, $x_1 = x_o+z$ are two Lebesgue points of h then $h(x_0) = h(x_1)$. Since the set of Lebesgue points of $h \in \mathscr{L}^1(I)$ has the full measure $L^1(I)$, $h(x) = \text{const}$ a.e. in I. This proves

Theorem 1.1.55

If $f \in W^{1,p}_{loc}(I)$, $p \geqslant 1$, then f can be modified on a set of measure zero in I in such a way that it is absolutely continuous and $f'_c = f'_d$ a.e. in I.

Let $x = (x^k, x'_k)$, $x'_k \in \boldsymbol{R}^{n-1}$, for $k = 1, \ldots, n$.

Morrey (1966) proves a stronger and more general theorem (Theorem 3.1.8): For every $f \in W^{1,p}(\Omega)$, and a bounded domain $\Omega \subset \boldsymbol{R}^n$, there is a function g such that $g = f$ a.e., $x^k \mapsto g(x^k, x'_k)$ is absolutely continuous on each segment in Ω with endpoints in $\bar{\Omega}$, for $L^{n-1}-$a.a. x'_k, g tends to limits as x tends to the endpoints of such a segment and, for every bi-Lipschitz mapping $h\colon \mathscr{U} \to \Omega$, $g \circ h$ has the properties of g listed above. $g(x)$ can be defined as the Lebesgue derivative of $\int_{\Omega} dL^n f$ at x, for every $x \in \Omega$ for which the Lebesgue derivative exists.

The set of functions with μ-Hölder continuous derivatives of orders $0 \leqslant |\alpha| \leqslant m$ endowed with the norm $\sup\{|D^\alpha u(x)| \mid x \in \Omega,\ 0 \leqslant |\alpha| \leqslant m\} + \sup\{|D^\alpha u(x) - D^\alpha u(y)|\ |x-y|^{-\mu} \mid x \neq y \in \Omega,\ 0 \leqslant |\alpha| \leqslant m\}$ is a Banach space $\mathscr{C}^{m,\mu}(\Omega)$.

Theorem 1.1.56 (Adams, 1975, Theorem 6.2)
For a bounded domain $\Omega \subset \boldsymbol{R}^n$ satisfying the strong local Lipschitz condition there are compact embeddings $W^{j+m,p}(\Omega) \to \mathscr{C}^{j,\mu}(\Omega)$ for integer $j \geqslant 0$, $m \geqslant 1$, $1 \leqslant p < \infty$, $mp > n \geqslant (m-1)p$, $0 < \mu < m - n/p$.

1.1.8 Some Facts from Non-Linear Functional Analysis

We shall recall the basic derivative notions in Banach spaces.

Let F be a function $\mathscr{U} \to \boldsymbol{R}$ defined on an open subset $\mathscr{U}$ of a Banach space $\mathscr{X}$.

Definition 1.1.57
The **variation of F along a vector** $h \in \mathscr{X}$:

$$\delta F(x; h) := \frac{d}{dt} F(x+th)|_{t=0}$$

whenever the right-hand side exists.

It is easy to see that $\delta F(x; th) = t\delta F(x; h)\ \forall\, t \in \boldsymbol{R}$. In general however $\delta F(x; h_1 + h_2) \neq \delta F(x; h_1) + \delta F(x; h_2)$ (cf. a counterexample in Vainberg (1956)).

Definition 1.1.58
The variation of F at a point $x \in \mathscr{X}$ defines the **Gâteaux derivative** $DF(x) \in \mathscr{X}^*$:

$$DF(x)[h] := \delta F(x; h)$$

provided the right-hand side is a continuous linear functional of $h \in \mathscr{X}$.

Definition 1.1.59
$DF(x)$ is the **Fréchet derivative** if the following estimate holds

$$|F(x+h) - F(x) - DF(x)[h]| = o(\|h\|).$$

Note that the estimate in Definition 1.1.59 is uniform with respect to the "direction" $|h|^{-1}h$ of h. This need not be the case for a Gâteaux derivative.

The above derivatives can be extended easily to arbitrary $F: \mathscr{X} \to \mathscr{Y}$ where $\mathscr{X}$, $\mathscr{Y}$ are Banach spaces.

If F is defined on a dense subset $\mathscr{U}$ of $\mathscr{X}$ then $DF(x)[h]$ is defined on a dense subset of X and can be extended to all $h \in X$.

Suppose that $F: \mathscr{U} \to \boldsymbol{R}$ is Gâteaux-differentiable at every point of a convex subset Ω of $\mathscr{U}$. By elementary calculus the **Lagrange formula**

$$F(x+h)-F(x) = DF(x+\vartheta h)[h] \quad \text{for some } \vartheta \in]0,1[$$

can be established. It is equally easy to see that a functional $F: \mathscr{U} \to \boldsymbol{R}$ which is Gâteaux-differentiable at every point in a nbhd of $x_o \in \mathscr{U}$ and has a local extremum at x_o satisfies the equation $DF(x_o)[h] = 0 \;\forall\, h \in \mathscr{X}$.

If $\delta F(x; h)$ exists for all x in a nbhd of x_o and is a continuous function of h at $(x_o, 0)$ and of x at (x_o, h) then $\delta F(x; h) = DF(x)[h]$.

If DF is continuous in x in a nbhd of x_o then DF is a Fréchet derivative (cf. Vainberg, 1972).

In general the norm need not be Gâteaux-differentiable. The $\mathscr{L}^p$ norms are however differentiable for $p > 1$:

$$D(||x||^p)[h] = p \int_\Omega dL^n(\xi)|x(\xi)|^{p-1}\operatorname{sgn} x(\xi)h(\xi) \quad \text{for } 1 < p < 2, p > 2$$

$$D(||x||)[h] = ||x||^{1-p} \int_\Omega dL^n(\xi)|x(\xi)|^{p-2}x(\xi)h(\xi) \quad \text{for } x \neq 0,$$

cf. Vainberg (1972).

We shall now consider the nonlinear superposition operators in Lebesgue spaces $\mathscr{L}^p$. We shall use the notation $q = \dfrac{p}{p-1}$.

Definition 1.1.60

Let Ω be a domain in $\boldsymbol{R}^n$. A function $f: \Omega \times \boldsymbol{R}^m$ defined for a.a. $x \in \Omega$ and all $u \in \boldsymbol{R}^m$, is said to be a **Carathéodory function** if

(i) $\mathbf{u} \mapsto f(x, \mathbf{u})$ is continuous for a.e. $x \in \Omega$;

(ii) $x \mapsto f(x, \mathbf{u})$ is measurable* for every $\mathbf{u} \in \boldsymbol{R}^m$;

Let f be a Carathéodory function $\Omega \times \boldsymbol{R}^m \to \boldsymbol{R}$. For any measurable function $\mathbf{u}: \Omega \to \boldsymbol{R}^m$ the function $f(\cdot, \mathbf{u}(\cdot)): \Omega \to \boldsymbol{R}$ is measurable. Indeed, let $\mathbf{u}_n(x) = \sum_j \mathbf{c}_j^n \chi_{E_j^n}(x)$ be a sequence of simple functions converging pointwise a.e.

* Henceforth we shall have in mind L^n-measurability.

to $\mathbf{u}(x)$. We have $f(x, \mathbf{u}_n(x)) \to f(x, \mathbf{u}(x))$ for a.e. x. The functions $f(x, \mathbf{u}_n(x)) = \sum_j f(x, \mathbf{c}_j^n)\chi_{E_j^n}(x)$ are measurable, hence $f(x, \mathbf{u}(x))$ is measurable.

Theorem 1.1.61

If f is a Carathéodory function satisfying the inequality

$$|f(x, \mathbf{u})| \leqslant a(x) + K|\mathbf{u}|^{p/r}, \quad p, r > 1, \quad a(\cdot) \in \mathscr{L}^r(\Omega) \tag{1.1.28}$$

then the **superposition operator** $\mathscr{F}\colon \mathbf{u}(\cdot) \mapsto f(\cdot, \mathbf{u}(\cdot))$ maps $\mathscr{L}^p(\Omega)$ into $\mathscr{L}^r(\Omega)$.

Proof

Let f be a Carathéodory function satisfying (1.1.28) and $\mathbf{u} \in \mathscr{L}^p$. $\mathscr{F}\mathbf{u}$ is measurable and

$$|f(x, \mathbf{u}(x))|^r \leqslant [|a| + K|\mathbf{u}(x)|^{p/r}]^r \leqslant 2^r \max\{|a|^r, K^r|\mathbf{u}(x)|^p\}$$
$$\leqslant 2^r[|a(x)|^r + K^r|\mathbf{u}(x)|^p] \text{ is integrable.}$$ □

Remark

Let $\mathbf{f}$ be a vector-valued Carathéodory function with values in $\boldsymbol{R}^k$ (i.e. each component is a Carathéodory function) satisfying (1.1.28). Then the thesis of Theorem 1.1.61 remains valid. Indeed,

$$\left(\sum_{j=1}^{k} f_j^2\right)^{r/2} \leqslant k^r \max\{|f_j|^r \mid j \leqslant k\} \leqslant k^r \sum_{j=1}^{k} |f_j|^r.$$

There is a theorem of Krasnoselskii to the effect that the substitution operator $\mathscr{F}\colon \mathscr{L}^p(\Omega) \to \mathscr{L}^r(\Omega)$ is continuous (Ékeland and Témam, 1976). We shall use another method of proving continuity of $\mathscr{F}$. First of all we shall recall a fairly well known result about convergence in the $\mathscr{L}^p$ norm.

Lemma 1.1.62

Let $1 \leqslant p \leqslant \infty$. If $u_n \to u$ in $\mathscr{L}^p(\Omega)$ then a subsequence $\{u_k'\}$ of $\{u_n\}$ converges a.e. to u.

Proof

Let $p < \infty$ (for $p = \infty$ there is nothing to be proved).

For arbitrary $\delta > 0$ and $v \in \mathscr{L}^p(\Omega)$

$$L^n(\mathscr{U}) := L^n(\{x \in \Omega \mid |v(x)| \geqslant \delta\}) \leqslant \frac{1}{\delta^p} ||v||_p{}^p \tag{1.1.29}$$

Indeed,

$$||v||_p{}^p \geqslant \int_{\mathscr{U}} dL^n\, |v|^p \geqslant \delta^p L^n(\mathscr{U}).$$

Let $u_{n'} \to u$ in $\mathscr{L}^p(\Omega)$, $p \geqslant 1$. It follows from (1.1.29) that $\{u_{n'}\}$ converges in the measure to u:

$$\lim_{n' \to \infty} L^n(\{x \in \Omega \mid |u_{n'}(x) - u(x)| \geqslant \delta\}) = 0 \ \forall \delta > 0.$$

For every $n' \in \boldsymbol{Z}_+$ we pick out an $r(n') \in \boldsymbol{Z}_+$ such that $\forall s, m \geqslant r(n')$ $L^n(\{x \in \Omega \mid |u_m(x) - u_s(x)| \geqslant 1/2^{n'}\}) < 1/2^{n'}$, and $r(n'+1) > r(n')$. Let $\mathscr{A}_{n'}$

$$:= \{x \in \Omega \mid |u_{r(n'+1)}(x) - u_{r(n')}(x)| \geqslant 1/2^{n'}\},\ \mathscr{B}_{n'} = \bigcup_{k=0}^{\infty} \mathscr{A}_{n'+k},\ \ \mathscr{B} := \bigcap_{n'} \mathscr{B}_{n'}.$$

Since $L^n(\mathscr{A}_{n'}) < 1/2^{n'}$, $L^n(\mathscr{B}_{n'}) \leqslant 2^{1-n'}$, and $L^n(\mathscr{B}) = 0$. Let $\mathscr{A}$ be the set of $x \in \Omega$ of full measure such that all the functions f_j are defined everywhere on $\mathscr{A}$. Let $x \in \mathscr{A} \setminus \mathscr{B} = \bigcup_{n'} (\mathscr{A} \setminus \mathscr{B}_{n'})$. Then $\exists\, n_o$ such that $x \in \mathscr{A} \setminus \mathscr{B}_{n_o}$ and $x \in \mathscr{A} \setminus \mathscr{A}_{n'}$ for all $n' \geqslant n_o$. Hence $|u_{r(n'+1)}(x) - u_{r(n')}(x)| \leqslant 1/2^{n'}$ for $n' \geqslant n_o$ and $\{u_{r(n')}\}$ converges in the measure and a.e. to a function u', hence it converges a.e. to u. □

Lemma 1.1.63

Let $u \in \mathscr{L}^1(\Omega)$. Then $\forall\, \varepsilon > 0\ \exists\ \delta > 0$ such that $L^n(\mathscr{A}) < \delta \mapsto \int_{\mathscr{A}} dL^n |u| < \varepsilon$.

Proof

For every $m \in \boldsymbol{Z}_+$ let

$$\psi_m(x) :\equiv \begin{cases} u(x) & \text{if } u(x) \leqslant m, \\ m & \text{otherwise.} \end{cases}$$

$\{\psi_m\}$ is a non-decreasing sequence of measurable functions converging pointwise to $|u(x)|$. By Fatou's lemma

$$\lim_{m \to \infty} \int_{\Omega} dL^n\, \psi_m(x) = \int_{\Omega} dL^n\, |u(x)|.$$

Let $m_o \in \boldsymbol{Z}_+$ be such that $\forall\, m \geqslant m_o$, $\int_{\Omega} dL^n\,(|u| - \psi_m) < \dfrac{\varepsilon}{2}$, $\delta := \dfrac{\varepsilon}{2m_o}$,

$L^n(\mathscr{A}) < \delta$. We have $\int_{\mathscr{A}} dL^n \psi_{m_o} \leqslant m_o L^n(\mathscr{A}) < \dfrac{\varepsilon}{2}$ and

$$\left| \int_{\mathscr{A}} dL^n\, u \right| \leqslant \int_{\mathscr{A}} dL^n\, |u| = \int_{\mathscr{A}} dL^n\, [|u| - \psi_{m_o}] + \int_{\mathscr{A}} dL^n\, \psi_{m_o} < \frac{\varepsilon}{2} + \frac{\varepsilon}{2}. \quad \square$$

Theorem 1.1.64 (Vitali)

Let $u_\nu \in \mathscr{L}^p(\Omega)$, $u_\nu \to u$ a.e. for $\nu \to \infty$.

Then $u \in \mathscr{L}^p(\Omega)$ and $u_\nu \to u$ in $\mathscr{L}^p(\Omega)$ iff the following two conditions are satisfied:

(i) $\forall\, \varepsilon > 0$ there is a bounded measurable $\Omega_\varepsilon \subset \Omega$ such that

$$\int_{\Omega \setminus \Omega_\varepsilon} dL^n\, |u_\nu|^p < \varepsilon \;\; \forall \nu \in \boldsymbol{Z}_+;$$

(ii) $\forall\, \varepsilon > 0$, $\exists\, \delta > 0$ such that, for every measurable $\mathscr{A} \subset \Omega$, $L^n(\mathscr{A}) < \delta$ entails the inequality $\int_{\mathscr{A}} dL^n\, |u_\nu|^p < \varepsilon \;\forall\, \nu \in \boldsymbol{Z}_+$.

Proof

(1) Necessity of (i).

Let $\varepsilon > 0$. Let $\nu_o \in \boldsymbol{Z}_+$ be such that $||u_\nu - u||_p < \dfrac{\varepsilon}{2} \forall\, \nu \geqslant \nu_o$. We can choose such $\Omega'_\varepsilon, \Omega''_\varepsilon \subset \Omega$ that $L^n(\Omega'_\varepsilon) < \infty$, $L^n(\Omega''_\varepsilon) < \infty$, $\int_{\Omega \setminus \Omega'_\varepsilon} dL^n\, |u_\nu|^p < \varepsilon^p$ for $\nu \leqslant \nu_o$, $\int_{\Omega \setminus \Omega''_\varepsilon} dL^n\, |u|^p < \left(\dfrac{\varepsilon}{2}\right)^p$ (Lemma 1.1.63).

Let $\Omega_\varepsilon := \Omega'_\varepsilon \cup \Omega''_\varepsilon$. For $\nu \geqslant \nu_o$

$$\int_{\Omega \setminus \Omega_\varepsilon} dL^n\, |u_\nu|^p = \int_{(\Omega \setminus \Omega'_\varepsilon) \cap (\Omega \setminus \Omega''_\varepsilon)} dL^n\, |u_\nu|^p \leqslant$$

$$\left\{||u_\nu - u||_p + \Big(\int_{\Omega / \Omega''_\varepsilon} dL^n\, |u|^p\Big)^{1/p}\right\}^p \leqslant \varepsilon^p.$$

(2) Necessity of (ii).

Let $||u_\nu - u||_p < \dfrac{\varepsilon}{2}$ for all $\nu \geqslant \nu_o$. Let $\delta > 0$ be such that $\int_{\mathscr{A}} dL^n\, |u|^p < \left(\dfrac{\varepsilon}{2}\right)^p$ whenever $L^n(\mathscr{A}) < \delta$. Hence

$$\int_{\mathscr{A}} dL^n\, |u_\nu|^p \leqslant \left\{||u_\nu - u||_p + \Big(\int_{\mathscr{A}} dL^n\, |u|^p\Big)^{1/p}\right\}^p \leqslant \varepsilon^p.$$

(3) Sufficiency of (i) and (ii).

Let $\varepsilon > 0$. On account of (i) there is a subset $\Omega_\varepsilon \subset \Omega$ of finite measure such that for all $\mu, \nu \in \boldsymbol{Z}_+$

$$\int_{\Omega} dL^n\, |u_\mu - u_\nu|^p = \int_{\Omega \setminus \Omega_\varepsilon} dL^n\, |u_\mu - u_\nu|^p + \int_{\Omega_\varepsilon} dL^n\, |u_\mu - u_\nu|^p$$

$$\leqslant 2^p \varepsilon + \int_{\Omega_\varepsilon} dL^n\, |u_\mu - u_\nu|^p.$$

Indeed, the first integral on the right-hand side is bounded by

$$\Big\{\Big[\int_{\Omega\setminus\Omega_\varepsilon} dL^n\,|u_\mu|^p\Big]^{1/p}+\Big[\int_{\Omega\setminus\Omega_\varepsilon} dL^n\,|u_\nu|^p\Big]^{1/p}\Big\}^p \leqslant 2^p\varepsilon,$$

using the Minkowski inequality.

Let us choose $\delta > 0$ according to (ii) for the given $\varepsilon > 0$. By the Egoroff theorem (Section 1.1) there is a measurable subset $\mathscr{A} \subset \Omega_\varepsilon$ such that $L^n(\mathscr{A}) < \delta$ and $u_\mu \to u$ uniformly on $\Omega_\varepsilon\setminus\mathscr{A}$, so that

$$\limsup_{\mu,\nu\to\infty}\int_{\Omega_\varepsilon} dL^n\,|u_\mu-u_\nu|^p \leqslant \limsup_{\mu,\nu\to\infty}\int_{\Omega_\varepsilon\setminus\mathscr{A}} dL^n\,|u_\mu-u_\nu|^p+2^p\varepsilon = 2^p\varepsilon.$$

Hence $\limsup_{\mu,\nu\to\infty}||u_\mu-u_\nu||_p{}^p \leqslant 2^{p+1}\varepsilon$ and $\{u_\mu\}$ converges in $\mathscr{L}^p(\Omega)$. Since $\mathscr{L}^p(\Omega)$ is complete, $u_\mu \to u_o \in \mathscr{L}^p(\Omega)$ and a subsequence $u'_\mu(x) \to u_o(x)$ for a.a. $x \in \Omega$. Hence $u_o(x) = u(x)$ for a.a. $x \in \Omega$.

□

Theorem 1.1.65

Let f be a Carathéodory function $\Omega\times\boldsymbol{R}^m \to \boldsymbol{R}$ satisfying (1.1.28). Then the corresponding superposition operator $\mathscr{F}$: $\mathscr{L}^p(\Omega) \to \mathscr{L}^r(\Omega)$ is bounded.

Proof

Let $u_\nu \to u$ in $\mathscr{L}^p(\Omega)$. A subsequence $u'_\mu \to u$ a.e. in Ω (Lemma 1.1.62). Hence $f(x, u'_\mu(x)) \to f(x, u(x))$ a.e. in Ω. By the Vitali theorem $\forall\, \varepsilon > 0\ \exists\, \Omega_\varepsilon \subset \Omega$ such that $L^n(\Omega_\varepsilon) < \infty$ and $\exists\, \delta > 0$ such that $L^n(\mathscr{A}) < \delta$, $\mathscr{A} \subset \Omega$, implies

$$\int_{\mathscr{A}} dL^n\,|u'_\mu(x)|^p < \varepsilon, \qquad \int_{\Omega\setminus\Omega_\varepsilon} dL^n\,|u'_\mu(x)|^p < \varepsilon.$$

Hence, choosing δ so small that $\int_{\mathscr{A}} dL^n\,|a(x)|^r < \varepsilon$ as well,

$$\int_{\mathscr{A}} dL^n\,|f(x, u'_\mu(x))|^r \leqslant \int_{\mathscr{A}} dL^n\,[a(x)+K|u'_\mu|^{p/r}]^r$$

$$\leqslant 2^r\Big[\int_{\mathscr{A}} dL^n\,|a(x)|^r+K^r\int_{\mathscr{A}} dL^n\,|u'_\mu(x)|^p\Big] \leqslant 2^r\varepsilon(1+K^r).$$

The same estimate can be repeated for $\Omega\setminus\Omega_\varepsilon$ instead of $\mathscr{A}$. By the Vitali theorem $\mathscr{F}u'_\mu \to \mathscr{F}u$.

□

Definition 1.1.66

A superposition operator $\mathscr{F}$: $\mathscr{L}^p(\Omega) \to \mathscr{L}^q(\Omega)$, $p > 1$, $q = p/(p-1)$, is called a **Nemytskii operator.**

Note that a Nemytskii operator maps $\mathscr{L}^p$ into its conjugate.

The superposition operator plays an important role in the functional-analytic formulation of constitutive relations. In the variational formulation of non-linear elastostatics the fundamental constitutive relation expresses the strain energy (or stored energy) as a function of deformation gradients. From the point of view of the variational calculus the notion of a normal integrand is a natural extension of the notion of a Carathéodory function.

Definition 1.1.67
A function $f\colon \Omega\times \boldsymbol{R}^m \to \bar{\boldsymbol{R}}_+$ is said to be a **normal integrand** if
(i) for a.a. x, $f(x, \cdot)$ is lower semicontinuous on $\boldsymbol{R}^m$;
(ii) there is a Borel function* $g\colon \Omega\times \boldsymbol{R}^m \to \bar{\boldsymbol{R}}_+$ such that $f(x, \cdot) = g(x, \cdot)$ for a.e. $x \in \Omega$.

A description of normal integrands in a sense close to Lusin's theorem for measurable functions is given in Chapter 8 of Ékeland and Témam (1976). If f is a normal integrand and $u(\cdot)$ is measurable then $x \mapsto f(x, u(x))$ is a measurable function with values in the extended reals (op. cit.).

Theorem 1.1.68 (Ékeland and Témam, 1976)
If $f\colon \Omega\times (\boldsymbol{R}^l\times \boldsymbol{R}^m) \to \bar{\boldsymbol{R}}_+$ is a normal integrand and
(i) there is a convex increasing lower semicontinuous function $g\colon \bar{\boldsymbol{R}}_+ \to \bar{\boldsymbol{R}}_+$ such that $f(x, \mathbf{v}, \mathbf{u}) \geqslant g(|\mathbf{u}|)$ and $\lim_{t\to\infty} t^{-1}g(t) = \infty$;
(ii) $f(x, \mathbf{v}, \cdot)$ is convex for a.a. x and all $\mathbf{v} \in \boldsymbol{R}^l$;
then for $\mathbf{u}_\mu \rightharpoonup \mathbf{u}$ in $\mathscr{L}^1(\Omega)$, measurable $\mathbf{v}_\mu \to \mathbf{v}$ a.e. in Ω,

$$\int_\Omega dL^n f(x, \mathbf{v}(x), \mathbf{u}(x)) \leqslant \liminf_{\mu\to\infty} \int_\Omega dL^n f(x, \mathbf{v}_\mu(x), \mathbf{u}_\mu(x)).$$

A possible choice for the function g is $g(t) \equiv t^p$, $t > 0$, for some $p > 1$.

Theorem 1.1.69 (Ekeland and Témam, 1976, Proposition 8.1.1)
Every Carathéodory function is a normal integrand.

1.2 FUNDAMENTALS OF THE THEORY OF ELASTICITY

1.2.1 Galilean Reference Frames

We deal with a special class of objects, called **bodies** (or **media**, in some contexts). A precise definition of these objects will be given in the following

* By definition f is a Borel function if for every open (closed) set $\mathscr{A}$ in the range of f the set $f^{-1}(\mathscr{A})$ is Borel.

sections. For the purpose of this section we need only know the following property of a body. A body is a set of elements, called **material points.** A specification of the physical state of a body involves assigning each material point its position, i.e. a point in a three-dimensional space.

The physical state of a body depends on the relative position of the material points in the space. More specifically, it is the distances between material points in the space that is relevant for the physical state of the body. Hence the description of the physical state of the body requires assigning each material point its position in a metric space. Such an assignment is called a **placement** of the body.

For the static theory of a body we shall consider its placements in a three-dimensional Euclidean space. This space can be identified with $\boldsymbol{R}^3$ endowed with the Riemannian metric δ_{ij} in suitable coordinates. In the dynamic theory of the body every material point describes a trajectory in space-time. At a given time the physical state of the body is partly specified by the distances between the material points in the space. We must however account for the inertial properties of the body and of the material points. This fact implies that we must specify a physical model of space-time. Throughout this book we shall assume the Newtonian model of space-time.

The Newtonian space-time is a four-dimensional manifold E of **events** y, endowed with an absolute time and a geometrical structure that will be described later. The existence of an absolute time can be expressed in the following form. There is a one-dimensional linearly ordered time manifold $\mathscr{T}$, diffeomorphic to $\boldsymbol{R}^1$, and a surjective mapping $t\colon E \to \mathscr{T}$. The point $t(y)$ of $\mathscr{T}$ is the **time** of (occurrence of) the event y. Two events y_1, y_2 are said to be **simultaneous** if $t(y_1) = t(y_2)$. The inequality $t(y_1) < t(y_2)$ will be interpreted as the statement that y_1 occurred earlier than y_2.

It is obvious that simultaneity is an equivalence relation on E. It is well known that the possibility of establishing simultaneity of two events hinges on the availability of signals propagating with arbitrarily large (or infinite) speeds. The fastest signals propagate with the speed of light. In physical terms our assumption means that all the speeds and velocities appearing in our model are negligibly small in comparison with the speed of light in vacuo.

The geometry of the space-time $(E, \mathscr{T}, t)$ is determined by the laws of Newtonian mechanics as well as by the availability of rigid bodies which include yardsticks for measuring spatial distances.

According to the first law of Newtonian mechanics there is a privileged class Γ of global coordinate systems $\boldsymbol{\varphi}\colon E \to \boldsymbol{R}^4$ (onto) and a class of curves in E, called **trajectories** of free material points with the following properties:

(i) for each $\varphi \in \Gamma$ there is an order preserving diffeomorphism $\psi: \mathscr{T} \to R^1$ such that $\varphi(y) = (x^0, x^1, x^2, x^3)$, $x^0 = \psi(t(y))$ for all $y \in E$;

(ii) every trajectory of a free material point is given in terms of the coordinates φ by an equation

$$x^k = v^k x^0 + a^k, \quad k = 1, 2, 3, \tag{1.2.1}$$

with constant v^k, a^k;

(iii) if $\varphi \in \Gamma$, then any curve in E of the form (1.2.1) is a trajectory of a free material point;

(iv) $\forall \varphi_1, \varphi_2 \in \Gamma \ \ \varphi_2 \circ \varphi_1^{-1} \in \mathscr{C}^2$.

Let $\varphi_1, \varphi_2 \in \Gamma$ and let $\phi = \varphi_2 \circ \varphi_1^{-1}$. On account of (i) $\phi(x^0, x^1, x^2, x^3) = (\phi_0(x^0), \mathbf{q}(x^1, x^2, x^3))$. We shall write henceforth $\mathbf{x}$ for (x^1, x^2, x^3). On account of (ii) and (iii) for any $\mathbf{a}$, $\mathbf{v} \in R^3$ there are such $\mathbf{b}$, $\mathbf{w} \in R^3$ such that

$$\mathbf{q}(x^0, \mathbf{a}+x^0\mathbf{v}) = \mathbf{b}+\phi_0(x^0)\mathbf{w} \quad \forall\, x^0 \in R. \tag{1.2.2}$$

Differentiating (1.2.2) twice with respect to x^0 and setting $\mathbf{x} = \mathbf{a}+x^0\mathbf{v}$, we have

$$(\mathbf{q}_{,x^0x^0} - \phi_0''\mathbf{w}) + 2\mathbf{q}_{,x^0\mathbf{x}}[v] + \mathbf{q}_{,\mathbf{xx}}[\mathbf{v}, \mathbf{v}] = 0.$$

In the limit $\mathbf{v}^2 \to 0$ we get the identity $\mathbf{q}_{,x^0x^0}(x^0, \mathbf{x}) = \phi_0''\mathbf{w}$. Since $\mathbf{w}$ is arbitrary, we conclude that $\phi_0'' = 0$, $\mathbf{q}_{,x^0x^0} = \mathbf{0}$.

Substituting the above result into equation (1.2.2), dividing (1.2.2) by $|\mathbf{v}|$ and letting $\mathbf{v}^2 \to 0$ again we conclude that $\mathbf{q}_{,x^0\mathbf{x}} = \mathbf{0}$. Hence also $\mathbf{q}_{,\mathbf{xx}} = \mathbf{0}$,

$$\phi_0(x^0) \equiv \alpha x^0 + \beta, \tag{1.2.3}$$

$$\mathbf{q}(x^0, \mathbf{x}) \equiv \mathbf{A}\mathbf{x} + x^0\mathbf{b} + \mathbf{c}. \tag{1.2.4}$$

In view of (i) $\alpha > 0$.

From (1.2.3) and (1.2.4) it follows easily that the privileged class Γ of coordinate systems induces an affine structure on E. The associated class of coordinate systems ψ on $\mathscr{T}$ induces an affine structure on $\mathscr{T}$. The construction of the associated groups of translations of E and $\mathscr{T}$ is an easy exercise in set-theoretical operations involving division by equivalence relations and projecting algebraic structures onto sets thus obtained (cf. Hanyga (1984) for details). With respect to the affine structures on E and $\mathscr{T}$ the projection t is a linear mapping, cf. (1.2.3). It is also a submersion ($dt(y) = 0$ at some $y \in E$ would entail that $d(\psi \circ t)(y) = d\phi_0(y) = 0$).

Every coordinate transformation (1.2.3) and (1.2.4) is a composition of a transformation $\phi^{(1)}$:

$$\phi_0^{(1)}(x^0) = x^0, \quad \mathbf{q}^{(1)}(x^0, \mathbf{x}) \equiv \mathbf{x} + x^0\mathbf{b} \tag{1.2.5}$$

and a transformation $\phi^{(2)}$:

$$\phi_0^{(2)}(x^0) = \alpha x^0 + \beta, \quad \mathbf{q}^{(2)}(x^0, \mathbf{x}) = \mathbf{A}\mathbf{x} + \mathbf{c}. \tag{1.2.6}$$

Let $\mathscr{G}_0$, $\mathscr{E}$ be the group of all the transformations of the form (1.2.5), (1.2.6), resp. We introduce the following equivalence relation on Γ:

$$\varphi_1 \sim \varphi_2 \quad \text{iff} \quad \varphi_2 \circ \varphi_1^{-1} \in \mathscr{E}.$$

An equivalence class $R = [\varphi]_\sim$ in Γ will be called a **reference frame.**

A reference frame R induces a mapping

$$E \to \mathscr{T} \times S_R, \tag{1.2.7}$$

where S_R is an affine three-dimensional space. The points of S_R are the equivalence classes $z = [y]_{\sim_R}$ with respect to the equivalence relation

$$y_1 \sim_R y_2 \quad \text{iff} \quad \mathbf{q}(y_1) = \mathbf{q}(y_2) \quad \text{for some} \quad \varphi = (\phi_0, \mathbf{q}) \in R.$$

If $\varphi_1 \circ \varphi^{-1} \in \mathscr{E}$, $\varphi_1 = (\phi_{01}, \mathbf{q}_1)$, then $\mathbf{q}_1(y_1) = \mathbf{q}_1(y_2)$ is equivalent to $\mathbf{q}(y_1) = \mathbf{q}(y_2)$. Every $\varphi \in R$ induces a unique mapping $\mathbf{r}_\varphi$: $S_R \to \boldsymbol{R}^3$ defined by the formula $\mathbf{r}_\varphi(z) = \mathbf{q}(y)$ if $y \in z$. The bijective mappings $\mathbf{r}_\varphi$ induce a differential structure on S_R by the requirement that all the $\mathbf{r}_\varphi$, where $\varphi \in R$, are $\mathscr{C}^2$ diffeomorphisms.

The mappings $\mathbf{r}_\varphi$ also induce an affine structure on S_R. Indeed, let us consider the following equivalence relation on $S_R \times S_R$:

$$(z_1, z_2) \sim (z_3, z_4) \quad \text{iff} \quad \mathbf{r}_\varphi(z_1) - \mathbf{r}_\varphi(z_2) = \mathbf{r}_\varphi(z_3) - \mathbf{r}_\varphi(z_4)$$

for some $\varphi \in R$. This definition does not depend on the choice of φ in R. The equivalence classes $\mathbf{w} = [(z_1, z_2)]_\sim$ form a vector space if the vector space operations are defined as follows:

$$\alpha[(z_1, z_2)]_\sim + [(z_3, z_4)]_\sim = [(z_5, z_6)]_\sim$$

iff

$$\alpha[\mathbf{r}_\varphi(z_1) - \mathbf{r}_\varphi(z_2)] + [\mathbf{r}_\varphi(z_3) - \mathbf{r}_\varphi(z_4)] = \mathbf{r}_\varphi(z_5) - \mathbf{r}_\varphi(z_6)$$

for some $\varphi \in R$. It is easy to see that this definition does not depend on the choice of $\varphi \in R$. We have thus defined the vector space S_R' of translations of S_R and the coordinate systems $\mathbf{r}_\varphi'$:

$$\mathbf{r}_\varphi'([(z_1, z_2)]_\sim) := \mathbf{r}_\varphi(z_1) - \mathbf{r}_\varphi(z_2).$$

The mappings $\mathbf{r}_\varphi'$: $S_R' \to \boldsymbol{R}^3$ are bijective and linear. They induce a differential structure on S_R'. If $\mathbf{v} = [(z_1, z_2)]_\sim$, then we shall write $z_1 = z_2 + \mathbf{v}$. The mapping $S_R \times S_R' \to S_R$ thus defined is continuous and $\mathbf{r}_\varphi(z_1) = \mathbf{r}_\varphi(z_2) + \mathbf{r}_\varphi'(\mathbf{v})$.

For every reference frame R (1.1.7) defines a decomposition of the affine space E into a direct sum of two affine spaces $\mathscr{T}$ and S_R. In physical terms S_R is the **space** associated with R. A material point P occupies a fixed point in S_R iff it **rests with respect to** R.

The notion of spatial distance in S_R is intimately connected with the notion of **rigid motion.** A set $\mathscr{B}$ of material points moves rigidly if the distance between any pair of material points of $\mathscr{B}$ does not vary in time. Let us choose any privileged coordinate system $\varphi \in \Gamma$ and let the **distance** between two **simultaneous** events $y_1, y_2 \in E$ be defined by the formula

$$d(y_1, y_2) := \sum_{k,l=1}^{3} g_{kl}(x_1{}^k - x_2{}^k)(x_1{}^l - x_2{}^l), \quad x_i{}^k = \varphi^k(y_i), \tag{1.2.8}$$

$i = 1, 2$, $g_{kl} = g_{lk}$ positive definite. Let φ_1 be another privileged coordinate system on E, with $\bar{x}^k = \varphi_1{}^k(y)$, $\varphi \circ \varphi_1^{-1} = (\phi_0, \mathbf{q})$. From equations (1.2.3) and (1.2.4)

$$\sum_{k,l} g_{kl}(x_1{}^k - x_2{}^k)(x_1{}^l - x_2{}^l) = \sum_{k,l,r,s} g_{kl} A^k{}_r A^l{}_s (\bar{x}_1{}^r - \bar{x}_2{}^r)(\bar{x}_1{}^s - \bar{x}_2{}^s).$$

Hence $d(y_1, y_2) = \sum_{k,l} g_{kl}(\bar{x}_1{}^k - \bar{x}_2{}^k)(\bar{x}_1{}^l - \bar{x}_2{}^l)$ provided $g_{kl} A^k{}_r A^l{}_s = g_{rs}$, i.e. provided $\mathbf{A}$ is an orthogonal transformation of $\boldsymbol{R}^3$ with respect to the Riemannian metric g_{kl} on $\boldsymbol{R}^3$. On the other hand we can always find an invertible linear mapping $\mathbf{A} \in L(\boldsymbol{R}^3, \boldsymbol{R}^3)$ such that $g_{kl} A^k{}_r A^l{}_s = \delta_{rs}$.

In view of these remarks we can pick out a class $\Gamma_0 \subset \Gamma$ of coordinate systems $\varphi\colon E \to \boldsymbol{R}^4;\ y \mapsto (x^0, \mathbf{x})$, such that $d(y_1, y_2)^2 = \sum_{k,l=1}^{3} \delta_{kl}(x_1{}^k - x_2{}^k)(x_1{}^l - x_2{}^l)$ for any two simultaneous events y_1, y_2. Any two coordinate systems $\varphi_1, \varphi_2 \in \Gamma_0$ are connected by transformations (1.2.3) and (1.2.4) with $\mathbf{A} \in \mathrm{SO}(3)$. Moreover every reference frame R contains a coordinate system $\varphi \in \Gamma_0$.

Let $R_0 := R \cap \Gamma_0$. It is easy to see that R_0 induces the structure of an Euclidean space on S_R, with the distance in S_R defined by the formula

$$d(z_1, z_2)^2 := [\mathbf{r}_\varphi(z_1) - \mathbf{r}_\varphi(z_2)]^2 = \sum_{k=1}^{3} (x_1{}^k - x_2{}^k)^2 \tag{1.2.9}$$

for arbitrary $\varphi \in R_0$, $z_1, z_2 \in S_R$.

1.2.2 Bodies, Their Motion and Configurations: Topological Aspects

In Sections 1.2.3 and 1.2.4 we gradually develop the notion of a body and its configurations. In this Section we discuss those aspects of this notion which are related to the intuitive concepts of consistency and continuity of the body (as opposed to the looseness of sand or the quality of being torn).

A body $\mathscr{B}$ is a set of material points P. With respect to a fixed reference frame R each material point P assumes a position in the associated space S_R

at every moment $t \in I \subset \boldsymbol{R}$. The position of P at time $t \in I$ is specified by a point $z \in S_R$:

$$z = h_R(P, t). \tag{1.2.10}$$

It is assumed that two different material points P, P' cannot appear at the same point $z \in S_R$ simultaneously. Hence for every $t \in I$ the mapping $h_R(\cdot\,, t)$: $\mathscr{B} \to S_R$ is injective.

The mapping h_R: $\mathscr{B} \times I \to S_R$ is called the **motion** of the body $\mathscr{B}$ **relative to the reference frame** R. It is convenient to express the motion in terms of a privileged coordinate system $\varphi \in R_0$. In this case the motion is specified by a mapping

$$\mathbf{h}\colon \mathscr{B} \times I \to \boldsymbol{R}^3. \tag{1.2.11}$$

In terms of any other coordinate system $\varphi_1 \in \Gamma_0$ the motion is given by the mapping $\mathbf{h}_1\colon \mathscr{B} \times I \to \boldsymbol{R}^3$,

$$\mathbf{h}_1(P, \alpha t + \beta) = \mathbf{A}\mathbf{h}(P, t) + t\mathbf{b} + \mathbf{c}, \tag{1.2.12}$$

provided $\varphi_1 \circ \varphi^{-1}$ is given by equations (1.2.3) and (1.2.4), $\mathbf{b}$, $\mathbf{c} \in \boldsymbol{R}^3$, $\mathbf{A} \in \mathrm{SO}(3)$, $\det \mathbf{A} = 1$, $\alpha > 0$.

In addition to the actual positions of the material points in the space $S_R \sim \boldsymbol{R}^3$ during the motion of the body $\mathscr{B}$ we shall be interested in their virtual positions in the abstract space $\boldsymbol{R}^3$:

$$\mathbf{g}\colon \mathscr{B} \to \boldsymbol{R}^3. \tag{1.2.13}$$

Intuitively, a mapping (1.2.13) is a *virtual* configuration iff it can be attained from a *real* configuration, e.g. $\mathbf{h}(\cdot\,, t)$, without destroying the continuity of the body. Let $\mathscr{K}$ denote the class of all the virtual configurations. We shall assume that all the real configurations $\mathbf{h}(\cdot\,, t)$ of $\mathscr{B}$ belong to $\mathscr{K}$. In practice we shall seek the solution of the equations of motion with prescribed initial configuration $\mathbf{h}(\cdot\,, 0)$ within the class $\mathscr{K}$ containing $\mathbf{h}(\cdot\,, 0)$. We shall also seek the equilibrium configuration of the body under specified loads within a class of virtual (i.e. accessible) configurations.

It is important to note that configurations serve two different practical purposes. The second of them consists in labelling the material points. This function will be fulfilled by a somewhat different class of configurations, called (local) **reference configurations.**

Let $\mathscr{K}_0$ be the class of all the mappings $\mathbf{g}\colon \mathscr{B} \to \boldsymbol{R}^3$, which include $\mathbf{h}(\cdot\,, 0) =: g_0(\cdot)$ and which have the property that $\mathbf{g} \circ \mathbf{g}_0^{-1}$ is a $\mathscr{C}^1$-diffeomorphism. The class $\mathscr{K}_0$ includes all the configurations $\mathbf{g}\colon \mathscr{B} \to \boldsymbol{R}^3$ which preserve merely

local continuity,* leaving open the question of global implementability of the deformation from $\mathbf{g}_0$ to $\mathbf{g}$.

In order to see this let us consider the following configurations $\mathbf{g}_n$, $n = 0, 1, 2, \ldots$, of a torus:

$$\begin{aligned} x^1 &= [2+r\cos(\phi+n\theta)]\cos\theta, \\ x^2 &= [2+r\cos(\phi+n\theta)]\sin\theta, \\ x^3 &= r\sin(\phi+n\theta) \end{aligned} \tag{1.2.14}$$

with $r \in [0, 1]$, θ, $\phi \in [0, 2\pi[$. The coordinates (r, θ, ϕ) label the material points of the torus. Since $\tan^{-1}(x^2/x^1) = \theta$, $\operatorname{sgn}\sin\theta = \operatorname{sgn} x^2$, $[(x^1)^2 + (x^2)^2]^{1/2} - 2 = r\cos(\phi+n\theta)$, $x^3 = r\sin(\phi+n\theta)$, it is clear that $\mathbf{g}_n$ is injective and $\mathbf{g}_n$, $\mathbf{g}_n^{-1}$ are smooth. Hence $\mathbf{g}_0 \in \mathscr{K}_0$ implies that $\mathbf{g}_n \in \mathscr{K}_0$ for $n = 1, 2, 3, \ldots$. On the other hand for $n \neq m$ there is no continuous deformation $(\mathbf{x}, s) \mapsto \mathbf{f}(\mathbf{x}, s)$, $s \in [0, 1]$, such that $\mathbf{f}(\mathbf{x}, 0) \equiv \mathbf{x}$, $\mathbf{f}(\mathbf{g}_n(r, \theta, \phi), 1) \equiv \mathbf{g}_m(r, \theta, \phi)$ (in topological terms the mappings $\mathbf{g}_n$, $\mathbf{g}_m$, $n \neq m$, are not homotopic).

The configurations $\mathbf{g}_n$, $n \geqslant 1$, can be attained from the configuration $\mathbf{g}_0$ by cutting the torus at $\theta = 0$, twisting it in the direction of the coordinate ϕ through an angle $2\pi n$, in such a way that the faces of the cut do not become distorted, and glueing them together. The act of glueing the cut restores the initial continuity of the torus in the neighbourhood of $\theta = 0$. The transition from $\mathbf{g}_0$ to $\mathbf{g}_n$, $n \geqslant 1$, however, requires a temporary destruction of continuity. The *same* transition can be achieved by a temporary cut at some $\theta \neq 0$ while the continuity at $\theta = 0$ is preserved. This indicates why the discontinuity of the transition becomes apparent only at the *global* level.

The configurations $\mathbf{g}_n$, $n \geqslant 1$, are obtained from $\mathbf{g}_0$ by introducing a special kind of **Volterra dislocations** (cf. Chapter 4) into the torus $\mathscr{B}$. A larger class of Volterra dislocations can be constructed by twisting the torus in such a way that the faces of the cut rotate through an angle $\alpha \neq 2\pi n$ with respect to each other. Let $k = \alpha/2\pi$. After glueing the faces of the cut the initial continuity at $\theta = 0$, 2π will not be restored. Let $\mathbf{g}_k$ be defined by equation (1.2.14) with the real number k replacing n. Let us consider θ, ϕ as two real variables ranging over $\boldsymbol{R}$. Since, by assumption, the configuration $\mathbf{g}_0 \in \mathscr{K}_0$ correctly labels the material points, it follows that (r, θ, ϕ) and $(r, \theta+2\pi n, \phi)$, $n \in \boldsymbol{Z}$, refer to the same material point. Hence the mapping $\mathbf{g}_k$ assigns a denumerable set of points in the space to every material point and $\mathbf{g}_k \bar{\in} \mathscr{K}_0$.

* The assumption that $\mathbf{g} \circ \mathbf{g}_0^{-1}$ is a homeomorphism would be sufficient to guarantee this property. The requirement that $\mathbf{g} \cdot \mathbf{g}_0^{-1}$ be a $\mathscr{C}^1$-diffeomorphism is motivated by the local dependence of stress on the configuration in the theory of (local) elasticity.

In order to avoid this difficulty we shall consider the multivalued "inverse" mapping $\tilde{\mathbf{g}}_{-k} := \mathbf{g}_0^{-1} \circ \mathbf{g}_{-k} \circ \mathbf{g}_0^{-1}$ from the range $\mathscr{U} \subset \boldsymbol{R}^3$ of $\mathbf{g}_0$ in the space onto the body $\mathscr{B}$. Let us introduce the coordinate system $(\bar{r}, \bar{\theta}, \bar{\phi}) = \mathbf{g}_0^{-1}(\mathbf{x})$ on $\mathscr{U}$. We shall identify $(\bar{\theta}, \bar{\phi})$ with $(\bar{\theta}+2\pi n, \bar{\phi}+2\pi m)$. In view of our assumptions there is a bijective correspondence between the points of $\mathscr{U}$ and the material points in the dislocated configuration. Hence the mapping $\mathbf{g}_{-k}$ assigns a denumerable set of numerical labels to every material point. In the coordinates $(\bar{r}, \bar{\theta}, \bar{\phi})$ $\mathbf{g}_0^{-1} \circ \mathbf{g}_{-k}(\bar{r}, \bar{\theta}, \bar{\phi}) = (r, \theta, \phi)$ with $r = \bar{r}$, $\theta = \bar{\theta}$, $\phi = \bar{\phi} - k\bar{\theta}$. The restriction of $\mathbf{g}_{-k}$ to a sector $\beta < \bar{\theta} < \gamma$, $0 < \gamma - \beta < 2\pi$, consists of a denumerable family of mappings $\mathbf{g}^n_{-k}$, corresponding to the sheets of $\mathbf{g}_{-k}$, $n = \ldots, -1, 0, 1, 2, \ldots$. Each mapping $\mathbf{g}^n_{-k}$ is univalued and invertible and the corresponding mapping $\tilde{\mathbf{g}}^n_{-k}$ can be used to label the material points which occupy the sector $\beta < \bar{\theta} < \gamma$ of $\mathscr{U}$ in the dislocated configuration.

The above example suggests the idea of labelling the material points of $\mathscr{B}$ by means of local coordinates on $\mathscr{B}$. We shall see how fruitful and natural this idea may be in some applications. Let $\mathscr{B}$ be the set of material points with the topology and differential structure induced by some virtual configuration $\mathbf{g}_0 \in \mathscr{K}_0$, e.g. $\mathbf{g}_0(\cdot) = \mathbf{h}(\cdot, 0)$. Let $\{\mathscr{U}_\iota | \iota \in \mathscr{I}\}$ be a family of open domains covering $\mathscr{B}$. Every index $\iota \in \mathscr{I}$ will be assigned an injective mapping $\boldsymbol{\varkappa}_\iota$: $\mathscr{U}_\iota \to \boldsymbol{R}^3$ such that $\mathbf{g}_0 \circ \boldsymbol{\varkappa}_\iota^{-1}$: $\boldsymbol{\varkappa}_\iota(\mathscr{U}_\iota) \to \boldsymbol{R}^3$, and $\boldsymbol{\varkappa}_\iota \circ \mathbf{g}_0^{-1}$: $\mathbf{g}_0(\mathscr{U}_\iota) \to \boldsymbol{R}^3$ are both of class $\mathscr{C}^1$ and have non-vanishing Jacobians. The family $\mathfrak{a}_0 = \{(\mathscr{U}_\iota, \boldsymbol{\varkappa}_\iota) | \iota \in \mathscr{I}\}$ induces a differential structure on $\mathscr{B}$ which is equivalent to the structure induced by $\mathbf{g}_0$. We can complete the family $\mathfrak{a}_0$ to obtain a maximal atlas of the **differential manifold** $\mathscr{B}$. The body $\mathscr{B}$ appears now as an abstract differential manifold which can be mapped diffeomorphically onto open subsets of $\boldsymbol{R}^3$. There is no restriction on the choice of coordinate systems on $\mathscr{B}$. The material points $P \in \mathscr{U}_\iota \subset \mathscr{B}$ can be identified by means of **local reference configurations** $\boldsymbol{\varkappa}_\iota$: $\mathscr{U}_\iota \to \boldsymbol{R}^3$. In the case of the torus we can take $\mathscr{U}_\iota =]\beta_\iota, \gamma_\iota[$, $\boldsymbol{\varkappa}_\iota = \mathbf{g}^{n_\iota}_{-k}$.

An example of the use of local reference configurations will be found in the theory of dislocations (Chapter 4). In this case the configurations $\boldsymbol{\varkappa}_\iota$ are chosen in such a way that $\mathscr{U}_\iota$ appears as a homogeneous piece of material which does not exhibit any stress in the configuration $\boldsymbol{\varkappa}_\iota$.* Such reference configurations are convenient in the computation of stress introduced by the dislocations.

* Intuitively, internal stress associated with the dislocations can be removed from the body $\mathscr{B}$ by cutting it into the pieces $\mathscr{U}_\iota$. When the stress in the piece $\mathscr{U}_\iota$ is relaxed it assumes the configuration $\boldsymbol{\varkappa}_\iota$.

1.2.3 Local Impenetrability Condition, Configuration Gradients and Mass

A notion of virtual configurations incorporating such global conditions as injectivity and homotopy can hardly provide a workable framework for solving the differential equations of motion or equilibrium of $\mathscr{B}$. The homotopy of the configurations $\mathbf{h}(\cdot, t)$ with the initial one $\mathbf{h}(\cdot, 0)$ is ensured by the motion $\mathbf{h}$ itself provided $\mathbf{h}$ is continuous. In the case of equilibrium conditions the homotopy of the solutions with a given configuration can be checked *a posteriori*.

The global condition of impenetrability (injectivity) is sometimes built into the unilateral boundary conditions and enters the final formulation of the problem without influencing the choice of the underlying solution class. In our general and abstract discussion of the initial and boundary value problems of elasticity the global condition of injectivity will be disregarded. We shall replace it by the local impenetrability condition which imposes a local restriction on the class of reference configurations and on the class of virtual configurations.

We shall assume that all the local reference configurations $\{\varkappa_\iota |\ \iota \in \mathscr{I}\}$ satisfy the condition

$$\det\nabla(\varkappa_{\iota'} \circ \varkappa_\iota^{-1})\big(\varkappa_\iota(P)\big) > 0 \quad \text{for } P \in \mathscr{U}_\iota \cap \mathscr{U}_{\iota'},\ \iota, \iota' \in \mathscr{I}. \qquad (1.2.15)$$

The mapping $\mathbf{g} \in \mathscr{K}_0$ belongs to the class $\mathscr{K}_+$ iff

$$\det\nabla(\mathbf{g} \circ \varkappa_\iota^{-1})\big(\varkappa_\iota(P)\big) > 0 \quad \text{for } P \in \mathscr{U}_\iota,\ \iota \in \mathscr{I}. \qquad (1.2.16)$$

The condition (1.2.15) implies that the body $\mathscr{B}$ is an oriented manifold while (1.2.16) says that a virtual configuration $\mathbf{g} \in \mathscr{K}_+$ preserves the orientation. We note that a configuration reversing the orientation cannot be obtained from a configuration preserving the orientation by a continuous deformation process.

We now make a new step towards a local description of configurations. For every $P \in \mathscr{B}$ we introduce the following equivalence relation on local reference configurations:

$$\varkappa_\iota \sim_P \varkappa_{\iota'} \quad \text{iff} \quad \nabla(\varkappa_{\iota'} \circ \varkappa_\iota^{-1})\big(\varkappa_\iota(P)\big) = \mathbf{E}.$$

Since (every restriction of) a virtual configuration $\mathbf{g} \in \mathscr{K}_+$ is a local reference configuration, the equivalence relation encompasses the virtual configurations $\mathbf{g} \in \mathscr{K}_+$.

The equivalence classes $\mathbf{K} = [\varkappa_\iota]_{\sim P}$ will be called **configuration gradients** at P. If $\varkappa_0$ is a local reference configuration defined on a nbhd $\mathscr{U}_0$ of P, $\mathbf{K}$ is a configuration gradient at P and $\mathbf{g}$, $\varkappa_\iota$, $\varkappa_{\iota'} \in \mathbf{K}$, then $\nabla(\mathbf{g} \circ \varkappa_0^{-1})\big(\varkappa_0(P)\big) = \nabla(\varkappa_\iota \circ \varkappa_0^{-1})\big(\varkappa_0(P)\big) = \nabla(\varkappa_{\iota'} \circ \varkappa_0^{-1})\big(\varkappa_0(P)\big) = \mathbf{F} \in L_o^+(R^3, R^3)$. $L_o^+(R^3, R^3)$

denotes the set of all the linear mappings of $\boldsymbol{R}^3$ onto $\boldsymbol{R}^3$ with positive determinant. The mapping $\mathbf{F}$ is uniquely determined by $\boldsymbol{\varkappa}_0$ and $\mathbf{K}$. Conversely, for every $(\boldsymbol{\varkappa}_0, \mathscr{U}_0)$ and $\mathbf{F}$ there is a local reference configuration $(\boldsymbol{\varkappa}', \mathscr{U}_0)$ such that $\nabla(\boldsymbol{\varkappa}' \circ \boldsymbol{\varkappa}_0{}^{-1})(\boldsymbol{\varkappa}_0(P)) = \mathbf{F}$, e.g. $\boldsymbol{\varkappa}'(P') \equiv \mathbf{F} \circ \boldsymbol{\varkappa}_0(P')$ for $P' \in \mathscr{U}_0$.

In terms of differential geometry the configuration gradients form a cone in each fiber of the principal frame bundle over $\mathscr{B}$. The standard fiber of the principal frame bundle is $L(\boldsymbol{R}^3, \boldsymbol{R}^3)$, and the cone corresponds to the subset $L_0^+(\boldsymbol{R}^3, \boldsymbol{R}^3)$ of the standard fiber.

In view of equations (1.2.15) and (1.2.16) every local reference configuration $\boldsymbol{\varkappa}_\iota$ (or virtual configuration $\mathbf{g}$) can be assigned the corresponding positive mass density:

$$\varrho_{\varkappa_\iota} : \boldsymbol{\varkappa}_\iota(\mathscr{U}_\iota) \to \boldsymbol{R}_+$$

in such a way that

$$\varrho_{\varkappa_\iota}/\varrho_{\varkappa_{\iota'}} \circ \boldsymbol{\varkappa}_{\iota'} \circ \boldsymbol{\varkappa}_\iota{}^{-1} = \det\nabla(\boldsymbol{\varkappa}_{\iota'} \circ \boldsymbol{\varkappa}_\iota{}^{-1}) > 0. \tag{1.2.17}$$

Clearly, $\varrho_{\varkappa_\iota}(\xi^\alpha) = \varrho_{\varkappa_{\iota'}}(\zeta^\alpha)$ if $\boldsymbol{\varkappa}_\iota{}^{-1}(\xi^\alpha) = \boldsymbol{\varkappa}_{\iota'}{}^{-1}(\zeta^\alpha) = P$ and $\boldsymbol{\varkappa}_\iota \sim_P \boldsymbol{\varkappa}_{\iota'}$. Hence for every $P \in \mathscr{U}_\iota$ the density of mass at P depends on the configuration gradient at P only.

We assume that the function $\varrho_{\varkappa_\iota}$ is measurable and bounded on $\boldsymbol{\varkappa}_\iota(\mathscr{U}_\iota)$, for every $\iota \in \mathscr{I}$. For every measurable subset $\mathscr{V} \subset \mathscr{U}_\iota$, $\iota \in \mathscr{I}$, we define

$$M(\mathscr{V}) := \int_{\varkappa_\iota(\mathscr{V})} dL^3 \varrho_{\varkappa_\iota}. \tag{1.2.18}$$

For $\mathscr{V} \subset \mathscr{U}_\iota \cap \mathscr{U}_{\iota'}$ equation (1.2.18) implies that

$$M(\mathscr{V}) = \int_{\varkappa_{\iota'}(\mathscr{V})} dL^3 \varrho_{\varkappa_{\iota'}}, \tag{1.2.19}$$

on account of (1.2.17). $M(\cdot)$ is a σ-additive function on Borel subsets of $\mathscr{U}_\iota$, for every $\iota \in \mathscr{I}$. It can obviously be extended to a σ-additive function on Borel subsets of $\mathscr{B}$, if (1.2.19) is taken into account. The measure M on $\mathscr{B}$ constructed in this way is the **mass.**

1.2.4 Sobolev Classes of Virtual Configurations

In the last sections of this chapter we derive a system of partial differential equations satisfied by the motion $\mathbf{h}$ of the body as well as a system of equations determining the conditions of equilibrium of a virtual configuration $\mathbf{g}$. The common experience with partial differential equations of these types indicates that the solutions should be sought in a class of virtual configurations which are locally integrable and belong to a Sobolev class. We shall now proceed to define the Sobolev classes $\mathscr{K}_p$ of virtual solutions.

A subset $\mathscr{A}$ of $\mathscr{B}$ is said to **have measure zero** if it can be covered by a countable family of sets $\mathscr{A}_\iota \subset \mathscr{U}_\iota$ such that $\boldsymbol{\varkappa}_\iota(\mathscr{A}_\iota)$ has Lebesgue measure zero in $\boldsymbol{R}^3$. In this connection we shall note that for $\mathscr{A} \subset \mathscr{U}_\iota \cap \mathscr{U}_{\iota'}$, $\mathbf{g}_{\iota'\iota} := \boldsymbol{\varkappa}_{\iota'} \circ \boldsymbol{\varkappa}_\iota^{-1}$: $\boldsymbol{\varkappa}_\iota(\mathscr{U}_\iota \cap \mathscr{U}_{\iota'}) \to \boldsymbol{R}^3$, $L^3(\boldsymbol{\varkappa}_\iota(\mathscr{A})) = 0$ implies that

$$L^3(\boldsymbol{\varkappa}_{\iota'}(\mathscr{A})) = \int_{\boldsymbol{\varkappa}_\iota(\mathscr{A})} dL^3\, |J(\mathbf{g}_{\iota'\iota})| = 0.$$

$J(\mathbf{g}_{\iota'\iota})$ denotes the Jacobian of the transformation $\mathbf{g}_{\iota'\iota}$.

We shall say that a statement $S(P)$ holds for almost all $P \in \mathscr{B}$ (or **almost everywhere** *a.e.*, on $\mathscr{B}$) if it is true for all $P \in \mathscr{B} \setminus \mathscr{A}$, where $\mathscr{A}$ is a set of measure zero in $\mathscr{B}$. A function $\mathbf{f}$: $\mathscr{B} \to \boldsymbol{R}^3$ defined a.e. on $\mathscr{B}$ is said to be **measurable,** if $\forall\, \iota \in \mathscr{I}$, $\mathbf{f} \circ \boldsymbol{\varkappa}_\iota^{-1}$ is measurable on $\boldsymbol{\varkappa}_\iota(\mathscr{U}_\iota)$ in the sense of Lebesgue. We note that $\mathbf{f} \circ \boldsymbol{\varkappa}_\iota^{-1} | \boldsymbol{\varkappa}_\iota(\mathscr{U}_\iota \cap \mathscr{U}_{\iota'})$ is measurable iff $\mathbf{f} \circ \boldsymbol{\varkappa}_{\iota'}^{-1} | \boldsymbol{\varkappa}_{\iota'}(\mathscr{U}_\iota \cap \mathscr{U}_{\iota'})$ is measurable.

A function $\mathbf{f}$: $\mathscr{B} \to \boldsymbol{R}^3$ belongs to $\mathscr{L}^p_{loc}$, $p \geqslant 1$, if every $P \in \mathscr{B}$ has a compact nbhd $\mathscr{K} \subset \mathscr{U}_\iota$, $\iota \in \mathscr{I}$, such that

$$\int_{\boldsymbol{\varkappa}_\iota(\mathscr{K})} dL^3 \left| \sum_{k=1}^{3} (f^k \circ \boldsymbol{\varkappa}_\iota^{-1})^2 \right|^{p/2} < \infty. \tag{1.2.20}$$

Since $J(\mathbf{g}_{\iota'\iota})$ is bounded on the compact set $\boldsymbol{\varkappa}_\iota(\mathscr{K})$, $\mathscr{K} \subset \mathscr{U}_\iota \cap \mathscr{U}_{\iota'}$, definition (1.2.20) does not depend on the choice of $\iota \in \mathscr{I}$. The Minkowski inequality implies that $\mathscr{L}^p_{loc}$ is a linear space. Its topology is defined by the semi-norms (1.2.20).

The space $W^{1,p}_{loc}$ can be defined in two equivalent ways. We shall say that $\mathbf{f} \in \mathscr{L}^p_{loc}$ belongs to $W^{1,p}_{loc}$, $p \geqslant 1$, if every $P \in \mathscr{B}$ has a compact nbhd $\mathscr{K} \subset \mathscr{U}_\iota$, $\iota \in \mathscr{I}$, such that $\mathbf{f} \circ \boldsymbol{\varkappa}_\iota^{-1} \in W^{1,p}(\boldsymbol{\varkappa}_\iota(\mathscr{K}))$. It follows that $\mathbf{f} \circ \boldsymbol{\varkappa}_{\iota'}^{-1} = \mathbf{f} \circ \boldsymbol{\varkappa}_\iota^{-1} \circ \mathbf{g}_{\iota\iota'} \in W^{1,p}_{loc}(\boldsymbol{\varkappa}_{\iota'}(\mathscr{K}))$ provided $\mathscr{K} \subset \mathscr{U}_\iota \cap \mathscr{U}_{\iota'}$, by Theorem 1.1.52. Alternatively, we can construct the linear topological space $W^{1,p}_{loc}$ by completing $\mathscr{C}^\infty(\mathscr{B})$ with respect to the topology defined by the semi-norms $\|\cdot\|_{p,\mathscr{K},\iota}$, for all compact $\mathscr{K} \subset \mathscr{U}_\iota$, $\iota \in \mathscr{I}$,

$$\|\mathbf{f}\|^p_{p,\mathscr{K},\iota} = \int_{\boldsymbol{\varkappa}_\iota(\mathscr{K})} dL^3 \left\{ \left| \sum_{k=1}^{3} (f^k \circ \boldsymbol{\varkappa}_\iota^{-1})^2 \right|^{p/2} + \left| \sum_{k,\mu=1}^{3} \left(\frac{\partial (f^k \circ \boldsymbol{\varkappa}_\iota^{-1})}{\partial \xi^\mu} \right)^2 \right|^{p/2} \right\}. \tag{1.2.21}$$

It is easy to see that the semi-norms $\|\cdot\|_{p,\mathscr{K},\iota}$ and $\|\cdot\|_{p,\mathscr{K},\iota'}$ are equivalent if $\mathscr{K} \subset \mathscr{U}_\iota \cap \mathscr{U}_{\iota'}$.

In the following we assume that $\mathscr{B}$ is a compact manifold. Let $\mathscr{I}' \subset \mathscr{I}$ be a finite set of indices such that $\{\mathscr{U}_\iota |\ \iota \in \mathscr{I}'\}$ is a covering of $\mathscr{B}$. Let

$$\|\mathbf{f}\|_{p,\mathscr{I}'} := \sum_{\iota\in\mathscr{I}'} \int_{\varkappa_\iota(\mathscr{U}_\iota)} dL^3 \left\{ \left| \sum_{k=1}^{3} (f^k \circ \varkappa_\iota^{-1})^2 \right|^{p/2} + \left| \sum_{k,\mu=1}^{3} \left(\frac{\partial (f^k \circ \varkappa_\iota^{-1})}{\partial \xi^\mu} \right)^2 \right|^{p/2} \right\}, \tag{1.2.22}$$

for every $\mathscr{C}^\infty$ mapping $\mathbf{f}\colon \mathscr{B} \to \boldsymbol{R}^3$.

Let $\mathscr{I}'' \subset \mathscr{I}$ be another finite subset of $\mathscr{I}$ such that $\{\mathscr{U}_\iota | \ \iota \in \mathscr{I}''\}$ covers $\mathscr{B}$. Let $\mathscr{V}_{\iota\iota'} := \mathscr{U}_\iota \cap \mathscr{U}_{\iota'}$ for $\iota \in \mathscr{I}'$, $\iota' \in \mathscr{I}''$, $\varkappa_{\iota\iota'} := \varkappa_\iota | \mathscr{V}_{\iota\iota'}$, $\varkappa'_{\iota\iota'} := \varkappa_{\iota'} | \mathscr{V}_{\iota\iota'}$. We now compare the norms $\|\cdot\|_{p,\mathscr{I}''}$, $\|\cdot\|'_p$, $\|\cdot\|''_p$ defined in terms of the finite coverings $\{(\mathscr{U}_\iota, \varkappa_\iota) | \ \iota \in \mathscr{I}''\}$, $\{(\mathscr{V}_{\iota\iota'}, \varkappa_{\iota\iota'}) | \ \iota \in \mathscr{I}', \ \iota' \in \mathscr{I}''\}$ and $\{(\mathscr{V}_{\iota\iota'}, \varkappa'_{\iota\iota'}) | \ \iota \in \mathscr{I}', \iota' \in \mathscr{I}''\}$ resp. Clearly

$$K_1 \|\mathbf{f}\|'_p \leqslant \|\mathbf{f}\|_{p,\mathscr{I}'} \leqslant \|\mathbf{f}\|'_p, \quad K_1 > 0,$$

$$K_2 \|\mathbf{f}\|''_p \leqslant \|\mathbf{f}\|_{p,\mathscr{I}''} \leqslant \|\mathbf{f}\|''_p, \quad K_2 > 0$$

with K_1, K_2 independent of $\mathbf{f}$. By a transformation of variables we establish that the norms $\|\cdot\|'_p$, $\|\cdot\|''_p$ are equivalent. Hence the norms $\|\cdot\|_{p,\mathscr{I}'}$ and $\|\cdot\|_{p,\mathscr{I}''}$ are equivalent and define the same topology on $\mathscr{C}^\infty$. Completing $\mathscr{C}^\infty$ with respect to any of these norms we obtain the topological vector space $W^{1,p}$. From a set-theoretical and topological point of view $W^{1,p}$ does not depend on the choice of $\mathscr{I}'$, although every norm $\|\cdot\|_{p,\mathscr{I}'}$ induces a different Banach space structure on $W^{1,p}$.

It is easy to see that $W^{1,p} = W^{1,p}_{loc} \subset \mathscr{L}^p_{loc} = \mathscr{L}^p$ for a compact $\mathscr{B}$. The identity mapping $W^{1,p} \to \mathscr{L}^p$ is a compact continuous injection (cf. Adams, 1975; Neĉas, 1968). It is possible to define a natural norm on $\mathscr{L}^p$, $p \geqslant 1$:

$$\|\mathbf{f}\|_p = \left\{ \int_{\mathscr{B}} dM \, |\mathbf{f}|^p \right\}^{1/p}, \tag{1.2.23}$$

using the mass M as the measure on $\mathscr{B}$. There is no such natural norm on $W^{1,p}$. Indeed, even if there is a global reference configuration $\varkappa$ defined on $\mathscr{U} = \mathscr{B}$, the norm (1.2.22) defined in terms of the one-element set $\mathscr{I} = \{1\}$ with $\mathscr{U}_1 = \mathscr{B}$, $\varkappa_1 = \varkappa$ depends on the choice of $\varkappa$ (one might use some $\varkappa' = \mathbf{r} \circ \varkappa$ instead). Replacing the Lebesgue measure by the mass does not remove this ambiguity.

If $\mathbf{f} \in W^{1,p}$, then $\nabla(\mathbf{f} \circ \varkappa_\iota^{-1}) \in \mathscr{L}^p_{loc}$ for every $\iota \in \mathscr{I}$. We shall now define the Sobolev class $\mathscr{K}_p$ of virtual configurations: $\mathbf{g} \in \mathscr{K}_p$ iff $\mathbf{g}\colon \mathscr{B} \to \boldsymbol{R}^3$ belongs to $W^{1,p}$ and

$$\forall \iota \in \mathscr{I} \ \det \nabla(\mathbf{g} \circ \varkappa_\iota^{-1}) > 0 \quad \text{a.e. on } \varkappa_\iota(\mathscr{U}_\iota). \tag{1.2.24}$$

On account of (1.2.15) the condition (1.2.24) is invariant with respect to the coordinate transformations $\mathbf{g}_{\iota\iota'}$ on $\boldsymbol{\varkappa}_{\iota'}(\mathscr{U}_\iota \cap \mathscr{U}_{\iota'})$.

$\mathscr{K}_p$ is a cone in the topological vector space $W^{1,p}$. The class $\mathscr{K}_+$ is dense in $\mathscr{K}_p$. As usual in the theory of Sobolev spaces, two mappings $\mathbf{g}_1, \mathbf{g}_2 \in \mathscr{K}_p$ define the same element of $\mathscr{K}_p$ if they coincide except on a set of measure zero in $\mathscr{B}$. Since the value of $\mathbf{g}$ at a fixed point $P \in \mathscr{B}$ may be changed at will the notion of injectivity of $\mathbf{g}$ is rather vague.* The condition of (global) impenetrability can sometimes be expressed in terms of boundary conditions.

Let $\mathbf{h}$ be the motion of $\mathscr{B}$ with respect to some $\varphi_0 \in \Gamma_0$, $\mathbf{h}(\cdot\,, t) \in \mathscr{K}_p$ for all t. The same motion expressed in terms of another coordinate system $\varphi_1 \in \Gamma_0$ assumes the form

$$\mathbf{h}'(P, t) \equiv \mathbf{A}(t)\mathbf{h}(P, t) + t\mathbf{b} + \mathbf{c},$$

$\det\mathbf{A}(t) = 1$. Hence $\mathbf{h}'(\cdot\,, t) \in \mathscr{K}_p$ too.

1.2.5 Balance Equations in Continuum Mechanics

Equations of motion of a body will be derived by combining the balance of momentum with the constitutive relations of an elastic material. Every balance equation states that the temporal variation of an extensive quantity A within a part $\mathscr{P}$ of $\mathscr{B}$ is balanced by the influx B of A through $\partial\mathscr{P}$ plus the internal supply C of A:

$$A(t_2, \mathscr{P}) - A(t_1, \mathscr{P}) - \int_{t_1}^{t_2} dt\, B(t, \partial\mathscr{P}) = \int_{t_1}^{t_2} dt\, C(t, \mathscr{P}). \tag{1.2.25}$$

A quantity A is said to be **extensive** if it is an additive function defined on all the Borel subsets $\mathscr{P}$,

$$A(t, \mathscr{P}_1 \cup \mathscr{P}_2) = A(t, \mathscr{P}_1) + A(t, \mathscr{P}_2) \tag{1.2.26}$$

for any two disjoint Borel sets $\mathscr{P}_1, \mathscr{P}_2 \subset \mathscr{B}$ and all t.

It is often true that an extensive quantity A satisfies an estimate

$$|A(t, \mathscr{P})| \leqslant K(t) L^3(\boldsymbol{\varkappa}(\mathscr{P})), \quad 0 < K(t) < \infty \tag{1.2.27}$$

for all $t \in [0, T]$ and some reference configuration $\boldsymbol{\varkappa}$. We shall see later that the property (1.2.27) is invariant with respect to the change of reference configuration.

* A Sobolev mapping $\boldsymbol{\chi}: \Omega \to \boldsymbol{R}^3$ satisfying the condition $\det\nabla\boldsymbol{\chi} > 0$ a.e. in Ω is not in general invertible. The invertibility properties of such mappings were studied by Ball (1980). If $\boldsymbol{\chi} \in W^{1,p}(\Omega)$, $p \geqslant 1$, coincides on $\partial\Omega$ with a homeomorphism $\boldsymbol{\chi}_0: \Omega \to U$, $\gamma\boldsymbol{\chi} = \boldsymbol{\chi}_0|\partial\mathscr{U}$ then $\boldsymbol{\chi}(\Omega) = \boldsymbol{\chi}_0(\Omega) = \mathscr{U}$ and for every $x \in \mathscr{U}$ $\boldsymbol{\chi}^{-1}(\{\mathbf{x}\})$ is a continuum. Under an additional integral condition the inverse mapping $\boldsymbol{\chi}^{-1}$ exists, is continuous and belongs to $W^{1,q}(\mathscr{U})$, $q^{-1} = 1 - p^{-1}$.

Let us assume that $\overline{\mathscr{B}}$ is compact. Equations (1.2.26) and (1.2.27) imply then that $A(t, \cdot)$ is σ-additive for all $t \in [0, T]$. Indeed, let $\mathscr{P} = \bigcup_{i=1}^{\infty} \mathscr{P}_i$, $\mathscr{P}_i$ being disjoint Borel subsets of $\mathscr{B}$. Equation (1.2.26) implies that

$$A(t, \bigcup_{i=1}^{n} \mathscr{P}_i) = \sum_{i=1}^{n} A(t, \mathscr{P}_i).$$

Since

$$\left|A(t, \mathscr{P}) - \sum_{i=1}^{n} A(t, \mathscr{P}_i)\right| = \left|A(t, \bigcup_{i=n+1}^{\infty} \mathscr{P}_i)\right| \leqslant K(t) L^3 \big(\varkappa(\bigcup_{i=n+1}^{\infty} \mathscr{P}_i)\big) \tag{1.2.28}$$

and $L^3(\varkappa(\mathscr{P})) = \sum_{i=1}^{\infty} L^3(\varkappa(\mathscr{P}_i)) < \infty$, the expression on the right-hand side tends to zero with $n \to \infty$. The Radon–Nikodym theorem implies that there is a locally integrable function $a(t, \cdot)$ such that

$$A(t, \mathscr{P}) = \int_{\varkappa(\mathscr{P})} dL^3 \, a(t, \xi), \tag{1.2.29}$$

$|a(t, \xi)| \leqslant K(t)$ for almost all $\xi \in \boldsymbol{\varkappa}(\mathscr{B})$, and all $t \in [0, T]$.

Analogous considerations can be applied to the rate of supply $C(t, \mathscr{P})$. For $\mathscr{B}$ we shall assume that it is a function defined on surface elements and satisfying some additivity and continuity conditions (with respect to the area). We shall see in the next section that under these assumptions $B(t, \partial\mathscr{P}) = \int_{\partial\varkappa(\mathscr{P})} dH^2 \mathbf{b}(t, \xi) \cdot \mathbf{n}$ ($\mathbf{n}$ denotes here the unit outer normal on $\partial\boldsymbol{\varkappa}(\mathscr{P})$. Hence (1.2.25) assumes the following form:

$$\int_{\varkappa(\mathscr{P})} dL^3 \, [a(t_2, \xi) - a(t_1, \xi)] = \int_{t_1}^{t_2} dt \int_{\partial\varkappa(\mathscr{P})} dH^2 \, \mathbf{b}(t, \xi) \cdot \mathbf{n} + \int_{t_1}^{t_2} dt \int_{\varkappa(\mathscr{P})} dL^3 \, c(t, \xi) \tag{1.2.30}$$

for every Borel set $\mathscr{P} \subset \mathscr{B}$.

In particular equation (1.2.30) holds for every cube $\boldsymbol{\varkappa}(\mathscr{P})$. Suppose that $\mathbf{b}$ is sufficiently regular so that the Gauss–Green theorem can be applied. In this case the surface integral is replaced by a volume integral of $\mathrm{div}_\xi \mathbf{b}$ and the Lebesgue differentiation theorem yields

$$a(t_2, \xi) - a(t_1, \xi) = \int_{t_1}^{t_2} dt \, \mathrm{div}_\xi \mathbf{b}(t, \xi) + \int_{t_1}^{t_2} dt \, c(t, \xi)$$

a.e. in $\varkappa(\mathscr{P})$. The function $a(\cdot\,,\xi)$ is absolutely continuous and for almost all t, ξ

$$a_{,t}(t,\xi) = \operatorname{div}_\xi \mathbf{b}(t,\xi) + c(t,\xi). \tag{1.2.31}$$

This sketchy argument involves an unjustified use of the Fubini theorem. It works only for smooth $\mathbf{b}$ and continuous a, c. In practice this is hardly ever the case. Moreover, the existence of the flux density $\mathbf{b}(t,\,\cdot)$ has not been proved. In the following sections we shall prove that under some natural hypotheses the balance equation (1.2.25) can actually be expressed in terms of some density functions a, $\mathbf{b}$, c which are defined almost everywhere and satisfy a differential equation (1.2.31) in the sense of distributions. The latter assertion for the balance of momentum amounts to the virtual work principle.

1.2.6 Preparatory Results

It will be more convenient to formulate the balance equation (1.2.25) in terms of volume and surface integrals in $\tilde{\mathscr{B}} = \mathscr{B}\times I$, $I = [0, T[$, $T<\infty$. Let $\tilde{\mathscr{P}} = \mathscr{P}\times[t_1, t_2]$ be a cylinder set in $\tilde{\mathscr{B}}$. The left-hand side of equation (1.2.25) can be treated as an additive function F of three-dimensional surface elements in $\tilde{\mathscr{B}}$, evaluated at $\partial\tilde{\mathscr{P}}$. The right-hand side is an additive function G defined on a class of four-dimensional sets including the cylinder sets $\tilde{\mathscr{P}}$. Equation (1.2.25) assumes the form:

$$F(\partial\tilde{\mathscr{P}}) = G(\tilde{\mathscr{P}}). \tag{1.2.32}$$

The underlying set $\tilde{\mathscr{B}} = \mathscr{B}\times I$ is a manifold. In the following we shall assume the continuity of the set functions with respect to some volume and area measures induced by the Lebesgue and Hausdorff measures in $\boldsymbol{R}^4$. For this purpose we shall fix a globally defined reference configuration $\varkappa$ on $\mathscr{B}$* and use the Lebesgue and Hausdorff measure on $\hat{\mathscr{B}} := \varkappa(\mathscr{B})\times I$.

Let $\varkappa'$ be another configuration, $\varkappa'\circ\overset{-1}{\varkappa} = \boldsymbol{\phi}\in\mathscr{C}^1(\varkappa(\mathscr{B}))$. The transformation from $\varkappa$ to $\varkappa'$ induces a transformation of coordinates g on $\tilde{\mathscr{B}}$: $t' = t$, $(\zeta^\alpha) = \boldsymbol{\phi}(\xi^\beta)$. We shall prove that for precompact $\mathscr{A}$, $\mathscr{S}\subset\hat{\mathscr{B}}$

(1) $L^4(g(\mathscr{A})) \leqslant KL^4(\mathscr{A})$,

(2) $H^3(g(\mathscr{S})) \leqslant MH^3(\mathscr{S})$

and K, M can be chosen to be constant for $\mathscr{A}$, $\mathscr{S}$ contained in an arbitrary fixed compact subset of $\hat{\mathscr{B}}$.

* In the case of local reference configurations $(\varkappa_\iota, \mathscr{U}_\iota)$ it is necessary to assume that the sets $\mathscr{A}$, $\mathscr{S}$, ... considered below are contained in some $\mathscr{U}_\iota$.

(1) follows immediately from the boundedness of the Jacobian $J(g)$ of g on compact sets.

For (2) we shall introduce the following notations: $x^0 = t$, $x^\alpha = \xi^\alpha$, $y^k = g^k(x^l)$, $\alpha, \beta = 1, 2, 3$, $k, l = 0, \ldots, 3$. Let $\mathscr{S} \subset \hat{\mathscr{B}}$ be a smooth piece of a surface given in a parametric form $x^k = \bar{x}^k(\eta^\alpha)$, $\alpha = 1, 2, 3$, $k = 0, \ldots, 3$, (η^α) ranging over a set $\Sigma \subset \boldsymbol{R}^3$. Let $\mathscr{S}$ be contained in a compact set $\mathscr{K}$. The Hausdorff measure of $g(\mathscr{S})$ is

$$H^3(g(\mathscr{S})) = \int_\Sigma dL^3(\eta)\gamma^{1/2}, \tag{1.2.33}$$

$$\gamma_{\alpha\beta} := \sum_{k,p,q=0}^{3} \frac{\partial y^k}{\partial x^p}\frac{\partial \bar{x}^p}{\partial \eta^\alpha}\frac{\partial y^k}{\partial x^q}\frac{\partial \bar{x}^q}{\partial \eta^\beta}, \qquad \gamma := \det\|\gamma_{\alpha\beta}\|.$$

Let

$$g_{pq} := \sum_{k=0}^{3} \frac{\partial y^k}{\partial x^p}\frac{\partial y^k}{\partial x^q}.$$

Since

$$\gamma = \frac{1}{6} A^{p_1 p_2 p_3} g_{p_1 q_1} g_{p_2 q_2} g_{p_3 q_3} A^{q_1 q_2 q_3}, \tag{1.2.34}$$

$$A^{p_1 p_2 p_3} := \epsilon^{\alpha\beta\gamma} \frac{\partial \bar{x}^{p_1}}{\partial \eta^\alpha}\frac{\partial \bar{x}^{p_2}}{\partial \eta^\beta}\frac{\partial \bar{x}^{p_3}}{\partial \eta^\gamma},$$

from the general inequality

$$G_{AB} w^A w^B \leqslant C \sum_{A,B} |w^A||w^B| \leqslant C\Big(\sum_A |w^A|\Big)^2$$

$$= C\Big(\sum_A |w^A|^2 + \sum_{A \neq B} |w^A w^B|\Big) \leqslant 2C \sum_A |w^A|^2,$$

$$C := \sup\{|G_{AB}(x)| \;\; x \in \mathscr{K}, A, B\},$$

it follows that

$$H^3(g(\mathscr{S})) \leqslant C \int_\Sigma dL^3(\eta)\gamma_o^{1/2}, \quad C > 0, \tag{1.2.35}$$

where γ_o is given by (1.2.34) with g_{pq} replaced by δ_{pq}. Hence the integral on the right-hand side of (1.2.35) is equal to $H^3(\mathscr{S})$. For Lipschitz continuous functions $\bar{x}^k$ (on $\bar{\Sigma}$) $A^{p_1 p_2 p_3}$ is defined L^3-a.e. on Σ and essentially bounded, but the assertion (2) remains valid.

1.2.7 Existence of Flux Densities

Going back to equation (1.2.32) we assume that F is defined on oriented plane Borel sets and equation (1.2.32) holds for all polyhedral subsets of $\hat{\mathscr{B}}$. From this fact we deduce the existence of densities f, g of F, G following a method due to Gurtin and Martins (1976). In Section 1.2.8 we shall proceed to determine a larger class of sets $\mathscr{P}$ for which (1.2.32) remains valid, following in part the paper of Antman and Osborn (1979).

We shall assume that G is absolutely continuous with respect to L^4:

$$|G(\mathscr{P})| \leqslant KL^4(\mathscr{P})$$

for every bounded polyhedron $\mathscr{P} \subset \hat{\mathscr{B}}$ (1.2.36). Equation (1.2.32) implies the following estimate for every bounded polyhedron

$$|F(\partial\mathscr{P})| \leqslant KL^4(\mathscr{P}). \tag{1.2.37}$$

In order to discuss the set function F we shall consider the class $\mathscr{R}$ of all Borel subsets of three-dimensional hyperplanes $\Pi \subset \boldsymbol{R}^4$, endowed with an external orientation specified by a unit normal $\mathbf{n}$. By a **plane set** we shall mean a subset of a three-dimensional hyperplane $\Pi \subset \boldsymbol{R}^4$. $\Sigma \subset \Pi$ is a **plane Borel set** if it is a Borel set in Π with respect to the topology of Π induced by the embedding $\Pi \subset \boldsymbol{R}^4$. A subset Σ of $\boldsymbol{R}^4$ is said to be a **polygonal set** if (i) Σ is a plane set, $\Sigma \subset \Pi$, (ii) Σ is the closure of an open subset of Π, (iii) the boundary of Σ in the topology of Π consists of two-dimensional planes in Π. A two-dimensional plane in $\Pi = \{x \in \boldsymbol{R}^4 |\ \mathbf{n}\cdot x = a\}$ is given by two linear equations $\mathbf{n}\cdot x = a,\ \ \mathbf{m}\cdot x = b,\ \ \mathbf{m} \neq \lambda\mathbf{n}$.

An **oriented plane set (o.p.s.)** $\mathscr{S}$ is a pair $(\Sigma, \mathbf{n})$ consisting of a plane set Σ and a unit normal n on Σ, specifying an external orientation of $\mathscr{S}$. If $S = (\Sigma, \mathbf{n})$ then $-S := (\Sigma, -\mathbf{n})$.

Let $\Sigma + y := \{x \in \boldsymbol{R}^4 | x = z + y, z \in \Sigma\}$ for any $\Sigma \subset \boldsymbol{R}^4$, $y \in \boldsymbol{R}^4$. We shall define $\mathscr{S} + y = (\Sigma + y, \mathbf{n})$ for any o.p.s. $\mathscr{S} = (\Sigma, \mathbf{n})$.

An o.p.s. $\mathscr{S}$ is called a **Borel** (or **polygonal**) **o.p.s.** if Σ is a plane Borel (or polygonal) set.*

The closure $\mathscr{A}$ of an open subset of $\boldsymbol{R}^4$ is called a **polyhedron** if its boundary $\partial\mathscr{A}$ is contained in a finite number of hyperplanes $\Pi_1, \ldots, \Pi_n$. A component of $\partial\mathscr{A} \cap \Pi_r$, $r \leqslant n$, is called a **face** of the polyhedron $\mathscr{A}$. We shall always assume that $\partial\mathscr{A}$ is oriented, the orientation being specified by the exterior unit normal $\mathbf{n}$. In an obvious sense the boundary $\partial\mathscr{A}$ of a polyhedron $\mathscr{A}$ can be regarded as the sum of its faces, which are polygonal o.p.s.

* Henceforth we shall use the symbol $\mathscr{S}$ to denote both the o.p.s. and the underlying set Σ, whenever there is no risk of confusion.

For reasons of convenience we shall assume that I is an open interval and $\hat{\mathscr{B}} := \varkappa(\mathscr{B}) \times I$ is an open subset of $\boldsymbol{R}^4$.

Let $\mathscr{S}$ be an o.p.s. By $\mathscr{D}(\mathscr{S})$ we shall denote the set of those $y \in \boldsymbol{R}^4$ for which $\mathscr{S} + y \subset \hat{\mathscr{B}}$.

By $\mathscr{R}(\mathscr{R}_o)$ we shall denote the set of all Borel (polygonal) o.p.s. contained in $\hat{\mathscr{B}}$.

Lemma 1.2.1
If $\mathscr{S} \in \mathscr{R}$ is compact then $\mathscr{D}(\mathscr{S})$ is open.

Proof
Let $\mathscr{S} + y_o \subset \hat{\mathscr{B}}$, $y \in \mathscr{S}$.

The mapping $(y, z) \mapsto y + z$ is continuous from $\boldsymbol{R}^4 \times \boldsymbol{R}^4$ to $\boldsymbol{R}^4$. Hence there are two positive numbers $r_y, \varrho_y > 0$, dependent on $y \in \mathscr{S}$, such that $\mathscr{K}(y_o, r_y) + \mathscr{K}(y, \varrho_y) \subset \hat{\mathscr{B}}$. Since $\mathscr{S}$ is compact, there is a finite set of y_i, $i = 1, \ldots, m$ such that $\mathscr{S} \subset \bigcup_{i=1}^{m} \mathscr{K}(y_i, \varrho_{y_i})$. Let $\mathscr{W} := \bigcap_{i=1}^{m} \mathscr{K}(y_o, r_{y_i})$. $\mathscr{W}$ is open and $z \in \mathscr{W}$ then $z + \mathscr{K}(y_i, \varrho_{y_i}) \subset \hat{\mathscr{B}}$ for $i = 1, \ldots, m$. Hence $W + \mathscr{S} \subset \hat{\mathscr{B}}$. □

We shall assume that

(i) G is an additive set function defined on all polyhedra $\mathscr{A} \subset \hat{\mathscr{B}}$ and satisfying (1.2.36);

(ii) for an arbitrary o.p.s. $\mathscr{S} \in \mathscr{R}_o$, $\mathscr{S} = (\Sigma, \mathbf{n})$, $F(\mathscr{S} + \lambda\mathbf{n})$ is defined for almost all real λ such that $\lambda\mathbf{n} \in \mathscr{D}(\mathscr{S})$;

(iii) if a Borel o.p.s. $\mathscr{S}' \subset \mathscr{S}$ and $F(\mathscr{S} + y)$ is defined then $F(\mathscr{S}' + y)$ is defined too;

(iv) if $\mathscr{S}_j = (\Sigma_j, \mathbf{n}) \in \mathscr{R}_o$, $j = 1, \ldots, k$, Σ_j are disjoint subsets of a hyperplane Π, $\mathscr{S} = (\bigcup_{j=1}^{k} \Sigma_j, \mathbf{n})$, then $F(\mathscr{S}) = \sum_{j=1}^{k} F(\mathscr{S}_j)$;

(v) $\exists\, M > 0$ such that

$$|F(\mathscr{S})| \leqslant M H^3(\mathscr{S}) \tag{1.2.38}$$

for every $\mathscr{S}$ in the domain of definition of F.

Let $\mathscr{S}$ be an oriented surface consisting of a finite number of o.p.s. $\mathscr{S}_j \in \mathscr{R}_o$, $j = 1, \ldots, k$, such that $F(\mathscr{S}_j)$ is defined for $j \leqslant k$. We define

$$F(\mathscr{S}) = \sum_{j=1}^{k} F(\mathscr{S}_j). \tag{1.2.39}$$

Our final assumption is

(vi) For every bounded polyhedron $\mathscr{A} \subset \hat{\mathscr{B}}$ such that $F(\partial\mathscr{A})$ is defined in the sense of (1.2.39)

$$F(\partial\mathscr{A}) = G(\mathscr{A}). \tag{1.2.40}$$

Suppose that G is the restriction of a set function $\tilde{G}$ defined on all Borel sets (or all open sets) contained in $\hat{\mathscr{B}}$ and satisfying (1.2.36). It follows that $\tilde{G}$ is σ-additive and Borel-regular. By an argument presented in Section 1.1.4 in connection with the definition of the Lebesgue measure $\tilde{G}$ is uniquely defined by G.

Let $\mathscr{R}_o^{\Pi,\mathbf{n}}$, $\mathscr{R}^{\Pi,\mathbf{n}}$ be the set of all the polygonal, or Borel, o.p.s. contained in the hyperplane Π and having the orientation $\mathbf{n}$. By a similar argument, if F is defined on $\mathscr{R}^{\Pi,\mathbf{n}}$ and satisfies (1.2.38), then it is determined by its values on $\mathscr{R}_o^{\Pi,\mathbf{n}}$.

We shall asusme that G is defined on Borel subsets of $\hat{\mathscr{B}}$ and replace $\mathscr{R}_o$ by $\mathscr{R}$ in (ii), (iii), (iv).

Theorem 1.2.2
There is a locally integrable function g: $\hat{\mathscr{B}} \to \boldsymbol{R}$ such that

$$\tilde{G}(\mathscr{P}) = \int_{\mathscr{P}} dL^4\, g(x) \tag{1.2.41}$$

for every Borel set $\mathscr{P} \subset \hat{\mathscr{B}}$ and

$$|g(x)| \leqslant K \quad \text{for a.a. } x \in \hat{\mathscr{B}}. \tag{1.2.42}$$

Proof

Equation (1.2.32) implies that $\tilde{G}$ is a σ-additive set function defined on Borel subsets of $\hat{\mathscr{B}}$ and absolutely continuous with respect to L^4. The thesis follows from the Radon–Nikodym theorem (Theorem 1.1.18). □

Lemma 1.2.3
For any Borel o.p.s. in the domain of F

$$F(-\mathscr{S}) = -F(\mathscr{S}). \tag{1.2.43}$$

Proof
First of all, we shall note that for every $y_o \in \mathscr{D}(\mathscr{S})$ and $\varepsilon > 0$ such that $\mathscr{K}(y_o, \varepsilon) \subset \mathscr{D}(\mathscr{S})$ there is a $\delta > 0$ such that $\delta < \varepsilon$ and $\mathscr{S} + y_o + \delta\mathbf{n}$ belongs to the domain of F (by (ii)).

Let $\mathscr{S} = (\Sigma, \mathbf{n})$ be a convex polygonal o.p.s., $\Sigma \subset \hat{\mathscr{B}} \cap \Pi$, such that $F(\mathscr{S})$ and $F(-\mathscr{S})$ are defined. Let $\varepsilon > 0$ be such that $\mathscr{K}(0, \varepsilon) \subset \mathscr{D}(\mathscr{S})$. We now construct another polygonal set Σ_ε by shifting the faces of Σ over a distance $< \varepsilon$ in the inward normal directions in Π. Let

$$\mathscr{P}_+^\delta := \bigcup_{\lambda \in [0,\delta]} (\Sigma_\varepsilon + \lambda\mathbf{n}), \quad \mathscr{P}_-^\delta := \bigcup_{\lambda \in [0,\delta]} (\Sigma_\varepsilon - \lambda\mathbf{n}), \quad \mathscr{P}^\delta := \mathscr{P}_+^\delta \cup \mathscr{P}_-^\delta,$$

$0 < \delta \leqslant \varepsilon$. Suppose that Σ_ε has been constructed in such a way that the lateral faces of $\mathscr{P}_+^\delta, \mathscr{P}_-^\delta$ belong to the domain of F (we have taken advantage of (ii)). For $\delta' \leqslant \delta$ the lateral faces of $\mathscr{P}_+^{\delta'}, \mathscr{P}_-^{\delta'}$ also belong to the domain of F. We choose δ in such a way that $\mathscr{S}_\varepsilon + \delta\mathbf{n}$, $\mathscr{S}_\varepsilon - \delta\mathbf{n}$ also belong to the domain of F, $\mathscr{S}_\varepsilon := (\Sigma_\varepsilon, \mathbf{n})$.

We have

$$F(\partial\mathscr{P}_+^\delta)+F(\partial\mathscr{P}_-^\delta)-F(\partial\mathscr{P}^\delta) = F(\mathscr{S}_\varepsilon)+F(-\mathscr{S}_\varepsilon). \tag{1.2.44}$$

The absolute value of the left-hand side of (1.2.44) is bounded by $K[L^4(\mathscr{P}_+^\delta)+ +L^4(\mathscr{P}_-^\delta)+L^4(\mathscr{P}^\delta)] \leqslant 4K\delta H^3(\Sigma_\varepsilon)$. The number $\delta > 0$ can be chosen arbitrarily small and

$$F(\mathscr{S}_\varepsilon)-F(\mathscr{S}) = 0(\varepsilon), \qquad F(-\mathscr{S}_\varepsilon)-F(-\mathscr{S}) = O(\varepsilon)$$

by (v). Hence follows the theorem for convex polygonal o.p.s.

A non-convex polygonal o.p.s. $\mathscr{S}$ can be decomposed into the union of a finite number of convex polygonal sets. By (iv) this implies the result for all polygonal o.p.s.

Every plane Borel set can be approximated in the measure by polygonal sets (joins of a finite number of intervals, see Section 1.1.4). Hence follows the required result. □

We shall now establish two estimates related to the continuity of the function $y \mapsto F(\mathscr{S}+y)$.

Let $\mathscr{S} = (\Sigma, \mathbf{n})$ be a polygonal o.p.s. and let $y_o \in \mathscr{D}(\mathscr{S})$, $(y-y_o)\cdot\mathbf{n} = 0$. The polygonal o.p.s. $\mathscr{S}+y$, $\mathscr{S}+y_o$ lie in the same hyperplane and the polygonal o.p.s. $\mathscr{S}+y$ can be obtained by translation of $\mathscr{S}+y_o$ along the vector $y-y_o$. In the course of the translation a face of Σ whose H^2-area is A sweeps over an H^3-area $\leqslant |y-y_o|A$. Hence

$$|F(\mathscr{S}+y)-F(\mathscr{S}+y_o)| \leqslant MH^3\big((\Sigma+y)\div(\Sigma+y_o)\big) \leqslant MP(\Sigma)|y-y_o|, \tag{1.2.45}$$

provided $\mathscr{S}+y$ and $\mathscr{S}+y_o$ lie in the domain of F. $P(\Sigma)$ denotes the perimeter of Σ and $\mathscr{A}\div\mathscr{B}$ denotes the symmetric difference of sets $\mathscr{A}$, $\mathscr{B}$ ($= (\mathscr{A}\setminus\mathscr{B})\cup \cup(\mathscr{B}\setminus\mathscr{A})$).

Let us now consider the case of a convex polygonal o.p.s. $\mathscr{S} = (\Sigma, \mathbf{n})$, $y_o \in \mathscr{D}(\mathscr{S})$, $y \in \mathscr{K}(y_o, r) \subset \mathscr{D}(\mathscr{S})$, $(y-y_o)\cdot\mathbf{n} > 0$. Let $\mathscr{P}$ be the prism

$$\mathscr{P} := \bigcup_{\vartheta\in[0,1]} (\Sigma_\varepsilon+\vartheta y+(1-\vartheta)y_o).$$

$\Sigma_\varepsilon \subset \Sigma$ has been obtained by shifting the faces of Σ in the hyperplane $\Pi \supset \Sigma$ over distances $< \varepsilon$ in such a way that the lateral faces of $\mathscr{P}$ lie in the domain of F. Clearly $L^4(\mathscr{P}) \leqslant H^3(\Sigma_\varepsilon)|y-y_o|$. Let $\mathscr{S}_\varepsilon := (\Sigma_\varepsilon, \mathbf{n})$. The boundary of $\mathscr{P}$

consists of the o.p.s. $\mathscr{S}_\varepsilon+y$, $-(\mathscr{S}_\varepsilon+y_o)$ and the lateral faces whose total H^3-area does not exceed $P(\Sigma_\varepsilon)|y-y_o|$. Hence, by (i), (iv), (v), (vi) and Lemma 1.2.3

$$|F(\mathscr{S}+y)-F(\mathscr{S}+y_o)| \leqslant [KH^3(\Sigma)+MP(\Sigma)+0(\varepsilon)]|y-y_o|. \qquad (1.2.46)$$

Since $\varepsilon > 0$ is arbitrarily small the term $0(\varepsilon)$ can be omitted.

Every Borel o.p.s, $\mathscr{S} \subset \Pi$ can be approximated in the measure L^3 on Π by a sequence of polygonal subsets $\mathscr{A}_n$ of Π (joins of finite numbers of closed intervals, cf. 1.1.4). L^3 on Π_- coincides with H^3 restricted to subsets of Π. In general, however, the perimeters $P(\mathscr{A}_n)$ may be unbounded for $n \to \infty$. Hence we shall consider a subclass of Borel o.p.s.

Definition 1.2.4

A Borel o.p.s. $\mathscr{S} = (\Sigma, \mathbf{n})$ is said to **have bounded perimeter** if there is a sequence $\{\mathscr{A}_n\}$ of polygonal sets $\mathscr{A}_n \subset \Pi$ such that $H^3(\Sigma \div \mathscr{A}_n) \to 0$ and $P(\mathscr{A}_n) < R$. $\forall\, n$ for some fixed $R < \infty$.

Lemma 1.2.5

The estimates (1.2.45) and (1.2.46) hold for every pair $\mathscr{S}+y$, $\mathscr{S}+y_o$ of Borel o.p.s. with bounded perimeter provided $P(\mathscr{S})$ is replaced by $\limsup_{n\to\infty} P(\mathscr{A}_n)$, where $\{\mathscr{A}_n\}$ is an approximating sequence for $\mathscr{S}$.

Let $\mathscr{S}$ be a Borel o.p.s. of bounded perimeter. It follows from Lemma 1.2.5 and (ii) that $y \mapsto F(\mathscr{S}+y)$ can be extended to a continuous function defined for all $y \in \mathscr{D}(\mathscr{S})$. We shall use the symbol F to denote the latter function.

Let $\mathscr{D}_r(\mathbf{y}, \mathbf{n})$ denote the disc $\{y \in \mathbf{R}^4 |\ \mathbf{n}\cdot(y-y_o) = 0,\ |y-y_o| \leqslant r\}$ with the orientation defined by the unit normal $\mathbf{n}$. $\mathscr{D}_r(y, \mathbf{n})$ is a Borel o.p.s. of bounded perimeter.

Let $\mathscr{D}_r$ be the set of $x \in \mathbf{R}^4$ such that $\mathscr{K}(x, r) \subset \hat{\mathscr{B}}$. Let

$$f_r(x, \mathbf{n}) := \frac{F\mathscr{D}_r((x, \mathbf{n}))}{H^3(\mathscr{D}_r(x, \mathbf{n}))}. \qquad (1.2.47)$$

Clearly

$$|f_r(x, \mathbf{n})| \leqslant M, \qquad (1.2.48)$$

$$f_r(x, -\mathbf{n}) = -f_r(x, \mathbf{n}). \qquad (1.2.49)$$

The continuity of $f_r(x, \mathbf{n})$ with respect to r and $x \in \mathscr{D}_r$ follows from (v) and equations (1.2.45) and (1.2.46).

We now proceed to establish an estimate displaying the continuity of $f_r(x, \cdot)$. Let $\mathbf{n}_1, \mathbf{n}_2$ be two unit vectors in $\mathbf{R}^4$ such that $\mathbf{n}_1\cdot\mathbf{n}_2 =: \cos\Delta \neq \pm 1$. We can choose Δ in such a way that $0 < \Delta \leqslant \pi/2$. Let $\mathscr{P}$ be the two-dimensional plane $\{y \in \boldsymbol{R}^4 | \; y = x + \alpha \mathbf{n}_1 + \beta \mathbf{n}_2, \alpha, \beta \in \boldsymbol{R}\}$. The discs $\mathscr{D}_r(x, \mathbf{n}_1)$, $\mathscr{D}_r(x, \mathbf{n}_2)$ divide the ball $\mathscr{K}(x, r) \subset \boldsymbol{R}^4$ into four parts $\mathscr{K}_1, \mathscr{K}_2, \mathscr{K}_3, \mathscr{K}_4$. We choose $\mathscr{K}_1, \mathscr{K}_2$ in such a way that $\mathscr{K}_1 \cap \mathscr{P}$ and $\mathscr{K}_2 \cap \mathscr{P}$ is a sector with the acute angle at x.

We introduce the spherical coordinates $(\varrho, \vartheta, \theta, \varphi) \mapsto y \in \boldsymbol{R}^4$

$$\begin{aligned} y^1 - x^1 &= \varrho \sin\vartheta \sin\theta \cos\varphi, \\ y^2 - x^2 &= \varrho \sin\vartheta \sin\theta \sin\varphi, \\ y^3 - x^3 &= \varrho \sin\vartheta \cos\theta, \\ y^4 - x^4 &= \varrho \cos\vartheta, \\ \varrho \geqslant 0, \quad 0 \leqslant \vartheta, &\quad \theta \leqslant \pi, \quad 0 \leqslant \varphi < 2\pi, \end{aligned}$$

in such a way that $\mathscr{D}_r(x, \mathbf{n}_1)$ lies in the plane $\varphi = 0$ and $\mathscr{D}_r(x, \mathbf{n}_2)$ lies in the plane $\varphi = \Delta$. The measure H^3 of $\partial \mathscr{K}_1 \cap \partial \mathscr{K}(x, r)$ is equal to

$$r^3 \int_0^\pi d\vartheta \int_0^\pi d\theta \, \sin^2\vartheta \sin\theta \int_0^\Delta d\varphi = \pi \Delta r^3,$$

while the four-dimensional volume of $\mathscr{K}_1$ is

$$\int_0^r d\varrho \; \varrho^3 \pi \Delta = \frac{\pi}{4} r^4 \Delta.$$

Hence

$$|F(\mathscr{D}_r(x, \mathbf{n}_1)) - F(\mathscr{D}_r(x, \mathbf{n}_2))| \leqslant \pi \Delta r^3 (M + \tfrac{1}{4} K r), \tag{1.2.50}$$

and

$$|f_r(x, \mathbf{n}_1) - f_r(x, \mathbf{n}_2)| \leqslant \frac{3}{4} \Delta (M + \tfrac{1}{4} K r). \tag{1.2.51}$$

Hence $f_r(x, \mathbf{n})$ is a continuous function of r, x and $\mathbf{n}$.

On account of (v) and the Radon–Nikodym theorem

$$\lim_{r \to 0} f_r(x, \mathbf{n}) =: f(x, \mathbf{n}) \tag{1.2.52}$$

exists for H^3-a.a. x in every hyperplane $\Pi \subset \boldsymbol{R}^4$ and

$$F(\mathscr{S}) = \int_{\mathscr{S}} dH^3 f(x, \mathbf{n}) \tag{1.2.53}$$

for every Borel o.p.s. $\mathscr{S} \subset \Pi$.

Let $\mathscr{D}(\mathbf{n})$ be the set of $x \in \hat{\mathscr{B}}$ such that $f(x, \mathbf{n})$, defined by equation (1.2.52), exists. There is an $r_o > 0$ such that $\mathscr{D}_{r_o} \neq \emptyset$ and $\mathscr{D}_r \subset \mathscr{D}_{r'}$ if $r' < r$. Since $r \mapsto f_r(x, \mathbf{n})$ is continuous, the limit in (1.2.52) exists iff the limit

$$\lim_{\substack{m\to\infty \\ m\in \mathbf{Z}_+}} f_{1/m}(x, \mathbf{n}) \tag{1.2.54}$$

exists. The limit (1.2.54) does not exist iff for some $\varepsilon > 0$ there are arbitrarily large positive integers m_1, m_2 such that

$$|f_{1/m_1}(x, \mathbf{n}) - f_{1/m_2}(x, \mathbf{n})| > \varepsilon.$$

Hence the set of such $x \in \mathscr{D}_{r_o}$ that $f(x, \mathbf{n})$ does not exist is

$$\mathscr{A} = \bigcup_{k\in \mathbf{Z}_+} \bigcap_{\substack{m\in \mathbf{Z}_+ \\ m^{-1} < r_o}} \bigcup_{\substack{\nu,\mu\in \mathbf{Z}_+ \\ \nu,\mu > m}} \left\{ x \in \mathscr{D}_{r_o} \middle| |f_{1/\nu}(x, \mathbf{n}) - f_{1/\mu}(x, \mathbf{n})| > \frac{1}{k} \right\}.$$

For $x \,\overline{\in}\, \mathscr{D}_{1/\nu}$ or $x \,\overline{\in}\, \mathscr{D}_{1/\mu}$ we shall assume that the inequality in the braces is satisfied trivially. The sets defined by the expression in the braces are open in $\mathscr{D}_{r_o}$ (since μ^{-1}, $\nu^{-1} < r_o$), hence $\mathscr{A}$, $\mathscr{D}(\mathbf{n}) \cap \mathscr{D}_{r_o}$ is a Borel subset of $\mathscr{D}_{r_o}$. For $r < r_o f_r(\cdot, \mathbf{n})$ is continuous on $\mathscr{D}_{r_o}$. Hence the set $\{x \in \mathscr{D}_{r_o} \cap \mathscr{D}(\mathbf{n}) |\ f(x, \mathbf{n}) > R\} = \bigcup_{\substack{m\in \mathbf{Z}_+ \\ m^{-1} \leqslant r_o}} \bigcap_{\substack{m'\in \mathbf{Z}_+ \\ m' \geqslant m}} \{\mathbf{x} \in \mathscr{D}_{r_o} \cap \mathscr{D}(\mathbf{n}) |\ f_{1/m'}(x, \mathbf{n}) > R\}$ is a Borel subset of $\mathscr{D}_{r_o}$ for every real R. The same is true for the set $f(x, \mathbf{n}) < R$. Hence $f(\cdot, \mathbf{n})$ is a Borel function on $\mathscr{D}_{r_o}$. Since $\hat{\mathscr{B}} = \bigcup_m \mathscr{D}_{1/m}$, the set $\mathscr{D}(\mathbf{n})$ is a Borel subset of $\hat{\mathscr{B}}$. We proved above that for every real a the set $\{x \in \hat{\mathscr{B}} \setminus \mathscr{D}(\mathbf{n}) |\ x^1 = a\}$ is of L^3 measure zero in $\{x |\ x^1 = a\}$. By the Fubini theorem $L^4(\hat{\mathscr{B}} \setminus \mathscr{D}(\mathbf{n})) = 0$. Since $f(\cdot, \mathbf{n})$ is measurable and $|f(x, \mathbf{n})| \leqslant M$ for $x \in \mathscr{D}(\mathbf{n})$, $f(\cdot, \mathbf{n}) \in \mathscr{L}^1_{loc}(\hat{\mathscr{B}})$.

Since the unit sphere $\mathbf{S}^3$ in $\boldsymbol{R}^4$ is compact, we can choose a dense countable set of points $\mathbf{n}_k \in \mathbf{S}^3$, $k = 1, 2, \ldots$ in $\mathbf{S}^3$. Let $\mathscr{D} := \bigcap_{k=1}^{\infty} \mathscr{D}_{\mathbf{n}_k}$. $\mathscr{D}$ is a Borel subset of $\hat{\mathscr{B}}$ and $\hat{\mathscr{B}} \setminus \mathscr{D}$ has L^4 measure zero.

For $x \in \mathscr{D}$, $\mathbf{n} \in S^3$, $\cos \Delta_k := \mathbf{n} \cdot \mathbf{n}_k$, $0 \leqslant \Delta_k \leqslant \pi/2$ we have

$$\begin{aligned} |f_r(x, \mathbf{n}) - f_s(x, \mathbf{n})| &\leqslant |f_r(x, \mathbf{n}) - f_r(x, \mathbf{n}_k)| + |f_s(x, \mathbf{n}) - f_s(x, \mathbf{n}_k)| \\ &\quad + |f_r(x, \mathbf{n}_k) - f_s(x, \mathbf{n}_k)| \\ &\leqslant \frac{3}{4} \Delta_k(M + Kr/4) + \frac{3}{4} \Delta_k(M + Ks/4) + \varepsilon/3, \end{aligned}$$

provided r, s are sufficiently small (smaller than some $\delta(\varepsilon, k)$). For given $\mathbf{n} \in \mathbf{S}^3$ we may choose $\mathbf{n}_k \in \mathbf{S}^3$ in such a way that Δ_k is arbitrarily small. Hence $\lim_{r\to 0} f_r(x, \mathbf{n})$ exists for $x \in \mathscr{D}$

We thus have

Theorem 1.2.6

There is a subset $\mathscr{D} \subset \hat{\mathscr{B}}$ of full measure L^4 in $\hat{\mathscr{B}}$ such that

(i) the limit (1.2.52) exists for all $(x, \mathbf{n}) \in \mathscr{D} \times S^3$,

(ii) $f(\cdot\,, \mathbf{n}) \in \mathscr{L}^1_{loc}(\hat{\mathscr{B}})$,

(iii) $f(x, \cdot\,)$ is continuous for $x \in \mathscr{D}$.

Proof

It remains to prove (iii). But (iii) follows from (1.2.51) in the limit $r \to 0$. □

Lemma 1.2.7 (Cauchy)

Suppose that F satisfies (i)–(vi). Let $f(\cdot\,, \cdot\,)$ be the density of F in the sense of eq. (1.2.53). Suppose that f is a continuous function of $x \in \hat{\mathscr{B}}$. Then there is a continuous function $\mathbf{f}\colon \hat{\mathscr{B}} \to R^4$ such that $f(\mathbf{x}, \mathbf{n}) \equiv \mathbf{f}(x) \cdot \mathbf{n}$.

Proof

Consider the polyhedron $\mathscr{P}$ with a vertex at $\mathbf{x}$ and the faces $\mathscr{S}_k$: $y^k = x^k$, $\sum_l n_l(y^l - x^l) \leqslant M$, $y^l \geqslant x^l$ for $l \neq k$, oriented by the coordinate vector $\mathbf{e}_k := \partial/\partial x^k$, $k = 0, 1, 2, 3$, $\mathscr{S}_4$: $\sum_{k=0}^{3} n_k(y^k - x^k) = M$, $y^k \geqslant x^k$ for $k = 0, 1, 2, 3$ oriented by the unit vector $\mathbf{n}$, $n_k > 0$ for $k = 0, 1, 2, 3$, $M > 0$.

Since

$$|F(\partial\mathscr{P})| \leqslant KL^4(\mathscr{P}) = O(M^4),$$

it follows that

$$\frac{F(\mathscr{S}_4)}{H^3(\mathscr{S}_4)} + \sum_{k=0}^{3} \frac{F(-\mathscr{S}_k)}{H^3(\mathscr{S}_4)} = O(M). \tag{1.2.55}$$

We shall calculate $H^3(\mathscr{S}_4)$ and $H^3(\mathscr{S}_k)$, $k = 0, \ldots, 3$. Consider a polygonal set $\mathscr{S}$: $n_k y^k = M$, $y^k \geqslant 0$, with $\sum_k n_k{}^2 = 1$, $n_0 > 0$, $M > 0$. Let $\nu_\alpha := -n_\alpha/n_0$, $\alpha = 1, 2, 3$. We can rewrite the equation of $\mathscr{S}$ in the following form:

$$y^0 = M/n_0 + \nu_\alpha y^\alpha, \quad y^0 \geqslant 0, \quad y^\alpha \geqslant 0 \quad \text{for } \alpha = 1, 2, 3.$$

The second fundamental form of $\mathscr{S}$ is

$$g_{\alpha\beta} = \delta_{\alpha\beta} + \nu_\alpha \nu_\beta$$

and det $((g_{\alpha\beta})) = 1+\mathbf{v}^2 = n_0{}^{-2}$. Hence the H^3-area of $\mathscr{S}$ is

$$\frac{1}{n_0}\int\limits_0^{M/n_1} dy^1 \int\limits_0^{(M-n_1y^1)/n_2} dy^2 \int\limits_0^{(M-n_1y^1-n_2y^2)/n_3} dy^3$$

$$= \frac{1}{2n_1n_2n_3n_0}\int\limits_0^{M} d\xi\,(M-\xi)^2 = \frac{M^3}{6n_0n_1n_2n_3}.$$

The area of $\mathscr{S}_0$: $y^0 = 0$, $n_\alpha y^\alpha \leqslant M$, $y^\alpha \geqslant 0$, $\alpha = 1, 2, 3$, is

$$\int\limits_0^{M/n_1} dy^1 \int\limits_0^{(M-n_1y^1)/n_2} dy^2 \int\limits_0^{(M-n_1y^1-n_2y^2)/n_3} dy^3 = \frac{M^3}{6n_1n_2n_3}.$$

Hence $H^3(\mathscr{S}_k)/H^3(\mathscr{S}_4) = n_k$ for $k = 0, 1, 2, 3$. In the limit $M \to 0$ equation (1.2.55) yields the identity $f(x, \mathbf{n}) - f(x, \mathbf{e}_k)n_k = 0$ (sum over $k = 0, 1, 2, 3$).

Let $f^k(x) := f(x, \mathbf{e}_k)$. □

Theorem 1.2.8 (Gurtin and Martins, 1976)
Suppose that F is a function of Borel o.p.s. satisfying (i)–(vi).

There is a locally H^3 integrable function $\mathbf{f}$: $\hat{\mathscr{B}} \to \boldsymbol{R}^4$ such that F is given by (1.2.53) with $f(x, \mathbf{n}) = \mathbf{f}(x) \cdot \mathbf{n}$.

Proof
We shall extend the function $f(\cdot\,,\,\cdot)$, defined so far on $\bigcup_{\mathbf{n}\in S^3} \mathscr{D}(\mathbf{n}) \times \{\mathbf{n}\}$ to $\boldsymbol{R}^4 \times S^3$ by setting

$$\bar{f}(x, \mathbf{n}) = \begin{cases} f(x, \mathbf{n}) & \text{if } x \in \mathscr{D}(\mathbf{n}), \\ 0 & \text{if } x \,\bar{\in}\, \mathscr{D}(\mathbf{n}). \end{cases}$$

Since $\mathscr{D}(\mathbf{n})$ is a Borel set, $\bar{f}(\cdot\,, \mathbf{n})$ is a Borel function for every $\mathbf{n}$. Let $\delta_\varepsilon \in \mathscr{C}_0^\infty(\boldsymbol{R}^4)$, $\delta_\varepsilon(x) \geqslant 0$, $\operatorname{supp}\delta_\varepsilon \subset \mathscr{K}(0, \varepsilon)$, $\int_{\boldsymbol{R}^4} dL^4\delta_\varepsilon = 1$ and

$$f^\varepsilon(\cdot\,, \mathbf{n}) := \delta_\varepsilon * \bar{f}(\cdot\,, \mathbf{n}). \tag{1.2.56}$$

The asterisk denotes the convolution: $(f*g)(x) = \int_{\boldsymbol{R}^4} dL^4(y) f(x-y)g(y)$. We have: $f^\varepsilon(\cdot\,, \mathbf{n}) \in \mathscr{C}^\infty$ (Adams, 1975, 2.17), $|f^\varepsilon(x, \mathbf{n})| \leqslant \int dL^4(y) M\delta_\varepsilon(y-x) = M$, $f^\varepsilon \to \bar{f}$ in $\mathscr{L}^1(\boldsymbol{R}^4)$ as $\varepsilon \to 0$ (op. cit.).

For any Borel o.p.s. $\mathscr{S} \subset \hat{\mathscr{B}}$ let

$$F^\varepsilon(\mathscr{S}) := \int\limits_\Sigma dH^3 f^\varepsilon(x, \mathbf{n}) \quad \text{for } \varepsilon < r_o,\ \mathscr{S} \subset \mathscr{D}_\varepsilon,\ \mathscr{S} = (\Sigma, \mathbf{n}) \tag{1.2.57}$$

in the notations of the proof of Theorem 1.2.6. Clearly $|F^\varepsilon(\mathscr{S})| \leqslant MH^3(\mathscr{S})$.

We now establish an auxiliary identity. Let $y \mapsto L(y)$ be L^3-measurable and locally integrable on every hyperplane $\Pi \subset \boldsymbol{R}^4$. The function

$$(y, x) \mapsto L(y)\,\delta_\varepsilon(y-x)\bar{f}(x, \mathbf{n})$$

is $(L^3 \times L^4)$-measurable on $\mathscr{S} \times \boldsymbol{R}^4$. Hence

$$\begin{aligned}\int_{\mathscr{S}} dH^3(y) L(y) f^\varepsilon(y, \mathbf{n}) &= \int_{\mathscr{S}} dH^3(y) L(y) \int_{\boldsymbol{R}^4} dL^4(x)\,\delta_\varepsilon(y-x)\bar{f}(x, \mathbf{n}) \\ &= \int_{\boldsymbol{R}^4} dL^4(x) \int_{\mathscr{S}} dH^3(y)\,\delta_\varepsilon(y-x) L(y)\bar{f}(x, \mathbf{n}) \\ &= \int_{\boldsymbol{R}^4} dL^4(z)\,\delta_\varepsilon(z) \int_{\mathscr{S}-\mathbf{z}} dH^3(y) L(y+z)\bar{f}(y, \mathbf{n}).\end{aligned}$$

For a closed polyhedron $\mathscr{P} \subset \mathscr{D}_{r_o}$ we have for $\varepsilon < r_o$

$$\int_{\partial\mathscr{P}} dH^3\, L(y) f^\varepsilon(y, \mathbf{n}) = \int_{\boldsymbol{R}^4} dL^4(z)\,\delta_\varepsilon(z) \int_{\partial\mathscr{P}-\mathbf{z}} dH^3(y) L(y+z)\bar{f}(y, \mathbf{n}).$$

In this formula $\mathbf{n}$ denotes the outer unit normal on $\partial\mathscr{P}$.

In particular for $L(\mathbf{y}) \equiv 1$, $\varepsilon < r_o$

$$\begin{aligned}\left|\int_{\partial\mathscr{P}} dH^3\, f^\varepsilon(y, \mathbf{n})\right| &\leqslant \sup\left\{\left|\int_{\partial\mathscr{P}-\mathbf{z}} dH^3\,\bar{f}(y, \mathbf{n})\right| \,\middle|\, |z| < \varepsilon\right\} \\ &\leqslant K\sup\left\{L^4(\mathscr{P}-z) \,\middle|\, |z| < \varepsilon\right\} = KL^4(\mathscr{P}).\end{aligned}$$

By Lemma 1.2.7 $f^\varepsilon(y, \mathbf{n}) = \mathbf{f}^\varepsilon(y)\cdot\mathbf{n}$, $f^{\varepsilon,k}(y) \equiv f^\varepsilon(y, \mathbf{e}_k) \to \bar{f}(y, \mathbf{e}_k)$, for $\varepsilon \to 0$, $k = 0, 1, 2, 3$, in $\mathscr{L}^1(\mathscr{D}_{r_o})$, since $\mathscr{D}_\varepsilon \supset \mathscr{D}_{r_o}$. Hence

$$\bar{f}(y, \mathbf{n}) = \sum_{k=0}^{3} \bar{f}(y, \mathbf{e}_k) n_k \tag{1.2.58}$$

for L^4-a.a. $\mathbf{y} \in \mathscr{D}_{r_o}$.

Now $\hat{\mathscr{B}} = \bigcup_{m \in \boldsymbol{Z}_+} \mathscr{D}_{1/m}$ and $\mathscr{D}_{1/m} \supset \mathscr{D}_{r_o}$ if $m^{-1} < r_o$. Hence (1.2.58) holds on some $\mathscr{D}^*(\mathbf{n})$ such that $L^4(\hat{\mathscr{B}} \setminus \mathscr{D}^*(\mathbf{n})) = 0$.

Let $\{\mathbf{n}_k\}$ be dense in $\boldsymbol{S}^3$, $\mathscr{D}^* = \bigcap_{k \in \boldsymbol{Z}_+} \mathscr{D}^*(\mathbf{n}_k) \cap \mathscr{D}$, $\mathscr{D}^*$ is an L^4-measurable set in $\boldsymbol{R}^4$ and $L^4(\hat{\mathscr{B}} \setminus \mathscr{D}^*) = 0$. Moreover it is easy to see by an argument used in the proof of Theorem 1.2.6 that $f(x, \mathbf{n}) = \mathbf{f}(x)\cdot\mathbf{n}$ on $\mathscr{D}^* \times \boldsymbol{S}^3$. □

1.2.8 Remarks about General Balance Equations

So far we have dealt successfully with balance equations (1.2.32) defined for a "dense" set of polyhedra in $\boldsymbol{R}^4$. We established existence of densities f^k,

g for F, G in the sense of equations (1.2.53), (1.2.41). In terms of these densities equation (1.2.32) assumes the form

$$\int_{\mathscr{P}} dL^4\, g(x) = \int_{\partial\mathscr{P}} dH^3\,(x)\mathbf{f}(x)\cdot\mathbf{n}, \tag{1.2.59}$$

valid for every polyhedron $\mathscr{P} \subset \hat{\mathscr{B}}$. *Prima facie*, since $\mathbf{f} \in \mathscr{L}^1_{loc}(\hat{\mathscr{B}})$, one would expect that the right-hand side of equation (1.2.59) is defined for a.e. polyhedron, on account of the Fubini theorem. Indeed, if $\mathscr{S}$ is L^3-measurable and I is an interval in $\boldsymbol{R}$, then $\mathscr{S}\times I$ is L^4-measurable and $\int_{\mathscr{S}} dH^3(x')f(x^0, x')$, $x' := (x^1, x^2, x^3)$, is defined for a.e. $x^0 \in I$. Equation (1.2.32) with (1.2.36) (or (1.2.59) with (1.2.42)) implies continuity of F with respect to translations so that (1.2.59) makes sense and is valid for *every* polyhedron $\mathscr{P} \subset \hat{\mathscr{B}}$. It is clear that the densities $\mathbf{f} \in \mathscr{L}^1_{loc}(\hat{\mathscr{B}})$ must have an additional property which allows for this fact. The difficulty is even more conspicuous if one tries to enlarge the class of sets $\mathscr{P}$ in (1.2.59).

Indeed, suppose that $f^k \in \mathscr{L}^1_{loc}(\hat{\mathscr{B}})$ merely. The right-hand side of (1.2.59) is defined for a.e. cell $\mathscr{P} = \{x|\ a^i \leqslant x^i \leqslant b^i\} \subset \hat{\mathscr{B}}$, i.e. for a.a. values of a^i, b^i, $i = 0, 1, 2, 3$. In Section 1.1 it was shown that a Borel set can be approximated in measure L^4 by joins of finite numbers of cells. Borel sets with bounded perimeter can perhaps be approximated with due account of the boundary as well. In Section 1.1 approximating sequences were constructed from rational cells (picking out rational values of a^i, b^i). But the set of rational numbers has measure zero in $\boldsymbol{R}$ and we cannot be sure that equation (1.2.59) has any meaning even for the approximating sequences.

In practice one is confronted with the problem of finding a class Γ of sets $\mathscr{P}$ and a class Φ of functions $\mathbf{f}$ such that (1.2.59) is meaningful (or valid) for all $g \in \mathscr{L}^1_{loc}(\hat{\mathscr{B}})$, $\mathbf{f} \in \Phi$, $\mathscr{P} \in \Gamma$, $\mathscr{P} \subset \hat{\mathscr{B}}$. The class Φ should be defined explicitly, preferably in functional-analytic terms. We expect that a suitable restriction on $\Phi \subset \mathscr{L}^1_{loc}(\hat{\mathscr{B}})$ will allow a reasonably large class Γ of sets $\mathscr{P}$.

Let us consider the following example: $\Phi = W^{1,p}_{loc}(\hat{\mathscr{B}})$, $p \geqslant 1$, $\Gamma =$ the class of all open subsets of $\hat{\mathscr{B}}$ with the strong local Lipschitz property (Section 1.1). In this case the trace theorem (Theorem 1.1.47 and the following remark) ensures that (1.2.59) is meaningful on $\mathscr{L}^1_{loc}(\hat{\mathscr{B}})\times\Phi\times\Gamma$ provided $\mathbf{f}$ is interpreted as $\gamma\mathbf{f}$. We now examine the balance equation for polyhedra more closely.

Let $p \geqslant 1$, $f \in \mathscr{L}^p(\hat{\mathscr{B}})$. Assume that $\hat{\mathscr{B}}$ is compact in $\boldsymbol{R}^4$. The Hölder inequalities imply that

$$||f||_1 := \int_{\hat{\mathscr{B}}} dL^4|f| \leqslant ||f||_p||1||_q = (L^4(\hat{\mathscr{B}}))^{1/q}||f||_p, \quad 1/q = 1-1/p.$$

Hence $f \in \mathscr{L}^1(\hat{\mathscr{B}})$. For unbounded $\hat{\mathscr{B}}$ we have the embedding $\mathscr{L}^p_{loc}(\hat{\mathscr{B}}) \subset \mathscr{L}^1_{loc}(\hat{\mathscr{B}})$.

Let $g \in W^{1,p}_{loc}(\hat{\mathscr{B}})$, $p \geqslant 1$, and let $\mathscr{S} \subset \hat{\mathscr{B}}$ be a bounded Borel subset of the hyperplane $x^0 = a$. We shall introduce the notation $x = (x^0, \mathbf{x}')$, $\mathbf{x}' = (x^1, x^2, x^3)$. Let $\tilde{\mathscr{S}} := \{x' \in \boldsymbol{R}^3 | \ (a, \mathbf{x}') \in \tilde{\mathscr{S}}\}$. We assume that $[a, c] \times \tilde{\mathscr{S}} \subset \hat{\mathscr{B}}$ for some $c > a$. Let $\theta \in \mathscr{C}^\infty(\boldsymbol{R}, \boldsymbol{R})$, $\theta(\xi) = 1$ for $a < \xi < b$, $\theta(\xi) = 0$ for $\xi \geqslant c > b$. Obviously $\tilde{g}(\mathbf{x}) : \equiv \theta(x^0) g(x^0, \mathbf{x}')$ belongs to $W^{1,p}(\hat{\mathscr{B}})$ and $\tilde{g}(x^0, \mathbf{x}') = 0$ for $x^0 \geqslant c$.

Since the distributional derivative $\partial \tilde{g}/\partial x^0 \in \mathscr{L}^p_{loc}(\hat{\mathscr{B}})$, it follows that $\partial \tilde{g}/\partial x^0 \in \mathscr{L}^1_{loc}(\hat{\mathscr{B}})$. By a theorem due to Nikodym (cf. Mizohata, 1969; Schwartz, 1957) for L^3-a.e. values of $\mathbf{x}'$ the function $x^0 \mapsto g(x^0, \mathbf{x}')$ is absolutely continuous. Since

$$\int_a^c dx^0 \int_{\tilde{\mathscr{S}}} dL^3(\mathbf{x}') \left| \frac{\partial \tilde{g}}{\partial x^0} \right| < \infty,$$

it follows by the Fubini theorem that

$$\begin{aligned} \int_{a+\varepsilon}^c dx^0 \int_{\tilde{\mathscr{S}}} dL^3(\mathbf{x}') \frac{\partial \tilde{g}}{\partial x^0} &= -\int_{\tilde{\mathscr{S}}} dL^3(\mathbf{x}') \tilde{g}(a+\varepsilon, \mathbf{x}') \\ &= -\int_{\tilde{\mathscr{S}}} dL^3(\mathbf{x}') g(a+\varepsilon, \mathbf{x}') \end{aligned}$$

exists for *every* $\varepsilon \geqslant 0$ such that $a+\varepsilon < b$ and depends *continuously* on ε. On the other hand

$$\begin{aligned} -\int_{\tilde{\mathscr{S}}} dL^3 g(a, \mathbf{x}') &= \int_{\tilde{\mathscr{S}}} dL^3(x') \int_a^c dx^0 \frac{\partial \tilde{g}}{\partial x^0}(x^0, \mathbf{x}') \\ &= -\int_{\mathscr{S}} dH^3 \gamma_{\mathscr{S}} g(x), \end{aligned} \tag{1.2.60}$$

where $\gamma_{\mathscr{S}}$ denotes the trace operator in $W^{1,p}_{loc}(\hat{\mathscr{B}})$ associated with $\mathscr{S}$ (cf. Mizohata, 1969, Lemma 3.5).

It follows that (1.2.59) remains valid for a class Γ of sets $\mathscr{P}$ containing all the polyhedra. We shall also note that for every $\mathscr{P}$ with the strong local Lipschitz property which is bounded and has a boundary $\partial\mathscr{P}$ of finite measure H^3 the Green formula is valid. Indeed, every $\mathbf{f} \in W^{1,p}_{loc}(\hat{\mathscr{B}})$ can be approximated by a sequence $\{\mathbf{f}_n\} \subset \mathscr{C}^\infty$. Since $\partial\mathscr{P}$ has the strong local Lipschitz property,

$$\int_{\partial\mathscr{P}} dH^3 f^k_n \nu_k = \int_{\mathscr{P}} dL^4 \, \partial f^k_n / \partial x^k$$

(ν is the outer unit normal, defined H^3-a.e. on $\partial\mathscr{P}$, cf. Nečas, 1966, Section 1.2.4). For $\mathbf{f}_n \to \mathbf{f}$ in $W^{1,p}_{loc}(\hat{\mathscr{B}})$ the integrand on the right-hand side converges in $\mathscr{L}^p_{loc}(\hat{\mathscr{B}}) \subset \mathscr{L}^1_{loc}(\hat{\mathscr{B}})$ to $\partial f^k/\partial x^k$ while the integrand on the left-hand side conv ergesto $\nu_k \gamma f^k$ in $W^{1-1p,p}_{loc}(\partial\hat{\mathscr{B}}) \to \mathscr{L}^1_{loc}(\partial\hat{\mathscr{B}})$ (the embeddings are continuous). Hence

$$\int_{\partial\mathscr{P}} dH^3 \gamma f^k \nu_k = \int_{\mathscr{P}} dL^4\, \partial f^k/\partial x^k. \tag{1.2.61}$$

Unfortunately the class $\Phi = W^{1,p}_{loc}(\hat{\mathscr{B}})$ is too small for practical applications. In Section 1.2.9 we shall reformulate balance equations in a more satisfactory form. In the case of the law of momentum conservation we shall arrive at the virtual work principle. In the language of mathematics the virtual work principle is a differential equation in the sense of distributions. On the other hand a balance equation is a convenient point of departure for the theory since it expresses the assumption that something is conserved. The next step involves derivation of differential equations from the conservation laws. These differential equations must, however, be interpreted in the distributional sense. This brings us back to the virtual work principle and its counterparts for all the conservation laws of the theory.

1.2.9 Conservation of Momentum and the Virtual Work Principle

We shall use a global reference configuration $\boldsymbol{\varkappa}\colon \mathscr{B} \to R^3$. Let $\boldsymbol{\chi}\colon \boldsymbol{\varkappa}(\mathscr{B})\times I \to R^3$ denote the motion of $\mathscr{B}$ referred to $\boldsymbol{\varkappa}$ and to the space coordinates of a fixed Galilean reference frame.

The conservation of momentum in an elastic body is postulated in the form of the following balance equations:

$$\int_{\varkappa(\mathscr{R})} dL^3\, \varrho_o[\chi^k{}_{,t}(\xi, t_2) - \chi^k{}_{,t}(\xi, t_1)] - \int_{\partial\varkappa(\mathscr{R})\times I} dH^2\, dt\, S_k{}^{\mu}(\xi, t)\nu_{\mu}$$

$$- \int_{\varkappa(\mathscr{R})\times I} dL^3\, dt\, \varrho_o(\xi) b_k(\xi, t) = 0, \tag{1.2.62}$$

$\varrho_o := \varrho_{\varkappa}$, valid for all the sets $\mathscr{R} \subset \mathscr{B}$, $I' = [t_1, t_2]$, for which the left-hand side of (1.2.62) is defined. In view of subsequent applications in Chapter 2 we assume that

$$\chi^k \in W^{1,p}_{loc}(\hat{\mathscr{B}}), \quad S_k^{\mu} \in \mathscr{L}^q_{loc}(\hat{\mathscr{B}}), \quad b_k \in \mathscr{L}^q_{loc}(\hat{\mathscr{B}}), \quad \varrho_o \in \mathscr{L}^\infty(\boldsymbol{\varkappa}(\mathscr{B})),$$
$$p \geqslant 2, \quad 1 \leqslant q \leqslant 2, \quad p^{-1} + q^{-1} = 1. \tag{1.2.63}$$

Here $\hat{\mathscr{B}} := \boldsymbol{\varkappa}(\mathscr{B})\times I$ and ν_{μ} denotes the outer unit normal on $\partial\boldsymbol{\varkappa}(\mathscr{R})$ in indicial notation.

For each $k = 1, 2, 3$ equation (1.2.62) is a special case of the balance equations considered in Sections 1.2.6 and 1.2.7 with

$$f^0 = \varrho_0 \chi^k{}_{,t}, \quad f^\mu = -S_k{}^\mu, \; \mu = 1, 2, 3. \tag{1.2.64}$$

Equation (1.2.62) introduces the **Piola–Kirchhoff stress tensor** $S_k{}^\mu$ for the reference configuration $\varkappa$. The functions b_k, $k = 1, 2, 3$, are the components of the **body force** vector field.

We assume that

$$\partial\varkappa(\mathscr{B}) = \bar{\Sigma}_1 \cup \bar{\Sigma}_2, \; \Sigma_1 \cap \Sigma_2 = \emptyset, \quad \Sigma_1 \text{ and } \Sigma_2 \text{ are open in } \partial\varkappa(\mathscr{B}), \tag{1.2.65}$$

Σ_1, Σ_2 and $\Gamma := \partial\Sigma_1 = \partial\Sigma_2$ (boundaries in the topology of $\partial\varkappa(\mathscr{B})$) are uniformly Lipschitz continuous, viz. $\exists\ \delta, K > 0$, $\nu \in \boldsymbol{Z}_+$ and a locally finite open covering $\{\mathscr{U}_i^\alpha |\ i = 1, \ldots, N_\alpha,\ \alpha = 0, 1, 2\}$, $N_\alpha \leqslant \infty$, of $\partial\varkappa(\mathscr{B})$, such that

(1) the intersection of more than ν sets $\mathscr{U}_i^1$ or $\mathscr{U}_i^2$ is empty;

(2) $\mathscr{U}_i^1 \cap \bar{\Sigma}_2 = \emptyset$ for $i \leqslant N_1$, $\mathscr{U}_i^2 \cap \bar{\Sigma}_1 = \emptyset$ for $i \leqslant N_2$, $\mathscr{U}_i^0 \cap \Gamma = \emptyset$, for $\alpha = 1, 2$, $\bigcup_i \mathscr{U}_i^\alpha \cup \bigcup_j \mathscr{U}_j^0 \supset \Sigma_\alpha$, $\bigcup_i \partial\mathscr{U}_i^\alpha \supset \Gamma$;

(3) each $\mathscr{U}_i^\alpha$ is endowed with a Cartesian coordinate system (z_1, z_2, z_3) such that

(a) $\varkappa(\mathscr{B}) \cap \mathscr{U}_i^0$ is given by the inequalities $\gamma(z_1, z_2) < z_3 < \gamma(z_1, z_2) + \delta$, $|z_k| < \delta$, $k = 1, 2$,

(b) $\varkappa(\mathscr{B}) \cap \mathscr{U}_i^\alpha$ is given by $|z_1| < \delta$, $|z_2 - \lambda(z_1)| < \delta$, $\gamma(z_1, z_2) < z_3 < \gamma(z_1, z_2) + \delta$,* for $\alpha = 1, 2$;

(4) γ is defined on $[-\delta, \delta] \times [-\delta, \delta]$, λ is defined on $[-\delta, \delta]$, $\gamma(0, 0) = 0$, $\lambda(0) = 0$, $\operatorname{Lip}\gamma \leqslant K$, $\operatorname{Lip}\lambda \leqslant K$, i.e.

$$|\gamma(z_1', z_2') - \gamma(z_1'', z_2'')| \leqslant K[(z_1' - z_1'')^2 + (z_2' - z_2'')^2]^{1/2}, \tag{1.2.66}$$

$$|\lambda(z_1') - \lambda(z_1'')| \leqslant K|z_1' - z_1''|. \tag{1.2.67}$$

Moreover, let

$$t_k \in \mathscr{L}^s_{loc}(\Sigma_2 \times I), \; s := \max\{1, 3q/4\}, \tag{1.2.68}$$

$$\chi_1{}^k \in W^{1-1p/1}_{loc}(\Sigma_1 \times I), \tag{1.2.69}$$

$$\chi_0{}^k \in W^{1-1p/1}_{loc}(\varkappa(\mathscr{B})), \tag{1.2.70}$$

$$v_0{}^k \in \mathscr{L}^p_{loc}(\varkappa(\mathscr{B})), \quad k = 1, 2, 3. \tag{1.2.71}$$

We assume that $\chi^k(\xi, t) = \chi_1^k(\xi, t)$ for $\xi \in \Sigma_1$ and $\chi^k(\xi, 0) = \chi_0^k(\xi)$ for $\xi \in \varkappa(\mathscr{B})$, in the sense of traces.

* The functions γ, λ depend on i, α.

We consider the following sets:

$$\mathscr{Q} := \{w \in \boldsymbol{R}^3 \mid |w_k| \leqslant a_k,\ k = 1, 2, 3\},$$
$$\mathscr{Q}_+ := \{w \in \boldsymbol{R}^3 \mid |w_k| \leqslant a_k,\ k = 1, 2, 0 \leqslant w_3 \leqslant a_3\},$$
$$\mathscr{P} := \{z \in \boldsymbol{R}^3 \mid |z_k| \leqslant a_k,\ k = 1, 2, 3\},$$
$$\mathscr{P}_+ := \{z \in \boldsymbol{R}^3 \mid |z_k| \leqslant a_k,\ k = 1, 2, \gamma(z_1, z_2) \leqslant z_3 \leqslant \gamma(z_1, z_2) + a_3\},$$
$$\mathscr{P}_{++} := \{z \in \boldsymbol{R}^3 \mid |z_1| \leqslant a_1, |z_2 - \lambda(z_1)| \leqslant a_2, \gamma(z_1, z_2) \leqslant z_3 \leqslant \gamma(z_1, z_2) + a_3\},$$

((z_k) are the coordinates defined on $\mathscr{U}_i^\alpha$),

$$\mathscr{P}_+ \cap \Gamma = \emptyset, \quad \mathscr{P}_{++} \cap \Gamma = \{z \in \mathscr{P}_{++} \mid z_2 = \lambda(z_1) - a_2\}.$$

Moreover, we define the mappings $f, f_1 \colon w \mapsto z$,

$$f(w) := (w_1, w_2, w_3 + \gamma(w_1, w_2)),$$
$$f_1(w) := (w_1, w_2 + \lambda(w_1), w_3 + \gamma(w_1, w_2)).$$

The mappings f, f_1 are Lipschitz continuous and their inverses

$$\overset{-1}{f}(z) = (z_1, z_2, z_3 - \gamma(z_1, z_2)),$$

$$\overset{-1}{f_1}(z) = (z_1, z_2 - \lambda(z_1), z_3 - \gamma(z_1, z_2 - \lambda(z_1)))$$

are also Lipschitz continuous. We note that

$$f(\mathscr{Q}_+) = \mathscr{P}_+, f_1(\mathscr{Q}_+) = \mathscr{P}_{++}.$$

In the nbhd of every point $\xi_o \in \overline{\varkappa(\mathscr{B})}$ we introduce local Cartesian coordinates z in such a way that ξ_o corresponds to $z = 0$. In terms of these coordinates we now define the sets $\mathscr{P}$, $\mathscr{P}_+$, $\mathscr{P}_{++}$ according to the position of ξ_o.

The family of sets $\mathscr{Q}$, $\mathscr{Q}_+$ can be parametrized by the three real numbers a_1, a_2, a_3. For every $\xi_o \in \overline{\varkappa(\mathscr{B})}$ we consider the sets $\mathscr{P}$ (if ξ_o lies inside $\varkappa(\mathscr{B})$), $\mathscr{P}_+$ (if ξ_o lies on Σ_1 or on Σ_2 away from Γ) or $\mathscr{P}_{++}$ (if ξ_o lies near Γ). These sets are also parametrized by a_1, a_2, a_3 (sufficiently small so that $\mathscr{P}, \mathscr{P}_+, \mathscr{P}_{++} \subset \overline{\varkappa(\mathscr{B})}$). Following the remarks made in Section 1.2.8 we shall assume that the balance equation (1.2.62) holds for

(1) $\xi_o \in \varkappa(\mathscr{B})$, $\varkappa(\mathscr{R}) = \mathscr{P} \subset \varkappa(\mathscr{B})$, for a.a. t_1, t_2, a_1, a_2, a_3, $t_2 > t_1$ in I, a_k sufficiently small,

(2) for $\xi_o \in \Sigma_1$, $\varkappa(\mathscr{R}) = \mathscr{P}_+$ or $\mathscr{P}_{++} \subset \varkappa(\mathscr{B})$, for a.a. (t_1, t_2) such that $t_2 > t_1$, $t_2, t_1 \in I$ and a.a. (a_1, a_2, a_3) sufficiently small.

For $\xi_o \in \Sigma_2$ we shall assume that

$$\int_{\mathscr{P}'} dL^3\, \varrho_o(\xi)[\chi^k{}_{,t}(\xi, t_2)-\chi^k{}_{,t}(\xi, t_1)] = \int_{t_1}^{t_2} dt \int_{\partial\mathscr{P}'\cap\Sigma_2} dH^2\, t_k(\xi, t)+$$

$$+\int_{t_1}^{t_2} dt \int_{\partial\mathscr{P}'\setminus\Sigma_2} dH^2\, S_k{}^{\mu}(\xi, t)\nu_\mu + \int_{[t_1, t_2]\times\mathscr{P}'} dt\, dL^3\, \varrho_o(\xi) b_k(\xi, t) \quad (1.2.72)$$

for a.a. $t_1, t_2 \in I$ such that $t_2 > t_1$ and a.a. $\mathscr{P}' = \mathscr{P}_+$ or $\mathscr{P}_{++} \subset \overline{\varkappa(\mathscr{B})}$.*

The modification (1.2.72) of the momentum conservation law accounts for the traction boundary condition. In a naive approach it would take the form $S_k{}^{\mu}\nu_\mu = t_k$ on Σ_2. Under the assumption (1.2.63) the last equation is meaningless.

Let $\mathscr{G}$ be the projection onto the (z_1, z_2)-plane of the sets $\{z \in \mathscr{P}' |\ z_3 = a_3+\gamma(z_1, z_2)\}$. Contracting $\mathscr{P}'$ to $\mathscr{P}'\cap\Sigma_2 = \{z \in \mathscr{P}' |\ z_3 = \gamma(z_1, z_2)\}$ we conclude that

$$\lim_{a_3\to 0_+} \int_{t_1}^{t_2}\int_{\mathscr{G}} dt\, dH^2\, S_k{}^{\mu}\big(z_1, z_2, a_3+\gamma(z_1, z_2)\big)\,\nu_\mu(z_1, z_2)$$

$$= \int_{t_1}^{t_2}\int_{\mathscr{G}} dt\, dH^2\, t_k(z_1, z_2)$$

for all t_1, t_2, a_1, a_2, a_3 for which (1.2.72) remains valid.**

The initial condition (1.2.71) cannot be interpreted in the sense of traces either. It will be incorporated in the balance equations by assuming that

$$\int_{\varkappa(\mathscr{R})} dL^3 \varrho_o[\chi^k{}_{,t}(\xi, t)-v_0^k(\xi)] = \int_{\partial\varkappa(\mathscr{R})\times[0, t_2]} dH^2\, dt\, S_k{}^{\mu}(\xi, t)\nu_\mu +$$

$$+ \int_{\varkappa(\mathscr{R})\times[0, t_2]} dL^3\, dt\, \varrho_o b_k(\xi, t) \quad \text{for a.a. } t_2,\ \varkappa(\mathscr{R}) = \mathscr{P},\ \mathscr{P}_+ \text{ or } \mathscr{P}_{++},$$

$$(1.2.73)$$

which implies that

$$\lim_{n\to\infty} \int_{\varkappa(\mathscr{R})} dL^3\, \varrho_o[\chi^k{}_{,t}(\xi, t_n)-v_0^k(\xi)] = 0$$

provided the sequence $t_n \to 0$ is chosen in such a way that the l.-h.s. is well defined for each t_n and the given $\varkappa(\mathscr{R})$.

* That is, for a.a. (a_1, a_2, a_3) such that $\mathscr{P}' \subset \overline{\varkappa(\mathscr{B})}$.

** Suppose that there is a set $\mathscr{A}$ of (t_1, t_2, a_1, a_2) of positive measure and a number $\varepsilon > 0$ such that (1.2.72) ceases to hold on $\mathscr{A}$ for $a_3 < \varepsilon$. Then (1.2.72) cannot hold for a.a. $(t_1, t_2, a_1, a_2, a_3)$.

Let $\mathscr{W}$ be the linear manifold spanned by the functions

$$\mathbf{u}(\xi, t) = \psi(\overset{-1}{\mathbf{k}}(\xi), t)\mathbf{c}, \quad \xi \in \overline{\varkappa(\mathscr{B})},\ t \in \bar{I}_+,\ \mathbf{c} \in \mathbf{R}^3 \tag{1.2.74}$$

where $\mathbf{k}$: $\mathbf{R}^3 \supset \overline{\mathscr{U}} \to \overline{\varkappa(\mathscr{B})}$ is a bi-Lipschitz mapping,* $\psi(w_1, w_2, w_3, t) \equiv \psi_0(t)\psi_1(w_1)\psi_2(w_2)\psi_3(w_3)$, $\operatorname{supp}\mathbf{u} \subset (\varkappa(\mathscr{B})\cup\Sigma_2\times)\bar{I}_+$, $I_+ := I\cap\mathbf{R}_+$. Let $\mathscr{V}^p$ be the closure of $\mathscr{W}$ in $W^{1,p}(\varkappa(\mathscr{B})\times I_+)$. It can be proved that

(i) for $p < \infty$, $\Sigma_1 = \partial\varkappa(\mathscr{B})$, and $\mathscr{V}^p = W_o^{1,p}(\varkappa(\mathscr{B})\times I_+)$,

(ii) for $p < \infty$, $\Sigma_2 = \partial\varkappa(\mathscr{B})$, and $\mathscr{V}^p = W^{1,p}(\varkappa(\mathscr{B})\times I_+)$

(Témam, 1973; cf. Section 1.1). Antman and Osborn (1979) conjectured that

$$\mathscr{V}^p = \{\mathbf{u} \in W^{1,p}(\varkappa(\mathscr{B})\times I_+)|\ \gamma_{\Sigma_1}\mathbf{u} = 0\}. \tag{1.2.75}$$

Under the assumptions formulated above we shall prove the **virtual work principle** (VWP):

For all $\mathbf{u} \in \mathscr{V}^p$ with compact support in $\overline{\varkappa(\mathscr{B})}\times \bar{I}_+$

$$\int\limits_{I_+\times\varkappa(\mathscr{B})} dt\, dL^3\, [S_k^{\mu}u^k{}_{,\xi^\mu} - \varrho_o b_k u^k - \varrho_o \chi^k{}_{,t} u^k{}_{,t}] - \int\limits_{I_+\times\partial\varkappa(\mathscr{B})} dt\, dH^2\, t_k u^k$$

$$= \int\limits_{\varkappa(\mathscr{B})} dL^3\, \varrho_o \sum_{k=1}^{3} v_0^k(\xi)u^k(\xi, 0). \tag{1.2.76}$$

The idea of the proof is quite straightforward although the proof looks cumbersome when written out in full detail. Therefore, we shall indicate the procedures, leaving the details to the reader (cf. also Antman and Osborn, 1979).

We shall consider the functions (1.2.74) supported by some set $\mathscr{P}$, $\mathscr{P}_+$, $\mathscr{P}_{++}$ around ξ_o:

(1) for $\xi_o \in \varkappa(\mathscr{B})$: $\psi_k \in W^{1,p}([-a_k^o, a_k^o])$, $\operatorname{supp}\psi_k \subset]-a_k^o, a_k^o[$, $\psi_k(-y) = \psi_k(y)$, $k = 1, 2, 3$; $\mathbf{k} = \mathrm{id}$;

(2) for $\xi_o \in \Sigma_1$: $\psi_k \in W^{1,p}([-a_k^o, a_k^o])$, $\operatorname{supp}\psi_k \subset]-a_k^o, a_k^o[$, $\psi_k(-y) = \psi_k(y)$, $k = 1, 2$; $\psi_3 \in W^{1,p}([0, a_3^o])$, $\operatorname{supp}\psi_3 \subset]0, a_3^o[$; $\mathscr{P}' = \mathscr{P}_+$ or $\mathscr{P}_{++}$, appropriate **f**, **k**;

(3) for $\xi_o \in \Sigma_2$: $\psi_k \in W^{1,p}([-a_k^o, a_k^o])$, $\operatorname{supp}\psi_k \subset]-a_k^o, a_k^o[$, $\psi_k(y) = \psi_k(-y)$, $k = 1, 2$; $\psi_3 \in W^{1,p}([0, a_3^o])$, $\operatorname{supp}\psi_3 \subset [0, a_3^o[$; $\mathscr{P}' = \mathscr{P}_+$ or $\mathscr{P}_{++}$, appropriate **f** and **k**.

We shall always assume that a_k^o are sufficiently small so that $\mathbf{k}(\bigtimes_{l=1}^{3}[-a_l^o, a_l^o]) \subset \overline{\varkappa(\mathscr{B})}$. For ψ_0 we assume that $\psi_0 \in W^{1,p}([0, t])$, $\operatorname{supp}\psi_0 \subset [0, t[$ for some $t \in I_+$.

* $\mathbf{k}$ is the composition of the coordinate transformations f: $w \mapsto z$ and $z \mapsto \xi$.

By Theorem 1.1.56 ψ_k, $k = 0, 1, 2, 3$, is Hölder continuous with exponent $1-1/p$. By Theorem 1.1.55 it is also absolutely continuous.*

Let us write out equations (1.2.62), (1.2.72), (1.2.73) for one of the sets $\mathscr{P}$, $\mathscr{P}_+$ or $\mathscr{P}_{++}$, with $t_1 = 0 < t_2 \leqslant t$, $a_k \leqslant a_k^o$ such that the equation is meaningful. We shall write τ for t_2 and use w for the integration variables. Now $dL^3 = |J(\mathbf{k})| dw_1 dw_2 dw_3$, $dH^2\boldsymbol{\nu} = \dfrac{\partial \mathbf{k}}{\partial w_1} \times \dfrac{\partial \mathbf{k}}{\partial w_2} dw_1 dw_2$, $dH^2 = \gamma^{1/2} dw_1 dw_2$ on the faces $w_3 = \text{const}$, $\gamma = \det[\gamma_{rs}] = \det \left[\sum_{l=1}^{3} \dfrac{\partial k_l}{\partial w_r} \dfrac{\partial k_l}{\partial w_s} \right]$.

The left-hand side of equation (1.2.62), (1.2.72) or (1.2.73) is a function $F(a_1, a_2, a_3, \tau)$ defined for a.a. (a_1, a_2, a_3, τ). We assume that it is locally integrable. We calculate the integrals

$$\frac{1}{4}\int_0^t d\tau \int_\alpha^{a_3^o} da_3 \int_{-a_1^o}^{a_1^o} da_1 \int_{-a_2^o}^{a_2^o} da_2\, \psi_0'(\tau)\psi_1'(a_1)\psi_2'(a_2)\psi_3'(a_3) F(a_1, a_2, a_3, \tau),$$

$$\alpha = 0 \text{ or } -a_3^o, \tag{1.2.77}$$

noting that for absolutely continuous ψ**

$$\int_0^{a^o} da\, \psi'(a) \int_0^a dw\, \omega(w) = \psi(a^o)\omega(a^o) - \int_0^{a^o} dw\, \omega(w)\psi(w),$$

$$\int_{-a^o}^{a^o} da\, \psi'(a) \int_{-a}^{a} dw\, \omega(w) = -\int_{-a^o}^{o} dw\, \psi(w)[\omega(w)+\omega(-w)]$$

$$= -2 \int_{-a^o}^{a^o} dw\, \psi(w)\omega(w)$$

provided $\psi(-w) = \psi(w)$, $0 \leqslant w \leqslant a^o$, $\psi(a^o) = 0$,

$$\int_0^{a^o} da\, \psi'(a) \cdot \text{const} = -\psi(0) \cdot \text{const}$$

provided $\psi(a_\cdot^o) = 0$.

The contribution of the terms

$$\mp \int_0^\tau ds \int_0^{a_3} dw_3 \left| \int_{-a_2}^{a_2} dw_2\, S_k^\mu(\mathbf{k}(\pm a_1, w_2, w_3), s) \left(\frac{\partial \mathbf{k}}{\partial w_2} \times \frac{\partial \mathbf{k}}{\partial w_3} \right) \cdot \mathbf{e}_\mu \right.$$

* More precisely $\psi_k \in W^{1,p}$ has an absolutely continuous and Hölder continuous representative. It is then assumed that the support of this representative has the properties specified in (1)–(3).

** Cf. Section 1.1.7.

in $F(a_1, a_2, a_3, \tau)$ to (1.2.77) is

$$-\int_0^t d\tau \int_{\mathscr{Q}_+} dL^3(w)\psi_0(\tau)\psi_1'(w_1)\psi_2(w_2)\psi_3(w_3)S_k^{\mu}(\mathbf{k}(w_1, w_2, w_3), \tau)\cdot$$
$$\cdot\left(\frac{\partial \mathbf{k}}{\partial w_2}\times\frac{\partial \mathbf{k}}{\partial w_3}\right)\cdot \mathbf{e}_{\mu}, \tag{1.2.78}$$

while the contributions of

$$-\int_0^{\tau} ds \int_{-a_1}^{a_1} dw_1 \int_{-a_2}^{a_2} dw_2\, S_k^{\mu}(\mathbf{k}(w_1, w_2, a_3), s)\left(\frac{\partial \mathbf{k}}{\partial w_1}\times\frac{\partial \mathbf{k}}{\partial w_2}\right)\cdot \mathbf{e}_{\mu}$$

and

$$-\int_0^{\tau} ds \int_{-a_1}^{a_1} dw_1 \int_{-a_2}^{a_2} dw_2\, t_k \gamma^{1/2}|_{w_3=0}$$

are

$$\int_0^t d\tau \int_{\mathscr{Q}_+} dL^3(w) S_k^{\mu}(\mathbf{k}(w_1, w_2, a_3), \tau)\left(\frac{\partial \mathbf{k}}{\partial w_1}\times\frac{\partial \mathbf{k}}{\partial w_2}\right)\cdot \mathbf{e}_{\mu}$$

and

$$\psi_3(0)\int_0^t d\tau \int_{-a_1^0}^{a_1^0} dw_1 \int_{-a_2^0}^{a_2^0} dw_2\, t_k \gamma^{1/2}|_{w_3=0}, \text{ etc.}$$

Contracting (1.2.78) with a vector $\mathbf{a} \in \boldsymbol{R}^3$ we obtain the expression

$$\sum_{k=1}^{3} a^k \int_0^t d\tau \int_{\mathscr{Q}_+} dL^3(w)\, \frac{\partial \psi(w, \tau)}{\partial w_1}\, S_k^{\ u}\left(\frac{\partial \mathbf{k}}{\partial w_2}\times\frac{\partial \mathbf{k}}{\partial w_3}\right)\cdot \mathbf{e}^{\mu}$$
$$= \sum_{k=1}^{3}\int_0^t d\tau \int_{\mathbf{k}(\mathscr{Q}_+)} dL^3\, \frac{\partial u^k}{\partial \xi^{\gamma}}\, S_k^{\mu}\, \frac{\partial k^{\gamma}}{\partial w_1}\, \epsilon_{\alpha\beta\mu}\, \frac{\partial k^{\alpha}}{\partial w_2}\, \frac{\partial k^{\beta}}{\partial w_3}\,.$$

The other expressions can be dealt with in a similar way. Hence, collecting all the terms of equation (1.2.62), (1.2.72), (1.2.73) contracted with a constant vector $\mathbf{a} \in \boldsymbol{R}^3$ and noting that

$$\frac{1}{2}\,\epsilon_{\alpha\beta\mu}\,\frac{\partial k^{\gamma}}{\partial w_r}\,\frac{\partial k^{\alpha}}{\partial w_s}\,\frac{\partial k^{\beta}}{\partial w_p}\,\epsilon^{rsp} = J(\mathbf{k})\,\delta^{\gamma}{}_{\mu}$$

we arrive at equation (1.2.76) for test functions $\mathbf{u}$ of the form (1.2.74). Since linear combinations of functions $\mathbf{u}$ are dense in $\mathscr{V}^p$, this implies the VWP.

The converse implication, leading from the VWP to the balance equations for a.a. sets $\mathscr{P}$, $\mathscr{P}_{+}$, $\mathscr{P}_{++}$ can also be proved. Note, however, that the proofs contain serious gaps, e.g. the conjecture (1.2.75).

The VWP and its modifications provide the basis for the rigorous formulation of boundary value problems of elastostatics. In elastodynamics we shall also use a class of functions and sets which is equally well adapted to the VWP and balance formulations (see Chapter 3, in particular the Appendix).

1.2.10 The Conservation of Moment of Momentum. Constitutive Equations

The law of conservation of moment of momentum in elasticity is expressed by the following balance equation:

$$\int_{\mathscr{P}} dL^3\, \varrho_o \chi^{[k},{}_t\chi^{l]}\Big|_{t_1}^{t_2} - \int_{\partial\mathscr{P}\times[t_1,t_2]} dH^2\, dt\, S_{[k}{}^{\mu}\nu_\mu \chi^{l]} = \int_{\mathscr{P}\times[t_1,t_2]} dL^4\, \varrho_o b_{[k}\chi^{l]} \tag{1.2.79}$$

for a.a. 4-cells $\tilde{\mathscr{P}} = \mathscr{P}\times[t_1, t_2]$ in $\hat{\mathscr{B}} := \boldsymbol{\varkappa}(\mathscr{B})\times]0, T[$. We assume that $\boldsymbol{\chi}$ is Lipschitz continuous:

$$|\boldsymbol{\chi}(y)-\boldsymbol{\chi}(y')| \leqslant K|y-y'|, \quad \forall y, y' \in \hat{\mathscr{B}},\ y = (\xi, t).$$

Setting $f_k{}^0 = \varrho_o v^k$, $f_k{}^\mu = -S_k{}^\mu$, k, $\mu = 1, 2, 3$, and taking into account equation (1.2.62) we can rewrite equation (1.2.79) in the form

$$\int_{\partial\tilde{\mathscr{P}}} dH^3 f_{[k}{}^{\alpha}\nu_\alpha \bar{\chi}^{l]} = \int_{\tilde{\mathscr{P}}} dL^4\, \varrho_o b_{[k}\bar{\chi}^{l]}; \quad \text{sum over } \alpha = 0, \ldots, 3 \tag{1.2.80}$$

with $\bar{\boldsymbol{\chi}}(\xi, t) := \boldsymbol{\chi}(\xi, t)-\boldsymbol{\chi}(\xi_o, t_o)$, $t_o = \dfrac{1}{2}(t_1+t_2)$, ξ_o = the centre of the 3-cell $\mathscr{P}$.

By the Lebesgue derivation theorem for a.a. $(\xi_o, t_o)\in\hat{\mathscr{B}}$ $\dfrac{1}{L^4(\tilde{\mathscr{P}})}\displaystyle\int_{\tilde{\mathscr{P}}} dL^4\, \varrho_o b_{[k}\bar{\chi}^{l]}$ $\to 0$ as $\tilde{\mathscr{P}}$ shrinks uniformly to (ξ_o, t_o), since $\bar{\boldsymbol{\chi}}(\xi_o, t_o) = \mathbf{0}$. Hence

$$\lim_{\tilde{\mathscr{P}}\searrow(\xi_o,t_o)} \frac{1}{L^4(\tilde{\mathscr{P}})} \int_{\partial\tilde{\mathscr{P}}} dH^3 f_{[k}{}^{\alpha}\nu_\alpha \bar{\chi}^{l]} = 0 \tag{1.2.81}$$

for a.e. $(\xi_o, t_o) \in \hat{\mathscr{B}}$.

For continuous $f_k{}^\alpha(\xi, t)$

$$\int_{\partial\tilde{\mathscr{P}}} dH^3\, \nu_\alpha f_{[k}{}^\alpha \bar{\chi}^{l]} = f_{[k}{}^\alpha(\xi_o, t_o) \int_{\partial\tilde{\mathscr{P}}} dH^3\, \bar{\chi}^{l]}\nu_\alpha$$

$$+ \int_{\partial\tilde{\mathscr{P}}} dH^3\, [f_{[k}{}^\alpha(\xi, t) - f_{[k}{}^\alpha(\xi_o, t_o)]\bar{\chi}^{l]}(\xi, t)\nu_\alpha. \tag{1.2.82}$$

Since $\boldsymbol{\chi}$ is Lipschitz continuous it has a total differential a.e. in $\hat{\mathscr{B}}$ (by Rademacher's theorem; Morrey, 1966, Theorem 3.1.6). We assume that $\hat{\mathscr{B}}$ satisfies the strong local Lipschitz condition. Hence the Green formula can be applied and the first integral on the right-hand side of (1.2.82) becomes

$$f_{[k}{}^\alpha(\xi_o, t_o) \int_{\tilde{\mathscr{P}}} dL^4\, \chi^{l]}{}_{,y^\alpha}$$

while the absolute value of the second integral is bounded by

$$K\sup\{|f_k{}^\alpha(\xi, t) - f_k{}^\alpha(\xi_o, t_o)|\,|\ (\xi, t) \in \tilde{\mathscr{P}},\ k = 1, 2, 3,\ \alpha = 0, \ldots, 3\}$$

$$\times \int_{\partial\tilde{\mathscr{P}}} dH^3\, |\xi - \xi_o|.$$

In the limit (1.2.81) only the first term on the right-hand side of (1.2.82) contributes and we get the formula $f_{[k}{}^\alpha \chi^{l]}{}_{,y^\alpha} = 0$ a.e. in $\hat{\mathscr{B}}$, i.e.

$$S_{[k}{}^\mu F^{l]}{}_\mu = 0, \quad \text{sum over } \mu = 1, 2, 3, \text{ a.e. in } \hat{\mathscr{B}}, \quad F^k{}_\mu = \chi^k{}_{,\xi^\mu}. \tag{1.2.83}$$

This result can be extended to $f_k^\alpha \in \mathscr{L}^1_{loc}(\hat{\mathscr{B}})$ by retracing the proof of Theorem 1.2.8 with $L\mathbf{f}^\alpha := \bar{\boldsymbol{\chi}} \times \mathbf{f}^\alpha$, $\mathbf{f}^\alpha := (f_k{}^\alpha)$ in place of $\mathbf{f}^\alpha$.

The VWP is equivalent to the system of three equations

$$\varrho_o \chi^k{}_{,tt} = S_k{}^\mu{}_{,\xi^\mu} + \varrho_o b_k, \quad k = 1, 2, 3 \tag{1.2.84}$$

in the weak sense of distribution theory. The body loads are often prescribed as functions $b_k(\xi, t)$, sometimes as functions of $\boldsymbol{\chi}$ and its derivatives at (ξ, t) as well. The basic idea is that the body loads b_k are not related to the material properties of the body in a unique fashion—they depend on the environment of the body, which is essentially arbitrary. The unknown functions in (1.2.84) are χ^k, $S_k{}^\mu$. Hence the system (1.2.84) is underdetermined. We shall remove this indeterminacy by substituting the **constitutive relations**

$$S_k{}^\mu = \hat{S}_k{}^\mu(\mathbf{F}, \mathbf{x}, \xi) \tag{1.2.85}$$

with $F^k{}_\mu = \chi^k{}_{,\xi^\mu}$, $x^k = \chi^k$ into (1.2.84). We shall assume that the **constitutive function** $\hat{S}_k{}^\mu$ satisfies (1.2.83) identically:

$$\hat{S}_{[k}{}^\mu(\mathbf{F}, \mathbf{x}, \xi) F^{l]}{}_\mu = 0 \quad \text{for all } \mathbf{F}, \mathbf{x}^\star \text{ and a.a. } \xi \in \boldsymbol{\varkappa}(\mathscr{B}). \tag{1.2.86}$$

⋆ More precisely, for all $(\mathbf{F}, \mathbf{x})$ in the domain of definition of $\hat{S}_k{}^\mu$.

The constitutive function $\hat{S}_k^{\mu}$ includes a complete description of the material properties of the elastic body $\mathscr{B}$. This additional information is necessary for predicting the behaviour of the body in response to given external conditions. Similarly the laws of conservation of momentum, energy and moment of momentum do not determine the outcome of a collision of balls (or material points) under given initial conditions: the exact mechanism of the collision must also be specified.

Suppose that there is a $\mathscr{C}^1$ function $(\mathbf{F}, \mathbf{x}) \mapsto W(\mathbf{F}, \mathbf{x}, \xi)$ for a.e. $\xi \in \varkappa(\mathscr{B})$ such that

$$\hat{S}_k^{\mu}(\mathbf{F}, \mathbf{x}, \xi) = \varrho_o(\xi) \frac{\partial W}{\partial F^k{}_{\mu}} (\mathbf{F}, \mathbf{x}, \xi) \quad \text{for a.a. } \xi \in \varkappa(\mathscr{B}). \tag{1.2.87}$$

Let $\boldsymbol{\chi} \in \mathscr{C}^2(\hat{\mathscr{B}})$ be a solution of equations (1.2.84) and (1.2.85). Equation (1.2.87) implies that $\boldsymbol{\chi}$ satisfies the equation of energy balance

$$\frac{\partial}{\partial t} \varrho_o \left[\frac{1}{2} \sum_{k=1}^{3} (\chi^k{}_{,t})^2 + W(\nabla\boldsymbol{\chi}, \boldsymbol{\chi}, \xi) \right] = \frac{\partial}{\partial \xi^{\mu}} [\hat{S}_k^{\mu}(\nabla\boldsymbol{\chi}, \boldsymbol{\chi}, \xi) \chi^k{}_{,t})] + \varrho_o \tilde{b}_k \chi^k{}_{,t} \tag{1.2.88}$$

with $\tilde{b}_k := b_k + W_{,x^k}$.* The vector field $-\hat{S}_k^{\mu}(\nabla\boldsymbol{\chi}, \boldsymbol{\chi}, \xi)\chi^k{}_{,t}$ represents the energy flux density.

Let us consider a homogeneous body $\mathscr{B}$ satisfying the constitutive equation (1.2.85) with $\hat{S}_k^{\mu}$ independent of ξ, $\boldsymbol{\chi}$, subject to a homogeneous deformation gradient $\mathbf{F}(t)$, $t \in [t_1, t_2]$. In a quasistatic process the time variation of kinetic energy $(1/2)\, \varrho_o \boldsymbol{\chi}_{,t} \cdot \boldsymbol{\chi}_{,t}$ can be neglected and the work done on the body by external agencies can be identified as

$$A = \int_{t_1}^{t_2} dt\, \hat{S}_k^{\mu}(\mathbf{F}) \dot{F}^k{}_{\mu}.$$

If there is a potential W such that $\hat{S}_k^{\mu}$ is given by equation (1.2.87) then $A = 0$ in every closed cycle $(\mathbf{F}(t_2) = \mathbf{F}(t_1))$. Conversely, if there is no potential W for $\hat{S}_k^{\mu}$, then there is at least one closed cycle $\mathbf{F}(t)$ such that $A \neq 0$. If $A > 0$ for this cycle, then $A < 0$ for the time-inverted cycle. Hence the body $\mathscr{B}$ can work as a perpetuum mobile of first kind, producing energy in a closed cycle.

If $\hat{S}_k^{\mu}$ satisfies (1.2.87), then $\frac{1}{2}\, \varrho_o \boldsymbol{\chi}_{,t}{}^2 + \varrho_o W(\nabla\boldsymbol{\chi}, \boldsymbol{\chi}, \xi)$ is the energy density and the total energy E of the body satisfies the differential equation

$$\frac{dE}{dt} = \int_{\varkappa(\mathscr{B})} dL^3\, [S_k^{\mu} \dot{F}^k{}_{\mu} + \varrho_o \boldsymbol{\chi}_{,t} \cdot \boldsymbol{\chi}_{,tt}].$$

* The total body force b_k consists of $\tilde{b}_k$ and a potential part $-W_{,x^k}$.

The first term represents the increase of the **stored energy** $\int_{\varkappa(\mathscr{B})} dL^3 W\varrho_0$, the second is equal to the rate of increase of the kinetic energy.

In addition to (1.2.85) the constitutive relations may include some relations involving the configurations or the configuration gradients only. An example is provided by **incompressibility**:

$$\det \mathbf{F} = f(\xi) \quad \text{for a.a. } \xi \in \varkappa(\mathscr{B}). \tag{1.2.89}$$

Constitutive relations of this kind are commonly called **internal constraints.** The adjective "internal" emphasizes the intuitive idea that the constraint is a material property and cannot be imposed or lifted at will. The distinction between internal and external constraints does not play any role in the mathematical framework of the theory—it is only a matter of interpretation.

The system of equations (1.2.84) and (1.2.85) supplemented by an internal constraint is overdetermined. This difficulty is, however, circumvented by adding an appropriate number of unknown functions, e.g. pressure in the case of incompressibility. In a variational theory of elasticity these additional unknowns appear quite naturally as Lagrange multipliers for the constraints (Chapter 2). In the case of a constraint

$$f(\mathbf{F}(\xi, t), \xi) \equiv A_k^{\mu}(\xi) F^k{}_{\mu} + a(\xi) = 0 \tag{1.2.90}$$

the extra unknowns represent some stress responsible for maintaining the constrained states. Loosely speaking, the total stress $S_k{}^{\mu}$ is now the sum of a derivative (1.2.87) tangent to the constraint manifold and an extra stress orthogonal to the constraint manifold. Only the former is given as a function of deformation.

1.2.11 Invariance of the Theory with Respect to Galilean Transformations and Changes of Reference Configuration

So far equations of motion (1.2.76), (1.2.85) of an elastic body $\mathscr{B}$ have been expressed (1) in terms of a fixed Galilean coordinate system on space-time, (2) in terms of a fixed reference configuration. It is therefore necessary to consider the effects of varying either of these. We shall establish invariance of all the equations hitherto considered with respect to changes of Galilean coordinates and Lagrange coordinates.

Suppose that $\boldsymbol{\chi}\colon \varkappa(\mathscr{B}) \to R^3$ is a solution of (1.2.76) (or (1.2.84)) and (1.2.85) and

$$\boldsymbol{\chi}'(\xi, t) := \mathbf{A}\boldsymbol{\chi}(\xi, t) + t\mathbf{b} + \mathbf{c}, \quad \mathbf{b}, \mathbf{c} \in R^3,\ \mathbf{A} \in \mathrm{SO}(3). \tag{1.2.91}$$

χ' satisfies the same equations with the transformed body forces b'_k,

$$\mathbf{b}'(\mathbf{A}\mathbf{x}+t\mathbf{b}+\mathbf{c}, \xi) = \mathbf{A}\mathbf{b}(\mathbf{x}, \xi) \tag{1.2.92}$$

if

$$\hat{S}_k^{\mu}(\mathbf{A}\mathbf{F}, \mathbf{A}\mathbf{x}+t\mathbf{b}+\mathbf{c}, \xi) = A^k{}_l \hat{S}_l^{\mu}(\mathbf{F}, \mathbf{x}, \xi) \quad \text{(sum over } l = 1, 2, 3). \tag{1.2.93}$$

We do not claim that (1.2.93) is a necessary condition. We shall however assume that $\hat{S}_k^{\mu}$ satisfies (1.2.93).

It follows from (1.2.93) that $\hat{S}_k^{\mu}$ does not depend on $\mathbf{x}$, and on account of (1.2.87)

$$W(\mathbf{F}, \mathbf{x}, \xi) = W_o(\mathbf{F}, \xi)+P(\mathbf{x}, \xi). \tag{1.2.94}$$

$\mathbf{W}_o$ is the **specific stored** (or strain) **energy** and P is the potential energy of potential body forces acting on the material point ξ.

Furthermore

$$\hat{S}_k^{\mu}(\mathbf{A}\mathbf{F}, \xi) = A^k{}_l \hat{S}_l^{\mu}(\mathbf{F}, \xi), \quad W_o(\mathbf{A}\mathbf{F}, \xi) = W_o(\mathbf{F}, \xi). \tag{1.2.95}$$

It follows from the theory of scalar invariant functions of a set of vectors that there is a function $\tilde{W}$ of symmetric 3×3 tensors and ξ such that

$$W(F^k{}_{\mu}, \xi) = \tilde{W}(C_{\mu\nu}, \xi), \quad C_{\mu\nu} = \sum_{k=1}^{3} F^k{}_{\mu} F^k{}_{\nu}. \tag{1.2.96}$$

This argument applies only to algebraic functions W_o (cf. Spencer, 1971). For an arbitrary function W_o we may apply the polar decomposition: $\mathbf{F} = \mathbf{R}\mathbf{V}$, $\mathbf{R} \in SO(3)$, $\mathbf{V}$ is symmetric positive definite, since $\det\mathbf{F} > 0$. From ($1.2.95_2$) it follows that $W_o(\mathbf{F}, \xi) = W_o(\mathbf{V}, \xi)$. Equation (1.2.96) follows from the fact that $\mathbf{V} = \mathbf{C}^{1/2}$.

The symmetric positive definite tensor $\mathbf{C}$ is known as the **right Cauchy–Green tensor.** $\mathbf{V}$ is often called the **right extension tensor.** $\mathbf{C}$ has six independent components. Hence we may view $\tilde{W}$ as a function of $C_{\mu\nu}$, $\mu \leqslant \nu$, or as a function of nine independent arguments $C_{\mu\nu}$, $\mu, \nu = 1, 2, 3$, satisfying the identity $\tilde{W}(C_{\mu\nu}, \xi) = \tilde{W}(C_{\nu\mu}, \xi)$. In the latter case

$$S_k^{\sigma} = \varrho_o \frac{\partial \tilde{W}}{\partial C_{\mu\nu}} [\delta^{\sigma}{}_{\mu} F^k{}_{\nu} + \delta^{\sigma}{}_{\nu} F^k{}_{\mu}] = 2\varrho_o \frac{\partial \tilde{W}}{\partial C_{\sigma\mu}} F^k{}_{\mu}. \tag{1.2.97}$$

It is a remarkable fact that (1.2.97) entails the identity (1.2.86) related to the conservation of moment of momentum.

Whenever the constitutive relations involve a constraint

$$f(\mathbf{F}, \mathbf{x}, \xi) = 0 \tag{1.2.98}$$

we shall require that the constraint equation is invariant under Galilean transformations, viz. $f(\mathbf{F}, \mathbf{x}, \xi) = 0$ entails $f(\mathbf{AF}, \mathbf{Ax}+\mathbf{b}, \xi) = 0 \ \ \forall\, \mathbf{A} \in SO(3)$, $\mathbf{b} \in \boldsymbol{R}^3$. For $\mathbf{A} = \mathbf{E}$ this implies that $\mathbf{x} \mapsto f(\mathbf{F}, \mathbf{x}, \xi)$ is identically zero if it vanishes for some $\mathbf{x}_1$ and the given $(\mathbf{F}, \xi)$. Since we are interested only in those values of $(\mathbf{F}, \xi)$ for which $\exists\, \mathbf{x}_1$ such that $f(\mathbf{F}, \mathbf{x}_1, \xi) = 0$, we can safely assume that f is a function of $\mathbf{F}$, ξ only.

Suppose now that $f(\mathbf{F}_1, \xi) = 0$ for some $\mathbf{F}_1 \in \mathscr{L}_+$, $\xi \in \boldsymbol{\varkappa}(\mathscr{B})$, and let $\mathbf{A}(t)$ be a $\mathscr{C}^1$ curve in SO(3), $\mathbf{W} = \dot{\mathbf{A}}(0)\mathbf{A}(0)^{-1} = -{}^t\mathbf{W}$. Differentiating the identity $f(\mathbf{A}(t)\mathbf{F}_1, \xi) \equiv 0$ we conclude that

$$\frac{\partial f}{\partial F^{[k}{}_{\mu}} (\mathbf{F}_1, \xi) F_1{}^{l]}{}_{\mu} = 0, \tag{1.2.99}$$

since $\mathbf{W}$ is an arbitrary skew-symmetric matrix. We have thus obtained a necessary condition of invariance: $f(\mathbf{F}_1, \xi) = 0$ implies (1.2.99).

For a sufficient condition we shall note that $f(\mathbf{F}_1, \xi) = 0$ iff $f(\mathbf{V}_1, \xi) = 0$, $\mathbf{F}_1 = \mathbf{RV}_1$, $\mathbf{R} \in SO(3)$, $\mathbf{V}_1 = {}^t\mathbf{V}_1$. Hence the zeros of $f(\cdot\,, \xi)$ coincide with those of a function $\mathbf{F} \mapsto \tilde{f}(C_{\mu\nu}, \xi)$. Hence we can replace (1.2.98) by

$$\tilde{f}(C_{\mu\nu}, \xi) = 0. \tag{1.2.100}$$

Incompressibility can be expressed in the form $\det \mathbf{C} = a(\xi)^2$.

We now proceed to examine the effects of varying the reference configuration in the momentum conservation law, the VWP and in the constitutive relations. The integrals appearing in these equations involve the volume measure $L^3(\boldsymbol{\varkappa}(\mathscr{P}))$ and the area measure $H^2(\boldsymbol{\varkappa}(\mathscr{S}))$ on $\mathscr{B}$ induced by the embedding $\boldsymbol{\varkappa}\colon \mathscr{B} \to \boldsymbol{R}^3$ of $\mathscr{B}$ in the Euclidean space $\boldsymbol{R}^3$ (with the metric $\delta_{\mu\nu}$).

Let $(\xi^\mu) = \boldsymbol{\varkappa}(P)$, $(\zeta^\varrho) = \boldsymbol{\varkappa}'(P)$, $\mathbf{g} = \boldsymbol{\varkappa} \circ \overset{-1}{\boldsymbol{\varkappa}'}\colon \boldsymbol{\varkappa}'(\mathscr{B}) \to \boldsymbol{\varkappa}(\mathscr{B})$. We shall denote the measures induced by $\boldsymbol{\varkappa}$, $\boldsymbol{\varkappa}'$ by the subscripts ξ, ζ, resp. On account of (1.2.17)

$$\varrho_{\varkappa} dL^3_{(\xi)} = \varrho_{\varkappa'} dL^3{}_{(\zeta)}. \tag{1.2.101}$$

Let $\mathscr{S}$ be a piece of smooth (or Lipschitz continuous) surface given in the parametric form $\xi^\mu(u, v)$ or $\zeta^\varrho(u, v)$.

$$\nu^{(\xi)}_{\mu} dH^2_{(\xi)} = \epsilon_{\mu\varkappa\lambda} \frac{\partial \xi^\varkappa}{\partial u} \frac{\partial \xi^\lambda}{\partial v} du\, dv = \epsilon_{\mu\varkappa\lambda} \frac{\partial \xi^\varkappa}{\partial \zeta^\varrho} \frac{\partial \xi^\lambda}{\partial \zeta^\sigma} \frac{\partial \zeta^\varrho}{\partial u} \frac{\partial \zeta^\sigma}{\partial v} du\, dv$$

$$= \epsilon_{\tau\varrho\sigma} J(\mathbf{g}) \frac{\partial \zeta^\tau}{\partial \xi^\mu} \frac{\partial \zeta^\varrho}{\partial u} \frac{\partial \zeta^\sigma}{\partial v} du\, dv = \nu^{(\zeta)}_{\tau} dH^2_{(\zeta)}\, J(\mathbf{g}) \frac{\partial \zeta^\tau}{\partial \xi^\mu}. \tag{1.2.101'}$$

Hence $dH^2 \nu_\mu S_k{}^\mu$ is invariant with respect to changes of reference configuration provided

$$S_k^{(\zeta)\varrho} = S_k^{(\xi)\mu} \frac{\partial \zeta^\varrho}{\partial \xi^\mu} J(\mathbf{g}). \tag{1.2.102}$$

Clearly

$$F^k_{(\zeta)\varrho} = F^k_{(\xi)\mu} \frac{\partial \xi^\mu}{\partial \zeta^\varrho}. \tag{1.2.103}$$

The transformation laws (1.2.102) and (1.2.103) are in accordance with the invariance of the strain energy W_o and equation (1.2.87),

$$W_o^{(\xi)}(F^k_{(\xi)\mu}, \xi) \equiv W_o^{(\xi)}\left(F^k_{(\zeta)\varrho} \frac{\partial \zeta^\varrho}{\partial \xi^\mu} \circ \mathbf{g}(\zeta),\quad \mathbf{g}(\zeta)\right) \equiv W_o^{(\zeta)}(F^k_{(\zeta)\varrho}, \zeta). \tag{1.2.104}$$

We shall also assume that the body loads b_k are invariant under transformations of the reference configuration. It is easy to see that the momentum and energy balance is invariant under the transformations (1.2.101)–(1.2.104). For the VWP it is enough to note that

$$dL^3 S_k^\mu \frac{\partial}{\partial \xi^\mu}, \quad dH^2 t_k$$

are invariant under changes of reference configuration. The test functions u^k are scalars with respect to these transformations.

The argument of Section 1.2.9 proves the VWP for test functions $\mathbf{u}$ with $\operatorname{supp} \mathbf{u} \subset \boldsymbol{\varkappa}_\iota(\mathscr{U}_\iota)$. In view of the boundary conditions it is necessary to consider $\overline{\mathscr{B}}$ as a manifold with boundary: some maps $\boldsymbol{\varkappa}_\iota$ map $\mathscr{U}_\iota \cap \overline{\mathscr{B}}$ onto sets $\{\xi \in \boldsymbol{R}^3 | \, |\xi^\alpha| < a^\alpha,\ \alpha = 1, 2;\ 0 \leqslant \xi^3 < a^3\}$. Let us now consider a test function $\mathbf{u}\colon \overline{\mathscr{B}} \to \boldsymbol{R}^3$ vanishing on Σ_1 and let $\mathbf{u}_\iota := \mathbf{u} \circ \overset{-1}{\boldsymbol{\varkappa}_\iota}\colon \boldsymbol{\varkappa}_\iota(\mathscr{U}_\iota) \to \boldsymbol{R}^3$. Let $\{q_\iota\}$ be a partition of unity subordinate to the cover $\{\mathscr{U}_\iota\}$ of $\overline{\mathscr{B}}$, $\tilde{q} := q_\iota \circ \overset{-1}{\boldsymbol{\varkappa}_\iota}$. We can define the invariant expression

$$\langle \mathbf{S}, \nabla \mathbf{u} \rangle := \sum_\iota \int_{\mathscr{B}} dL^3_{(\iota)}\, S_k^{(\iota)\mu} \frac{\partial}{\partial \xi^\mu_{(\iota)}} (\tilde{q}_\iota u_\iota^k), \quad \xi_{(\iota)} = \boldsymbol{\varkappa}_\iota(P), \tag{1.2.105}$$

where $L^3_{(\iota)}(\mathscr{P}) := L^3(\boldsymbol{\varkappa}_\iota(\mathscr{P}))$, $S_k^{(\iota)\mu}$ is the Piola–Kirchhoff stress tensor for the reference $\boldsymbol{\varkappa}_\iota$. For $\operatorname{supp} \mathbf{u} \subset \mathscr{U}_\iota$ equation (1.2.105) reduces to

$$\int_{\mathscr{U}_\iota} dL^3_{(\iota)}\, S_k^{(\iota)\mu} \frac{\partial}{\partial \xi^\mu_{(\iota)}} u_\iota^k$$

since $\sum_\iota \tilde{q}_\iota = 1$ and $dL^3_{(\iota)} S_k^{(\iota)\mu} \frac{\partial}{\partial \xi^\mu_{(\iota)}}$ is invariant on overlaps $\mathscr{U}_\iota \cap \mathscr{U}_{\iota'}$.

In terms of local reference configurations the VWP assumes the following form:

$$\int_0^T dt \int_{\mathscr{B}} dM\, [\boldsymbol{\chi}_{,t} \cdot \mathbf{u}_{,t} + \mathbf{b} \cdot \mathbf{u}]$$

$$= \int_0^T dt\, \langle \mathbf{S}, \nabla \mathbf{u} \rangle - \int_0^T dt \int_{\Sigma_2} dH^2\, t_k u^k - \int_{\mathscr{B}} dM\, \boldsymbol{\chi}_{,t} \cdot \mathbf{u}|_{t=0}, \tag{1.2.106}$$

where the second integral on the right-hand side should be interpreted as $\sum_\iota \int_0^T dt \int_{\Sigma_2} dH^2_{(\iota)} \tilde{q}_\iota t_k^{(\iota)} u_\iota^k$.

We close this section with a remark on a possible geometrical approach to the VWP and momentum balance.* The role of the stress tensor consists essentially in assigning tractions to each surface $\mathscr{S} \subset \mathscr{B}$. The tractions are defined by an integral over the oriented surface $\mathscr{S}$. Since there is no natural area measure in $\mathscr{B}$ we have had to exploit the Hausdorff measure on some $\varkappa(\mathscr{B}) \subset \boldsymbol{R}^3$. Such a procedure is artificial and cumbersome from a geometrical point of view. It would be more satisfactory to exploit the invariant concept of a vector-valued two-form on $\mathscr{B}$.

Indeed, let us write

$$S_k^{\ \mu} \nu_\mu dH^2 = S_k^{\ \mu} \in_{\mu\varrho\sigma} \frac{\partial \xi^\varrho}{\partial u} \frac{\partial \xi^\sigma}{\partial v} du dv$$

$$= S_{k\varrho\sigma} \frac{\partial \xi^\varrho}{\partial u} \frac{\partial \xi^\sigma}{\partial v} du dv.$$

The first equation makes sense provided the coordinates (ξ^ϱ) on $\varkappa(\mathscr{B}) \subset \boldsymbol{R}^3$ are Cartesian. The tractions on $\mathscr{S}$ are now given by

$$\int_{\mathscr{S}} dH^2\, S_k^{\ \mu} \nu_\mu = \int_{\mathscr{S}} S_k, \quad S_k := S_{k\varrho\sigma} d\xi^\varrho \wedge d\xi^\sigma \tag{1.2.107}$$

and S_k is an invariant geometrical object independent of the choice of Lagrangian coordinates. There is no need to interprete (ξ^ϱ) as Cartesian coordinates on $\varkappa(\mathscr{B}) \subset \boldsymbol{R}^3$, since Hausdorff measure does not appear any more.

We also note that

$$d(S_{k\varrho\sigma} d\xi^\varrho \wedge d\xi^\sigma) = S_{k\varrho\sigma,\xi^\mu} d\xi^\mu \wedge d\xi^\varrho \wedge d\xi^\sigma = \in^{\mu\varrho\sigma} S_{k\varrho\sigma,\xi^\mu} dL^3_{(\xi)}$$

$$= 2 S_k^{\ \mu}{}_{,\xi^\mu} dL^3_{(\xi)},$$

* For a detailed analysis of the geometric meaning of the VWP and of the constitutive equations see Hanyga and Seredyńska (1984).

$$du^k \wedge S_k = \frac{\partial u^k}{\partial \xi^\mu} S_{k\varrho\sigma} d\xi^\mu \wedge d\xi^\varrho \wedge d\xi^\sigma = 2S_k{}^\mu \frac{\partial u^k}{\partial \xi^\mu} dL^3_{(\xi)}.$$

For the invariance of the VWP and of the momentum balance it is necessary to replace $dH^2 t_k$ by an invariant expression. Hence we shall define a vector measure τ_k on Borel $\mathscr{S} \subset \Sigma_2$. The VWP can be rewritten in the invariant form

$$\int_0^T dt \int_{\mathscr{B}} dM\, [\boldsymbol{\chi}_{,t} \cdot \mathbf{u}_{,t} + \mathbf{b} \cdot \mathbf{u}] = \int_0^T dt \int_{\mathscr{B}} du^k \wedge S_k - \\ - \int_0^T dt \int_{\Sigma_2} d\tau_k\, u^k - \int_{\mathscr{B}} dM\, \boldsymbol{\chi}_{,t} \cdot \mathbf{u}|_{t=0} \tag{1.2.108}$$

for $\mathbf{u}$ vanishing at $t = T$ and on Σ_1 for $0 \leqslant t \leqslant T$. The momentum conservation law can also be rewritten in terms of S_k.

Chapter 2

Elastostatics

2.1 FUNDAMENTAL BOUNDARY-VALUE PROBLEMS

2.1.1 Introduction

Elastostatics is concerned with elastic bodies in equilibrium.

Definition 2.1
An elastic body $\mathscr{B}$ is said to be **in equilibrium** if it rests in a Galilean reference frame.

In some cases it is preferable to define equilibrium in terms of a larger class of reference frames. For example one is tempted to speak about the equilibrium of a rotating body. In this case it is essential that the body *does not deform* in time, i.e. it performs a rigid motion. Hence it rests with respect to an Euclidean frame. Euclidean frames are obtained by applying coordinate transformations $'\mathbf{x} = \mathbf{A}(t)\mathbf{x}+\mathbf{b}(t)$, $\mathbf{A}(t) \in \mathrm{SO}(3)\ \forall t$, to Galilean coordinate systems and identifying space points with trajectories $'\mathbf{x} = \text{const}$ in space-time. Since constitutive equations of elasticity happen to be invariant with respect to Euclidean transformations

$$\boldsymbol{\chi}(\xi, t) \mapsto \mathbf{A}(t)\boldsymbol{\chi}(\xi, t)+\mathbf{b}(t), \quad \mathbf{A}(t) \in \mathrm{SO}(3),\ \mathbf{b}(t) \in \boldsymbol{R}^3,$$

the problem of equilibrium of a rotating body reduces to the problem of equilibrium under body loads equal to the inertial forces. For non-elastic bodies the possibility of such a reduction hinges on the validity of the principle of material frame indifference (cf. Truesdell and Noll, 1965), which need not be universally true (this point is discussed in some detail by Hanyga (1981)).

For simplicity we shall stick to Definition 2.1.

Let $\boldsymbol{\chi}(\xi)$ be the motion of $\mathscr{B}$ expressed in terms of Lagrangian coordinates ξ ranging over $\Omega = \boldsymbol{\varkappa}(\mathscr{B}) \subset \boldsymbol{R}^3$ and referred to a Galilean coordinate system φ such that $\mathscr{B}$ rests with respect to the reference frame $[\varphi]_\sim$. Since $\boldsymbol{\chi}_{,t} \equiv \mathbf{0}$, the virtual work principle assumes the following form

$$\int_{\Omega} dL^3\, S_k^{\alpha} w^k{}_{,\xi\alpha} = \int_{\Omega} dL^3\, \varrho_0 b_k w^k + \int_{\partial\Omega} dH^2\, S_k^{\alpha}\nu_{\alpha} w^k. \tag{2.1.1}$$

If $\boldsymbol{\chi}(\xi)$ is prescribed on $\Gamma_1 \subset \partial\Omega$, $\partial\Omega = \overline{\Gamma}_1 \cup \overline{\Gamma}_2$, $\Gamma_1 \cap \Gamma_2 = \emptyset$, then we set $w^k = 0$ on Γ_1 (virtual displacements $\mathbf{w}$ must vanish on Γ_1) and (2.1.1) reduces to a special case of (1.2.76), viz.

$$\int_{\Omega} dL^3\, S_k^{\alpha} w^k{}_{,\xi\alpha} = \int_{\Omega} dL^3\, \varrho_0 b_k w^k + \int_{\Gamma_2} dH^2\, s_k w^k \quad \forall \mathbf{w}: \gamma_{\Gamma_1}\mathbf{w} = \mathbf{0}. \tag{2.1.2}$$

The problem (2.1.2), (1.2.85) can be expected to be well-posed provided the tractions s_k per unit area of Γ_2 (measured in the reference configuration) are prescribed on Γ_2. This observation suggests the most representative **boundary-value problem (BVP)** of elastostatics:

$$\boldsymbol{\chi}(\xi) = \boldsymbol{\chi}_1(\xi) \quad \text{for } \xi \in \Gamma_1, \tag{2.1.3}$$

$$S_k^{\alpha}\nu_{\alpha} = s_k \quad \text{for } \xi \in \Gamma_2 \tag{2.1.4}$$

$S_k{}^{\alpha}{}_{,\xi\alpha} + \varrho_0 b_k = 0$, $k = 1, 2, 3$ in the weak sense (set $\mathbf{w} \in \mathscr{C}_0^{\infty}$ in (2.1.2), (1.2.85)). Equation (2.1.4) is implicit in (2.1.2). In fact equation (2.1.4) is meaningless for Sobolev configurations $\boldsymbol{\chi} \in W^{1,p}(\Omega)$ and equation (2.1.2) provides one of its acceptable formulations.

Definition 2.1.1
The problem defined by (2.1.2), (1.2.85), (2.1.3) is called a **mixed BVP.**

The surface tractions s_k on Γ_2 can be prescribed as functions of $(\boldsymbol{\chi}(\xi), \xi)$, $\xi \in \Gamma_2$ (**configuration-dependent loads**) or of $(\boldsymbol{\chi}_{,t}(\xi, t), \xi)$, $\xi \in \Gamma_2$ (**dissipative surface loads**). One can conjecture that the boundary condition (BC) (2.1.3) is equivalent to a configuration-dependent load which vanishes when $\boldsymbol{\chi} \equiv \boldsymbol{\chi}_1$ on Γ_1 and for each deviation $\boldsymbol{\chi} \not\equiv \boldsymbol{\chi}_1$ from (2.1.3) provides exactly the traction needed to prevent the deviation.

Conceptually unilateral constraints fall into the category of configuration-dependent (and possibly dissipative) surface loads. A representative example is the case of a rigid support on the part Γ_1 of the boundary. In loose terms, let us imagine a surface $\mathscr{R}$ rigidly fixed in space. For simplicity, assume that $\mathscr{R}$ is smooth. The BC can be formulated as follows: (1) $\boldsymbol{\chi}(\Gamma_1)$ cannot move beyond the support $\mathscr{R}$; (2) if there is contact between a point $\xi_0 \in \Gamma_1$ and $\mathscr{R}$, i.e. $\boldsymbol{\chi}(\xi_0) \in \mathscr{R}$, then the support exerts a surface force directed along the normal to $\mathscr{R}$ into $\boldsymbol{\chi}(\Omega)$; (3) if $\boldsymbol{\chi}(\xi_0) \bar{\in} \mathscr{R}$, $\xi_0 \in \Gamma_1$, then $s_k = 0$ at ξ_0.

In the case of small displacements or geometrically simple shapes of $\mathscr{R}$, Ω it is easy to express these BC in terms of some inequalities. Notwithstanding their conceptual similarity different kinds of BC are tractable by different methods. One may feel confident that any reasonable BVP of elastostatics is solvable if mixed configuration-dependent BVP have this property. Since we are interested in the foundations of elastostatics more than in specific problems we shall restrict our attention to mixed configuration-dependent BVPs.

In Chapter 2 we shall be concerned with existence and uniqueness theorems for mixed BVPs of elastostatics. Validity of such theorems hinges on certain assumptions concerning the functionals (2.1.2) on the set of functions $\boldsymbol{\chi}$ satisfying (2.1.3) and det $\nabla\boldsymbol{\chi} > 0$ a.e. in Ω. These hypotheses can often be replaced by somewhat stronger hypotheses which bear on the constitutive functions $\hat{S}_k^{\alpha}$, W and do not involve the specific BC. The latter hypotheses will be referred to as **constitutive assumptions** It is our aim to prove existence and related theorems on the basis of constitutive assumptions alone. We shall begin with considerations which shed some light on the non-uniqueness and related problems.

2.1.2 Non-Uniqueness and Necessary Conditions for Existence of Solutions to Mixed BVPs

It follows from the Galilean invariance of the equations of elastostatics that solutions to certain mixed BVPs cannot be unique. If $\boldsymbol{\chi}$ is a solution to (2.1.2) and (2.1.3) with given $\mathbf{b}$, $\mathbf{s}$, $\boldsymbol{\chi}_1$, then $\boldsymbol{\chi}' = \mathbf{A}\boldsymbol{\chi}+\mathbf{a}$, $\mathbf{A} \in \mathrm{SO}(3)$, $\mathbf{a} \in \boldsymbol{R}^3$, is a solution to (2.1.2) and (2.1.3) with $\mathbf{Ab}$, $\mathbf{As}$ and $\mathbf{A}\boldsymbol{\chi}_1+\mathbf{a}$ replacing $\mathbf{b}$, $\mathbf{s}$, $\boldsymbol{\chi}_1$. In order to see this it is enough to substitute $\boldsymbol{\chi}'$ and $\mathbf{Aw}$ in place of $\boldsymbol{\chi}$ and $\mathbf{w}$ in (2.1.2) and (2.1.3) and make use of (1.2.95).

Let us consider the set $\mathscr{R}$ of rigid displacements of $\mathscr{B}$ consistent with the BC (2.1.3):

$$\mathscr{R} = \{(\mathbf{A}, \mathbf{a}) \in \mathrm{SO}(3)\times \boldsymbol{R}^3 |\ \mathbf{A}\chi_1(\xi)+\mathbf{a} = \boldsymbol{\chi}_1(\xi)\forall\boldsymbol{\xi} \in \Gamma_1\}. \tag{2.1.4}$$

If $\Gamma_1 = \emptyset$ then $\mathscr{R} = \mathrm{SO}(3)\times \boldsymbol{R}^3$. Suppose now that $\Gamma_1 \neq \emptyset$. For every $(\mathbf{A}, \mathbf{a}) \in \mathscr{R}$, $\xi \in \Gamma_1$

$$(\mathbf{A}-\mathbf{E})\boldsymbol{\chi}_1(\xi)+\mathbf{a} = \mathbf{0}. \tag{2.1.5}$$

If $\mathbf{A} = \mathbf{E}$, $\mathbf{a} \neq \mathbf{0}$ belongs to $\mathscr{R}$, then $\Gamma_1 = \emptyset$, in contradiction with our hypothesis. If $\mathbf{A} \neq \mathbf{E}$ then dim ker $(\mathbf{A}-\mathbf{E}) = 1$. Indeed, suppose that there are two linearly independent vectors $\mathbf{p}$, $\mathbf{n} \in \boldsymbol{R}^3$ such that $\mathbf{Ap} = \mathbf{p}$, $\mathbf{An} = \mathbf{n}$, and let $\mathbf{m} := \mathbf{p}\times\mathbf{n}$. Clearly, $\mathbf{m} \neq \mathbf{0}$ and $\mathbf{Am} = \mathbf{m}$. Hence $\mathbf{A} = \mathbf{E}$, which contradicts our hypothesis. We conclude that dim ker$(\mathbf{A}-\mathbf{E}) \leqslant 1$. Every $\mathbf{A} \in \mathrm{SO}(3)$ can be written in the form $\mathbf{A} = \exp(\mathbf{W})$, ${}^t\mathbf{W} = -\mathbf{W}$. Let the vector $\mathbf{p}$ be defined by $p^i = \dfrac{1}{2}\epsilon^{ijk}W_{jk}$ so that $W_{ij} = \epsilon_{ijk}p^k$ and $\mathbf{Wp} = 0$. Hence $\mathbf{Ap} = \mathbf{p}$ and we conclude that dim ker$(\mathbf{A}-\mathbf{E}) = 1$.

Suppose that $(\mathbf{A},\mathbf{a})\in\mathscr{R}$, $\mathbf{y}=\boldsymbol{\chi}_1(\xi)$, $\xi\in\Gamma_1$, $\mathbf{Ap}=\mathbf{p}$, $\mathbf{Am}_i\neq\mathbf{m}_i$ for $i=1,2$, and $\mathbf{p}$, $\mathbf{m}_1$, $\mathbf{m}_2$ are linearly independent. Let $\mathbf{y}=\sum_{i=1}^{2}\mu^i\mathbf{m}_i+\lambda\mathbf{p}$, for some $\mu^1,\mu^2,\lambda\in\boldsymbol{R}$. It follows from (2.1.5) that

$$\mathbf{a}=-(\mathbf{A}-\mathbf{E})\sum_{i=1}^{2}\mu^i\mathbf{m}_i,\quad \mathbf{y}=-(\mathbf{A}-\mathbf{E})^{-1}\mathbf{a}+\lambda\mathbf{p}$$

($(\mathbf{A}-\mathbf{E})^{-1}$ denotes the inverse of $(\mathbf{A}-\mathbf{E})$ restricted to $\mathbf{p}^{\perp}$). Hence if $(\mathbf{A},\mathbf{a})\in\mathscr{R}$ and $\mathbf{A}\neq\mathbf{E}$, then $\boldsymbol{\chi}_1(\Gamma_1)$ is a subset of a straight line parallel to $\mathbf{p}$. Also, if $(\mathbf{A}_1,\mathbf{a}_1)$, $(\mathbf{A}_2,\mathbf{a}_2)\in\mathscr{R}$, $\mathbf{A}_1,\mathbf{A}_2\neq\mathbf{E}$, and $\ker(\mathbf{A}_1-\mathbf{E})\neq\ker(\mathbf{A}_2-\mathbf{E})$, then $\boldsymbol{\chi}_1(\Gamma_1)$ lies on the intersection of two non-parallel straight lines and is either empty or a singleton.

On the other hand, if $\boldsymbol{\chi}_1(\Gamma_1)$ contains two different points $\mathbf{y}_1$, $\mathbf{y}_2$ and $(\mathbf{A},\mathbf{a})\in\mathscr{R}$, $\mathbf{A}\neq\mathbf{E}$, then

$$\mathbf{A}(\mathbf{y}_1-\mathbf{y}_2)=\mathbf{y}_1-\mathbf{y}_2. \tag{2.1.6}$$

If $\mathbf{A}=\exp(\mathbf{W})$, ${}^t\mathbf{W}=-\mathbf{W}$, then $\mathbf{W}(\mathbf{y}_1-\mathbf{y}_2)=0$ and the property (2.1.6) is shared by all the rotations $\exp(s\mathbf{W})$, $s\in\boldsymbol{R}$, around the axis through $\mathbf{y}_1,\mathbf{y}_2$.

If $\boldsymbol{\chi}_1(\Gamma_1)$ contains three non-collinear points $\mathbf{y}_1,\mathbf{y}_2,\mathbf{y}_3$, then clearly $\mathscr{R}=\{(\mathbf{E},\mathbf{0})\}$. Hence, from a group-theoretical point of view the mixed BCs fall into four classes:

(1) $\boldsymbol{\chi}_1(\Gamma_1)=\emptyset$, $\mathscr{R}=\mathrm{SO}(3)\times\boldsymbol{R}^3$;

(2) $\boldsymbol{\chi}_1(\Gamma_1)=\{\mathbf{y}_0\}$, $\mathscr{R}=\mathrm{SO}(3)\times\{\mathbf{0}\}$ (all the rotations leaving $\mathbf{y}_0$ fixed);

(3) $\boldsymbol{\chi}_1(\Gamma_1)$ contains two different points and is contained in a straight line Σ, $\mathscr{R}=\{\exp(s\mathbf{W})|\ s\in\boldsymbol{R}\}\times\{\mathbf{0}\}$ (we have chosen the origin $(0,0,0)$ to lie on Σ), for some fixed $\mathbf{W}$;

(4) $\boldsymbol{\chi}_1(\Gamma_1)$ contains three non-collinear points, $\mathscr{R}=\{(\mathbf{E},\mathbf{0})\}$.

In the case (1) we shall substitute $\mathbf{w}=\exp(s\mathbf{W})\boldsymbol{\chi}+t\mathbf{a}$ in (2.1.2) and differentiate the resulting identity with respect to s, t. There result the following *necessary conditions for the existence of solutions*:

$$\int_{\Omega}dL^3\,\varrho_0\mathbf{b}+\int_{\Gamma_2}dH^2\,\mathbf{s}=\mathbf{0}, \tag{2.1.7}$$

$$\int_{\Omega}dL^3\,S_k^{\ \alpha}\,W^k{}_l\,\chi^l{}_{,\xi\alpha}=\int_{\Omega}dL^3\,\varrho_0 b_k\,W^k{}_l\,\chi^l+\int_{\Gamma_2}dH^2\,s_k\,W^k{}_l\,\chi^l. \tag{2.1.8}$$

If we substitute $S_k^{\ \alpha}=\hat{S}_k^{\ \alpha}(\nabla\boldsymbol{\chi},\xi)$, then $S_k^{\ \alpha}\chi^l{}_{,\xi\alpha}$ is symmetric in k, l and equation (2.1.8) reduces to

$$\int_{\Omega}dL^3\,\varrho_0\mathbf{b}\times\boldsymbol{\chi}+\int_{\Gamma_2}dH^2\,\mathbf{s}\times\boldsymbol{\chi}=\mathbf{0}. \tag{2.1.9}$$

Equations (2.1.7) state the condition of equilibrium of the applied loads, while equation (2.1.9) is the condition of equilibrium of their moments. In case (2) we obtain only the condition of equilibrium of moments of loads with respect to the fixed point $\mathbf{y}_0$:

$$\int_{\Omega} dL^3\, \varrho_0 \mathbf{b} \times (\boldsymbol{\chi} - \mathbf{y}_0) + \int_{\Gamma_2} dH^2\, \mathbf{s} \times (\boldsymbol{\chi} - \mathbf{y}_0) = \mathbf{0}. \qquad (2.1.10)$$

In case (3) the necessary condition of existence of solutions reads as follows:

$$\Big\{\int_{\Omega} dL^3\, \varrho_0 b_k (\chi^l - y_0{}^l) + \int_{\Gamma_2} dH^2\, s_k (\chi^l - y_0{}^l)\Big\} W^k{}_l = 0 \qquad (2.1.11)$$

for some arbitrarily chosen $\mathbf{y}_0 \in \boldsymbol{\chi}_1(\Gamma_1)$ and a fixed skew-symmetric $\mathbf{W}$. Equation (2.1.11) expresses the condition of equilibrium of the moment of the applied loads with respect to the axis through $\boldsymbol{\chi}_1(\Gamma_1)$.

All the necessary conditions presented above are invariant with respect to the translations and orthogonal transformations of the coordinate system (x^k). For equation (2.1.9) this follows from equation (2.1.7).

Let $\mathscr{R}_0 := \{(\mathbf{A}, \mathbf{a}) \in \mathscr{R} |\ \mathbf{Ab} = \mathbf{b}$ in Ω, $\mathbf{As} = \mathbf{s}$ on $\Gamma_2\}$. The invariance properties of the BVP clearly imply that $\boldsymbol{\chi}' = \mathbf{A}\boldsymbol{\chi} + \mathbf{a}$ is a solution of this problem if $\boldsymbol{\chi}$ is and $(\mathbf{A}, \mathbf{a}) \in \mathscr{R}_0$. It is desirable to reformulate the problem in such a way as to remove this trivial non-uniqueness, e.g. by considering the solutions modulo $\mathscr{R}_0$.

Let $\mathscr{V}$ be the space of displacements $\mathbf{u}\colon \Omega \to \boldsymbol{R}^3$ of class $\mathscr{C}^2$ satisfying the BC $\mathbf{u}|\Gamma_1 = \mathbf{0}$. For Γ_1 sufficiently regular (e.g. without cusps) and $\boldsymbol{\chi}_1 \in \mathscr{C}^2(\Gamma_1)$ we can find such a configuration $\boldsymbol{\chi}_o \in \mathscr{C}^2\colon \Omega \to \boldsymbol{R}^3$ such that $\boldsymbol{\chi}_o|\Gamma_1 = \boldsymbol{\chi}_1$ (by Whitney's extension theorem, cf. Narasimhan, 1968). The linear manifold $\mathscr{W}$ of configurations $\boldsymbol{\chi} \in \mathscr{C}^2$ satisfying (2.1.3) can be expressed in terms of the subspace $\mathscr{V}$:

$$\mathscr{W} = \boldsymbol{\chi}_o + \mathscr{V} := \{\boldsymbol{\chi}_o + \mathbf{u}|\ \mathbf{u} \in \mathscr{V}\}.$$

We define the following equivalence relation in $\mathscr{W}$: $\boldsymbol{\chi} \sim \boldsymbol{\chi}'$ if $\exists\, (\mathbf{A}, \mathbf{a}) \in \mathscr{R}_0$ such that $\boldsymbol{\chi}'(\xi) = \mathbf{A}\boldsymbol{\chi}(\xi) + \mathbf{a}$ everywhere in Ω. Since $\boldsymbol{\chi}'_o = \mathbf{A}\boldsymbol{\chi}_o + \mathbf{a}$ satisfies (2.1.3) for every $(\mathbf{A}, \mathbf{a}) \in \mathscr{R}_0$, we shall define the corresponding equivalence relation in $\mathscr{V}$: $\mathbf{w}' \sim \mathbf{w}$ in $\mathscr{V}$ if $\mathbf{w}'(\xi) = \mathbf{Aw}(\xi)$ for all $\xi \in \Omega$ and some $(\mathbf{A}, \mathbf{a}) \in \mathscr{R}_0$.

Let $\mathscr{V}_\sim$ denote the set of equivalence classes in $\mathscr{V}$. Equations (2.1.7) and (2.1.9) for BCs of class (1), equation (2.1.10) for BCs of class (2) and equation (2.1.11) for BCs of class (3) imply that the right-hand side of equation (2.1.2) is in fact a function of equivalence classes $[\mathbf{w}]_\sim \in \mathscr{V}_\sim$. The left-hand side of equation (2.1.2) is invariant under simultaneous transformations of $\boldsymbol{\chi} \in \mathscr{W}$ and $\mathbf{w} \in \mathscr{V}$ by $\mathscr{R}_0$. Hence we shall consider the left-hand side of

(2.1.2) as a function of the equivalence classes $[\chi]_\sim \in \mathscr{W}_\sim$ and $[\mathbf{w}]_\sim \in \mathscr{V}_\sim$. By a solution of the BVP we shall mean an equivalence class $[\chi]_\sim \in \mathscr{W}_\sim$ satisfying equation (2.1.2) for all $[\mathbf{w}]_\sim \in \mathscr{V}_\sim$.

For BCs of class (4) $\mathscr{R}_0$ is a semi-direct product $\mathscr{G} \times \boldsymbol{R}^3$. For the remaining classes of BCs $\mathscr{R}_0 = \mathscr{G} \times \{\mathbf{0}\}$. In either case $\mathscr{G}$ is a subgroup of SO(3). There is no natural linear structure on $\mathscr{V}_\sim$ except when $\mathscr{G} = \{\mathbf{E}\}$ and $\mathscr{V} = \mathscr{V}_\sim$.

2.1.3 Classification of the BVPs in the Case of Sobolev Configurations

In practice the solutions to a BVP will be sought in a subset $\mathscr{C}$ of $W^{1,p}(\Omega)$, $p > 1$, defined by the BC (2.1.3) and by the condition $\det\nabla\chi(\xi) > 0$ L^3-a.e. in Ω. In some treatments the last condition will be neglected.

On account of the trace theorems the BVPs can be divided into two classes:

$$(1')\ H^2(\Gamma_1) > 0, \quad \mathscr{R} = \{(\mathbf{E}, 0)\},$$

$$(2')\ H^2(\Gamma_1) = 0, \quad \mathscr{R} = \mathrm{SO}(3) \times \boldsymbol{R}^3.$$

The closure in $W^{1,p}$ of all the smooth functions $\Omega \to \boldsymbol{R}^3$ satisfying a BC of class (2) or (3), as defined in Section 2.1.2, coincides with $W^{1,p}(\Omega)$. Hence the case of a body supported by a sharp pivot or suspended at a single point of its boundary has to be reformulated in such a way that allowance is made for a finite area of the clamped boundary.

Let $\mathscr{W}$ be the subset of $W^{1,p}(\Omega)$ defined by the BC (2.1.3) in the sense of traces and let $\mathscr{V}$ be the subspace of $W^{1,p}(\Omega)$ defined by the BC $\gamma_{\Gamma_1}\mathbf{u} = \mathbf{0}$. Suppose that $\partial\Omega$ has the strong Lipschitz property. For a BC of class (1′) and $\chi_1 \in W^{1/q,p}(\Gamma_1)$, $q^{-1} = 1-p^{-1}$, we can find a configuration $\chi_o \in W^{1,p}(\Omega)$ such that $\gamma_{\Gamma_1}\chi_o = \chi_1$ and $\mathscr{W} = \chi_o + \mathscr{V}$.

For a BC of class (2′) $\mathscr{W} = \mathscr{V} = W^{1,p}(\Omega)$. Let $\mathscr{R}_0 := \{(\mathbf{A}, \mathbf{a}) \in \mathscr{R} |\ \mathbf{Ab} = \mathbf{b} L^3\text{-a.e. in } \Omega,\ \mathbf{As} = \mathbf{s}\ H^2\text{-a.e. on } \partial\Omega\}$. $\mathscr{R}_0$ is a semidirect product $\mathscr{G} \times \boldsymbol{R}^3$, with $\mathscr{G} \subset \mathrm{SO}(3)$. We define the following equivalence relations on $\mathscr{W}$ and $\mathscr{V}$: $\chi \sim \chi'$ if $\chi'(\xi) = \mathbf{A}\chi(\xi) + \mathbf{a}$ L^3-a.e. in Ω for some $(\mathbf{A}, \mathbf{a}) \in \mathscr{R}_0$, $\mathbf{w} \sim \mathbf{w}'$ if $\mathbf{w}'(\xi) = \mathbf{A}\mathbf{w}(\xi)$ L^3-a.e. in Ω for some $(\mathbf{A}, \mathbf{a}) \in \mathscr{R}_0$. We note that the condition $\det\nabla\chi > 0$ a.e. in Ω is preserved by the equivalence relation on $\mathscr{W}$. The set $\mathscr{W}_\sim$ of equivalence classes in $\mathscr{W}$ will serve as the solution space for our BVP.

$\mathscr{W}$ is a Banach space with a norm $\|\cdot\|_{\mathscr{W}}$, e.g. $\|\cdot\|_{\mathscr{W}} = \|\cdot\|_{1,p}$. If $\mathscr{G} = \{\mathbf{E}\}$, then the set $\mathscr{V}_\sim$ of equivalence classes in $\mathscr{V}$ coincides with $\mathscr{V}$ and there is a natural linear structure on $\mathscr{W}_\sim$:

$$\lambda[\chi_1]_\sim + [\chi_2]_\sim := [\lambda\chi_1 + \chi_2]_\sim \quad \forall \lambda \in \boldsymbol{R}, \quad \chi_1, \chi_2 \in \mathscr{W}. \tag{2.1.12}$$

Indeed, $\chi_1 \sim \chi_1'$ and $\chi_2 \sim \chi_2'$ entail that $\lambda\chi_1' + \chi_2' \sim \lambda\chi_1 + \chi_2$. In this case we have $[0]_\sim = [\chi - \chi]_\sim = [\chi]_\sim - [\chi]_\sim = \mathbf{0} \in \mathscr{W}_\sim$. Moreover $\|\tilde{\chi}\|_\sim$

$:= \inf\{\|\chi\|_{\mathscr{W}} | \chi \in \tilde{\chi}\}$ is a norm on $\mathscr{W}$. Indeed, $\|\cdot\|_\sim$ satisfies the triangle inequality and $\|\tilde{\chi}\|_\sim = 0$ implies that $0 \in \chi$ (since $\tilde{\chi}$ is a closed subset of $\mathscr{W}$), and hence $\tilde{\chi} = \mathbf{0} \in \mathscr{W}_\sim$.

Equations (2.1.7) and (2.1.9) ensure that the right-hand side of equation (2.1.2) is a function of $[\mathbf{w}]_\sim \in \mathscr{V}_\sim$ only. The left-hand side of (2.1.2) is invariant with respect to simultaneous transformations of $\chi \in \mathscr{W}$ and $\mathbf{w} \in \mathscr{V}$ by $(\mathbf{A}, \mathbf{a}) \in \mathscr{R}_0$. It follows easily that every $\chi' \sim \chi$ is a solution of (2.1.2) if χ is a solution of this equation.

The right-hand side of (2.1.2) is continuous with respect to the topology defined by $\|\cdot\|_\sim$. Indeed, the right-hand side of (2.1.2) can be estimated by $K\|\mathbf{w}'\|_{\mathscr{V}}$ for arbitrary $\mathbf{w}' \sim \mathbf{w}$ and some $K > 0$, hence by $K\|[\mathbf{w}]_\sim\|_\sim$. For $\mathscr{G} = \{\mathbf{E}\}$ the right-hand side of equation (2.1.2) is a continuous linear functional on $\mathscr{V}_\sim = \mathscr{V}$.

In practice it is often more convenient to reduce the non-uniqueness of the BVP by imposing a subsidiary condition on the solutions which picks out exactly one configuration $\chi \in \mathscr{W}$ from each equivalence class. This additional condition enters the definition of the solution space. In Section 2.2 this condition will be used in connection with the Poincaré lemma.

For BVPs of class (1′) the BC on Γ_1 combined with the Friedrichs lemma plays an analogous role.

2.2 ENERGETIC EXTENSIONS OF MIXED BVPs

2.2.1 Introduction

It is our objective in this section to cast the mixed BVPs in the operator form

$$A\mathbf{u} = \mathbf{f}, \quad \mathbf{u} = \chi - \tilde{\chi}_1 \in \mathscr{V}, \tag{2.2.1}$$

where $\tilde{\chi}_1$ is a fixed configuration and $\mathscr{V}$ is an appropriate Banach space. The condition $\det\nabla\chi > 0$ will be disregarded in this approach to the BVPs.

If the Dirichlet data $\chi = \chi_1$ are prescribed on a part Γ_1 of $\partial\Omega$ such that $H^2(\Gamma_1) > 0$, then $\tilde{\chi}_1 \in W^{1,p}(\Omega)$ denotes an arbitrary extension of χ_1 to Ω. Otherwise $\tilde{\chi}_1 \in W^{1,p}(\Omega)$ can be arbitrary. The existence of such extensions was discussed in the preceding section.

We shall consider somewhat more general BCs on $\Gamma_2 = \Gamma_2' \cup \Gamma_3$ than hitherto discussed:

$$S_k{}^\alpha \nu_\alpha = s_k(\xi) \quad \text{on } \Gamma_2', \tag{2.2.2}$$

$$S_k{}^\alpha \nu_\alpha + \lambda(\xi)|\mathbf{u}|^{p-2}u^k = s_k(\xi) \quad \text{on } \Gamma_3, \tag{2.2.3}$$

with

$$\bar{\Gamma}_1 \cup \bar{\Gamma}_2 = \partial\Omega, \quad H^2(\Gamma_1 \cap \Gamma_2') = H^2(\Gamma_2' \cap \Gamma_3) = H^2(\Gamma_1 \cap \Gamma_3) = 0.$$

The sets $\Gamma_1, \Gamma_2', \Gamma_3$ are assumed to be H^2-measurable.

We consider two classes of BVPs separately:

(1) $H^2(\Gamma_1 \cup \Gamma_3) > 0$;

(2) $H^2(\Gamma_1 \cup \Gamma_3) = 0$.

For the BVPs we construct the solution space $\mathscr{V}^\dagger$ by imposing an extra constraint rather than by dividing $\mathscr{V}$ by an equivalence. For simplicity of notation we shall use the symbol $\mathscr{V}$ to denote $\mathscr{V}^\dagger$ in this case, or, in other words, the constraint will enter the definition of $\mathscr{V}$.

2.2.2 Construction of the Operator A: A Simple BVP of Class (1)

We consider an operator of the form

$$A = L^\dagger \circ S_0 \circ L, \tag{2.2.4}$$

where $L: \mathscr{V} \to \mathscr{Y}$ is a linear isometry, $L^\dagger : \mathscr{Y}^* \to \mathscr{V}^*$ is the adjoint isometry, $\mathscr{V}, \mathscr{Y}, \mathscr{V}^*, \mathscr{Y}^*$ are Banach spaces and their conjugates with strong topology, S_0 is an operator $\mathscr{Y} \to \mathscr{Y}^*$, to be specified below. S_0 involves the non-linear Nemytskii operator $\tilde{S}: \mathscr{L}^p \to \mathscr{L}^q$,

$$\tilde{S}(\mathbf{F})(\xi) \equiv [\hat{S}_k^{\alpha}(\mathbf{F}(\xi), \xi)], \tag{2.2.5}$$

$1/p + 1/q = 1$, $p > 1$. The Carathéodory function $\hat{S}_k^{\alpha}$ satisfies the estimate

$$|\hat{S}_k^{\alpha}(\mathbf{F}, \xi)| \leqslant a(\xi) + b|\mathbf{F}|^{p-1}, \tag{2.2.6}$$

$a \in \mathscr{L}^q(\Omega)$. This estimate is unrealistic for $\mathbf{F} \to \mathbf{0}$, but we shall ignore this difficulty in Sections 2.2 and 2.3. $\mathscr{V}$ is a reflexive Banach space, typically a subspace of $W^{1,p}(\Omega)$, $p > 1$ specified by the boundary conditions on Γ_1 or by the extra constraint in the case of a BVP of class (2).

For a comparison with the classical differential operator

$$(E\mathbf{u})_k = -\frac{\partial}{\partial \xi^{\alpha}} \hat{S}_k^{\alpha}(\nabla \mathbf{u}(\xi), \xi) \tag{2.2.7}$$

we shall assume additionally that $\hat{S}_k^{\alpha}(\cdot\,, \cdot) \in \mathscr{C}^1$ and construct a Hilbert space $\mathscr{H}$,

$$\mathscr{V} \subset \mathscr{H} \subset \mathscr{V}^* \tag{2.2.8}$$

with a dense and continuous embedding $\mathscr{V} \to \mathscr{H}$. Typically $\mathscr{H}$ is (or involves) the space $\mathscr{L}^2(\Omega)$. Since we identify $\mathscr{H}$ with $\mathscr{H}^*$ in (2.2.8), the duality form

$\langle \cdot | \cdot \rangle$ for $\mathscr{V}$, $\mathscr{V}^*$ is the integral $\int_\Omega dL^3 \mathbf{u} \cdot \mathbf{v}$, cf. Section 1.1. For a bounded Ω the first embedding in (2.2.8) implies that $p \geqslant 2$.

For a BVP of class (1) with $\Gamma_3 = \emptyset$, $\Gamma_2' = \Gamma_2$ we define $\mathscr{V} := \{\mathbf{u} \in W^{1,p}(\Omega) |\ \gamma_{\Gamma_1}\mathbf{u} = \mathbf{0}\}$, with $p > 1$ fixed by (2.2.6). The space $\mathscr{V}$ will be endowed with the norm $||\mathbf{u}|| := ||\nabla \mathbf{u}||_p$, where $\nabla = \left[\frac{\partial}{\partial \xi^\alpha}\right]$ denotes the distributional gradient with respect to ξ^α. On account of the Friedrichs lemma the norm $||\cdot||$ is equivalent to the $W^{1,p}$ norm $||\cdot||_{1,p}$ on $\mathscr{V}$. Let $L\colon \mathscr{V} \to \mathscr{L}^p(\Omega) = \mathscr{Y}$ be the extension by continuity of the gradient operator $\nabla\colon \mathscr{C}^1(\Omega) \to \mathscr{L}^p(\Omega)$ in the norm $||\cdot||$. L is a linear isometry, i.e. $||L\mathbf{u}||_p = ||\mathbf{u}||$.

Since L is continuous and $\mathscr{D}(L) = \mathscr{V}$, we can consider the adjoint $L^\dagger\colon \mathscr{L}^q(\Omega) \to \mathscr{V}$, $q^{-1} = 1 - p^{-1}$,

$$\langle L\mathbf{S}^\dagger | \mathbf{v} \rangle = \langle \mathbf{S} | L\mathbf{v} \rangle_0,$$

where $\langle S | \mathbf{F} \rangle_0 = \int_\Omega dL^3 S_k{}^\alpha F^k{}_\alpha$, $\mathbf{S} \in \mathscr{L}^q(\Omega)$, $\mathbf{F} \in \mathscr{L}^p(\Omega)$.

We shall note that $L^\dagger$ is surjective since $\langle \mathbf{S} | L\mathbf{v} \rangle_0 = \langle L^\dagger \mathbf{S} | \mathbf{v} \rangle = 0$ for all $\mathbf{S} \in L^q(\Omega)$ implies $\mathscr{L}\mathbf{v} = 0$ in $L^p(\Omega)$ and hence $\mathbf{v} = \mathbf{0}$ in $\mathscr{V}$. For $\mathbf{S} \in \mathscr{Y}^* \cap \mathscr{C}^1(\bar{\Omega})$, $\mathbf{u} \in \mathscr{V} \cap \mathscr{C}^1(\bar{\Omega})$:

$$\langle L^\dagger \mathbf{S} | \mathbf{u} \rangle = \langle \mathbf{S} | L\mathbf{u} \rangle_0 = \int_\Omega dL^3\, S_k{}^\alpha \frac{\partial u^k}{\partial \xi^\alpha}$$

$$= \int_{\Gamma_2} dH^2\, S_k{}^\alpha u^k \nu_\alpha - \int_\Omega dL^3\, u^k \frac{\partial S_k{}^\alpha}{\partial \xi^\alpha}.$$

For $\mathbf{S} \in Y^* \cap \mathscr{C}_0^1(\Omega)$ this yields the identity

$$L^\dagger \mathbf{S} = -\left[\frac{\partial S_k{}^\alpha}{\partial \xi^\alpha}\right]$$

since $\mathscr{C}^1(\bar{\Omega}) \cap \mathscr{V}$ is dense in $\mathscr{V}$.

Let

$$S_0(\mathbf{F}) = \tilde{S}(\mathbf{F} + L\tilde{\chi}). \tag{2.2.9}$$

For $p \geqslant 2$ we have $\mathscr{V} \subset W^{1,p}(\Omega) \subset \mathscr{L}^2(\Omega) = \mathscr{H}$. Since $\mathscr{C}_0^2(\Omega) \subset \mathscr{V}$ and $\mathscr{C}_0^2(\Omega)$ is dense in $\mathscr{H}$, the embedding $\mathscr{V} \subset \mathscr{H}$ is dense. By the Friedrichs lemma (Theorem 1.1.49) the embedding $\mathscr{V} \subset \mathscr{H}$ is also continuous. By Theorem 1.1.31 the embeddings (2.2.8) are continuous and dense.

Let us consider the problem

$$A\mathbf{u} = \mathbf{f} \tag{2.2.10}$$

with $\mathbf{f} \in \mathscr{H} \subset \mathscr{V}^*$. For $\mathbf{v} \in \mathscr{V}$ we have

$$\langle A\mathbf{u}|\mathbf{v}\rangle = \langle L^\dagger \circ S_0 \circ L\mathbf{u}|\mathbf{v}\rangle = \langle S_0 \circ L\mathbf{u}|L\mathbf{v}\rangle_0$$
$$= \int_\Omega dL^3\, \hat{S}_k^{\ \alpha}(\nabla\tilde{\boldsymbol{\chi}}(\xi)+\nabla\mathbf{u}(\xi), \xi)v^k_{\ ,\xi\alpha}.$$

Hence (2.2.10) is equivalent to the VWP

$$\int_\Omega dL^3\, \hat{S}_k^{\ \alpha}(\nabla\boldsymbol{\chi}(\xi), \xi)v^k_{\ ,\xi\alpha} = \int_\Omega dL^3\, f_k v^k \ \forall \mathbf{v} \in \mathscr{V} \tag{2.2.11}$$

for the BVP under consideration provided $f_k(\xi) = \varrho_0(\xi)b_k(\xi)$. It is possible to replace $\mathbf{v} \in \mathscr{V}$ by $\mathbf{v}$ ranging over the dense subset

$$\mathscr{V}\cap\mathscr{C}^1(\bar{\Omega}) = \{\mathbf{v} \in \mathscr{C}^1(\bar{\Omega})|\ \mathbf{v}|\Gamma_1 = \mathbf{0}\}.$$

Let us now compare A with the classical operator (2.2.7) defined on

$$\mathscr{D}(E) = \{\mathbf{u} \in \mathscr{C}^2(\bar{\Omega})|\ \mathbf{u}|\Gamma_1 = 0,$$
$$\hat{S}_k^{\ \alpha}(\nabla\boldsymbol{\chi}(\xi)+\nabla\mathbf{u}(\xi), \xi)\nu_\alpha = 0 \text{ on } \Gamma_2\}.$$

Clearly $\mathscr{D}(E) \subset \mathscr{V}$. Since $\hat{S}_k^{\ \alpha} \in \mathscr{C}^1$ by assumption and Ω is bounded, $\mathscr{R}(E) \subset \mathscr{H}$. For $\mathbf{u} \in \mathscr{D}(E)$, $\mathbf{v} \in \mathscr{V}\cap\mathscr{C}^1(\bar{\Omega})$ we have

$$\langle A\mathbf{u}|\mathbf{v}\rangle = \langle S_0 \circ L\mathbf{u}|L\mathbf{v}\rangle_0 = \int_\Omega dL^3\, \hat{S}_k^{\ \alpha}(\nabla\boldsymbol{\chi}(\xi)+\nabla\mathbf{u}(\xi), \xi)v^k_{\ ,\xi\alpha}$$
$$= \int_\Omega dL^3\, E\mathbf{u}\cdot\mathbf{v}.$$

Since the set $\mathscr{V}\cap\mathscr{C}^1(\bar{\Omega})$ is dense in $\mathscr{V}$,

$$A\mathbf{u} = E\mathbf{u}\ \forall \mathbf{u} \in \mathscr{D}(E). \tag{2.2.12}$$

In Section 2.2.5 we shall also prove that

$$\mathscr{D}(E) = \{\mathbf{u} \in \mathscr{V}\cap\mathscr{C}^2(\bar{\Omega})|\ A\mathbf{u} \in \mathscr{H}\}. \tag{2.2.13}$$

In view of the properties (2.2.8), (2.2.12) and (2.2.13) the operator A is called the **energetic extension** of E (cf. Gajewski *et al.*, 1974). We shall note that the relation between the operators A and E is a secondary matter, since the problem (2.2.10) is directly related to the VWP and the operator E may not exist in general (e.g. if $\hat{S}_k^{\ \alpha}$ is not differentiable). From the classical formulation of the BVP one may read off the BCs more easily.

2.2.3 Construction of the Operator A for a BVP of Class (1), Continued

We assume that Ω has strong Lipschitz property, Γ_1, Γ_2' and Γ_3 are disjoint H^2-measurable subsets of $\partial\Omega$, $H^2(\Gamma_1\cup\Gamma_3) > 0$, $p > 1$ and $\tilde{S}$ is an arbitrary Nemytskii operator (2.2.5) satisfying (2.2.6). We shall set $q = p/(p-1)$.

In order to deal with the BC (2.2.2) and (2.2.3) we define a configuration $\tilde{\boldsymbol{\chi}} \in W^{1,p}$ in such a way that $\gamma_{\Gamma_1}\tilde{\boldsymbol{\chi}} = \boldsymbol{\chi}_1$ H^2-a.e. on Γ_1, $\gamma_{\Gamma_3}\tilde{\boldsymbol{\chi}} = 0$ H^2-a.e. on Γ_3. The extension exists if $\boldsymbol{\chi}_1 \in W^{1/q,p}(\Gamma_1)$, i.e. for $\boldsymbol{\chi}_1$ ranging over a dense subset of $\mathscr{L}^p(\Gamma_1)$. Let $\mathbf{u} := \boldsymbol{\chi}-\tilde{\boldsymbol{\chi}}$.

The space $\mathscr{V} := \{\mathbf{u} \in W^{1,p}(\Omega) |\ \gamma_{\Gamma_1}\mathbf{u} = \mathbf{0}\}$ will be endowed with the norm

$$||\mathbf{u}|| := \left[||\nabla\mathbf{u}||_p{}^p + \int_{\Gamma_3} dH^2 |\gamma_{\Gamma_3}\mathbf{u}|^p\right]^{1/p}. \tag{2.2.14}$$

Since $H^2(\Gamma_1 \cup \Gamma_3) > 0$, the norm $||\cdot||$ is equivalent to the norm $||\cdot||_{1,p}$.

Let $\mathscr{Y} := \mathscr{L}^p(\Omega)\times\mathscr{L}^p(\Gamma_3)$, $H := \mathscr{L}^2(\Omega)$, and define the operator $L: \mathscr{V} \to \mathscr{Y}$; $\mathbf{u} \mapsto (\nabla\mathbf{u}, \gamma_{\Gamma_3}\mathbf{u})$.

Then L is a linear isometry since

$$||L\mathbf{u}||_{\mathscr{Y}}{}^p = ||\nabla\mathbf{u}||_p{}^p + \int_{\Gamma_3} dH^2\, |\gamma_{\Gamma_3}\mathbf{u}|^p = ||\mathbf{u}||^p \ \forall \mathbf{u} \in \mathscr{V}.$$

Let $s_k \in \mathscr{L}^q(\Gamma_2)$. Let $g \in \mathscr{Y}^*$ be a continuous extension of the functional $g_0: \mathscr{R}(L) \to \mathbf{R}$,*

$$g_0: L\mathbf{v} \mapsto \int_{\Gamma_2} dH^2\, s_k \gamma_{\Gamma_2} v^k \ \ \forall \mathbf{v} \in \mathscr{V}.$$

The functional g_0 is well-defined and continuous since the trace γ_{Γ_2}: $\mathscr{V} \to \mathscr{L}^p(\mathscr{L}_2)$ is continuous and L is an isometry. We shall define S_0 by the formula

$$S_0(\mathbf{F}, \mathbf{w}) \equiv (\tilde{S}(\mathbf{F}), \nu|\mathbf{w}|^{p-2}w^k) - g, \tag{2.2.15}$$

$S_0: \mathscr{Y} \to \mathscr{Y}^*$. Note that

$$\int_{\Gamma_3} dH^2\, |\mathbf{w}|^{(p-1)q} = \int_{\Gamma_3} dH^2\, |\mathbf{w}|^p.$$

We calculate

$$\begin{aligned}\langle A\mathbf{u}|\mathbf{v}\rangle - \langle \mathbf{f}|\mathbf{v}\rangle &\equiv \langle S_0 \circ L\mathbf{u}|L\mathbf{v}\rangle - \langle \mathbf{f}|\mathbf{v}\rangle \\ &\equiv \int_{\Omega} dL^3\, \hat{S}_k^{\alpha}(\nabla\tilde{\boldsymbol{\chi}}+\nabla\mathbf{u}, \xi)v^{k}{}_{,\xi\alpha} - \\ &- \int_{\Gamma_3} dH^2 [s_k v^k - \nu|\mathbf{u}|^{p-1}\mathbf{u}\cdot\mathbf{v}] - \int_{\Gamma_2'} dH^2\, s_k v^k - \int_{\Omega} dL^3\, f_k v^k.\end{aligned} \tag{2.2.16}$$

Hence $\langle A\mathbf{u}-\mathbf{f}|\mathbf{v}\rangle = 0$ is equivalent to the VWP.

* It exists by the Hahn–Banach theorem.

2.2.4 Construction of A for Mixed BVPs of Class (2)

Let us now consider a general mixed BVP of class (2), with $\tilde{\boldsymbol{\chi}} = \mathbf{0}$. We shall make the usual assumptions about $\partial\Omega$ granting the validity of the Poincaré lemma. Let $\hat{S}_k^\alpha$ be a Carathéodory function satisfying (2.2.6) with $p > 1$.

In order to remove the arbitrary rigid translations admitted by the BC we incorporate an extra constraint in the definition of $\mathscr{V}$:

$$\mathscr{V} = \left\{\mathbf{u} \in W^{1,p}(\Omega) \middle| \int_\Omega dL^3\mathbf{u} = \mathbf{0}\right\}.$$

The norm $\|\cdot\|$, $\|\mathbf{u}\| := \|\nabla\mathbf{u}\|_p$, on $\mathscr{V}$ is equivalent to the $W^{1,p}$ norm in view of the Poincaré lemma, Section 1.1.

We define $L\colon \mathscr{V} \to \mathscr{Y}$, $\mathscr{Y} := \mathscr{L}^p(\Omega)$, as the continuous extension of ∇. L is a linear isometry. Let

$$g_0(L\mathbf{v}) :\equiv \int_{\partial\Omega} dH^2\, s_k\gamma v^k + \int_\Omega dL^3\, f_k v^k. \tag{2.2.17}$$

The right-hand side of (2.2.17) is a continuous linear functional of $\mathbf{v} \in \mathscr{V}$. We extend g_0 to a continuous functional g on $\mathscr{Y}$, so that $g \in \mathscr{Y}^*$, $\mathscr{Y}^* = \mathscr{L}^q(\Omega)$, $q^{-1} = 1-p^{-1}$.

Let $S_0\colon \mathscr{Y} \to \mathscr{Y}^*$ be defined by

$$S_0(\mathbf{F}) = \tilde{S}(\mathbf{F})-g. \tag{2.2.18}$$

We introduce the space

$$\mathscr{H} = \left\{\mathbf{u} \in \mathscr{L}^2(\Omega) \middle| \int_\Omega dL^3\,\mathbf{u} = 0\right\}.$$

It is a closed subspace of $\mathscr{L}^2(\Omega)$, corresponding to the projection operator

$$P\colon \mathbf{u} \mapsto \mathbf{u}-L^3(\Omega)^{-1}\int_\Omega dL^3\mathbf{u}$$

hence a Hilbert space in its own right. Its orthogonal complement is the space of all rigid translations.

Clearly $\mathscr{V} \subset \mathscr{H}$ provided $p \geqslant 2$ and Ω is bounded. The embedding is continuous on account of the Poincaré lemma and a remark on p. 46. We show that $\mathscr{V}\cap\mathscr{C}_0^2(\Omega)$ (and hence $\mathscr{V}$) is dense in $\mathscr{H}$. Let $\mathbf{h} \in \mathscr{H}$, $\mathbf{v}_n \to \mathbf{h}$ in $\mathscr{L}^2(\Omega)$, $\mathbf{v}_n \in \mathscr{C}_0^2(\Omega)$. Let $\varphi \in \mathscr{C}_0^2(\Omega)$ be such that $\int_\Omega dL^3\varphi = 1$. We set $\mathbf{h}_n := \mathbf{v}_n - \varphi\int_\Omega dL^3\mathbf{v}_n$. Clearly

$$\mathbf{h}_n \in V\cap\mathscr{C}_0^2(\Omega), \quad \mathbf{h}_n \in \mathscr{H},$$

$$\|\mathbf{h}_n-\mathbf{h}\|_2 \leqslant \|\mathbf{v}_n-\mathbf{h}\|_2+\|\varphi\|_2\left|\int_\Omega dL^3(\mathbf{v}_n-\mathbf{h})\right|,$$

since $\int_\Omega dL^3\mathbf{h} = \mathbf{0}$.

By the Schwartz inequality

$$\|\mathbf{h}_n-\mathbf{h}\|_2 \leqslant [1+\|\varphi\|_2 L^3(\Omega)]^{1/2}\|\mathbf{v}_n-\mathbf{h}\|_2 \to 0,$$

hence $\mathscr{V}\cap\mathscr{C}_0^2(\Omega)$ is dense in $\mathscr{H}$. From Theorem 1.1.31 it follows that the embeddings (2.2.8) are all continuous and dense.

Let $\mathbf{u} = \boldsymbol{\chi}$ be a solution of $A\mathbf{u} = \mathbf{0}$, and let $\mathbf{v}\in\mathscr{V}$. We have

$$\langle A\mathbf{u}|\mathbf{v}\rangle = \int_\Omega dL^3\, \hat{S}_k^\alpha(\nabla\boldsymbol{\chi}(\xi),\xi)v^k{}_{,\xi^\alpha} - \int_\Omega dL^3\, f_k v^k - \int_{\partial\Omega} dH^2\, s_k v^k, \tag{2.2.19}$$

in accordance with the VWP.

2.2.5 Comparison with Classical Differential Operators

In this subsection we shall assume that $p \geqslant 2$, Ω is bounded and $\hat{S}_k^\alpha\in\mathscr{C}^1$ with respect to both arguments. The spaces $\mathscr{H}$ have been defined in Sections 2.2.3 and 2.2.4 for either class of BC.

In the first place we shall compare the operator A defined in Section 2.2.3 with the operator (2.2.7) restricted to the domain

$$\begin{aligned}\mathscr{D}(E) := \{\mathbf{u}\in\mathscr{C}^2(\bar{\Omega})|\mathbf{u}|\Gamma_1 = 0,\\ \hat{S}_k^\alpha(\nabla\tilde{\boldsymbol{\chi}}(\xi)+\nabla\mathbf{u}(\xi),\xi)\nu_\alpha = s_k \text{ on } \Gamma_2',\\ \nu|\mathbf{u}|^{p-2}u^k+\hat{S}_k^\alpha(\nabla\tilde{\boldsymbol{\chi}}(\xi)+\nabla\mathbf{u}(\xi),\xi)\nu_\alpha = s_k \text{ on } \Gamma_3\}.\end{aligned}$$

Clearly $\mathscr{D}(E)\subset\mathscr{V}$ and $\mathscr{R}(E)\subset\mathscr{H}$ (i.e. $E\mathbf{u}\in\mathscr{H}$ for $\mathbf{u}\in\mathscr{D}(E)$). For $\mathbf{u}\in\mathscr{D}(E)$, $\mathbf{v}\in\mathscr{V}\cap\mathscr{C}^1(\bar{\Omega})$ we have

$$\begin{aligned}\langle A\mathbf{u}|\mathbf{v}\rangle &= \int_\Omega dL^3\, \hat{S}_k^\alpha(\nabla\tilde{\boldsymbol{\chi}}(\xi)+\nabla\mathbf{u}(\xi),\xi)v^k{}_{,\xi^\alpha} + \int_{\Gamma_3} dH^2\,\nu|\mathbf{u}|^{p-2}\mathbf{u}\cdot\mathbf{v}\\ &\quad - \int_{\Gamma_2} dH^2\, s_k v^k = \int dL^3\, E\mathbf{u}\cdot\mathbf{v}.\end{aligned}$$

Since $\mathscr{V}\cap\mathscr{C}^1(\bar{\Omega})$ is dense in $\mathscr{V}$,

$$A\mathbf{u} = E\mathbf{u}\ \ \forall\mathbf{u}\in\mathscr{D}(E).$$

Hence on $\mathscr{D}(E)$ the equation $A\mathbf{u} = \mathbf{f}$, $\mathbf{f}\in H$, is equivalent to the classical differential equation $E\mathbf{u} = \mathbf{f}$.

Let $\mathbf{u}\in\mathscr{V}\cap\mathscr{C}^2(\bar{\Omega})$, $A\mathbf{u}\in\mathscr{H}$, $\mathbf{v}\in\mathscr{V}\cap\mathscr{C}^1(\bar{\Omega})$. From equations (2.2.16) and (2.2.7) we have

$$\begin{aligned}\int_\Omega dL^3\,(A\mathbf{u}-E\mathbf{u})\cdot\mathbf{v} &= \int_{\Gamma_3} dH^2\,[\nu|\mathbf{u}|^{p-2}u^k+\hat{S}_k^\alpha(\nabla\tilde{\boldsymbol{\chi}}+\nabla\mathbf{u},\xi)\,\nu_\alpha - s_k]\,v^k +\\ &\quad + \int_{\Gamma_2'} dH^2\,[\hat{S}_k^\alpha(\nabla\tilde{\boldsymbol{\chi}}(\xi)+\nabla\mathbf{u}(\xi),\xi)\nu_\alpha - s_k]\gamma v^k.\end{aligned} \tag{2.2.20}$$

For $\mathbf{v} \in \mathscr{V} \cap \mathscr{C}_0^1(\Omega)$ this implies

$$\int_{\Omega} dL^3 \,(A\mathbf{u} - E\mathbf{u}) \cdot \mathbf{v} = 0. \tag{2.2.21}$$

By continuity this equation extends to all $\mathbf{v} \in \mathscr{H}$ (note that $A\mathbf{u} \in \mathscr{H}$ by assumption). But $\mathscr{V} \subset \mathscr{H}$, hence (2.2.20) implies that

$$\int_{\Gamma_3} dH^2 \,[\nu|\mathbf{u}|^{p-2}u^k + \hat{S}_k^{\alpha}(\nabla\tilde{\boldsymbol{\chi}} + \nabla\mathbf{u}, \xi) - s_k]\gamma_{\Gamma_3} v^k + \\ + \int_{\Gamma_2'} dH^2 \,[\hat{S}_k^{\alpha}(\nabla\tilde{\boldsymbol{\chi}} + \nabla\mathbf{u}, \xi)\nu_\alpha - s_k]\gamma_{\Gamma_2'} v^k = 0 \tag{2.2.22}$$

for all $\mathbf{v} \in \mathscr{V} \cap \mathscr{C}^1(\bar{\Omega})$ and hence for all $\mathbf{v} \in \mathscr{V}$.

The trace operators map $\mathscr{V}$ onto $W^{1-1/p,p}(\Gamma_2')$, $W^{1-1/p,p}(\Gamma_3)$ and the latter spaces are dense subsets of $\mathscr{L}^p(\Gamma_2')$, $\mathscr{L}^p(\Gamma_3)$, resp. Hence equation (2.2.22) remains valid if we substitute arbitrary $w^k \in \mathscr{L}^p(\Gamma_2')$, $w^k \in \mathscr{L}^p(\Gamma_3)$ for $\gamma_{\Gamma_2'} v^k$, $\gamma_{\Gamma_3} v^k$. Equation (2.2.22) implies then that the expressions in the brackets in the integrands vanish H^2-a.e. Hence $\mathbf{u} \in \mathscr{D}(E)$. We have thus proved (2.2.13) for the operator A constructed in Section 2.2.3.

The operator A constructed in Section 2.2.4 is now compared with the classical differential operator E_f: $E_f\mathbf{u} \equiv E\mathbf{u} - \mathbf{f}$, $\mathbf{f} \in \mathscr{H} \cap \mathscr{C}^0(\bar{\Omega})$. Let

$$\mathscr{D}(E_f) := \{\mathbf{u} \in \mathscr{C}^2(\bar{\Omega}) | \int_{\Omega} dL^3 \,\mathbf{u} = 0,\ \mathscr{S}_k^{\alpha}(\nabla\mathbf{u}, \xi) = s_k \text{ on } \partial\Omega\}.$$

Clearly $\mathscr{D}(E_f) \subset \mathscr{V}$. Moreover for $\mathbf{u} \in \mathscr{D}(E_f)$

$$\int_{\Omega} dL^3 \,(E_f\mathbf{u})_k = \int_{\Omega} dL^3 \,[(E\mathbf{u})_k - f_k] \\ = -\int_{\partial\Omega} dH^2 \,\hat{S}_k^{\alpha}(\nabla\mathbf{u}, \xi)\nu_\alpha - \int_{\Omega} dL^3 f_k = -\int_{\Omega} dL^3 f_k - \int_{\partial\Omega} dH^2 s_k.$$

Hence $\mathscr{R}(E_f) \subset \mathscr{H}$ provided the necessary condition of existence of solutions

$$\int_{\Omega} dL^3 \mathbf{f} + \int_{\partial\Omega} dH^2 \mathbf{s} = \mathbf{0}$$

is satisfied.

For $\mathbf{u} \in \mathscr{D}(E_f)$, $\mathbf{v} \in \mathscr{V} \cap \mathscr{C}^1(\bar{\Omega})$

$$\langle A\mathbf{u}|\mathbf{v}\rangle = \int_{\Omega} dL^3 \,[\hat{S}_k^{\alpha}(\nabla\mathbf{u}(\xi), \xi)v^k{}_{,\xi\alpha} - f_k v^k] - \int_{\partial\Omega} dH^2 \,s_k v^k \\ = \int_{\Omega} dL^3 \,(E_f\mathbf{u})_k v^k.$$

Hence

$$A\mathbf{u} = E_f\mathbf{u} \quad \text{for all } \mathbf{u} \in \mathscr{D}(E_f)$$

and $A\mathbf{u} = 0$ is equivalent to $E\mathbf{u} = \mathbf{f}$ for every $\mathbf{u} \in \mathscr{D}(E_f)$.

For $\mathbf{u} \in \mathscr{C}^2(\bar{\Omega}) \cap \mathscr{V}$, $A\mathbf{u} \in \mathscr{H}$, $\mathbf{v} \in \mathscr{V} \cap \mathscr{C}^1(\bar{\Omega})$

$$\langle (A - E_f)\mathbf{u} | \mathbf{v} \rangle = \int_{\partial\Omega} dH^2 \, [\hat{S}_k^{\alpha}(\nabla \mathbf{u}(\xi), \xi) \nu_\alpha - s_k] v^k. \tag{2.2.23}$$

Since $\mathscr{V} \cap \mathscr{C}_0^2 \subset \mathscr{V} \cap \mathscr{C}^1 \subset \mathscr{H}$ densely, we conclude that

$$\langle (A - E_f)\mathbf{u} | \mathbf{v} \rangle = 0 \quad \text{for } \mathbf{v} \in \mathscr{V} \cap \mathscr{C}_0^2(\Omega)$$

and therefore

$$\langle (A - E_f)\mathbf{u} | \mathbf{v} \rangle = 0 \quad \text{for all } \mathbf{v} \in \mathscr{H}. \tag{2.2.24}$$

Since $\mathscr{V} \subset \mathscr{H}$, equations (2.2.23) and (2.2.24) imply that

$$\int_{\partial\Omega} dH^2 \, [\hat{S}_k^{\alpha}(\nabla \mathbf{u}, \xi) \nu_\alpha - s_k] \gamma v^k = 0 \tag{2.2.25}$$

for every $\mathbf{v} \in \mathscr{V} \cap \mathscr{C}^1(\bar{\Omega})$ and hence for every $\mathbf{v} \in \mathscr{V}$. Like before, we conclude that the expression in the brackets vanishes, $\mathbf{u} \in \mathscr{D}(E_f)$, and hence

$$\mathscr{D}(E_f) = \{\mathbf{u} \in \mathscr{C}^2(\bar{\Omega}) \cap \mathscr{V} \,|\, A\mathbf{u} \in \mathscr{H}\}. \tag{2.2.13$'$}$$

Hence A is an energetic extension of E_f.

2.2.6 Linearized BVPs of Elastostatics

Let $x^p = \chi_0{}^p(\xi^\alpha)$ be a virtual equilibrium configuration of the body corresponding to the Dirichlet data on $\Gamma_1 \subset \partial\Omega$ and the loads $b_k{}^{(0)}$ in Ω, $s_k{}^{(0)}$ on Γ_2. Assuming that $\boldsymbol{\chi}_0$ is a bi-Lipschitz mapping we can use $\boldsymbol{\chi}_0 \circ \boldsymbol{\varkappa}$ as a reference configuration for nearby Sobolev configurations $y^k = \chi^k(\xi^\alpha, t) \equiv \delta^k{}_p x^p + u^k(\mathbf{x}, t)$, possibly time-dependent and satisfying the Dirichlet BC

$$\chi^k(\xi^\alpha, t) = \chi_0{}^k(\xi^\alpha) \;\; \forall \xi \in \Gamma_1, \tag{2.2.26}$$

i.e.

$$\mathbf{u}(\mathbf{x}, t) = 0 \;\; \forall \mathbf{x} \in \Sigma_1 := \boldsymbol{\chi}_0(\Gamma_1). \tag{2.2.27}$$

Let $W(\mathbf{F}, \cdot)$ be a $\mathscr{C}^2$ function of $\mathbf{F}$ for almost every $\xi \in \Omega$ and $\varrho_1 := \varrho_0(\det \mathbf{F}_0)^{-1}$, $\mathbf{F}_0 = \nabla \boldsymbol{\chi}_0$. Then the Piola–Kirchhoff stress tensor for the configuration $\boldsymbol{\chi} \circ \boldsymbol{\varkappa}$ and the reference configuration $\boldsymbol{\chi}_0 \circ \boldsymbol{\varkappa}$ is

$$\begin{aligned} S_k{}^p = \varrho_1 \frac{\partial W(F^k{}_p F_0{}^p{}_\alpha, \xi)}{\partial F^k{}_p} &= \varrho_1 \frac{\partial W}{\partial F^k{}_\alpha}(\mathbf{F}, \xi) F_0{}^p{}_\alpha \\ &= \varrho_1 \frac{\partial W}{\partial F^k{}_\alpha}(\mathbf{F}_0, \xi) F_0{}^p{}_\alpha \\ &\quad + \varrho_1 \frac{\partial^2 W}{\partial F^k{}_\alpha \partial F^l{}_\beta}(\mathbf{F}_0, \xi) \mathbf{F}_0{}^p{}_\alpha F_0{}^q{}_\beta H^l{}_q \\ &\quad + O(|\mathbf{H}|^2) = T_0{}^{kp} + B_k{}^p{}_l{}^q(\mathbf{x}) H^l{}_q + \mathbf{0}(|\mathbf{H}|^2), \end{aligned}$$

with $F^k{}_p = \dfrac{\partial y^k}{\partial x^p} = \delta^k{}_p + u^k{}_{,x^p}$, $H^k{}_p = u^k{}_{,x^p}$, $T_0{}^{kp}$ = the Cauchy stress tensor in the configuration $\chi_0 \circ \varkappa$,

$$B_k{}^p{}_l{}^q(\mathbf{x}) :\equiv \varrho_1(\mathbf{x}) A_k{}^\alpha{}_l{}^\beta(\mathbf{F}_0(\xi), \xi)) F_0{}^p{}_\alpha(\xi) F_0{}^q{}_\beta(\xi) \equiv B_l{}^q{}_k{}^p(\mathbf{x}). \quad (2.2.28)$$

The VWP for the configurations χ, χ_0 in terms of the reference configuration χ_0 yields the equations

$$\int_{\mathscr{B}_0} dL^3(\mathbf{x}) S_k{}^p v^k{}_{,x^p} = \int_{\mathscr{B}_0} dL^3(\mathbf{x}) \varrho_1 b_k v^k + \int_{\Sigma_2} dH^2\, s_k \gamma v^k, \qquad (2.2.29)$$

$$\int_{\mathscr{B}_0} dL^3(\mathbf{x}) T_0{}^{kp} v^k{}_{,x^p} = \int_{\mathscr{B}_0} dL^3(\mathbf{x}) \varrho_1 b_k{}^{(0)} v^k + \int_{\Sigma_2} dH^2\, s_k{}^{(0)} \gamma v^k \qquad (2.2.30)$$

$\forall \mathbf{v}$: $\gamma_{\Gamma_1}\mathbf{v} = \mathbf{0}$, $\mathscr{B}_0 := \chi_0(\varkappa(\mathscr{B}))$. Hence, setting $b_k{}^{(1)} := b_k - b_k{}^{(0)}$, $s_k{}^{(1)} := s_k - s_k{}^{(0)}$, we get the equation

$$\int_{\mathscr{B}_0} dL^3(\mathbf{x}) B_k{}^p{}_l{}^q u^l{}_{,x^q} v^k{}_{,x^p} = \int_{\mathscr{B}_0} dL^3(\mathbf{x}) \varrho_1 b_k{}^{(1)} v^k + \int_{\Sigma_2} dH^2\, s_k{}^{(1)} \gamma v^k, \qquad (2.2.31)$$

if the terms of order $\mathbf{0}(|\mathbf{H}|^3)$ can be neglected on the left-hand side. We shall not justify this assumption here.

If $B_k{}^p{}_l{}^q \in \mathscr{L}^\infty(\mathscr{B}_0)$, then the left-hand side of (2.2.31) is a symmetric bilinear form $B(\mathbf{u}, \mathbf{v})$ on $H^1 \times H^1$,* continuous with respect to either argument. The right-hand side is a continuous linear functional $\mathbf{v} \mapsto f(\mathbf{v})$ on $H^1(\mathscr{B}_0)$ provided $\varrho_1 \in \mathscr{L}^\infty(\mathscr{B}_0)$, $b_k{}^{(1)} \in \mathscr{L}^2(\mathscr{B}_0)$, $s_k{}^{(1)} \in \mathscr{L}^2(\Gamma_2)$. Equation (2.2.31) assumes the form

$$B(\mathbf{u}, \mathbf{v}) = f(\mathbf{v}) \;\; \forall \mathbf{v} \in \mathscr{V}, \qquad (2.2.32)$$

$\mathscr{V} := \{\mathbf{v} \in H^1(\mathscr{B}_0) | \gamma_{\Sigma_1}\mathbf{v} = \mathbf{0}\} = \overline{\{\mathbf{v} \in \mathscr{C}^1(\overline{\mathscr{B}}_0) | \mathbf{v}|\Sigma_1 = \mathbf{0}\}}$ (closure in the H^1-topology; for the last identity see Adams (1975), Nečas (1968), Fichera (1972)).

In the absence of **prestress** $\mathbf{T}_0$ (or in the presence of a hydrostatic prestress $\mathbf{T}_0 = -p\mathbf{E}$, p: $\mathscr{B}_0 \to \boldsymbol{R}$) the tensor $B_k{}^p{}_l{}^q$ has an additional symmetry

$$B_k{}^p{}_l{}^q = B_k{}^p{}_q{}^l \quad (= B_p{}^k{}_l{}^q). \qquad (2.2.33)$$

Indeed, $S_k{}^p(\delta^l{}_p + H^l{}_p)$ is symmetric on account of the law of conservation of moment of momentum. Hence

$$T_0{}^{kl} + T_0{}^{kp} H^l{}_p + B_k{}^l{}_m{}^q H^m{}_q + B_k{}^p{}_m{}^q H^m{}_q H^l{}_p + \ldots$$

* $\mathscr{H}^m(\Omega) := W^{m,2}(\Omega)$.

is symmetric in k, l. The first term is symmetric too. Replace $\mathbf{H}$ by $\varepsilon\mathbf{H}$, $\varepsilon \neq 0$, and divide the sum of the remaining terms by ε. Letting $\varepsilon \to 0$ we conclude that

$$T_0{}^{kp}H^l{}_p + B_k{}^l{}_m{}^q H^m{}_q \tag{2.2.34}$$

is symmetric in k, l. Hence (2.2.33) follows when $\mathbf{T}_0 = \mathbf{0}$.

Equation (2.2.33) can also be derived from the Galilean invariance of W: $W(\mathbf{AF}, \xi) \equiv W(\mathbf{F}, \xi)\ \forall \mathbf{A} \in SO(3)$. Hence $\dfrac{\partial W}{\partial F^k{}_\alpha}(\mathbf{AF}, \xi) A^k{}_p \equiv \dfrac{\partial W}{\partial F^p{}_\alpha}(\mathbf{F}, \xi)$. Let $\mathbf{A}(t)$ be a curve in SO(3), $\mathbf{A}(0) = \mathbf{E}$, $\dot{\mathbf{A}}(0) = \mathbf{W} = -{}^t\mathbf{W}$. We conclude that

$$\frac{\partial^2 W}{\partial F^k{}_\alpha \partial F^l{}_\beta}(\mathbf{F}, \xi)\, W^l{}_q F^q{}_\beta + \frac{\partial W}{\partial F^l{}_\alpha}(\mathbf{F}, \xi)\, W^l{}_p = 0,$$

and, setting $\mathbf{F} = \mathbf{F}_0$,

$$B_k{}^p{}_l{}^q W^l{}_q + T_0{}^{lp} W^l{}_k = 0 \tag{2.2.35}$$

for arbitrary skew-symmetric $\mathbf{W}$. In particular

$$B_k{}^p{}_l{}^q W^k{}_p W^l{}_q = -T_0{}^{lp} W^k{}_p W^k{}_p \tag{2.2.36}$$

vanishes if $\mathbf{T}_0 = \mathbf{0}$.

Equation (2.2.36) shows that the bilinear form $B_k{}^p{}_l{}^q H^k{}_p H^l{}_q$ is not in general positive definite. We shall however assume the (uniform) **strong ellipticity (SE)**:

$$\operatorname{Re}(B_k{}^p{}_l{}^q \bar{a}^k a^l k_p k_q) \geqslant K|\mathbf{a}|^2\mathbf{k}^2 \quad \forall\, \mathbf{a} \in C^3, \quad \mathbf{k} \in R^3 \tag{2.2.37}$$

for some fixed positive constant K. In view of the fact that $B_k{}^p{}_l{}^q$ are real it is possible to consider only real $\mathbf{a} \in R^3$ and drop "Re".

There is a standard method of dealing with the elliptic problems in variational form (2.2.32) of arbitrary order, cf. Lions and Magenes (1968), Nečas (1968), Fichera (1972). The first step consists in proving that the form B is $\mathscr{V}$-coercive.

In what follows we shall consider complex-valued solutions. Let $\mathscr{V}$ be a complex Hilbert subspace of $\mathscr{H}^m(\mathscr{B}_0)$,

$$B(\mathbf{u}, \mathbf{v}) = \int_{\mathscr{B}_0} dL^3 \sum_{|i|,|j|=0}^{m} a_{ij} D^i\bar{u} D^j v,$$

$$|i| := i_1 + i_2 + i_3, \quad D^i := \frac{\partial^{|i|}}{\partial x_1{}^{i_1} \partial x_2{}^{i_2} \partial x_3{}^{i_3}}.$$

Definition 2.2.1
A bilinear continuous form B on $\mathscr{V}\times\mathscr{V}\subset\mathscr{H}^m\times\mathscr{H}^m$ is said to be $\mathscr{V}$**-coercive** if there is a real number λ_0 and a positive number K such that

$$\operatorname{Re} B(\mathbf{v},\mathbf{v})+\lambda_0\|\mathbf{v}\|_0^2 \geqslant K\|\mathbf{v}\|_m^2 \ \forall \mathbf{v}\in\mathscr{V}. \tag{2.2.38}$$

By $\|\cdot\|_m$, $\|\cdot\|_0$ we denote the usual norms on $\mathscr{H}^m$, $\mathscr{L}^2$, resp. Although in practice the form B is defined on $\mathscr{H}^m\times\mathscr{H}^m$, it is coercive on $\mathscr{V}\times\mathscr{V}$.

We shall use the triple $\mathscr{V}\subset\mathscr{H}\subset\mathscr{V}^*$, $\mathscr{H}=\mathscr{L}^2(\mathscr{B}_0)$, with continuous and dense embeddings, and denote by $\langle\cdot|\cdot\rangle_0$ the duality between $\mathscr{V}$ and $\mathscr{V}^*$, which reduces to the $\mathscr{L}^2$ scalar product $(\cdot|\cdot)_0$ on $\mathscr{V}\times\mathscr{H}\subset\mathscr{V}\times\mathscr{V}^*$.

The bilinear form

$$\tilde{B}(\mathbf{u},\mathbf{v}) :\equiv B(\mathbf{u},\mathbf{v})+\lambda_0\langle\mathbf{u}|\mathbf{v}\rangle_0 \tag{2.2.39}$$

on $\mathscr{V}\times\mathscr{V}$ satisfies the hypotheses of the Lax–Milgram lemma (Section 1.1). Hence for every $\mathbf{f}\in\mathscr{V}^*$ there is a unique $\mathbf{u}=\tilde{G}\mathbf{f}\in\mathscr{V}$ such that

$$\tilde{B}(\mathbf{u},\mathbf{v}) = \langle\mathbf{f}|\mathbf{v}\rangle_0 \ \forall\mathbf{v}\in\mathscr{V}. \tag{2.2.40}$$

Indeed, $\mathscr{V}^*$ is isomorphic to $\mathscr{V}$ by the Riesz theorem and $\langle\mathbf{f}|\mathbf{v}\rangle_0=(\mathbf{u}_f|\mathbf{v})$ for some $\mathbf{u}_f\in\mathscr{V}$. $(\cdot|\cdot)$ denotes the scalar product in $\mathscr{V}$ induced by the scalar product in $\mathscr{H}^m$. Moreover $\|\mathbf{f}\|_{\mathscr{V}^*}=\sup\{|\langle\mathbf{f}|\mathbf{v}\rangle_0| \ \|\mathbf{v}\|_{\mathscr{V}}\leqslant 1\}=\|\mathbf{u}_f\|_{\mathscr{V}}$. Hence $\|\mathbf{u}\|_{\mathscr{V}}\leqslant K^{-1}\|\mathbf{u}_f\|_{\mathscr{V}}=K^{-1}\|\mathbf{f}\|_{\mathscr{V}^*}$. For $\mathbf{f}\in\mathscr{H}\subset V^*$ we have $\|\mathbf{u}\|_{\mathscr{V}}\leqslant K'\|\mathbf{f}\|_0$ for some $K'>0$.

Let $G_0:=\tilde{G}|H\colon \mathscr{H}\to\mathscr{V}$. We shall assume that $\mathscr{B}_0$ is properly regular. Since the embedding $j\colon\mathscr{V}\to\mathscr{H}$ is completely continuous and $\|\tilde{G}_0\|\leqslant K'$, the linear operator $G:=j\circ\tilde{G}_0\colon\mathscr{H}\to\mathscr{H}$ is also completely continuous.

Consider the problem

$$\mathbf{v}-\lambda G\mathbf{v}=\mathbf{f}, \quad \mathbf{v}\in H,\ \lambda\in\mathbf{C} \tag{2.2.41}$$

for given $f\in\mathscr{H}$. For simplicity we shall consider $\mathscr{V}$ as a subset of $\mathscr{H}$, so that $\mathscr{R}(G)\subset\mathscr{V}\subset\mathscr{H}$. Let $\mathbf{u}=G\mathbf{v}$, $\mathbf{w}\in\mathscr{V}$. Since $\tilde{B}(\mathbf{u},\mathbf{w})=\langle\mathbf{v}|\mathbf{w}\rangle_0$, equation (2.2.41) is equivalent to

$$\tilde{B}(\mathbf{u},\mathbf{w})-\lambda\langle\mathbf{u}|\mathbf{w}\rangle_0=\langle\mathbf{f}|\mathbf{w}\rangle_0 \ \forall\mathbf{w}\in\mathscr{V} \tag{2.2.42}$$

(note that $\mathscr{V}$ is dense in H).

From the Riesz–Fredholm theory (Section 1.1) it follows that the spectrum of the eigenvalue problem

$$\mathbf{v}-\lambda G\mathbf{v}=\mathbf{0} \tag{2.2.43}$$

is at most countable with no accumulation points except at infinity. The space of eigensolutions corresponding to an eigenvalue λ_n, $n=1,2,\ldots$, has at most finite dimensions. If λ is not an eigenvalue of (2.2.43), then equation (2.2.41), and hence also equation (2.2.42), has a unique solution. Suppose now that $\lambda = \lambda_n$ is an eigenvalue of (2.2.43). Then $\bar{\lambda}_n$ is an eigenvalue of the adjoint problem

$$\boldsymbol{\varphi} - \mu G^{\dagger}\boldsymbol{\varphi} = \mathbf{0}, \quad \boldsymbol{\varphi} \in H. \tag{2.2.44}$$

Let $\boldsymbol{\varphi}_1, \ldots, \boldsymbol{\varphi}_r$ be a complete set of solutions of (2.2.44) for $\mu = \bar{\lambda}_n$. For $\lambda = \lambda_n$ equation (2.2.41) (and hence equation (2.2.42)) has a solution iff $(\mathbf{f}|\boldsymbol{\varphi}_j)_0 = 0$ for $j = 1, \ldots, r$. The solution is determined modulo arbitrary linear combinations of the solutions $\boldsymbol{\psi}_1, \ldots, \boldsymbol{\psi}_s$ of (2.2.43) with $\lambda = \lambda_n$.

If B is symmetric on $\mathscr{V} \times \mathscr{V}$, i.e. $B(\mathbf{u}, \mathbf{v}) = \overline{B(\mathbf{v}, \mathbf{u})}$, then for arbitrary $\mathbf{u}$, $\mathbf{v} \in \mathscr{H}$

$$(\mathbf{u}|G\mathbf{v})_0 = \tilde{B}(G\mathbf{u}, G\mathbf{v}) = \overline{\tilde{B}(G\mathbf{v}, G\mathbf{u})} = \overline{(\mathbf{v}|G\mathbf{u})_0} = (G\mathbf{u}|\mathbf{v})_0,$$

i.e. $G = G^{\dagger}$. Moreover $(\mathbf{v}|G\mathbf{v})_0 = \tilde{B}(G\mathbf{v}, G\mathbf{v}) \geqslant K\|G\mathbf{v}\|_m{}^2 \geqslant 0$ and $(\mathbf{v}|G\mathbf{v})_0 = 0$ implies that $G\mathbf{v} = 0$ in $\mathscr{V}$, $\mathscr{H}$, $\langle \mathbf{v}|\mathbf{w}\rangle_0 = \tilde{B}(G\mathbf{v}, \mathbf{w}) = 0$ for all $\mathbf{w} \in \mathscr{V}$. Hence $\mathbf{v} = 0$ in $\mathscr{H}$. Hence for symmetric B we have a positive self-adjoint Green operator G. In this case, and notably in elastostatics, the eigenvalues of (2.2.43) are real, positive and can be arranged into a sequence $\lambda_n \to \infty$. Moreover it follows from the spectral theory of self-adjoint operators that the eigensolutions span $\mathscr{H}$.

2.2.7 Existence, Uniqueness and Continuous Dependence on Input Data in Linearized Elastostatics: Some Examples

The "classical" linearized BVP

$$\begin{aligned}
&[B_k{}^p{}_l{}^q(\mathbf{x})u^l{}_{,x^q}]_{,x^p} + \varrho_1 b_k = 0 \quad \text{in } \mathscr{B}_0,\\
&B_k{}^p{}_l{}^q(\mathbf{x})u^l{}_{,x^q} n_p = s_k(\mathbf{x}) \quad \text{on } \Sigma_2',\\
&B_k{}^p{}_l{}^q(\mathbf{x})u^l{}_{,x^q} n_p + \nu(\mathbf{x})u^k = s_k(\mathbf{x}) \quad \text{on } \Sigma_3,\\
&\mathbf{u}|\Sigma_1 = \mathbf{0} \quad \text{on } \Sigma_1
\end{aligned} \tag{2.2.45}$$

can be put in the more general form (2.2.32), with

$$B(\mathbf{u}, \mathbf{v}) := \int_{\mathscr{B}_0} dL^3\, B_k{}^p{}_l{}^q(\mathbf{x})u^k{}_{,x^p}v^l{}_{,x^q} + \int_{\Sigma_3} dH^2\, \nu\gamma_{\Sigma_3}\mathbf{u}\cdot\gamma_{\Sigma_3}\mathbf{v},$$

$$f(\mathbf{v}) := \int_{\mathscr{B}_0} dL^3\, \varrho_1 b_k v^k + \int_{\Sigma_2} dH^2\, s_k\gamma_{\Sigma_2}v^k,$$

$$V := \{\mathbf{u} \in \mathscr{H}^1(\mathscr{B}_0)|\ \gamma_{\Sigma_1}\mathbf{u} = \mathbf{0}\}.$$

In the classical formulation we must assume that $B_k{}^p{}_l{}^q$ are differentiable and $\mathbf{u} \in \mathscr{C}^2(\overline{\mathscr{B}_0})$, whereas in (2.2.32) $B_k{}^p{}_l{}^q \in \mathscr{L}^\infty(\mathscr{B}_0)$, $s_k \in \mathscr{L}^2(\Sigma_2)$, $\nu \in \mathscr{L}^\infty(\Sigma_3)$.

The BC on Σ_1 can be expressed in terms of traces, while the natural BCs in (2.2.45) cannot be taken at face value. They have been accounted for in (2.2.32) by an appropriate definition of B and f.

The natural BCs are sometimes called **unstable** for the following reason. Suppose that $s_k \equiv 0$, $\nu \equiv 0$ and f ranges over $\mathscr{H}$ in eq. (2.2.32). In this case the solution $\mathbf{u}$ of (2.2.32) ranges over a dense subset of $\mathscr{V}$ (Nečas, 1968, Theorem 3.4).

We shall establish the relation between (2.2.32) and (2.2.45). For $\mathbf{u} \in \mathscr{C}^2(\overline{\mathscr{B}_0})$ and $\mathbf{v} \in \mathscr{V} \cap \mathscr{C}^1(\overline{\mathscr{B}_0})$ we have the following identities:

$$\int_{\mathscr{B}_0} dL^3\, B_k{}^p{}_l{}^q u^k{}_{,x^p}\, v^l{}_{,x^q} = \int_{\Sigma_2} dH^2\, B_k{}^p{}_l{}^q u^k{}_{,x^p}\, n_q v^l - \int_{\mathscr{B}_0} dL^3\, [B_k{}^p{}_l{}^q u^k{}_{,x^p}]_{,x^q}\, v^l$$

(we have assumed differentiability of $B_k{}^p{}_l{}^q$). Substituting arbitrary $\mathbf{v} \in \mathscr{C}^1_0(\mathscr{B}_0)$ in (2.2.32) we recover the first equation (2.2.45). Substituting this result into (2.2.32) we get an equation involving surface integrals. This equation remains valid for arbitrary $\mathbf{v} \in \mathscr{V}$ provided appropriate trace operators are inserted where necessary. For $\mathbf{v}$ ranging over $\mathscr{V}$ the traces range over dense subsets of $\mathscr{L}^2(\Sigma_2')$ and $\mathscr{L}^2(\Sigma_3)$. Hence the natural BCs specified in equation (2.2.45) follow.

For the Dirichlet problem $\gamma_{\partial\mathscr{B}_0}\mathbf{u} = 0$ we set $\mathscr{V} = H^1_0(\mathscr{B}_0)$ and $\mathscr{V}$-coercivity of B is granted by (2.2.37) and the Gårding lemma (Section 1.1). In the case of symmetry (2.2.37) the **Korn inequalities** can be used:

$$\int_{\mathscr{B}_0} dL^3 \sum_{i,j} [e_{ij}(u)]^2 \geqslant K\|\mathbf{u}\|_1{}^2 \quad \text{for } \mathbf{u} \in \mathscr{H}^1_0(\mathscr{B}_0), \tag{2.2.46}$$

$$\int_{\mathscr{B}_0} dL^3 \sum_{i,j} [e_{ij}(\mathbf{u})]^2 + \lambda_0\|\mathbf{u}\|_0{}^2 \geqslant K'\|\mathbf{u}\|_1{}^2 \quad \forall \mathbf{u} \in \mathscr{H}^1(\mathscr{B}_0), \tag{2.2.47}$$

with

$$e_{ij}(\mathbf{u}) := \frac{1}{2}(u^i{}_{,x^j} + u^j{}_{,x^i}), \tag{2.2.48}$$

$K, K' > 0$. Equation (2.2.47) is valid provided $\mathscr{B}_0$ satisfies the restricted cone hypothesis (Section 1.1). The first of the Korn inequalities is readily verified by integrating by parts:

$$\int_{\mathscr{B}_0} dL^3 \sum_{i,j} [e_{ij}(\mathbf{u})]^2 = \frac{1}{2}\int_{\mathscr{B}_0} dL^3\, [u^i{}_{,x^j} u^i{}_{,x^j} + u^i{}_{,x^j} u^j{}_{,x^i}]$$

$$= \frac{1}{2}\int_{\mathscr{B}_0} dL^3\, [u^i{}_{,x^j} u^i{}_{,x^j} + (\operatorname{div} \mathbf{u})^2] \geqslant K\|\mathbf{u}\|_1{}^2$$

on account of the Friedrichs inequality. The proof of the second Korn inequality is much more complicated (Fichera, 1972; Nečas, 1968). There is a vast literature on the Korn inequalities, including optimal estimates of K, K' for given $\mathscr{B}_0$.

$\mathscr{V}$-coercivity can be deduced from the Korn inequalities under somewhat more stringent conditions on $B_k{}^p{}_l{}^q$ than SE, viz.

$$B_k{}^p{}_l{}^q H^k{}_p H^l{}_q \geqslant M \sum_{k,p} (H^k{}_p)^2, \quad M > 0 \tag{2.2.49}$$

for every symmetric $H^k{}_p = H^p{}_k$. For arbitrary $\mathbf{a}, \mathbf{k} \in \boldsymbol{R}^3$

$$B_k{}^p{}_l{}^q a^k a^l k_p k_q = B_k{}^p{}_l{}^q a_{(k} k_{p)} a_{(l} k_{q)} \geqslant M \sum_{k,} a_{(k} k_{p)}{}^2$$
$$= \frac{M}{2} [\mathbf{a}^2 \mathbf{k}^2 + (\mathbf{a} \cdot \mathbf{k})^2] \geqslant \frac{M}{2} \mathbf{a}^2 \mathbf{k}^2.$$

Hence (2.2.48) implies SE. For the converse we shall note that SE implies that

$$B_k{}^p{}_l{}^q a_{(k} k_{p)} a_{(l} k_{q)} \geqslant M \mathbf{a}^2 \mathbf{k}^2 \geqslant \frac{M}{2} [\mathbf{a}^2 \mathbf{k}^2 + (\mathbf{a} \cdot \mathbf{k})^2].$$

Hence (2.2.48) follows for $H^k{}_p = a_{(k} k_{p)}$ but not for general symmetric $H^k{}_p$. In fact, for an isotropic medium without prestress (2.2.49) becomes

$$\lambda (H^k{}_k)^2 + 2\mu H^k{}_l H^l{}_k \geqslant M \sum_{k,l} (H^k{}_l)^2 \tag{2.2.49a}$$

is equivalent to $\mu > 0$ (set $H^k{}_k = 0$) and $3\lambda + 2\mu > 0$ (set $\mathbf{H} = \mathbf{E}$) while SE is equivalent to

$$\lambda (\mathbf{k} \cdot \mathbf{a})^2 + 2\mu \sum_{i,j} k_{(i} a_{j)}{}^2 \geqslant M \mathbf{a}^2 \mathbf{k}^2, \tag{2.2.50}$$

or $\mu > 0$ (let $\mathbf{k} \cdot \mathbf{a} = 0$), $\lambda + 2\mu > 0$ (let $\mathbf{a} = \mathbf{k}$). Now $3(\lambda + 2\mu) = (3\lambda + 2\mu) + 4\mu > 0$ provided $3\lambda + 2\mu > 0$, $\mu > 0$ but it is possible that $\lambda + 2\mu > 0$, $\mu > 0$ and $\lambda < 0$, $3\lambda + 2\mu \leqslant 0$, at least in theory. In practice $\lambda < 0$ has never been observed and hence (2.2.49) can be applied safely to isotropic media without prestress.

Under the hypotheses (2.2.33), (2.2.49) existence and uniqueness for the Dirichlet problem $\gamma \mathbf{u} = 0$ on $\partial \mathscr{B}_0$ follows directly from the first Korn inequality and the Lax–Milgram lemma. Estimates displaying continuous dependence on $\mathbf{f}$ are also obtained. For mixed BVPs (2.2.33), (2.2.49) imply the $\mathscr{V}$-coercivity of B.

We also note that $\mathbf{f} \in \mathscr{H}$ provided the BC on Σ_2 are homogeneous, $s_k = 0$. For non-homogeneous BC on Σ_2 the operator G should be regarded as a (completely continuous) mapping $\mathscr{V}^* \to \mathscr{V}^*$. For the homogeneous Neumann problem: $\Sigma_1 = \emptyset$, $s_k = 0$, $\varrho_0 b_k \in \mathscr{H}$ we must investigate the eigenvalue problem (2.2.43) with $\lambda = \lambda_0 :=$ the constant appearing in (2.2.47). The latter problem is equivalent to

$$B(\mathbf{u}, \mathbf{v}) = 0 \ \forall \mathbf{v} \in \mathscr{V}. \tag{2.2.51}$$

We assume that B satisfies (2.2.48), (2.2.33). Setting $\mathbf{u} = \mathbf{v}$ we conclude that $e_{ij}(\mathbf{u}) = 0$ in $\mathscr{H}$. Conversely, $e_{ij}(\mathbf{u}) = 0$ in $\mathscr{H}$ implies (2.2.51).

Lemma 2.2.2

Let $e_{ij}(\mathbf{u}) = 0$ in the sense of distributions. Then

$$\mathbf{u}(\mathbf{x}) \equiv \mathbf{a} + \mathbf{b} \times \mathbf{x} \equiv \mathbf{a} + \mathbf{W}\mathbf{x}, \quad {}^t\mathbf{W} = -\mathbf{W}. \tag{2.2.52}$$

Proof

Since $u^i{}_{,x^i} = e_{ii} = 0$ and $u^k{}_{,x^k x^l} + u^l{}_{,x^k x^k} = 0$, it follows that $\mathbf{u}$ is harmonic. By the well-known regularity theorems for elliptic (or harmonic) functions $\mathbf{u} \in \mathscr{C}^2$ it follows that $\mathbf{u}$ is also real-analytic.

Let $\varphi(\mathbf{x}) :\equiv \sum_{i=1}^{3} x^i u^i(\mathbf{x})$. Since $\partial\varphi/\partial x^k = u^k - x^l \partial u^l/\partial x^k$, it follows that φ satisfies the equation $x^k \partial\varphi/\partial x^k = 2\varphi - [\partial(x^l u^l)/\partial x^k] x^k$, i.e. $\varphi = x^k \partial\varphi/\partial x^k$. Hence φ is a homogeneous function of first degree, $\varphi = \sum_{k=1}^{3} a_k x^k$, $a_k = \text{const}$ for $k = 1, 2, 3$. Hence $u^k - x^l \partial u^k/\partial x^l = a^k$ and $w^k = u^k - a_k$ is an analytic and homogeneous function of first degree. Hence $u^k = W^k{}_l x^l + a_k$. Since

$$\sum_{k=1}^{3} u^k x^k \equiv \sum_{k=1}^{3} a_k x^k, \quad \text{it follows that} \quad W^k{}_l = -W^l{}_k. \qquad \square$$

The displacements (2.2.52) are **infinitesimal rigid** since $\mathbf{W}$ generates a one-parameter subgroup $t \mapsto \exp(t\mathbf{W})$ of SO(3). Consider a rigid displacement with the rotation $\exp(t\mathbf{W})$ followed by a translation $t\mathbf{b}$. A point $\mathbf{x} = \chi_0(\xi)$ goes to $\mathbf{x} + \mathbf{u}(\mathbf{x}, t) \equiv \exp(t\mathbf{W})x + t\mathbf{b}$. For small t this becomes $\mathbf{x} + t\mathbf{W}\mathbf{x} + t\mathbf{b}$ $\equiv: \mathbf{x} + t\mathbf{u}(\mathbf{x})$.

Since B is symmetric, the problem is self-adjoint. The homogeneous Neumann problem with non-zero body force

$$B(\mathbf{u}, \mathbf{v}) = \langle \mathbf{f} | \mathbf{v} \rangle_0 \ \forall \mathbf{v} \in \mathscr{V}, \tag{2.2.53}$$

where $\mathbf{f} \in H$, has a solution $\mathbf{u} \in \mathscr{V}$ provided

$$\langle \mathbf{f} | \mathbf{v} \rangle_0 = 0 \tag{2.2.54}$$

for every infinitesimal rigid displacement $\mathbf{v}$, i.e. provided the following conditions are satisfied:

$$\int_{\mathscr{B}_0} dL^3\, \varrho_1 \mathbf{b}^{(1)} = \mathbf{0}, \tag{2.2.55}$$

$$\int_{\mathscr{B}_0} dL^3\, \varrho_1 \mathbf{b}^{(1)} \times \mathbf{x} = \mathbf{0}. \tag{2.2.56}$$

The solution is defined modulo arbitrary infinitesimal rigid diplacements.

We now outline an alternative approach to the Neumann problem with $B_k{}^p{}_l{}^q$ satisfying (2.2.33), (2.2.49). Since

$$B(\mathbf{u}, \mathbf{v}) \equiv \int_{\mathscr{B}_0} dL^3\, B_k{}^p{}_l{}^q e_{kp}(\mathbf{u}) e_{lq}(\mathbf{v}),$$

equation (2.2.53) has a solution only if the condition (2.2.54) (or, equivalently (2.2.55) and (2.2.56)) is satisfied. Suppose then that (2.2.54) is satisfied. Let R be the linear subspace of infinitesimal rigid displacements in the real Hilbert space H^1 and let $R^\perp$ be its orthogonal complement, $H^1 = \mathscr{R} \oplus \mathscr{R}^\perp$. We shall prove that

$$\sum_{p,q} \langle e_{pq}(\mathbf{u}) | e_{pq}(\mathbf{u}) \rangle_0 \geqslant K ||\mathbf{u}||_1{}^2 \quad \forall \mathbf{u} \in \mathscr{R}^\perp. \tag{2.2.57}$$

Suppose the contrary. Then there is a sequence $\{\mathbf{u}_n\} \subset \mathscr{R}^\perp$ such that $||\mathbf{u}_n||_1 = 1$ while

$$\sum_{p,q} \langle e_{pq}(\mathbf{u}_n) | e_{pq}(\mathbf{u}_n) \rangle_0 < \frac{1}{n}. \tag{2.2.58}$$

By virtue of the Rellich–Kondrachov theorem (assuming that $\mathscr{B}_0$ satisfies the cone condition) a subsequence $\{\mathbf{u}_{n_k}\} = \{\mathbf{u}'_k\}$ of $\{\mathbf{u}_n\}$ converges in $\mathscr{L}^2(\mathscr{B}_0)$. The second Korn inequality implies that

$$K||\mathbf{u}'_k - \mathbf{u}'_l||_1{}^2 \leqslant ||\mathbf{u}'_k - \mathbf{u}'_l||_0{}^2 + \sum_{p,q} \langle e_{pq}(\mathbf{u}'_k - \mathbf{u}'_l) | e_{pq}(\mathbf{u}'_k - \mathbf{u}'_l) \rangle_0$$
$$\leqslant ||\mathbf{u}'_k - \mathbf{u}'_l||_0{}^2 + 1/k + 1/l + 2(kl)^{-1/2}.$$

Hence $\{\mathbf{u}'_k\}$ converges in $H^1(\mathscr{B}_0)$, $\mathbf{u}'_k \to \mathbf{u}_0 \in \mathscr{R}^\perp$, $||\mathbf{u}_0||_1 = 1$. On the other hand (2.2.58) implies that $e_{ij}(\mathbf{u}_0) = \lim_{k\to\infty} e_{ij}(\mathbf{u}'_k) = 0$ and $\mathbf{u}_0 \in \mathscr{R}$. Hence $\mathbf{u}_0 = 0$, which contradicts with $||\mathbf{u}_0||_1 = 1$. □

Equation (2.2.49) and (2.2.57) imply that the form B restricted to $\mathscr{R}^\perp$ satisfies the hypotheses of the Lax–Milgram lemma. Hence a unique solution exists in $\mathscr{R}^\perp$ provided (2.2.54) holds.

We note that $\mathbf{u} \in \mathscr{R}^\perp$ iff the following constraints are satisfied:

$$\int_{\mathscr{B}_0} dL^3\, \mathbf{u} = \mathbf{0}, \quad \int_{\mathscr{B}_0} dL^3\, [\mathbf{x} \times \mathbf{u} + \operatorname{curl} \mathbf{u}] = \mathbf{0} \tag{2.2.59}$$

The above methods based on the Riesz–Fredholm theory clearly cover the case of infinitesimal harmonic motions $\mathbf{u}(\mathbf{x}, t) = \mathbf{u}(\mathbf{x})\exp(i\omega t)$ under mixed BC. The eigensolutions of equation (2.2.43) correspond in part to the free vibrations of the body with $\omega^2 = \lambda_n - \lambda_0$.

2.2.8 Remarks about Regularity and Transmission Problems in Linearized Elastostatics

There are several formulations of linear elliptic BVPs in the language of functional analysis. The most natural one for elasticity (and for many other theories of continuous media) is the variational one, with $\mathbf{u} \in H^1(\mathscr{B}_0)$ and $B_k{}^p{}_l{}^q \in \mathscr{L}^\infty(\mathscr{B}_0)$, since it follows directly from the VWP. Alternative approaches are possible if $B_k{}^p{}_l{}^q$ are sufficiently smooth (at least $\mathscr{C}^1$). We may consider strong solutions $\mathbf{u} \in \mathscr{D} \subset H^2(\mathscr{B}_0)$ of the equation $A\mathbf{u} :\equiv \left(\dfrac{\partial}{\partial x^p} B_k{}^p{}_l{}^q(\mathbf{x}) u^l{}_{,x^q}\right)$ $= -(\varrho_1 b_k^{(1)})$, with $\mathscr{D} := \{\mathbf{u} \in H^2 |\ \gamma_{\Sigma_1}\mathbf{u} = \mathbf{u}_0,\ \nu_p B_k{}^p{}_l{}^q u^l{}_{,x^q} + \nu u^k = s_k \text{ on } \Sigma_2\}$, or even distributional solutions in $H^s(\mathscr{B}_0)$, $s < 2$ (cf. Peetre, 1961; Babuška and Aziz, 1972).

A link between these solutions is provided by the regularity theorems. The mere ellipticity condition, in our case $\det[B_k{}^p{}_l{}^q(\mathbf{x}) k_p k_q] \neq 0$ for every $\mathbf{k} \in R^3 \setminus \{\mathbf{0}\}$, guarantees the following interior regularity property (cf. Fichera, 1972; Nečas, 1968):

Theorem 2.2.3
If $A\mathbf{u} = \mathbf{f} \in H^s(\mathscr{B}_0)$ in the weak sense, i.e.

$$\langle A\mathbf{v}|\mathbf{u}\rangle_0 = \langle \mathbf{v}|\mathbf{f}\rangle_0 \quad \forall\, \mathbf{v} \in \mathscr{C}_0^\infty(\mathscr{B}_0)^\star$$

$B_k{}^p{}_l{}^q \in \mathscr{C}^\infty(\mathscr{B}_0)$ and $\mathbf{x}_0 \in \mathscr{B}_0$, then $\exists\ r > 0$ such that

$$\mathbf{u}|\mathscr{K}(\mathbf{x}_0, r) \in H^{s+2}(\mathscr{K}(\mathbf{x}_0, r)),$$

$$||\mathbf{u}||_{s+2,\mathbf{x}_0,r}{}^2 \leqslant c(\mathbf{x}_0)[||\mathbf{f}||_{s,\mathbf{x}_0,2r}{}^2 + ||\mathbf{u}||_{s+1,\mathbf{x}_0,2r}{}^2]. \tag{2.2.59}$$

Here $||\cdot||_{m,\mathbf{x}_0,r}$ denotes the norm in $H^m(\mathscr{K}(\mathbf{x}_0, r))$ applied to a suitable restriction of $\mathbf{u}$.

The important point in the interior regularity theory is that if we neglect the (smooth) $\mathbf{x}$-dependence of $B_k{}^p{}_l{}^q$, then a formal solution by the Fourier transformation has the following form:

$$\hat{u}^k(\mathbf{k}) \equiv \frac{D^k{}_l(\mathbf{k})\hat{f}^l(\mathbf{k})}{\det[(B_k{}^p{}_l{}^q k_p k_q)]} \tag{2.2.60}$$

with a non-vanishing denominator for $\mathbf{k} \neq 0$.

* Note that the differential operator A is formally self-adjoint.

If $s > 3/2$ ($3 = \dim\mathscr{B}_0$), then $\mathbf{u} \in \mathscr{C}^2$. If $\mathbf{f} \in \mathscr{C}^\infty$ then $\mathbf{u} \in \mathscr{C}^\infty$. In fact $\mathbf{u} \in \mathscr{C}^2$ if $\mathbf{f}$ is at least Hölder continuous.

Regularity at the boundary $\partial\mathscr{B}_0$ is sensitive to the boundary conditions (cf. Lions and Magenes (1968), Nečas (1968) for the definition of **covering systems of BC**. Compatibility problems at $\partial\Sigma_1 = \partial\Sigma_2$ appear in the case of mixed BVPs. A typical estimate for the solution $\mathbf{u}$ of a variational elliptic problem with $B_k{}^p{}_l{}^q \in \mathscr{C}^\infty$, $\partial\mathscr{B}_0 \in \mathscr{C}^\infty$, near $\mathbf{x}_0 \in \partial\mathscr{B}_0$ is

$$||\mathbf{u}||_{s+2,\mathbf{x}_0,r,\mathbf{n}}^2 \leqslant c(r)[||\mathbf{f}||_s{}^2 + ||\mathbf{u}||_s{}^2] \tag{2.2.61}$$

(cf. Fichera, 1972). Here $||\cdot||_{m,\mathbf{x}_0,r,\mathbf{n}}$ denotes the $\mathscr{H}^m(\mathscr{K}_+(\mathbf{x}_0, r))$ norm, $\mathscr{K}_+(\mathbf{x}_0, r) := \mathscr{K}(\mathbf{x}_0, r) \cap \mathscr{B}_0 \approx \mathscr{K}(\mathbf{x}_0, r) \cap \{\mathbf{x}|\mathbf{n}\cdot(\mathbf{x}-\mathbf{x}_0) < 0\}$, while $||\cdot||_m$ denotes the $\mathscr{H}^m(\mathscr{B}_0)$ norm and $\mathbf{n}$ is the outer normal to $\partial\mathscr{B}_0$ at $\mathbf{x}_0$. The estimate is based on the assumption of ellipticity and some additional assumptions on the solution space $\mathscr{V}$ and on the regularity properties of $\partial\mathscr{B}_0$ near $\mathbf{x}_0$.

Although our starting point was the VWP and the resulting variational formulation of the BVP we must admit that the regularity properties of a solution involve a great deal of physics. This is particularly conspicuous if we consider a BVP for the equations

$$\frac{\partial}{\partial x^p}[B_k{}^p{}_l{}^q(\mathbf{x})u^l{}_{,x^q}] + \omega^2 u^k = -f_k \tag{2.2.62}$$

governing infinitesimal harmonic motions under harmonic loads $f_k e^{i\omega t}$. Suppose that $\mathscr{B}_0$ can be subdivided into domains $\mathscr{B}_j$, $j = 1, \ldots, N$, with sufficiently regular boundaries, $\mathscr{B}_j \cap \mathscr{B}_i = \emptyset$, $\bigcup_{i=1}^{N} \overline{\mathscr{B}}_i = \overline{\mathscr{B}}_0$, in such a way that $B_k{}^p{}_l{}^q$ is smooth on each $\mathscr{B}_j$ and either $B_k{}^p{}_l{}^q$ or its derivatives have jump discontinuities at $\partial\mathscr{B}_j$. The variational approach contains the interface conditions (continuity of u^k and $B_k{}^p{}_l{}^q u^l{}_{,x^q}$ on $\partial\mathscr{B}_i \cap \partial\mathscr{B}_j$) in implicit form. It is possible to derive these interface conditions by a judicious choice of the test functions $\mathbf{v} \in X^1 \cap \mathscr{C}^\infty$ with supports near the interface. Hence it becomes apparent that the reflection-transmission phenomena are related to regularity.

2.3 NON-LINEAR ELASTOSTATICS AND MONOTONE OPERATORS

2.3.1 Monotone Operators in Elasticity. Theorems about Existence and Uniqueness

Let $\mathscr{V}$ be a reflexive and separable Banach space.

Definition 2.3.1
An operator $A: \mathscr{V} \to \mathscr{V}^*$ is said to be **monotone** if

$$\langle A\mathbf{u} - A\mathbf{v} | \mathbf{u} - \mathbf{v} \rangle \geqslant 0 \ \forall \mathbf{u}, \mathbf{v} \in \mathscr{V}, \tag{2.3.1}$$

strictly monotone if

$$\langle A\mathbf{u} - A\mathbf{v} | \mathbf{u} - \mathbf{v} \rangle > 0 \ \forall \mathbf{u}, \mathbf{v} \in \mathscr{V},\ \mathbf{u} \neq \mathbf{v}, \tag{2.3.2}$$

coercive if

$$\langle A\mathbf{u} | \mathbf{u} \rangle \geqslant \gamma(||\mathbf{u}||)\, ||\mathbf{u}|| \ \forall \mathbf{u} \in \mathscr{V} \tag{2.3.3}$$

with $\gamma(s) \to \infty$ for $s \to \infty$.

In Definition 2.3.1 and throughout Section 2.3 $\langle \cdot | \cdot \rangle$ denotes the duality between $\mathscr{V}$ and $\mathscr{V}^*$ and $|| \cdot ||$ is the norm in $\mathscr{V}$.

Definition 2.3.2
An operator $A: \mathscr{V} \to \mathscr{V}^*$ is said to be **hemicontinuous** if it is weakly continuous on straight lines in $\mathscr{V}$, i.e. if $s \mapsto \langle A(\mathbf{u}+s\mathbf{v}) | \mathbf{w} \rangle$ is continuous $\forall\, \mathbf{u}, \mathbf{v}, \mathbf{w} \in \mathscr{V}$, **radially continuous** if the same property holds for all $\mathbf{v}, \mathbf{u} \in \mathscr{V}$ and $\mathbf{w} = \mathbf{v}$, **demicontinuous** if $\mathbf{u}_n \to \mathbf{u}$ in $\mathscr{V}$ entails $A\mathbf{u}_n \to A\mathbf{u}$ in $\mathscr{V}^*$.

Since $\mathscr{V}$ is reflexive, the last convergence is equivalent to the convergence in the weak star topology on $\mathscr{V}$ (cf. Yosida, 1965).

The Lax–Milgram lemma will now be replaced by the Minty–Browder theorem.

Theorem 2.3.3 (Minty–Browder)
Let $A: \mathscr{V} \to \mathscr{V}^*$ be radially continuous, monotone and coercive. Then the set of solutions of

$$A\mathbf{u} = \mathbf{f} \tag{2.3.4}$$

for arbitrary $\mathbf{f} \in \mathscr{V}^*$ is non-empty, convex and weakly closed.

It is clear that strong monotonicity implies uniqueness. Indeed, let $A\mathbf{u} = \mathbf{f}$, $A\mathbf{v} = \mathbf{f}$, $\mathbf{u} \neq \mathbf{v}$. Then $\langle A\mathbf{u} - A\mathbf{v} | \mathbf{u} - \mathbf{v} \rangle = 0$ in contradiction of (2.3.2). Actually we have

Theorem 2.3.4
Let $A: \mathscr{V} \to \mathscr{V}^*$ be radially continuous, strictly monotone and coercive. Then A is surjective and $A^{-1}: \mathscr{V}^* \to \mathscr{V}$ is strictly monotone, bounded and demicontinuous.

We prove the most important assertion of Theorem 2.3.3 in Section 2.3.3. In this subsection we examine the applicability of Theorems 2.3.3 and 2.3.4 in non-linear elastostatics.

Let us consider the case when S_0 is the Nemytskii operator $\tilde{S}$, equation (2.2.8). The operator S_0 is demicontinuous (*a fortiori* radially continuous). Hence A is radially continuous. A is monotone provided

$$\langle A\mathbf{u}-A\mathbf{v}|\mathbf{u}-\mathbf{v}\rangle \equiv \int_{\mathscr{B}_0} dL^3\, [\hat{S}_k^{\alpha}(\nabla\boldsymbol{\chi}_0(\xi)+\nabla\mathbf{u}(\xi),\xi))- \\ -\hat{S}_k^{\alpha}(\nabla\boldsymbol{\chi}_0(\xi)+\nabla\mathbf{v}(\xi),\xi))][u^k{}_{,\xi^\alpha}-v^k{}_{,\xi^\alpha}] \geqslant 0. \tag{2.3.5}$$

The inequality (2.3.5) is satisfied if $\hat{S}_k^{\alpha}(\mathbf{F},\xi)$ satisfies the pointwise inequality

$$[\hat{S}_k^{\alpha}(\mathbf{F}_1,\xi)-\hat{S}_k^{\alpha}(\mathbf{F}_2,\xi)][F_1{}^k{}_\alpha-F_2{}^k{}_\alpha] \geqslant 0 \tag{2.3.6}$$

for almost every $\xi\in\Omega$ and all $\mathbf{F}_1,\mathbf{F}_2\in\mathscr{L}_+$.*

The inequality (2.3.6) cannot be satisfied for all physically interesting values of $\mathbf{F}_1,\mathbf{F}_2$. Indeed, let us consider $\mathbf{F}_2=\mathbf{Q}\mathbf{F}_1$, $\mathbf{Q}\in SO(3)$. Since $\hat{\mathbf{S}}(\mathbf{QF},\xi)\equiv\mathbf{Q}\hat{\mathbf{S}}(\mathbf{F},\xi)$, equation (2.3.6) implies that

$$\operatorname{tr}[(\mathbf{Q}-\mathbf{E})\hat{\mathbf{T}}(\mathbf{F}_1,\xi)({}^t\mathbf{Q}-\mathbf{E})] \\ =(\det\mathbf{F}_1)^{-1}\operatorname{tr}\{[\hat{\mathbf{S}}(\mathbf{F}_2,\xi)-\hat{S}(\mathbf{F}_1,\xi)][{}^t\mathbf{F}_2-{}^t\mathbf{F}_1]\}\geqslant 0. \tag{2.3.7}$$

$\hat{\mathbf{T}}(\mathbf{F},\xi)$ denotes the Cauchy stress. Choose the coordinates x^1, x^2, x^3 in such a way that the principal axes of $\hat{\mathbf{T}}(\mathbf{F}_1,\xi)$ coincide with the coordinate vectors $\partial/\partial x^i$, $i=1,2,3$. Let $\mathbf{Q}=\operatorname{diag}\{-1,-1,1\}$. The inequality (2.3.7) yields immediately the inequality $t_1+t_2\geqslant 0$ for the principal stresses t_i (= eigenvalues of $\mathbf{T}$). Analogously we deduce the inequalities $t_1+t_3\geqslant 0$, $t_3+t_2\geqslant 0$. Hence at least two principal stresses are non-negative and if one is negative then the other two are positive. This condition exclude pure compression, in particular the hydrostatic compression $\mathbf{T}=-p\mathbf{E}$, $p>0$ at any $\xi\in\Omega$. In the case of a *homogeneous* body made of the material taken at ξ we can achieve $\mathbf{T}=-p\mathbf{E}$ throughout the body by imposing the BC $\mathbf{Tn}=-p\mathbf{n}$ and zero body loads ($\mathbf{n}$ denotes the normal to $\partial\boldsymbol{\chi}(\Omega)$).

Let us now consider a weaker hypothesis, viz.

$$\int_\Omega dL^3\,[\hat{S}_k^{\alpha}(\mathbf{QF},\xi)-\hat{S}_k^{\alpha}(\mathbf{F},\xi)][(\mathbf{QF})^k{}_\alpha-F^k{}_\alpha]\geqslant 0$$

$\forall\,\mathbf{F}(\cdot)\in\mathscr{L}^p(\Omega)$ (monotonicity of $\tilde{S}$), i.e.

$$\int_\Omega dL^3\,[2\hat{S}_k^{\alpha}(\mathbf{F},\xi)F^k{}_\alpha-(\mathbf{Q}+{}^t\mathbf{Q})^k{}_l\hat{S}_l^{\alpha}(\mathbf{F},\xi)F^k{}_\alpha]\geqslant 0$$

* According to the definition of monotonicity the constitutive function $\hat{S}_k^{\alpha}$ should be extended to non-physical deformation gradients with $\det\mathbf{F}\leqslant 0$ in such a way that (2.3.6) holds for all $\mathbf{F}$.

$\forall\, \mathbf{F}(\cdot) \in \mathscr{L}^p(\Omega)$ or for $\mathbf{F} = \nabla\boldsymbol{\chi}_0 + \nabla\mathbf{u},\ \mathbf{u} \in \mathscr{V}$. For $\mathbf{Q} = \exp(t\mathbf{W})$, ${}^t\mathbf{Q} = \exp(-t\mathbf{W}), -t^2 \int_\Omega dL^3\, \hat{T}^{kl}(\mathbf{F}, \xi)\, W^k{}_p W^p{}_l + 0(t^3) \geqslant 0$. Now $W^k{}_p = \in_{kpr} w_r$,

$$W^k{}_p W^p{}_l = -\delta^k{}_l \mathbf{w}^2 + w^k w^l.$$

Hence

$$\int_\Omega dL^3\, [T^{kk}\mathbf{w}^2 - T^{kl} w^k w^l] \geqslant 0.$$

Let $\mathbf{w}$ be an eigenvector of $\overline{T}^{kl} := \int_\Omega dL^3\, T^{kl} / L^3(\Omega)$, corresponding to the eigenvalue $\bar{t}_1$. It follows that

$$\sum_{r=1}^{3} \bar{t}_r - \bar{t}_1 \geqslant 0.$$

Hence again at most one average principal stress $\bar{t}_i$ is negative. This excludes some BC (e.g. the one discussed above).

It may however be the case that the relative deformation $\mathbf{Q}$ is forbidden by the BC defining $\mathscr{V}$: $\mathbf{u}(\xi) \equiv \boldsymbol{\chi}(\xi) - \boldsymbol{\chi}_0(\xi) = \mathbf{0}$ for $\xi \in \Gamma_1$. Let $\mathbf{QF} = \nabla\mathbf{v}(\xi) + \nabla\boldsymbol{\chi}_0(\xi)$, $\mathbf{F} = \nabla\mathbf{u}(\xi) + \nabla\boldsymbol{\chi}_0(\xi)$, $\mathbf{v}(\xi) = \mathbf{Qu}(\xi) + (\mathbf{Q}-\mathbf{E})\boldsymbol{\chi}_0(\xi)$ and $\mathbf{v}(\xi) = (\mathbf{Q}-\mathbf{E})\boldsymbol{\chi}_0(\xi)$ for $\xi \in \Gamma_1$. Hence $\boldsymbol{\chi}_0(\xi) = \lambda(\xi)\mathbf{w}$ and $\boldsymbol{\chi}_0(\Gamma_1)$ lies on a straight line.* If the Dirichlet data $\boldsymbol{\chi}_0$ can be extended to a bi-Lipschitz mapping $\boldsymbol{\chi}_0$, then the above conclusion is in contradiction with the assumption that $H^2(\Gamma_1) > 0$. Hence in the case of $H^2(\Gamma_1) > 0$ there is no apparent contradiction between monotonicity of A and rotational invariance of the equations.

Let us now consider the case of $\Gamma_1 = \emptyset$, $\int_\Omega dL^3\, \mathbf{u} = \mathbf{0} = \int_\Omega dL^3\, \mathbf{v}$. We can choose $\boldsymbol{\chi}_0$ in such a way that $\mathbf{a} = \int_\Omega dL^3\, \boldsymbol{\chi}_0 = \mathbf{0}$** and hence $\int_\Omega dL^3\, \mathbf{v} = \mathbf{Q}\int_\Omega dL^3\, \mathbf{u} = \mathbf{0}$ for all $\mathbf{Q} \in SO_+(3)$. Hence in this case monotonicity of A is not ensured by the definition of $\mathscr{V}$.

Coercivity amounts to

$$||\nabla\mathbf{u}||_p^{-1} \int_\Omega dL^3\, \hat{S}_k^\alpha(\nabla\boldsymbol{\chi}_0(\xi) + \nabla\mathbf{u}(\xi), \xi)\, u^k{}_{,\xi^\alpha} \to \infty$$

for $||\nabla\mathbf{u}||_p \to \infty$. The integral is $\mathbf{0}(||\nabla\mathbf{u}||_p{}^p)$. It is also non-negative if the monotonicity assumption is satisfied and $\hat{S}_k^\alpha(\nabla\boldsymbol{\chi}_0(\xi), \xi) = 0$ a.e. in Ω.

* The possibility of a **variable** $\mathbf{Q}(\xi)$ can be excluded by the argument of Section 4.2.1.

** Alternatively, let $\mathbf{v} + \boldsymbol{\chi}_0$ differ from $\mathbf{u} + \boldsymbol{\chi}_0$ by a rigid displacement, $\mathbf{v} + \boldsymbol{\chi}_0 = \mathbf{Q}(\mathbf{u} + \boldsymbol{\chi}_0) + \mathbf{b}$. We have then

$$\int_\Omega dL^3 \mathbf{v} = \mathbf{Q} \int_\Omega dL^3 \mathbf{u} + (\mathbf{Q}-\mathbf{E}) \int_\Omega dL^3 \boldsymbol{\chi}_0 + L^3(\Omega)\mathbf{b}$$

For every $\mathbf{Q} \in SO(3)$ there is a $\mathbf{b} \in \boldsymbol{R}^3$ such that $\mathbf{u} \in \mathscr{V}$ entails $\mathbf{v} \in \mathscr{V}$

Let us now consider the case (2.2.2) and (2.2.3), $p > 1$. The mapping Φ: $\mathbf{w} \mapsto |\mathbf{w}|^{p-2}\mathbf{w}$ is a duality mapping $\mathscr{L}^p(\Gamma_3) \to \mathscr{L}^q(\Gamma_3)$, $q = \dfrac{p}{p-1}$, viz.

(i) $\| \, |\mathbf{w}|^{p-2}\mathbf{w}\|_q = \left[\int_{\Gamma_3} dH^2 \, |w|^p\right]^{1/q} = \|\mathbf{w}\|_p^{p/q} = \|\mathbf{w}\|_p^{p-1}$ is an increasing function of $\|\mathbf{w}\|_p$, growing from 0 to ∞;

(ii) $\langle |\mathbf{w}|^{p-2}\mathbf{w}|\mathbf{w}\rangle = \int_{\Gamma_3} dH^2 \, |\mathbf{w}|^p = \|\mathbf{w}\|_p^p = \|\mathbf{w}\|_p \|\mathbf{w}\|_p^{p/q}$

$= \|\mathbf{w}\|_p \| \, |\mathbf{w}|^{p-2}\mathbf{w}\|_q$.

Hence

$$\begin{aligned}
&\langle \Phi(\mathbf{w}_1) - \Phi(\mathbf{w}_2)|\mathbf{w}_1 - \mathbf{w}_2\rangle \\
&\geqslant \|\Phi(\mathbf{w}_1)\|_q \|\mathbf{w}_1\|_p + \|\Phi(\mathbf{w}_2)\|_q \|\mathbf{w}_2\|_p - \|\Phi(\mathbf{w}_1)\|_q \|\mathbf{w}_2\|_p \\
&- \|\Phi(\mathbf{w}_2)\|_q \|\mathbf{w}_1\|_p \\
&= [\|\Phi(\mathbf{w}_1)\|_q - \|\Phi(\mathbf{w}_2)\|_q][\|\mathbf{w}_1\|_p - \|\mathbf{w}_2\|_p] \geqslant 0.
\end{aligned}$$

It follows immediately that the operator S_0, and hence A, is monotone provided $\tilde{S}$ is monotone in the sense of equation (2.3.5).

The duality mapping Φ is demicontinuous. More generally, we have

Lemma 2.3.5

Let $\mathscr{V}$ be a reflexive Banach space with a strictly convex dual $\mathscr{V}^*$.* Let Φ: $\mathscr{V} \to \mathscr{V}^*$ be a duality mapping, viz.

(i′) $\|\Phi(\mathbf{u})\|_{\mathscr{V}^*} = \phi(\|\mathbf{u}\|_{\mathscr{V}})$, with a continuous strictly increasing ϕ, $\phi(0) = 0$, $\phi(s) \to \infty$ for $s \to \infty$;

(ii′) $\langle \Phi(\mathbf{u})|\mathbf{u}\rangle = \|\Phi(\mathbf{u})\|_{\mathscr{V}^*} \|\mathbf{u}\|_{\mathscr{V}}$.

Then Φ is demicontinuous.

Proof

Let $\mathbf{u}_n \to \mathbf{u}_0$, $\|\mathbf{u}_n\|_{\mathscr{V}} \leqslant M$. It follows that $\|\Phi(\mathbf{u}_n)\|_{\mathscr{V}^*} \leqslant \phi(M)$ and a subsequence $\Phi(\mathbf{u}_{n_k})$ converges weakly to some $\mathbf{v} \in \mathscr{V}^*$,

$$\langle \Phi(\mathbf{u}_{n_k})|\mathbf{u}_{n_k}\rangle \to \langle \mathbf{v}|\mathbf{u}_0\rangle \leqslant \|\mathbf{v}\|_{\mathscr{V}^*} \|\mathbf{u}_0\|_{\mathscr{V}}.$$

The left-hand side is equal to $\|\Phi(\mathbf{u}_{n_k})\|_{\mathscr{V}^*} \|\mathbf{u}_{n_k}\|_{\mathscr{V}}$ and converges to the limit $\phi(\|\mathbf{u}_0\|_{\mathscr{V}}) \|\mathbf{u}_0\|_{\mathscr{V}}$. Hence

$$\phi(\|\mathbf{u}_0\|_{\mathscr{V}}) \leqslant \|\mathbf{v}\|_{\mathscr{V}^*}, \quad \langle \mathbf{v}|\mathbf{u}_0\rangle = \phi(\|\mathbf{u}_0\|_{\mathscr{V}}) \|\mathbf{u}_0\|_{\mathscr{V}}.$$

Also, for every $\mathbf{w} \in \mathscr{V}$,

$$\langle \Phi(\mathbf{u}_{n_k})|\mathbf{w}\rangle \to \langle \mathbf{v}|\mathbf{w}\rangle$$

and

$$\langle \Phi(\mathbf{u}_{n_k})|\mathbf{w}\rangle \leqslant \phi(\|\mathbf{u}_0\|_{\mathscr{V}}) \|\mathbf{w}\|_{\mathscr{V}}.$$

* This means that $\|\mathbf{u}\|_{\mathscr{V}} \leqslant 1$, $\|\mathbf{v}\|_{\mathscr{V}} \leqslant 1$, $\mathbf{u} \neq \mathbf{v}$ imply that $\|\mathbf{u}+\mathbf{v}\|_{\mathscr{V}} < 2$.

Hence $||\mathbf{v}||_{\mathscr{V}^*} \leqslant \phi(||\mathbf{u}_0||_{\mathscr{V}})$ and the results of the preceding paragraph imply that $||\mathbf{v}||_{\mathscr{V}^*} = \phi(||\mathbf{u}_0||_{\mathscr{V}})$.

Since $\langle \mathbf{v}|\mathbf{u}_0\rangle = ||\mathbf{v}||_{\mathscr{V}^*}\phi(||\mathbf{u}_0||_{\mathscr{V}})$, it follows that $\mathbf{v} = \Phi(\mathbf{u}_0)$. Indeed, if $\mathbf{v} \neq \Phi(\mathbf{u}_0)$, then

$$||\mathbf{u}_0||_{\mathscr{V}}||\mathbf{v}+\Phi(\mathbf{u}_0)||_{\mathscr{V}^*} \geqslant \langle \mathbf{v}+\Phi(\mathbf{u}_0)|\mathbf{u}_0\rangle = [||\mathbf{v}||_{\mathscr{V}^*}+\phi(||\mathbf{u}_0||_{\mathscr{V}})]||\mathbf{u}_0||_{\mathscr{V}}$$

and $||\mathbf{v}+\Phi(\mathbf{u}_0)||_{\mathscr{V}^*} \geqslant ||\mathbf{v}||_{\mathscr{V}^*}+||\Phi(\mathbf{u}_0)||_{\mathscr{V}^*}$. Hence $\mathbf{v} = \Phi(\mathbf{u}_0)$ on account of the strict convexity of $\mathscr{V}^*$. □

Let $\mathscr{V} = \mathscr{L}^p$, $p > 1$. Then $\mathscr{V}^* = \mathscr{L}^q$ is strictly convex by a sharpened version of the Minkowski inequality.

Coercivity amounts to

$$\int_\Omega dL^3\, \hat{S}_k^{\alpha}(\nabla\boldsymbol{\chi}_0+\nabla\mathbf{u},\xi)u^k{}_{,\xi^\alpha} + \int_{\Gamma_3} dH^2\, \nu|\mathbf{u}|^p - \int_{\Gamma_2} dH^2\, \mathbf{s}\cdot\gamma\mathbf{u} \geqslant \phi(||\mathbf{u}||)||\mathbf{u}||.$$

On account of (2.2.14) the latter inequality is satisfied for $p > 1$ and $\nu > 0$ H^2-a.e. on Γ_3. For $\Gamma_3 = \emptyset$ the proof of coercivity is even simpler.

In nonlinear elastostatics a complete set of BC neednot determine a unique solution. This is true even if we define the solution modulo rigid displacements. Let $\boldsymbol{\chi}_1$ be an equilibrium configuration under the given BC. If the data are momentarily perturbed in such a way that the body leaves the original equilibrium and then is allowed to relax under the original BC it may go to a different equilibrium configuration. (We have tacitly assumed that a mechanism restoring the equilibrium acts in the body although no such mechanism is accounted for by the theory of elasticity). Also, if the loads or Dirichlet BC vary continuously a hitherto stable equilibrium configuration may lose its stability while another kind of equilibrium configurations becomes stable. This phenomenon occurs at the bifurcation points in the space of solutions and load parameters. Hence it is not always justified to require uniqueness of elastic equilibrium configurations. At bifurcation points continuous dependence on the loads also breaks down. On account of Theorem 2.3.4 this implies that strict monotonicity may be untenable even if monotonicity is acceptable.

2.3.2 Monotone Potential Operators in Elastostatics

Definition 2.3.6

An operator $A\colon \mathscr{D} \to \mathscr{V}^*$, defined on a dense linear subset $\mathscr{D}$ of a Banach space $\mathscr{V}$, is said to be **potential** if there is a functional $f\colon \mathscr{D} \to \boldsymbol{R}$ such that

$$\left.\frac{df(\mathbf{u}+t\mathbf{v})}{dt}\right|_{t=0} = \langle A\mathbf{u}|\mathbf{v}\rangle \;\forall \mathbf{u}, \mathbf{v} \in \mathscr{D}. \qquad (2.3.8)$$

Equation (2.3.8) states that f has a Gâteaux derivative $Df(\mathbf{u})[\mathbf{v}]$, which is a (linear and) continuous functional of $\mathbf{v} \in \mathscr{D}$. The right-hand side of (2.3.8) is defined for all $\mathbf{v} \in \mathscr{V}$, hence $A\mathbf{u}$ is the unique extension of $Df(\mathbf{u})$ by continuity.

Let us assume that $W(\cdot\,,\,\cdot)$ is a Carathéodory function. Since $\hat{S}_k{}^\alpha(\mathbf{F},\xi) = \varrho_0 \dfrac{\partial W(\mathbf{F},\xi)}{\partial F^k{}_\alpha}$, it follows by a theorem of Vainberg (Vainberg, 1956, Section 21) that $\tilde{\mathbf{S}}$ is a potential operator. Also

$$\varrho_0 |W(\mathbf{F},\xi) - W(\mathbf{F}_0,\xi)| = \varrho_0 \left| \int_0^1 dt \frac{d}{dt} W(\mathbf{F}_0 + t\mathbf{H},\xi) \right|$$

$$= \varrho_0 \left| \int_0^1 dt\, H^k{}_\alpha \frac{\partial W}{\partial F^k{}_\alpha} (\mathbf{F}_0 + t\mathbf{H},\xi) \right| \leqslant \varrho_0 |\mathbf{H}| \left| a + K \int_0^1 dt\, t^{p-1} |\mathbf{H}|^{p-1} \right|$$

$$\leqslant \varrho_0 \left[a|\mathbf{H}| + \frac{K}{p} |\mathbf{H}|^p \right],$$

on account of (2.2.6). For $p \geqslant 1$ and $\xi \mapsto W(\mathbf{F}_0(\xi),\xi)$ in $\mathscr{L}^1(\Omega)$ it implies that $\varrho_0 W(\mathbf{F}(\cdot),\,\cdot) \in \mathscr{L}^1(\Omega)$ provided $\mathbf{H} \in \mathscr{L}^p(\Omega)$.

The potential $f(\mathbf{F}(\cdot))$ is given by the expression

$$\int_\Omega dL^3\, \varrho_0\, W(\mathbf{F}(\xi),\xi).$$

Indeed,

$$I := \frac{1}{t} \int_\Omega dL^3\, \{\varrho_0(\xi)[W(\mathbf{F}_0(\xi) + t\mathbf{H}(\xi),\xi) - W(\mathbf{F}_0(\xi),\xi)]$$

$$- t\hat{S}_k{}^\alpha(\mathbf{F}_0(\xi),\xi) H^k{}_\alpha\}$$

$$= \int_\Omega dL^3 \int_0^1 ds\, [\hat{S}_k{}^\alpha(\mathbf{F}_0(\xi) + st\mathbf{H}(\xi),\xi) - \hat{S}_k{}^\alpha(\mathbf{F}_0(\xi),\xi)] H^k{}_\alpha.$$

The function $s\mathbf{H}(\xi)$ and hence the integrand is a measurable function of (s,ξ). Hence by the Fubini theorem we obtain another expression for I

$$\int_0^1 ds \int_\Omega dL^3\, [\hat{S}_k{}^\alpha(\mathbf{F}_0(\xi) + st\mathbf{H}(\xi),\xi) - \hat{S}_k{}^\alpha(\mathbf{F}_0(\xi),\xi)] H^k{}_\alpha$$

$$= \int_0^1 ds\, \langle \tilde{S}(\mathbf{F}_0 + st\mathbf{H}) - \tilde{S}(\mathbf{F}_0) | \mathbf{H} \rangle.$$

In view of the radial continuity of $\tilde{S}$ the integrand is a continuous function of s and tends to zero boundedly for $t \to 0$. (Since $\tilde{S}$ maps bounded sets onto bounded sets, $|\langle\tilde{S}(\mathbf{F}_0+st\mathbf{H})-S(\mathbf{F}_0)|\mathbf{H}\rangle| \leqslant M\|\mathbf{H}\|_p$, with M independent of s, t for $0 \leqslant s$, $t \leqslant 1$.) This proves our claim.

It is easy to see that $A = L^\dagger \circ \tilde{S} \circ L$ is also a potential operator. Indeed,

$$\lim_{t\to 0}\frac{1}{t}[f(L\mathbf{u}+\mathbf{F}_0+tL\mathbf{v})-f(L\mathbf{u}+\mathbf{F}_0)] = Df(\mathbf{F}_0+L\mathbf{u})[L\mathbf{v}]$$

$$= \langle\tilde{S}(L\mathbf{u}+\mathbf{F}_0)|L\mathbf{v}\rangle = \langle A\mathbf{u}|\mathbf{v}\rangle.$$

The potential for A is

$$f(\mathbf{F}_0+L\mathbf{u}) \equiv \int_\Omega dL^3\,\varrho_0(\xi)\,W(\mathbf{F}_0(\xi)+\nabla\mathbf{u}(\xi),\,\xi). \tag{2.3.9}$$

It is not difficult to construct potentials for $H^2(\Gamma_3) > 0$, $g \neq 0$. The potential for $\mathbf{w}$: $\mathbf{w} \mapsto |\mathbf{w}|^{p-2}\mathbf{w}$, $\mathbf{w} \in \mathscr{L}^p(\Gamma_3)$, is

$$\frac{1}{p}\int_{\Gamma_3} dH^2\,|\mathbf{w}|^p.$$

We shall now prove this fact. Let $J(\mathbf{u}) :\equiv \|\mathbf{u}\|_p^{2-p}|\mathbf{u}|^{p-2}\mathbf{u}$. We have then

$$\langle J(\mathbf{u})|\mathbf{u}\rangle = \|\mathbf{u}\|_p^2 = \|J(\mathbf{u})\|_q^2.$$

Hence

$$\langle J(\mathbf{u})|\mathbf{u}-\mathbf{v}\rangle \geqslant \|\mathbf{u}\|_p^2-\|\mathbf{u}\|_p\|\mathbf{v}\|_p \geqslant \|\mathbf{u}\|_p^2-\frac{1}{2}(\|\mathbf{u}\|_p^2+\|\mathbf{v}\|_p^2)$$

$$= \frac{1}{2}\|\mathbf{u}\|_p^2-\frac{1}{2}\|\mathbf{v}\|_p^2 = \frac{1}{2}(\|\mathbf{u}\|_p^2+\|\mathbf{v}\|_p^2)-\|\mathbf{v}\|_p^2$$

$$\geqslant \|\mathbf{u}\|_p\|\mathbf{v}\|_p-\|\mathbf{v}\|_p^2 \geqslant \langle J(\mathbf{v})|\mathbf{u}-\mathbf{v}\rangle,$$

i.e. J is monotone. Two of the above chain of inequalities yield for $\mathbf{u} = \mathbf{v}+t\mathbf{w}$ the inequality

$$t\langle J(\mathbf{v})|\mathbf{w}\rangle \leqslant \frac{1}{2}\|\mathbf{v}+t\mathbf{w}\|_p^2-\frac{1}{2}\|\mathbf{v}\|_p^2 \leqslant t\langle J(\mathbf{v}+t\mathbf{w})|\mathbf{w}\rangle.$$

For $t > 0$

$$\langle J(\mathbf{v})|\mathbf{w}\rangle \leqslant \frac{1}{t}\left[\frac{1}{2}\|\mathbf{v}+t\mathbf{w}\|_p^2-\frac{1}{2}\|\mathbf{v}\|_p^2\right] \leqslant \langle J(\mathbf{v}+t\mathbf{w})|\mathbf{w}\rangle.$$

The operator J is radially continuous and in the limit $t \to 0$ the right-hand side of the above inequality tends to the left-hand side; hence

$$\lim_{t\to 0}\frac{1}{t}\left[\frac{1}{2}\|\mathbf{v}+t\mathbf{w}\|_p^2-\frac{1}{2}\|\mathbf{v}\|_p^2\right] = \langle J(\mathbf{v})|\mathbf{w}\rangle, \quad J(\mathbf{v}) \in \mathscr{L}^q(\Gamma_3),$$

and therefore $D\left(\frac{1}{p}\,||\mathbf{v}||_p{}^p\right)[\mathbf{w}] = \langle J(\mathbf{v})|\mathbf{w}\rangle$. Working out the last formula we have

$$D\left(\frac{1}{p}\,||\mathbf{v}||_p{}^p\right)[\mathbf{w}] = \langle|\mathbf{v}|^{p-2}\mathbf{v}|\mathbf{w}\rangle. \tag{2.3.10}$$

The potential of the operator $A = L^{\dagger}\circ S_0 \circ L$, equation (2.2.15), is given by

$$P(\mathbf{u}) = \int_{\Omega} dL^3\, \varrho_0\, W(\mathbf{F}_0+\nabla\mathbf{u},\, \xi)+\int_{\Gamma_3} dH^2\, \nu|\gamma\mathbf{u}|^p - \int_{\Gamma_2} dH^2\, s_k\gamma u^k. \tag{2.3.11}$$

Indeed, $DP(\mathbf{u})[\mathbf{v}] = \langle A\mathbf{u}|\mathbf{v}\rangle$. Hence equation (2.3.4) is equivalent to the assertion that $\mathbf{u}$ is a stationary point of

$$P_{\mathbf{f}}(\mathbf{u}) := P(\mathbf{u})-\langle\mathbf{f}\,|\mathbf{u}\rangle.$$

In view of the symmetry of $B(\mathbf{u}, \mathbf{v})$ it is also clear that equation (2.2.32) expresses the condition that $\mathbf{u}$ is a stationary point of the functional

$$\mathbf{u}\mapsto\frac{1}{2}\,B(\mathbf{u}, \mathbf{u})-f(\mathbf{u}).$$

Monotone potential operators have the following remarkable property

Theorem 2.3.7
Every monotone potential operator is demicontinuous.

For the proof see Vainberg (1972) or Gajewski *et al.* (1974), Lemma 3.4.12. We do not need this theorem here since we have proved the demicontinuity of A in a different way. In view of Theorem 2.3.9 below it shows however how strong the theory of monotone potential operators is in comparison with the variational theory presented in Section 2.4.

Definition 2.3.8
A functional f on a linear space $\mathscr{V}$ (or on a convex subset $\mathscr{D}$ of a linear space $\mathscr{V}$) is said to be **convex** if $f(\nu\mathbf{u}+(1-\nu)\mathbf{v}) \leqslant \nu f(\mathbf{u})+(1-\nu)f(\mathbf{v})$ for $\nu\in[0, 1]$, and **strictly convex** if

$$f(\nu\mathbf{u}+(1-\nu)\mathbf{v}) < \nu f(\mathbf{u})+(1-\nu)f(\mathbf{v}) \quad \text{for } \nu\in\,]0, 1[,\ \mathbf{u}\neq\mathbf{v}.$$

Theorem 2.3.9
A potential operator A is monotone (strictly monotone) iff its potential f is convex (strictly convex).

Proof
Let A be a monotone potential operator. Then

$$\begin{aligned}\varDelta &:= \nu f(\mathbf{u})+(1-\nu)f(\mathbf{v})-f(\nu\mathbf{u}+(1-\nu)\mathbf{v})\\ &= \nu[f(\mathbf{u})-f(\nu\mathbf{u}+(1-\nu)\mathbf{v})]+(1-\nu)[f(\mathbf{v})-f(\nu\mathbf{u}+(1-\nu)\mathbf{v})].\end{aligned}$$

Applying the Lagrange formula as well as the definition of the Gâteaux derivative we find that

$$\begin{aligned}\varDelta &= \nu\langle A(\nu\mathbf{u}+(1-\nu)\mathbf{v}-\lambda_1(1-\nu)\mathbf{v}-\mathbf{u})|(1-\nu)(\mathbf{u}-\mathbf{v})\rangle\\ &\quad+(1-\nu)\langle A(\nu\mathbf{u}+(1-\nu)\mathbf{v}+\lambda_2\nu(\mathbf{v}-\mathbf{u}))|\nu(\mathbf{v}-\mathbf{u})\rangle\end{aligned}$$

for some $0 < \lambda_1, \lambda_2 < 1$ Hence

$$\gamma\varDelta = \nu(1-\nu)\langle A(\mathbf{x})-A(\mathbf{y})|\mathbf{x}-\mathbf{y}\rangle \geqslant 0$$

for $\mathbf{x} = \nu\mathbf{u}+(1-\nu)\mathbf{v}-\lambda_1(1-\nu)(\mathbf{v}-\mathbf{u})$, $\mathbf{y} = \nu\mathbf{u}+(1-\nu)\mathbf{v}+\lambda_2\nu(\mathbf{v}-\mathbf{u})$, $\gamma = \lambda_1(1-\nu)+\lambda_2\nu > 0$. Thus monotonicity of A implies convexity of f and strict monotonicity of A implies strict convexity of f (note that $\mathbf{v} \neq \mathbf{u}$ implies that $\mathbf{x} \neq \mathbf{y}$, since $\lambda_1(1-\nu) > 0$, $\lambda_2\nu > 0$ for $0 < \nu < 1$).

Conversely, convexity (or strict convexity) of f implies that $\varphi(\nu) := f(\mathbf{u}+\nu(\mathbf{v}-\mathbf{u}))$, $0 \leqslant \nu \leqslant 1$, $\mathbf{v} \neq \mathbf{u}$, is convex (or strictly convex, resp.). It is also differentiable and its derivative φ' is non-decreasing (or increasing, resp.). Since

$$\varphi'(\nu) = \langle A(\mathbf{u}+\nu(\mathbf{v}-\mathbf{u}))|\mathbf{v}-\mathbf{u}\rangle,$$

$$\langle A(\mathbf{v})-A(\mathbf{u})|\mathbf{v}-\mathbf{u}\rangle = \varphi'(1)-\varphi'(0) \geqslant 0 \quad (\text{or } > 0).$$ □

2.3.3 Proof of the Minty–Browder Theorem

Lemma 2.3.10
Let $A: \mathscr{V} \to \mathscr{V}^*$ be monotone.

Then $\forall\, \mathbf{u} \in \mathscr{V}\ \exists\, \varepsilon > 0$, $M > 0$ such that $||\mathbf{v}-\mathbf{u}||_{\mathscr{V}} \leqslant \varepsilon$ entails $||A\mathbf{v}||_{\mathscr{V}^*} \leqslant M$.

Proof
Suppose that the thesis is false. Then $\exists\, \mathbf{u} \in \mathscr{V}$ such that for $\varepsilon = 1/n$ arbitrarily small there are elements $\mathbf{u}_n$ of $\mathscr{V}$ such that $||\mathbf{u}-\mathbf{u}_n||_{\mathscr{V}} < 1/n$ but $||Au_n||_{\mathscr{V}^*}$ is arbitrarily large. We now construct a sequence $\{\mathbf{u}_n\}$ such that $||\mathbf{u}-\mathbf{u}_n||_{\mathscr{V}} < 1/n$ and $||A\mathbf{u}_n||_{\mathscr{V}^*} \to \infty$. Let $\lambda_n := 1+||A\mathbf{u}_n||_{\mathscr{V}^*}||\mathbf{u}_n-\mathbf{u}||_{\mathscr{V}}$. Since A is monotone, we have for arbitrary $\mathbf{w} \in \mathscr{V}$

$$\langle A(\mathbf{u}_n)-A(\mathbf{u}+\mathbf{w})|\mathbf{u}_n-\mathbf{u}-\mathbf{w}\rangle \geqslant 0.$$

Hence

$$\frac{1}{\lambda_n}\langle A(\mathbf{u}_n)|\mathbf{w}\rangle \leqslant \frac{1}{\lambda_n}\Big\{\langle A(\mathbf{u}_n)|\mathbf{u}_n-\mathbf{u}\rangle-\langle A(\mathbf{u}+\mathbf{w})|\mathbf{u}_n-\mathbf{u}-\mathbf{w}\rangle\Big\}$$

$$\leqslant 1+\frac{1}{\lambda_n}\,||A(\mathbf{u}+\mathbf{w})||_{\mathscr{V}^*}\{||\mathbf{u}_n-\mathbf{u}||_{\mathscr{V}}+||\mathbf{w}||_{\mathscr{V}}\} \leqslant K_1(\mathbf{u},\mathbf{w}).$$

Replacing $\mathbf{w}$ by $-\mathbf{w}$ we obtain a lower estimate

$$\lambda_n^{-1}\langle A\mathbf{u}_n|\mathbf{w}\rangle \geqslant K_2(\mathbf{u},\mathbf{w}).$$

Hence

$$\left|\frac{1}{\lambda_n}\langle A\mathbf{u}_n|\mathbf{w}\rangle\right| \leqslant M(\mathbf{u},\mathbf{w}) \quad \text{for all } \mathbf{w}\in\mathscr{V}.$$

By the Banach–Steinhaus (or resonance) theorem

$$\frac{1}{\lambda_n}\,||A\mathbf{u}_n||_{\mathscr{V}^*} \leqslant M(\mathbf{u}) \quad \text{for some } M(\mathbf{u})>0.$$

Hence

$$||A\mathbf{u}_n||_{\mathscr{V}^*} \leqslant M(\mathbf{u})[1+||A\mathbf{u}_n||_{\mathscr{V}^*}||\mathbf{u}-\mathbf{u}_n||_{\mathscr{V}}].$$

Choose n_o so large that for $n \geqslant n_o M||\mathbf{u}-\mathbf{u}_n||_{\mathscr{V}} < 1$. Hence for $n \geqslant n_o ||A\mathbf{u}_n||_{\mathscr{V}^*} \leqslant 2M(\mathbf{u}) < \infty$ which contradicts our hypothesis that $||A\mathbf{u}_n||_{\mathscr{V}^*} \to \infty$. □

Lemma 2.3.11

If A is monotone and radially continuous then it is demicontinuous.

Proof (in three steps)

(1) We prove that $\langle \mathbf{f}-A\mathbf{w}|\mathbf{u}-\mathbf{w}\rangle \geqslant 0 \;\; \forall\, \mathbf{w}\in\mathscr{V}$ implies that $A\mathbf{u}=\mathbf{f}$.

Set $\mathbf{w}(t)=\mathbf{u}-t\mathbf{w}$, $t>0$. Assuming the premises of the implication above, we have $\langle \mathbf{f}-A\mathbf{w}(t)|\mathbf{w}\rangle \geqslant 0$. Letting $t\to 0$ we conclude from radial continuity that $\langle \mathbf{f}-A\mathbf{u}|\mathbf{w}\rangle \geqslant 0$. Since $\mathbf{w}$ is arbitrary, $A\mathbf{u}=\mathbf{f}$, q.e.d.

(2) We shall prove that $\mathbf{u}_n \rightharpoonup \mathbf{u}$ in V, $A\mathbf{u}_n \to \mathbf{f}$ in $\mathscr{V}^*$ and $\limsup_{n\to\infty}\langle A\mathbf{u}_n|\mathbf{u}_n\rangle \leqslant \langle \mathbf{f}|\mathbf{u}\rangle$ imply $A\mathbf{u}=\mathbf{f}$.

NB. This implication is referred to as the **property** (M) of A.

Let the premises of the above implication be satisfied and let $\mathbf{w}\in\mathscr{V}$. Then

$$\begin{aligned}\langle \mathbf{f}-A\mathbf{w}|\mathbf{u}-\mathbf{w}\rangle &= \langle \mathbf{f}|\mathbf{u}\rangle-\langle \mathbf{f}|\mathbf{w}\rangle-\langle A\mathbf{w}|\mathbf{u}-\mathbf{w}\rangle \\ &\geqslant \limsup_{n\to\infty}[\langle A\mathbf{u}_n|\mathbf{u}_n\rangle-\langle \mathbf{f}|\mathbf{w}\rangle-\langle A\mathbf{w}|\mathbf{u}-\mathbf{w}\rangle] \\ &\geqslant \limsup_{n\to\infty}[\langle A\mathbf{u}_n|\mathbf{u}_n\rangle-\langle A\mathbf{u}_n|\mathbf{w}\rangle-\langle A\mathbf{w}|\mathbf{u}_n-\mathbf{w}\rangle] \\ &= \limsup_{n\to\infty}\langle A\mathbf{u}_n-A\mathbf{w}|\mathbf{u}_n-\mathbf{w}\rangle \geqslant 0.\end{aligned}$$

From (1) it follows that $A\mathbf{u}=\mathbf{f}$.

(3) We prove finally that A is demicontinuous.

Let $\mathbf{u}_n \to \mathbf{u}$ in $\mathscr{V}$. By Lemma 2.3.10 $||A\mathbf{u}_n||_{\mathscr{V}^*}$ is bounded and a subsequence $\{A\mathbf{u}_{n_k}\}$ converges weakly to some $\mathbf{f} \in \mathscr{V}^*$. Hence

$$\lim_{k\to\infty} \langle A\mathbf{u}_{n_k}|\mathbf{u}_{n_k}\rangle = \langle \mathbf{f}|\mathbf{u}\rangle .$$

From (2) it follows that $A\mathbf{u} = \mathbf{f}$, i.e. $A\mathbf{u}_{n_k} \to A\mathbf{u}$. We have proved that (a) there is at least one subsequence $\{\mathbf{u}_{n_k}\}$ of $\{\mathbf{u}_n\}$ such that $\{A\mathbf{u}_{n_k}\}$ is weakly convergent; (b) every weakly convergent sequence $\{A\mathbf{u}_{n_k}\}$ converges weakly to $A\mathbf{u}$. It remains to prove that $\{A\mathbf{u}_n\}$ converges weakly to $A\mathbf{u}$. If $\{A\mathbf{u}_n\}$ does not converge weakly to $A\mathbf{u}$, then for some subsequence $\{\mathbf{u}'_{n_k}\}$ of $\{\mathbf{u}_n\}$ and $\mathbf{w} \in \mathscr{V}$, $\varepsilon > 0$

$$|\langle A\mathbf{u}'_{n_k}|\mathbf{w}\rangle - \langle A\mathbf{u}|\mathbf{w}\rangle| > \varepsilon . \tag{2.3.12}$$

Since $\{A\mathbf{u}'_{n_k}\}$ is bounded, a subsequence of this sequence is weakly convergent and hence converges weakly to $A\mathbf{u}$, in contradiction of (2.3.12). □

Lemma 2.3.12

Let $\mathbf{B}: \boldsymbol{R}^n \to \boldsymbol{R}^n$ be a continuous mapping such that there is a positive number $r > 0$ with the property that

$$\langle \mathbf{Bx}|\mathbf{x}\rangle \geqslant 0 \quad \text{for } |\mathbf{x}| = r.$$

Then $\exists\, \mathbf{x}_o \in \boldsymbol{R}^n$ such that $|\mathbf{x}_o| \leqslant r$, $\mathbf{Bx}_o = \mathbf{0}$.

Proof

Suppose that $\mathbf{Bx} \neq \mathbf{0}$ for all $|\mathbf{x}| \leqslant r$. Then $\mathbf{x} \mapsto -r|\mathbf{Bx}|^{-1}\mathbf{Bx}$ is a continuous mapping of $\mathscr{K}(\mathbf{0}, r)$ into $\mathscr{K}(\mathbf{0}, r)$. By the Brouwer fixed point theorem there is an $\mathbf{x} \in \mathscr{K}(\mathbf{0}, r)$ such that

$$\mathbf{x} = -r|\mathbf{Bx}|^{-1}\mathbf{Bx}, \qquad |\mathbf{x}| = r,\ \langle \mathbf{Bx}|\mathbf{x}\rangle = -|\mathbf{Bx}|r^{-1}\langle \mathbf{x}|\mathbf{x}\rangle < 0,$$

a contradiction. □

Lemma 2.3.13

Let $A: \mathscr{V} \to \mathscr{V}^*$ be monotone, $||\mathbf{u}||_{\mathscr{V}} \leqslant M_1$, $\langle A\mathbf{u}|\mathbf{u}\rangle \leqslant M_2$ for every $\mathbf{u} \in \mathscr{V}$. Then $||A\mathbf{u}||_{\mathscr{V}^*} \leqslant M\ \forall\ \mathbf{u} \in \mathscr{V}$.

Proof

By Lemma 2.3.10 $\exists\ \varepsilon > 0$, $M_3 > 0$, such that $||\mathbf{u}||_{\mathscr{V}} < \varepsilon$ implies $||A\mathbf{u}||_{\mathscr{V}^*} < M_3$. Hence for $\mathbf{u} \in \mathscr{V}$

$$\begin{aligned}
||A\mathbf{u}||_{\mathscr{V}^*} &= \sup\left\{\frac{1}{\varepsilon}\langle A\mathbf{u}|\mathbf{v}\rangle|\ \ ||\mathbf{v}||_{\mathscr{V}} \leqslant \varepsilon\right\} \\
&\leqslant \sup\left\{\frac{1}{\varepsilon}\left[\langle A\mathbf{u}|\mathbf{u}\rangle - \langle A\mathbf{v}|\mathbf{v}\rangle - \langle A\mathbf{v}|\mathbf{u}\rangle\right]|\ ||\mathbf{v}||_{\mathscr{V}} \leqslant \varepsilon\right\} \\
&\leqslant \frac{1}{\varepsilon}(M_2 + M_3\varepsilon + M_3 M_1) = M,
\end{aligned}$$

using monotonicity. □

Theorem 2.3.14 (Minty–Browder)*
Let $A: \mathscr{V} \to \mathscr{V}^*$ be monotone, radially continuous and coercive, $\mathbf{f} \in \mathscr{V}^*$. Then $\exists\, \mathbf{u} \in \mathscr{V}$ such that $A\mathbf{u} = \mathbf{f}$.

Proof
Let $\{v_n\}$ be a basis of $\mathscr{V}$, assumed to be *separable*, and let $\mathscr{V}_n$ be the subspace spanned by $\{\mathbf{v}_1, \mathbf{v}_2, \ldots, \mathbf{v}_n\}$.

Let $\mathbf{C}_n$ be the linear mapping $\mathbf{R}^n \to \mathscr{V}_n$ defined by

$$\{x^1, \ldots, x^n\} \mapsto \sum_{j=1}^{n} x^j \mathbf{v}_j.$$

Since all the norms on a finite-dimensional space are equivalent,

$$|\mathbf{x}| \leqslant K_n \|\mathbf{C}_n \mathbf{x}\|_{\mathscr{V}}.$$

Let $\mathbf{B}\mathbf{x} := \{\langle A\mathbf{C}_n\mathbf{x} - \mathbf{f} | \mathbf{v}_i \rangle | i = 1, 2, \ldots, n\} \in \mathbf{R}^n$. A is radially continuous and monotone, hence demicontinuous. Hence $\mathbf{B}$ is continuous. The coercivity of A implies that $\exists\, R > 0$ such that

$$\frac{\langle A\mathbf{u}|\mathbf{u}\rangle}{\|\mathbf{u}\|_{\mathscr{V}}} \geqslant \|\mathbf{f}\|_{\mathscr{V}^*} \quad \text{for } \|\mathbf{u}\|_{\mathscr{V}} > R.$$

For $\mathbf{u}_n = \mathbf{C}_n\mathbf{x}$, $\mathbf{x} \in \mathbf{R}^n$, $|\mathbf{x}| = RK_n$ we have $\|\mathbf{u}_n\|_{\mathscr{V}} \geqslant R$ and

$$\langle \mathbf{B}\mathbf{x}|\mathbf{x}\rangle = \sum_{j=1}^{n} \langle A\mathbf{C}_n\mathbf{x} - \mathbf{f}|\mathbf{v}_j\rangle x^j = \langle A\mathbf{C}_n\mathbf{x} - \mathbf{f}|\mathbf{C}_n\mathbf{x}\rangle$$

$$\geqslant \left[\frac{\langle A\mathbf{u}_n|\mathbf{u}_n\rangle}{\|\mathbf{u}_n\|_{\mathscr{V}}} - \|\mathbf{f}\|_{\mathscr{V}^*}\right] \|\mathbf{u}_n\|_{\mathscr{V}} \geqslant 0.$$

By Lemma 2.3.11 $\exists\, \mathbf{x}_o \in \mathbf{R}^n$ $\mathbf{B}\mathbf{x}_o = \mathbf{0}$, i.e.

$$\langle A\mathbf{u}_n - \mathbf{f}|\mathbf{v}_j\rangle = 0, \quad j = 1, 2, \ldots, n. \tag{2.3.13}$$

Taking the linear combination of (2.3.13) with the coefficients x^j we get the formula $\langle A\mathbf{u}_n|\mathbf{u}_n\rangle = \langle \mathbf{f}|\mathbf{u}_n\rangle$. Hence

$$\frac{\langle A\mathbf{u}_n|\mathbf{u}_n\rangle}{\|\mathbf{u}_n\|_{\mathscr{V}}} \leqslant \|\mathbf{f}\|_{\mathscr{V}^*}.$$

By the coercivity $\gamma(\|\mathbf{u}_n\|_{\mathscr{V}}) \leqslant \|\mathbf{f}\|_{\mathscr{V}^*}$ and $\|\mathbf{u}_n\|_{\mathscr{V}} \leqslant M_2$. Hence $\langle A\mathbf{u}_n|\mathbf{u}_n\rangle \leqslant \|\mathbf{f}\|_{\mathscr{V}^*} M_2 = M_1$ with $M_1 > 0$ independent of n. By Lemma 2.3.13 $\|A\mathbf{u}_n\|_{\mathscr{V}^*} \leqslant M$ for some $M > 0$ and all n. Hence there is a subsequence $\{\mathbf{u}_{n_k}\}$

* We prove only the existence of solutions. The proof of the remaining assertions of this theorem can be found in many references, e.g. Gajewski *et al.* (1974).

such that $\mathbf{u}_{n_k} \rightharpoonup \mathbf{u}$, $A(\mathbf{u}_{n_k}) \rightharpoonup \boldsymbol{\chi}$. Since $\{A\mathbf{u}_n\}$ is bounded in $\mathscr{V}^*$ and $\lim_{n\to\infty} \langle A\mathbf{u}_n|\mathbf{w}\rangle = \langle \mathbf{f}|\mathbf{w}\rangle$ for a dense set of $\mathbf{w} \in \mathscr{V}$ on account of (2.3.13), it follows by a theorem of functional analysis (Yosida, 1965, Theorem 5.1.3) that $A\mathbf{u}_n \rightharpoonup \mathbf{f} \in \mathscr{V}^*$.

Since $\{||\mathbf{u}_n||_{\mathscr{V}}\}$ is bounded, a subsequence $\{\mathbf{u}_{n_k}\}$ converges weakly to some $\mathbf{u} \in \mathscr{V}$ (since $\mathscr{V}$ is weakly closed),

$$\lim_{k\to\infty}\langle A\mathbf{u}_{n_k}|\mathbf{u}_{n_k}\rangle = \lim_{k\to\infty}\langle \mathbf{f}|\mathbf{u}_{n_k}\rangle = \langle \mathbf{f}|\mathbf{u}\rangle .$$

By the property (M), cf. the assertion (2) in the proof of Lemma 2.3.11, $A\mathbf{u} = f$, q.e.d. □

Remark

Monotonicity is relevant only for the proofs of demicontinuity and the property (M). This observation suggests the possibility of replacing monotone radially continuous operators by a larger class of demicontinuos operators with the property (M). An important subclass of this class is the class of **operators of variational calculus** (Lions 1969). An operator of variational calculus has the form $A\mathbf{u} = A_1(\mathbf{u}, \mathbf{u})$ with $A_1(\mathbf{u}, \cdot)$, $A_1(\cdot, \mathbf{u})$ hemicontinuous, $A_1(\mathbf{u}, \cdot)$ monotone and some additional properties. In practice the second argument of A_1 represents the dependence on the highest order derivatives of $\mathbf{u}$ (on $\mathbf{F}$ in the case of elastostatics), while the lower order derivatives of $\mathbf{u}$ are represented by the first argument of A_1. The requirement of monotonicity affects the dependence of A on the highest order derivatives only. In the context of elastostatics we might consider $S_k^{\alpha} = \hat{S}_k^{\alpha}(\mathbf{F}, \boldsymbol{\chi}, \xi)$ with the usual conditions on $\hat{S}_k^{\alpha}(\cdot, \boldsymbol{\chi}, \xi)$.

2.4 VARIATIONAL THEORY OF ELASTOSTATICS. PRELIMINARIES

2.4.1 Introduction

It follows from Section 2.3 that the solutions of elastostatic BVPs are stationary points of certain Gâteaux-differentiable functionals, representing the energy of the strains plus the energy of the loads Φ:

$$\Phi(\boldsymbol{\chi}) := -\int_{\Omega} dL^3\, \varrho_0\, b_k \chi^k - \int_{\Gamma_2} dH^2\, s_k \gamma \chi^k .$$

In Chapter 3 we shall see that the energy of an elastic body in motion $E(t) = K(t) + \mathscr{W}(t) + \Phi(t)$, with a non-negative kinetic energy $K(t)$ vanishing only at rest, is a non-increasing function of time. Hence we can expect that under some circumstances the body can approach some static configuration corresponding to a minimum of $\mathscr{W}+\Phi$. A static configuration which corresponds to a saddle point, maximum or inflexion of $\mathscr{W}+\Phi$ is presumably an unstable equilibrium. A transition of the body from a stable equilibrium (= proper minimum of $\mathscr{W}+\Phi$) to another static configuration is possible only if sufficient energy is supplied to ensure a temporary increase of $\mathscr{W}+\Phi$ and a $K(0) > 0$.

The basic idea of this section is to develop a more acceptable mathematical framework of elastostatics by a recourse to direct variational methods. We shall thus be able to get rid of the unrealistic assumptions of monotonicity (convexity of the energy functional) and boundedness of the Nemytskii operator $\tilde{S}$ at $\mathbf{F} \to \mathbf{0}$. We shall also be able to impose the condition $\det \mathbf{F} > 0$ a.e. *a priori*. The BVPs of the preceding section will be replaced by a more general problem of finding the local minima of the potential $\mathscr{W}+\Phi$ on a conical subset of the space $\mathscr{V}$. The existence of an Euler–Lagrange differential operator annihilating a minimizer of $\mathscr{W}+\Phi$ is an open problem. In addition to the realistic constitutive assumptions on which this framework is based we are offered another advantage. We know in advance that the equilibrium configurations obtained by the direct methods correspond to stable equilibria.

We shall also consider more general potential loads:

$$\Phi(\boldsymbol{\chi}) = \int_{\Omega} dL^3\, f(\boldsymbol{\chi}(\xi), \xi) + \int_{\Gamma_2} dH^2\, g(\boldsymbol{\chi}(\xi), \xi). \tag{2.4.1}$$

We begin with some elementary remarks. Let F be a functional defined on an open subset $\mathscr{D}$ of a normed vector space $\mathscr{V}$. Suppose that F has a local minimum (maximum) at $\mathbf{u}_0 \in \mathscr{D}$. Hence $t \mapsto F(\mathbf{u}_0 + t\mathbf{w})$ has a local minimum (maximum) at $t = 0$. If F is Gâteaux-differentiable at $\mathbf{u}_0$ then

$$DF(\mathbf{u}_0)[\mathbf{w}] = 0 \ \forall\, \mathbf{w} \in \mathscr{V}. \tag{2.4.2}$$

Suppose now that F is convex and Gâteaux-differentiable in a convex nbhd $\mathscr{U} \subset \mathscr{D}$ of $\mathbf{u}_0$. In this case (2.4.2) is a necessary and sufficient condition for $\mathbf{u}_0$ to be a local minimum. For the proof, consider

$$\varphi(\mathbf{v}) :\equiv F(\mathbf{v}+\mathbf{u}_0) - F(\mathbf{u}_0).$$

Clearly, $\varphi(\mathbf{0}) = 0$. Suppose that $\mathbf{0}$ is not a minimum of φ in $\mathscr{U} - \mathbf{u}_0$, i.e. $\exists\, \mathbf{u}_1 \neq \mathbf{u}_0$, $\mathbf{u}_1 = \mathbf{u}_0 + \mathbf{v}_1 \in \mathscr{U}$ such that $\varphi(\mathbf{v}_1) < 0$. Since φ is convex

$$\varphi(\vartheta \mathbf{v}_1 + (1-\vartheta)\mathbf{0}) \leqslant \vartheta \varphi(\mathbf{v}_1) \ \forall\, \vartheta \in \,]0, 1[.$$

Hence $\vartheta^{-1}\varphi(\vartheta \mathbf{v}_1) \leqslant \varphi(\mathbf{v}_1)$ and in the limit $\vartheta \to 0_+$ $D\varphi(\mathbf{0})[\mathbf{v}_1] \leqslant \varphi(\mathbf{v}_1) < 0$, $D\varphi(\mathbf{0}) \neq \mathbf{0}$, contradicting the hypothesis.

Let us now work out the condition (2.4.2) for

$$F = \Phi + \mathscr{W} = \int_\Omega dL^3\, \varrho_0\, W(\nabla \boldsymbol{\chi}_0 + \nabla \mathbf{u}, \xi) + \int_\Omega dL^3\, f(\boldsymbol{\chi}_0 + \mathbf{u}, \xi)$$

$$+ \int_{\Gamma_2} dH^2 g(\boldsymbol{\chi}_0 + \mathbf{u}, \xi).$$

We first calculate $DW(\boldsymbol{\chi}_0)$

$$\frac{1}{t}\int_\Omega dL^3\, \varrho_0 [W(\nabla \boldsymbol{\chi}_0 + t\nabla \mathbf{u}, \xi) - W(\nabla \boldsymbol{\chi}_0, \xi)]$$

$$= \int_\Omega dL^3\, \hat{S}_k^{\alpha}(\nabla \boldsymbol{\chi}_0(\xi) + \vartheta(\xi) t \nabla \mathbf{u}(\xi), \xi) u^k{}_{,\xi^\alpha}, \quad 0 \leqslant \vartheta \leqslant 1,\star$$

since W is a continuous function of $\mathbf{F}$ for a.e. $\xi \in \Omega$. Since $\tilde{S}$ is radially continuous, in the limit $t \to 0$ we have

$$DW(\boldsymbol{\chi}_0)[\mathbf{u}] = \int_\Omega dL^3\, \hat{S}_k^{\alpha}(\nabla \boldsymbol{\chi}_0(\xi), \xi) u^k{}_{,\xi^\alpha}.$$

Suppose now that the functions f, g are differentiable with respect to the first argument and their derivatives are Nemytskii operators $\mathscr{L}^p(\Omega) \to \mathscr{L}^q(\Omega)$, $\mathscr{L}^p(\Gamma_2) \to \mathscr{L}^q(\Gamma_2)$. The same argument yields

$$D\Phi(\boldsymbol{\chi}_0)[\mathbf{u}] = \int_\Omega dL^3 \frac{\partial f}{\partial x^k}(\boldsymbol{\chi}_0(\xi), \xi) u^k(\xi)$$

$$+ \int_{\Gamma_2} dH^2 \frac{\partial g}{\partial x^k}(\gamma \boldsymbol{\chi}_0(\xi), \xi) \gamma u^k(\xi).$$

Hence (2.4.2) is equivalent to

$$\int_\Omega dL^3\, \hat{S}_k^{\alpha}(\nabla \boldsymbol{\chi}_0(\xi), \xi) u^k{}_{,\xi^\alpha} + \int_\Omega dL^3 \frac{\partial f}{\partial x^k}(\boldsymbol{\chi}_0(\xi), \xi) u^k(\xi)$$

$$+ \int_{\Gamma_2} dH^2 \frac{\partial g}{\partial x^k}(\gamma \boldsymbol{\chi}_0(\xi), \xi) \gamma u^k(\xi) \quad \forall \mathbf{u} \in \mathscr{V} \subset W^{1,p}(\Omega) \tag{2.4.3}$$

and we can identify $\dfrac{\partial f}{\partial x^k}$, $\dfrac{\partial g}{\partial x^k}$ as $-\varrho_0(\xi) b_k(\mathbf{x}, \xi)$, $-s_k(\mathbf{x}, \xi)$, resp. Setting $\boldsymbol{\chi}_0(\xi) = \boldsymbol{\chi}_1(\xi) + \mathbf{u}_0(\xi)$ with $\boldsymbol{\chi}_1 \in W^{1,p}(\Omega)$ prescribed on Γ_1 and extended to $\bar{\Omega}$,

$$\mathscr{V} = \{\mathbf{u} \in W^{1,p}(\Omega) |\ \gamma_{\Gamma_1} \mathbf{u} = \mathbf{0}\} \quad \text{if } H^2(\Gamma_1) > 0,$$

★ This is a heuristic calculus ($\vartheta(\cdot)$) need not be measurable). For a rigorous argument, see Section 2.3.2.

or

$$\mathscr{V} = \left\{\mathbf{u} \in W^{1,p}(\Omega) \middle| \int_{\Omega} dL^3\, \mathbf{u} = \mathbf{0}\right\} \quad \text{if } H^2(\Gamma_1) = 0,$$

$$C = \{\mathbf{u} \in \mathscr{V} \mid \det(\nabla\boldsymbol{\chi}_0 + \nabla\mathbf{u}) > 0 \text{ a.e.}\}$$

we arrive at the variational formulation

$$\mathscr{W} + \Phi \to \min \quad \text{on} \quad \mathscr{C}$$

of the problem

$$S_k^{\alpha}{}_{,\xi^\alpha} + \varrho_0 b_k(\mathbf{x}, \xi) = 0,$$
$$S_k^{\alpha} \nu_\alpha = s_k(\mathbf{x}, \xi) \quad \text{on } \Gamma_2,$$
$$\boldsymbol{\chi} = \boldsymbol{\chi}_1 \quad \text{on } \Gamma_1. \tag{2.4.4}$$

In the following section we shall review some concepts related to the direct method. In Section 2.5 these methods will be refined and adapted to the specific problems of elasticity.

2.4.2 Fundamental Notions and Theorems for the Direct Methods of Variational Calculus

Definition 2.4.1

A functional $f\colon \mathscr{V} \to \mathbf{R}$ is said to be **weakly lower semicontinuous (wlsc)*** if for every sequence $\mathbf{u}_n \rightharpoonup \mathbf{u}_0$, $n = 1, 2, \ldots$ in $\mathscr{V}$

$$f(\mathbf{u}_0) \leqslant \liminf_{n\to\infty} f(\mathbf{u}_n). \tag{2.4.5}$$

Note that a real function f is continuous at $\mathbf{u}_0$ if $\mathbf{u}_n \to \mathbf{u}_0$ implies $\lim_{n\to\infty} f(\mathbf{u}_n) = f(\mathbf{u}_0)$, or if $\forall \varepsilon > 0\ \exists \delta > 0$ such that $|\mathbf{u}-\mathbf{u}_0| < \delta \Rightarrow |f(\mathbf{u})-f(\mathbf{u}_0)| < \varepsilon$. For lower semicontinuity it is only required that $|\mathbf{u}-\mathbf{u}_0| < \delta \Rightarrow f(\mathbf{u})-f(\mathbf{u}_0) > -\varepsilon$. If $\mathbf{u}_n \to \mathbf{u}_0$, then for $n \geqslant n_0$ sufficiently large $f(\mathbf{u}_n) > f(\mathbf{u}_0)-\varepsilon$ and therefore $\liminf_{n\to\infty} f(\mathbf{u}_n) \geqslant f(\mathbf{u}_0)$. Suppose now that (2.4.5) is satisfied, but $\exists\, \varepsilon > 0\ \forall\, \delta > 0$ $\exists\, \mathbf{u}$ such that $|\mathbf{u}-\mathbf{u}_0| < \delta$ and $f(\mathbf{u}) < f(\mathbf{u}_0)-\varepsilon$. Let us choose $\{\mathbf{u}_n\}$ satisfying the above inequalities with $\delta = 1/n$. Hence $\mathbf{u}_n \to \mathbf{u}_0$ and for the sequence $\{\mathbf{u}_n\}$ $\liminf_{n\to\infty} f(\mathbf{u}_n) < f(\mathbf{u}_0)-\varepsilon$, a contradiction.

Lower semicontinuity is a natural property of functionals appearing in the variational calculus. For example the length of curves joining two

* The correct term is: **sequentially weakly lower semicontinuous.** We shall not use any other notion of weak lower semicontinuity and therefore we can drop the word "sequentially" throughout this section

fixed points is a lower semicontinuous (but not continuous) functional in the $\mathscr{C}^0$ topology. Curves lying in a $\mathscr{C}^0$ nbhd of a given curve $\mathscr{S}$ joining the same two points can be much longer (if they oscillate densely) but not much shorter than $\mathscr{S}$. This fact is in close relation with the fact that in general there are curves of minimal but not maximal length.

Lemma 2.4.2

Let $\mathscr{V}$ be a reflexive Banach space and suppose that $f\colon \mathscr{D} \to \boldsymbol{R}$ is wlsc on a non-empty bounded closed convex set $\mathscr{D} \subset \mathscr{V}$. Then f assumes a minimum value on $\mathscr{D}$.

Proof

Let $\{\mathbf{u}_n\} \subset \mathscr{D}$ be a minimizing sequence for f, i.e.

$$\lim_{n\to\infty} f(\mathbf{u}_n) = \inf\{f(\mathbf{u}) |\ \mathbf{u} \in \mathscr{D}\} =: f_o.$$

Since $\mathscr{D}$ is bounded and $\mathscr{V}$ is reflexive, some subsequence $\{\mathbf{u}_{n_k}\}$ converges weakly to some $\mathbf{u} \in \mathscr{V}$ (Dunford and Schwartz, 1958; Sections 2.3.28 and 2.3.29. Yosida, 1965; Theorems 5.2.1 and 5.1.7). Since $\mathscr{D}$ is closed and convex in a Banach space $\mathscr{V}$, it is weakly closed (Dunford and Schwartz, 1958, Section 5.3.13) and $\mathbf{u} \in \mathscr{D}$. Since $\lim_{k\to\infty} f(\mathbf{u}_{n_k}) = f_o$ and f is wlsc, $f(\mathbf{u}) \leqslant f_o$ and hence $f(\mathbf{u}) = f_o$. □

It is more usual to deal with an unbounded domain $\mathscr{D}$ and a **coercive** functional f:

$$f(\mathbf{u}) \to \infty \quad \text{for } ||\mathbf{u}|| \to \infty. \tag{2.4.6}$$

Theorem 2.4.3

Let $\mathscr{D}$ be a non-empty closed convex subset of a reflexive Banach space $\mathscr{V}$, and let $f\colon \mathscr{D} \to \boldsymbol{R}$ be wlsc and coercive. Then f assumes its minimum on $\mathscr{D}$.

Proof

Let $\mathbf{u}_o \in \mathscr{D}$. For some $r > 0$ the inequality $||\mathbf{u}|| > r$ entails the inequality $f(\mathbf{u}) > f(\mathbf{u}_o)$. By Lemma 2.4.2 f assumes its minimum on the bounded closed convex set $\{\mathbf{u} \in \mathscr{D} |\ ||\mathbf{u}|| \leqslant r\}$. This minimum is obviously a minimum of f on $\mathscr{D}$. □

Theorem 2.4.4

Let $A\colon \mathscr{V} \to \mathscr{V}^*$ be a potential operator with the potential $f\colon \mathscr{V} \to \boldsymbol{R}$.

Then f is convex iff $f(\mathbf{v}) - f(\mathbf{u}) \geqslant \langle A\mathbf{u} | \mathbf{v} - \mathbf{u}\rangle$ for all $\mathbf{v}, \mathbf{u} \in \mathscr{V}$ iff A is monotone.

Proof

Let $\varphi(s) :\equiv f(\mathbf{u}+s\mathbf{w})$, $\mathbf{w} := \mathbf{v}-\mathbf{u}$, $\mathbf{v}, \mathbf{u} \in \mathscr{V}$. The derivative $\varphi'(s) = \langle A(\mathbf{u}+s\mathbf{w})|\mathbf{w}\rangle$ is everywhere defined. If f is convex, then φ' is non-decreasing and A is monotone.

Assume that A is monotone. Since φ' is everywhere defined on [0, 1], finite and monotone, φ is Lipschitz continuous and

$$\varphi(1)-\varphi(0) = \int_0^1 ds\, \varphi'(s) \geqslant \varphi'(0),^\star$$

which amounts to the inequality

$$f(\mathbf{v})-f(\mathbf{u}) \geqslant \langle A\mathbf{u}|\mathbf{v}-\mathbf{u}\rangle. \tag{2.4.7}$$

Finally, we prove that equation (2.4.7) implies convexity of f. For

$$\mathbf{w}(s) := s\mathbf{u}+(1-s)\mathbf{v}, \quad 0 \leqslant s \leqslant 1,$$

we have the inequalities

$$f(\mathbf{u}) \geqslant f\big(\mathbf{w}(s)\big)+\langle A\mathbf{w}(s)|\mathbf{u}-\mathbf{w}(s)\rangle$$

and

$$f(\mathbf{v}) \geqslant f\big(\mathbf{w}(s)\big)+\langle A\mathbf{w}(s)|\mathbf{v}-\mathbf{w}(s)\rangle.$$

We shall multiply these inequalities by the non-negative numbers s and $1-s$, resp., and add them up. Since $s[\mathbf{u}-\mathbf{w}(s)]+(1-s)[\mathbf{v}-\mathbf{w}(s)] = 0$, we get the inequality

$$sf(\mathbf{u})+(1-s)f(\mathbf{v}) \geqslant f\big(\mathbf{w}(s)\big).$$ □

Corollary 2.4.5

Let f: $\mathscr{V} \to \boldsymbol{R}$ be a convex Gâteaux-differentiable functional on a Banach space $\mathscr{V}$. Then f is wlsc.

Proof

Let $\mathbf{u}_n \rightharpoonup \mathbf{u}$ in $\mathscr{V}$. Then

$$f(\mathbf{u}) \leqslant f(\mathbf{u}_n)+\langle A\mathbf{u}|\mathbf{u}-\mathbf{u}_n\rangle$$

and

$$f(\mathbf{u}) \leqslant \liminf_{n\to\infty} f(\mathbf{u}_n)$$

in the limit $n \to \infty$. □

$\star$ A reader unfamiliar with such arguments can use an alternative argument based on Lemma 2.5.2 below.

Remark

Gâteaux-differentiability is not essential for the validity of the corollary although it makes the proof nearly trivial. Cf. Vainberg (1972), Theorem 3.8.10.

In view of these results it is interesting to prove the existence theorem for monotone potential operators using variational methods. In Section 2.3.2 we noted that a monotone potential operator is demicontinuous and hence the Browder–Minty theorem can be applied. Another possible approach is illustrated by the proof of the following theorem.

Theorem 2.4.6

Let $A: \mathscr{V} \to \mathscr{V}^*$ be a monotone coercive potential operator with a potential φ. Then

$$\forall \mathbf{f} \in \mathscr{V}^* \quad \exists \mathbf{u} \in \mathscr{V} \quad A\mathbf{u} = \mathbf{f}.$$

Proof

We shall take advantage of the fact that A is demicontinuous, and hence radially continuous, without proving it. For the proof see Gajewski *et al.* (1974).

A potential for A can be calculated from the formula

$$\varphi(\mathbf{v}) = \int_0^1 ds \, \langle A(s\mathbf{v})|\mathbf{v}\rangle .$$

Indeed,

$$\lim_{\Delta\to 0} \frac{\varphi\big((s+\Delta)\mathbf{v}\big)-\varphi(s\mathbf{v})}{\Delta} = D\varphi(s\mathbf{v})[\mathbf{v}] = \langle A(s\mathbf{v})|\mathbf{v}\rangle .$$

Hence $\varphi(\mathbf{v})-\varphi(0) = \int_0^1 ds \langle A(s\mathbf{v})|\mathbf{v}\rangle$.

Let $\psi(\mathbf{v}) \equiv \varphi(\mathbf{v})-\langle \mathbf{f}|\mathbf{v}\rangle$. Since the operator $\mathbf{v} \mapsto A\mathbf{v}-\mathbf{f}$ is monotone, ψ is convex and Gâteaux-differentiable and hence wlsc. Moreover

$$\psi(\mathbf{u}) = \int_0^1 ds \, \langle A(s\mathbf{u})|\mathbf{u}\rangle - \langle \mathbf{f}|\mathbf{u}\rangle$$

$$= \int_0^1 ds \, s^{-1}\langle A(s\mathbf{u})-A(\mathbf{0})|s\mathbf{u}\rangle - \langle \mathbf{f}-A(\mathbf{0})|\mathbf{u}\rangle .$$

The last integrand is non-negative, hence

$$\psi(\mathbf{u}) \geqslant \int_{1/2}^{1} ds\,\langle A(s\mathbf{u})-A(\mathbf{0})|\mathbf{u}\rangle-\langle \mathbf{f}-A(\mathbf{0})|\mathbf{u}\rangle$$

$$\equiv \int_{1/2}^{1} ds\,\langle A(s\mathbf{u})|\mathbf{u}\rangle-\langle \mathbf{f}-\tfrac{1}{2}A(\mathbf{0})|\mathbf{u}\rangle .$$

Since $\varphi'(su)$ is non-decreasing, so is $s\mapsto\langle A(s\mathbf{u})|\mathbf{u}\rangle$. Hence

$$\psi(\mathbf{u}) \geqslant \langle A(\tfrac{1}{2}\mathbf{u})|\tfrac{1}{2}\mathbf{u}\rangle-\langle \mathbf{f}-\tfrac{1}{2}A(\mathbf{0})|\mathbf{u}\rangle$$

$$\geqslant \frac{1}{2}\,||\mathbf{u}||\gamma\,(\tfrac{1}{2}||\mathbf{u}||)-|\mathbf{f}-\tfrac{1}{2}A(\mathbf{0})||_{\mathscr{V}^*}||\mathbf{u}|| \to \infty$$

for $||\mathbf{u}|| \to \infty$. Hence ψ assumes its minimum on $\mathscr{V}$ at a point $\mathbf{u}_o\in\mathscr{V}$,

$$D\psi(\mathbf{u}_o)[\mathbf{w}] = 0 \quad \forall\in\mathscr{V}$$

$A\mathbf{u}_o = \mathbf{f}$, q.e.d. □

We note that for a monotone potential operator A the equation $A\mathbf{u}_o = \mathbf{f}$ entails that $\psi(\mathbf{u}) \to \min$ at $\mathbf{u}_o$. Indeed,

$$\psi(\mathbf{v})-\psi(\mathbf{u}_o) = \varphi(\mathbf{v})-\varphi(\mathbf{u}_o)+\langle\mathbf{f}|\mathbf{v}-\mathbf{u}_o\rangle$$

$$= \varphi(\mathbf{v})-\varphi(\mathbf{u}_o)+\langle A\mathbf{u}_o|\mathbf{v}-\mathbf{u}_o\rangle \geqslant 0$$

$\forall\,\mathbf{v}\in\mathscr{V}$ by Theorem 2.4.4.

Let us consider the potential operator $\tilde{A}\colon \mathscr{V}\to\mathscr{V}^*$ extending DF, $F=\mathscr{W}+$ $+\Phi$, in the case of potential loads which neednot be dead. The convexity of F (or monotonicity of A) amounts to

$$DF(\mathbf{u})[\mathbf{u}-\mathbf{v}]-DF(\mathbf{v})[\mathbf{u}-\mathbf{v}] \geqslant 0 \;\; \forall \mathbf{u},\mathbf{v}\in\mathscr{V},$$

or

$$\int_{\Omega} dL^3\,[\hat{S}_k^{\alpha}(\nabla\chi_0+\nabla\mathbf{u},\xi)-\hat{S}_k^{\alpha}(\nabla\chi_0+\nabla\mathbf{v},\xi)][u^k{}_{,\xi^\alpha}-v^k{}_{,\xi^\alpha}]$$

$$-\int_{\Omega} dL^3\,\varrho_0(\xi)[b_k(\mathbf{u},\xi)-b_k(\mathbf{v},\xi)][u^k-v^k]$$

$$-\int_{\Gamma_2} dH^2\,[s_k(\mathbf{u},\xi)-s_k(\mathbf{v},\xi)]\cdot[\gamma u^k-\gamma v^k] \geqslant 0. \tag{2.4.8}$$

In the above arguments wlsc entails the existence of minima while convexity essentially ensures the wlsc of the potential. However, convexity brings us back to the difficulties encountered in connection with monotonicity in elastostatics, Section 2.3. Moreover, equation (2.4.8) places unduly restrictive monotonicity conditions on the loads. It turns out however that these conditions can be lifted.

Let

$$\langle A_1(\mathbf{u}, \mathbf{v})|\mathbf{w}\rangle := \int_\Omega dL^3\, [\hat{S}_k{}^\alpha(\nabla\mathbf{u}, \mathbf{v}, \xi)\, w^k{}_{,\xi^\alpha} - \varrho_0 b_k(\mathbf{v}, \xi)\, w^k]$$

$$- \int_{\Gamma_2} dH^2\, s_k(\gamma\mathbf{v}, \xi)\gamma w^k \qquad \forall \mathbf{u}, \mathbf{v}, \mathbf{w} \in \mathscr{V}.$$

As noted in Section 2.3.3, existence is ensured by the property of $A\mathbf{u} = A_1(\mathbf{u}, \mathbf{u})$ being a coercive operator of variational calculus, with A_1 hemicontinuous with respect to either argument. Monotonicity is then replaced by an analogous condition on $A_1(\cdot\,, \mathbf{v})$. It can be proved (Lions, 1969, Chapter 2) that A is an operator of variational calculus if

$$[\hat{S}_k{}^\alpha(\mathbf{F}_1, \mathbf{x}, \xi) - \hat{S}_k{}^\alpha(\mathbf{F}_2, \mathbf{x}, \xi)][F_1{}^k{}_\alpha - F_2{}^k{}_\alpha] > 0$$

for every $\mathbf{x}, \mathbf{F}_1 \neq \mathbf{F}_2$ and a.e. $\xi \in \Omega$. This is however too strong a condition for elasticity. It can nevertheless be deduced from Theorem 8.2.1 of Ekeland and Témam (1976) that wlsc follows from suitable growth properties and mere convexity of $W(\cdot\,, \xi)$ for arbitrary $f(\cdot\,, \xi)$, $g(\cdot\,, \xi)$ continuous for a.e. $\xi \in \Omega$, Γ_2, resp.

2.5 DIRECT VARIATIONAL METHODS AND CONSTITUTIVE ASSUMPTIONS

2.5.1 Introduction

In Sections 2.2–2.4 we observed two important facts:

(1) the BVPs of elastostatics can be expressed in terms of potential operators (Section 2.3) or equivalent variational problems (Section 2.4),

(2) the potential operators of elastostatics are monotone (and the potentials are convex) at best for some specific BVPs with $H^2(\Gamma_1) > 0$. Monotonicity can be satisfied only on a set of deformations with quite special stress fields.

Another shortcoming of the methods developed in Sections 2.3–2.4 is the upper bound on $\hat{S}_k{}^\alpha$, which is inconsistent with the expectation that stress grows to infinity for $\det\mathbf{F} \to 0$. Finally, we have not taken into account the *a priori* condition $\det\mathbf{F} > 0$ a.e. (although it may turn out to be satisfied *a posteriori*.

In Section 2.5 we adopt a somewhat different approach developed essentially by J. M. Ball. The BVP will be presented in the variational form $F \to \min$

and the solution of the latter problem will be sought by a recourse to direct variational methods. The convexity assumption will be replaced by an acceptable constitutive assumption. The exposition is somewhat simplified in comparison with the paper of Ball (1977) at the cost of some loss of generality.

To give an idea of the direct methods we shall use in this section, suppose that $a := \inf\{\mathscr{W}(\mathbf{u}) | \ \mathbf{u} \in \mathscr{V}\} > -\infty$. Let $\{\mathbf{u}_n\} \subset \mathscr{V}$ be a minimizing sequence for $\mathscr{W}$, i.e. $\mathscr{W}(\mathbf{u}_n) \to a$ for $n \to \infty$. A growth condition $W(\mathbf{F}_0 + \mathscr{H}, \xi) \geqslant b(\xi) + k|\mathscr{H}|^p$, $k > 0, p > 1$, implies that the sequence $||\mathbf{u}_n||_p$ is bounded. By the Friedrichs or Poincaré lemma the sequence $\{\mathbf{u}_n\}$ is bounded in $W^{1,p}$ and contains a weakly convergent subsequence $\{\mathbf{u}_{n_k}\}$, $\mathbf{u}_{n_k} \to \mathbf{u}_o$. If $\mathscr{V}$ is sequentially weakly closed then $\mathbf{u}_o \in \mathscr{V}$. If $\mathscr{W}$ is wlsc then $\mathscr{W}(\mathbf{u}_o) \leqslant \lim\limits_{n\to\infty} \mathscr{W}(\mathbf{u}_n) = a$ and hence $\mathscr{W}(\mathbf{u}_o) = a$. The inequality $\det(\nabla\boldsymbol{\chi}_0 + \nabla\mathbf{u}_o) > 0$ a.e. in Ω follows from the fact that $\mathscr{W}(\mathbf{u}_o) = \infty$ if this inequality is not satisfied (on account of the growth conditions).

It is desirable to find constitutive assumption on the function W that would ensure that F is a wlsc functional for all the fundamental BVPs of elastostatics. Some guidelines for this undertaking are provided by the following observations of Ball:

(1) the very existence of a sufficiently regular ($\mathscr{C}^1$) minimizer $\mathbf{u}_o$ entails some local conditions on W, satisfied on the graph of $\nabla\mathbf{u}_o$;

(2) some of these conditions (QC, SE, LH) can be extrapolated to the entire domain of definition of W, if they ensure that $\mathscr{W}$ is wlsc;

(3) other conditions (PC) are obtained by a slight strengthening of previously obtained conditions (QC etc.);

(4) growth conditions are suggested by physical insight and by their utility in the proof of existence;

(5) all the constitutive assumptions on W should be insensitive to the transformations of the integrand $\mathscr{W}$ which do not influence the content of the problems $F \to \min$ (gauge transformations of the Lagrangian $\mathscr{W}$).

2.5.2 The Legendre–Hadamard Condition

Suppose that $\boldsymbol{\chi}_0$ minimizes $F = \mathscr{W} + \Phi$ in $\boldsymbol{\chi}_1 + \mathscr{V}$ ($\mathscr{V}$ is a subspace of $W^{1,p}$). For every $\mathbf{u} \in \mathscr{V} \cap \mathscr{C}_o(\Omega)$

$$\frac{d}{dt} F(\boldsymbol{\chi}_0 + t\mathbf{u})|_{t=0} = DF(\boldsymbol{\chi}_0)[\mathbf{u}] = 0, \quad \forall \mathbf{u} \in \mathscr{V}, \tag{2.5.1}$$

$$\frac{d^2}{dt^2} F(\boldsymbol{\chi}_0 + t\mathbf{u})|_{t=0} \geqslant 0. \quad \forall \mathbf{u} \in \mathscr{V}. \tag{2.5.2}$$

Equation (2.5.1) turns out to be the weak form of the Euler–Lagrange equation for the minimizer χ_0. Suppose that $W(\cdot, \xi)$ has continuous derivatives up to second order in a nbhd of $\mathbf{F}_0(\xi) := \nabla\chi_0(\xi)$ for almost every $\xi \in \Omega$,

$$|\varrho_0 W_{,F^k{}_\alpha}| = |\hat{S}_k{}^\alpha(\mathbf{F}_1+\mathbf{H}, \xi)| \leqslant a(\xi)+b|\mathbf{H}|^{p-1},$$
$$|\phi_{,x^k}(\chi_1+\mathbf{u}, \xi)| \leqslant \bar{a}(\xi)+\bar{b}|\mathbf{u}|^{p-1}, \quad p > 1,\ a, a \in \mathcal{L}^q(\Omega),$$
$$q = p/(p-1),\ b, \bar{b} > 0,\ \mathbf{F}_1 := \nabla\chi_1,$$
$$|\varrho_0 W_{,F^k{}_\alpha F^l{}_\beta}(\mathbf{F}_1+\mathbf{H}, \xi)| \leqslant c(\xi)+d|\mathbf{H}|^{p'-1},$$
$$|\phi_{,x^k x^l}(\chi_1+\mathbf{u}, \xi)| \leqslant \bar{c}(\xi)+\bar{d}|\mathbf{u}|^{p'-1}, \quad p' > 1,\ c, \bar{c} \in \mathcal{L}^{q'}(\Omega),$$
$$q' = p'/(p'-1), \quad d, \bar{d} > 0$$

(one might expect that $p' = p/2$). We have

$$\frac{d^2}{dt^2} F(\chi_0+t\mathbf{u})|_{t=0} = \int_\Omega dL^3\, \varrho_0(\xi)\, W_{,F^k{}_\alpha F^l{}_\beta}(\nabla\chi_0(\xi), \xi) u^k{}_{,\xi^\alpha} u^l{}_{,\xi^\beta}$$
$$+ \int_\Omega dL^3\, \phi_{,x^k x^l}(\chi_0(\xi), \xi) u^k u^l.$$

Surface integrals do not contribute since $\mathbf{u} \in \mathscr{C}_o(\Omega)$. Differentiation under the integral sign is justified by the Lebesgue dominated convergence theorem, since the difference quotients are bounded uniformly with respect to $t \leqslant t_o$ by integrable functions, e.g. for $t \leqslant t_o$

$$|\varrho_0 W_{,F^k{}_\alpha F^l{}_\beta}(\mathbf{F}_0(\xi)+t(\xi)\nabla\mathbf{u}(\xi), \xi)|$$
$$\leqslant c(\xi)+d\Big\{\sum_{k,\alpha} (F_0{}^k{}_\alpha+t(\xi)u^k{}_{,\xi^\alpha})^2\Big\}^{(p'-1)/2}$$
$$\leqslant c(\xi)+4^{(p'-1)/2}\{|\mathbf{F}_0|^{p'-1}+t(\xi)^{p'-1}|\nabla\mathbf{u}|^{p'-1}\}$$
$$\leqslant c(\xi)+2^{(p'-1)}\{|\mathbf{F}_0|^{p'-1}+|\nabla\mathbf{u}|^{p'-1} t_o{}^{p'-1}\},$$

since $0 \leqslant t(\xi) \leqslant t \leqslant t_o$.* Let $\mathbf{u}(\xi) = \mathbf{v}(R\xi)$. Then

$$\frac{d^2}{dt^2} F(\chi_0+t\mathbf{u})|_{t=0} = R^2\int_\Omega dL^3\, \varrho_0(\xi)\, W_{,F^k{}_\alpha F^l{}_\beta}(\nabla\chi_0(\xi), \xi) v^k{}_{,\xi^\alpha}(R\xi) v^l{}_{,\xi^\beta}(R\xi)$$
$$+ \int_\Omega dL^3\, \phi_{,x^k x^l}(\chi_0(\xi), \xi) v^k v^l.$$

* Note that

$$\Big[\sum_{k,\alpha} (F_0{}^k{}_\alpha+tu^k{}_{,\xi^\alpha})^2\Big]^{(p'-1)/2} \leqslant \Big\{2\sum_{k,\alpha} [(F_0{}^k{}_\alpha)^2+t^2(u^k{}_{,\xi^\alpha})^2]\Big\}^{(p'-1)/2},$$

while for $\alpha > 0$, $x_i > 0$

$$\Big(\sum_{i=1}^N x_i\Big)^\alpha \leqslant [N \max\{x_i|\ i = 1, ..., N\}]^\alpha \leqslant N^\alpha \sum_{i=1}^N x_i{}^\alpha.$$

Since we can make R arbitrarily large, it follows that

$$\int_{\Omega} dL^3\, \varrho_0(\xi)\, W_{,F^k{}_\alpha F^l{}_\beta}(\nabla\chi_0(\xi), \xi) u^k{}_{,\xi^\alpha} u^l{}_{,\xi^\beta} \geqslant 0 \tag{2.5.3}$$

for all $\mathbf{u} \in \mathscr{V} \cap \mathscr{C}_o^\infty(\Omega)$.

Theorem 2.5.1 (Hadamard)
Suppose that Ω is bounded, $A_k{}^\alpha{}_l{}^\beta = A_l{}^\beta{}_k{}^\alpha$ is a continuous function of $\xi \in \bar{\Omega}$

$$\int_{\Omega} dL^3\, A_k{}^\alpha{}_l{}^\beta u^k{}_{,\xi^\alpha} u^l{}_{,\xi^\beta} \geqslant 0 \tag{2.5.3'}$$

for every $\mathbf{u} \in \mathscr{C}_o^\infty(\Omega)$. Then the **Legendre–Hadamard inequality** (LH)

$$A_k{}^\alpha{}_l{}^\beta(\xi) a^k a^l k_\alpha k_\beta \geqslant 0 \;\; \forall \xi \in \bar{\Omega} \;\; \forall \mathbf{a}, \mathbf{k} \in \mathbf{R}^3 \tag{2.5.4}$$

holds.

Proof (Fichera)
Suppose that $\xi_0 \in \Omega$,

$$A_k{}^\alpha{}_l{}^\beta(\xi_0) a^k a^l k_\alpha k_\beta \leqslant -\varepsilon \mathbf{k}^2 \mathbf{a}^2, \tag{2.5.5}$$

$\varepsilon > 0$, $\mathbf{k} \neq 0$, $\mathbf{a} \neq \mathbf{0}$. Let $\delta > 0$ be such that $|\xi - \xi_0| < \delta$ implies the inequality

$$|A_k{}^\alpha{}_l{}^\beta(\xi) - A_k{}^\alpha{}_l{}^\beta(\xi_0)| < \frac{\varepsilon}{18}.$$

Let $\mathbf{u} \in \mathscr{C}_o^\infty(\Omega)$ be complex-valued, supp $\mathbf{u} \subset \mathscr{K}(\xi_0, \delta) \cap \Omega$.
Clearly,

$$\left| \int_{\Omega} dL^3\, [A_k{}^\alpha{}_l{}^\beta(\xi) - A_k{}^\alpha{}_l{}^\beta(\xi_0)] \bar{u}^k{}_{,\xi^\alpha} u^l{}_{,\xi^\beta} \right| \leqslant \frac{\varepsilon}{18} \left| \int_{\Omega} dL^3 \sum_{k,l,\alpha,\beta} \bar{u}^k{}_{,\xi^\alpha} u^l{}_{,\xi^\beta} \right|$$

$$\leqslant \frac{\varepsilon}{18} \left| \int_{\Omega} dL^3\, \frac{1}{2} \sum_{k,l,\alpha,\beta} [|u^k{}_{,\xi^\alpha}|^2 + |u^l{}_{,\xi^\beta}|^2] \right|$$

$$= \frac{\varepsilon}{2} \int_{\Omega} dL^3 \sum_{k,\alpha} |u^k{}_{,\xi^\alpha}|^2.$$

The inequality (2.5.3′) implies that

$$\int_{\Omega} dL^3\, A_k{}^\alpha{}_l{}^\beta(\xi_0) \bar{u}^k{}_{,\xi^\alpha} u^l{}_{,\xi^\beta} \geqslant -\frac{\varepsilon}{2} ||\nabla \mathbf{u}||_2{}^2. \tag{2.5.6}$$

Substituting $\mathbf{u}(\xi) = \varphi(\xi) \exp(i\lambda k_\alpha \xi^\alpha) \mathbf{a}$, supp$\varphi \subset \mathscr{K}(\xi_0, \delta) \cap \Omega$, into (2.5.6) we have

$$\lambda^2 A_k{}^\alpha{}_l{}^\beta(\xi_0) a^k a^l k_\alpha k_\beta ||\varphi||_2{}^2 + O(\lambda) \geqslant -\frac{\varepsilon}{2} \lambda^2 \mathbf{a}^2 \mathbf{k}^2 ||\varphi||_2{}^2 + O(\lambda).$$

For $\lambda \to \infty$ we get the inequality

$$A_k{}^\alpha{}_l{}^\beta(\xi_0)a^k a^l k_\alpha k_\beta \geqslant -\frac{\varepsilon}{2}\mathbf{a}^2\mathbf{k}^2,$$

which contradicts equation (2.5.5).

According to Theorem 2.5.1 □

$$W_{,F^k{}_\alpha F^l{}_\beta}\nabla^k(\chi_0(\xi),\xi)a^k a^l k_\alpha k_\beta \geqslant 0 \tag{2.5.7}$$

provided $W_{,\mathbf{FF}}$ is a continuous function of ξ, $\mathbf{F}$ and the minimizer $\chi_0 \in \mathscr{C}^1(\bar{\Omega})$. Suppose now that the functional $F = \mathscr{W}+\Phi$ satisfies the hypotheses listed at the beginning of this subsection and is convex. The same argument implies now that (2.5.7) holds for every $\chi_0 \in (\chi_1+\mathscr{V})\cap\mathscr{C}^1(\bar{\Omega})$, provided $W_{,\mathbf{FF}}$ is a continuous function of ξ, $\mathbf{F}$. Note, however, that $C = \{\chi \in \chi_1+\mathscr{V} \mid \det\nabla\chi > 0 \text{ a.e.}\}$ is not convex. The converse implication: (2.5.7) → (2.5.3) is not true, as shown by the counterexample 5 of Fichera (1972a).

Following Fichera, let us consider the case of an isotropic body

$$B_{kplq} = B_{lqkp} = B_{pklq},$$

$$B_{kplq}(\mathbf{x})u^k{}_{,x^p}u^l{}_{,x^q} = \mu(r)\sum_{k,p} u_{(k,p)}{}^2 + \frac{1}{2}\lambda(r)\sum_k u_{k,k}{}^2,$$

$$r = |\mathbf{x}|, \quad u_{k,l} := u^k{}_{,x^l}, \quad \mu > 0,\ \lambda+2\mu > 0 \text{ for all } r \in \boldsymbol{R}_+.$$

As we know $B_{kplq}a^k a^l k^p k^q > 0$ for $\mathbf{a}$, $\mathbf{k} \neq 0$ in this case.

Let μ, λ be continuous functions of r. Let

$$\Omega = \{\mathbf{x} \in \boldsymbol{R}^3 \mid \pi < r < 2\pi\}, \quad \varphi \in \mathscr{C}_0^\infty(]\pi, 2\pi[), \quad \mathbf{u}(\mathbf{x}) = r^{-1}\varphi(r)\mathbf{x}.$$

$$\int_\Omega dL^3\, B_{kplq}(\mathbf{x})u_{k,p}u_{l,q} = \int_\Omega dL^3 \left[\mu\sum_{k,l} u_{(k,l)}{}^2 + \frac{1}{2}\lambda\sum_k u_{k,k}{}^2\right]$$

$$= \frac{1}{2}\int_\pi^{2\pi} dr\,[(\lambda+2\mu)r^2\varphi'^2 + 4\lambda r\varphi\varphi' + 4(\lambda+\mu)\varphi^2].$$

For $\lambda+2\mu = r^{-2}$, $\mu = \dfrac{3}{4}r^{-2} + \dfrac{1}{2}\ln\dfrac{2\pi}{r}$ we have

$$\int_\Omega dL^3\, B_{kplq}(\mathbf{x})u_{k,p}u_{l,q} = \frac{1}{2}\int_\pi^{2\pi} dr(\varphi'^2 - 2\varphi^2).$$

Now for $\varphi = \sin r$ the right-hand side equals $-\pi/4$, which is negative.

It is plausible that we can assume **LH**:

$$W_{,F^k{}_\alpha F^l{}_\beta}(\mathbf{F},\xi)a^k a^l k_\alpha k_\beta \geqslant 0 \quad \forall \mathbf{F} \in \mathscr{L}_+, \quad \forall \mathbf{a}, \mathbf{k} \in \boldsymbol{R}^3,$$

without running into the difficulties associated with convexity. The effective treatment of linearized problems in Section 2.2 requires a stronger constitutive assumption, viz. **SE**:

$$\varrho_0(\xi)\, W_{,F^k{}_\alpha F^l{}_\beta}(\mathbf{F}, \xi) a^k a^l k_\alpha k_\beta > 0$$
$$\text{for all} \quad \mathbf{F} \in \mathscr{L}_+, \mathbf{a}, \mathbf{k} \neq 0, \ \xi \in \overline{\Omega}. \tag{2.5.4a}$$

Indeed, the inequality $\varrho_0 W_{,F^k{}_\alpha F^l{}_\beta}(\nabla\boldsymbol{\chi}_0(\xi), \xi)\ a^k a^l k_\alpha k_\beta > 0$ for some $\boldsymbol{\chi}_0 \in \boldsymbol{\chi}_1 + \mathscr{V}$, all $\mathbf{a}$, $\mathbf{k} \neq 0$ and a.a. $\xi \in \Omega$ ensures existence, uniqueness and continuous dependence on the data of solutions to BVPs linearized around $\boldsymbol{\chi}_0$. Such solutions represent small perturbations of $\boldsymbol{\chi}_0$.

In the proof of the Hadamard theorem we used test functions with supports contained in sufficiently small balls $\mathscr{K}(\xi_0, \delta)$. Hence the thesis remains true if we assume that (2.5.3) holds only for $\operatorname{supp}\mathbf{u} \subset \mathscr{K}(\xi, \delta) \cap \Omega$, $\xi \in \Omega$, for $\delta > 0$ sufficiently small. SE turns out to be equivalent to the inequality

$$\inf\left\{ \int_\Omega dL^3\, \varrho_0 W_{,F^k{}_\alpha F^l{}_\beta}(\mathbf{F}_0(\xi), \xi) u^k{}_{,\xi^\alpha} u^l{}_{,\xi^\beta} \middle|\ \operatorname{supp}\mathbf{u} \subset \mathscr{K}(\xi, \delta), \ \|\nabla\mathbf{u}\|_2 = 1 \right\} > 0$$

or some sufficiently small $\delta > 0$ (Fichera, 1972a).

2.5.3 Some Theorems on Convex Functions

We shall need some lemmas on convex and strictly convex functions. Recall that a function f defined on a convex subset $\mathscr{U}$ of $\boldsymbol{R}^n$ is said to be **convex** if

$$f(\nu\mathbf{x} + (1-\nu)\mathbf{y}) \leqslant \nu f(\mathbf{x}) + (1-\nu) f(\mathbf{y}) \quad \forall \mathbf{x}, \mathbf{y} \in \mathscr{U}, \ \nu \in [0, 1].$$

Lemma 2.5.2

Suppose that f is a differentiable function on the interval $I \subset \boldsymbol{R}$. Then the following assertions are equivalent:

(i) f is convex on I;

(ii) $f(x) - f(y) \geqslant f'(y)(x-y)\ \forall x, y \in I$;

(iii) f' is non-decreasing on I.

Proof

(a) Let f be convex and let $x > y$ for definiteness. Then

$$f'(y) = \lim_{\nu \to 0} \frac{f(y + \nu(x-y)) - f(y)}{x-y} \leqslant \frac{f(x) - f(y)}{x-y}.$$

Hence (i) implies (ii).

(b) Let f satisfy (ii). Then

$$f(x)-f(y) \geqslant f'(y)(x-y),$$
$$f(y)-f(x) \geqslant f'(x)(y-x).$$

Let $x > y$ for definiteness. It follows that

$$f'(x) \geqslant \frac{f(x)-f(y)}{x-y} \geqslant f'(y).$$

Hence (ii) implies (iii).

(c) Suppose that f' is non-decreasing. Let $y > x_o$, $y, x_o \in I$. Since f' is everywhere defined and bounded on $[x_o, y]$ by the numbers $f'(x_0)$ and $f'(y)$, both finite, we conclude that f is Lipschitz continuous (by Rolle's theorem, Bourbaki, 1960) on $[x_o, y]$. Hence

$$f(y) = \int_{x_o}^{y} dx\, f'(x)+f(x_o).$$

For $\nu \in [0, 1]$

$$(1-\nu)f(y)+\nu f(x)-f\big(\nu x+(1-\nu)y\big)$$
$$= \nu(x-y)\int_0^1 d\lambda\,[f'(y+\lambda(x-y))-f'(y+\nu\lambda(x-y))] \geqslant 0.$$

Hence (iii) implies (i). □

Remark

The following can be proved:

(i) A convex function f: $I \to \boldsymbol{R}$ is continuous and has a left derivative $f'(x-0)$ and a right derivative $f'(x+0)$ at every $x \in I$, $f'(x-0) \leqslant f'(x+0)$ and the functions $x \mapsto f'(x-0)$ and $x \mapsto f'(x+0)$ are non-decreasing.

(ii) A function f: $I \to \boldsymbol{R}$ is convex iff the following two conditions are satisfied: (a) f is continuous and has a derivative at every point of I except for a countable set of points; (b) for any numbers x, $y \in I$ such that $x \leqslant y$ and the derivatives $f'(x)$, $f'(y)$ exist we have the inequality $f'(x) \leqslant f'(y)$. For the proof, see Bourbaki (1960, Section 1.4).

Corollary 2.5.3

Suppose that the function f: $\mathscr{U} \to \boldsymbol{R}$ is differentiable at every point of the open convex subset $\mathscr{U}$ of $\boldsymbol{R}^n$. Then the following assertions are equivalent:

(i) f is convex on $\mathscr{U}$;
(ii) $f(\mathbf{x})-f(\mathbf{y}) \geqslant \langle f'(\mathbf{y})|\mathbf{x}-\mathbf{y}\rangle$;
(iii) $\langle f'(\mathbf{x})-f'(\mathbf{y})|\mathbf{x}-\mathbf{y}\rangle \geqslant 0$.

If f'' exists at every point of $\mathscr{U}$, then (i) is equivalent to
(iv) $\langle f''(\mathbf{x})(\mathbf{y}-\mathbf{x})|\mathbf{y}-\mathbf{x}\rangle \geqslant 0 \;\forall \mathbf{x}, \mathbf{y} \in \mathscr{U}$.

Proof

Convexity of f on $\mathscr{U}$ is equivalent to convexity of the function $t \mapsto g(t) \equiv f(y+ +t(\mathbf{x}-\mathbf{y}))$ for arbitrary $\mathbf{x}, \mathbf{y} \in \mathscr{U}$. □

In the above corollary differentiability is meant in the sense of Gâteaux.

Definition 2.5.4

A function f: $\mathscr{U} \to \boldsymbol{R}$, $\mathscr{U}$ convex in $\boldsymbol{R}^n$, is **strictly convex** if

$$f(\mathbf{x}+\nu(\mathbf{x}-\mathbf{y})) < \nu f(\mathbf{x})+(1-\nu)f(\mathbf{y})$$
$$\forall \mathbf{x}, \mathbf{y} \in \mathscr{U},\; \mathbf{x} \neq \mathbf{y},\; \nu \in]0, 1[.$$

Lemma 2.5.5

Let f: $\mathscr{U} \to \boldsymbol{R}$ be continuous and differentiable (in the sense of Gâteaux) at every point of $\mathscr{U}$. Then the following are equivalent:
(i) f is strictly convex;
(ii) $f(\mathbf{x})-f(\mathbf{y}) > \langle f'(\mathbf{y})|\mathbf{x}-\mathbf{y}\rangle$ for every $\mathbf{x} \neq \mathbf{y}, \mathbf{x}, \mathbf{y} \in \mathscr{U}$;
(iii) f' is **strictly monotone**:

$$\langle f'(\mathbf{x})-f'(\mathbf{y})|\mathbf{x}-\mathbf{y}\rangle > 0 \quad \text{for every pair } \mathbf{x} \neq \mathbf{y},\; \mathbf{x},\; \mathbf{y} \in \mathscr{U}.$$

Proof

(A) Let $n = 1$, $y > x$,

$$\varphi(\xi) := f(x)+\frac{f(y)-f(x)}{y-x}\,\xi-f(x+\xi).$$

Obviously $\varphi(0) = 0 = \varphi(y-x)$. Suppose that f is strictly convex. Then $\varphi(\xi) > 0$ for $0 < \xi < y-x$ and φ has a proper maximum in $]0, y-x[$, at ξ_0, say. From the very definition of the derivative it follows that $\varphi'(\xi_0) = 0$, $\varphi(\xi_0) > 0$. Hence

$$\frac{f(y)-f(x)}{y-x} = f'(x+\xi_0), \quad \xi_0 < y-x.$$

Suppose that $f'(y) = \dfrac{f(y)-f(x)}{y-x}$. Since f' is non-decreasing it is constant on $[x+\xi_0, y]$. Hence f is a linear function over a finite interval and cannot be strictly convex. Consequently

$$f(y)-f(x) > f'(x)(y-x),$$
$$f'(x) < \frac{f(y)-f(x)}{y-x} < f'(y) \quad \text{for } y > x.$$

If f' is (strictly) increasing then the argument (c) of the proof of Lemma 2.5.2 shows that f is strictly convex.

(B) Let $f\colon \mathscr{U} \to \boldsymbol{R}$, $\mathscr{U} \subset \boldsymbol{R}^n$ be convex, $g(t) :\equiv f(\mathbf{x}+t(\mathbf{y}-\mathbf{x}))$, $\mathbf{y} \neq \mathbf{x}$. It is enough to note that $g(1)-g(0) > g'(0)$ is equivalent to $f(\mathbf{y})-f(\mathbf{x}) > \langle f'(\mathbf{x})|\mathbf{y}-\mathbf{x}\rangle$. Also, $g'(1)-g'(0) > 0$ is equivalent to strict monotonicity of f'. □

Corollary 2.5.6
If f is strictly convex then f' is one-to-one.

Proof
The thesis follows from the strict monotonicity of f'. □

Lemma 2.5.7
Let $\mathscr{U} \subset \boldsymbol{R}^n$ be open and convex and suppose that $f \in \mathscr{C}^1$ and f has second-order derivatives everywhere in $\mathscr{U}$. Then f is convex iff f'' is positive everywhere in $\mathscr{U}$. If f'' is positive definite everywhere in in $\mathscr{U}$, then f is strictly convex.

Proof
For $n = 1$ monotonicity of f' implies that f'' exists a.e. on $\mathscr{U} = I$ and $f''(x) \geqslant 0\ \forall\, x \in I$. Hence for $n \geqslant 1$ monotonicity of the mapping $f'\colon \mathscr{U} \to \boldsymbol{R}^n$ implies that the matrix $f''(\mathbf{x})$ is positive for every $\mathbf{x} \in \mathscr{U}$: $f''(\mathbf{x})[\mathbf{v}, \mathbf{v}] \geqslant 0$ $\forall\, \mathbf{v} \in \boldsymbol{R}^n$.

Suppose that the matrix $f''(\mathbf{x})$ of second-order derivatives of f exists at every $\mathbf{x} \in \mathscr{U}$ and is positive (or positive definite). Then $\langle f'(\mathbf{x})-f'(\mathbf{y})|\mathbf{x}-\mathbf{y}\rangle = \langle \mathbf{x}-\mathbf{y}|f''(\mathbf{z})(\mathbf{x}-\mathbf{y})\rangle$ for some $\mathbf{z} \in \mathscr{U}$, by Rolle's theorem, and the mapping f' is (strictly) monotone. □

Lemma 2.5.8
Suppose that the function $f\colon I \to \boldsymbol{R}$, defined on an interval $I \subset \boldsymbol{R}$, is continuous and has a second-order derivative at every $x \in I$. Then f is strictly convex iff $f''(x) \geqslant 0\ \forall x \in I$ and $f''(x) > 0$ for x ranging over a dense subset of I.

Proof
See Bourbaki (1960).

Lemma 2.5.9 (Ball, 1980)
Let $f\colon \mathscr{U} \to \boldsymbol{R}$ be defined on a convex subset $\mathscr{U}$ of $\boldsymbol{R}^n$, $f \in \mathscr{C}^1(\mathscr{U})$. Then f is strictly convex iff the following two conditions are satisfied:

(i) the mapping $f'\colon \mathscr{U} \to \boldsymbol{R}^n$ is locally one-to-one;

(ii) $\exists\, \mathbf{x}_o \in \mathscr{U}\ \exists\, \varepsilon > 0\ f(\mathbf{x})-f(\mathbf{x}_o) \geqslant \langle f'(\mathbf{x}_o)|\mathbf{x}-\mathbf{x}_o\rangle\ \forall\, \mathbf{x} \in \mathscr{K}(\mathbf{x}_o, \varepsilon)\cap\mathscr{U}$.

For a general convex function f on an open convex $\mathscr{U} \subset \boldsymbol{R}^n$ continuity of f at a point $\mathbf{x}_o \in \mathscr{U}$ follows from local boundedness of f near $\mathbf{x}_o$.

Remark
Every convex finite-valued function f on a convex open $\mathscr{U} \subset \boldsymbol{R}^n$ is continuous.

For the proof note that f is bounded from above on open sets of the form

$$\left\{\sum_{i=1}^{n+1} \nu_i \mathbf{z}_i \middle| \; \nu_i \geqslant 0 \text{ for } i = 1, \ldots, n+1, \; \sum_{i=1}^{n+1} \nu_i = 1\right\},$$

where $\mathbf{z}_i \in \mathscr{U}$ and no $\mathbf{z}_i$ is a convex combination of the remaining $\mathbf{z}_j$. The bound is $\max\{f(\mathbf{z}_i) | \; i = 1, \ldots, n+1\}$.

It is enough to consider the case of a function f satisfying the conditions $\mathbf{0} \in \mathscr{U}$, $f(\mathbf{0}) = 0$ and prove continuity at $\mathbf{0}$. From the preceding paragraph we know that $f(\mathbf{z}) < a$ for all $\mathbf{z}$ in a nbhd $\mathscr{V}$ of $\mathbf{0}$. Let $\nu \in \,]0, 1[$. For $\mathbf{z} \in \nu\mathscr{V}$ we have that $f(\mathbf{z}) \leqslant (1-\nu)f(\mathbf{0}) + \nu f(\mathbf{z}/\nu)$. Since $(1+\nu)^{-1}\mathbf{z} + \nu(1+\nu)^{-1}(-\nu^{-1}\mathbf{z}) = 0$, we have for $\mathbf{z} \in -\nu\mathscr{V}$ the inequality $f(\mathbf{z}) \geqslant (1+\nu)f(\mathbf{0}) - \nu f(-\mathbf{z}/\nu)$. Hence for $\mathbf{z} \in (\nu\mathscr{V}) \cap (-\nu\mathscr{V})$ we have that $|f(\mathbf{z})| \leqslant \nu a$, which proves continuity at $\mathbf{0}$. □

Theorem 2.5.10
Let $\mathscr{U}$ be a convex open subset of $\boldsymbol{R}^n$.

The function f: $\mathscr{U} \to \boldsymbol{R}$ is convex iff $\forall \, \mathbf{x} \in \mathscr{U} \; \exists \mathbf{g}(\mathbf{x}) \in \boldsymbol{R}^n$ such that

$$f(\mathbf{y}) - f(\mathbf{x}) \geqslant \langle \mathbf{g}(\mathbf{x}) | \mathbf{y} - \mathbf{x} \rangle \;\; \forall \, \mathbf{y} \in \mathscr{U}. \tag{2.5.8}$$

Proof
Necessity follows from Proposition 1.5.2 of Ékeland and Témam (1976) (set $f(\mathbf{y}) = \infty$ for $\mathbf{y} \,\overline{\in}\, \mathscr{U}$). Sufficiency follows from the formula

$$f(\mathbf{y}) = \sup\{f(\mathbf{x}) + \langle \mathbf{g}(\mathbf{x}) | \mathbf{y} - \mathbf{x} \rangle | \; \mathbf{x} \in \mathscr{U}\} \;\; \forall \, \mathbf{y} \in \mathscr{U}.$$

Since f is the supremum of a family of affine functions, it is convex by Proposition 1.2.2 in Ékeland and Témam (op. cit.). □

2.5.4 Rank-One Convexity

Definition 2.5.11
The Carathéodory function W: $L_+ \times \Omega \to \overline{\boldsymbol{R}}$ is said to be **rank-one convex** if

$$W(\mathbf{F}_0 + \nu \mathbf{a} \otimes \mathbf{k}, \xi) \leqslant \nu W(\mathbf{F}_0 + \mathbf{a} \otimes \mathbf{k}, \xi) + (1-\nu) W(\mathbf{F}_0, \xi) \tag{2.5.9}$$

or all $\mathbf{F}_0 \in L_+$, $\nu \in [0, 1]$, all $\mathbf{a}$, $\mathbf{k} \in \boldsymbol{R}^3$, such that

$$1 + \overset{-1}{F_0}{}^{\alpha}{}_{i} a^i k_\alpha > 0 \tag{2.5.10}$$

and almost all $\xi \in \Omega$.

The Carathéodory function $W: L_+ \times \Omega \to \boldsymbol{R}$ is said to be **strictly rank-one convex** if

$$W(\mathbf{F}_0+\nu\mathbf{a}\otimes\mathbf{k}, \xi) < \nu W(\mathbf{F}_0+\mathbf{a}\otimes\mathbf{k}, \xi)+ (1-\nu) W(\mathbf{F}_0, \xi) \qquad (2.5.11)$$

for all $\mathbf{F}_0 \in L_+, \nu \in]0, 1[$, $\mathbf{a}, \mathbf{k} \neq \mathbf{0}$ satisfying (2.5.10) and almost all $\xi \in \Omega$.

Equation (2.5.10) is equivalent to $\det(\mathbf{F}_0+\nu\mathbf{a}\otimes\mathbf{k}) > 0 \ \forall \nu \in [0, 1]$.

Lemma 2.5.12

Let $W: \mathscr{L}_+ \times \Omega \to \boldsymbol{R}$ be a Carathéodory function and let $W(\cdot\,, \xi) \in \mathscr{C}^2(\mathscr{L}_+)$ for a.e. $\xi \in \Omega$. The condition SE, equation (2.5.4a), is satisfied iff W is strictly rank-one convex. The condition LH, equation (2.5.4), is satisfied iff W is rank-one convex.

Proof

It is sufficient to note that $\nu \mapsto W(\mathbf{F}_0+\nu\mathbf{a}\otimes\mathbf{k}, \xi)$ is strictly convex iff equation (2.5.4a) holds for $\mathbf{F} = \mathbf{F}_0+\nu\mathbf{a}\otimes\mathbf{k}$ for a dense set of $\nu \in [0, 1]$. By the continuity of the left-hand side of equation (2.5.4a) with respect to $\mathbf{F}$ equation (2.5.4a) is then satisfied for all ν, including $\nu = 0$. The thesis follows from the fact that $\mathbf{F}_0 \in \mathscr{L}_+$ is arbitrary.

The considerations concerning equation (2.5.4) and LH are analogous. □

Suppose that ϱ_0 and $\hat{S}_k{}^\alpha$ are continuous functions of $\xi \in \bar{\Omega}$, $\nabla\boldsymbol{\chi}_1 \in \mathscr{C}^0(\bar{\Omega})$, $\mathbf{u} \in \mathscr{C}^0(\bar{\Omega})\cap\mathscr{V}$ is a minimizer of $F = \mathscr{W}+\Phi$ and $\nabla\mathbf{u}$ has a jump discontinuity at a sufficiently smooth surface $\Sigma \subset \Omega$. (A precise definition of jump discontinuities can be found in Section 3.1. It follows then that the limits $\nabla\mathbf{u}_+(\xi_0)$, $\nabla\mathbf{u}_-(\xi_0)$ of $\nabla\mathbf{u}(\xi)$ for $\xi \to \xi_0$ from either side of Σ satisfy a relation

$$\nabla\mathbf{u}_-(\xi_0)-\nabla\mathbf{u}_+(\xi_0) = \mathbf{a}(\xi_0)\otimes\mathbf{n}(\xi_0), \qquad (2.5.12)$$

where $\mathbf{n}(\xi_0)$ denotes the unit normal to Σ at ξ_0 pointing into $\Omega_+ \subset \Omega$. For all $\mathbf{v} \in \mathscr{V}$ vanishing near $\partial\Omega$

$$\int_\Omega dL^3\,\hat{S}_k{}^\alpha(\nabla\boldsymbol{\chi}_1+\nabla\mathbf{u}, \xi)v^k{}_{,\xi^\alpha}+\int_\Omega dL^3\,\phi_{,x^k}v^k = 0 \qquad (2.5.13)$$

hence $S_k{}^\alpha(\xi) \equiv \hat{S}_k{}^\alpha(\nabla\boldsymbol{\chi}_1(\xi)+\nabla\mathbf{u}(\xi), \xi)$ satisfies the equation

$$S_k{}^\alpha{}_{,\xi^\alpha} = \phi_{,x^k}$$

in the sense of distributions. Since $S_k{}^\alpha(\cdot)$ is continuous except for a jump discontinuity at Σ, we shall apply the calculus in the BV class of functions (Section 3.2) to conclude that

$$[\hat{S}_k{}^\alpha(\nabla\boldsymbol{\chi}_1(\xi_0)+\nabla\mathbf{u}_-(\xi_0), \xi_0)-\hat{S}_k{}^\alpha(\nabla\boldsymbol{\chi}_1(\xi_0)+\nabla\mathbf{u}_+(\xi_0), \xi_0)]n_\alpha(\xi_0) = 0,$$
$$k = 1, 2, 3 \qquad (2.5.13a)$$

for H^2—almost all $\xi_0 \in \Sigma$.

Alternatively, one can note that equation (2.5.13) has the form of a VWP and conclude by the method outlined in Section 1.2 that the conservation law

$$\int_{\partial\mathscr{P}} dH^2\, S_k^{\alpha}\nu_\alpha = \int_{\mathscr{P}} dL^3\, \phi_{,x^k}, \quad k = 1, 2, 3,$$

holds for a large collection of subsets $\mathscr{P}$ of Ω. Taking $\mathscr{P}$ to be narrow slices containing a part Σ_0 of Σ and contracting $\mathscr{P}$ to Σ_0 one readily arrives at (2.5.13a).

Setting $\mathbf{F} = \nabla\boldsymbol{\chi}_1(\xi_0)+\nabla\mathbf{u}_+(\xi_0)$ we have from (2.5.12) and (2.5.13a) the following equation

$$\left[\frac{\partial W}{\partial F^k{}_\alpha}(\mathbf{F}+\mathbf{a}(\xi_0)\otimes\mathbf{n}(\xi_0), \xi_0)-\frac{\partial W}{\partial F^k{}_\alpha}(\mathbf{F}, \xi_0)\right] n_\alpha(\xi_0)a^k = 0. \quad (2.5.14)$$

Suppose that $\mathbf{a}(\xi_0) \neq \mathbf{0}$. Equation (2.5.14) implies that the function $\mathbf{a} \mapsto W(\mathbf{F}+\mathbf{a}\otimes\mathbf{n}, \xi_0)$ cannot be strictly convex, in view of Lemma 2.5.5 (iii), and hence W cannot be strictly rank-one convex. It follows that strict rank-one convexity excludes continuous piecewise $\mathscr{C}^1$ solutions $\mathbf{u} \in \mathscr{V}$ from $F \to \min$ with non-trivial jump discontinuities of $\nabla\mathbf{u}$.*

Since $W^{1,p}$, $p \geqslant 1$, is contained in the class BV, defined in Section 3.2, this conclusion extends to $\nabla\mathbf{u} \in W^{1,p}(\Omega)$ with jump discontinuities in the sense of Section 3.2.

Ball (1980) proved that a function W independent of ξ is strictly rank-one convex iff the following two conditions are satisfied:

(i) for an arbitrary plane $\Sigma \subset \boldsymbol{R}^3$, all weak solutions of equilibrium equations $S_k{}^\alpha{}_{,\xi^\alpha} = 0$ which are affine in either component of $\Omega\setminus\Sigma$, are $\mathscr{C}^1$ in Ω;

(ii) $\exists\, \mathbf{F}_0 \in \mathscr{L}_+$, $\mathbf{k} \in \boldsymbol{R}^3$, $\varepsilon > 0$, such that $\forall\, \mathbf{a} \in \boldsymbol{R}^3$ the inequality $|\mathbf{a}| < \varepsilon$ implies that

$$W(\mathbf{F}_0+\mathbf{a}\otimes\mathbf{k})-W(\mathbf{F}_0) \geqslant \frac{\partial W}{\partial F^i{}_\alpha}(\mathbf{F}_0)a^i k_\alpha.$$

The condition (i) refers to some arbitrary affine structure on $\mathscr{B}$ defined by its embedding $\boldsymbol{\varkappa}\colon \mathscr{B} \to \boldsymbol{R}^3$. In order to give a more satisfactory formulation to this condition let us assume that $x^p = \chi_1{}^p(\xi^\alpha)$ is a $\mathscr{C}^1$ configuration of the body. We shall use it as a reference configuration and assume that $W(F^k{}_\alpha, \xi) = \overline{W}(G^k{}_p)$ for $F^k{}_\alpha = G^k{}_p\chi_1{}^p{}_{,\xi^\alpha}$, $\varrho_1(\mathbf{x}) :\equiv \varrho_0(\xi)\ \det\nabla\boldsymbol{\chi}_1(\xi)^{-1} = \text{const}$. It is readily seen that $W(\cdot\,,\,\cdot)$ is strictly rank-one convex iff $\overline{W}(\cdot)$ is strictly rank-one convex.

* Except perhaps for solutions with special surfaces of discontinuity.

Let $y^k = x^k + u^k(\mathbf{x})$ denote the actual configuration corresponding to zero body loads: $\phi_{,y^k} \equiv 0$. The condition (i) states that there are no piecewise affine displacements

$$u^k(\mathbf{x}) \equiv \begin{cases} H^k{}_p x^p + h^k & \text{for } n_p x^p > \lambda,\ \mathbf{x} \in \chi_1(\Omega), \\ (H^k{}_p + a^k n_p) x^p + g^k & \text{for } n_p x^p < \lambda,\ \mathbf{x} \in \chi_1(\Omega) \end{cases} \tag{2.5.15}$$

(with arbitrary $\lambda \in \boldsymbol{R}$, $\mathbf{h}, \mathbf{a}, \mathbf{n}, \mathbf{g} \in \boldsymbol{R}^3$, $\mathbf{H} \in \mathcal{M}_{3\times 3}$), satisfying the equilibrium equations

$$\frac{\partial}{\partial x^p} \hat{S}_k{}^p(\mathbf{E} + \nabla \mathbf{u}) = 0, \quad \hat{S}_k{}^p := \varrho_1 \frac{\partial \overline{W}}{\partial G^k{}_p},$$

in the weak sense, for which $\mathbf{a} \neq \mathbf{0}$.

The condition (ii) can also be readily rewritten in terms of $\overline{W}$.

In a configuration (2.5.15) the body consists of two parts with a constant stress in either of them. Phase transformations of the first kind involve similar situations. In order to consider a phase transformation of the first kind it is, however, necessary to take into account the (stable and natural) BCs as well as the behaviour of $F = \mathscr{W} + \overline{\Phi}$ in a nbhd of the configuration under consideration.

2.5.5 Quasiconvexity

The theory of linearized static BVPs perturbed around a solution of a nonlinear BVP (Section 2.2) as well as nonlinear elastodynamics (Chapter 3) are essentially* based on the assumption of SE, although occasionally somewhat stronger assumptions were used in Section 2.2. Unfortunately SE is not a sufficiently strong assumption to ensure existence or uniqueness in nonlinear elastostatics. In order to find a more satisfactory constitutive assumption we shall again start from an observation about the behaviour of the function W in a nbhd of the graph of a $\mathscr{C}^1$ minimizer and try to extrapolate this behaviour to the whole domain of definition of W. The effectiveness of the assumption in the existence proofs will provide a test of its "correctness".

Definition 2.5.13 (Morrey, 1952, 1966; Meyers, 1965)
The Carathéodory function W: $\mathscr{L}_+ \times \Omega \to \boldsymbol{R}$ is said to be **quasiconvex at** $(\mathbf{F}_0, \xi_0) \in \mathscr{L}_+ \times \Omega$ if for every bounded open set $\mathscr{U} \subset \boldsymbol{R}^3$ and for every $\mathscr{C}_0^\infty$ function $\mathbf{u}$: $\mathscr{U} \to \boldsymbol{R}^3$, satisfying the inequality $\det(\mathbf{F}_0 + \nabla \mathbf{u}(\zeta)) > 0 \ \forall \xi \in \mathscr{U}$ the following inequality holds:

$$\int_{\mathscr{U}} dL^3(\xi)\, W(\mathbf{F}_0 + \nabla \mathbf{u}(\xi), \xi_0) \geqslant \int_{\mathscr{U}} dL^3\, W(\mathbf{F}_0, \xi_0) = L^3(\mathscr{U})\, W(\mathbf{F}_0, \xi_0). \tag{2.5.16}$$

* In nonlinear elastodynamics growth conditions also play an important role.

Physically (2.5.16) amounts to the statement that a homogeneous elastic body $\mathscr{B}'$ which has the considered constitutive properties (relative to a homogeneous reference configuration $\boldsymbol{\varkappa}'\colon \mathscr{B}' \to \mathscr{U}$) and which is subject to zero body loads and fixed Dirichlet BCs on $\partial\mathscr{U}$ compatible with the homogeneous deformation $\mathbf{F}_0$ (relative to the reference $\boldsymbol{\varkappa}'$) attains its lowest energy precisely in the homogeneous deformation $\mathbf{F}_0$.

It is reasonable to expect that a local inhomogeneous perturbation of an nitially homogeneous stress increases the strain energy of the body.

In the case of a function W depending on the derivatives of $\boldsymbol{\chi}$ of order $\leqslant m$ quasiconvexity at a jet$(\xi_0, \mathbf{x}_0, \mathbf{d}^\alpha)$, $|\alpha| \leqslant m$, bears on the dependence of W on the m-th order derivatives at fixed $(\xi_0, \mathbf{x}_0, \mathbf{d}^\alpha)$, $|\alpha| < m$.

Let

$$\mathscr{A} := \{\boldsymbol{\chi} \in W^{1,1}_{loc}(\Omega) |\ \det\nabla\boldsymbol{\chi} > 0 \text{ a.e. in } \Omega,\ \mathscr{W}(\boldsymbol{\chi}) < \infty\}.$$

Theorem 2.5.14 (Ball, 1977)
Suppose that the function W is defined and continuous on $\mathscr{L}_+ \times \bar{\Omega}$, $\xi_0 \in \Omega$, $\tilde{\boldsymbol{\chi}}_0 \in \mathscr{A}$ and

$$\mathscr{W}(\tilde{\boldsymbol{\chi}}_0) \leqslant \mathscr{W}(\tilde{\boldsymbol{\chi}}_0 + \mathbf{u}) \tag{2.5.17}$$

for all $\mathbf{u} \in \mathscr{C}_0^\infty(\Omega)$ such that $\tilde{\boldsymbol{\chi}}_0 + \mathbf{u} \in \mathscr{A}$ and $\sup\{|\mathbf{u}(\xi)|\ |\ \xi \in \Omega\}$ is sufficiently small. Suppose that there is a mapping $\boldsymbol{\chi}_0\colon \Omega \to \boldsymbol{R}^3$ such that $\boldsymbol{\chi}_0 = \tilde{\boldsymbol{\chi}}_0$ a.e. and $\nabla\boldsymbol{\chi}_0 = \nabla\tilde{\boldsymbol{\chi}}_0$ a.e., $\nabla\boldsymbol{\chi}_0$ is continuous at ξ_0 and $\det\nabla\boldsymbol{\chi}_0(\xi_0) > 0$. Then W is quasiconvex at $(\nabla\boldsymbol{\chi}_0(\xi_0), \xi_0)$.

Proof
Let $\varepsilon > 0$, $\mathbf{v} \in \mathscr{C}_0^\infty(\mathscr{U})$, with $\mathscr{U}$ open and bounded in $\boldsymbol{R}^3$, $\det(\mathbf{F}_0 + \nabla\mathbf{v}(\zeta)) > 0$ $\forall \zeta \in \mathscr{U}$, $\mathbf{F}_0 := \nabla\boldsymbol{\chi}_0(\xi_0)$. We set

$$\mathbf{u}_\varepsilon(\xi) :\equiv \begin{cases} \varepsilon\mathbf{v}\big(\varepsilon^{-1}(\xi - \xi_0)\big) & \text{for } \varepsilon^{-1}(\xi - \xi_0) \in \mathscr{U}, \\ \mathbf{0} & \text{otherwise,} \end{cases}$$

$$\boldsymbol{\chi}_\varepsilon(\xi) :\equiv \boldsymbol{\chi}_0(\xi) + \mathbf{u}_\varepsilon(\xi).$$

For $\varepsilon > 0$ small enough $\operatorname{supp}\mathbf{u}_\varepsilon \subset \xi_0 + \varepsilon\mathscr{U}$ and hence $\mathbf{u}_\varepsilon \in \mathscr{C}_0^\infty(\Omega)$. Let $\zeta := \varepsilon^{-1}(\xi - \xi_0)$. The function $\xi \mapsto \big|W(\nabla\boldsymbol{\chi}_\varepsilon(\xi), \xi)\big| \equiv \big|W(\nabla\boldsymbol{\chi}_0(\xi) + \nabla\mathbf{v}(\xi), \xi)\big|$ is bounded on $\xi_0 + \varepsilon\mathscr{U}$ uniformly with respect to $\varepsilon \leqslant \varepsilon_o$, $\varepsilon_o > 0$, provided ε_o is sufficiently small. Indeed, for some $\varepsilon_o > 0$ the set $\bigcup_{\varepsilon \leqslant \varepsilon_o} (\xi_0 + \varepsilon\mathscr{U})$ is contained in a small compact nbhd of ξ_0. Since $\nabla\boldsymbol{\chi}_0$ is continuous at ξ_0, $|\nabla\boldsymbol{\chi}_0(\xi) - F_0|$ $< \delta$ and $|\nabla\mathbf{u}_\varepsilon| < M$ for ξ in this set, with both constants independent of ε. Moreover $\nabla\mathbf{u}_\varepsilon(\xi) = \nabla\mathbf{v}(\zeta)$, hence $\boldsymbol{\chi}_\varepsilon \in \mathscr{A}$ and $\mathbf{u}_\varepsilon \in \mathscr{C}_0^\infty(\Omega)$ for sufficiently small ε. On account of our hypotheses

$$\int_{\mathscr{U}} dL^3(\zeta)\, W(\nabla\chi_0(\xi_0+\varepsilon\zeta)+\nabla\mathbf{v}(\zeta),\ \xi_0+\varepsilon\zeta)$$

$$\geqslant \int_{\mathscr{U}} dL^3(\zeta)\, W(\nabla\chi_0(\xi_0+\varepsilon\zeta),\ \xi_0+\varepsilon\zeta).$$

Since both integrands are bounded uniformly with respect to $\varepsilon \leqslant \varepsilon_0$, we can go to the limit $\varepsilon \to 0$ using the Lebesgue dominated convergence theorem. As a result we obtain equation (2.5.16). □

It follows from Theorem 2.5.14 that equation (2.5.16) holds for every bounded open subset $\mathscr{U} \subset \boldsymbol{R}^3$ provided it holds for some special bounded set $\mathscr{U}_0 \subset \boldsymbol{R}^3$. Indeed, set $\Omega = \mathscr{U}_0$, $\chi_0{}^k(\xi) \equiv F_0{}^k{}_\alpha \xi^\alpha$. By Theorem 2.5.14 W is quasiconvex at $(\mathbf{F}_0, \xi_0)$ if equation (2.5.16) is satisfied for $\mathscr{U} = \mathscr{U}_0$.

Definition 2.5.15

A Carathéodory function $W\colon \mathscr{L}_+ \times \bar{\Omega} \to \boldsymbol{R}$ is said to be **quasiconvex** (QC) if it is quasiconvex at every point of $\mathscr{L}_+ \times \bar{\Omega}$.

Quasiconvexity is an interesting candidate for a constitutive assumption. As pointed out by Ball (1977a) QC does not follow from existence of $\mathscr{C}^1$-minimizers. It does follow however from existence of $\mathscr{C}^1$-minimizers in the following special case: (a) $\mathscr{B}$ is a homogeneous elastic cube in the reference configuration $\boldsymbol{\varkappa}$; (b) body loads vanish; (c) $\mathscr{B}$ is subject to Dirichlet boundary conditions which allow for a homogeneous deformation $\mathbf{F}_0$. Then W is quasiconvex at $\mathbf{F}_0$ (Ball, 1977). This conclusion is valid for arbitrary $\mathbf{F}_0 \in \mathscr{L}_+$ if the BCs vary.

The latter observation of Ball deserves some comments. Our primitive notion of a body incorporates two notions which in technological applications* are separated and regarded as independent of each other; viz. the notion of a material specified by its constitutive equations and the notion of a sample of arbitrary size and shape made of any material. From this point of view we are entitled to imagine samples of arbitrary size and shape for a given constitutive equation independent of ξ. A constitutive equation is acceptable if it ensures existence of solutions for BVPs for arbitrary samples of the corresponding material.

As regards sufficiency of QC for existence theorems, Moriey (1952) and Meyers (1965) proved that the functional $\mathscr{W}$ is sequentially wlsc on $W^{1,p}(\Omega)$

* The situation is less clear in the case of rocks and minerals.

if the function W is continuous on $\mathscr{L}_+ \times \bar{\Omega}$, quasiconvex, bounded from below by an integrable function $\psi\colon \Omega \to \boldsymbol{R}$ and if it satisfies the inequalities:

$$|W(\mathbf{F}+\mathbf{H}, \xi) - W(\mathbf{F}, \xi)| \leqslant K[1+(|\mathbf{F}|+|\mathbf{H}|)^{p-\gamma}]|\mathbf{H}|^{\gamma} \tag{2.5.18}$$

for some $\gamma \in]0, 1]$,

$$|W(\mathbf{F}, \xi+\zeta) - W(\mathbf{F}, \xi)| \leqslant K_1[1+|\mathbf{F}|^p]\eta(|\zeta|), \tag{2.5.19}$$

with $\eta(0) = 0$, η continuous and increasing.

For existence theorems we need a coercivity assumption as well, e.g.

$$W(\mathbf{F}, \xi) \geqslant K_2|\mathbf{F}|^p + a(\xi), \quad K_2 > 0,\ a \in \mathscr{L}^1(\Omega). \tag{2.5.20}$$

Conditions (2.5.18) and (2.5.19) are physically unrealistic in a nbhd of $\mathbf{F} = \mathbf{0}$ since we expect that $W \to \infty$ as $\det\mathbf{F} \to 0$. Another difficulty with QC is the fact that it involves integration and hence is difficult to verify. In Section 2.5.6 we shall replace QC by a pointwise inequality which is acceptable for all $\mathbf{F} \in \mathscr{L}_+$. We note, however, the following implication.

Theorem 2.5.16 (Ball, 1977)
If W is quasiconvex at $(\mathbf{F}_0, \xi_0) \in \mathscr{L}_+ \times \Omega$, then W is rank-one convex at $(\mathbf{F}_0, \xi_0)$, viz.

$$W(\mathbf{F}_0, \xi_0) \leqslant \nu W\big(\mathbf{F}_0 - (1-\nu)\mathbf{a}\otimes\mathbf{k}, \xi_0\big) + (1-\nu)\, W(\mathbf{F}_0 + \nu\mathbf{a}\otimes\mathbf{k}, \xi_0). \tag{2.5.21}$$

Remark
The relationship between rank-one convexity at $\mathbf{F}_0$ and rank-one convexity becomes apparent if one substitutes $\mathbf{F} = \mathbf{F}_0 - (1-\nu)\mathbf{a}\otimes\mathbf{k}$, $\mathbf{G} = \mathbf{F}_0 + \nu\mathbf{a}\otimes\mathbf{k}$, $\mathbf{F}_0 = \nu\mathbf{F} + (1-\nu)\mathbf{G}$ in (2.5.21).

Proof of the theorem (in two steps)

(1) Suppose that W is quasiconvex at $(\mathbf{F}_0, \xi_0)$ and $\operatorname{supp}\mathbf{u} \subset \mathscr{U}$, $\mathbf{u} \in W_0^{1,\infty}(\mathscr{U})$ for some bounded $\mathscr{U} \subset \boldsymbol{R}^3$. Clearly $\mathbf{F}_0 + \nabla\mathbf{u}(\xi) \in \mathscr{K}$ for almost all $\xi \in \mathscr{U}$, with $\mathscr{K}$ compact in $\mathscr{L}_+$. Since $\mathscr{C}_0^\infty(\mathscr{U})$ is dense in $W_0^{1,\infty}(\mathscr{U})$, we can find a sequence $\{\mathbf{u}_n\} \subset \mathscr{C}_0^\infty(\mathscr{U})$ such that $\mathbf{u}_n \to \mathbf{u}$ in $W_0^{1,\infty}(\mathscr{U})$. Since $\nabla\mathbf{u}_n$ converge in the ess sup norm, there is a compact $\mathscr{K}_1 \subset \mathscr{L}_+$ and an integer n_o such that for $n > n_0$ we have $\mathbf{F}_0 + \Delta\mathbf{u}_n(\xi)$ for all $\xi \in \mathscr{U}$. Since

$$\int_{\mathscr{U}} dL^3\, W\big(\mathbf{F}_0 + \nabla\mathbf{u}_n(\xi), \xi_0\big) \geqslant L^3(\mathscr{U})\, W(\mathbf{F}_0, \xi_0)$$

and $W(\cdot\,, \xi_0)$ is bounded on $\mathscr{K}_1$, the Lebesgue dominated convergence theorem implies that

$$\int_{\mathscr{U}} dL^3\, W(\mathbf{F}_0+\nabla\mathbf{u}(\xi), \xi_0) \geqslant L^3(\mathscr{U})\, W(\mathbf{F}_0, \xi_0)$$

in the limit $n \to \infty$, $n > n_o$, i.e. the inequality (2.5.11) remains valid for $\mathbf{u} \in W_0^{1,\infty}(\mathscr{U})$.

(2) We consider $\mathscr{U}$ as a subset of the *Euclidean* space $\boldsymbol{R}^3$ with the metric $\delta_{\alpha\beta}$. Let $\{\mathbf{i}_k |\ k = 1,2,3\}$ be an orthonormal basis in $\boldsymbol{R}^3$ and $y_k := (\boldsymbol{\xi}-\boldsymbol{\xi}_0)\cdot \mathbf{i}_k = \sum_{\alpha=1}^{3} (\xi^\alpha - \xi_0{}^\alpha) i_k{}^\alpha$.

Let

$$\mathscr{D} := \{\mathbf{y} \in \boldsymbol{R}^3 |\ -h < y_1 < H, |y_k| < \varrho \text{ for } k = 2, 3\}, h, H, \varrho > 0,$$

$$\mathbf{F}_1^- = \{\mathbf{y} \in \boldsymbol{R}^3 |\ y_1 = -h, |y_k| \leqslant \varrho \text{ for } k = 2, 3\},$$

$$\mathbf{F}_1^+ = \{\mathbf{y} \in \boldsymbol{R}^3 |\ y_1 = H, |y_k| \leqslant \varrho\}, \quad F_k^\mp = \{\mathbf{y} \in \bar{\mathscr{D}} |\ y_k = \mp\varrho\}$$

for $k = 2, 3$. Note that $\mathbf{0} \in \mathscr{D}$. Let $\Pi_k^\pm$ be the pyramid with the vertex at $\mathbf{0}$ and the base $F_k^\pm$, $k = 1, 2, 3$. Substitute into (2.5.11) the function $\mathbf{u} \in W_o^{1,\infty}(\boldsymbol{R}^3)$ with $\operatorname{supp}\mathbf{u} \subset \mathscr{D}$, defined by the conditions

(i) $\mathbf{u}|\Pi_k^\pm$ is affine for $k = 1, 2, 3$;

(ii) $\mathbf{u}(\mathbf{0}) = \mathbf{a} \in \boldsymbol{R}^3$;

(iii) $\mathbf{u}(\mathbf{y}) = \mathbf{0}$ for $\mathbf{y} \in F_k^\pm$, $\quad k = 1, 2, 3$.

(i), (ii) and (iii) imply that $\nabla\mathbf{u}(\mathbf{y}) = \lambda_k^\pm \mathbf{a}\otimes\mathbf{i}_k$ on $\Pi_k^\pm$, with $\lambda_1^- = -h^{-1}$, $\lambda_1^+ = H^{-1}$, $-\lambda_k^- = \varrho^{-1} = \lambda_k^+$ for $k = 2, 3$. Since

$$\int_{\mathscr{D}} dL^3(\mathbf{y})\, W(\mathbf{F}_0+\nabla\mathbf{u}(\mathbf{y}), \mathbf{0}) \geqslant W(\mathbf{F}_0, \mathbf{0})(2\varrho)^2(h+H)$$

in terms of the coordinates $\mathbf{y}$, we have that

$$\frac{4h\varrho^2}{3} W\left(\mathbf{F}_0 - \frac{1}{h}\mathbf{a}\otimes\mathbf{i}_1, \mathbf{0}\right) + \frac{4H\varrho^2}{3} W\left(\mathbf{F}_0 + \frac{1}{H}\mathbf{a}\otimes\mathbf{i}_1, \mathbf{0}\right)$$
$$+ \frac{2(H+h)\varrho^2}{3} \sum_{k=2}^{3} \left[W\left(\mathbf{F}_0 - \frac{1}{\varrho}\mathbf{a}\otimes\mathbf{i}_k, \mathbf{0}\right) + W\left(\mathbf{F}_0 + \frac{1}{\varrho}\mathbf{a}\otimes\mathbf{i}_k, \mathbf{0}\right)\right]$$
$$\geqslant W(\mathbf{F}_0, \mathbf{0})(2\varrho)^2(h+H).$$

Dividing both sides of the above inequality by $\varrho^2(h+H)$ and letting $\varrho \to \infty$, we conclude on account of the continuity of $W(\cdot\,, \mathbf{0})$ that

$$\frac{h}{h+H} W\left(\mathbf{F}_0 - \frac{1}{h}\mathbf{a}\otimes\mathbf{i}_1, \mathbf{0}\right) + \frac{H}{h+H} W\left(\mathbf{F}_0 + \frac{1}{H}\mathbf{a}\otimes\mathbf{i}_1, \mathbf{0}\right)$$
$$\geqslant W(\mathbf{F}_0, \mathbf{0}).$$

For $\nu = h(h+H)^{-1}$, $\mathbf{k} = [(h+H)/hH]\mathbf{i}_1$ we obtain (2.5.21). In view of the arbitrariness of $\mathbf{i}_1$, h, H this proves the theorem. □

2.5.6 Polyconvexity and Existence of Minimizers

Following Ball (1977, 1977a) we replace QC by a stronger "pointwise" constitutive assumption about W, viz. the polyconvexity (PC). PC is both easier to be verified and ensures existence under physically acceptable growth and continuity conditions on W.

Definition 2.5.17

$W(\cdot\,,\cdot)$ is said to be **polyconvex** if there is a convex function $g(\cdot\,,\xi)$ defined on a convex subset M of $\mathscr{L}\times\mathscr{L}\times\boldsymbol{R}_+ \subset \boldsymbol{R}^{19}$ such that

$$W(\mathbf{F},\xi) = g(\mathbf{F},\operatorname{adj}\mathbf{F},\det\mathbf{F},\xi) \tag{2.5.22}$$

for every $\mathbf{F}\in\mathscr{L}_+$ and a.e. $\xi\in\Omega$.

Here

$$(\operatorname{adj}\mathbf{F})^{\gamma}{}_{k} := \frac{1}{6}\in^{\alpha\beta\gamma}\in_{ijk} F^{i}{}_{\alpha}F^{j}{}_{\beta} = \frac{1}{3}(\det\mathbf{F})\overset{-1}{F}{}^{\gamma}{}_{k} \tag{2.5.23}$$

(read "the **adjugate of F**").

Remark

Let us define the mapping T: $\mathscr{L}_+ \to M$; $\mathbf{F}\mapsto(\mathbf{F},\operatorname{adj}\mathbf{F},\det\mathbf{F})$. The range $T(\mathscr{L}_+)$ of T is not convex. Hence the function g on the right-hand side of (2.5.22) has to be interpreted as the restriction to $T(\mathscr{L}_+)$ of a convex function defined on the convex hull $\operatorname{Co}T(\mathscr{L}_+)$ of $T(\mathscr{L}_+)$ and we can set $M = \operatorname{Co}T(\mathscr{L}_+)$. It can be shown (Ball, 1977a, pp. 208–209) that $M = \mathscr{L}\times\mathscr{L}\times\boldsymbol{R}_+$. If g is known to be the restriction to $T(\mathscr{L}_+)$ of a convex function, then a possible extension of g to a convex function on M is

$$\tilde{g}(\mathbf{z}) = \inf\Big\{\sum_{i=1}^{n}\nu_i g(\mathbf{z}_i) \Big| \ \mathbf{z}_i\in T(\mathscr{L}_+),\ \nu_i \geqslant 0,$$

$$i = 1,\dots n,\ \sum_{i=1}^{n}\nu_i = 1\Big\},\quad n = 19.$$

A necessary and sufficient condition for a function f defined on a set $\mathscr{U}$ with open $\operatorname{Co}\mathscr{U}\subset\boldsymbol{R}^n$ to be the restriction of a convex function is the existence of functions h_i, $i = 1,\dots,n$, on $\mathscr{U}$, such that

$$f(\mathbf{z}) \geqslant f(\mathbf{z}_0) + \sum_{i=1}^{n} h_i(\mathbf{z}_0)(\mathbf{z}^i - \mathbf{z}_0^i)\quad \forall z, z_0\in\mathscr{U}$$

(Busemann, Ewald and Shephard, 1963).

Theorem 2.5.18
$W(\cdot\,,\xi)$ is polyconvex iff

$$W(\mathbf{F}_0+\mathbf{H},\xi) \geqslant W(\mathbf{F}_0,\xi)+a_i^\alpha(\mathbf{F}_0,\xi)H^i{}_\alpha + b_\alpha^i(\mathbf{F}_0,\xi)(\operatorname{adj}\mathbf{H})_i^\alpha + c(\mathbf{F}_0,\xi)\det\mathbf{H} \quad \forall \mathbf{F}_0,\mathbf{F}_0+\mathbf{H}\in\mathscr{L}_+, \tag{2.5.24}$$

or iff

$$W(\mathbf{F},\xi) \geqslant W(\mathbf{F}_0,\xi)+A_i^\alpha(\mathbf{F}_0,\xi)(F^i{}_\alpha-F^i_{0\alpha}) + B_\alpha^i(\mathbf{F}_0,\xi)[(\operatorname{adj}\mathbf{F})^\alpha{}_i-(\operatorname{adj}\mathbf{F}_0)^\alpha{}_i]+C(\mathbf{F}_0,\xi)[\det\mathbf{F}-\det\mathbf{F}_0] \quad \forall \mathbf{F},\mathbf{F}_0\in\mathscr{L}_+. \tag{2.5.25}$$

Proof
Equivalence of (2.5.24) and (2.5.25) can be established by expressing $\operatorname{adj}\mathbf{F}-\operatorname{adj}\mathbf{F}_0$ in terms of $\operatorname{adj}(\mathbf{F}-\mathbf{F}_0)$, $\mathbf{F}-\mathbf{F}_0$ etc.

Equation (2.5.25) follows from the convexity of the extension $\tilde{g}$ of g and Theorem 2.5.10. Conversely, repeating the argument used in the proof of Theorem 2.5.10 we can define the convex function $\tilde{g}(\mathbf{F},\mathbf{H},\Delta,\xi)$ as the supremum of the affine functions

$$W(\mathbf{F}_0,\xi)+A_i^\alpha(\mathbf{F}_0,\xi)(F^i{}_\alpha-F^i_{0\alpha})+B_\alpha^i(\mathbf{F}_0,\xi)\left(H^\alpha{}_i-(\operatorname{adj}\mathbf{F}_0)^\alpha{}_i\right) + C(\mathbf{F}_0,\xi)\,(\Delta-\det\mathbf{F}_0)$$

of $(\mathbf{F},\mathbf{H},\Delta)$. Clearly $W(\mathbf{F},\xi)=\tilde{g}(\mathbf{F},\operatorname{adj}\mathbf{F},\det\mathbf{F},\xi)$ for $\mathbf{F}\in\mathscr{L}_+$ and a.a. $\xi\in\Omega$. Since $\tilde{g}$ is convex and bounded on $T(\mathscr{L}_+)$, it is bounded on $M = \operatorname{Co}T(\mathscr{L}_+)$. Hence (2.5.25) implies that $W(\cdot\,,\xi)$ is polyconvex. □

Theorem 2.5.19 (Morrey, 1952)
If W is polyconvex on $\mathscr{D}\times\Omega$, $\mathscr{D}\subset\mathscr{L}_+$, and $\operatorname{Co}T(\mathscr{D})$ is open, then W is quasiconvex on $\mathscr{D}\times\Omega$.

Remark
If $\mathscr{D}=\mathscr{L}_+$ then $\operatorname{Co}T(\mathscr{L}_+)=\mathscr{L}\times\mathscr{L}\times\boldsymbol{R}_+$ is open in $\boldsymbol{R}^{19}$.

Proof
Let $g(\cdot\,,\xi)$ be a convex function defined on $\operatorname{Co}T(\mathscr{D})$ and satisfying equation (2.5.22) on $T(\mathscr{D})$. For every $(\mathbf{F},\mathbf{H},\Delta)\in\operatorname{Co}T(\mathscr{D})$ there are such $\mathbf{F}_i\in\mathscr{D}$ and $\theta_i\geqslant 0$ that $\sum_i\theta_i=1$ and

$$g(\mathbf{F},\mathbf{H},\Delta,\xi)\leqslant\sum_i\theta_i g(\mathbf{F}_i,\operatorname{adj}\mathbf{F}_i,\det\mathbf{F}_i,\xi)\equiv\sum_i\theta_i W(\mathbf{F}_i,\xi).$$

Since W is finite, so is g on $\operatorname{Co}T(\mathscr{D})$. Hence g and W are continuous functions on $\operatorname{Co}T(\mathscr{D})$ and $\mathscr{D}$, resp. (see a remark in Section 2.5.3).

Let $\mathscr{U}$ be open and bounded in $\boldsymbol{R}^3$, $\mathbf{F}_0 \in \mathscr{D}$. Let $\mathbf{u} \in \mathscr{C}_0^\infty(\mathscr{U})$ be such that $\mathbf{F}(\xi) = \mathbf{F}_0 + \nabla\mathbf{u}(\xi) \in \mathscr{D}$. We note that $\mathbf{F}(\xi) - \mathbf{F}_0$, $\mathrm{adj}\mathbf{F} - \mathrm{adj}\mathbf{F}_0$, $\det\mathbf{F} - \det\mathbf{F}_0$ are divergences. Indeed,

$$\mathrm{adj}(\mathbf{F}_0+\nabla\mathbf{u})^\gamma{}_k = (\mathrm{adj}\,\mathbf{F}_0)^\gamma{}_k + \left(\frac{1}{6}\,\epsilon^{\alpha\beta\gamma}{}_{ijk} F_{0\alpha}^i u^j\right)_{,\xi^\beta}$$

$$\left(\frac{1}{6}\,\epsilon^{\alpha\beta\gamma}\,\epsilon_{ijk}\,u^i F_0{}^j{}_\beta\right)_{,\xi^\alpha} + \left(\frac{1}{6}\,\epsilon^{\alpha\beta\gamma}\,\epsilon_{ijk}\,u^i{}_{,\xi^\alpha} u^j\right)_{,\xi^\beta},$$

$$\det(\mathbf{F}_0+\nabla\mathbf{u}) = \det\mathbf{F}_0 + \left(\frac{1}{2}\,\epsilon^{\alpha\beta\gamma}\,\epsilon_{ijk}\,u^i F_0{}^j{}_\beta F_0{}^k{}_\gamma\right)_{,\xi^\alpha}$$

$$+\left(\frac{1}{2}\,\epsilon^{\alpha\beta\gamma}\,\epsilon_{ijk}\,u^i{}_{,\xi^\alpha} u^j F_0{}^k{}_\gamma\right)_{,\xi^\beta} + \left(\frac{1}{6}\,\epsilon^{\alpha\beta\gamma}\,\epsilon_{ijk} u^i{}_{,\xi^\alpha} u^i{}_{,\xi^\alpha} u^k\right)_{,\xi^\gamma}.$$

The result follows from this observation and (2.5.25). □

There are examples of quasiconvex functions W which are not polyconvex (Ball, 1977).

So far we have dealt with $\mathrm{adj}\,\nabla\mathbf{u}$, $\det\nabla\mathbf{u}$ for $\mathbf{u} \in \mathscr{C}^1(\bar{\Omega})$. In the following we shall deal with $\mathrm{adj}\,\nabla\mathbf{u}$, $\det\nabla\mathbf{u}$ for Sobolev functions $\mathbf{u}$. The distributions

$$(\mathrm{Adj}\,\nabla\mathbf{u})^\alpha{}_i := \left(\frac{1}{6}\,\epsilon^{\alpha\beta\gamma}\,\epsilon_{ijk}\,u^j u^k{}_{,\xi^\gamma}\right)_{,\xi^\beta}, \tag{2.5.26}$$

$$\mathrm{Det}\nabla\mathbf{u} := 3[u^1(\mathrm{Adj}\,\nabla\mathbf{u})^\alpha{}_1]_{,\xi^\alpha} = [u^i(\mathrm{Adj}\,\nabla\mathbf{u})^\alpha{}_i]_{,\xi^\alpha} \tag{2.5.27}$$

are well-defined provided $\mathbf{u}\otimes\nabla\mathbf{u}$, $\mathbf{u}\otimes\mathrm{Adj}\,\nabla\mathbf{u}$ are distributions, resp. Note that $(\mathrm{Adj}\,\nabla\mathbf{u})^\alpha{}_{i,\xi^\alpha} \equiv 0$. We shall be interested in the case where these distributions belong to suitable subspaces of $\mathscr{L}^1_{loc}(\Omega)$. Therefore we shall consider the maps $\mathbf{u} \mapsto \mathrm{adj}\,\nabla\mathbf{u}$, $\mathbf{u} \mapsto \det\nabla\mathbf{u}$ from the point of view of functional analysis.

Lemma 2.5.20

(i) If $\boldsymbol{\chi} \in W^{1,2}(\Omega)$, then $\mathrm{adj}\,\nabla\boldsymbol{\chi} \in \mathscr{L}^1(\Omega)$ and

$$(\mathrm{adj}\,\nabla\boldsymbol{\chi})^\alpha_i = \frac{1}{6}\,(\epsilon_{ijk}\,\epsilon^{\alpha\beta\gamma}\,\chi^j\chi^k{}_{,\xi^\gamma})_{,\xi^\beta} \text{ in } \mathscr{D}'(\Omega). \tag{2.5.28}$$

(ii) If $p \geqslant 2$, $\boldsymbol{\chi} \in W^{1,p}(\Omega)$ and $\mathrm{adj}\,\nabla\boldsymbol{\chi} \in \mathscr{L}^q(\Omega)$, $q = \dfrac{p}{p-1}$, then $\det\,\nabla\boldsymbol{\chi} \in \mathscr{L}^1(\Omega)$ and

$$\det\nabla\boldsymbol{\chi} = [\chi^i(\mathrm{adj}\,\nabla\boldsymbol{\chi})^\alpha{}_i]_{,\xi^\alpha} \quad \text{in } \mathscr{D}'(\Omega). \tag{2.5.29}$$

The left-hand side of equations (2.5.28) and (2.5.29) are defined by the formulae

$$(\operatorname{adj}\nabla\boldsymbol{\chi})^{\alpha}{}_{i} := \frac{1}{6}\,\epsilon_{ijk}\,\epsilon^{\alpha\beta\gamma}\,\chi^{j}{}_{,\xi^{\beta}}\,\chi^{k}{}_{,\xi^{\gamma}},$$

$$\det\nabla\boldsymbol{\chi} := \frac{1}{6}\,\epsilon_{ijk}\,\epsilon^{\alpha\beta\gamma}\chi^{i}{}_{,\xi^{\alpha}}\,\chi^{j}{}_{,\xi^{\beta}}\,\chi^{k}{}_{,\xi^{\gamma}}$$

in terms of pointwise multiplication, while the derivatives on the right-hand side are understood in the sense of distributions.

Proof

(i) The mapping $\boldsymbol{\chi} \mapsto \operatorname{adj}\nabla\boldsymbol{\chi}$ from $\mathscr{C}^1(\bar{\Omega})$ into $\mathscr{L}^1(\Omega)$ is continuous with respect to the $W^{1,2}(\Omega)$-topology on $\mathscr{C}^1(\bar{\Omega})$ and can be defined on $W^{1,2}(\Omega)$ by continuous extension. For any $\phi \in \mathscr{C}_0^{\infty}(\Omega)$, $\boldsymbol{\chi} \in \mathscr{C}^1(\Omega)$

$$\int_{\Omega} dL^3\,\phi(\operatorname{adj}\nabla\boldsymbol{\chi})^{\alpha}{}_{i} = -\frac{1}{6}\int_{\Omega} dL^3\,\epsilon_{ijk}\,\epsilon^{\alpha\beta\gamma}\chi^{j}\chi^{k}{}_{,\xi^{\gamma}}\phi_{,\xi^{\beta}} \tag{2.5.30}$$

and both sides of equation (2.5.30) are continuous functionals of $\boldsymbol{\chi} \in W^{1,2}(\Omega)$. Hence equation (2.5.28) holds for $\boldsymbol{\chi} \in W^{1,2}(\Omega)$ in the sense of distributions.

(ii). Let $p \geqslant 2$, $\boldsymbol{\chi} \in W^{1,p}(\Omega)$, $\operatorname{adj}\nabla\boldsymbol{\chi} \in \mathscr{L}^q(\Omega)$. By the Hölder inequality we can extend (2.5.30) from $\boldsymbol{\chi} \in \mathscr{C}_0^{\infty}(\bar{\Omega})$ to the set of all $\boldsymbol{\chi} \in W^{1,p}(\Omega)$, $p \geqslant 2$, by continuity.

For $\boldsymbol{\chi} \in \mathscr{C}^{\infty}(\bar{\Omega})$ $(\operatorname{adj}\nabla\boldsymbol{\chi}^{\alpha}{}_{i})_{,\xi^{\alpha}} = 0$ and therefore

$$\int_{\Omega} dL^3\,(\operatorname{adj}\nabla\boldsymbol{\chi})^{\alpha}{}_{i}\phi_{,\xi^{\alpha}} = 0 \;\; \forall\phi \in \mathscr{C}_0^{\infty}(\Omega). \tag{2.5.31}$$

By continuity this formula extends to $\boldsymbol{\chi} \in W^{1,p}(\Omega)$*. Let

$$\varrho \in \mathscr{C}_0^{\infty}(\boldsymbol{R}^3), \quad \operatorname{supp}\varrho \subset \mathscr{K}(\mathbf{0}, 1), \quad \varrho \geqslant 0, \quad \int dL^3\varrho = 1,$$

and

$$\varrho_{\varepsilon}(\xi) :\equiv \varepsilon^{-3}\varrho(\varepsilon^{-1}\xi), \quad \operatorname{supp}\varrho_{\varepsilon} \subset \mathscr{K}(\mathbf{0}, \varepsilon) \quad \text{for } \varepsilon > 0.$$

Let

$$w^{\alpha}{}_{i}(\xi) := \begin{cases} (\operatorname{adj}\nabla\boldsymbol{\chi})^{\alpha}{}_{i}(\xi) & \text{for } \xi \in \bar{\Omega}, \\ 0 & \text{for } \xi \bar{\in} \bar{\Omega}. \end{cases}$$

Clearly the convolutions $\varrho_{\varepsilon} * w^{\alpha}{}_{i} \in \mathscr{C}_0^{\infty}(\boldsymbol{R}^3)$, $\varrho_{\varepsilon} * w^{\alpha}{}_{i} \to w^{\alpha}{}_{i} \in \mathscr{L}^q(\Omega)$ as $\varepsilon \to 0$.

Let $\phi \in \mathscr{C}_0^{\infty}(\Omega)$. Since $\partial\Omega$ and $\operatorname{supp}\phi$ are compact disjoint sets, $\operatorname{dist}(\operatorname{supp}\phi, \partial\Omega) > \varepsilon_0$ for $\varepsilon_0 > 0$ sufficiently small. Hence for $\xi \in \operatorname{supp}\phi$, $\varepsilon < \varepsilon_0$

$$\frac{\partial}{\partial\xi^{\alpha}}\,\varrho_{\varepsilon} * w^{\alpha}{}_{i}(\xi) = \int_{\Omega} dL^3\,(\zeta)\,\frac{\partial\varrho_{\varepsilon}}{\partial\xi^{\alpha}}\,(\xi - \zeta)\,w^{\alpha}{}_{i}(\zeta) = 0 \tag{2.5.32}$$

on account of (2.5.31).

* The same formula follows from (2.5.30) by substituting $\phi_{,\xi^{\alpha}}$ for ϕ.

Let $\chi_n \in \mathscr{C}^\infty(\bar{\Omega})$, $\chi_n \to \chi$ in $W^{1,p}(\Omega)$. Equation (2.5.32) implies that

$$\int_\Omega dL^3\, \phi \chi_n{}^i{}_{,\xi^\alpha}\, \varrho_\varepsilon * w^\alpha{}_i = \int_{\partial\Omega} dH^2\, \phi \chi_n{}^i \varrho_\varepsilon * w_i{}^\alpha \nu_\alpha - \int_\Omega dL^3\, \chi_n{}^i \varrho_\varepsilon * w^\alpha{}_i \phi_{,\xi^\alpha}.$$

The first integral on the right-hand side vanishes and in the limits $\varepsilon \to 0$, $n \to \infty$ we have

$$\int_\Omega dL^3\, \chi^i{}_{,\xi^\alpha}\, w^\alpha{}_i \phi = -\int_\Omega dL^3\, \chi^i w^\alpha{}_i \phi_{,\xi^\alpha},$$

i.e.

$$\det \nabla\chi \equiv \chi^i{}_{,\xi^\alpha} (\operatorname{adj} \nabla\chi)^\alpha{}_i = (\chi^i w^\alpha{}_i)_{,\xi^\alpha}$$

in the sense of distributions. □

Theorem 2.5.21 (Ball, 1977a)

Let $p \geqslant 2$, $p^{-1}+q^{-1} = 1$.

(i) If $\chi_n \rightharpoonup \chi$ in $W^{1,p}(\Omega)$, then $\operatorname{adj}\nabla\chi_n \rightharpoonup \operatorname{adj}\nabla\chi$ in $\mathscr{D}'(\Omega)$.

(ii) If $\chi_n \rightharpoonup \chi$ in $W^{1,p}(\Omega)$, $\operatorname{adj}\nabla\chi_n \rightharpoonup \operatorname{adj}\nabla\chi$ in $\mathscr{L}^q(\Omega)$, then $\det\nabla\chi_n \to \det\nabla\chi$ in $\mathscr{D}'(\Omega)$.

Proof

Ad (i). Let $\phi \in \mathscr{C}_0^\infty(\Omega)$. By Lemma 2.5.20(i)

$$\int_\Omega dL^3\, (\operatorname{adj}\nabla\chi_n)^\alpha{}_i \phi = -\frac{1}{6}\int_\Omega dL^3\, \epsilon_{ijk}\, \epsilon^{\alpha\beta\gamma} \chi_n{}^j \chi_n{}^k{}_{,\xi^\gamma} \phi_{,\xi^\beta}.$$

By Theorem 1.1.42 there is a domain Ω' such that

$$\operatorname{supp}\phi \subset \Omega' \subset \bar{\Omega}' \subset \Omega$$

and Ω' has the cone property. Since $\chi_n \rightharpoonup \chi$ in $W^{1,p}(\Omega')$, by the Rellich–Kondrachov theorem (Section 1.1) $\chi_n \to \chi$ in $\mathscr{L}^2(\Omega')$ (note that $p \geqslant 2$). Hence $\chi_n{}^j \chi_n{}^k{}_{,\xi^\gamma} \rightharpoonup \chi^j \chi^k{}_{,\xi^\gamma}$ in $\mathscr{L}^1(\Omega')$*

and

$$\int_\Omega dL^3\, (\operatorname{adj}\nabla\chi_n)^\alpha{}_i \phi \to \int_\Omega dL^3\, (\operatorname{adj}\nabla\chi)^\alpha{}_i \phi \quad \forall \phi \in \mathscr{C}_0^\infty(\Omega).$$

The assertion (ii) is proved similarly by means of Lemma 2.5.20(ii). □

$$\left|\int_{\Omega'} dL^3\, [\chi_n{}^j \chi_n{}^k{}_{,\xi^\gamma} - \chi^j \chi^k{}_{,\xi^\gamma}]\phi\right|$$
$$\leqslant \int_{\Omega'} dL^3\, |\phi \chi_n{}^k{}_{,\xi^\gamma}|\, |\chi_n{}^j - \chi^j| + \left|\int_{\Omega'} dL^3\, \chi^j \phi [\chi_n{}^k{}_{,\xi^\gamma} - \chi^k{}_{,\xi^\gamma}]\right|.$$

* The first integral $\leqslant \|\chi_n - \chi\|_{2,\Omega'} \|\phi\|_\infty \sup\{\|\chi_{n,\xi^\gamma}\|_{2,\Omega'} | \, n = 1, 2, \dots\}$ and the second tends to 0 since $\chi_n{}^k{}_{,\xi^\gamma} \rightharpoonup \chi^k{}_{,\xi^\gamma}$ in $\mathscr{L}^2(\Omega')$.

Corollary 2.5.22 (Reshetnyak, 1968; Ball, 1977)

(i) If $p > 2$, then the map $\chi \mapsto \operatorname{adj}\nabla\chi$ from $W^{1,p}(\Omega)$ into $\mathscr{L}^{p/2}(\Omega)$ is sequentially weakly continuous.

(ii) If $p > 3$, then the map $\chi \mapsto \det\nabla\chi$; $W^{1,p}(\Omega) \to \mathscr{L}^{p/3}(\Omega)$ is sequentially weakly continuous.

Proof

Ad (i). If $\chi_n \rightharpoonup \chi$ in $W^{1,p}(\Omega)$, then the sequence $\{\operatorname{adj}\nabla\chi_n\}$ is bounded in $\mathscr{L}^{p/2}(\Omega)$ (Theorem 1.1.20). By Theorem 2.5.21(i) and Theorem 1.1.23 the latter sequence converges weakly to $\operatorname{adj}\nabla\chi$ in $\mathscr{L}^{p/2}(\Omega)$.

Ad (ii) If $\chi_n \rightharpoonup \chi$ in $W^{1,p}(\Omega)$, $p > 3$, then $\operatorname{adj}\nabla\chi_n \rightharpoonup \operatorname{adj}\nabla\chi$ in $\mathscr{L}^{p/2}(\Omega)$. Since $p/2 > 3/2 > q$, it follows that $\operatorname{adj}\nabla\chi_n \rightharpoonup \operatorname{adj}\nabla\chi$ in $\mathscr{L}^{q}(\Omega)$. By Theorem 2.5.21(ii) as well as by Theorem 1.1.20, the Hölder inequality and Theorem 1.1.23 $\det\nabla\chi_n \rightharpoonup \det\nabla\chi$ in $\mathscr{L}^{p/3}(\Omega)$. □

We now formulate the hypotheses of the existence theorems of this section.

(PC) $W(\mathbf{F}, \xi) \equiv g(\mathbf{F}, \operatorname{adj}\mathbf{F}, \det\mathbf{F}, \xi)$ is polyconvex and g is a **Carathéodory function** (or a **normal integrand** in the sense of Ékeland and Témam (1976));

(GC_1) **(growth conditions)** $\exists\, p \geqslant 2,\ s \geqslant q = \dfrac{p}{p-1},\ r > 1,\ K > 0$ such that

$$g(\mathbf{F}, \mathbf{H}, \Delta, \xi) \geqslant c + K[|\mathbf{F}|^p + |\mathbf{H}|^s + |\Delta|^r]$$

for almost all $\xi \in \Omega$, and all $(\mathbf{F}, \mathbf{H}, \Delta) \in M$;

(GC_2) $g(\mathbf{F}, \mathbf{H}, \Delta, \xi) \to \infty$ if $(\mathbf{F}, \mathbf{H}, \Delta)$ tends to the boundary ∂M of M.

In the case of $\mathscr{C} = \{\mathbf{u} \in \mathscr{V} \mid \det(\nabla\chi_0 + \nabla\mathbf{u}) > 0 \text{ a.e. in } \Omega\}$, $\mathscr{D} = \mathscr{V}_+$ this becomes equivalent to $\Delta \to 0 \mapsto g \to \infty$.

We assume that g is a normal integrand $M \times \Omega \to \bar{\boldsymbol{R}}$ and $g(\mathbf{F}, \mathbf{H}, \Delta, \xi) = \infty$ for $(\mathbf{F}, \mathbf{H}, \Delta) \bar{\in} M$.

The remaining assumptions are

(P) $\phi\colon \boldsymbol{R}^3 \to \boldsymbol{R}_+$ is continuous;

(SL) Ω is open, bounded and has the strong Lipschitz property, $\partial\Omega = \bar{\Gamma}_1 \cup \bar{\Gamma}_2$, $\Gamma_1 \cap \Gamma_2 = \emptyset$, $H^2(\Gamma_1) > 0$.

We now define the solution set

$$\mathscr{C} := \{\chi \in W^{1,p}(\Omega) \mid \operatorname{adj}\nabla\chi \in \mathscr{L}^s(\Omega),\ \det\nabla\chi \in \mathscr{L}^r(\Omega),\ \det\nabla\chi(\xi) > 0 \text{ a.e. in } \Omega,\ \gamma\chi = \chi_1\ H^2\text{-a.e. on } \Gamma_1\}.$$

Let

$$F := \mathscr{W} + \Phi,\ \Phi(\chi) := \int_\Omega dL^3\, \phi(\chi(\xi)).$$

Theorem 2.5.23 (Ball, 1977a)

Suppose that for some $\chi_2 \in C$ $F(\chi_2) < \infty$. Then there exists $\chi_0 \in \mathscr{C}$ which gives a minimum value to F on $\mathscr{C}$.

Proof

From the Friedrichs inequality

$$\|\chi\|_{1,p}{}^{p} \leqslant K\{\|\nabla\chi\|_{p}{}^{p}+\|\gamma\chi\|_{\Gamma_1,p}{}^{p}\}$$

and (GC_1) we conclude that for $\chi \in \mathscr{C}$

$$F(\chi) \geqslant cL^3(\Omega)+K_1\|\chi\|_{1,p}{}^{p}+K[\|\mathrm{adj}\,\nabla\chi\|_s{}^s+\|\det\nabla\chi\|_r{}^r] \qquad (2.5.33)$$

with $K, K_1 > 0$. Hence F is bounded from below on $\mathscr{C}$.

Let $a = \inf\{F(\chi) |\ \chi \in \mathscr{C}\}$. We have that $a > -\infty$. Let $\{\chi_n\}$ be a minimizing sequence for F on $\mathscr{C}$, i.e. $F(\chi_n) < \infty$, $F(\chi_n) \to a$. From (2.5.33) the sequence $\{\chi_n\}$ is bounded in $W^{1,p}(\Omega)$, $\{\mathrm{adj}\nabla\chi_n\}$ is bounded in $\mathscr{L}^s(\Omega)$, $\{\det\nabla\chi_n\}$ is bounded in $\mathscr{L}^r(\Omega)$. Hence there is a subsequence $\{\chi_k'\}$ of $\{\chi_n\}$ such that $\chi_k' \rightharpoonup \chi_0$ in $W^{1,p}(\Omega)$, $\mathrm{adj}\nabla\chi_k' \rightharpoonup \mathbf{H}_0$ in $\mathscr{L}^s(\Omega)$, $\det\nabla\chi_k' \rightharpoonup \Delta_0$ in $\mathscr{L}^r(\Omega)$. By Theorem 2.5.21 $\mathbf{H}_0 = \mathrm{adj}\nabla\chi_0$, $\Delta_0 = \det\nabla\chi_0$, since both sides of either of these two equations coincide (as functionals) on $\mathscr{C}_0^\infty(\Omega)$, $s \geqslant q$. By the Rellich–Kondrachov theorem and (SL) $\chi_k' \to \chi_0$ in $\mathscr{L}^p(\Omega)$, hence $\chi_k'(\xi) \to \chi_0(\xi)$ a.e. in Ω. By the trace theorem $\gamma\chi_k' \to \gamma\chi_0$ in $\mathscr{L}^1(\Gamma_1)$ and hence $\gamma\chi_0(\xi) = \chi_1(\xi)$ a.e. on Γ_1.

In particular

$$(\nabla\chi_k', \mathrm{adj}\nabla\chi_k', \det\nabla\chi_k') \to (\nabla\chi_0, \mathrm{adj}\nabla\chi_0, \det\nabla\chi_0) \quad \text{in } \mathscr{L}^1(\Omega).$$

Since $g(\cdot\,,\,\cdot\,,\,\cdot\,,\xi)$ is convex and satisfies (GC_1) with $p, r, s > 1$, we can apply Theorems 1.1.68 and 1.1.69 and conclude that

$$F(\chi_0) \leqslant \liminf_{n\to\infty} F(\chi_n) = a,$$

hence $F(\chi_0) = a$.

We have proved that $\gamma\chi_0 = \chi_1$ a.e. on Γ_1. Since $F(\chi_0) < \infty$, it follows that $\nabla\chi_0(\xi) \bar{\in} T^{-1}(\partial M)$, i.e. $\det\nabla\chi_0(\xi) > 0$ a.e. in Ω. Hence $\chi_0 \in \mathscr{C}$. □

Note that the property (GC_2) forces the solution χ_0 into the set $\mathscr{C}$ in $\mathscr{V}$. Hence we neednot require that $\mathscr{C}$ is weakly closed.

2.5.7 Existence Theorem for Incompressible Elastic Media

Theorem 2.5.23 can be easily adapted to the case of incompressible elastic bodies. In this case we make the following assumptions:

(PC′) $W(\mathbf{F}, \xi) = g(\mathbf{F}, \mathrm{adj}\mathbf{F}, \xi)\ \forall\, \mathbf{F} \in \mathrm{SU}(3)$, and a.a. $\xi \in \Omega$; g is a Carathéodory function (continuous of $\mathbf{F}$, $\mathrm{adj}\mathbf{F}$), convex on $\mathscr{L}\times\mathscr{L}$ for a.e. Ω.

(GC′) $\exists K > 0$, c, $p \geqslant 2$, $s \geqslant q = \dfrac{p}{p-1}$ such that $g(\mathbf{F}, \mathbf{H}, \xi) \geqslant c_+$

$+K[|\mathbf{F}|^p+|\mathbf{H}|^s]$ for all $(\mathbf{F},\mathbf{H})\in\mathscr{L}\times\mathscr{L}$, a.a. $\xi\in\Omega$. We assume (SL) and (P) as before.

Let $H^2(\Gamma_1)>0$,

$$\mathscr{C}_1 := \{\boldsymbol{\chi}\in W^{1,p}(\Omega)|\ \operatorname{adj}\nabla\boldsymbol{\chi}\in\mathscr{L}^s(\Omega), \det\nabla\boldsymbol{\chi}=1 \text{ a.e. in } \Omega,$$
$$\gamma_{\Gamma_1}\boldsymbol{\chi}=\boldsymbol{\chi}_1\ H^2 \text{ a.e. on } \Gamma_1\},\quad F=\mathscr{W}+\Phi \text{ as above.}$$

Theorem 1.1.24 (Ball, 1977a)
Suppose that $\exists\,\boldsymbol{\chi}_2\in\mathscr{C}_1$ such that $F(\boldsymbol{\chi}_2)<\infty$. Then $\exists\,\boldsymbol{\chi}_0\in\mathscr{C}_1$ such that $F(\boldsymbol{\chi}_0)=\inf\{F(\boldsymbol{\chi})|\ \boldsymbol{\chi}\in\mathscr{C}_1\}$.

Proof
Let $F(\boldsymbol{\chi}_n)\to a=\inf\{F(\boldsymbol{\chi})|\ \boldsymbol{\chi}\in\mathscr{C}_1\}$. On account of (GC′) and the hypothesis of the theorem $-\infty<a<\infty$.

Since $\det\nabla\boldsymbol{\chi}_n(\xi)=1$ a.e. in Ω, $\{\det\nabla\boldsymbol{\chi}_n\}$ is bounded in $\mathscr{L}^2(\Omega)$. (GC′) implies that $\{\boldsymbol{\chi}_n\}$ is bounded in $W^{1,p}(\Omega)$, $\{\operatorname{adj}\nabla\boldsymbol{\chi}_n\}$ is bounded in $\mathscr{L}^s(\Omega)$. Hence there is a subsequence $\{\boldsymbol{\chi}_k'\}$ of $\{\boldsymbol{\chi}_n\}$ such that $\boldsymbol{\chi}_k'\rightharpoonup\boldsymbol{\chi}_0$ in $W^{1,p}(\Omega)$, $\operatorname{adj}\nabla\boldsymbol{\chi}_k'\rightharpoonup H_0$ in $\mathscr{L}^s(\Omega)$, $\det\nabla\chi_k'\rightharpoonup\Delta_0$ in $\mathscr{L}^2(\Omega)$, $\Delta_0=1$ a.e. On account of Theorem 2.5.21 and the inequality $s\geqslant q$ we conclude that $\det\nabla\boldsymbol{\chi}_0(\xi)=1$ a.e. in Ω, $\mathbf{H}_0=\operatorname{adj}\nabla\boldsymbol{\chi}_0$, $\boldsymbol{\chi}_0\in\mathscr{C}_1$. The thesis follows from the wlsc of F, Theorem 1.1.68. □

It can be shown (Ball, 1977a, Theorem 3.2) that a mapping $\Psi:\boldsymbol{\chi}\mapsto\psi\circ\nabla\boldsymbol{\chi}$, with $\psi\in\mathscr{C}^0$, $|\psi(\mathbf{F})|\leqslant K[1+|\mathbf{F}|^p]$, $p\geqslant 1$, is sequentially weakly continuous from $W^{1,p}(\Omega)$ into $\mathscr{L}^1(\Omega)$ iff

$$\psi(\mathbf{F})\equiv a+b^{\alpha}{}_iF^i{}_{\alpha}+c^i{}_{\alpha}(\operatorname{adj}\mathbf{F})^{\alpha}{}_i+d\det\mathbf{F}. \tag{2.5.34}$$

Suppose that the constraint $\Psi(\boldsymbol{\chi})=0$ enters the definition of the solution set $\mathscr{C}_1$, as above. The fact that $\Psi(\boldsymbol{\chi}_n)=0$ for the minimizing sequence $\{\boldsymbol{\chi}_n\}$ as well as the sequential weak continuity entail that $\Psi(\boldsymbol{\chi}_0)=0$ for the minimizer $\boldsymbol{\chi}_0$. This is a crucial step in the proof that $\boldsymbol{\chi}_0\in\mathscr{C}_1$. Unfortunately sequential weak continuity, viz. (2.5.34), and the requirement that

$$\psi(\mathbf{AF})\equiv\psi(\mathbf{F})\ \forall\mathbf{A}\in\mathrm{SO}(3),\quad \mathbf{F}\in\mathscr{L}_+$$

are satisfied by the incompressibility constraint $\det\mathbf{F}=$ const only. Other constraints are not tractable by the above method.

An elegant method of taking into account the pointwise constraints Ψ_j: $W^{1,p}(\Omega)\to\mathscr{L}^{s_j}(\Omega)$, $\Psi_j(\boldsymbol{\chi})(\xi)\equiv\psi_j(\boldsymbol{\chi}(\xi),\nabla\boldsymbol{\chi}(\xi),\xi)=0$ involves the **Lagrange multipliers.** For a pointwise constraint Ψ_j: $W^{1,p}(\Omega)\to\mathscr{L}^{s_j}(\Omega)$ $s_j\geqslant 1$ the Lagrange multiplier is a function $\mu_j\in\mathscr{L}^{t_j}(\Omega)$, $t_j^{-1}+s_j^{-1}=1$, $1<t_j\leqslant\infty$. Roughly, a necessary condition for $F(\boldsymbol{\chi}_0)=\inf\{F(\boldsymbol{\chi})|\ \boldsymbol{\chi}\in\mathscr{V}+\boldsymbol{\chi}_1,\ \Psi_j(\boldsymbol{\chi})=0,\ j=1,\dots,N\}$, $\boldsymbol{\chi}_0\in\mathscr{V}+\boldsymbol{\chi}_1$, requires that there is a set

of functions $\mu_j \in \mathscr{L}^{t_j}(\Omega)$, $j = 1, \dots, N$, such that $D\tilde{F}(\boldsymbol{\chi}_0)[\mathbf{v}] = 0 \ \forall \mathbf{v} \in \mathscr{V}$, $\tilde{F}(\boldsymbol{\chi}) :\equiv F(\boldsymbol{\chi}) + \sum_j \int_\Omega dL^3 \mu_j \Psi_j(\boldsymbol{\chi})$ (cf. Gel'fand and Fomin, 1961; Lavrentiev and Lyusternik, 1935). Physically the Lagrange multipliers can be identified as forces (if ψ_j depends on $\boldsymbol{\chi}$ only) or stresses (if ψ_j depends on $\nabla\boldsymbol{\chi}$ only) sustaining the constraints.

We now apply this method to the incompressibility constraint $\det\nabla\boldsymbol{\chi} - 1 = 0$. The associated Lagrange multiplier is pressure $p(\cdot)$, as can be readily verified from the Euler–Lagrange equations. Since F is expected to assume its minimum on the manifold $\det\nabla\boldsymbol{\chi} = 1$ at $\boldsymbol{\chi}_0$, we have $\frac{d}{dt}F(\boldsymbol{\chi}_0 + t\mathbf{v})|_{t=0} = 0$ for all $\mathbf{v} \in \mathscr{V}$ satisfying the equation $\det(\nabla\boldsymbol{\chi}_0 + t\nabla\mathbf{v}) = 1$. Hence $DF(\boldsymbol{\chi}_0)[\mathbf{v}] = 0$ for all $\mathbf{v} \in \mathscr{V}$ satisfying the constraint $(\operatorname{adj}\nabla\boldsymbol{\chi}_0)^\alpha{}_i(\xi)v^i{}_{,\xi^\alpha}(\xi) = 0$ a.e. in Ω. Since $\det\nabla\boldsymbol{\chi}_0(\xi) = 1$ a.e. in Ω, this equation is equivalent to $\overset{-1}{F_0}{}^\alpha{}_i v^i{}_{,\xi^\alpha} = 0$ a.e. in Ω, $\mathbf{F}_0 = \nabla\boldsymbol{\chi}_0$.

Suppose that $\operatorname{adj}\nabla\boldsymbol{\chi}_0 \in \mathscr{L}^\infty(\Omega)$ and let $p = 2$, $\mathscr{V} \subset W^{1,2}(\Omega)$. Let $G\colon \mathscr{V} \to \mathscr{L}^2(\Omega)$ be the bounded linear operator $G(\mathbf{v})(\xi) :\equiv (\operatorname{adj}\nabla\boldsymbol{\chi}_0)^\alpha{}_i v^i{}_{,\xi^\alpha}$. We shall seek a $\boldsymbol{\chi}_0 \in \mathscr{V} + \boldsymbol{\chi}_1$ such that $DF(\boldsymbol{\chi}_0)[\mathbf{v}] = 0$ for all $\mathbf{v} \in \ker G$. Since $\ker G$ is closed in $\mathscr{V}$, we can decompose the Hilbert space $\mathscr{V}$ into $\ker G$ and its orthogonal complement: $\mathscr{V} = \ker G \otimes (\ker G^\perp)$. Analogously, $\mathscr{L}^2(\Omega) = \overline{\mathscr{R}(G)} \otimes \mathscr{R}(G)^\perp$, $\overline{\mathscr{R}(G)} = \mathscr{R}(G)^{\perp\perp}$, $\mathscr{R}(G) := G(\mathscr{V})$. The superscript" $\perp$" denotes the orthogonal complement of a linear manifold in the Hilbert space $W^{1,2}(\Omega)$ or $\mathscr{L}^2(\Omega)$, resp.

Suppose that $\mathscr{R}(G)$ is closed. It is clear that $G_1 := G|(\ker G)^\perp : (\ker G)^\perp \to \mathscr{R}(G)$ is bijective. Hence by Theorem 1.1.29 $G_1{}^{-1}$ exists, is linear, bounded and bijective. By Riesz' theorem $DF(\boldsymbol{\chi}_0)[\mathbf{v}] = \langle \mathbf{r}|\mathbf{v}\rangle_1$, $\mathbf{r} \in \mathscr{V}$. $\langle\cdot|\cdot\rangle_1$ denotes the scalar product in $\mathscr{V}$.

Let $\mathbf{v} \in (\ker G)^\perp$. Then $\mathbf{v} = G_1{}^{-1}\mathbf{f}$, $\mathbf{f} \in \mathscr{R}(G)$, and $\langle \mathbf{r}|\mathbf{v}\rangle_1 = \langle \mathbf{r}|G_1{}^{-1}\mathbf{f}\rangle_1$ is a linear continuous functional on $\mathscr{R}(G)$. Hence, by Riesz' theorem there is a $\phi(\mathbf{r}) \in \mathscr{R}(G)$ such that

$$\langle \mathbf{r}|\mathbf{v}\rangle_1 = \langle \mathbf{r}|G_1{}^{-1}\mathbf{f}\rangle_1 = \langle \phi(\mathbf{r})|\mathbf{f}\rangle_0 = \langle \phi(\mathbf{r})|G\mathbf{v}\rangle_0 = \langle G^\dagger\phi(\mathbf{r})|\mathbf{v}\rangle_1.$$

Suppose now that $\mathbf{v} \in \ker G$. Then $\langle \mathbf{r}|\mathbf{v}\rangle_1 = 0$ by our hypothesis and $\langle G^\dagger\phi(\mathbf{r})|\mathbf{v}\rangle_1 = \langle \phi(\mathbf{r})|G\mathbf{v}\rangle_0 = 0$. Hence

$$DF(\boldsymbol{\chi}_0)[\mathbf{v}] = \langle \mathbf{r}|\mathbf{v}\rangle_1 = \langle G^\dagger\phi(\mathbf{r})|\mathbf{v}\rangle_1 = \langle \phi(\mathbf{r})|G\mathbf{v}\rangle_0 \quad \text{for all } \mathbf{v} \in \mathscr{V}$$

and $\phi(\mathbf{r})$ can be identified as the Lagrange multiplier.

Note that we have assumed without proof that (i) $\operatorname{adj}\mathbf{F}_0 \in \mathscr{L}^\infty(\Omega)$, (ii) $\mathscr{R}(G)$ is closed.* Although it is conceptually attractive the Lagrange multiplier method can be quite difficult to apply to elasticity.

* In the case of Banach spaces $W^{1,p}(\Omega)$, $\mathscr{L}^p(\Omega)$ the existence of a decomposition of these spaces into a direct sum of $\mathscr{R}(G)$ or $\ker G$ and its complement is another moot point.

2.5.8 Existence Theorems for Non-Vanishing Surface Loads. The Neumann Problem

Since the main steps in the proofs of the theorems below are quite similar to those of Theorems 2.5.23 and 2.5.24, we shall concentrate on the necessary modifications associated with estimating the surface integral in F. We shall use the notation

$$W(\mathbf{F},\xi)+\phi(\chi,\xi) = g(\mathbf{F},\mathbf{H},\Delta,\chi,\xi), \quad \mathbf{H}:=\operatorname{adj}\mathbf{F}, \quad \Delta := \det\mathbf{F}. \tag{2.5.35}$$

For dead surface loads

$$F(\chi) = \int_\Omega dL^3\, g(\nabla\chi(\xi), \operatorname{adj}\nabla\chi(\xi), \det\nabla\chi(\xi), \chi(\xi), \xi) - \int_{\Gamma_2} dH^2\, s_k(\xi)\gamma\chi^k(\xi) \tag{2.5.36}$$

with $s_k \in \mathscr{L}^\sigma(\Gamma_2)$ for $k = 1, 2, 3$, and some $\sigma > 1$.

In order to estimate the surface integral in (2.5.36) we need the inequality

$$ab \leqslant \frac{1}{p}a^p + \frac{1}{q}b^q \quad \text{for } p > 1, \quad p^{-1}+q^{-1} = 1, \quad a, b \geqslant 0 \tag{2.5.37}$$

(equality is possible only for $a = b$).

For $b = 0$ equation (2.5.37) is trivial. For $b > 0$ let $x := a/b^{q/p}$, $f(x) :\equiv x^p/p+1/q-x$. We shall prove that $f(x) \geqslant 0$ for $x \geqslant 0$, $f(x) = 0$ only for $x = 1$. Indeed, $f(0) = 1/q > 1$, $\lim_{x\to\infty} f(x) = \infty$, $f'(x) \equiv x^{p-1}-1 = 0$ has exactly one solution $x = 1$, $f(1) = 0$. Hence $f(x) > f(1) = 0$ for $x \neq 1$.

For $\tau^{-1} = 1-\sigma^{-1}$, $\varepsilon > 0$ we conclude from (2.5.37) that

$$\left|\int_{\Gamma_2} dH^2\, s_k\gamma\chi^k\right| \leqslant \|\mathbf{s}\|_{\sigma,\Gamma_2}\|\gamma\chi\|_{\tau,\Gamma_2} \leqslant \frac{1}{\varepsilon}\|\mathbf{s}\|_{\sigma,\Gamma_2}\,\varepsilon M\|\chi\|_{1,p,\Omega}$$

$$\leqslant \frac{1}{q\varepsilon^q}\|\mathbf{s}\|_{\sigma,\Gamma_2}{}^q + \varepsilon^p M^p\|\chi\|_{1,p,\Omega}^p, \quad q = \frac{p}{p-1} \tag{2.5.38}$$

provided $1 < p < 3$ and $\tau^{-1} \geqslant p^{-1}-1/2q$, $\tau \geqslant 1$, i.e. $\sigma \geqslant \max\{1, 2q/3\}$, on account of the trace theorems (Section 1.1). For $p = 3$ the same estimate is valid for arbitrary $\sigma > 1$.

Suppose that g satisfies (GC) with $p \geqslant 2$, $s \geqslant q$, $r > 1$. For $\gamma_{\Gamma_1}\chi = \chi_1$, $H^2(\Gamma_1) > 0$,

$$F(\chi) \geqslant cL^3(\Omega)+K[\|\nabla\chi\|_p{}^p+\|\operatorname{adj}\nabla\chi\|_s{}^s+\|\det\nabla\chi\|_r{}^r] - \frac{1}{q\varepsilon^q}\|\mathbf{s}\|_\sigma{}^q - \frac{1}{p}\varepsilon^p M^p L[\|\nabla\chi\|_p{}^p+\|\chi_1\|_{\Gamma_1,p}{}^p], \quad K, M, L > 0$$

on account of (2.5.38) and the Friedrichs lemma. Let $\varepsilon > 0$ be such that $K - p^{-1}\varepsilon^p M^p L > 0$. Assume that $\partial\Omega$ is strongly Lipschitz and $H^2(\Gamma_1) > 0$. Let $a := \inf\{F(\boldsymbol{\chi}) | \ \boldsymbol{\chi} \in \mathscr{C}\}$. Clearly $a > -\infty$. If $F(\boldsymbol{\chi}_2) < \infty$ for some $\boldsymbol{\chi}_2 \in \mathscr{C}$, then $a < \infty$. In this case $F(\boldsymbol{\chi}_n) \to a$ implies that $\boldsymbol{\chi}_n \rightharpoonup \boldsymbol{\chi}_0 \in W^{1,p}(\Omega)$, $\boldsymbol{\chi}_n \to \boldsymbol{\chi}_0$ a.e. in Ω, $\operatorname{adj} \nabla \boldsymbol{\chi}_n \to \mathbf{H}_0$ in $\mathscr{L}^s(\Omega)$, $\det \nabla \boldsymbol{\chi}_n \rightharpoonup \Delta_0$ in $\mathscr{L}^r(\Omega)$ by the Friedrichs lemma and the Rellich–Kondrachov theorem. By Theorem 2.5.21 $\operatorname{adj} \nabla \boldsymbol{\chi}_n \to \operatorname{adj} \nabla \boldsymbol{\chi}_0$, $\det \nabla \boldsymbol{\chi}_n \to \det \nabla \boldsymbol{\chi}_0$ in $\mathscr{D}'(\Omega)$ and hence also in the weak topologies of $\mathscr{L}^s(\Omega)$ and $\mathscr{L}^r(\Omega)$, resp. Since γ_{Γ_1}: $W^{1,p}(\Omega) \to \mathscr{L}^1(\Gamma_1)$ is completely continuous, $\boldsymbol{\chi}_1(\xi) = \gamma_{\Gamma_1}\boldsymbol{\chi}_n(\xi) \to \gamma_{\Gamma_1}\boldsymbol{\chi}_0(\xi)$ H^2-a.e. and $\gamma_{\Gamma_1}\boldsymbol{\chi}_0 = \boldsymbol{\chi}_1$. If $\sigma > 2q/3$, i.e. $\tau^{-1} > p^{-1} - 1/2q$, then the trace operator γ_{Γ_2}: $W^{1,p}(\Omega) \to \mathscr{L}^\tau(\Gamma_2)$ is completely continuous and the surface integral in (2.5.36) is weakly continuous. If g satisfies (PC), then the volume integral in (2.5.36) is wlsc and $F(\boldsymbol{\chi}_0) = a$. (GC$_2$), with $g = \infty$ outside M, implies that $\boldsymbol{\chi}_0 \in C$.

We have thus proved

Theorem 2.5.25

Suppose that C is the set defined in Section 2.5.6, $H^2(\Gamma_1) > 0$ and Ω satisfies (SL). Let F be given by equation (2.5.36) and suppose that the normal integrand g satisfies (PC), (P), (GC$_1$), (GC$_2$) and is infinite outside M, $\mathbf{s} \in \mathscr{L}^\sigma(\Gamma_2)$, $r > 1$, $s \geqslant q$, and either $2 \leqslant p < 3$, $\sigma > \max\{1, 2q/3\}$ or $p = 3$, $\sigma > 1$. If $\exists \boldsymbol{\chi}_2 \in C$ $F(\boldsymbol{\chi}_2) < \infty$, then the problem $F \to \min$ on C has a solution $\boldsymbol{\chi}_0$.

Remark

Ball (1977) proved a slightly different and more general theorem with the alternative assumptions $2 \leqslant p < 3$, $\sigma > 2q/3$ or $p = 3$, $\sigma > 1$, or $p > 3$, $\sigma = 1$.

The case of incompressible materials can be treated by an analogous modification of Theorem 2.5.24 (cf. Ball, 1977).

The case of pure tractions needs some comments. Let F be given by (2.5.36) with $\Gamma_2 = \partial\Omega$. The control of $||\boldsymbol{\chi}_n||_{1,p}$ can be ensured by a recourse to the Poincaré lemma provided one of the following assumptions is made:

(1) the solution set is $\mathscr{C}_* := \{\boldsymbol{\chi} \in W^{1,p}(\Omega) | \operatorname{adj} \nabla \boldsymbol{\chi} \in \mathscr{L}^s(\Omega), \det \nabla \boldsymbol{\chi} \in \mathscr{L}^r(\Omega)$, $\det \nabla \boldsymbol{\chi}(\xi) > 0$ a.e. in Ω, $\int_\Omega dL^3\, \boldsymbol{\chi}(\xi) = \mathbf{k}\}$ for some fixed $\mathbf{k} \in \boldsymbol{R}^3$, while the normal integrand g satisfies (PC), (P), (GC$_1$), (GC$_2$) and $g(\mathbf{F}, \mathbf{H}, \Delta, \xi) = \infty$ if $(\mathbf{F}, \mathbf{H}, \Delta) \,\overline{\in}\, M$;

(2) the solution set is C, while the normal integrand g satisfies (PC) and the following two conditions: $g = \infty$ for $(\mathbf{F}, \mathbf{H}, \Delta) \,\overline{\in}\, M$,

(GC$_*$) $g(\mathbf{F}, \mathbf{H}, \Delta, \boldsymbol{\chi}, \xi) \geqslant c + K[|\mathbf{F}|^p + |\mathbf{H}|^s + |\Delta|^r + |\boldsymbol{\chi}|^\lambda]$, with $K > 0$, $r > 1$, $p \geqslant 2$, $s \geqslant q$, $\lambda \geqslant 1$.

In either case we assume that Ω satisfies (SL) and $s \in \mathscr{L}^\sigma(\partial\Omega)$.

In the first case we have that for every $\boldsymbol{\chi} \in \mathscr{C}_*$

$$||\boldsymbol{\chi}||_{1,p}{}^p \leqslant M[||\nabla\boldsymbol{\chi}||_p{}^p + |\mathbf{k}|^p], \quad M > 0,$$

and a minimizing sequence $\{\boldsymbol{\chi}_n\} \subset \mathscr{C}_*$ is bounded in $W^{1,p}(\Omega)$. Suppose that $\exists \boldsymbol{\chi}_2 \in \mathscr{C}_* F(\boldsymbol{\chi}_2) < \infty$, $2 \leqslant p < 3$, $r > 1$, $s \geqslant q$ and $s \in \mathscr{L}^\sigma(\partial\Omega)$. If $\sigma > 2q/3$, then γ: $W^{1,p}(\Omega) \to \mathscr{L}^\tau(\partial\Omega)$ is completely continuous (cf. the remarks following Theorem 1.1.47) and the surface integral in F is weakly continuous. If $\boldsymbol{\chi}_0$ is the weak limit of $\{\boldsymbol{\chi}_n\}$, then $F(\boldsymbol{\chi}_0) \leqslant a < \infty$ and $\det\nabla\boldsymbol{\chi}_0(\xi) > 0$ a.e. in Ω. By the Rellich–Kondrachov theorem $\boldsymbol{\chi}_n \to \boldsymbol{\chi}_0$ in $\mathscr{L}^1(\Omega)$, hence $\int_\Omega dL^3\, \boldsymbol{\chi}_0(\xi) = \mathbf{k}$, $\boldsymbol{\chi}_0 \in \mathscr{C}_*$ and $F(\boldsymbol{\chi}_0) = a$.

In the second case there is an obvious control over $||\boldsymbol{\chi}_n||_{1,p}$ if $\lambda \geqslant p$. If $1 \leqslant \lambda < p$, then $\boldsymbol{\chi}_n \rightharpoonup \boldsymbol{\chi}_0 \in W^{1,\lambda}(\Omega)$ on account of (GC$_*$), $\boldsymbol{\chi}_n(\xi) \to \boldsymbol{\chi}_0(\xi)$ a.e. in Ω as well as in $\mathscr{L}^1_{loc}(\Omega)$. Hence $\nabla\boldsymbol{\chi}_n \to \nabla\boldsymbol{\chi}_0$ in $\mathscr{D}'(\Omega)$. Also $\nabla\boldsymbol{\chi}_n \rightharpoonup \mathbf{F}_0$ in $\mathscr{L}^p(\Omega)$ and hence $\nabla\boldsymbol{\chi}_n \to \mathbf{F}_0$ in $\mathscr{D}'(\Omega)$. Since $\mathbf{F}_0$, $\nabla\boldsymbol{\chi}_0 \in \mathscr{L}^1_{loc}(\Omega)$, it follows that $\mathbf{F}_0 = \nabla\boldsymbol{\chi}_0$ a.e., $\nabla\boldsymbol{\chi}_n \rightharpoonup \nabla\boldsymbol{\chi}_0$ in $\mathscr{L}^p(\Omega)$.* Since $p \geqslant 2$, we can repeat the reasoning of the proof of Theorem 2.5.23.

Theorem 2.5.26

Suppose that Ω satisfies (SL), the normal integrand g satisfies (PC), $p \geqslant 2$, $s \geqslant q$, $r > 1$, $\sigma > \max\{1, 2q/3\}$, $g = \infty$ outside M.

If g satisfies (GC$_*$), then the problem $F \to \min$ in $\mathscr{C}$ has a solution provided

$$\exists \boldsymbol{\chi}_2 \in \mathscr{C} \quad F(\boldsymbol{\chi}_2) < \infty.$$

If g satisfies (GC) and (P), then $F \to \min$ in $\mathscr{C}_*$ has a solution provided $\exists \boldsymbol{\chi}_2 \in \mathscr{C}_*\ F(\boldsymbol{\chi}_2) < \infty$.

We shall finally prove existence for a mixed BVP with $\partial\Omega$ consisting of the components Γ_j, $j = 1, \ldots, N$, $\gamma_{\Gamma_1}\boldsymbol{\chi} = \boldsymbol{\chi}_1$ on Γ_1 and pressure loads on Γ_j, $j = 2, \ldots, N$. The pressure loads may be different on different components. Let $p \in W^{1,\infty}(\Omega)$ be such that $\gamma_{\Gamma_j} p$ is the pressure on Γ_j, $j = 2, \ldots, N$ and $\gamma_{\Gamma_1} p = 0$. The potential for the pressure loads is

$$\Psi(\boldsymbol{\chi}) = \int_\Omega dL^3\, [p \det\nabla\boldsymbol{\chi} + p_{,\xi^\alpha} (\operatorname{adj}\nabla\boldsymbol{\chi})^\alpha{}_k \chi^k],$$

cf. Section 3.5. We assume that $H^2(\Gamma_1) > 0$. Consider the problem $F = \mathscr{W} + \Phi + \Psi \to \min$ on $\mathscr{C}$, defined above. Suppose that g satisfies (PC), (P) and (GC).

* Boundedness and weak convergence of $\{\boldsymbol{\chi}_n\}$ in $W^{1,p}(\Omega)$ follows from the Poincaré inequality: $||\boldsymbol{\chi}||_{1,p}{}^p \leqslant K[||\nabla\boldsymbol{\chi}||_p{}^p + ||\boldsymbol{\chi}||_1{}^p] \leqslant K||\nabla\boldsymbol{\chi}||_p{}^p + R||\boldsymbol{\chi}||_\lambda{}^p$.

For the proof of existence we have to estimate Ψ. The first term of Ψ is easily handled since the lower estimate of F involves a term

$$K\|\det\nabla\chi\|_r^r - \|p\|_\infty\|\det\nabla\chi\|_1 \geqslant K[\|\nabla\chi\|_r^{r-1} - N'\|p\|_\infty]\|\nabla\chi\|_r \geqslant K'\|\nabla\chi\|_r$$

for $\|\nabla\chi\|_r^{r-1} \geqslant N'\|p\|_\infty + \delta$, $\delta > 0$, $K' = K\delta > 0$, $r > 1$. For the second term of Ψ we have

$$\left|\int_\Omega dL^3\, p_{,\xi^\alpha}(\operatorname{adj}\nabla\chi)^\alpha{}_k\chi^k\right| \leqslant \|\nabla p\|_\infty\|\operatorname{adj}\nabla\chi\|_s\|\chi\|_t \leqslant \|\nabla p\|_\infty\left[\frac{\varepsilon^s}{s}\|\operatorname{adj}\nabla\chi\|_s^s + \frac{1}{t\varepsilon^t}\|\chi\|_t^t\right],$$

$$t^{-1} + s^{-1} = 1. \tag{2.5.39}$$

By the Friedrichs lemma the lower estimate of $\mathscr{W} + \Phi$ involves an expression

$$\frac{K}{2}\|\nabla\chi\|_p^p + \frac{K}{2L}\|\chi\|_p^p - \frac{K}{2}\|\chi_1\|_{\Gamma_1,p}^p.$$

The second term of this expression can be combined with the second term of the right-hand side of (2.5.39):

$$A := \frac{K}{2L}\|\chi\|_p^p - \|\nabla p\|_\infty\frac{1}{t\varepsilon^t}\|\chi\|_t^t.$$

Fix $\varepsilon > 0$ in such a way that $K > \|\nabla p\|_\infty\varepsilon^s/s$. For $t < p$ (i.e. $s > q$) and $\|\chi\|_p > [2LMK^{-1}\|\nabla p\|_\infty(t\varepsilon^t)^{-1}]^{1/(p-t)}$

$$A \geqslant \left[\frac{K}{2L}\|\chi\|_p^{p-t} - M\|\nabla p\|_\infty\frac{1}{t\varepsilon^t}\right]\|\chi\|_p^t \geqslant 0.$$

It follows that for any sequence $\{\chi_n\}$ minimizing F the sequences $\{\|\chi_n\|_{1,p}\}$, $\{\|\operatorname{adj}\nabla\chi_n\|_s\}$, $\{\|\det\nabla\chi_n\|_r\}$ are bounded. Hence we have

Theorem 2.5.27
Suppose that $\partial\Omega$ satisfies (SL), and g satisfies (PC), (P) and (GC) with $p \geqslant 2$, $r > 1$, $s > q$. If $\exists\, \chi_2 \in C$ such that $F(\chi_2) < \infty$ then the problem $F = \mathscr{W} + \Phi + \Psi$ min on $\mathscr{C}$ has a solution χ_0.

Similar theorems can be found in the paper of Ball (1977).

2.5.9 Generalizations of Existence Theorems

Ball (1977) proved existence theorems for a substantially larger class of configurations χ and under somewhat weaker growth conditions (GC). He replaced the functions $\operatorname{adj}\nabla\chi$, $\det\nabla\chi \in \mathscr{L}^1_{loc}(\Omega)$ by the distributions $\operatorname{Adj}\nabla\chi$,

$\operatorname{Det}\nabla\boldsymbol{\chi}$. The latter are represented by locally integrable functions under significantly weaker assumptions than those of Lemma 2.5.20 and Theorem 2.5.21. If $\boldsymbol{\chi}$ is assumed to belong to a suitable Orlicz space then Lemma 2.5.20 and Theorem 2.5.21 remain valid for $\operatorname{Adj}\nabla\boldsymbol{\chi}$, $\operatorname{Det}\nabla\boldsymbol{\chi}$, *mutatis mutandis*. The existence Theorems 2.5.23, 2.5.24 also remain valid provided $\boldsymbol{\chi}$, $\operatorname{Adj}\nabla\boldsymbol{\chi}$, $\operatorname{Det}\nabla\boldsymbol{\chi}$ are locally integrable and belong to appropriate Orlicz spaces.

For future reference we shall consider an example of a function with $\det\nabla\boldsymbol{\chi} \neq \operatorname{Det}\nabla\boldsymbol{\chi}$, $\nabla\boldsymbol{\chi} \in \mathscr{L}^1_{loc}(\Omega)$. Let (x^k) be an orthonormal Cartesian coordinate system in the space. We shall use $x^k = \chi_0{}^k(\xi^\alpha)$ as our reference configuration with the Lagrange coordinates (x^p) and consider a configuration $y^k = \delta^k{}_p x^p + u^k(\mathbf{x}) = \chi^k(\mathbf{x})$ referred to the same coordinate system in the space. In other words the material point (ξ^α) that would assume the position (x^k) in the space in the reference configuration $\boldsymbol{\chi}_0$ assumes the position (y^k) in the configuration $\boldsymbol{\chi}$, with respect to the same spatial coordinates.

Let

$$u^k(\mathbf{x}) \equiv \lambda \frac{x^k}{r}, \qquad r := |\mathbf{x}|, \ \lambda \in \boldsymbol{R}_+ .$$

Let $\mathbf{x}$ range over $\Omega = \mathscr{K}(0, 1)$. At $\mathbf{x} = \mathbf{0}$ we have $|\mathbf{y}| = \lambda$, i.e. the body exhibits a hole of radius $\lambda > 0$ in the configuration $\boldsymbol{\chi}$. It is clear that $\boldsymbol{\chi}, \mathbf{u} \in \mathscr{L}^\infty(\Omega)$. Since $|\nabla\mathbf{u}| = (\sum_{k,p} (u^k{}_{,x^p})^2)^{1/2} = K/r$ for some constant $K > 0$, $\boldsymbol{\chi}, \mathbf{u} \in W^{1,p}(\Omega)$ for $p < 3$ and $\operatorname{adj}\nabla\mathbf{u}$, $\operatorname{adj}\nabla\boldsymbol{\chi} \in \mathscr{L}^s(\Omega)$ for $s < 3/2$. Note that $q > 3/2 > s$.

Let us now calculate $\nabla\mathbf{u}$. For $r > 0$

$$u^k{}_{,x^p} = r^{-1}\delta^k{}_p - r^{-3}x^k x^p$$

and for an arbitrary $\varepsilon > 0$, $\varphi \in \mathscr{C}_0^\infty(\Omega)$,

$$\int_{\mathscr{K}(0,\varepsilon)} dL^3\, u^k{}_{,x^p}\varphi = -\int_{\mathscr{K}(0,\varepsilon)} dL^3\, u^k \varphi_{,x^p} + \int_{r=\varepsilon} dH^2\, u^k \frac{x^p}{r}\varphi .$$

The last integral is equal to

$$\lambda\varepsilon^2 \int_{r=1} dH^2\, x^k x^p \varphi(\varepsilon\mathbf{x}) \to 0 \qquad \text{for } \varepsilon \to 0.$$

Hence the distributional derivative $u^k{}_{,x^p}$ is represented by the locally integrable function $u^k{}_{,x^p}$ calculated above and extended to $r \geqslant 0$. We can calculate $\det\nabla\boldsymbol{\chi}$ by pointwise multiplication of $\nabla\boldsymbol{\chi}$:

$$\begin{aligned}\det\nabla\boldsymbol{\chi} &= \tfrac{1}{6}\, \epsilon_{ijk}\, \epsilon^{pqr}(\delta^i{}_p + u^i{}_{,x^p})(\delta^j{}_q + u^j{}_{,x^q})(\delta^k{}_r + u^k{}_{,x^r}) \\ &= \left(1 + \frac{1}{r}\right)^2 \equiv: \Delta(\mathbf{x}), \qquad \Delta \in \mathscr{L}^1_{loc}(\Omega).\end{aligned}$$

Let us now calculate the distribution $\operatorname{Det}\nabla\chi$ defined by (2.5.26) and 2.5.27). Let $\varphi \in \mathscr{C}_0^\infty(\Omega)$. Omitting some details* we have

$$\int_\Omega \varphi \operatorname{Det}\nabla\chi = -\int_\Omega dL^3\, \chi^i (\operatorname{Adj}\nabla\chi)^p{}_i \varphi_{,x^p}$$

$$= -\frac{1}{6}\int_\Omega dL^3\, \epsilon_{ijk}\, \epsilon^{pqr}\left(x^i+\frac{x^i}{r}\right)$$

$$\times\left[\left(x^j+\frac{x^j}{r}\right)\left(x^k+\frac{x^k}{r}\right)_{,x^r}\right]_{,x^q}\varphi_{,x^p}$$

$$= -\frac{1}{6}\int_\Omega dL^3\, \epsilon_{ijk}\, \epsilon^{pqr} x^i\left(1+\frac{1}{r}\right)\left(x^j+\frac{x^j}{r}\right)_{,x^q}\left(x^k+\frac{x^k}{r}\right)_{,x^r}\varphi_{,x^p}$$

$$= -\frac{4\pi}{3}\int_0^1 dr\,(r+1)^3\frac{d\tilde{\varphi}}{dr} = \frac{4\pi}{3}\,\tilde{\varphi}(0)+4\pi\int_0^1 dr\,(r+1)^2\tilde{\varphi}(r),$$

where

$$\tilde{\varphi}(r) :\equiv \frac{1}{4\pi}\int dH^2\, \varphi(r,\theta,\phi),$$

$$\frac{d\tilde{\varphi}}{dr} = \frac{1}{4\pi}\int dH^2\, \varphi_{,r}(r,\theta,\phi)$$

in spherical coordinates (r,θ,ϕ). Hence $\operatorname{Det}\nabla\chi = \frac{4\pi}{3}\delta+\Delta$, with δ = the Dirac delta supported by $\mathbf{0}$.

2.5.10 Comments on Regularity

So far we have established the existence of equilibrium configurations for some BVPs defining them merely as minimizers of the associated variational problems. Under the assumptions (GC) the Nemytskii operator $\tilde{S}$ does not exist and $\mathscr{W}$ need not be Gâteaux-differentiable. Hence we cannot derive a weak equilibrium differential equation in the form of a VWP along the lines of Section 2.3.

In Chapter 3 we shall see that the energy of an elastic body in motion plus the potential energy of the loads under given BCs does not increase in

* Green's formula should be applied on $\mathscr{K}(\mathbf{0},1)\setminus\mathscr{K}(\mathbf{0},\varepsilon)$, $\varepsilon > 0$. However the surface integral arising from integration by parts is $O(\varepsilon)$.

time and generically it decreases after some time. In general however the motion $\boldsymbol{\chi}(\xi, t)$ neednot be asymptotically convergent (for $t \to \infty$) to an equilibrium configuration. In linear elastodynamics the energy is conserved and the asymptotic state (if it exists) may be a periodic vibration. Asymptotic convergence to equilibrium states is possible if some lossy mechanisms (viscosity, heat conduction, lossy loads $\mathbf{b}(\boldsymbol{\chi}_{,t})$, $\mathbf{s}(\boldsymbol{\chi}_{,t})$) attenuate the vibrations. Nevertheless it is possible to define a stable equilibrium as the minimizer of $F = \mathscr{W} + \Phi$. Indeed, a body in stable equilibrium will not start to move until some energy is supplied to it bringing it to a configuration with higher potential and some kinetic energy. Such an event implies however that the BCs or the loads have been perturbed.

If one stands by the traditional interpretation of equilibrium configurations as solutions of some differential equations, the following questions arise: (1) is a minimizer a weak solution of the appropriate differential equations? (2) if so, is it a strong solution (in $W^{2,p}(\Omega)$) or even a regular one (in $\mathscr{C}^2(\bar{\Omega})$)?

Concerning the first question, it is possible to prove that a minimizer $\boldsymbol{\chi}_0$ is a weak solution of the Euler–Lagrange equations:

$$\int_{\Omega} dL^3 \left[\varrho_0 \frac{\partial W}{\partial F^k{}_{\alpha}} v^k{}_{,\xi^\alpha} + \frac{\partial \phi}{\partial x^k} v^k \right] = 0 \ \forall \mathbf{v} \in \mathscr{C}_0^\infty(\Omega) \qquad (2.5.40)$$

under some additional assumptions, viz. $p \geqslant s \geqslant r$,*

(R) $W(\cdot\,, \xi) \in \mathscr{C}^1(\mathscr{L}_+)$ $\forall \xi \in \Omega$ and $\forall d > 0$ $\exists c(d) \in \boldsymbol{R}$ such that

$$\frac{\partial W}{\partial \mathbf{F}}(\mathbf{F}, \xi) \leqslant c(d)\,[1 + |\mathbf{F}|^p + |\mathrm{adj}\,\mathbf{F}|^s + |\det \mathbf{F}|^r] \qquad (2.5.41)$$

$$\forall \mathbf{F} \in \mathscr{L}_+ \quad \text{with } \det \mathbf{F} > d,$$

(POT) ϕ: $\boldsymbol{R}^3 \to \boldsymbol{R}_+$ is $\mathscr{C}^1$ and for $p \leqslant 3$ $\exists \gamma$, K such that

(i) if $p < 3$ then $1 \leqslant \gamma \leqslant \dfrac{3p}{3-p}$,

(ii) if $p = 3$ then $\gamma \geqslant 1$, and

$$\left| \frac{\partial \phi}{\partial \mathbf{x}} \right| \leqslant K[1 + |\boldsymbol{\chi}|^\gamma]. \qquad (2.5.42)$$

If $p > 3$ then the condition (2.5.42) drops out.

One of the hypotheses of Theorem 2.5.28 is stated in terms of a property of the minimizer $\boldsymbol{\chi}_0$ that can be verified *a posteriori*. Hence we do not know

* Note that $p \geqslant q$ under the hypotheses of the existence theorems above.

any constitutive asumptions that would guarantee *a priori* that the minimizer is a weak solution.

We shall also note that equation (2.5.40) amounts to the statement that F has a Gâteaux derivative at $\boldsymbol{\chi}_0$ and that this derivative vanishes.

Theorem 2.5.28

Suppose that the hypotheses of Theorem 2.5.23 as well as (R) and (POT) are satisfied with $p \geqslant s \geqslant r$. Let $\boldsymbol{\chi}_0 \in \mathscr{C}$, $F(\boldsymbol{\chi}_0) = \inf\{F(\boldsymbol{\chi}) | \ \boldsymbol{\chi} \in \mathscr{C}\}$.

Let Ω' be an open subset of Ω and suppose that $\forall \mathbf{v} \in \mathscr{C}_0^\infty(\Omega')\ \exists d > 0$, $\varepsilon > 0$ such that

$$\det \nabla \left(\boldsymbol{\chi}_0(\xi) + t\mathbf{v}(\xi)\right) \geqslant d \quad \text{a.e. in } \Omega' \ \forall |t| < \varepsilon. \tag{2.4.43}$$

Then $\boldsymbol{\chi}_0$ satisfies equation (2.5.40) for every $\mathbf{v}$ such that $\operatorname{supp}\mathbf{v} \subset \Omega'$.

Proof

Let $\mathbf{v} \in \mathscr{C}_o^\infty(\Omega')$, $\boldsymbol{\chi}_t := \boldsymbol{\chi}_0 + t\mathbf{v}$. Since $p \geqslant s \geqslant r$, it is clear that $\operatorname{adj}\nabla\boldsymbol{\chi}_t \in \mathscr{L}^s(\Omega)$, $\det\nabla\boldsymbol{\chi}_t \in \mathscr{L}^r(\Omega)$. Now

$$\frac{1}{t}[F(\boldsymbol{\chi}_t) - F(\boldsymbol{\chi}_0)] = \int_{\Omega'} dL^3 \frac{1}{t}\phi[(\boldsymbol{\chi}_t(\xi)) - \phi(\boldsymbol{\chi}_0(\xi))]$$

$$+ \int_{\Omega'} dL^3 \frac{1}{t}\, \varrho_0 W[(\nabla\boldsymbol{\chi}_t(\xi), \xi) - W(\nabla\boldsymbol{\chi}_0(\xi), \xi)]. \tag{2.5.44}$$

Let $|t| < t_o$ for some fixed $t_o > 0$. On account of (R) the second integrand is bounded by the expression

$$c(d)\,[1 + |\nabla\boldsymbol{\chi}_\tau|^p + |\operatorname{adj}\nabla\boldsymbol{\chi}_\tau|^s + |\det\nabla\boldsymbol{\chi}_\tau|^r],$$

with $\nabla\boldsymbol{\chi}_\tau(\xi) :\equiv \nabla\boldsymbol{\chi}_0(\xi) + \tau(\xi)\nabla\mathbf{v}(\xi)$ for some $0 \leqslant \tau(\xi) \leqslant t \leqslant t_o$,

$$|\nabla\boldsymbol{\chi}_\tau(\xi)|^p \leqslant |\nabla\boldsymbol{\chi}_0(\xi)|^p + t_o{}^p|\nabla\mathbf{v}(\xi)|^p,$$

$$|\operatorname{adj}\nabla\boldsymbol{\chi}_\tau(\xi)|^s \leqslant |\operatorname{adj}\nabla\boldsymbol{\chi}_0(\xi)|^s + t_o{}^s K_1|\nabla\mathbf{v}(\xi)|^s|\nabla\boldsymbol{\chi}_0(\xi)|^s$$

$$+ K_2 t_o{}^{2s}|\nabla\mathbf{v}(\xi)|^{2s} \leqslant |\operatorname{adj}\nabla\boldsymbol{\chi}_0|^s + t_o{}^s K_1'|\nabla\boldsymbol{\chi}_0|^s + K_2' t_o{}^{2s} \quad \text{etc.}$$

Hence, by the Lebesgue dominated convergence theorem, the second integral tends to the first term of (2.5.40) as $t \to 0$.

For the first integral we note that $p > 3$ implies that $\boldsymbol{\chi}_0 \in \mathscr{C}^0(\Omega')$, while for $p < 3$ $\boldsymbol{\chi}_0 \in \mathscr{L}^{3p/(3-p)}(\Omega')$, for $p = 3$ $\boldsymbol{\chi}_0 \in \mathscr{L}(\Omega')$ for any $\gamma > 1$. Using (POT) and repeating the argument of the preceding paragraph we conclude that the first integral in (2.5.44) tends to the second term of (2.5.40). □

It is clear that (R) cannot be sharpened because of (GC) and a lower bound for $\det \nabla \boldsymbol{\chi}_0$ is necessary.

We now turn to the regularity of equilibrium configurations.

Regularity of weak solutions of elliptic systems has been investigated by many authors in connection with the 19th problem posed by Hilbert. The first positive results for nonlinear elliptic systems were obtained by Morrey (1940) (second-order systems, two independent variables), De Giorgi and Nash (De Giorgi, 1957) (n independent variables, one second-order equation). Subsequently examples of irregular solutions for systems in $n \geqslant 3$ independent variables and analytic data ($\hat{S}_k{}^{\alpha}$, $\boldsymbol{\chi}_1$, $\mathbf{f} = -\varrho_0 b_k$ in our case) were produced (De Giorgi, 1968; Giusti and Miranda, 1968). They have essentially the form x^i/r.

The existence of an irregular solution is particularly disturbing if it is the only solution in a class including all the regular solutions. Nečas (1977) and Nečas, John and Stará (1980) proved the existence of solutions which are irregular and have the form $u^{ij} = x^i x^j/r$ or the related form for $n \geqslant 5$ and $n \geqslant 3$ resp. In these papers the irregular minimizer is *uinque*. The irregular solution of Giusti and Miranda is also unique provided n is sufficiently large. The counterexample of De Giorgi can easily be adapted to the case of nonlinear elasticity with appropriately chosen constitutive equations.

For elliptic systems with smooth data the set of singular points of a weak solution has zero H^{n-1} measure (Almgren, 1968; Morrey, 1968; Giusti, 1969).

It should also be remarked that the above mentioned papers assumed Višik ellipticity, viz.

$$\frac{\partial^2 W}{\partial F^k{}_\alpha \, \partial F^l{}_\beta} H^k{}_\alpha H^l{}_\beta \geqslant c(\mathbf{F})|\mathbf{H}|^2, \quad c(\mathbf{F}) > 0, \ \forall \mathbf{H} \in \mathscr{L} \tag{2.5.45}$$

in the case of elasticity.

It is a common feature of the counterexamples constructed by De Giorgi, Giusti, Miranda, Nečas *et al.* that $\hat{S}_k{}^{\alpha}$ exhibits linear or quadratic growth with $\mathbf{F}$ and hence W exhibits quadratic or cubic growth in $\mathbf{F}$. Ball (1977a) conjectured that non-regular solutions with a hole in the center may be ruled out by appropriate growth conditions on W.

A physically plausible growth condition is

$$W(\mathbf{F}, \xi) \geqslant a(\xi) + b|\mathbf{F}|^p, \quad b > 0, \ p > 3. \tag{2.5.46}$$

Indeed, consider a one-parameter family of homogeneous cubes with the strain energy $W(\mathbf{F}) \equiv W(\mathbf{F}, \xi_0)$ independent of ξ, of side λ^{-1}, subject to a homogeneous deformation gradient $\lambda\mathbf{F}$, with $\mathbf{F} \in \mathscr{L}_+$ fixed and λ ranging over $\boldsymbol{R}_+$. The cubes have identical sizes and material properties but for $\lambda \to \infty$ they exhibit increasing strain. We expect that the energy $\lambda^{-3}W(\lambda\mathbf{F}, \xi_0)$ grows unboundedly for $\lambda \to \infty$,

$$\lambda^{-3}W(\lambda\mathbf{F}, \xi_0) \to \infty \quad \text{for } \lambda \to \infty, \tag{2.5.47}$$

which suggests (2.5.46).

Following Ball (1978, unpublished) we consider the following deformation with a hole of radius $\lambda > 0$

$$\mathbf{y} = \boldsymbol{\chi}_\lambda(\mathbf{x}) \equiv (r^3+\lambda^3)^{1/3}\frac{\mathbf{x}}{r}, \quad r < 1$$

in an incompressible material.* In fact

$$\begin{aligned}\det(\partial\mathbf{y}/\partial\mathbf{x}) &= \det\{(r^3+\lambda^3)^{-2/3}\mathbf{x}\otimes\mathbf{x}+(r^3+\lambda^3)^{1/3}r^{-3}(r^2\mathbf{E}-\mathbf{x}\otimes\mathbf{x})\}\\ &= \det[r^{-1}\mathbf{E}(r^3+\lambda^3)-\lambda^3\mathbf{x}\otimes\mathbf{x}]\,(r^3+\lambda^3)^{-2} = 1.\end{aligned}$$

Now $\mathbf{F} = \mathbf{AV}$, with $\mathbf{A} \in \mathrm{SO}(3)$ and $\mathscr{V}$ symmetric and positive definite, hence $W(\mathbf{F}, \xi) = W(\mathbf{V}, \xi)$. In the case of a homogeneous isotropic material, which we assume, $W = W(\mathbf{F}, \xi) = W(\mathbf{AFB}, \xi)$, $\mathbf{A}, \mathbf{B} \in \mathrm{SO}(3)$** and $W = g(v_1, v_2, v_3)$ for $\mathbf{V} = \mathbf{B}^{-1}\mathrm{diag}\{v_1, v_2, v_3\}\mathbf{B}$, $\mathbf{B} \in \mathrm{SO}(3)$. For $\boldsymbol{\chi}_\lambda$ the principal stretches v_i are $v_1 = \dfrac{d|\boldsymbol{\chi}_\lambda|}{dr} = \dfrac{d|\mathbf{y}|}{dr}$, $v_2 = \dfrac{|\mathbf{y}|}{r} = v_3 =: v$. Hence

$$v_1 = v^{-2},\ r^3 = \frac{\lambda^3}{v^3-1},\ r^2dr = -\lambda^3v^2(v^3-1)^{-2}dv,$$

$$\begin{aligned}\mathscr{W}(\boldsymbol{\chi}) &= 4\pi\int_0^1 dr\, r^2 g(v_1, v_2, v_3)\\ &= 4\pi\lambda^3\int_{(1+\lambda^3)^{1/3}}^{\infty} dv\,\frac{v^2}{(v^3-1)^2}\,g(v^{-2}, v, v).\end{aligned}$$

If $\boldsymbol{\chi}_{\lambda_0}$ minimizes $F(\boldsymbol{\chi}) \equiv \mathscr{W}(\boldsymbol{\chi})-P(1+\lambda^3)^{1/3}$, i.e. under a normal load P at the external boundary, then $\dfrac{dF(\boldsymbol{\chi}_\lambda)}{d\lambda} = 0$ at $\lambda = \lambda_0$. Working out this equation and integrating by parts we get the equation

$$4\pi\lambda^2\left[\int_{(1+\lambda^3)^{1/3}}^{\infty} dv\,\frac{1}{v^3-1}\,\frac{dg(v^{-2}, v, v)}{dv} - \frac{P}{(1+\lambda^3)^{1/3}}\right] = 0. \tag{2.5.48}$$

* The solution considered here is a universal static solution for arbitrary isotropic incompressible elastic bodies (Wang and Truesdell, 1973, Section 4.4.). It is not universal for isotropic compressible bodies in view of Ericksen's theorem (op. cit.).

** See Chapter 4.

Equation (2.5.48) determines the load P which produces the cavity of radius λ provided the integral in (2.5.48) is convergent. For a model of rubber-like materials suggested by Ogden (1972)

$$g(v_1, v_2, v_3) = v_1^{\alpha}+v_2^{\alpha}+v_3^{\alpha}-3$$

and the integral converges iff $-3/2 < \alpha < 3$. It can also be shown that W is polyconvex if $\alpha \geqslant 1$ (Ball, 1977a). The growth condition (2.5.46) is not satisfied. For $\alpha = 2$ the solutions with holes appear at the minimum load $P = 5$.

Physically, there is an additional obstacle to the creation of holes: the loads must provide the energy of the free surface at $|\mathbf{y}| = \lambda$. Roughly, the loads should supply an amount of energy of order $\lambda d\lambda$ for an increase $d\lambda$ of the hole radius for the surface energy and $d(1+\lambda^3) = 3\lambda^2 d\lambda$ for the volume energy, cf. equation (2.5.48) for λ not too large. At small λ the expenditure of energy for the surface of the cavity dominates and cannot be neglected.

2.5.11 Invariance of Constitutive Assumptions

Constitutive assumptions must satisfy several *a priori* conditions. The most obvious one is the invariance with respect to the transformations of the spatial and Lagrange coordinates.

The invariance of the constitutive assumptions with respect to the Galilean transformations follows from the invariance of W with respect to the left action of SO(3) on $\mathscr{L}_+$: $\mathbf{F} \mapsto \mathbf{AF}$. Since the latter invariance is not explicit in the formulae of this section it is advisable to check the invariance of constitutive assumptions by inspection. The invariance of pointwise constitutive assumptions is apparent from their scalar character with respect to the left action of SO(3) on $\mathscr{L}_+$. For the invariance of (PC) note that the associated action of SO(3) on M is linear: $(\mathbf{F}, \mathbf{H}, \Delta) \mapsto (\mathbf{AF}, \mathbf{H}^t\mathbf{A}, \Delta)$. Convexity of $g(\cdot\,, \xi)$ is preserved by such transformations. The other constitutive assumptions such as (QC) are also manifestly invariant with respect to spatial rotations and translations. It is worth noting that non-constitutive assumptions of monotonicity and coercivity considered in Section 2.3 are also invariant with respect to these transformations.

The invariance of pointwise constitutive assumptions such as (LH), (SE), rank-one convexity and strict rank-one convexity with respect to transformations of the Lagrange coordinates is a consequence of their invariance with respect to the right action of SL(3) on $\mathscr{L}_+$: $\mathbf{F} \mapsto \mathbf{FL}$. The associated action of SL(3) on M is linear: $(\mathbf{F}, \mathbf{H}, \Delta) \mapsto (\mathbf{FL}, (\det\mathbf{L})\mathbf{L}^{-1}\mathbf{H}, (\det\mathbf{L})\Delta)$. Hence (PC) is also invariant with respect to such transformations. The invariance of (QC) under a.e. differentiable transformations $\xi = \boldsymbol{\chi}(\zeta)$ of Lagrange coordinates follows immediately from the definition (2.5.16) and the invariance of W. Note that the associated transformation of the integration variables in equation (2.5.16) is $\xi^\alpha = \chi^\alpha(\zeta_0) + \frac{\partial \chi^\alpha}{\partial \zeta^\rho}(\zeta_0)(\zeta^\rho - \zeta_0^\rho)$.* On account of the remark following the proof of Theorem 2.5.14 (QC) can be expressed in terms of local Lagrange coordinates. Upper bounds, e.g. (2.5.18) and (2.5.19), are preserved by transformations $\boldsymbol{\chi} \in W^{1,\infty}$ or $\boldsymbol{\chi} \in \mathrm{Lip}$, provided the constants are suitably readjusted. Since all the norms on a finite-dimensional space are equivalent it is convenient to work with the norm $|\mathbf{F}| := \sup\{|F^k{}_\alpha| \mid k, \alpha = 1, 2, 3\}$ at this point.

Coercivity and growth conditions, equation (2.5.42), are obviously invariant under $\mathscr{C}^1$-smooth transformations $\boldsymbol{\chi}$ satisfying the condition $\det\nabla\boldsymbol{\chi} > 0$. For a bi-Lipschitz transformation $\boldsymbol{\chi}$ the following inequalities for $\mathbf{H} = \nabla\boldsymbol{\chi}$ will be useful:

$$|\mathbf{H}| \leqslant L, \quad 3 \leqslant \sum_{\alpha, k} |H^k{}_\alpha| |\overset{-1}{H}{}^\alpha{}_k| \leqslant K \sum_{\alpha, k} |H^k{}_\alpha| \leqslant 9K|\mathbf{H}|$$

with $K := \sup\{\operatorname{ess\,sup}|\overset{-1}{H}{}^\alpha{}_k(\cdot)| \mid \alpha, k = 1, 2, 3\}$.

We now turn to some less obvious invariance requirements.

A Lagrangian $\Lambda(\nabla\boldsymbol{\chi}, \boldsymbol{\chi}, \xi) = \varrho_o W(\nabla\boldsymbol{\chi}, \xi) + f(\boldsymbol{\chi}, \xi)$ admits certain gauge transformations

$$\int_\Omega dL^3 \Lambda' = \int_\Omega dL^3 \Lambda + \text{terms which are constant under the given BCs on } \Gamma_1.$$

In particular we consider the gauge

$$\Lambda'(\nabla\mathbf{u}, \mathbf{u}, \xi) - \Lambda(\nabla\mathbf{u}, \mathbf{u}, \xi) \equiv \frac{d}{d\xi^\alpha} X^\alpha(\mathbf{u}(\xi), \xi) + c(\xi) \tag{2.5.49}$$

with $X^\alpha(\mathbf{u}, \xi)\nu_\alpha \equiv 0$ for $\xi \in \Gamma_2$.** The variational problem $F = \mathscr{W} + \Phi \to \min$ is clearly invariant under such gauge transformations. The Euler–Lagrange operators for Λ and Λ' coincide:

$$\frac{\partial \Lambda'}{\partial u^k} - \frac{d}{d\xi^\alpha}\frac{\partial \Lambda'}{\partial F^k{}_\alpha} \equiv \frac{\partial \Lambda}{\partial u^k} - \frac{d}{d\xi^\alpha}\frac{\partial \Lambda}{\partial F^k{}_\alpha}$$

identically.

* In geometrical terms, (QC) involves integration over a subset of a fiber of $T\mathscr{B}$.

** For the variational problem $F \to \min$ with $F = \mathscr{W} + \Phi$ and Φ given by equation (2.4.1) it is possible to consider the gauge (2.5.49) jointly with $g \to g' = g - X^\alpha\nu_\alpha$ on Γ_2.

At this stage we shall leave aside the question whether the trace of $X^\alpha\nu_\alpha$ on Γ_1 is a H^2-integrable function of the trace of $\mathbf{u}$ on Γ_1.

If Λ, Λ' are functions of $\mathbf{F}$, ξ only ($f = 0$) then

$$\frac{\partial X^\alpha}{\partial u^k} = c_k{}^\alpha(\xi), \quad X^\alpha(\mathbf{u}, \xi) = c_k{}^\alpha(\xi) u^k + Y^\alpha(\xi),$$

$$\frac{dX^\alpha}{d\xi^\alpha} \equiv \frac{\partial Y^\alpha}{\partial \xi^\alpha} + \frac{\partial c_k{}^\alpha}{\partial \xi^\alpha} u^k + c_k{}^\alpha u^k{}_{,\xi^\alpha} = b(\nabla \mathbf{u}, \xi).$$

Hence

$$X^\alpha \equiv c_k{}^\alpha u^k + Y^\alpha(\xi), \quad c_k{}^\alpha = \text{const},$$

and the most general gauge (2.5.49) of $\Lambda = \varrho_0 W$ that does not affect the loads is

$$\frac{dX^\alpha}{d\xi^\alpha} = c_k{}^\alpha u^k{}_{,\xi^\alpha} + \frac{\partial Y^\alpha}{\partial \xi^\alpha}(\xi).$$

Another interesting class of gauges of $\Lambda = \varrho_0 W$ has the following form:

$$\Lambda' - \Lambda = G(\mathbf{F}), \quad G \in \mathscr{C}^0(\mathscr{L}) \tag{2.5.50}$$

with

$$\int_\Omega dL^3\, G(\nabla \boldsymbol{\chi} + \nabla \mathbf{u}) \equiv \int_\Omega dL^3\, G(\nabla \boldsymbol{\chi}) \forall \boldsymbol{\chi} \in \mathscr{C}^1(\bar{\Omega}), \quad \mathbf{u} \in \mathscr{C}_0^\infty(\Omega), \tag{2.5.51}$$

or, equivalently

$$\int_\Omega dL^3\, G\big(\mathbf{F}_0 + \nabla \mathbf{u}(\xi)\big) \equiv G(\mathbf{F}_0) L^3(\Omega). \tag{2.5.52}$$

It is obvious that (2.5.52) follows from (2.5.51). On the other hand it follows from (2.5.52) that both G and $-G$ are quasiconvex. By Theorem 2.5.16 G and $-G$ are both rank-one convex.

If $G \in \mathscr{C}^2(\mathscr{L})$ satisfies (2.5.52) then both G and $-G$ satisfy (LH), i.e.

$$A_k{}^\alpha{}_l{}^\beta a^k a^l k_\alpha k_\beta = 0 \;\; \forall \mathbf{a}, \mathbf{k} \in \mathbf{R}^3 \tag{2.5.53}$$

with

$$A_k{}^\alpha{}_l{}^\beta := \frac{\partial^2 G}{\partial F^k{}_\alpha \partial F^l{}_\beta}(\mathbf{F}). \tag{2.5.54}$$

Let $g^{\alpha\beta} = A_k{}^\alpha{}_l{}^\beta a^k a^l$. By polarization $g^{\alpha\beta} k_\alpha l_\beta = -g^{\alpha\beta} l_\alpha k_\beta$. Hence

$$A_k{}^\alpha{}_l{}^\beta = A_l{}^\beta{}_k{}^\alpha = -A_k{}^\beta{}_l{}^\alpha = -A_l{}^\alpha k^\beta.$$

Since $A_1{}^1{}_1{}^1 = A_2{}^2{}_2{}^2 = A_3{}^3{}_3{}^3 = 0$, G is a linear function of each of the arguments $F^1{}_1$, $F^2{}_2$, $F^3{}_3$, i.e.

$$G(\mathbf{F}) = a(\tilde{\mathbf{F}}) F^1{}_1 F^2{}_2 F^3{}_3 + b_1(\tilde{\mathbf{F}}) F^2{}_2 F^3{}_3 + \text{cyclic} + \sum_{j=1}^3 c_j(\tilde{\mathbf{F}}) F^j{}_j + d(\tilde{F}),$$

where $\tilde{\mathbf{F}}$ denotes the off-diagonal entries of $\mathbf{F}$.

Since $A_k{}^k{}_l{}^k = 0 = A_k{}^k{}_k{}^l$ for all $k \neq l$,

$$\frac{\partial a}{\partial F^k{}_l} = \frac{\partial b_j}{\partial F^k{}_l} = \frac{\partial c_k}{\partial F^k{}_l} = \frac{\partial d}{\partial F^k{}_l} = \frac{\partial b_j}{\partial F^l{}_k} = \frac{\partial c_k}{\partial F^l{}_k} = 0$$

for $j \neq k$, $k \neq l$, hence a, b_j, $d = \text{const}$, while $c_1 = \bar{c}_1(F^2{}_3, F^3{}_2)$ etc.

Since $A_1{}^1{}_2{}^2 = -A_1{}^2{}_2{}^1$ etc., it follows that G has the form

$$G = a + B_i^\alpha F^i{}_\alpha + C_\alpha{}^i H^\alpha{}_i + d \det \mathbf{F}, \tag{2.5.55}$$

$$\mathbf{H} = \operatorname{adj} \mathbf{F} = (\det \mathbf{F}) \overset{-1}{\mathbf{F}},$$

For $G \in \mathscr{C}^0(\mathscr{L})$ let $G_\varepsilon := \varrho_\varepsilon * G$, $\varrho \in \mathscr{C}_0^\infty(\mathscr{L})$, $\varrho \geqslant 0$, $\int dL^9 \varrho = 1$, $\varrho_\varepsilon(F) :\equiv \varepsilon^{-9} \varrho(\varepsilon^{-1}\mathbf{F})$. Note that we have identified the set $\mathscr{L}$ with $\boldsymbol{R}^9$, endowed with the Lebesgue measure $\mathscr{L}^9$. We assume that G and $-G$ are both rank-one convex. Hence G_ε and $-G_\varepsilon$ are both rank-one convex and G_ε satisfies equations (2.5.53) and (2.5.54). We conclude that

$$G_\varepsilon(\mathbf{F}) = a(\varepsilon) + B_i^\alpha(\varepsilon) F^i{}_\alpha + C_\alpha{}^i(\varepsilon) H^\alpha{}_i + d(\varepsilon) \det \mathbf{F}.$$

Since $G \in \mathscr{C}^0(\mathscr{L})$, it follows that $G_\varepsilon \to G$ for $\varepsilon \to 0$ uniformly on compact subsets of $\mathscr{L}$.

Fix i, α and let $F^j{}_\beta = 0$ unless $j = i$ and $\beta = \alpha$, $F^i{}_\alpha = z$. Then $G_\varepsilon(\mathbf{F}) \equiv a(\varepsilon) + B_i^\alpha(\varepsilon) z \to g(z)$ uniformly on compact sets of $z \in \boldsymbol{R}$. Hence $g(z) = a + B_i^\alpha z$, $a(\varepsilon) \to a$, $B_i^\alpha(\varepsilon) \to B_i^\alpha$. Using this result we conclude that $C_\alpha{}^i(\varepsilon) H^\alpha{}_i + d(\varepsilon) \det \mathbf{F}$ converges uniformly on compact subsets of $\mathscr{L}$. Choose $\mathbf{F} \in L$ so that $H^\beta{}_j = 0$ unless $j = i$ and $\beta = \alpha$, while $H^\alpha{}_i = z$.* Since $\det \mathbf{H} = (\det \mathbf{F})^2$, we have that $\det \mathbf{F} = 0$. Again we conclude that $C_\alpha{}^i(\varepsilon) \to C_\alpha{}^i$ and finally $d(\varepsilon) \to d$. This proves that a function $G \in \mathscr{C}^0(\mathscr{L})$ satisfying (2.5.52) has the form (2.5.55).

From (2.5.55) it is clear that

$$G(\nabla \boldsymbol{\chi}) = a + B^\alpha{}_i (\nabla \boldsymbol{\chi})^i{}_\alpha + C_\alpha{}^i (\operatorname{adj} \nabla \boldsymbol{\chi})^\alpha{}_i + d \det \nabla \boldsymbol{\chi}$$

is a divergence $dX^\alpha / d\xi^\alpha$ and for $\mathbf{u} \in \mathscr{C}_0^\infty(\Omega)$ $X^\alpha(\boldsymbol{\chi} + \mathbf{u}, \nabla \boldsymbol{\chi} + \nabla \mathbf{u}) = X^\alpha(\boldsymbol{\chi}, \nabla \boldsymbol{\chi})$ on $\partial\Omega$. Hence follows (2.5.51).

We have thus established equivalence of (2.5.51), (2.5.52) and (2.5.55). for arbitrary $G \in \mathscr{C}^0(\mathscr{L})$.

Gauge transformations (2.5.50) do not affect the problem $F \to \min$ under the given Dirichlet boundary conditions on $\partial\Omega$. Hence it is reasonable to expect that constitutive assumptions ensuring existence are invariant under gauge transformations (2.5.50). The invariance of (QC), (PC), (LH) and (SE) follows from (2.5.51), (2.5.55), (2.5.53) and (2.5.54). For (PC) use (2.5.25).

* For $i = 1$, $\alpha = 2$, $z > 0$ let $\mathbf{F}^1{}_3 = -F^3{}_2 = z^{1/2}$, the remaining entries of $\mathbf{F}$ being zero. For $i = 1 = \alpha$ let $\mathbf{F} = \operatorname{diag}\{0, z^{1/2}, z^{1/2}\}$.

We have proved that any $G \in \mathscr{C}^0(\mathscr{L})$ such that both G and $-G$ are quasiconvex has the form (2.5.55). By Theorem 2.5.19 any function $G: \mathscr{L} \to \boldsymbol{R}$ such that G and $-G$ satisfy (PC) has the form (2.5.55). This fact does not follow trivially from the definition of polyconvex functions, since the convex functions g, $\bar{g}$: $M \to \boldsymbol{R}$ corresponding to G and $-G$ according to Definition 2.5.17 neednot satisfy the equation $\bar{g} = -g$. Equation (2.5.24) can be used for a more direct proof. Indeed, if G and $-G$ are polyconvex, then

$$a + a_i^\alpha H_\alpha^i + b_\alpha^i (\operatorname{adj}\mathbf{H})_i^\alpha + c \det \mathbf{H}$$

$$\leqslant G(\mathbf{F}_0 + \mathbf{H}) \leqslant -\bar{a} - \bar{a}_i^\alpha H_\alpha^i - \bar{b}_\alpha^i (\operatorname{adj}\mathbf{H})_i^\alpha - \bar{c} \det \mathbf{H},$$

$$a = -\bar{a} = G(\mathbf{F}_0).$$

Substituting $\mathbf{H} = \varepsilon \mathbf{H}'$, dividing both sides of the inequality

$$a_\alpha{}^i H^i{}_\alpha + b^\alpha{}_i (\operatorname{adj}\mathbf{H})^\alpha{}_i + c \det \mathbf{H} \leqslant -\bar{a}_i^\alpha H^i{}_\alpha - \bar{b}_\alpha^i (\operatorname{adj}\mathbf{H})^\alpha{}_i - \bar{c} \det \mathbf{H}$$

by $\varepsilon > 0$ and letting $\varepsilon \to 0$ we arrive at the inequality $a_i^\alpha H \leqslant -\bar{a}_i^\alpha H^i{}_\alpha$. Substituting $-\mathbf{H}$ for $\mathbf{H}$ we get the opposite inequality. Hence $a_i^\alpha = -\bar{a}_i^\alpha$. Proceeding this way we obtain the identities $b_\alpha^i = -\bar{b}_\alpha^i$, $c = -\bar{c}$. Hence G has the form (2.5.55).*

A general study of the functions G satisfying (2.5.51), generalized to arbitrary dimensions and orders of derivatives, can be found in the paper of Ball, Currie and Olver (1981).

2.6 PROLEGOMENA TO THE COMPLEMENTARY ENERGY PRINCIPLES

2.6.1 Introduction. The Castigliano-Menabrea Principle in Linear Elasticity

The complementary energy principles owe their popularity to the following two reasons. Firstly, they allow solving elastostatic BVPs directly in terms of stress and stress is often the only object of interest in the problem. Secondly, in favorable circumstances a complementary energy principle used in conjunction with the principle of minimum energy allows estimates of error of approximate solutions.

The second application hinges on the following property of the complementary principles. While the principle of minimum potential energy has the form $F(\mathbf{u}) \to \min$ on a subspace $\mathscr{V}$ of $W^{1,p}(\Omega)$, the principle of maximum

* Note that this argument remains valid for G defined on $\mathscr{L}_+$.

complementary energy assumes the form $F^c(\mathbf{S}) \to \max$ on a set Σ **of statically admissible stress fields**

$$\Sigma := \{S_k^{\ \alpha} \in \mathscr{L}^q(\Omega) |\ S_k^{\ \alpha}{}_{,\xi\alpha} + \varrho_0 b_k = 0,\ S_k^{\ \alpha}\nu_\alpha = s_k \text{ on } \Gamma_2\},\star$$

$q^{-1} = 1 - p^{-1}$. Moreover, if $F(\mathbf{u}) \to \min$ in $\mathscr{V}$ has a solution $\mathbf{u}_o \in \mathscr{V}$ then $F^c(\mathbf{S}) \to \max$ has a solution $\mathbf{S}_o \in \Sigma$, which in addition satisfies the equations $S_{ok}^{\ \alpha}(\xi) = \hat{S}_k^{\ \alpha}(\mathbf{F}_o + \nabla\mathbf{u}, \xi)$ and $F^c(\mathbf{S}_o) = F(\mathbf{u}_o)$. Let $\mathbf{u}_1$, $\mathbf{S}_1$ be two approximate solutions to the dual problems $F \to \min$, $F^c \to \max$. Then

$$F^c(\mathbf{S}_1) \leqslant F(\mathbf{u}_o) = F^c(\mathbf{S}_o) \leqslant F(\mathbf{u}_1). \tag{2.6.1}$$

Equation (2.6.1) provides two-sided bounds for (the energy of) the exact solution.

A formal derivation of the complementary principle and other variational principles related to the original one, suggested by Friedrichs in 1929, is based on the application of Lagrange multipliers (Funk, 1962). Let us write $F(\mathbf{u}) = \mathscr{W}(\mathbf{u}) + \Phi(\mathbf{u}) = f(L\mathbf{u}) + \Phi(\mathbf{u}) \to \min$ in $\mathscr{V}$ in the form of a variational principle with constraints:

$$f(\mathbf{F}) + \Phi(\mathbf{u}) \to \min, \quad \mathbf{F}(\xi) = \mathbf{F}_o(\xi) + \nabla\mathbf{u}(\xi), \quad \mathbf{u}|\Gamma_1 = \mathbf{v}_1 .$$

For simplicity we assume that $\Phi(\mathbf{u}) = -\int_{\Gamma_2} dH^2 s_k \gamma u^k - \int_\Omega dL^3 \varrho_0 b_k u^k$, although this assumption does not entail any loss of generality. By means of the Lagrange multipliers $S_k'^{\alpha}$, s_k' we arrive at the variational principle

$$f(\mathbf{F}) - \int_\Omega dL^3\, S_k'^{\alpha}(F^k{}_\alpha - F_o{}^k{}_\alpha - u^k{}_{,\xi\alpha})$$
$$- \int_{\Gamma_1} dH^2\, s_k'(\gamma_{\Gamma_1} u^k - v_1^k) + \Phi(\mathbf{u}) \to \text{stat}. \tag{2.6.2}$$

Taking the variation with respect to $\mathbf{F}$, $\mathbf{S}'$, $\mathbf{u}$ and $\mathbf{s}'$ we get the equations:

$$\varrho_0(\xi) \frac{\partial W}{\partial F^k{}_\alpha}(\mathbf{F}, \xi) = S_k'^{\alpha}(\xi), \tag{2.6.3}$$

$$\mathbf{F} = \mathbf{F}_o(\xi) + \nabla\mathbf{u}(\xi), \tag{2.6.4}$$

$$S_k'^{\alpha}{}_{,\xi\alpha} + \varrho_0 b_k = 0 \quad (\text{take } \mathbf{u} \in \mathscr{C}_0^\infty(\Omega)), \quad S_k'^{\alpha}\nu_\alpha = s_k \quad \text{on } \Gamma_2, \tag{2.6.5}$$

$$S_k'^{\alpha}\nu_\alpha = s_k' \quad \text{on } \Gamma_1, \tag{2.6.6}$$

$$\gamma_{\Gamma_1} u^k = v_1{}^k \quad \text{on } \Gamma_1. \tag{2.6.7}$$

If equation (2.6.3) can be solved for $\mathbf{F}$,

$$F^k{}_\alpha = \hat{F}^k{}_\alpha(\mathbf{S}', \xi) \tag{2.6.8}$$

$\star$ The differential equation in this definition is understood in the distributions sense. For the interpretation of the BC see below.

being the unique solution, then substituting equations (2.6.5), (2.6.8) as well as s_k' from equation (2.6.6) in equation (2.6.2) and integrating by parts we get the variational principle

$$-f^c(\mathbf{S}) := f(\hat{\mathbf{F}}(\mathbf{S}')) - \int_\Omega dL^3 S_k'^\alpha [\hat{F}^k{}_\alpha(\mathbf{S}') - F_o{}^k{}_\alpha]$$

$$- \int_{\Gamma_1} dH^2 S_k'^\alpha \nu_\alpha v_1{}^k \to \text{stat on } \Sigma,$$

$$\hat{\mathbf{F}}(\mathbf{S}')(\xi) :\equiv \hat{\mathbf{F}}(\mathbf{S}'(\xi), \xi). \tag{2.6.9}$$

Note that the first two terms of (2.6.9) resemble a Legendre transformation of $f(\mathbf{F})$.

We shall restate (2.6.9) in a more acceptable form. Only the last term needs some manipulation. Let $\mathbf{u}_o \in W^{1,p}(\Omega)$ be such that $\gamma_{\Gamma_1}\mathbf{u}_o = \mathbf{v}_1$. Using Green's formula and the definition of Σ we have the following identity:

$$-\int_{\Gamma_1} dH^2\, S_k'^\alpha \nu_\alpha v_1{}^k = \int_{\Gamma_2} dH^2\, s_j \gamma_{\Gamma_2} u_o{}^j + \int_\Omega dL^3\, \varrho_0 b_k u_o{}^k - \int_\Omega dL^3\, S_k'^\alpha u_o{}^k{}_{,\xi^\alpha}.$$

Hence finally

$$-f^c(\mathbf{S}') = f(\hat{\mathbf{F}}(\mathbf{S}')) - \int_\Omega dL^3\, S_k'^\alpha [\hat{F}^k{}_\alpha(\mathbf{S}') - F_o{}^k{}_\alpha]$$

$$- \int_\Omega dL^3\, S_k'^\alpha u_o{}^k{}_{,\xi^\alpha} + \text{const}. \tag{2.6.9'}$$

The proof of (2.6.9) is only heuristic. What is even worse, equation (2.6.3) is not invertible. Indeed, if $\hat{S}_k{}^\alpha(\mathbf{F}_1, \xi_1) = 0$ for some deformation gradient $\mathbf{F}_1$ at ξ_1—i.e. if there is at least one stress-free configuration gradient at ξ_1—then $\hat{\mathbf{S}}(\mathbf{A}\mathbf{F}_1, \xi_1) = \mathbf{A}\hat{\mathbf{S}}(\mathbf{F}_1, \xi_1) = \mathbf{0}$ for all $\mathbf{A} \in \mathrm{SO}(3)$ and hence $\hat{\mathbf{S}}(\cdot\,, \xi_1)$ cannot be one-to-one. This gap can however be easily filled in linear elasticity.

Indeed, let us consider a linear elastic body satisfying condition (2.2.49). In equations (2.6.2–9) we replace $S_k{}^\alpha$ by T^{kl}, $\mathbf{F}$ by $\nabla\mathbf{u}$, $\mathbf{S} = \hat{\mathbf{S}}(\mathbf{F}, \xi)$ by

$$T^{kl} = B_k{}^l{}_m{}^p(\mathbf{x}) u^m{}_{,x^p},$$

$\mathscr{W}(\mathbf{F})$ by

$$\int_\Omega dL^3\, \tfrac{1}{2} B_k{}^l{}_m{}^p(\mathbf{x}) u^k{}_{,x^l} u^m{}_{,x^p},$$

etc. $\mathbf{F}_o$ will be omitted. If $C^m{}_p{}^k{}_l(\mathbf{x})$ is the inverse of $B_k{}^l{}_m{}^p(\mathbf{x})$:

$$B_k{}^l{}_m{}^p(\mathbf{x}) C^m{}_p{}^r{}_s(\mathbf{x}) = \delta^r{}_k \delta^l{}_s,$$

then $\mathbf{F} = \hat{\mathbf{F}}(\mathbf{S}, \xi)$ should be replaced by

$$u^m{}_{,x^p} = C^m{}_p{}^k{}_l(\mathbf{x}) T^{kl}.$$

Let $\mathscr{V} = \{\mathbf{u} \in W^{1,2}(\Omega) | \gamma_{\Gamma_1}\mathbf{u} = \mathbf{0}\}$ and let $\mathscr{H}$ be the space of symmetric tensor fields $T^{kl} \in \mathscr{L}^2(\Omega)$ endowed with the inner product

$$\int_\Omega dL^3 \, C^{m\ k}_{\ p\ l}(\mathbf{x}) T^{mp} T^{kl}.$$

We assume as usual that $B_{k\ m}^{\ l\ p}$ and $C^{m\ k}_{\ p\ l}$ belong to $\mathscr{L}^\infty(\Omega)$. Furthermore, let $\Sigma_c := \{T^{kl} \in \mathscr{L}^2(\Omega) | \exists\, \mathbf{u} \in \mathscr{V}\ T^{kl} = B_{k\ m}^{\ l\ p} u^m{}_{,x^p}\}$ and let $\Sigma_c^\perp$ be the orthogonal complement of Σ_c in $\mathscr{H}$. If $\mathbf{T} \in \Sigma_c^\perp$ then $0 = \int_\Omega dL^3 T^{rs} C^{r\ k}_{\ s\ l} B_{k\ m}^{\ l\ p} u^m{}_{,x^p}$ $= \int_\Omega dL^3 T^{rs} u^r{}_{,x^s}\ \forall\, \mathbf{u} \in \mathscr{V}$, and *vice versa.*

We define the set of **statically admissible stress fields**

$$\Sigma = \left\{\mathbf{T} \in \mathscr{H} | \int_\Omega dL^3 \, T^{kl} u^k{}_{,x^l} = \int_\Omega dL^3 \, \varrho_0 b_k u^k + \int_{\Gamma_1} dH^2 \, s_k \gamma u^k \ \forall \mathbf{u} \in \mathscr{V}\right\}. \tag{2.6.10}$$

Since $T^{kl} = T^{lk}$, the left-hand side of the equation defining Σ vanishes for $\mathbf{u} = \mathbf{a} + \mathbf{b} \times \mathbf{x}$. If $H^2(\Gamma_1) = 0$ such $\mathbf{u} \in \mathscr{V}$ and we must assume that equations (2.1.7) and (2.1.9) are satisfied.

Up to a constant the complementary energy equals

$$f^c(\mathbf{T}) = -\frac{1}{2} \int_\Omega dL^3 \, C^{k\ r}_{\ l\ s} T^{kl} T^{rs} + \int_\Omega dL^3 \, T^{kl} u_o{}^k{}_{,x^l}, \tag{2.6.11}$$

where $\gamma_{\Gamma_1} \mathbf{u}_o = \mathbf{u}_1$ if $H^2(\Gamma_1) > 0$, $\mathbf{u}_o = \mathbf{0}$ if $H^2(\Gamma_1) = 0$ and $\mathbf{u}_1$ is the BC for $\mathbf{u}$.

Let $\mathbf{u} = \mathbf{u}_o + \mathbf{w}$, $\mathbf{w} \in \mathscr{V}$, $\tilde{T}^{kl}(\mathbf{u}) := B_{k\ m}^{\ l\ p} u^m{}_{,x^p}$. Suppose that $\mathbf{u}$ is a solution of $F(\mathbf{u}) \to \min$ satisfying the BC $\gamma_{\Gamma_1}\mathbf{u} = \mathbf{u}_1$. Then $\tilde{\mathbf{T}}(\mathbf{u}) \in \Sigma$, $\tilde{\mathbf{T}}(\mathbf{u}) = \tilde{\mathbf{T}}(\mathbf{u}_o) + \tilde{\mathbf{T}}(\mathbf{w})$, $\tilde{\mathbf{T}}(\mathbf{w}) \in \Sigma_c$. Let $\mathbf{T} \in \Sigma$. Then $\mathbf{T} - \tilde{\mathbf{T}}(\mathbf{u}) \in \Sigma_c^\perp$ and $-f^c(\mathbf{T}) = \frac{1}{2} ||\mathbf{T} - \tilde{\mathbf{T}}(\mathbf{u}_o)||_{\mathscr{H}}^2 - \frac{1}{2} ||\tilde{\mathbf{T}}(\mathbf{u}_o)||_{\mathscr{H}}^2 = \frac{1}{2} ||\mathbf{T} - \tilde{\mathbf{T}}(\mathbf{u})||_{\mathscr{H}}^2 + \frac{1}{2} ||\tilde{\mathbf{T}}(\mathbf{w})||_{\mathscr{H}}^2 - \frac{1}{2} ||\tilde{\mathbf{T}}(\mathbf{u}_o)||_{\mathscr{H}}^2 \to \min$ has a unique solution $\mathbf{T} = \tilde{\mathbf{T}}(\mathbf{u})$ (note that $\mathbf{w} = \mathbf{u} - \mathbf{u}_o$ is fixed here).

Hence we have proved

Theorem 2.6.1 (principle of Castigliano–Menabrea).
Let $\mathbf{u}$ be a solution of $F(\mathbf{u}) \to \min$ on $\mathscr{V} + \mathbf{u}_o$. Then the dual problem $f^c(\mathbf{T}) \to \max$ has a solution $\mathbf{T}(\mathbf{u})$.*

The method of the above proof is known as the **hypercircle method** of Synge.

* Both solutions are unique (see Section 2.2.7).

2.6.2 Remarks about the Principle of Complementary Energy in Non-Linear Elastostatics

We need a few facts from convex analysis (see. Ékeland and Témam, 1976, for example).

Let $\mathscr{V}$, $\mathscr{V}^*$ be two Banach spaces dual with respect to the product $\langle \cdot | \cdot \rangle$, endowed with the weak topologies induced by $\langle \cdot | \cdot \rangle$. Let $\bar{\boldsymbol{R}} = \boldsymbol{R} \cup \{+\infty, -\infty\}$. We shall say that $w \in \mathscr{V}^*$ is a **subgradient** of $f: \mathscr{V} \to \bar{\boldsymbol{R}}$ at u or $w \in \partial f(u)$, if

$$f(v) - f(u) \geqslant \langle w | v - u \rangle \ \forall v \in \mathscr{V}. \tag{2.6.12}$$

The **subdifferential** $\partial f(u)$ is the set of all the subgradients of f at u, possibly empty. We shall note a few obvious facts.

$$f(u) = \inf\{f(v) | \ v \in \mathscr{V}\} \quad \text{iff } 0 \in \partial f(u).$$

If $f'(u)$ is the Gâteaux derivative of f at u and f is convex*, then $\partial f(u) = \{f'(u)\}$.

The **Fenchel transform** $f \mapsto f^*$ is defined by the formula

$$f^*(u^*) := \sup\{\langle u^* | u \rangle - f(u) | \ u \in \mathscr{V}\} \quad \text{for all } u^* \in \mathscr{V}^*. \tag{2.6.13}$$

f^* is convex since it is the upper envelope of a family of affine functions. Obviously $f^*(u^*) \geqslant \langle u^* | u \rangle - f(u)$, i.e.

$$f(u) + f^*(u^*) - \langle u^* | u \rangle \geqslant 0 \ \forall u \in \mathscr{V}, \ \forall u^* \in \mathscr{V}^*. \tag{2.6.14}$$

If $u^* \in \partial f(u)$ then

$$\langle u^* | u \rangle - f(u) \geqslant \langle u^* | v \rangle - f(v) \forall v \in \mathscr{V},$$

hence $\langle u^* | u \rangle - f(u) = f^*(u^*)$, i.e.

$$f(u) + f^*(u^*) - \langle u^* | u \rangle = 0. \tag{2.6.15}$$

Conversely, equation (2.6.15) implies that $u^* \in \partial f(u)$, $u \in \partial f^*(u^*)$.

It is easy to see from the definition (2.6.13) that $f^{**} := (f^*)^*$ is the pointwise lowest upper bound of affine functions $\leqslant f$. Hence f^{**} is convex and

$$f^{**} \leqslant f. \tag{2.6.16}$$

f^{**} is called the **Γ-regularization** of f. By a straightforward argument $f^{**}(u) = f(u)$ if $\partial f(u) \neq \varnothing$. Also $f^{***} = f^*$.

* Note that the notion of a derivative is strictly local while the notion of a subgradient is global. Consequently they can coincide only for special classes of functions.

Suppose that $g\colon \Omega\times \boldsymbol{R}^m \to \boldsymbol{R}$ is a Carathéodory function such that the corresponding superposition operator (recall Section 1.1.7) $\mathscr{G}\colon \mathbf{u}(\cdot) \mapsto g(\cdot, \mathbf{u}(\cdot))$ is continuous from $\mathscr{L}^p(\Omega)$ into $\mathscr{L}^r(\Omega)$, $1 \leqslant p, r < \infty$. Let

$$G(\mathbf{u}) := \int_\Omega dL^n\, g\big(x, \mathbf{u}(x)\big). \tag{2.6.17}$$

Then

$$G^*(\mathbf{u}^*) = \int_\Omega dL^n\, g^*\big(x, \mathbf{u}^*(x)\big)\ \forall \mathbf{u}^* \in \mathscr{L}^q(\Omega), \quad q^{-1} = 1-p^{-1}, \tag{2.6.18}$$

where $g^*(x, \cdot)$ is the Fenchel transform of $g(x, \cdot)$ for a.a. $x \in \Omega$ (Ékeland and Témam, 1976, Proposition 4.1.2).

Let g be a non-negative normal integrand $g(\cdot, \cdot)$ (Definition 1.1.67) such that the functional G, defined by (2.6.17), is defined on $\mathscr{L}^p(\Omega)$ and $G(\mathbf{u}_o) < \infty$ for some $\mathbf{u}_o \in \mathscr{L}^\infty(\Omega)$. In this case

$$G^*(\mathbf{u}^*) = \int_\Omega dL^n\, g^*\big(x, \mathbf{u}^*(x)\big)\ \forall \mathbf{u}^* \in \mathscr{L}^q(\Omega), \quad q^{-1} = 1-p^{-1} \tag{2.6.19}$$

(Ékeland and Témam, 1976, Proposition 9.2.1).

Integrating the Fenchel inequality

$$g\big(x, \mathbf{u}(x)\big)+g^*\big(x, \mathbf{u}^*(x)\big)-\langle \mathbf{u}^*(x)|\mathbf{u}(x)\rangle \geqslant 0 \quad \text{for a.a. } x \in \Omega$$

we recover the Fenchel inequality for G, G^*:

$$G(\mathbf{u})+G^*(\mathbf{u}^*)-\langle \mathbf{u}^*|\mathbf{u}\rangle \geqslant 0. \tag{2.6.20}$$

If $G(\mathbf{u})+G^*(\mathbf{u}^*)-\langle \mathbf{u}^*|\mathbf{u}\rangle = 0$ then

$$g\big(x, \mathbf{u}(x)\big)+g^*\big(x, \mathbf{u}^*(x)\big)-\langle \mathbf{u}^*(x)|\mathbf{u}(x)\rangle = 0 \quad \text{for a.a. } x \in \Omega \tag{2.6.21}$$

and $\mathbf{u}^*(x) \in \partial g\big(x, \mathbf{u}(x)\big)$, $\mathbf{u}(x) \in \partial g\big(x, \mathbf{u}^*(x)\big)$ for a.a. $x \in \Omega$.

Let $\mathscr{V} = \{\mathbf{u} \in W^{1,p}(\Omega)|\ \gamma_{\Gamma_1}\mathbf{u} = 0\}$, $p > 1$, $q^{-1} = 1-p^{-1}$, $\varrho_0 b_k \in \mathscr{L}^q(\Omega)$, $s_k \in W^{-1/q,q}(\Gamma_2)$. $\mathscr{L}^p(\Omega)$, $\mathscr{L}^q(\Omega)$ are dual with respect to the form $\int_\Omega dL^n \mathbf{u}\cdot\mathbf{v}$, while $W^{-1/q,q}(\Gamma_2)$ is a distribution space dual to $W^{1-1/p,p}(\Gamma_2)$ with respect to a bilinear form which is an extension of the integral $\int_{\Gamma_2} dH^2 uv$.

We shall define the set of statically admissible stress fields $\Sigma = \{\mathbf{S} \in \mathscr{L}^q(\Omega)|\ S_k{}^\alpha{}_{,\xi^\alpha}+\varrho_0 b_k = 0$ in the distributions sense, $S_k{}^\alpha \nu_\alpha = s_k \in W^{-1/q,q}(\Gamma_2)\}$. The BC on Γ_2 requires some comment. Applying formally Green's formula we have

$$\int_\Omega dL^3\, S_k{}^\alpha{}_{,\xi^\alpha} u^k + \int_\Omega dL^3\, S_k{}^\alpha u^k{}_{,\xi^\alpha} = \int_{\Gamma_2} dH^2\, S_k{}^\alpha \nu_\alpha \gamma_{\Gamma_2} u^k \tag{2.6.22}$$

for all $\mathbf{u} \in \mathscr{V}$. Note that $S_k{}^\alpha{}_{,\xi\alpha} = -\varrho_0 b_k \in \mathscr{L}^q(\Omega)$, $S_k{}^\alpha \in \mathscr{L}^q(\Omega)$ and the left-hand side is a continuous function of $\mathbf{u} \in \mathscr{V}$. If $\mathbf{u} \in \mathscr{V}$ and $\gamma_{\Gamma_2}\mathbf{u} = \mathbf{0}$ then $\mathbf{u} \in W_0^{1,p}(\Omega)$ and the left-hand side of (2.6.22) vanishes (approximate $\mathbf{u}$ by $\mathbf{u}_n \to \mathbf{u}$, $\mathbf{u}_n \in \mathscr{C}_0^\infty(\Omega)$ and recall the definition of a distributional derivative). Hence it is a linear function of $\gamma_{\Gamma_2}\mathbf{u}$. The trace operator γ_{Γ_2} maps $\mathscr{V}$ **onto** $W^{1-1/p,p}(\Gamma_2)$. By the Banach open mapping theorem (Theorem 1.1.29) γ_{Γ_2} is an open mapping. Let us write the left-hand side of (2.6.22) in the form $F \circ \gamma_{\Gamma_2}(\mathbf{u})$. For every open interval I the set $F^{-1}(I) = \gamma_{\Gamma_2}((F \circ \gamma_{\Gamma_2})^{-1}(I))$ is open. Hence F is linear and continuous and can be identified with an element $S_k{}^\alpha \nu_\alpha$ of $W^{-1/q,q}(\Gamma_2)$ by means of the duality form mentioned in the preceding paragraph. We have thus defined $S_k{}^\alpha \nu_\alpha$.

Alternatively, we might have defined Σ as follows

$$\Sigma = \Big\{\mathbf{S} \in \mathscr{L}^q(\Omega) \Big| \int_\Omega dL^3\, S_k{}^\alpha u^k{}_{,\xi\alpha}$$

$$= \int_\Omega dL^3\, \varrho_0 b_k u^k + \int_{\Gamma_2} dH^2\, s_k \gamma_{\Gamma_2} \mathbf{u}^k \;\; \forall \mathbf{u} \in \mathscr{V}\Big\}. \tag{2.6.23}$$

Let $\mathbf{S}_o \in \Sigma$. Every $\mathbf{S} \in \Sigma$ has the form $\mathbf{S} = \mathbf{S}_o + \bar{\mathbf{S}}$ with

$$\int_\Omega dL^3\, \bar{S}_k{}^\alpha u^k{}_{,\xi\alpha} = 0 \;\; \forall \mathbf{u} \in \mathscr{V}. \tag{2.6.24}$$

We define the functionals

$$f(\mathbf{F}) = \int_\Omega dL^3\, \varrho_0\, W(\mathbf{F}(\xi), \xi), \quad \mathscr{W}(\mathbf{u}) = f(\mathbf{F}_o + L\mathbf{u}) \tag{2.6.25}$$

of $\mathbf{F}(\cdot) \in \mathscr{L}^p(\Omega)$, $\mathbf{u} \in \mathscr{V}$, resp. Let f^* be the Fenchel transform of f. Consider the non-negative functional on $\mathscr{V} \times \Sigma$:

$$I(\mathbf{u}, \mathbf{S}) := f(\mathbf{F}_o + L\mathbf{u}) + f^*(\mathbf{S}_o + \bar{\mathbf{S}}) - \langle \mathbf{S}_o + \bar{\mathbf{S}} | \mathbf{F}_o + \nabla \mathbf{u}\rangle = \tilde{F}(\mathbf{u}) - F_c(\bar{\mathbf{S}}) \tag{2.6.26}$$

by (2.6.24), if we introduce the notations

$$\tilde{F}(\mathbf{u}) = f(\mathbf{F}_o + L\mathbf{u}) - \langle \mathbf{S}_o | \mathbf{F}_o + L\mathbf{u}\rangle \equiv F(\mathbf{u}) - \langle \mathbf{S}_o | \mathbf{F}_o\rangle,$$

$$F_c(\bar{\mathbf{S}}) := -f^*(\mathbf{S}_o + \bar{\mathbf{S}}) + \langle \bar{\mathbf{S}} | \mathbf{F}_o\rangle. \tag{2.6.27}$$

Suppose that $I(\mathbf{u}_1, \mathbf{S}_1) = 0$, $\mathbf{S}_1 = \mathbf{S}_o + \bar{\mathbf{S}}_1 \in \Sigma$, $\mathbf{u}_1 \in \mathscr{V}$. I attains its minimum at $(\mathbf{u}_1, \mathbf{S}_1)$ and in view of (2.6.26)

$$F(\mathbf{u}_1) = \inf\{F(\mathbf{u}) |\; \mathbf{u} \in \mathscr{V}\}, \; F_c(\bar{\mathbf{S}}_1) = \sup\{F_c(\bar{\mathbf{S}}) |\; \bar{\mathbf{S}} \in \Sigma - \mathbf{S}_o\}, \tag{2.6.28}$$

$$\tilde{F}(\mathbf{u}_1) = F_c(\bar{\mathbf{S}}_1). \tag{2.6.29}$$

On account of (2.6.24) equation (2.6.29) is equivalent to $f(\mathbf{F}_o+L\mathbf{u}_1)+f^*(\mathbf{S}_1) = \langle \mathbf{S}_1|\mathbf{F}_o+L\mathbf{u}_1\rangle$. Since

$$f^*(\mathbf{S}) = \int_\Omega dL^3\, \varrho_0(\xi)\, W^*(\mathbf{S}(\xi),\xi), \tag{2.6.30}$$

we must have that

$$\mathbf{S}_1(\xi) \in \partial_1\left[\varrho_0(\xi)\, W(\mathbf{F}_o(\xi)+L\mathbf{u}_1(\xi),\, \xi)\right] \quad \text{a.e. in } \Omega, \, ^\star$$

i.e. $\mathbf{S}_1$ and $\mathbf{u}_1$ satisfy a generalized form of constitutive equation.

Conversely, if $F(\mathbf{u}) \to \min$, $F_c(\bar{\mathbf{S}}) \to \max$ have the solutions $\mathbf{u}_1, \bar{\mathbf{S}}_1 = \mathbf{S}_1 - \mathbf{S}_o$, then I achieves its minimum at $(\mathbf{u}_1, \mathbf{S}_1)$. It does not follow that $I(\mathbf{u}_1, \mathbf{S}_1) = 0$ however and

$$\tilde{F}(\mathbf{u}_1) \geqslant F_c(\bar{\mathbf{S}}_1). \tag{2.6.31}$$

For approximate solutions $\mathbf{u}_1'$, $\mathbf{S}_1'$ of $F \to \min$, $F_c \to \max$ we have

$$F(\mathbf{u}_1') \geqslant F(\mathbf{u}_1) \geqslant F_c(\mathbf{S}_1') + \langle \mathbf{S}_o|\mathbf{F}_o\rangle. \tag{2.6.32}$$

Hence we still have two-sided bounds on energy.

A rigorous theory of dual variational problems is presented in the book of Ékeland and Témam (1976). According to this reference the dual problem to the variational problem $F(\mathbf{u}) \equiv f(\mathbf{F}_o+L\mathbf{u})+\Phi(\mathbf{u}) \to \min$ on $\mathscr{V}$ has the form $F^c(\mathbf{S}) = \langle \mathbf{S}|\mathbf{F}_o\rangle - f^*(\mathbf{S}) - \Phi^*(-L^\dagger\mathbf{S}) \to \max$. In our case

$$\Phi(\mathbf{u}) = -\int_\Omega dL^3\, \varrho_0 b_k u^k - \int_{\Gamma_2} dH^2\, s_k \gamma_{\Gamma_2} u^k,$$

and

$$\varrho_0 b_k \in \mathscr{L}^q(\Omega), \; s_k \in W^{-1/q,\,q}(\Gamma_2)$$

$$\Phi^*(-L^\dagger\mathbf{S}) = \sup\Big\{-\int_\Omega dL^3\, S_k^{\alpha} u^k{}_{,\xi\alpha} + \int_\Omega dL^3\, \varrho_0 b_k u^k$$

$$+\int_{\Gamma_2} dH^2\, s_k \gamma_{\Gamma_2} u^k \,\Big|\; \mathbf{u} \in \mathscr{V}\Big\}.$$

The linear functional in the braces is unbounded unless it vanishes identically, i.e. unless $\mathbf{S} \in \Sigma$, equation (2.6.23). Hence

$$\Phi^*(-L^\dagger\mathbf{S}) = \begin{cases} 0 & \text{if } \mathbf{S} \in \Sigma, \\ \infty & \text{otherwise,} \end{cases}$$

i.e. $\mathbf{S} \mapsto \Phi^*(-L^\dagger\mathbf{S})$ is the indicator function of the set Σ. Hence $F^c(\mathbf{S}) \to \max$ is equivalent to $\mathbf{S} \in \Sigma$ and $\langle \mathbf{S}|\mathbf{F}_o\rangle - f^*(\mathbf{S}) \to \max$.

From Theorem 3.4.1 of Ékeland and Témam we deduce the following

⋆ The subscript "1" indicates that the subdifferential is to be calculated at fixed ξ.

Theorem 2.6.2
If f is convex on $\mathscr{L}^p(\Omega)$, $f(L\mathbf{u}_o+\mathbf{F}_o)+\Phi(\mathbf{u}_o) < \infty$ and f is continuous at $L\mathbf{u}_o+\mathbf{F}_o$ for some $\mathbf{u}_o \in \mathscr{V}$, then $\inf\{F(\mathbf{u})|\ \mathbf{u} \in \mathscr{V}\} = \sup\{F^c(\mathbf{S})|\ \mathbf{S} \in \mathscr{L}^q(\Omega)\} = \sup\{\langle \mathbf{S}|\mathbf{F}_o\rangle - f^*(\mathbf{S})|\ \mathbf{S} \in \Sigma\}$, and $F^c(\mathbf{S}) \to \max$ on $\mathscr{L}^q(\Omega)$ has at least one solution $\mathbf{S}_1 \in \Sigma$.

If the conditions of existence of a unique solution $\mathbf{u}_1$ of $F(\mathbf{u}) \to \min$ on $\mathscr{V}$ are satisfied (Section 2.5) then

$$F(\mathbf{u}_1) = F^c(\mathbf{S}_1) = \langle \mathbf{S}_1|\mathbf{F}_o\rangle - f^*(\mathbf{S}) \tag{2.6.33}$$

and we have equality in (2.6.31). Hence $I(\mathbf{u}_1, \mathbf{S}_1) = 0$ and for a.e. ξ $\mathbf{S}_1(\xi) \in \partial[\varrho_o(\xi)\, W(\mathbf{F}_o(\xi)+\nabla\mathbf{u}_1(\xi), \xi)]$, i.e. $\mathbf{S}_1$ satisfies the constitutive equations with $\mathbf{F} = \mathbf{F}_o+\nabla\mathbf{u}_1$. We also have the identity

$$f(\mathbf{F}_o+L\mathbf{u}_1)+f^*(\mathbf{S}) = \int_\Omega dL^3\, \varrho_0 b_k u^k + \int_{\Gamma_2} dH^2\, s_k \gamma_{\Gamma_2} u^k + \langle \mathbf{S}_1|\mathbf{F}_o\rangle. \tag{2.6.34}$$

The main difficulty in this approach is the assumption of convexity of f, which amounts to the monotonicity of the BVP. In this connection we mention a theorem of Rockafellar (1968, 1971) to the effect that f is convex if $W(\cdot\,, \xi)$ is a lsc convex function for a.e. $\xi \in \Omega$.

In elastostatics the assumption of convexity of $W(\xi, \cdot)$ is untenable. In order to avoid the restrictive monotonicity hypothesis we take advantage of the existence theorems proved in Section 2.5. With this in view we assume that W is a nonnegative normal integrand defined on $\Omega \times M_{3\times 3}$ and $W(\xi, \mathbf{F}) = \infty$ if $\det \mathbf{F} \leqslant 0$.* We consider f and $F = \mathscr{W}+\Phi$, equation (2.6.25), as functionals defined on $\mathscr{L}^p$ and on $\mathscr{V} := \{\mathbf{u} \in W^{1,p}|\ \gamma_{\Gamma_1}\mathbf{u} = \mathbf{0}\}$, resp.

Suppose that $F(\mathbf{u}_o) = \inf F := \inf\{F(\mathbf{u})|\ \mathbf{u} \in \mathscr{V}\}$, and f is finite in a nbhd of $\mathbf{F}_o' = \mathbf{F}_o+L\mathbf{u}_o'$, $\mathbf{u}_o' \in \mathscr{V}$. We claim that $F^{**}(\mathbf{u}_o) = F(\mathbf{u}_o) = \inf F = \inf F^{**}$.

Indeed, $F^{**}(\mathbf{u}_o) \leqslant F(\mathbf{u}_o)$ on account of (2.6.16). Since $\mathbf{u}_o$ gives an absolute minimum to F, the constant function $F(\mathbf{u}_o)$ is a continuous affine minorant of F on $\mathscr{V}$, and hence $F^{**}(\mathbf{u}) \geqslant F(\mathbf{u}_o)\ \forall\, \mathbf{u} \in \mathscr{V}$. This proves our claim.

We shall apply Theorem 2.6.2 to the convex functional F^{**}. Since F^{**} is the pointwise supremum of all the continuous affine minorants of F, L is a linear isometry and Φ is a continuous linear functional on $\mathscr{V}$, we have $F^{**}(\mathbf{u}) = f^{**}(\mathbf{F}_o+L\mathbf{u})+\Phi(\mathbf{u})$. Since f^{**} is the supremum of a family of continuous affine functions on $\mathscr{L}^p$ and the operation sup preserves lower semicontinuity and convexity, f^{**} is lsc and convex on $\mathscr{L}^p$. On account of (2.6.16) f^{**} is finite in a nbhd of $\mathbf{F}_o'$. By a theorem of convex analysis (Corollary

* Note that this extension of the Carathéodory function W cannot be convex.

1.2.5 in Ekeland and Témam, 1976) every lsc convex function on a Banach space is continuous on every open set on which it does not assume the value $+\infty$. Hence f^{**} is continuous in a nbhd of $\mathbf{F}_o'$.*

Hence Theorem 2.6.2 can be applied to F^{**}. Since $F^{***} = F^*$ and $F^{**}(\mathbf{u}_o) = F(\mathbf{u}_o)$, we have proved

Theorem 2.6.3

Suppose that there are $\mathbf{u}_o$, $\mathbf{u}_o' \in \mathscr{V}$ such that $F(\mathbf{u}_o) = \inf\{F(\mathbf{u})|\ \mathbf{u} \in \mathscr{V}\} < \infty$ and f is finite in a nbhd of $\mathbf{F}_o' = \mathbf{F}_o + L\mathbf{u}_o'$ in $\mathscr{L}^p$.

Then

$$\sup\{F^c(\mathbf{S})|\ \mathbf{S} \in \Sigma\} = F^c(\mathbf{S}_1) = F(\mathbf{u}_o) \quad \text{for some } \mathbf{S}_1 \in \Sigma.$$

Note that

$$\langle \mathbf{S}_1|\mathbf{F}_o\rangle - f^*(\mathbf{S}_1) = f(\mathbf{F}_o + L\mathbf{u}_o) + \Phi(\mathbf{u}_o) \quad \text{and} \quad \Phi(\mathbf{u}_o) = -\langle \mathbf{S}_1|L\mathbf{u}_o\rangle$$

for $\mathbf{S}_1 \in \Sigma$, $\mathbf{u}_o \in \mathscr{V}$. Hence

$$f(\mathbf{F}_o + L\mathbf{u}_o) + f^*(\mathbf{S}_1) - \langle \mathbf{S}_1|\mathbf{F}_o + L\mathbf{u}_o\rangle = 0$$

and $\mathbf{S}_1 \in \partial f(\mathbf{F}_o + L\mathbf{u}_o)$.

Suppose that $W \geqslant 0$ is a normal integrand. Proposition 9.2.1 of Ekeland and Témam (1976), quoted above, and the Fenchel inequality for W, W^* immediately imply that

$$\varrho_o(\xi)\, W(\xi, \mathbf{F}_o(\xi) + \nabla\mathbf{u}(\xi)) + \varrho_0(\xi)\, W^*(\xi, \mathbf{S}_1(\xi)) - S_{1k}^{\alpha}(\xi)\big(F_o{}^k{}_\alpha(\xi) + u_o{}^k{}_{,\xi^\alpha}(\xi)\big) = 0$$

L^3-a.e. in Ω and hence $\mathbf{S}_1(\xi) \in \varrho_0(\xi)\,\partial_{\mathbf{F}} W(\xi, \mathbf{F}_o(\xi) + \nabla\mathbf{u}_o(\xi))$ L^3-a.e. in Ω. The latter conclusion is equivalent to the constitutive equations.

It should be noted that our assumption that $\exists\, \mathbf{u}_o \in \mathscr{V}$ such that $F(\mathbf{u}_o) = \inf\{F(\mathbf{u})|\ \mathbf{u} \in \mathscr{V}\}$ is stronger than the assertions of the existence theorems of Section 6.5. The latter assertions have the form: $\exists\, \mathbf{u}_o' \in \mathscr{C}$ such that $F(\mathbf{u}_o') = \inf\{F(\mathbf{u})|\mathbf{u} \in \mathscr{C}\}$ and $\mathscr{C}$ is a proper subset of $\mathscr{V}$. The proofs of the existence theorems hinge on the fact that the minimizers satisfy the conditions

* The existence theorems of Section 2.5 for vanishing loads ($\Phi = 0$) imply that $f(\mathbf{F}_o'') = \inf\{f(\mathbf{F}_o + L\mathbf{u})|\ \mathbf{u} \in \mathscr{V}\}$ for some $\mathbf{F}_o'' = \mathbf{F}_o + L\mathbf{u}_o''$, $\mathbf{u}_o'' \in \mathscr{V}$. It is easy to deduce henceforth that the restriction of f^{**} to the hyperplane $\mathbf{F}_o + L(\mathscr{V})$ is continuous at $\mathbf{F}_o''$. Obviously, $\liminf_{n\to\infty} f^{**}(\mathbf{F}_n) = f^{**}(\mathbf{F}_o'')$ for every sequence $\{\mathbf{F}_n\} \subset \mathbf{F}_o + L(\mathscr{V})$ converging to $\mathbf{F}_o''$ in $\mathscr{L}^p$. Suppose that $\limsup_{n\to\infty} f^{**}(\mathbf{F}_n) = a > f^{**}(\mathbf{F}_o'')$. Then there is a subsequence $\{\overline{\mathbf{F}}_n\} \subset \{\mathbf{F}_n\}$ such that $f^{**}(\mathbf{F}_n) \to a > f^{**}(\mathbf{F}_o'')$, in contradiction with the lower semicontinuity of f^{**}.

Unfortunately, the continuity of $f^{**}|\mathbf{F}_o + L(\mathscr{V})$ at a point is not sufficient for the validity of Theorem 2.6.3.

$\operatorname{adj}\nabla\boldsymbol{\chi}_n \in \mathscr{L}^s$, $\det\nabla\boldsymbol{\chi}_n \in \mathscr{L}^r(\Omega)$, $\boldsymbol{\chi}_n = \boldsymbol{\chi}_o + \mathbf{u}_n$, in accordance with the definition of $\mathscr{C}$. It is not advisable to incorporate these conditions in the definition of F (by setting $F = \infty$ if they fail to be satisfied) since in this case F and, more significantly, F^* are no longer expressible in integral form. As a result the argument of the preceding paragraph leading to the constitutive equations would break down.

Theorem 2.6.3 can readily be extended to unilateral boundary value problems, Hanyga and Seredyńska (1982). In this case the linear manifolds $\mathscr{V}$, Σ are replaced appropriate cones ($\gamma_{\Gamma_1}\mathbf{u}\cdot\mathbf{n} \leqslant 0$ and $\mathbf{n}\cdot\mathbf{Tn} \leqslant 0$ respectively), while the duality relation $F^c(\mathbf{S}_1) = F(\mathbf{u}_o)$ implies the relation $(\mathbf{n}\cdot\mathbf{Tn})\gamma_{\Gamma_1}\mathbf{u}\cdot\mathbf{n} = 0$ a.e. on Γ_1.

2.6.3 Closing Remarks

Let us return to the definition of W^*:

$$\varrho_0 W^*(\mathbf{S}, \xi) = \sup\{S_k^{\alpha}F^k{}_{\alpha} - \varrho_0 W(\mathbf{F}, \xi) |\ \mathbf{F} \in M_{3\times 3}\} \quad \text{for a.e. } \xi \in \Omega. \tag{2.6.35}$$

If W is a differentiable function of $\mathbf{F}$ for a.e. $\xi \in \Omega$, then

$$\varrho_0 W^*(\mathbf{S}, \xi) = S_k^{\alpha}F^k{}_{\alpha} - \varrho_0 W(\mathbf{F}, \xi) \tag{2.6.36}$$

for $S_k^{\alpha} = \dfrac{\partial\varrho_0 W(F, \xi)}{\partial F^k{}_{\alpha}}$. If S_k^{α} does not lie in the range of $\hat{S}_k^{\alpha} = \partial(\varrho_0 W)/\partial F^k{}_{\alpha}$, then $W^*(\mathbf{S}, \xi) = \infty$. It is clear that $\mathbf{S}_1$ cannot be a solution of $F^c \to \max$ unless $\mathbf{S}_1(\xi)$ is in the range of $\hat{\mathbf{S}}(\cdot\,, \xi)$ for a.e. $\xi \in \Omega$.

The Legendre transformation (2.6.35) is an attractive method of calculating W^* and f^*. In general there are several values of $\mathbf{F}$ for which $\mathbf{S} = \hat{\mathbf{S}}(\mathbf{F}, \xi)$. Although this fact does not influence the validity of (2.6.35)* a vast literature has been dedicated to the problem of inverting the stress—strain relations. The usual starting point is a convenient choice of conjugate strain and stress measures, e.g. the right Cauchy–Green tensor $C_{\alpha\beta}$ and the Kirchhoff tensor

$$\Sigma^{\alpha\beta} = 2\varrho_0 \frac{\partial\tilde{W}}{\partial C_{\alpha\beta}} = S_k^{\beta}\overset{-1}{F}{}^{\alpha}{}_k. \tag{2.6.37}$$

The assumption that equation (2.6.37) is invertible does not contradict rotational invariance. Applying the Legendre transformation of $\tilde{W}$ with respect to $C_{\alpha\beta}$, $\Sigma^{\alpha\beta}$ and reverting to the Piola–Kirchhoff stress S_k^{α} yields a complemen-

* In this case take a value of $\mathbf{F}$ for which (2.6.36) is maximal.

tary energy function W^c of **S** and some additional variables. Fraeijs de Veubeke (1972), Dill (1977) and others expressed the complementary energy as a function of **S** and the rotational part of the deformation gradient $\mathbf{R} = \overset{-1}{\mathbf{C}^{1/2}}\mathbf{F}$. Quite a different approach to the complementary energy principle was suggested by Lee and Shield (1980).

For other variational principles the interested reader is referred to the classic book by Washizu (1968).

Chapter 3

Waves in non-linear hyperelastic media

3.1 PRELIMINARIES. SIMPLE AND SHOCK WAVES

3.1.1 Introduction

Existence of solutions $\chi \in \mathscr{C}^1$ to the equations of motion of a non-linear elastic medium can be proved at best for short intervals of time. For an unbounded medium and smooth initial data such a theorem will be quoted in Section 3.1.8. It is our objective in this chapter to investigate solutions χ with discontinuous derivatives. Such solutions can be expected to exist for arbitrarily large time provided the constitutive properties of the medium are realistic. The existence of the hyperelastic strain energy W (Section 1.2) as well as the SE condition (Chapter 2) are crucial for the theory of dynamic elasticity.

In order to get a deeper insight into the problems involved we shall consider some elementary non-linear waves—simple waves and isolated shock waves (Section 3.1). An elementary analysis of a single first-order equation shows the relevance of solutions with shock waves for non-linear hyperbolic equations governing the motion of a continuous medium (Sections 3.1.1 and 3.1.2). In Sections 3.2.4–8 a class of globally defined solutions of the Cauchy problem for general strictly hyperbolic systems will be introduced. The methods developed in Sections 3.1 and 3.2 will be extended in (Section 3.3. to deal with plane waves in non-linear elasticity. Some of them will also be applied to bounded three-dimensional bodies.

In Section 3.4 convex extensions will be applied to study non-linear hyperbolic conservation laws and in particular equations of motion of elastic bodies. The equations of non-linear elastodynamics will be considered against the background of general non-linear hyperbolic systems in order to exhibit the generality of the methods applied as well as their physical meaning.

In Section 3.5 some non-existence theorems for weak solutions to the basic problems of dynamics of non-linear elastic bounded bodies are presented.

They are formulated in a framework which bears some relation to the BVPs of elastostatics discussed in Chapter 2. The results of Section 3.5 indicate the importance of positive definiteness of strain energy and coercivity.

3.1.2 Quasilinear Hyperbolic Systems of First Order

For the purposes of this chapter it will be convenient to rewrite the equations of elastodynamics in the form of a first-order system of conservation laws, Section 3.3. In Sections 3.1 and 3.2 a general theory of hyperbolic systems of first order will be developed.

Let

$$\boldsymbol{R}^{m+1}_{+} := \{x \in \boldsymbol{R}^{m+1} | \; x^0 > 0\}.$$

For $\mu = 0, 1, \dots, m$ let $\mathbf{u} \mapsto \mathbf{A}^{\mu}(\mathbf{u}, x)$ be a $\mathscr{C}^q$ mapping of an open subset $\mathscr{H} \times \mathscr{D}$ of $\boldsymbol{R}^n \times \boldsymbol{R}^{m+1}$ into the set M of $n \times n$ matrices, $q \geqslant 1$. We shall consider quasilinear systems of equations of first order

$$\mathbf{A}^{\mu}(\mathbf{u}, x)\mathbf{u}_{,x^{\mu}} = \mathbf{b}(\mathbf{u}, x) \tag{3.1.1}$$ ⋆

for an unknown mapping $\mathbf{u}\colon \mathscr{D} \to \mathscr{H}$, defined on an open subset $\mathscr{D}$ of $\boldsymbol{R}^{m+1}$ (usually $\mathscr{D} = \boldsymbol{R}^{m+1}_{+}$). We identify the variable x^0 with the time t. Since we want the initial value problems

$$\mathbf{u}(0, x^1, \dots, x^m) = \mathbf{u}_0(x^1, \dots, x^m) \tag{3.1.2}$$

to be solvable it is reasonable to require that the matrix $\mathscr{A}^0(\mathbf{u}, x)$ be invertible for every $(\mathbf{u}, x)$ in $\mathscr{H} \times \mathscr{D}$.

Definition 3.1.1

The quasilinear first-order system (3.1.1) is said to be **hyperbolic** if

(i) the matrix $\mathscr{A}^0(\mathbf{u}, x)$ is invertible for every $(\mathbf{u}, x)$ in $\mathscr{H} \times \mathscr{D}$;

(ii) for every covector $\mathbf{p} = [p_k, k = 1, \dots, m]$ and for every $(\mathbf{u}, x) \in \mathscr{H} \times \mathscr{D}$ there are n linearly independent vectors $\mathbf{r}_s(\mathbf{u}, x, \mathbf{p}) \in \boldsymbol{R}^n$, $s = 1, \dots, n$, such that

$$p_k \mathbf{A}^k(\mathbf{u}, x)\mathbf{r}_s = \lambda_s \mathbf{A}^0(\mathbf{u}, x)\mathbf{r}_s \tag{3.1.3}$$

for some real $\lambda_s = \lambda_s(\mathbf{u}, x, \mathbf{p})$.

Definition 3.1.2

The system (3.1.1) is said to be **strictly hyperbolic** if the eigenvalue problem (3.1.3) has n district eigenvalues λ_s, $s = 1, \dots, n$ for every $(\mathbf{u}, x) \in \mathscr{H} \times \mathscr{D}$, $\mathbf{p} \in \boldsymbol{R}^m$, $\mathbf{p} \neq \mathbf{0}$:

$$\lambda_1(\mathbf{u}, x, \mathbf{p}) < \dots < \lambda_n(\mathbf{u}, x, \mathbf{p}) \tag{3.1.4}$$

⋆ We often use the Einstein summation convention for the dummy indices $\mu, \nu = 0, 1, \dots, m$ and $k, l = 1, 2, \dots, m$.

Definition 3.1.3

The system (3.1.1) is said to be **symmetric hyperbolic** if the matrices $\mathbf{A}^{\mu}(\mathbf{u}, x)$ are symmetric for $\mu = 0, 1, \ldots, m$, $(\mathbf{u}, x) \in \mathscr{H} \times \mathscr{D}$, and the matrix $\mathscr{A}^{0}(\mathbf{u}, x)$ is uniformly positive definite (i.e. $\exists\, c \in \boldsymbol{R}_{+}\ \forall\, (\mathbf{u}, x) \in \mathscr{H} \times \mathscr{D}\ \ \forall\, \mathbf{v} \in \boldsymbol{R}^{n}$ $\langle \mathbf{v}, \mathscr{A}^{0}(\mathbf{u}, x)\mathbf{v}\rangle \geqslant c\langle \mathbf{v}, \mathbf{v}\rangle$).

It follows from the spectral theory of matrices and bilinear forms that every strictly hyperbolic or symmetric hyperbolic system is hyperbolic. Some hyperbolic systems can be converted into symmetric hyperbolic ones merely by multiplying them on the left by a suitable matrix function of $(\mathbf{u}, x)$ and changing the coordinates in $\mathscr{H}$ (see Section 3.4). Strictly hyperbolic and symmetric hyperbolic systems are of particular importance because the Cauchy problem with smooth initial data is well-posed for their linear counterparts (Mizohata, 1969). For quasilinear symmetric hyperbolic systems we have the following important result (Kato, 1975):

Theorem 3.1.4

Suppose that the system (3.1.1) is symmetric hyperbolic.

Let $s > (m/2)+1$ and let $\mathscr{V}$ be an open subset of $\mathscr{H}^{s}(\boldsymbol{R}^{m})$. Under certain assumptions of boundedness and Lipschitz continuity of the mappings $\mathscr{V} \ni \mathbf{v}(\cdot) \mapsto \mathbf{A}^{\mu}(\mathbf{v}(\cdot), t, \cdot)$ the Cauchy problem (3.1.1), (3.1.2) with $\mathbf{u}_{0} \in \mathscr{V}$ has a unique solution

$$\mathbf{u} \in \mathscr{C}^{0}([0, T]; \mathscr{V}) \cap \mathscr{C}^{1}([0, T]; \mathscr{H}^{s-1}(\boldsymbol{R}^{m})), \tag{3.1.5}$$

where $T > 0$ depends on $\mathbf{u}_{0}$ and can be chosen constant for a sufficiently small nbhd of every point of $\mathscr{V}$.

Note that $\mathscr{H}^{s}$ is continuously embedded in the space $\mathscr{C}^{1}_{o}$ of bounded functions with bounded continuous first-order derivatives. For details, see Kato (1975). It will be shown in Section 3.4.5 that Theorem 3.1.4 can be applied in elastodynamics only for plane wave solutions ($m = 1$). For $m > 1$ Theorem 3.1.30 has to be used.

Any hyperbolic system (3.1.1) can be rewritten in the following equivalent form

$$\mathbf{u}_{t} + \tilde{\mathbf{A}}^{k}(\mathbf{u}, x)\mathbf{u}_{,x^{k}} = \tilde{\mathbf{b}}(\mathbf{u}, x) \tag{3.1.6}$$

with $\tilde{\mathbf{A}}^{k} := (\mathbf{A}^{0})^{-1}\mathbf{A}^{k}$, $\tilde{\mathbf{b}} = (\mathbf{A}^{0})^{-1}\mathbf{b}\mathbf{u}_{,t} = \mathbf{u}_{,x^{0}}$.

3.1.3 Geometry of Strictly Hyperbolic Systems

Let (3.1.1) be a strictly hyperbolic system whose eigenvalues satisfy (3.1.4), and let $\mathbf{A}^{0} = \mathbf{E} := \mathrm{id}_{\boldsymbol{R}^{n}}$. Henceforth we omit the argument x in the functions $\mathbf{A}^{k}$, $\mathbf{b}$ in order to simplify the notations.

Let $\mathscr{H}$ be a simply connected domain of $\boldsymbol{R}^n$. The eigenvalues λ_α, $\alpha = 1, \ldots, n$ satisfy the algebraic equation of n-th degree

$$\det\left(p_k \mathbf{A}^k(\mathbf{u}) - \lambda \mathbf{E}\right) = 0. \tag{3.1.7}$$

Let $\mathbf{D}(\mathbf{u}, \mathbf{p}, \lambda) :\equiv p_k \mathbf{A}^k(\mathbf{u}) - \lambda \mathbf{E}$ and let $\mathbf{C}(\mathbf{u}, \mathbf{p}, \lambda)$ be the cofactor matrix of $\mathbf{D}$:

$$C_r^{\ s}(\mathbf{u}, \mathbf{p}, \lambda) := \frac{1}{n!} \in_{ra_1 \ldots a_{n-1}} \in^{sb_1 \ldots b_{n-1}} D^{a_1}{}_{b_1} \ldots D^{a_{n-1}}{}_{b_{n-1}} \tag{3.1.8}$$

so that

$$D^p{}_s C_r^{\ s} = (\det \mathbf{D})\, \delta^p{}_r . \tag{3.1.9}$$

It is easy to see that

$$\frac{\partial \det \mathbf{D}}{\partial \lambda} = -n \operatorname{tr} \mathbf{C} = -n C_r^{\ r}. \tag{3.1.10}$$

Under the similarity transformation $\mathbf{D}' = \mathbf{R}^{-1}\mathbf{D}\mathbf{R}$ the cofactor matrix transforms according to the formula $\mathbf{C}' = {}^t\mathbf{R}\mathbf{C}^t\mathbf{R}^{-1}$. Let $\mathbf{R}$ be the invertible matrix whose columns are the eigenvectors $\mathbf{r}_1, \ldots, \mathbf{r}_n$, i.e. $R^k{}_\alpha = (\mathbf{r}_\alpha)^k$. Since $\mathbf{D}' = \mathbf{R}^{-1}\mathbf{D}\mathbf{R} = \operatorname{diag}\{\lambda_\alpha - \lambda \mid \alpha = 1, \ldots, n\}$ and $\mathbf{C}' = \operatorname{diag}\left\{ \prod_{\beta=1, \neq \alpha}^{n}{}' (\lambda_\beta - \lambda) \mid \alpha = 1, \ldots, n\right\}$, it follows that $\operatorname{tr}\mathbf{C} = \operatorname{tr}\mathbf{C}' = \sum_{\alpha=1}^{n} \prod_{\beta \neq \alpha} (\lambda_\beta(\mathbf{u}, \mathbf{p}) - \lambda)$. In view of (3.1.4) $\operatorname{tr}\mathbf{C} \neq 0$ at $\lambda = \lambda_\alpha(\mathbf{u}, \mathbf{p})$ and equation (3.1.10) implies that for $\alpha = 1, \ldots, n$ the eigenvalue λ_α is a $\mathscr{C}^1$ function of $(\mathbf{u}, x, \mathbf{p}) \in \mathscr{H} \times \mathscr{D} \times \boldsymbol{R}^m$. More precisely, $(\mathbf{u}, x) \mapsto \lambda_\alpha(\mathbf{u}, x, \mathbf{p})$ is a $\mathscr{C}^q$ function whereas $\mathbf{p} \mapsto \lambda_\alpha(\mathbf{u}, x, \mathbf{p})$ is $\mathscr{C}^\infty$. In view of (3.1.1) the latter function is also positively homogeneous of degree one.

Since the eigenvalue λ_α is simple, every point $(\mathbf{u}_0, x_0, \mathbf{p}_0)$ in $\mathscr{H} \times \mathscr{D} \times \boldsymbol{R}^m$ has a nbhd $\mathscr{U}$ such that for some fixed $r, s \in \{1, \ldots, n\}$ and, for every $(\mathbf{u}, x, \mathbf{p}) \in \mathscr{U}$, $C_{rs}(\mathbf{u}, x, \mathbf{p}, \lambda_\alpha(\mathbf{u}, x, \mathbf{p})) \neq 0$. Equivalently, for every $(\mathbf{u}_0, x_0, \mathbf{p}_0) \in \mathscr{H} \times \mathscr{D} \times \boldsymbol{R}^m$ there is a nbhd $\mathscr{U}$ of $(\mathbf{u}_0, x_0, \mathbf{p}_0)$ and a vector $\mathbf{v} \in \boldsymbol{R}^n$ such that

$$\mathbf{C}\left(\mathbf{u}, x, \mathbf{p}, \lambda_\alpha(\mathbf{u}, x, \mathbf{p})\right)\mathbf{v} \neq \mathbf{0} \ \forall (\mathbf{u}, x, \mathbf{p}) \in \mathscr{U}.$$

In view of (3.1.7) and (3.1.9)

$$\mathbf{C}\left(\mathbf{u}, x, \mathbf{p}, \lambda_\alpha(\mathbf{u}, x, \mathbf{p})\right)\mathbf{v} = k\mathbf{r}_\alpha(\mathbf{u}, x, \mathbf{p}) \tag{3.1.11}$$

for some real k, which may depend on $(\mathbf{u}, x, \mathbf{p})$.

Choose the eigenvectors $\mathbf{r}_\alpha$ to be unit vectors in $\boldsymbol{R}^n$ with respect to the metric δ_{rs}. Fix the direction of $\mathbf{r}_\alpha$ at some fixed point $(\mathbf{u}_0, x_0, \mathbf{p}_0)$ and extend it by continuity to the whole $\mathscr{H} \times \mathscr{D} \times \boldsymbol{R}^m$. In view of (3.1.11) the functions so defined are of class $\mathscr{C}^\infty$ and positively homogeneous of degree zero with respect to the argument $\mathbf{p}$. For fixed $\mathbf{p} \in \boldsymbol{R}^m$, $x \in \mathscr{D}$ the mapping $\mathbf{u} \mapsto r_\alpha(\mathbf{u}, x, \mathbf{p})$ defines a vector field of class $\mathscr{C}^q$ on $\mathscr{H}$. Let $\Phi_\alpha(\cdot\,, \cdot\,, x, \mathbf{p})$ be the associated flow $\mathscr{D}' \to \mathscr{H}$, where $\mathscr{D}'$ is an open subset of $\boldsymbol{R} \times \mathscr{H}$, projecting onto $\mathscr{H}$.

Definition 3.1.5
The curves $s \mapsto \Phi_\alpha(s, \mathbf{u}, x, \mathbf{p}) \in \mathscr{H}$ will be called α-**hodographs**.

We shall henceforth assume, throughout this section, that the matrices $\mathscr{A}^\mu$ do not depend on x. Let $\mathbf{v}\colon \mathscr{I} \mapsto \mathscr{H}$ be a $\mathscr{C}^1$ function defined on an interval $\mathscr{I}$ of $\boldsymbol{R}$ and let φ be a $\mathscr{C}^1$ mapping from an open subset $\mathscr{W}$ of $\boldsymbol{R}_+ \times \boldsymbol{R}$ to $\mathscr{I}$ such that

$$\mathbf{u}(x^\mu) \equiv \mathbf{v} \circ \varphi(t, p_k x^k) \tag{3.1.12}$$

is a solution of equation (3.1.1) with $\mathbf{b} \equiv \mathbf{0}$. Substitution of (3.1.12) into equation (3.1.1) yields the condition

$$\{\varphi_t \mathbf{A}^0(\mathbf{v}) + \varphi_\xi p_k \mathbf{A}^k(\mathbf{v})\}\, \mathbf{v}' = \mathbf{0} \tag{3.1.13}$$

where $\xi := p_k x^k$. We shall assume that $\varphi_\xi \neq 0$. Equation (3.1.13) implies that

$$-\frac{\varphi_t}{\varphi_\xi} = \lambda_\alpha\big(\mathbf{v}(\varphi), \mathbf{p}\big), \tag{3.1.14}$$

$$\mathbf{v}' = k\mathbf{r}_\alpha(\mathbf{v}, \mathbf{p}) \tag{3.1.15}$$

for some *fixed* $\alpha \in \{1, \ldots, n\}$ (in view of (3.1.14)) and for fixed $\mathbf{p}$. We can take $s \mapsto \mathbf{v}(s)$ to be an arc of an α-hodograph with the parametrization defined above. Equation (3.1.14) implies then that the curves φ = const on $\mathscr{W}$ are straight lines $\xi = \bar{\xi}(t)$ of constant slope $\dfrac{d\bar{\xi}}{dt} = \lambda_\alpha\big(\mathbf{v}(\varphi), \mathbf{p}\big)$. Since $\mathbf{v}' = \mathbf{r}_\alpha \neq \mathbf{0}$, these lines cannot intersect in $\mathscr{W}$.

Suppose that the intercepts $\zeta = \bar{\xi}(0)$ of the lines $\bar{\xi}(t)$ on $t = 0$ vary over an interval $\mathscr{J}$ and $\varphi(0, \xi) \equiv \phi(\xi)$, $\phi'(\xi) \neq 0$ for $\xi \in \mathscr{J}$. In this case

$$\varphi = \phi\big(\xi - \lambda_\alpha\big(\mathbf{v}(\varphi), \mathbf{p}\big)t\big). \tag{3.1.16}$$

In the other extreme case $\mathscr{W} \subset \boldsymbol{R}^{m+1}_+$ and all the curves $\bar{\xi}(\cdot)$ originate at a single point $\xi = \xi_0$ on $t = 0$,

$$\xi = \lambda_\alpha\big(\mathbf{v}(\varphi), \mathbf{p}\big)t + \xi_0. \tag{3.1.17}$$

In the latter case φ has the special form $\varphi(t, \xi) \equiv \psi\big((\xi - \xi_0)t^{-1}\big)$ for $t > 0$ and $(\xi - \xi_0)t^{-1}$ varying over an interval, with

$$d\lambda_\alpha[\mathbf{r}_\alpha]\psi' > 0, \tag{3.1.18}$$

where $d\lambda[\mathbf{r}] := \dfrac{\partial \lambda}{\partial u^s}\, r^s(\mathbf{u})$.

In the case (3.1.16) the straight lines $\varphi = \text{const}$ converge as the time $t \geqslant 0$ grows provided

$$\phi' d\lambda_\alpha[\mathbf{r}_\alpha] < 0 \tag{3.1.19}$$

and diverge if

$$\phi' d\lambda_\alpha[\mathbf{r}_\alpha] > 0. \tag{3.1.20}$$

In the former case the solution φ of equation (3.1.16) ceases to be smooth on the locus $\mathscr{E}$ of points (ξ, t) defined by equation (3.1.16) and

$$d\lambda_\alpha[\mathbf{r}_\alpha]\phi' t + 1 = 0. \tag{3.1.21}$$

The set $\mathscr{E}$ is the envelope of the family of straight lines $\varphi = \text{const}$, viz.

$$\xi - \lambda_\alpha\big(\mathbf{v}(\phi(\zeta)), \mathbf{p}\big)t - \zeta = 0, \quad \zeta \in \mathscr{I}. \tag{3.1.22}$$

The behaviour of φ in a nbhd of $\mathscr{E}$ will be illustrated by some examples in Sections 3.2.1 and 3.2.2. The solutions given by equations (3.1.12), (3.1.16) will be called **α-simple waves**. The α-simple waves satisfying (3.1.17) will be called **α-centered simple waves** (abbreviation: α-c.s.w.).

For every $\alpha \in \{1, \ldots, n\}$ we shall find $n-1$ independent solutions J of the equation

$$dJ[\mathbf{r}_\alpha] = 0 \tag{3.1.23}$$

and denote them by the symbols J_α^k, $k = 1, \ldots, n-1$. By definition the one-forms dJ_α^k, $k = 1, \ldots, n-1$, are linearly independent at every point $\mathbf{u} \in \mathscr{H}$. The existence of J_α^k is granted by our assumptions concerning $\mathscr{H}$, $\mathbf{r}_\alpha \neq \mathbf{0}$, in view of a well-known theorem of differential geometry (cf. Boothby, 1975, Theorem 3.14). Every α-hodograph is given by the equations

$$J_\alpha^k(\mathbf{u}) = c_k, \quad k = 1, \ldots, n-1 \tag{3.1.24}$$

which exhibit it as a one-dimensional submanifold of $\mathscr{H}$. Every α-simple wave $\mathbf{u}(t, \mathbf{p}\cdot\mathbf{x})$ satisfies equation (3.1.24) in the domain of its definition.

The functions J_α^k, $k = 1, \ldots, n-1$, are called the **Lax–Riemann invariants of the αth kind**.

3.1.4 Isolated Discontinuity Waves

General discontinuous solutions to the integral conservation laws

$$\int_{\partial\mathscr{P}} dH^m\, \nu_\mu \mathbf{f}^\mu\big(x, \mathbf{u}(x)\big) = \int_{\mathscr{P}} dL^{m+1}\, \mathbf{h}\big(x, \mathbf{u}(x)\big) \tag{3.1.25}$$

(to be satisfied for every set $\mathscr{P}$ belonging to a sufficiently rich class of subsets of $\boldsymbol{R}_+ \times \mathscr{B} \subset \boldsymbol{R}^{m+1}_+$) will be considered in Section 3.2. In this section our attention will be focused on isolated discontinuities. We assume that $\mathbf{f} \in \mathscr{C}^1$, $\mathbf{h} \in \mathscr{C}^0$ are defined on $\mathscr{H}$. Before embarking on the main subject we note that every $\mathscr{C}^1$ function $\mathbf{u}: \mathscr{D} \to \mathscr{H}$ which satisfies equations (3.1.25) also satisfies the differential equation

$$\partial_\mu \mathbf{f}^\mu(x, \mathbf{u}(x)) = \mathbf{h}(x, \mathbf{u}(x)) \tag{3.1.25a}$$

Here and throughout Chapter 3

$$\partial_\mu \mathbf{f}^\mu(x, \mathbf{u}) := \frac{\partial}{\partial x^\mu} \mathbf{f}^\mu + \frac{\partial \mathbf{f}^\mu}{\partial u_s} \frac{\partial u_s}{\partial x^\mu}.$$

Let $\mathbf{u}$ be a function defined on a domain $\mathscr{D} \subset \boldsymbol{R}^{m+1}$ with values in $\mathscr{H}$.

Definition 3.1.6

A surface $\Sigma \subset \mathscr{D}$ is said to be a **Lipschitzian isolated discontinuity** of the function $\mathbf{u}$ if every point $x \in \Sigma$ has a nbhd $\mathscr{U} \subset \mathscr{D}$ with the following properties:

(i) there is a $\mathscr{C}^1$ diffeomorphism $x \mapsto y$, $y = \bar{y}(x)$, of $\mathscr{U}$ onto a cube $|y^\varrho| < \delta$, $\delta > 0$, $\varrho = 0, 1, \ldots, m$;

(ii) the set $\bar{y}(\Sigma \cap \mathscr{U})$ is given by the equations

$$y^0 = \varphi(y^1, \ldots, y^m), \quad \text{for } |y^k| < \delta,\ k = 1, \ldots, m$$

$$\text{and} \quad \sup|\varphi| \leqslant \delta;$$

(iii) the function φ is Lipschitz continuous;

(iv) the functions $\mathbf{u}|\mathscr{U}_+$, $\mathbf{u}|\mathscr{U}_-$ are continuous, with

$$\mathscr{U}_+ := \{x|\ y^0 > \varphi(y^1, \ldots, y^m)\}, \quad \mathscr{U}_- := \{x|\ y^0 < \varphi(y^1, \ldots, y^m)\},$$

and the limits

$$\begin{aligned} \mathbf{u}_+(y^1, \ldots, y^m) &= \lim_{\substack{y' \to y \\ y' \in \mathscr{U}_+}} \mathbf{u}(\bar{x}(y')), \\ \mathbf{u}_-(y^1, \ldots, y^m) &:= \lim_{\substack{y' \to y \\ y' \in \mathscr{U}_-}} \mathbf{u}(\bar{x}(y')) \end{aligned} \tag{3.1.26}$$

exist for $y = (\varphi(y^1, \ldots, y^m), y^1, \ldots, y^m)$, $|y^k| < \delta$, $k = 1, \ldots, m$;

(v) $\mathbf{u}_+(y^1, \ldots, y^m) \neq \mathbf{u}_-(y^1, \ldots, y^m)$ for $|y^k| < \delta$, $k = 1, \ldots, m$.

The limits $\mathbf{u}_+$, $\mathbf{u}_-$ are continuous on Σ. Indeed, let $x_1 \in \Sigma \cap \mathscr{U}$ and $x \in \mathscr{U}_+$. For any $\varepsilon > 0$ there is an $\eta > 0$ such that $|x - x_1| < \eta$ implies $|\mathbf{u}(x) - \mathbf{u}(x_1)| < \varepsilon$. Let $x_2 \in \Sigma \cap \mathscr{U}$ satisfy the inequality $|x_2 - x_1| < \eta$. Letting $x \to x_2$ through $\mathscr{U}_+$ we establish the inequality $|\mathbf{u}_+(x_2) - \mathbf{u}_+(x_1)| < \varepsilon$.

A Lipschitzian isolated discontinuity of a function $\mathbf{u}$ satisfying (3.1.25) will be called an **isolated discontinuity wave.**

Let $\mathbf{a}^\varrho = \mathbf{f}^\varrho(x, \mathbf{u}(x))$, $\mathbf{b} = \mathbf{h}(x, \mathbf{u}(x))$, $\varrho = 0, 1, \ldots, m$ and let Σ be a Lipschitzian isolated discontinuity of a function $\mathbf{u}$ satisfying equation (3.1.25) for every subset $\mathscr{P} \subset \boldsymbol{R}^{m+1}$ which is measurable and has a piecewise Lipschitzian boundary. For $\varepsilon \in]0, \delta[$ let $\mathscr{P}_\varepsilon$ be the set

$$\{x \in \mathscr{U} |\ \varphi(y^1, \ldots, y^m) - \varepsilon < y^0 < \varphi(y^1, \ldots, y^m) + \varepsilon\}.$$

Its boundary $\partial\mathscr{P}_\varepsilon$ consists of the following surfaces

$$\Sigma_\varepsilon^\pm = \{x |\ y^0 = \varphi(y^1, \ldots, y^m) \pm \varepsilon,\ |y^k| < \delta \text{ for } k = 1, \ldots, m\},$$

$$\mathscr{S}_\varepsilon = \bigcup_{k=1}^{m} \{x |\ y^k = \pm\delta,\ y^l \in [-\delta, \delta] \text{ for } l \neq k,\ l \leqslant m,$$

$$\varphi(y^1, \ldots, y^m) - \varepsilon \leqslant y^0 \leqslant \varphi(y^1, \ldots, y^m) + \varepsilon\}.$$

Let ν_ϱ be the outward unit normal on $\partial\mathscr{P}_\varepsilon$. It exists almost everywhere on $\partial\mathscr{P}_\varepsilon$ in the Hausdorff measure H^m. Since

$$\nu_\varrho(y) dH^m(y) = \iota\, \epsilon_{\mu\nu\lambda\varrho} \frac{\partial x^\mu}{\partial y^2} \frac{\partial x^\nu}{\partial y^3} \frac{\partial x^\lambda}{\partial y^0} dy^2 dy^3 dy^0, \tag{3.1.27}$$

$\iota := \operatorname{sgn}\det\left[\dfrac{\partial \bar{x}^\mu}{\partial y^\varrho}\right]$, on the surfaces $y^1 = \pm\delta$, we have

$$\int_{\mathscr{S}_\varepsilon} \mathbf{a}^\varrho(y) \nu_\varrho(y) dH^m(y) \leqslant 4\varepsilon m \sup|\mathbf{a}^\varrho| K = O(\varepsilon). \tag{3.1.28}$$

Similarly

$$\int_{\mathscr{P}_\varepsilon} \mathbf{b}\, dL^{m+1} = O(\varepsilon). \tag{3.1.29}$$

In the limit $\varepsilon \to 0$ equation (3.1.25) becomes

$$\int_{\mathscr{U} \cap \Sigma} \nu_\varrho [[\mathbf{a}^\varrho]] dH^m = 0, \tag{3.1.30}$$

where

$$[[\mathbf{a}^\varrho]] := \mathbf{a}_-^\varrho - \mathbf{a}_+^\varrho \tag{3.1.31}$$

and ν_ϱ is the unit normal on Σ pointing into $\mathscr{U}_+$. The integrand in equation (3.1.30) is locally integrable in the Lebesgue measure L^m, hence, by the Lebesgue derivation theorem we obtain in the limit $\delta \to 0$

$$(\nu_\varrho [[\mathbf{a}^\varrho]])(y^1, \ldots, y^m) = 0 \tag{3.1.32}$$

for almost all $(y^1, \ldots, y^m) \in \boldsymbol{R}^m$, $|y^k| < \delta$.

Note also that an inequality

$$\int_{\mathscr{P}} dH^m \nu_\mu \mathbf{f}^\mu(x, \mathbf{u}(x)) \geqslant \int_{\mathscr{P}} dL^{m+1} \mathbf{h}(x, \mathbf{u}(x)) \tag{3.1.33}$$

would analogously entail that

$$\left(\nu_\mu[[\mathbf{f}^\mu(\cdot, \mathbf{u}(\cdot))]]\right)(y^1, \dots, y^m) \leqslant 0 \tag{3.1.34}$$

for almost all $(y^1, \dots, y^m) \in \boldsymbol{R}^m$, $|y^k| < \delta$.

3.1.5 Isolated Discontinuity Waves in Strictly Hyperbolic Systems of Conservation Laws

Definition 3.1.7

Let $\mathbf{f}^\mu\colon \mathscr{H} \to \boldsymbol{R}^n$ be a mapping of class $\mathscr{C}^1$ for $\mu = 0, \dots, m$, and let $\mathbf{h}\colon \mathscr{H} \to \boldsymbol{R}^n$ be $\mathscr{C}^0$. A first-order system of PDE of the form

$$\frac{\partial}{\partial x^\mu}\mathbf{f}^\mu\left(x, \mathbf{u}(x)\right) = \mathbf{h}\left(x, \mathbf{u}(x)\right) \tag{3.1.35}$$

is said to be a **system of conservation laws.**

Equations (3.1.35) can be rewritten in the form (3.1.1), in so far as $\mathscr{C}^1$ solutions are concerned, with

$$\mathbf{A}^\mu(x, \mathbf{u})\mathbf{v} \equiv d\mathbf{f}^\mu(\mathbf{u})[\mathbf{v}], \quad \mathbf{b}(x, \mathbf{u}) \equiv \mathbf{h}(x, \mathbf{u}) - \frac{\partial}{\partial x^\mu}\mathbf{f}^\mu(x, \mathbf{u}).$$

We shall say that equation (3.1.35) is a **(strictly) hyperbolic system of conservation laws** if the associated equation (3.1.1) is (strictly) hyperbolic.

Definition 3.1.8

A function $\mathbf{u}\colon \mathscr{D} \to \mathscr{H}$, defined on a domain $\mathscr{D}$ of $\boldsymbol{R}^{m+1}$ is said to be **piecewise** $\mathscr{C}^1$ if $\mathscr{D}$ is a join of an (at most countable) set of Lipschitzian isolated discontinuity surfaces for $\mathbf{u}$ and open sets $\mathscr{W} \subset \mathscr{D}$ such that $\mathbf{u}|\mathscr{W} \in \mathscr{C}^1(\mathscr{W})$.

We shall say that a piecewise $\mathscr{C}^1$ function $\mathbf{u}\colon \mathscr{D} \to \mathscr{H}$ is a **solution** of equation (3.1.35) if equation (3.1.25) is satisfied for every Borel bounded set with a piecewise Lipschitzian boundary.

Every isolated discontinuity Σ of a piecewise $\mathscr{C}^1$ solution $\mathbf{u}$ to equation (3.1.35) satisfies the **Rankine-Hugoniot equations:**

$$\nu_\varrho[[\mathbf{f}^\varrho(\cdot, \mathbf{u}(\cdot))]] = 0 \quad \text{a.e. on } \Sigma \text{ in } H^m. \tag{3.1.36}$$

We assume that

$$\sum_{k=1}^{m} {\nu_k}^2 > 0. \tag{3.1.37}$$

(this inequality will be justified later by the extra conditions (L) imposed on the solution). The (normal) **propagation velocity** of the discontinuity Σ is, by definition, a vector $\mathbf{w} \in \boldsymbol{R}^m$ with the components

$$w^k = -\nu_0 \left(\sum_{l=1}^{m} \nu_l^2 \right)^{-1} \nu_k, \quad k = 1, \ldots, m. \tag{3.1.38}$$

Let $\mathbf{n} \in \boldsymbol{R}^m$,

$$n_k := \frac{\nu_k}{\left(\sum_l \nu_l^2 \right)^{1/2}}, \quad k = 1, \ldots, m, \tag{3.1.39}$$

$$w := -\frac{\nu_0}{\left(\sum_l \nu_l^2 \right)^{1/2}}. \tag{3.1.40}$$

Then, w is called the **normal propagation speed of** Σ.

Under the assumption (3.1.37) equation (3.1.36) can be rewritten in the following form

$$-w[[\mathbf{f}^0(\cdot, \mathbf{u}(\cdot))]] + n_k[[\mathbf{f}^k(\cdot, \mathbf{u}(\cdot))]] = \mathbf{0} \quad \text{a.e. on } \Sigma. \tag{3.1.41}$$

The function $\mathbf{f}^0(x, \cdot)$ is a local diffeomorphism and has locally in $\mathscr{H} \times \boldsymbol{R}^n$ a unique inverse. For simplicity we assume henceforth that the functions $\mathbf{f}^\iota$ do not depend on x and the function $\mathbf{f}^0$ is a global diffeomorphism of $\mathscr{H}$ onto an open subset of the affine space $\boldsymbol{R}^n$. This assumption does not restrict the generality of the results since the theorems we obtain below refer to a fixed point x and it is easy to restate them in the more general case when $\mathbf{f}^0$ is not globally invertible. Changing the coordinates on $\mathscr{H}$ we obtain a simpler form of equation (3.1.41):

$$-w[[\mathbf{u}]] + n_k[[\mathbf{f}^k \circ \mathbf{u}]] = \mathbf{0} \quad \text{a.e. on } \Sigma. \tag{3.1.42}$$

Here $\mathscr{H}$ is an open subset of the affine space $\boldsymbol{R}^n$.

Consider a fixed point x_0 on Σ, such that $\mathbf{n} = \mathbf{n}(x_0)$, $w(x_0)$ are well-defined. Let $\mathbf{u}_+(x_0) =: \mathbf{u}_0$. For fixed $\mathbf{u}_0, \mathbf{n}$ we investigate the set of those $(w, \mathbf{u}) \in \boldsymbol{R} \times \mathscr{H}$, $\mathbf{u} := \mathbf{u}_-(x_0)$, $w = w(x_0)$, which satisfy equation (3.1.42). The assumption of strict hyperbolicity is essential here (Hanyga, 1976).

Equation (3.1.42) in the form

$$-w(\mathbf{u}-\mathbf{u}_0) + \mathbf{f}(\mathbf{u}) - \mathbf{f}(\mathbf{u}_0) = \mathbf{0}, \quad \mathbf{f}(\mathbf{u}) :\equiv n_k \mathbf{f}^k(\mathbf{u}), \tag{3.1.43}$$

has a trivial solution $\mathbf{u} = \mathbf{u}_0$ for arbitrary $w \in \boldsymbol{R}$. The non-trivial solutions $(\mathbf{u} \neq \mathbf{u}_0)$, whenever they exist, form one-dimensional submanifolds $\mathbf{u} = \mathbf{v}(w)$ of $\boldsymbol{R} \times \mathscr{H}$, except possibly in a nbhd of every solution $(w_1, \mathbf{u}_1)$ with $w_1 = \lambda_\alpha(\mathbf{u}_1, \mathbf{n})$. Theorems 3.1.10 and 3.1.11 give a topological and analytic description of the set $\mathscr{H}'_{\mathbf{u}_0}$ of non-trivial solutions $(w, \mathbf{u})$ of equation (3.1.43).

Let $\mathbf{r}_\alpha(\mathbf{u}, \mathbf{n})$, $\alpha = 1, \ldots, n$ be defined as in Section 3.1.3. The matrix $\mathbf{A}(\mathbf{u}) := n_k \mathbf{A}^k(\mathbf{u})$ has n linearly independent left eigenvectors $\mathbf{l}^\alpha(\mathbf{u}, \mathbf{n})$. They are defined by the equations ${}^t\mathbf{A}\,{}^t l^\alpha = \lambda_\alpha {}^t\mathbf{l}^\alpha$ in matrix notation (for convenience the left eigenvectors will be regarded as covectors or row matrices). In view of hyperbolicity $\mathbf{A} = \mathbf{R}\Lambda\mathbf{R}^{-1}$, where $\Lambda = \operatorname{diag}\{\lambda_\alpha |\ \alpha = 1, \ldots, n\}$, and we can set ${}^t\mathbf{l}^\alpha := {}^t\mathbf{R}^{-1}\mathbf{e}^\alpha$, where the vector $\mathbf{e}^\alpha$ has the components $\delta^\alpha{}_\beta$, $\beta = 1, \ldots, n$. The left eigenvectors have the smoothness properties of the right eigenvectors $\mathbf{r}_\alpha$ and it is easy to see that

$$\langle \mathbf{l}^\alpha(\mathbf{u}, \mathbf{n}), \mathbf{r}_\beta(\mathbf{u}, \mathbf{n})\rangle = \delta^\alpha{}_\beta \tag{3.1.44}$$

(the left-hand side is equal to $\mathbf{l}^\alpha \mathbf{r}_\beta$ in matrix notation).

Theorem 3.1.9

Let $\mathbf{n} \in \boldsymbol{R}^m$ be a fixed unit vector, $\mathbf{f}(\mathbf{u}) := n_k \mathbf{f}^k(\mathbf{u})$, $\mathbf{A}(\mathbf{u}) := d\mathbf{f}(\mathbf{u})$.

Suppose that

(i) $\mathbf{f} \in \mathscr{C}^2$ and the eigenvalues of $\mathbf{A}(\mathbf{u})$ are simple for all $\mathbf{u} \in \mathscr{H}$;

(ii) for every point $(\lambda_\alpha(\mathbf{u}), \mathbf{u}) \in \mathscr{H}'_{\mathbf{u}_0}$, $\alpha \in \{1, \ldots, n\}$ we have that $\langle \mathbf{l}^\alpha(\mathbf{u}), \mathbf{u} - \mathbf{u}_0\rangle \neq 0$.

Then

(1) $\mathscr{H}'_{\mathbf{u}_0} \neq \emptyset$;

(2) the set $\mathscr{H}_{\mathbf{u}_0} := \mathscr{H}'_{\mathbf{u}_0} \cup \bigcup_{\alpha=1}^{n} \{(\lambda_\alpha(\mathbf{u}_0), \mathbf{u}_0)\}$ is a one-dimensional $\mathscr{C}^1$ submanifold of $\boldsymbol{R} \times \mathscr{H}$ and contains one-dimensional connected submanifolds $\mathscr{H}^\alpha_{\mathbf{u}_0}$ of $\boldsymbol{R} \times \mathscr{H}$ with the properties (a)–(d):

(a) $(\lambda_\alpha(\mathbf{u}_0), \mathbf{u}_0) \in \mathscr{H}^\alpha_{\mathbf{u}_0}$,

(b) for every $\alpha \in \{1, \ldots, n\}$ either of the two arcs of $\mathscr{H}^\alpha_{\mathbf{u}_0}$ beginning at $\lambda_\alpha((\mathbf{u}_0), \mathbf{u}_0)$ satisfies one of the following conditions:

(A) it runs to infinity or to a point of $\partial(\boldsymbol{R} \times \mathscr{H})$;

(B) it contains a point $(\lambda_\beta(\mathbf{u}_0), \mathbf{u}_0)$ with $\beta \neq \alpha$,

(c) if $\mathbf{f} \in \mathscr{C}^3$ in a nbhd of $\mathbf{u}_0$, then the projection $\tilde{\mathscr{H}}^\alpha_{\mathbf{u}_0}$ of $\mathscr{H}^\alpha_{\mathbf{u}_0}$ onto $\mathscr{H}$ osculates the α-hodograph through $\mathbf{u}_0$ to second order,

(d) if $\mathbf{f} \in \mathscr{C}^3$ in a nbhd of $\mathbf{u}_0$, then the derivative of w with respect to an arbitrary parameter s on $\mathscr{H}^\alpha_{\mathbf{u}_0}$

$$\frac{dw}{ds} = \frac{1}{2}\frac{d\lambda_\alpha}{ds} \quad \text{at } \mathbf{u}_0.$$

Remarks

(1) If $\langle \mathbf{l}^\alpha(\mathbf{u}), \mathbf{u} - \mathbf{u}_0\rangle = 0$ at some point $(\lambda_\alpha(\mathbf{u}), \mathbf{u}) \in \mathscr{H}'_{\mathbf{u}_0}$ then two different one-dimensional submanifolfs of $\boldsymbol{R} \times \mathscr{H}$ contained in $\mathscr{H}'_{\mathbf{u}_0}$ can meet at $(\lambda_\alpha(\mathbf{u}), \mathbf{u})$. An explicit example of such a situation appears in isotropic elasticity, Section 3.3.

(2) The behaviour corresponding to (B) in (2b) can be illustrated by the strictly hyperbolic system

$$u_t + (3\ln u + v)_x = 0,$$

$$v_t + \left(\frac{2}{u}\right)_x = 0$$

for $u > 0$. Here $\mathscr{H}^1_{\mathbf{u}_0} = \mathscr{H}^2_{\mathbf{u}_0}$ is a closed eight-shaped curve (Borovikov, 1969, for example).

Proof

Let $\mathbf{F}(w, \mathbf{y}) :\equiv w\mathbf{y} - \{\mathbf{f}(\mathbf{u}_0 + \mathbf{y}) - \mathbf{f}(\mathbf{u}_0)\}$, $\mathbf{y} = \mathbf{u} - \mathbf{u}_0$, $\mathbf{A}_0 := \mathbf{A}(\mathbf{u}_0)$. Since the implicit function theorem breaks down only at the points $(w, \mathbf{u}) = (\lambda_\alpha(\mathbf{u}_0), \mathbf{u}_0)$ of the trivial solution, we shall investigate the solutions of $\mathbf{F} = \mathbf{0}$ in some nbhd of each point $(\lambda_\alpha(\mathbf{u}_0), \mathbf{u}_0)$. Let $\lambda_0 := \lambda_\alpha(\mathbf{u}_0)$, $\mathbf{r}_0 := \mathbf{r}_\alpha(\mathbf{u}_0)$, $\mathbf{l}_0 := \mathbf{l}^\alpha(\mathbf{u}_0)$ for some fixed $\alpha \in \{1, \ldots, n\}$, and $\xi := w - \lambda_0$, $\mathbf{y} = s\mathbf{r}_0 + \boldsymbol{\gamma}$, $\langle \mathbf{l}_0, \boldsymbol{\gamma} \rangle = 0$. We decompose $T_{\mathbf{u}_0}\mathscr{H}$ into two complementary subspaces $\{k\mathbf{r}_0 |\ k \in \boldsymbol{R}\}$ and $\mathbf{l}_0^\perp := \{\mathbf{z} \in T_{\mathbf{u}_0}\mathscr{H} |\ \langle \mathbf{l}_0, \mathbf{z} \rangle = 0\}$, with the corresponding projections $\mathbf{P}$, $\mathbf{P}'$. Applying $\mathbf{P}'$ to the equation $\mathbf{F} = \mathbf{0}$ we obtain an equation for the unknown $\boldsymbol{\gamma}$:

$$w\boldsymbol{\gamma} - \mathbf{P}'\{\mathbf{f}(\mathbf{u}_0 + s\mathbf{r}_0 + \boldsymbol{\gamma}) - \mathbf{f}(\mathbf{u}_0)\} = 0. \tag{3.1.45}$$

The Fréchet derivative $w\mathbf{E} - \mathbf{P}'\mathbf{A}_0\mathbf{P}'$ of the left-hand side of equation (3.1.45) with respect to $\boldsymbol{\gamma}$ at $\boldsymbol{\gamma} = \mathbf{0}$, $s = 0$ is invertible for w near λ_0. Hence equation (3.1.45) has a unique solution

$$\boldsymbol{\gamma} = \boldsymbol{\gamma}(\xi, s) \tag{3.1.46}$$

satisfying the condition $\boldsymbol{\gamma}(\xi, 0) = \mathbf{0}$ and defined in a nbhd of $s = 0$, $\xi = 0$. Substitution of $\mathbf{y} = s\mathbf{r}_0 + \boldsymbol{\gamma}(\xi, s)$ into the equation $\mathbf{PF} = \mathbf{0}$ yields the *bifurcation equation*:

$$s(\lambda_0 + \xi)\mathbf{r}_0 + \mathbf{P}\{\mathbf{f}(\mathbf{u}_0 + s\mathbf{r}_0 + \boldsymbol{\gamma}(\xi, s)) - \mathbf{f}(\mathbf{u}_0)\} = \mathbf{0}, \tag{3.1.47}$$

which has to be solved somehow for s, ξ. Lemma 15.4 in the book of Prodi and Ambrosetti (1973) implies that equation (3.1.47) has a solution $\xi = \psi(s)$, $\psi \in \mathscr{C}^1$, defined in a nbhd of $s = 0$, $\xi = 0$. We have thus obtained a parametric representation of $\mathscr{H}^\alpha_{\mathbf{u}_0}$ in a nbhd of $(\lambda_\alpha(\mathbf{u}_0), \mathbf{u}_0)$:

$$w = \lambda_0 + \psi(s), \tag{3.1.48a}$$

$$\mathbf{u} = \mathbf{u}_0 + s\mathbf{r}_0 + \boldsymbol{\gamma}(\psi(s), s), \tag{3.1.48b}$$

where

$$\langle \mathbf{l}_0, \boldsymbol{\gamma}(\xi, s) \rangle = 0, \tag{3.1.49}$$

$$\boldsymbol{\gamma}(\xi, 0) = \mathbf{0} \tag{3.1.50}$$

and $\boldsymbol{\gamma}$, $\psi \in \mathscr{C}^1$. Since $|s| \leqslant |\mathbf{u}-\mathbf{u}_0|$, the set defined by equations (3.1.48) is a submanifold of $\boldsymbol{R}\times\mathscr{H}$. More explicit forms of the solution (3.1.48) can be found in Hanyga (1976).

Using two theorems on bifurcation from simple eigenvalues (Rabinowitz, 1971; Turner, 1971) we can also conclude that:

(1) for every sufficiently small nbhd $\mathscr{U}$ of $(\lambda_0, \mathbf{u}_0)$ which does not contain any point $(\lambda_\beta(\mathbf{u}_0), \mathbf{u}_0)$ with $\beta \neq \alpha$ the set $\mathscr{H}^{\alpha}_{\mathbf{u}_0}\cap\partial\mathscr{U}$ consists of exactly two points;

(2) for any sphere $\mathscr{S}_R = \mathscr{S}((\lambda_0, \mathbf{u}_0), R) = \{(w, \mathbf{u})|\ (w-\lambda_0)^2+(\mathbf{u}-\mathbf{u}_0)^2 = R^2\}$, $R > 0$, $\mathscr{H}^{\alpha}_{\mathbf{u}_0}$ is a connected closed set in $\mathscr{H}'_{\mathbf{u}_0}$ joining $(\lambda_0, \mathbf{u}_0)$ either to $\mathscr{S}_R$ or to some point $(\lambda_\beta(\mathbf{u}_0), \mathbf{u}_0)$ with $\beta \neq \alpha$, lying inside $\mathscr{S}_R$;

(3) if for every $\beta \neq \alpha$ the point $(\lambda_\beta(\mathbf{u}_0), \mathbf{u}_0)$ does not lie inside $\mathscr{S}_R$, then $\mathscr{H}^{\alpha}_{\mathbf{u}_0}\cap\mathscr{S}_R$ consists either of two points or of one point which is of multiplicity two for equation (3.1.43).

Consider the equation

$$\dot{w}(\mathbf{u}-\mathbf{u}_0)+(w\mathbf{E}-\mathbf{A}(\mathbf{u}))\dot{\mathbf{u}} = \mathbf{v} \tag{3.1.51}$$

for arbitrary fixed $\mathbf{v} \in \boldsymbol{R}^n$ and unknown $(\dot{w}, \dot{\mathbf{u}}) \in \boldsymbol{R}\oplus T_{\mathbf{u}}\mathscr{H}$. If $w \neq \lambda_\beta(\mathbf{u})$ for $\beta = 1, \ldots, n$, then equation (3.1.51) can be solved for the unknown $\dot{\mathbf{u}}$ for arbitrarily chosen $\dot{w}$. If $w = \lambda_\beta(\mathbf{u})$ for some $(w, \mathbf{u}) \in \mathscr{H}_{\mathbf{u}_0}$, $\beta \in \{1, \ldots, n\}$, then choose $\dot{w}$ so that $\mathbf{v}-\dot{w}(\mathbf{u}-\mathbf{u}_0) \in \mathbf{l}^{\beta\perp} := \{\mathbf{r} \in T_{\mathbf{u}}\mathscr{H}|\ \langle\mathbf{l}^\beta(\mathbf{u}), \mathbf{r}\rangle = 0\}$ (using (ii)), and then solve the equation $(\lambda_\beta\mathbf{E}-\mathbf{A})\dot{\mathbf{u}} = \mathbf{v}-\dot{w}(\mathbf{u}-\mathbf{u}_0)$ for $\dot{\mathbf{u}} \in \mathbf{l}^{\beta\perp}$, which is feasible in view of strict hyperbolicity. Hence we have proved that the mapping

$$\Phi\colon (w, \mathbf{u}) \mapsto w(\mathbf{u}-\mathbf{u}_0)-\{\mathbf{f}(\mathbf{u})-\mathbf{f}(\mathbf{u}_0)\} \in \boldsymbol{R}^n \tag{3.1.52}$$

is a submersion except for the bifurcation points $(\lambda_\beta(\mathbf{u}_0), \mathbf{u}_0)$.

The set (3.1.48) contains a point $(w, \mathbf{u}) \in \mathscr{H}'_{\mathbf{u}_0}$ lying on a $\mathscr{C}^1$ submanifold $\mathscr{H}^{\alpha}_{\mathbf{u}_0}$ through $(\lambda_\alpha(\mathbf{u}_0), \mathbf{u}_0)$. In view of the results of the preceding paragraph $\mathscr{H}'_{\mathbf{u}_0} \neq \emptyset$ is a $\mathscr{C}^1$ submanifold of $\boldsymbol{R}\times\mathscr{H}$ of dimension one.

The projection $\tilde{\mathscr{H}}_{\mathbf{u}_0}$ of the submanifold $\mathscr{H}_{\mathbf{u}_0}$ onto $\mathscr{H}$ consists of $\mathscr{C}^1$ curves. Since $\mathscr{H}'_{\mathbf{u}_0}$ is a submanifold of $\boldsymbol{R}\times\mathscr{H}$, it follows from equation (3.1.43) that these curves intersect only at $\mathbf{u}_0$. Let the projection $\tilde{\mathscr{H}}^{\alpha}_{\mathbf{u}_0}$ of $\mathscr{H}^{\alpha}_{\mathbf{u}_0}$ be represented by the equation $\mathbf{u} = \mathbf{d}(s)$ with $\mathbf{u}_0 = \mathbf{d}(0)$. Differentiating (3.1.43) with respect to s at $s = 0$ we find that

$$(\mathbf{A}(\mathbf{u}_0)-w(0)\mathbf{E})\dot{\mathbf{d}}(0) = \mathbf{0}, \quad \dot{\mathbf{d}}(s) := \frac{d\mathbf{d}(s)}{ds}. \tag{3.1.53}$$

Since $w(0) = \lambda_\alpha(\mathbf{u}_0)$ we conclude that $\dot{\mathbf{d}}(0) = k\mathbf{r}_\alpha(\mathbf{u}_0)$ and $k \neq 0$ (see equation (3.1.48b)). The parametrization of $\mathscr{H}^{\alpha}_{\mathbf{u}_0}$ can be changed locally in such a way that $k = \pm 1$. In fact it can be done in an arbitrarily small nbhd of $s = 0$ by setting $s = \sigma + \psi(\sigma)$,

$$\psi(\sigma) :\equiv \begin{cases} c\sigma \exp[-q(\varepsilon^2 - \sigma^2)^{-1}] & \text{for } |\sigma| \leqslant \varepsilon, \\ 0 & \text{for } |\sigma| \geqslant \varepsilon, \end{cases}$$

$q > 0$, $k^{-1} - 1 = c\exp(-q\varepsilon^{-2})$, $1 + \psi'(0) = k^{-1}$. Clearly, either $\psi'(\sigma) > 0$ or $\psi'(\sigma) \geqslant \psi'(0) = k^{-1} - 1 > -1$, so that $\psi'(\sigma) > -1$.*

Suppose that $k = 1$, $\dot{\mathbf{d}}(0) = \mathbf{r}_\alpha(\mathbf{u}_0)$. Differentiating equation (3.1.43) twice with respect to s we get for $s = 0$

$$2\dot{w}(0)\mathbf{r}_\alpha(\mathbf{u}_0) + \lambda_\alpha(\mathbf{u}_0)\ddot{\mathbf{d}}(0) = \dot{\mathbf{A}}\mathbf{r}_\alpha(\mathbf{u}_0) + \mathbf{A}(\mathbf{u}_0)\ddot{\mathbf{d}}(0) \tag{3.1.54}$$

with $\dot{\varrho}(s) := \dfrac{d}{ds}\varrho \circ \mathbf{d}(s)$ for any differentiable function ϱ on $\mathscr{H}$. Contracting equation (3.1.54) with $\mathbf{l}^\alpha(\mathbf{u}_0)$ we obtain the identity

$$2\dot{w}(0) = \langle \mathbf{l}^\alpha(\mathbf{u}_0), \dot{\mathbf{A}}\mathbf{r}_\alpha(\mathbf{u}_0)\rangle$$

$$= \frac{d}{ds}\bigg|_{s=0} \lambda_\alpha \circ \mathbf{d} - \left\langle \frac{d\mathbf{l}^\alpha \circ \mathbf{d}}{ds}, \mathbf{A}\mathbf{r}_\alpha \right\rangle\bigg|_{s=0} - \left\langle \mathbf{l}^\alpha, \mathbf{A}\frac{d\mathbf{r}_\alpha \circ \mathbf{d}}{ds} \right\rangle\bigg|_{s=0}$$

$$= d\lambda_\alpha[\dot{\mathbf{d}}] - \lambda_\alpha(\mathbf{u}_0)\frac{d}{ds}\langle \mathbf{l}^\alpha, \mathbf{r}_\alpha\rangle|_{s=0} = d\lambda_\alpha[\dot{\mathbf{d}}(0)] = \dot{\lambda}_\alpha(0).$$

Comparing (3.1.54) with the identities

$$\mathbf{A}\dot{\mathbf{r}}_\alpha + \dot{\mathbf{A}}\mathbf{r}_\alpha = \lambda_\alpha\dot{\mathbf{r}}_\alpha + \dot{\lambda}_\alpha\mathbf{r}_\alpha \tag{3.1.55}$$

and $2\dot{w}(0) = \dot{\lambda}_\alpha(0)$ we conclude that

$$\big(\mathbf{A}(\mathbf{u}_0) - \lambda_\alpha(\mathbf{u}_0)\mathbf{E}\big)\big(\ddot{\mathbf{d}}(0) - \dot{\mathbf{r}}_\alpha(0)\big) = \mathbf{0}, \tag{3.1.56}$$

whence $\ddot{\mathbf{d}}(0) = \dot{\mathbf{r}}_\alpha(0) + k\mathbf{r}_\alpha(\mathbf{u}_0)$. Let $s = \sigma + \varphi(\sigma)$, with $\varphi \in \mathscr{C}^\infty$, $\operatorname{supp}\varphi \subset]-\varepsilon, \varepsilon[$, $\varepsilon > 0$, $\varphi(0) = 0$, $\varphi'(0) = 0$, $\varphi'(\sigma) > -1$ and set $\mathbf{e}(\sigma) :\equiv \mathbf{d}\big(\sigma + \varphi(\sigma)\big)$. We have then

$$\frac{d^2\mathbf{e}}{d\sigma^2}(0) = \frac{d^2\mathbf{d}}{ds^2}(0) + \varphi''(0)\frac{d\mathbf{d}}{ds}(0) = \dot{\mathbf{r}}_\alpha(\mathbf{0})$$

provided $\varphi''(0) = -k$. We can take

$$\varphi(\sigma) :\equiv \begin{cases} c\sigma^2 \exp\big(-q(\varepsilon^2 - \sigma^2)^{-1}\big) & \text{for } |\sigma| < \varepsilon, \\ 0 & \text{for } |\sigma| \geqslant \varepsilon, \end{cases}$$

with $q > 0$, $-2c\exp(-q\varepsilon^{-2}) = k$, $|2\varepsilon c| < 1$, ε arbitrarily small. □

* The required parametrization is already provided by equation (3.1.48). The technique of local reparametrization presented here will, however, be useful later.

Remarks

(1) For any function $\varrho \in \mathscr{C}^2(\mathscr{H}, \boldsymbol{R}^1)$ we define

$$\grave{\varrho}(\sigma) := \frac{d}{d\sigma}\varrho \circ \mathbf{c}(\sigma), \quad \grave{\grave{\varrho}}(\sigma) := \frac{d^2}{d\sigma^2}\varrho \circ \mathbf{c}(\sigma), \quad \text{where } \sigma \mapsto \mathbf{c}(\sigma)$$

is the α-hodograph through $\mathbf{u}_0$. Theorem 3.1.9 implies that for any function $\varrho \in \mathscr{C}^2$ on $\mathscr{H}$ $\grave{\varrho}(0) = \dot{\varrho}(0)$ and $\grave{\grave{\varrho}}(0) = \ddot{\varrho}(0)$. In particular $\grave{\lambda}_\alpha(0) = \dot{\lambda}_\alpha(0) = d\lambda_\alpha[\mathbf{r}_\alpha](\mathbf{u}_0)$ provided the curve $\mathscr{H}^\alpha_{\mathbf{u}_0}$ is parametrized in the way indicated in the proof of Theorem 3.1.9.

(2) Each arc $\mathscr{J}$ of $\mathscr{H}_{\mathbf{u}_0}$ is a one-dimensional $\mathscr{C}^1$ submanifold of $\boldsymbol{R}\times\mathscr{H}$. Hence it is possible to choose the parametrization of $\mathscr{J}: s \mapsto (w(s), \mathbf{d}(s)) \in \mathscr{J}$ in such a way that in a nbhd of every point of $\mathscr{J}$ the parameter s coincides with the value of the first coordinate of a local coordinate system $\phi: \mathscr{U} \to \boldsymbol{R}^{n+1}$, the other coordinates being constant on $\mathscr{U}\cap\mathscr{J}$. A parametrization satisfying the above condition will be called **regular.** If $s \mapsto (w(s), \mathbf{d}(s))$ is a regular parametrization of $\mathscr{J}$ and $s = \sigma+\psi(\sigma)$, $\psi \in \mathscr{C}^1$, $\psi'(\sigma) > -1$, $\operatorname{supp}\psi \subset [\sigma_0, \sigma_1]$, then the parametrization of $\mathscr{J}$ by σ is also regular.

The parametrization of $\mathscr{H}^\alpha_{\mathbf{u}_0}$ defined by equations (3.1.48) in a nbhd of $(\lambda_\alpha(\mathbf{u}_0), \mathbf{u}_0)$ is regular. Moreover, if $\dot{w}(0) \neq 0$, then the coordinate transformation ϕ: $(w, \mathbf{u}) \mapsto (s, \mathbf{u}_0)$ has the required properties. Indeed, the Jacobian of ϕ has a non-zero determinant

$$\det\begin{bmatrix} \dot{w}(0) & d\lambda_\alpha(\mathbf{u}_0) \\ \mathbf{r}_\alpha(\mathbf{u}_0) & \mathbf{E}_{n\times n} \end{bmatrix} = \dot{w}(0) - d\lambda_\alpha[\mathbf{r}_\alpha](\mathbf{u}_0) = -\tfrac{1}{2}\dot{w}(0)$$

at $(\lambda_\alpha(\mathbf{u}_0), \mathbf{u}_0)$. If $\dot{w}(0) = 0$, then $d\lambda_\alpha[\mathbf{r}_\alpha](\mathbf{u}_0) = 0$ and ϕ is not a coordinate transformation. In fact it may happen that the arcs $\mathscr{H}^\alpha_{\mathbf{u}_0}$ cover only the subset $w \geqslant \lambda_\alpha(\mathbf{u}_0)$ of the nbhd of $(\lambda_\alpha(\mathbf{u}_0), \mathbf{u}_0)$ (see Theorem 3.1.10 below).

Theorem 3.1.10

Suppose that $\mathbf{f} \in \mathscr{C}^{k+1}$ in a nbhd of $\mathbf{u}_0 \in \mathscr{H}$ and $\left[\left(\mathbf{r}_\alpha(\mathbf{u})\cdot\frac{\partial}{\partial \mathbf{u}}\right)^l \lambda_\alpha\right](\mathbf{u}_0) = 0$ for $l = 1, \ldots, k$, with $1 \leqslant k \leqslant \infty$. Then the parametrization $s \mapsto \mathbf{d}(s)$ of $\tilde{\mathscr{H}}^\alpha_{\mathbf{u}_0}$ can be adjusted in an arbitrarily small nbhd of $s = 0$ in such a way that

$$w^{(l)}(0) = 0 = \lambda_\alpha^{(l)}(0)$$

and

$$\mathbf{d}^{(l+1)}(0) = \left(\mathbf{r}_\alpha(\mathbf{u})\cdot\frac{\partial}{\partial \mathbf{u}}\right)^l \mathbf{r}_\alpha(\mathbf{u})\Big|_{\mathbf{u}=\mathbf{u}_0} \qquad \text{for } l \leqslant k,\ l \geqslant 1,$$

$$w^{(k+1)}(0) = \frac{1}{k+2}\left(\mathbf{r}_\alpha(\mathbf{u})\cdot\frac{\partial}{\partial \mathbf{u}}\right)^{k+1}\Big|_{\mathbf{u}=\mathbf{u}_0} = \frac{1}{k+2}\lambda_\alpha^{(k+1)}(0),$$

where $\varphi^{(l)}(s) := \dfrac{d^l}{ds^l}\varphi \circ \mathbf{d}(s)$.

Proof

The proof of this theorem becomes a routine task after reading the proofs of Theorems 3.1.9, 3.1.16 and 3.1.17, and is left to the reader. □

Suppose that for some integer $k \geqslant 1$ $\mathbf{f} \in \mathscr{C}^{k+1}$ and $\left(\mathbf{r}_\alpha(\mathbf{u}) \cdot \frac{\partial}{\partial \mathbf{u}}\right)^k \lambda_\alpha \Big|_{\mathbf{u}=\mathbf{u}_0} \neq 0.$

In this case $w(s) \neq \lambda_\beta \circ \mathbf{d}(s)$ for $\beta \in \{1, \ldots, n\}$ and sufficiently small s. The case where all the existing derivatives of $s \mapsto \lambda_\alpha \circ \mathbf{c}(s)$ vanish at $s = 0$ (including the case where $\mathbf{f} \in \mathscr{C}^\infty$ locally and all the derivatives of $s \mapsto \lambda_\alpha \circ \mathbf{c}(s)$ vanish at $s = 0$) is discussed in the paper by Hanyga (1975).

Let $s \mapsto (w(s), \mathbf{d}(s))$ be a regular parametrization of an arc $\mathscr{I} \subset \mathscr{H}_{\mathbf{u}_0}$, with s running over $[s_1, s_2]$, $s_1 < s_2$. Assume that $\mathbf{f} \in \mathscr{C}^1$. It follows that $\mathbf{d} \in \mathscr{C}^1$, $w \in \mathscr{C}^1$. From (3.1.43)

$$\big(\mathbf{A}(\mathbf{d}(s_0)) - w(s_0)\mathbf{E}\big)\dot{\mathbf{d}}(s_0) = \dot{w}(s_0)\big(\mathbf{d}(s_0) - \mathbf{u}_0\big) \tag{3.1.57}$$

for every $s_0 \in [s_1, s_2]$. If $\dot{w}(s_0) = 0$ for some $s_0 \in [s_1, s_2]$, then $\dot{\mathbf{d}}(s_0) \neq \mathbf{0}$, since $\mathscr{I}$ is a submanifold with regular parametrization. Equation (3.1.57) implies immediately that $\exists\, \beta \in \{1, \ldots, n\}$ such that

$$w(s_0) = \lambda_\beta \circ \mathbf{d}(s_0), \tag{3.1.58}$$

$$\dot{\mathbf{d}}(s_0) = k\mathbf{r}_\beta \circ \mathbf{d}(s_0), \quad k \neq 0. \tag{3.1.59}$$

Suppose now that $\dot{w}(s) \equiv 0$ for all $s \in [s_1, s_2]$. Then the arc $s \mapsto \mathbf{d}(s)$, $s \in [s_1, s_2]$, coincides with an arc of a β-hodograph, $\beta \in \{1, \ldots, n\}$, up to parametrization, and $w(s) = \lambda_\beta \circ \mathbf{d}(s) = \text{const}\ \forall\, s \in [s_1, s_2]$.

Conversely, consider the mapping

$$\mathbf{u} \mapsto G(\mathbf{u}) :\equiv \lambda_\alpha(\mathbf{u})(\mathbf{u} - \mathbf{u}_0) - \{\mathbf{f}(\mathbf{u}) - \mathbf{f}(\mathbf{u}_0)\}. \tag{3.1.60}$$

If $G(\mathbf{u}_1) = 0$, then $(\lambda_\alpha(\mathbf{u}_1), \mathbf{u}_1) \in \mathscr{H}_{\mathbf{u}_0}$. Suppose that

$$\langle \mathbf{l}^\alpha(\mathbf{u}_1), \mathbf{u}_0 - \mathbf{u}_1 \rangle \neq 0.$$

A non-zero vector $\mathbf{y} \in \ker DG(\mathbf{u}_1)$ iff $\mathbf{y} = k\mathbf{r}_\alpha(\mathbf{u}_1)$ and $d\lambda_\alpha(\mathbf{u}_1)[\mathbf{r}_\alpha(\mathbf{u}_1)] = 0$. Consequently, it follows from $w(s) \equiv \lambda_\alpha \circ \mathbf{d}(s)$ and

$$\langle \mathbf{l}^\alpha(\mathbf{d}(s)), \mathbf{d}(s) - \mathbf{u}_0 \rangle \neq 0 \tag{3.1.61}$$

for all $s \in [s_1, s_2]$ that the arc $s \mapsto \mathbf{d}(s)$ coincides with an arc of an α-hodograph and $w(s) = \lambda_\alpha \circ \mathbf{d}(s) = \text{const}$ for $s \in [s_1, s_2]$.

Finally, if $\dot{\mathbf{d}}(s) = k(s)\mathbf{r}_\alpha \circ \mathbf{d}(s)$ and equation (3.1.61) holds for all $s \in [s_1, s_2]$, then it follows easily by the same kind of argument that $w(s) \equiv \lambda_\alpha \circ \mathbf{d}(s)$ $\forall\, s \in [s_1, s_2]$. We have thus

Lemma 3.1.11
Let $s \mapsto (w(s), \mathbf{d}(s))$, $s \in [s_1, s_2]$, $s_2 > s_1$, be an arc of $\mathscr{H}_{\mathbf{u}_0}$. Then

(i) $w(s) \equiv$ const implies that $s \mapsto \mathbf{d}(s)$ coincides with an arc of an α-hodograph and $w(s) \equiv \lambda_\alpha \circ \mathbf{d}(s)$ for some $\alpha \in \{1, \ldots, n\}$;

(ii) $w(s) \equiv \lambda_\alpha \circ \mathbf{d}(s)$ and equation (3.1.61) for all $s \in [s_1, s_2]$ imply that $s \mapsto \mathbf{d}(s)$ coincides with an arc of an α-hodograph, $w(s) \equiv$ const, $d\lambda_\alpha[\mathbf{r}_\alpha](\mathbf{u}) = 0$ for $\mathbf{u} = \mathbf{d}(s)$, $s \in [s_1, s_2]$.

(iii) $\dot{\mathbf{d}}(s) = k(s)\mathbf{r}_\alpha \circ \mathbf{d}(s)$, $k \neq 0$, and equation (3.1.61) imply that $w(s) \equiv \lambda_\alpha \circ \mathbf{d}(s) \equiv$ const.

Definition 3.1.12
We shall say that the α-mode is **linearly degenerate** on an open subset $\mathscr{U}$ of $\mathscr{H}$ if $d\lambda_\alpha[\mathbf{r}_\alpha](\mathbf{u}) = 0 \; \forall \, \mathbf{u} \in \mathscr{U}$.

Definition 3.1.13
An arc $s \mapsto \big(w(s), \mathbf{d}(s)\big)$, $s \in [s_1, s_2]$ is said to be **exceptional** if $s \mapsto \mathbf{d}(s)$ is an arc of an α-hodograph and $w(s) \equiv \lambda_\alpha \circ \mathbf{d}(s)$.

Remarks

(1) If $\mathbf{u}_1$ lies on an exceptional arc of $\tilde{\mathscr{H}}_{\mathbf{u}_0}$ containing $\mathbf{u}_0$, then $\mathbf{u}_0$ lies on an exceptional arc of $\tilde{\mathscr{H}}_{\mathbf{u}_1}$.

(2) If the α-mode is linearly degenerate on $\mathscr{U} \subset \mathscr{H}$ and the image of an α-simple wave $\mathbf{u}$ lies in $\mathscr{U}$, then the hyperplanes $\mathbf{u} =$ const in $\boldsymbol{R}^{m+1}$ are parallel.

There is another way of looking at the solutions of equation (3.1.43), which will occasionally be useful. Theorems 3.1.9, 3.1.10, 3.1.16 and 3.1.17 give a meaning to the term "weak shock" for every fixed $\mathbf{u}_0$. A weak shock enjoys some properties which are important for the validity of numerous statements about the solutions. It is sometimes necessary to deduce some facts from the assumption that the oscillation of the solution around a "mean value" is small enough (Section 3.4). For this reason it is useful to consider the set $\tilde{\mathscr{H}}$, defined below, of solutions $(w, \mathbf{u}, \mathbf{u}_0)$ of equation (3.1.43) with $\mathbf{u}_0$ varying.

Let $\mathscr{H}_0 := \boldsymbol{R} \times \mathscr{H} \times \mathscr{H} \setminus \{(w, \mathbf{u}, \mathbf{u}_0) | \; \mathbf{u} = \mathbf{u}_0\}$. $\mathscr{H}_0$ is an open submanifold of $\boldsymbol{R} \times \mathscr{H} \times \mathscr{H}$. Let $\tilde{\mathscr{H}}_0$ be the set of all the points $(w, \mathbf{u}, \mathbf{u}_0) \in \mathscr{H}_0$ satisfying equation (3.1.43). We make the following assumption:

(Z) if $(w, \mathbf{u}, \mathbf{u}_0) \in \tilde{\mathscr{H}}_0$ and $w = \lambda_\alpha(\mathbf{u}) = \lambda_\beta(\mathbf{u}_0)$ for some α, $\beta \in \{1, \ldots, n\}$, then either $\langle \mathbf{l}^\alpha(\mathbf{u}), \mathbf{u} - \mathbf{u}_0 \rangle \neq 0$ or $\langle \mathbf{l}^\beta(\mathbf{u}_0), \mathbf{u} - \mathbf{u}_0 \rangle \neq 0$.

Assumption (Z) is weaker than the hypothesis (ii) of Theorem 3.1.9. With this assumption it is easy to check that the mapping G of $\boldsymbol{R} \times \mathscr{H} \times \mathscr{H}$ into $\boldsymbol{R}^n$ defined by the r.h.s. of (3.1.52) is a submersion on $\mathscr{H}_0$. The set

$G^{-1}(\{\mathbf{0}\})\cap\mathscr{H}_0$ is non-empty (Theorem 3.1.9). Hence $\tilde{\mathscr{H}}_0$ is a non-empty $\mathscr{C}^1$ submanifold of $\mathscr{H}_0$ of dimension $n+1$.

Suppose that $\tilde{\mathscr{H}}_0 \ni (w_k, \mathbf{u}_k, \mathbf{u}_k') \to (w, \mathbf{u}, \mathbf{u}')$ for $k \to \infty$. If $\mathbf{u} \neq \mathbf{u}'$ then $(w, \mathbf{u}, \mathbf{u}') \in \tilde{\mathscr{H}}_0$. If $\mathbf{u} = \mathbf{u}'$, then $(w_k, \mathbf{u}_k, \mathbf{u}) \to (w, \mathbf{u}, \mathbf{u})$ and, by Theorem 3.1.9, $w = \lambda_\alpha(\mathbf{u})$ and $\mathbf{u}_k - \mathbf{u} = \varepsilon_k \mathbf{r}_\alpha(\mathbf{u}) + \varepsilon_k \mathbf{q}(\mathbf{u}, \varepsilon_k)$, $\mathbf{q}(\mathbf{u}, \varepsilon_k) = o(\varepsilon_k^0)$ for some $\alpha \in \{1, \dots, n\}$, $\varepsilon_k \to 0$, provided k is sufficiently large. Let $\tilde{\mathscr{H}}$ be the closure of $\tilde{\mathscr{H}}_0$ in $\boldsymbol{R}\times\mathscr{H}\times\mathscr{H}$, and $\partial\tilde{\mathscr{H}} := \tilde{\mathscr{H}}\setminus\tilde{\mathscr{H}}_0$. Then $\partial\tilde{\mathscr{H}} = \{(\lambda_\alpha(\mathbf{u}), \mathbf{u}, \mathbf{u}) \mid \mathbf{u} \in \mathscr{H}\}$ is an n-sheeted n-dimensional submanifold of $\boldsymbol{R}\times\mathscr{H}\times\mathscr{H}$. Let $\Pi_{\mathbf{u}_0} := \{(w, \mathbf{u}, \mathbf{u}_0) \mid (w, \mathbf{u}) \in \boldsymbol{R}\times\mathscr{H}\}$. Then $\tilde{\mathscr{H}}\cap\Pi_{\mathbf{u}_0} = \mathscr{H}_{\mathbf{u}_0}$ is a one-dimensional submanifold of $\boldsymbol{R}\times\mathscr{H}\times\mathscr{H}$ and hence of $\tilde{\mathscr{H}}$.

At every point $(w, \mathbf{u}, \mathbf{u}_0) \in \tilde{\mathscr{H}}_0$ we have that $\mathbf{u} \neq \mathbf{u}_0$. Let $u^1 \neq u_0^1$. Then $w = \{f^1(\mathbf{u}) - f^1(\mathbf{u}_0)\}/\{u^1 - u_0^1\}$ locally. Consequently w may be regarded as a continuous, differentiable function of $(\mathbf{u}, \mathbf{u}_0)$ for $(w, \mathbf{u}, \mathbf{u}_0) \in \tilde{\mathscr{H}}_0$.

3.1.6 Sonic Points. Admissible Discontinuities

Definition 3.1.14

A point $(\lambda_\alpha(\mathbf{u}_1), \mathbf{u}_1) \in \mathscr{H}'_{\mathbf{u}_0}$ which does not lie on an exceptional arc $s \mapsto (\lambda_\alpha(\mathbf{d}(s)), \mathbf{d}(s))$ of $\mathscr{H}'_{\mathbf{u}_0}$ is called a **sonic point of first kind** (for the α-mode).

A point $(\lambda_\alpha(\mathbf{u}_0), \mathbf{u}_1) \in \mathscr{H}'_{\mathbf{u}_0}$ which does not lie on an exceptional arc $s \mapsto (\lambda_\alpha(\mathbf{u}_0), \mathbf{d}(s))$ of $\mathscr{H}'_{\mathbf{u}_0}$ is called a **sonic point of second kind** (for the α-mode).

The set of all the sonic points of first (second) kind in $\tilde{\mathscr{H}}'_{\mathbf{u}_0}$ will be denoted by $\mathscr{S}_1$ ($\mathscr{S}_2$ resp.).

Definition 3.1.15

A point $(w(s_0), \mathbf{d}(s_0))$ lying on an arc $s \mapsto (w(s), \mathbf{d}(s))$ of $\mathscr{H}_{\mathbf{u}_0}$ is called a **Chapman–Jouguet point** if the function $s \mapsto w(s)$ has a local extremum at s_0.

Let $\mathscr{C}\mathscr{J}$ be the set of all Chapman–Jouguet points on $\mathscr{H}_{\mathbf{u}_0}$.

We make the following assumptions:

(A) if $\mathbf{u}_1 \in \mathscr{S}_1 (\mathbf{u}_1 \in \mathscr{S}_2)$, then $\mathbf{u}_1$ is not an accumulation point of the set $\mathscr{S}_1$ (or $\mathscr{S}_2$). $\mathbf{u}_0$ is not an accumulation point of $\mathscr{S}_1 \cup \mathscr{S}_2$.

(B) equation (3.1.61) is satisfied for every point $(\lambda_\alpha(\mathbf{u}_1), \mathbf{u}_1) \in \mathscr{H}'_{\mathbf{u}_0}$ with $\mathbf{u}_1 \neq \mathbf{u}_0$.

Here is a counterexample for the hypothesis (A): $n = 1, f(u) :\equiv u^5(\sin u^{-1} - 1)$, $u_0 = 2\pi^{-1}$, $u_1 = 0$. Note that $f \in \mathscr{C}^2$.

In the following we assume that the parametrization $s \mapsto (w(s), \mathbf{d}(s))$ of the arc $\mathscr{J} \subset \mathscr{H}_{\mathbf{u}_0}$ under consideration is regular. Let

$$\bar{\lambda}_\alpha(s) :\equiv \lambda_\alpha \circ \mathbf{d}(s), \quad \bar{\mathbf{A}}(s) :\equiv n_k \mathbf{A}^k \circ \mathbf{d}(s), \quad \bar{\mathbf{r}}_\alpha(s) :\equiv \mathbf{r}_\alpha \circ \mathbf{d}(s),$$

$$\bar{\mathbf{l}}^\alpha(s) :\equiv \mathbf{l}^\alpha \circ \mathbf{d}(s), \quad \varphi^{(k)}(s) = \frac{d^k\varphi}{ds^k} \text{ or } = \frac{d^k\bar{\varphi}}{ds^k} \quad \dot{\varphi} = \varphi^{(1)} \tag{3.1.62}$$

for $\alpha \in \{1, \ldots, n\}$.

Theorem 3.1.16

Let $(\lambda_\alpha(\mathbf{u}_1), \mathbf{u}_1) \in \mathscr{S}_1, \mathbf{u}_1 = \mathbf{d}(s_0) \ \mathbf{u}_1 \in \mathscr{I} \subset \mathscr{H}_{\mathbf{u}_0}$.

Suppose that $\mathbf{f} \in \mathscr{C}^{k+1}$ in a nbhd of $\mathbf{u}_1$ and the assumption (B) is satisfied. Let

$$w(s) - \bar{\lambda}_\alpha(s) = \frac{1}{k!}(w - \bar{\lambda}_\alpha)^{(k)}(s_0)(s - s_0)^k + O\left((s - s_0)^{k+1}\right) \tag{3.1.63}$$

and

$$(w - \bar{\lambda}_\alpha)^{(k)}(s_0) \neq 0 \tag{3.1.64}$$

for some $k \geqslant 1$.

Then the parametrization of $\mathscr{I}$ can be modified in an arbitrarily small nbhd of s_0 in such a way that the following equations hold:

$$w^{(l)}(s_0) = 0 \quad \text{for } l = 1, \ldots, k, \tag{3.1.65}$$

$$\bar{\lambda}_\alpha^{(l)}(s_0) = 0 \quad \text{for } l = 1, \ldots, k-1, \quad \bar{\lambda}_\alpha^{(k)}(s_0) \neq 0. \tag{3.1.66}$$

$$\mathbf{d}^{(l+1)}(s_0) = \iota \bar{\mathbf{r}}_\alpha^{(l)}(s_0), \tag{3.1.67}$$

$$d\lambda_\alpha[\mathbf{r}_\alpha] \circ \mathbf{d}(s) = \iota \frac{1}{(k-1)!} \bar{\lambda}_\alpha^{(k)}(s_0)(s - s_0)^{k-1} + O\left((s - s_0)^k\right), \tag{3.1.68}$$

where $\iota = +1$ or $\iota = -1$ in all the formulae.

Proof

Equation (3.1.64) and strict hyperbolicity imply that $(\lambda_\alpha(\mathbf{u}_1), \mathbf{u}_1)$ is not an accumulation point of $\mathscr{S}_1$.

Since $\mathbf{f} \in \mathscr{C}^{k+1}$ in a nbhd of $\mathbf{u}_1$ and the parametrization is regular, it follows from (B) and Theorem 3.1.9 that $w \in \mathscr{C}^k$, $\mathbf{d} \in \mathscr{C}^k$ in a nbhd of s_0. Equations (3.1.57) and (3.1.63) imply that $\dot{w}(s_0)\langle \bar{\mathbf{l}}^\alpha(s_0), \mathbf{d}(s_0) - \mathbf{u}_0\rangle = \langle \bar{\mathbf{l}}^\alpha(s_0), (\bar{\mathbf{A}}(s_0) - w(s_0)\mathbf{E})\dot{\mathbf{d}}(s_0)\rangle = 0$, whence $\dot{w}(s_0) = 0$ in view of (B). Equation (3.1.57) implies then that $\dot{\mathbf{d}}(s_0) = \varkappa \bar{\mathbf{r}}_\alpha(s_0)$ for some $\varkappa \neq 0$.

It is possible to modify the parametrization of $\mathscr{I}$ in a nbhd $\mathscr{I}$ of s_0 so that (1) no point $(w(s), \mathbf{d}(s))$ with $s \in \mathscr{I}$ lies in $\mathscr{S}_1 \cup \{\mathbf{u}_0\}$; (2) the new parametrization is regular; in the new parametrization

$$\dot{\mathbf{d}}(s_0) = \iota \bar{\mathbf{r}}_\alpha(s_0) \tag{3.1.69}$$

with $\iota = \operatorname{sgn} \varkappa$. If $k = 1$, then $\dot{w}(s_0) = 0$, $\dot{\bar{\lambda}}_\alpha(s_0) \neq 0$ and on account of (3.1.69) we are through with the proof.

Suppose now that $k > 1$, $w^{(j)}(s_0) = 0$, $\mathbf{d}^{(j)}(s_0) = \iota\bar{\mathbf{r}}_\alpha^{(j-1)}(s_0)$ for $j \leqslant l$, where $l < k$. Equation (3.1.63) implies that

$$\lambda_\alpha^{(j)}(s_0) = w^{(j)}(s_0) = 0 \tag{3.1.70}$$

for $j \leqslant l$. Substituting $w = w(s)$, $\mathbf{u} = \mathbf{d}(s)$ in equation (3.1.43), differentiating the resulting identity $l+1$ times with respect to s at s_0 and using the inductive hypotheses we arrive at the following formula:

$$\bar{\lambda}_\alpha(s_0)\mathbf{d}^{(l+1)}(s_0) + w^{(l+1)}(s_0)\big(\mathbf{d}(s_0) - \mathbf{u}_0\big) -$$

$$-\sum_{j=0}^{l-1}\binom{l}{j}\mathbf{A}^{(l-j)}(s_0)\bar{\mathbf{r}}_\alpha^{(j)}(s_0) - \bar{\mathbf{A}}(s_0)\mathbf{d}^{(l+1)}(s_0) = 0 \tag{3.1.71}$$

Differentiating the identity

$$\bar{\mathbf{A}}(s)\bar{\mathbf{r}}_\alpha(s) = \bar{\lambda}_\alpha(s)\bar{\mathbf{r}}_\alpha(s) \tag{3.1.72}$$

l times at $s = s_0$ and using (3.1.70) we get the identity

$$\bar{\lambda}_\alpha(s_0)\bar{\mathbf{r}}_\alpha^{(l)}(s_0) = \bar{\mathbf{A}}(s_0)\bar{\mathbf{r}}_\alpha^{(l)}(s_0) + \sum_{j+1}^{l}\binom{l}{j}\bar{\mathbf{A}}^{(j)}(s_0)\bar{\mathbf{r}}_\alpha^{(l-j)}(s_0). \tag{3.1.73}$$

Contracting equation (3.1.73) with $\bar{\mathbf{l}}^\alpha(s_0)$ we have

$$\Big\langle \bar{\mathbf{l}}^\alpha(s_0), \sum_{j=1}^{l}\binom{l}{j}\bar{\mathbf{A}}^{(j)}(s_0)\bar{\mathbf{r}}_\alpha^{(l-j)}(s_0)\Big\rangle = 0. \tag{3.1.74}$$

Contracting equation (3.1.71) with $\bar{\mathbf{l}}^\alpha(s_0)$ and using equation (3.1.74) we get the formula

$$w^{(l+1)}(s_0)\langle \bar{\mathbf{l}}^\alpha(s_0), \mathbf{d}(s_0) - \mathbf{u}_0\rangle = 0,$$

which implies that $w^{(l+1)}(s_0) = 0$, with $l+1 \leqslant k$. For $l+1 < k$ equation (3.1.63) implies that $\lambda_\alpha^{(l+1)}(s_0) = 0$; for, $l+1 = k$ we get $\lambda_\alpha^{(k)}(s_0) \neq 0$.

For $l+1 \leqslant k$ equations (3.1.71) and (3.1.73) imply that

$$\big(\bar{\mathbf{A}}(s_0) - \bar{\lambda}_\alpha(s_0)\mathbf{E}\big)\mathbf{d}^{(l+1)}(s_0) = \iota\{\bar{\mathbf{A}}(s_0) - \bar{\lambda}_\alpha(s_0)\mathbf{E}\}\bar{\mathbf{r}}_\alpha(s_0),$$

so that

$$\mathbf{d}^{(l+1)}(s_0) = \iota\bar{\mathbf{r}}_\alpha^{(l)}(s_0) + \mu\bar{\mathbf{r}}_\alpha(s_0) \tag{3.1.75}$$

A local change of parametrization in a nbhd of s_0: $s = \sigma + \psi(\sigma)$, $\psi^{(j)}(s_0) = 0$ for $j = 0, 1, \ldots, l$, $\psi^{(l+1)}(s_0) = -\mu$ leads to the formula (3.1.67).

For $l < k$ and $s = s_0$

$$(d\lambda_\alpha[\mathbf{r}_\alpha])^{(l)}(s_0) = \sum_{j=0}^{l} \binom{l}{j} \widehat{d\lambda_\alpha}^{(l-j)}[\hat{\mathbf{r}}_\alpha^{(j)}](s_0) =$$

$$= \iota \sum_{j=0}^{l} \binom{l}{j} \widehat{d\lambda_\alpha}^{(l-j)}[\mathbf{d}^{(j+1)}](s_0) = \iota (d\lambda_\alpha[\dot{\mathbf{d}}])^{(j)}(s_0) = \iota \lambda_\alpha^{(j+1)}(s_0),$$

where $\hat{\varphi}(s) := \varphi \circ \mathbf{e}(s)$ and $s \mapsto \mathbf{e}(s)$ is the α-hodograph with $\mathbf{e}(s_0) = \mathbf{d}(s_0)$. □

Theorem 3.1.17

Suppose that $\mathbf{f} \in \mathscr{C}^k$ in a nbhd of $\mathbf{u}_1 = \mathbf{d}(s_0)$, w, $\mathbf{d} \in \mathscr{C}^k$ in a nbhd of s_0, for some integer $k > 1$, and

$$w^{(l)}(s_0) = 0 \quad \text{for } l = 1, \ldots, k-1, \tag{3.1.76}$$

$$w^{(k)}(s_0) \neq 0. \tag{3.1.77}$$

Then $\exists\ \alpha \in \{1, \ldots, n\}$ such that $w(s_0) = \bar{\lambda}_\alpha(s_0)$ and

$$\lambda_\alpha^{(l)}(s_0) = 0 \quad \text{for } l = 1, \ldots, k-2, \tag{3.1.78}$$

$$\operatorname{sgn} \lambda_\alpha^{(k-1)}(s_0) = \iota \operatorname{sgn}\big(w^{(k)}(s_0)\langle \bar{\mathbf{l}}^\alpha(s_0), \mathbf{d}(s_0) - \mathbf{u}_0\rangle\big), \tag{3.1.79}$$

$$\mathbf{d}^{(l+1)}(s_0) = \iota \mathbf{r}_\alpha^{(l)}(s_0) \quad \text{for } l = 0, \ldots, k-2, \tag{3.1.80}$$

$\iota = +1$ or $= -1$.

Remark

Equations (3.1.76)–(3.1.79) imply that

$$w - \bar{\lambda}_\alpha = -\lambda_\alpha^{(k-1)}(s_0) \frac{1}{(k-1)!}(s - s_0)^{k-1} + \ldots, \tag{3.1.81}$$

which can be compared with the hypotheses (3.1.63) and (3.1.64) of Theorem 3.1.16. Equations. (3.1.80) and (3.1.67) are then equivalent.

Proof

There is a nbhd $\mathscr{I}$ of s_0 such that $\forall\ s \in \mathscr{I}\ \dot{w}(s) \neq 0$ unless $s = s_0$. On account of (3.1.65) $\forall\ s \in \mathscr{I} (w(s), \mathbf{d}(s)) \,\bar{\in}\, \mathscr{C}\mathscr{J} \cup \mathscr{S}_1$ except for $s = s_0$. This fact will allow us to change the parametrization in $\mathscr{I}$ without affecting the parametrization at other points of $\mathscr{C}\mathscr{J} \cup \mathscr{S}_1$.

Equations (3.1.57) and (3.1.76) imply again that $w(s_0) = \bar{\lambda}_\alpha(s_0)$, $\dot{\mathbf{d}}(s_0) = \varkappa \bar{\mathbf{r}}_\alpha(s_0)$ for some $\alpha \in \{1, \ldots, n\}$, $\varkappa \neq 0$. After a local change of parametrization we get equation (3.1.69) with $\iota = \operatorname{sgn} \varkappa \neq 0$.

Substituting $w = w(s)$, $\mathbf{u} = \mathbf{d}(s)$ into equation (3.1.43) and differentiating the resulting identity twice at s_0 we get the formula

$$\bar{\mathbf{A}}(s_0)\ddot{\mathbf{d}}(s_0) + \dot{\mathbf{A}}(s_0)\dot{\mathbf{d}}(s_0) = \ddot{w}(s_0)\big(\mathbf{d}(s_0) - \mathbf{u}_0\big) + \bar{\lambda}_\alpha(s_0)\ddot{\mathbf{d}}(s_0).$$

Hence $\dot{\lambda}_\alpha(s_0) = \langle \bar{\mathbf{l}}^\alpha(s_0), \ \dot{\mathbf{A}}(s_0)\bar{\mathbf{r}}_\alpha(s_0)\rangle = \iota\ddot{w}(s_0)\langle \bar{\mathbf{l}}^\alpha(s_0), \mathbf{d}(s_0)-\mathbf{u}_0\rangle$. If $k = 2$, we are through. If $k > 2$, we conclude that $\dot{\lambda}_\alpha(s_0) = 0$.

Suppose that $k \geqslant 2$, $\lambda_\alpha^{(j)}(s_0) = 0$ for $1 \leqslant j \leqslant l-1$ and $\mathbf{d}^{(j+1)}(s_0) = \iota\mathbf{r}_\alpha^{(j)}(s_0)$ for $0 \leqslant j \leqslant l-1$ for, some $l < k$. Differentiating the identity (3.1.72) l times at $s = s_0$ and using the inductive hypotheses we find that

$$\bar{\lambda}_\alpha \mathbf{r}_\alpha^{(l)} + \lambda_\alpha^{(l)}\bar{\mathbf{r}}_\alpha = \bar{\mathbf{A}}\mathbf{r}_\alpha^{(l)} + \sum_{j=0}^{l-1}\binom{l}{j}\mathbf{A}^{(l-j)}\mathbf{r}_\alpha^{(j)} \quad \text{at } s = s_0. \tag{3.1.82}$$

Contracting equations (3.1.82) and (3.1.71) (the latter holds on account of the inductive hypotheses) with $\bar{\mathbf{l}}^\alpha$ and combining them we find that

$$\lambda_\alpha^{(l)}(s_0) = \iota w^{(l+1)}(s_0)\langle \bar{\mathbf{l}}^\alpha(s_0), \mathbf{d}(s_0)-\mathbf{u}_0\rangle = 0 \tag{3.1.83}$$

if $l < k-1$, and

$$\lambda_\alpha^{(k-1)}(s_0) = \iota w^{(k)}(s_0)\langle \bar{\mathbf{l}}^\alpha(s_0), \mathbf{d}(s_0)-\mathbf{u}_0\rangle \neq 0. \tag{3.1.84}$$

Hence $\lambda_\alpha^{(l)}(s_0) = 0$ for $l < k-1$, but $\lambda_\alpha^{(k-1)}(s_0) \neq 0$. Equations (3.1.82), (3.1.83) and (3.1.71) imply then equation (3.1.75) for $l < k-1$ and a local modification of the parametrization leads to equation (3.1.80). This finishes the proof. □

Definition 3.1.18

A sonic point of first kind (a Chapman–Jouguet point) is said to be **regular** if it satisfies the hypotheses of Theorem 3.1.16 (of Theorem 3.1.17, resp.). The number k defined by (3.1.63) and (3.1.64) is called the **order of the sonic** (or **Chapman- Jouguet) point.**

Remarks

(1) A sonic point of first kind with $w - \bar{\lambda}_\alpha = O((s-s_0)^{k+1/2})$ (exactly) is not regular.

(2) It follows from the proofs of Theorems 3.1.16 and 3.1.17 that a regular sonic point of first kind (or a regular Chapman–Jouguet point) is not an accumulation point of $\mathcal{S}_1$ or $\mathcal{CJ}$. A sonic point of first kind (or a Chapman–Jouguet point) which is not regular can however be an accumulation point of $\mathcal{S}_1$. For instance, let $f(u) :\equiv \dfrac{u^6}{u-2/\pi}(\sin u^{-1}-1)$, $u_0 = 2/\pi$. The points $u = 0$ and $u_\nu = 2/(2\nu+1)\pi, \nu = 1, 2, \ldots$, are sonic points of first and second kind as well as Chapman–Jouguet points. At their accumulation point $u = 0$ $w^{(1)} = w^{(2)} = 0$, but $w^{(3)}$ does not exist. The functions $w - \lambda$ and $\lambda = f'$ exhibit dense oscillations near $u = 0$ and change sign at the points u_ν. For $u < 2/\pi$ we have that $w \geqslant 0$.

(3) The isolated Chapman–Jouguet point $u = 0$ for $f(u) :\equiv \exp(-|u|^{-1})(u-a)$, $u_0 = a$, is not regular but $w - \lambda = -(u|u|)^{-1}(u-a)$ changes sign at $u = 0$.

(4) For analytic $\mathbf{f}$ the statements of the theorems can be considerably simplified since all the sonic points are regular in this case (cf. Kuznetsov and Tupchiyev, 1975).

Corollary 3.1.19

(i) Every regular Chapman–Jouguet point is a sonic point of first kind.

(ii) Every regular sonic point of first kind and of an odd order is a Chapman–Jouguet point.

(iii) In a nbhd of a regular Chapman–Jouguet point for the α-mode the function $s \mapsto \overline{\lambda}_\alpha(s)$ is monotone.

(iv) In a nbhd of a regular sonic point of first kind $(\overline{\lambda}_\alpha(s_0), \mathbf{d}(s_0))$ which is not a Chapman–Jouguet point the function $s \mapsto \overline{\lambda}_\alpha(s)$ attains an extremum and $d\lambda_\alpha[\mathbf{r}_\alpha] \circ \mathbf{d}(s_0) = 0$.

Corollary 3.1.20

(i) The function $w - \overline{\lambda}_\alpha$ changes sign at every regular Chapman–Jouguet point $(\overline{\lambda}_\alpha(s), \mathbf{d}(s))$.

(ii) If the function $w - \overline{\lambda}_\alpha$ has an isolated zero of kth order at s_0, $\mathbf{d} \in \mathscr{C}^k$, $\mathbf{f} \in \mathscr{C}^{k+1}$ in a nbhd of s_0, $\mathbf{d}(s_0)$ resp., and $w - \overline{\lambda}_\alpha$ changes sign at s_0, then $(\overline{\lambda}_\alpha(s_0), \mathbf{d}(s_0))$ is a regular Chapman–Jouguet point.

(iii) Let $C = (\lambda_\alpha(\mathbf{u}_0), \mathbf{d}(s_0))$ be an isolated sonic point of second kind. Then there are three possibilities:

(1) $w - \lambda_\alpha(\mathbf{u}_0)$ changes sign at s_0;

(2) C is an irregular Chapman–Jouguet point;

(3) C is a regular Chapman–Jouguet point and $w - \overline{\lambda}_\beta$ for some $\beta \in \{1, \ldots, n\}$, changes sign at s_0.

Theorem 3.1.21

Suppose that $\mathbf{f} \in \mathscr{C}^2$.

Let $s \mapsto (w(s), \mathbf{d}(s))$ be a regularly parametrized arc $\mathscr{J}$ of $\mathscr{H}_{\mathbf{u}_0}$. Suppose that $\mathscr{J}$ contains two regular Chapman–Jouguet points $(\overline{\lambda}_\alpha(s_1), \mathbf{d}(s_1))$ and $(\overline{\lambda}_\alpha(s_2), \mathbf{d}(s_2))$ of respective orders r_1, r_2 and has the following properties:

(i) every sonic point of first kind for the α-mode on the arc $\mathscr{J}'$: $]s_1, s_2[\ni s \mapsto (w(s), \mathbf{d}(s))$ is regular;

(ii) there are no Chapman–Jouguet points for the α-mode on $\mathscr{J}'$;

(iii) $\langle \overline{\mathbf{l}}^\alpha(s), \mathbf{d}(s) - \mathbf{u}_0 \rangle \neq 0$. (3.1.85)

Then $\exists\, s_0 \in]s_1, s_2[$ such that $d\lambda_\alpha[\mathbf{r}_\alpha] \circ \mathbf{d}$ changes sign at s_0.

Proof
It follows from Theorems 3.1.16 and 3.1.17 that $w-\overline{\lambda}_\alpha$ does not change sign on $]s_1, s_2[$ and

$$w-\overline{\lambda}_\alpha = -\frac{1}{r_i!}\lambda_\alpha^{(r_i)}(s_i)(s-s_i)^{r_i}+ \ldots \quad \text{for } s \in]s_1, s_2[$$

close to s_1, s_2 resp., $\lambda_\alpha^{(r_i)}(s_i) \neq 0$. The integers r_1, r_2 are odd. Hence $\lambda_\alpha^{(r_1)}(s_1)\,\lambda_\alpha^{(r_2)}(s_2) < 0$.

Let ι_1, ι_2 be the corresponding values of ι in equation (3.1.80) at s_1, s_2. Equation (3.1.57) implies that

$$\big(w(s)-\overline{\lambda}_\alpha(s)\big)\langle\overline{\mathbf{l}}^\alpha(s), \dot{\mathbf{d}}(s)\rangle+\dot{w}(s)\langle\overline{\mathbf{l}}^\alpha(s), \mathbf{d}(s)-\mathbf{u}_0\rangle = 0. \tag{3.1.86}$$

In view of equations (3.1.77) and (3.1.81) the functions $\dot{w}$ and $w-\overline{\lambda}_\alpha$ vanish to the same order at s_1 and also at s_2. Hence from equation (3.1.85) $\langle\overline{\mathbf{l}}^\alpha(s_i), \dot{\mathbf{d}}(s_i)\rangle \neq 0$, $i = 1, 2$. Since $w-\overline{\lambda}_\alpha$ and $\dot{w}$ do not change sign on $]s_1, s_2[$, equations (3.1.85), (3.1.86) and (3.1.80) imply that $\iota_1 = \iota_2$. Equation (3.1.68) implies the thesis of the theorem. □

Theorem 3.1.22
Let $\mathscr{I} \ni s \mapsto (w(s), \mathbf{d}(s))$ be an arc of $\mathscr{H}_{\mathbf{u}_0}$ with the property that $\mathbf{d}(0) = \mathbf{u}_0$, $w(0) = \lambda_\alpha(\mathbf{u}_0)$ and suppose that the hypotheses of Theorem 3.1.10 are satisfied. Let the parametrization be chosen in accordance with Theorem 3.1.10 in a nbhd of $s = 0$.

Let $s_0 \in \mathscr{I}$ be such that

(i) $w(s_0) = \overline{\lambda}_\alpha(s_0)$ and $\big(w(s_0), \mathbf{d}(s_0)\big)$ is a regular Chapman–Jouguet point of order l;

(ii) the arc $\mathscr{I}' \ni s \mapsto \big(w(s), \mathbf{d}(s)\big)$ does not contain either regular Chapman–Jouguet points or irregular sonic points of first kind.
Here $\mathscr{I}' :=]0, s_0[$ if $s_0 > 0$ and $\mathscr{I}' :=]s_0, 0[$ if $s_0 < 0$.

(iii) $\langle\overline{\mathbf{l}}^\alpha(s), \mathbf{d}(s)-\mathbf{u}_0\rangle \neq 0$ for $s \in \mathscr{I}'$.

Under these hypotheses there is a point $s_1 \in \mathscr{I}'$ such that $d\lambda_\alpha[\mathbf{r}_\alpha] \circ \mathbf{d}$ changes sign at s_1.

Proof
Let $\iota_0 := \operatorname{sgn} \lambda_\alpha^{(k+1)}(0)$. It follows from Theorem 3.1.10 that for sufficiently small $s \in \mathscr{I}'$

$$\operatorname{sgn}\big(w(s)-\overline{\lambda}_\alpha(s)\big) = -\iota_0(\operatorname{sgn} s_0)^{k+1}. \tag{3.1.87}$$

The left-hand side of equation (3.1.87) does not change sign on $\mathscr{I}'$. Hence, using the expansion (3.1.81) near $s = s_0$ we conclude that

$$\operatorname{sgn} \lambda_\alpha^{(l)}(s_0)(-\operatorname{sgn} s_0)^l = \iota_0(\operatorname{sgn} s_0)^{k+1}. \tag{3.1.88}$$

We also have that $d\lambda_\alpha[\mathbf{r}_\alpha] \circ \mathbf{d}(s) = \iota_0(\operatorname{sgn} s_0)^k$ for $s \in \mathscr{I}'$ close to 0 and $\operatorname{sgn} d\lambda_\alpha[\mathbf{r}_\alpha] \circ \mathbf{d}(s) = \iota \operatorname{sgn}(\lambda_\alpha^{(l)}(s_0))(-\operatorname{sgn} s_0)^{l-1}$ for $s \in \mathscr{I}'$ close to s_0 (see equation (3.1.68)), with ι given by the formula (3.1.69). On account of equation (3.1.86) $\langle \mathbf{l}^\alpha, \dot{\mathbf{d}} \rangle$ has constant sign on $\mathscr{I}'$ and $\iota = 1$. Equation (3.1.88) implies that $\operatorname{sgn} d\lambda_\alpha[\mathbf{r}_\alpha] \circ \mathbf{d}(s) = -\iota_0(\operatorname{sgn} s_0)^k$, whence follows the thesis. □

Remarks

(1) Theorems 3.1.21 and 3.1.22 indicate the importance of the hypersurfaces $\Sigma_\alpha^m \subset \mathscr{H}$, at which $d\lambda_\alpha[\mathbf{r}_\alpha]$ changes sign. The assumption of **genuine non-linearity** (Lax, 1956),

$$(\mathrm{GN}_\alpha)\, d\lambda_\alpha[\mathbf{r}_\alpha] \neq 0 \text{ on } \mathscr{H},$$

which is so often made to simplify the deductions essentially amounts to excluding Chapman–Jouguet points for the α-mode on $\mathscr{H}^\alpha_{\mathbf{u}_0}$.

(2) We have often assumed the inequality (3.1.85). It is not essential except at the sonic points. The assumption (3.1.85) in Theorems 3.1.21 and 3.1.22 can be dropped provided the assertion "there is a point s_1 such that $d\lambda_\alpha[\mathbf{r}_\alpha] \circ \mathbf{d}$ changes sign at s_1" is replaced by the alternative "either $d\lambda_\alpha[\mathbf{r}_\alpha] \circ \mathbf{d}$ or $\langle \mathbf{l}^\alpha, \mathbf{d}(s) - \mathbf{u}_0 \rangle$ changes sign at some s_1". Moreover, the number of points $s_1 \in \mathscr{I}'$ at which either of these changes sign is an odd integer.

3.1.7 Admissible Shock Waves. The Riemann Problem

In this section we continue to restrict our attention to piecewise $\mathscr{C}^1$ solutions of equation (3.1.25). Following the intuitive arguments of Lax (1957), Jeffrey and Taniuti (1964), Hanyga (1974) and others, we consider only those solutions $\mathbf{u}(x)$ which have the following property:

(L) Every isolated discontinuity Σ for $\mathbf{u}(x)$ with the normal $(-w, \mathbf{n}) \in \boldsymbol{R}^{m+1}$ and the states $\mathbf{u}_+(y^1, \ldots, y^m)$, $\mathbf{u}_-(y^1, \ldots, y^m)$, as defined in Section 3.1.4, satisfies the Lax inequalities

$$\begin{aligned} &\lambda_\alpha(\mathbf{u}_+, \mathbf{n}) \leqslant w \leqslant \lambda_\alpha(\mathbf{u}_-, \mathbf{n}), \\ &\lambda_{\alpha-1}(\mathbf{u}_-, \mathbf{n}) < w < \lambda_{\alpha+1}(\mathbf{u}_+, \mathbf{n}) \end{aligned} \tag{3.1.89}$$

for some $\alpha \in \{1, \ldots, n\}$ and for almost all $(y^1, \ldots, y^m)$. Here we have set $\lambda_{n+1} \equiv \infty$, $\lambda_0 \equiv -\infty$.

We say that the discontinuity Σ **satisfies** $(\mathbf{L}_\alpha)$ at a point $(y^1, \ldots, y^m)$ if the values of w, $\mathbf{u}_+$, $\mathbf{u}_-$ at this point satisfy equations (3.1.89) for a given $\alpha \in \{1, \ldots, n\}$.

Remarks

The normal $(-w, \mathbf{n})$ points into $\mathscr{U}_+$ (cf. Sections 3.1.4 and 3.1.5). We say that the function $\mathbf{u}_+$ is the **right state** and the function $\mathbf{u}_-$ is the **left state** of the discontinuity Σ.

Definition 3.1.23
A discontinuity satisfying (L_α) at almost every point $(y^1, \ldots, y^m)$ will be called an **α-shock.**

Definition 3.1.24
Let $\mathbf{n}$ be a fixed unit vector in $\boldsymbol{R}^m$, $\mathbf{u}_L, \mathbf{u}_R \in \mathscr{H}$.

The Riemann problem $(\mathbf{n}, \mathbf{u}_L, \mathbf{u}_R)$ for equation (3.1.25) is defined as the ollowing initial-value problem for equation (3.1.25):

$$\mathbf{u}(0, x^1, \ldots, x^m) \equiv \begin{cases} \mathbf{u}_L & \text{for } n_k x^k < 0, \\ \mathbf{u}_R & \text{for } n_k x^k > 0. \end{cases} \tag{3.1.90}$$

For simplicity we define $y := n_k x^k$.

We look for piecewise $\mathscr{C}^1$ solutions

$$\mathbf{u} = \mathbf{u}(t, y) \tag{3.1.91}$$

to the Riemann problem (3.1.90).

If $\mathbf{u}(t, y)$ is a solution to (3.1.90), so is $\mathbf{u}(ct, cy)$ for every $c > 0$. Since we lack a uniqueness theorem for the general Riemann problem (3.1.90) and (3.1.91) we cannot assert that the solution is self-similar:

$$\mathbf{u} = \mathbf{v}(yt^{-1}) \quad \text{for } t > 0. \tag{3.1.92}$$

We shall however seek only self-similar solutions (3.1.92) to (3.1.90).

The following lemma is proved in Hanyga (1975, 1976):

Lemma 3.1.25
Suppose that $\mathbf{f} := n_k \mathbf{f}^k \in \mathscr{C}^2$.

Let $\boldsymbol{R}^2_+ := \{(t, y) |\ t > 0\}$ and let $\mathbf{u}$ be a piecewise $\mathscr{C}^1$ self-similar solution to the Riemann problem (3.1.90).

Then

$$\boldsymbol{R}^2_+ = \bigcup_{i=1}^{p} \Omega_i \cup \bigcup_{j=1}^{r} \Sigma_j \cup \bigcup_{j=1}^{s} \Sigma'_j,$$

where the sets

$$\Omega_i := \{(t, y) \in \boldsymbol{R}^2_+ |\ \zeta_i^- < yt^{-1} < \zeta_i^+\}, \quad \zeta_i^- < \zeta_i^+,$$
$$\Sigma_j := \{(t, y) \in \boldsymbol{R}^2_+ |\ yt^{-1} = \eta_j\}, \quad \eta_j \in \boldsymbol{R},$$
$$\Sigma'_j := \{(t, y) \in \boldsymbol{R}^2_+ |\ yt^{-1} = \eta'_j\}, \quad \eta'_j \in \boldsymbol{R},$$

do not intersect one another, $p, r, s < \infty$, $\mathbf{u}|\Omega_i$ is an α-centered simple wave or a constant function, $\alpha \in \{1, \ldots, n\}$, Σ_j is a discontinuity of $\mathbf{u}(t, y) = \mathbf{v}(yt^{-1})$ and Σ_j' is a discontinuity of $\mathbf{v}'$ such that $\mathbf{v}$ is continuous across Σ_j'.

If Σ_j' is not a boundary between a constant function $\mathbf{u}|\Omega_k$ and an α-c.s.w. $\mathbf{u}|\Omega_l$, then $d\lambda_\alpha[\mathbf{r}_\alpha] \circ \mathbf{u} = 0$ at Σ_j' and Σ_j' separates two α-c.s. waves for some $\alpha \in \{1, \ldots, n\}$.

If $d\lambda_\alpha[\mathbf{r}_\alpha] \circ \mathbf{v}(\eta) = 0$ and $\mathscr{L}_\eta := \{yt^{-1} = \eta\}$ is not a discontinuity for $\mathbf{v}$, $\mathbf{v}'$, then $\mathscr{L}_\eta$ lies in the domain Ω_k of a constant function or of a β-c.s.w., $\beta \neq \alpha$.

Remark
The assumption that $\mathbf{f} \in \mathscr{C}^2$ is indispensable for the validity of the theorem, cf. Hanyga (1975, 1976).

Definition 3.1.26
Every α-shock or α-c.s.w. will be called an α-**wave.**

Definition 3.1.27
We say that an α-wave **lies to the left of** a β-wave (or, equivalently, that the β-wave **lies to the right of** the α-wave) if the respective domains of the waves

$$\Omega_\alpha = \{yt^{-1} \in [\zeta_-, \zeta_+]\}, \quad \Omega_\beta = \{yt^{-1} \in [\zeta_-', \zeta_+']\},$$

satisfy one of the following conditions:

(i) $\zeta_+ \leqslant \zeta_-'$ and either $\zeta_- < \zeta_+$ or $\zeta_-' < \zeta_+'$;

(ii) $\zeta_- = \zeta_+ < \zeta_-' = \zeta_+'$.

Since the vector $\mathbf{n}$ is fixed in the problem we shall not mention it explicitly.

Lemma 3.1.28 (cf. Hanyga, 1975, 1976)
Suppose that $\mathbf{f} \in \mathscr{C}^2$ and the system (3.1.35) is strictly hyperbolic. Let $\mathbf{u} = \mathbf{v}(yt^{-1})$ be a piecewise $\mathscr{C}^1$ solution to the Riemann problem (3.1.90) satisfying the condition (L).

Then for $1 \leqslant \alpha < \beta \leqslant n$ every α-wave lies to the left of every β-wave.

Proof
Let $1 \leqslant \alpha < \beta \leqslant n$. It is enough to prove the lemma for an α-wave and a β-wave which are separated merely by a constant solution $\mathbf{u}|\Omega \equiv \mathbf{u}_0 \in \boldsymbol{R}^n$. Indicating the fact that a quantity is evaluated at $\mathbf{u} = \mathbf{u}_0$ by the zero superscript, we note from the inequality $\lambda_\alpha^0 < \lambda_\beta^0$ that an α-c.s.w. always lies to the left of a β-c.s.w. Suppose now that a β-shock lies to the left of an α-shock. Let w_α, w_β be the corresponding normal propagation speeds. We have, by assumption, $w_\beta < w_\alpha$ and also $w_\beta \geqslant \lambda_\beta^0$, hence $w_\alpha > \lambda_\beta^0 \geqslant \lambda_{\alpha+1}^0$, which is impossible.

If a β-shock lies to the left of an α-c.s.w., then $w_\beta \leqslant \lambda_\alpha^0 < \lambda_\beta^0$, which is impossible. Suppose now that a β-c.s.w. lies to the left of an α-shock. We have then $\lambda_\beta \leqslant w_\alpha < \lambda_{\alpha+1} \leqslant \lambda_\beta$, hence a contradiction. □

Remark

Lemma 3.1.28 implies that (i) $\boldsymbol{R}_+^2 = \bigcup_{\alpha=1}^{n} \tilde{\Omega}_\alpha \cup \bigcup_{\alpha=1}^{n} \hat{\Omega}_\alpha$; (ii) each $\hat{\Omega}_\alpha$ is either a sector $\zeta_- < yt^{-1} < \zeta_+$ or a line $yt^{-1} = \zeta_0$, and each $\hat{\Omega}_\alpha$ is a sector; (iii) the order of the sets $\hat{\Omega}_\alpha, \tilde{\Omega}_\alpha$ from the left to the right (i.e. in the sense of increasing values of yt^{-1}) is $\hat{\Omega}_0, \tilde{\Omega}_1, \hat{\Omega}_1, \tilde{\Omega}_2, \ldots, \hat{\Omega}_n$; (iv) $\mathbf{u}|\hat{\Omega}_\alpha = \text{const} =: \mathbf{u}_\alpha$; (v) $\tilde{\Omega}_\alpha$ is the join of the domains of all the α-waves in the solution and of some constant solutions.*

Definition 3.1.29
The function $\mathbf{u}|\tilde{\Omega}_\alpha$ will be referred to as the α-**group** of $\mathbf{u}$.

We now impose another condition on the discontinuities of the solution $\mathbf{u}$. Let $\mathbf{u}_-$, $\mathbf{u}_+$, w be the left, right states of an α-shock and its speed at a point x. Let $s \mapsto (w(s), \mathbf{d}(s))$ be the parametric representation of $\mathscr{H}_{\mathbf{u}_-}$, $\mathbf{d}(0) = \mathbf{u}_-$. The condition reads:

(E) Every α-shock in the solution $\mathbf{u}$ satisfies the following conditions:
(E_1) $(w, \mathbf{u}_+) \in \mathscr{H}_{\mathbf{u}_-}^{\alpha}$,
(E_2) if $(w, \mathbf{u}_+) = (w(s_1), \mathbf{d}(s_1))$, then for every s in the interval $]0, s_1[$ (or $]s_1, 0[$) $w(s) \geqslant w(s_1)$.

Suppose that for a fixed $\mathbf{u}_{\alpha-1} \in \mathscr{H}$ the value of $\mathbf{u}_\alpha$ lies on a curve in $\mathscr{H}$:

$$\mathbf{u}_\alpha = \mathbf{z}_\alpha(s; \mathbf{u}_{\alpha-1}), \quad s \in \mathscr{I}_\alpha \subset \boldsymbol{R}. \tag{3.1.93}$$

The function $\mathbf{z}_\alpha$ will be constructed below for all the possible α-groups satisfying (L) and (E). Before doing it we shall note that the solutions of the Riemann problem subject to (L) and (E) satisfy the equation

$$\begin{aligned}\mathbf{u}_{\mathscr{R}} = \mathbf{u}_n &= \mathbf{z}_n(s_n; \mathbf{z}_{n-1}(s_{n-1}; \ldots \mathbf{z}_1(s_1; \mathbf{u}_L) \ldots) \\ &:\equiv \mathbf{z}(s_1, \ldots, s_n; \mathbf{u}_L).\end{aligned} \tag{3.1.94}$$

Equation (3.1.94) determines a unique solution $\mathbf{u}$ provided

(1) equation (3.1.94) is uniquely solvable for the unknowns $s_1, s_2, \ldots, s_n$;
(2) every pair $(\mathbf{u}_{\alpha-1}, s_\alpha) \in \mathscr{H} \times \mathscr{I}_\alpha$ determines a unique α-group $\mathbf{u}|\hat{\Omega}_\alpha$.
It can be seen from our construction of the function $\mathbf{z}_\alpha$ below that

$$\frac{d\mathbf{z}_\alpha}{ds}(0; \mathbf{u}_{\alpha-1}) = \mathbf{r}_\alpha(\mathbf{u}_{\alpha-1}). \tag{3.1.95}$$

Hence the first condition is expected to be satisfied if $\max\{|\mathbf{u}_\alpha - \bar{\mathbf{u}}| \,|\, \alpha = 0, \ldots, n\}$ is small enough for some $\bar{\mathbf{u}} \in \mathscr{H}$. This condition is likely to be satisfied provided

* The domain of an α-shock is by definition a half-line in $\boldsymbol{R}^2$.

$|\mathbf{u}_L - \mathbf{u}_R|$ is small enough. Condition (2) is also satisfied for sufficiently weak α-groups. For arbitrarily large oscillations $|\mathbf{u}_L - \mathbf{u}_R|$ there is no uniqueness even in the class of continuous solutions—see Borovikov (1969), Rozhdestvenskii and Yanenko (1978) (for 3 equations), Dyachenko (1963) (for a strictly hyperbolic system of 3 conservation laws). Worse than that, it may happen that every nbhd of a solution satisfying (L) and (E) contains another solution (Tupchiyev, 1972, 1973).

We now proceed to construct the curves $\mathbf{z}_\alpha(\cdot\,; \mathbf{u}_0)$, $\mathbf{u}_0 \in \mathscr{H}$, taking into account (L), (E) and using the results of sections 3.1.5 and 3.1.6. For the sake of saving space we perform only the initial steps of the construction. The reader can complete the construction by induction. It will be assumed throughout this section that all the sonic points are regular.

Let $s \mapsto \mathbf{e}(s)$ be the α-hodograph through $\mathbf{u}_0$, parametrized in such a way that $\mathbf{e}(0) = \mathbf{u}_0$, $\mathbf{e}'(s) = \mathbf{r}_\alpha \circ \mathbf{e}(s)$. Let $s \mapsto \mathbf{d}(s)$ be a regular parametrization of $\mathscr{H}_{\mathbf{u}_0}$ satisfying $\mathbf{d}(0) = \mathbf{u}_0$, $\mathbf{d}'(0) = \mathbf{r}_\alpha(\mathbf{u}_0)$. We use the notations $\hat{\varphi} = \varphi \circ \mathbf{e}$, $\overline{\varphi} = \varphi \circ \mathbf{d}$ for every function φ defined in a nbhd of $s \mapsto \mathbf{e}(s)$ or of $s \mapsto \mathbf{d}(s)$. We consider the following three cases successively:

(i) $\hat{\lambda}_\alpha$ is a decreasing function in a nbhd of $s = 0$;

(ii) $\hat{\lambda}_\alpha$ has a local minimum at $s = 0$;

(iii) $\hat{\lambda}_\alpha$ has a local maximum at $s = 0$.

The case of an increasing $\hat{\lambda}_\alpha$ is equivalent to (i).

The case (i) corresponds to the hypotheses of Theorem 3.1.10 holding for some even integer $k > 0$ and

$$\left(\mathbf{r}_\alpha(\mathbf{u}) \cdot \frac{\partial}{\partial \mathbf{u}}\right)^{k+1} \lambda_\alpha(\mathbf{u}_0) < 0.$$

In this case we shall set

$$\mathbf{z}_\alpha(s; \mathbf{u}_0) = \begin{cases} \mathbf{e}(s) & \text{for } s < 0, \\ \mathbf{d}(s) & \text{for } s > 0 \end{cases} \tag{3.1.96}$$

for s close to 0. Theorem 3.1.10 implies that $\overline{\lambda}_\alpha(s) < w(s) < \lambda_\alpha(\mathbf{u}_0)$, $\lambda_{\alpha-1}(\mathbf{u}_0) < w(s) < \overline{\lambda}_{\alpha+1}(s)$ for $s > 0$ sufficiently small. Consequently for small $s > 0$ the α-group corresponding to (3.1.96) consists of a single α-shock. Since $w(\cdot)$ is monotone in a nbhd of $s = 0$, the condition (E) is also satisfied.

In the case (ii) we set

$$\mathbf{z}_\alpha(s; \mathbf{u}_0) = \mathbf{e}(s) \quad \text{for sufficiently small } s. \tag{3.1.97}$$

In the case (iii) let

$$\mathbf{z}_\alpha(s; \mathbf{u}_0) = \mathbf{d}(s) \quad \text{for sufficiently small } s. \tag{3.1.98}$$

The weak α-shocks corresponding to equation (3.1.98) satisfy the conditions (E) and (L_α) in view of Theorem 3.1.10.

Equation (3.1.96$_2$) or (3.1.98) can be extended up to a point s_1, beyond which one of the conditions (E), (L) breaks down. This will happen if one of the following three functions:

$$(1)\ w-\lambda_{\alpha-1}(\mathbf{u}_0); \quad (2)\ w-\bar{\lambda}_{\alpha+1}; \quad (3)\ w-\bar{\lambda}_\alpha$$

changes sign at s_1. In the cases (1) and (2) the condition (L_α) breaks down and the construction stops.

In the case (3) $(w(s_1), \mathbf{d}(s_1))$ is a regular Chapman–Jouguet point and $w(s_1) = \bar{\lambda}_\alpha(s_1)$ is a local minimum of w. In this case both (L_α) and (E) break down for the α-shocks with $s > s_1 > 0$ (or possibly $s < s_1 < 0$ in the case (iii)). For simplicity we assume that $s_1 > 0$. Let $s \mapsto \mathbf{e}_1(s)$ be the α-hodograph satisfying $\mathbf{e}_1'(s) = \mathbf{r}_\alpha \circ \mathbf{e}_1(s)$, $\mathbf{e}_1(s_1) = \mathbf{d}(s_1)$, and suppose that the parametrization of $s \mapsto \mathbf{d}(s)$ near s_1 has been chosen according to Theorem 3.1.17. It follows from the closing remark of the preceding section that $d\lambda_\alpha[\mathbf{r}_\alpha] \circ \mathbf{d}$ has changed sign k times on $]0, s_1[$ and $s \mapsto \langle \bar{\mathbf{l}}^\alpha(s), \mathbf{d}(s) - \mathbf{u}_0 \rangle$ has changed sign l times on $]0, s_1[$, with $k+l$ odd.

There are two possibilities,

(a) k is odd, l is even,

(b) k is even, l is odd.

In the case (a) $d\lambda_\alpha[\mathbf{r}_\alpha] \circ \mathbf{e}_1(s) > 0$ for $s \neq s_1$ close to s_1 and $\iota = 1$ (Theorems 3.1.16 and 3.1.17). We set

$$\mathbf{z}_\alpha(s; \mathbf{u}_0) = \mathbf{e}_1(s) \quad \text{for } s > s_1 \text{ sufficiently small.} \tag{3.1.99}$$

In the case (b) $d\lambda_\alpha[\mathbf{r}_\alpha] \circ \mathbf{e}_1(s) < 0$ for $s \neq s_1$ close to s_1, but $\iota = -1$ in equation (3.1.80). We set

$$\mathbf{z}_\alpha(s; \mathbf{u}_0) = \mathbf{e}_1(2s_1 - s) \quad \text{for } s > s_1 \text{ small enough.} \tag{3.1.100}$$

In the cases corresponding to equations (3.1.99) and (3.1.100) the α-group consists of an α-shock satisfying $w = \lambda_\alpha(\mathbf{u}_+)$ followed on the right by an α-c.s.w. The condition (E) is obviously satisfied. The curve $\mathbf{z}_\alpha$ is at least $\mathscr{C}^1$ in a nbhd of s_1 (Theorem 3.1.17).

We now consider the case where an α-hodograph $\mathscr{J}: s \mapsto \mathbf{e}(s)$ appearing as a part of the curve $\mathbf{z}_\alpha(\cdot\,; \mathbf{u}_0)$ attains a hypersurface Σ_α at which $d\lambda_\alpha[\mathbf{r}_\alpha]$ changes sign. We shall assume that

$$\left(\mathbf{r}_\alpha(\mathbf{u}) \cdot \frac{\partial}{\partial \mathbf{u}}\right)^l \lambda_\alpha = 0 \quad \text{for } l = 1, \ldots, k-1 \tag{3.1.101}$$

and

$$\left(\mathbf{r}_\alpha(\mathbf{u}) \cdot \frac{\partial}{\partial \mathbf{u}}\right)^k \lambda_\alpha \neq 0 \tag{3.1.102}$$

on $\Sigma_\alpha \cap \mathscr{J}$, for some even integer $k \geqslant 2$. Equation (3.1.102) ensures transversality of $\mathscr{J}$ and Σ_α.

In the following we assume that $k = 2$ for simplicity. In this case equations (3.1.101) and (3.1.102) ensure that Σ_α is a regular hypersurface in a nbhd of $\Sigma_\alpha \cap \mathscr{J}$. To fix the ideas we assume that $s < 0$, $\hat{\lambda}_\alpha := \lambda_\alpha \circ \mathbf{e}$ is a decreasing function on $]s_2, 0[$, $s_2 < 0$, and $\mathbf{e}(s_2) \in \Sigma_\alpha$. Our task consists of extending $\mathbf{z}_\alpha(\cdot\,; \mathbf{u}_0)$ to $s < s_2$ in such a way that $s \mapsto \mathbf{z}_\alpha(s; \mathbf{u}_0)$ crosses Σ_α at $s = s_2$. It is not possible to achieve this by a further increase of the strength of the α-c.s.w. on the extreme right of the α-group. The only possibility is to find α-shocks satisfying (E) and connecting some points $\mathbf{e}(\sigma)$, $\sigma \in]s_2, 0[$ to some points on the other side of Σ_α. Obviously the speed of such a shock $w = \hat{\lambda}_\alpha(\sigma)$ (in view of (L_α)) and its right state $\mathbf{z}_\alpha$ lies on some $\tilde{\mathscr{H}}^\alpha_{\mathbf{e}(\sigma)}$. Equation (3.1.102) and Theorem 3.1.10 imply that at least for $\mathbf{e}(\sigma)$ close enough to $\mathbf{e}(s_2)$ the curve $\tilde{\mathscr{H}}_{\mathbf{e}(\sigma)}$ is transversal to Σ_α. On the arc $\tau \mapsto \big(w(\tau, \sigma), \mathbf{d}(\tau, \sigma)\big)$ of $\tilde{\mathscr{H}}^\alpha_{\mathbf{e}(\sigma)}$ extending from $\mathbf{e}(\sigma)$ towards Σ_α we have the inequality $\lambda_\alpha \circ \mathbf{d}(\tau) > w(\tau, \sigma) > \hat{\lambda}_\alpha(\sigma)$. Between the point $\big(\hat{\lambda}_\alpha(\sigma), \mathbf{e}(\sigma)\big)$ and the point $\big(\hat{\lambda}_\alpha(\sigma), \mathbf{z}\big) \in \mathscr{H}^\alpha_{\mathbf{e}(\sigma)}$ the function $\tau \mapsto w(\tau, \sigma)$ attains a maximum $w\big(\tau_1(\sigma), \sigma\big)$ and in view of Theorems 3.1.17, 3.1.22 we can expect that $w - \lambda_\alpha \circ \mathbf{d}$ changes sign at $\tau = \tau_1(\sigma)$. Consequently, at $\mathbf{z}$, $w = \hat{\lambda}_\alpha(\sigma) > \lambda_\alpha(\mathbf{z})$. Assume that the arc of $\tilde{\mathscr{H}}^\alpha_{\mathbf{e}(\sigma)}$ between $\mathbf{e}(\sigma)$ and $\mathbf{z}$ does not intersect a hypersurface $\Sigma_{\alpha+1}$, at which $d\lambda_{\alpha+1}[\mathbf{r}_{\alpha+1}]$ changes sign. Then $w < \lambda_{\alpha+1}(\mathbf{z})$. Consequently the α-shock connecting the left state $\mathbf{e}(\sigma)$ with the right state $\mathbf{z}$ satisfies the conditions (L_α) and (E). Clearly, if the inequality $w < \lambda_{\alpha+1}(\mathbf{z})$ were violated the construction of $\mathbf{z}_\alpha$ would stop.

Let

$$\varphi(\tau, \mu) :\equiv w(\tau, s_2+\mu) - \hat{\lambda}_\alpha(s_2+\mu). \tag{3.1.103}$$

We assume that the parametrization of $\mathscr{H}^\alpha_{\mathbf{e}(\sigma)}$ is regular and $\mathbf{f} \in \mathscr{C}^3$. Considering $w(\cdot\,,\,\cdot)$ as a function on $\tilde{\mathscr{H}}_0$ it is easy to see that $w \in \mathscr{C}^2$ on $[0, \delta[\times$ $\times [0, \varepsilon[$ for some $\delta > 0$, $\varepsilon > 0$, i.e. $\varphi \in \mathscr{C}^2$ on a nbhd of this set. Since $\varphi(0, \sigma) \equiv 0$,

$$\varphi(\tau, \mu) \equiv \tau\psi(\tau, \mu), \quad \psi \in \mathscr{C}^1. \tag{3.1.104}$$

Moreover $\varphi_{,\tau}(0, \mu) \to 0$, $\varphi_{,\tau\tau}(0, \mu) \to \frac{1}{3}\hat{\lambda}''_\alpha(s_3) \neq 0$, $\varphi_{,\mu\tau}(0, \mu) = \frac{1}{2}\hat{\lambda}''_\alpha(s_3+\mu) \to \frac{1}{2}\hat{\lambda}''_\alpha(s_3) \neq 0$ as $\mu \to 0$. Hence $\psi(0, 0) = 0$, $\psi_{,\tau}(0, 0) \neq 0$, $\psi_{,\mu}(0, 0) \neq 0$.

We must find a $\tau \neq 0$ and μ such that $\varphi(\tau, \mu) = 0$, or equivalently,

$\psi(\tau, \mu) = 0$. The latter equation has a unique solution $\tau = \zeta(\mu)$, $\zeta \in \mathscr{C}^1$, in a nbhd of $(0, 0)$, satisfying $\zeta(0) = 0$. The function ζ is defined for $\mu \geqslant 0$ small enough. Since $\zeta'(0) = -\frac{1}{3}$, we also have $\zeta'(\mu) < 0$ for μ small enough. We set

$$\mathbf{z}_\alpha(s; \mathbf{u}_0) := \mathbf{d}(\zeta(\mu), s_2+\mu) \quad \text{for } s = s_2 - \frac{2}{3}\mu < s_2 \tag{3.1.105}$$

and for μ small enough. Since $\mathbf{d}(0, \sigma) \equiv \mathbf{e}(\sigma)$,

$$\frac{d\mathbf{z}_\alpha}{ds}(s_2; \mathbf{u}_0) = \frac{1}{3}\mathbf{d}_{,\tau}(0, s_2) - \mathbf{d}_{,\sigma}(0, s_2) = -\mathbf{r}_\alpha(\mathbf{e}(s_2)) \tag{3.1.106}$$

and the arc (3.1.105) actually extends the arc $\mathbf{z}_\alpha(s; \mathbf{u}_0) \equiv \mathbf{e}(s)$, $s > s_2$, $\mathscr{C}^1$-smoothly.

The solution $\tau = \zeta(\mu) = \tilde{\zeta}(s)$ of $\varphi = 0$ can be continued until the inequality $\frac{\partial w}{\partial \tau}(\tau, \sigma) \neq 0$ breaks down. The construction of the arc (3.1.105) can be followed through indefinitely unless there is a point $s = s_3 < s_2$ such that for $\tau = \tau_3 = \zeta(\frac{3}{2}(s_2 - s_3))$, $\sigma = \sigma_3 = s_2 + \frac{3}{2}(s_2 - s_3)$ one of the following is true:

(A) $\frac{\partial w}{\partial \tau}(\tau, \sigma) = 0$;

(B) $\mathbf{d}(0, \sigma) = \mathbf{u}_0$.

Let s_3 be the maximal value of $s < s_2$ enjoying one of these properties.

We note that the condition (L_α) can break down only if one of the functions $\tau \mapsto w(\tau, \sigma_3) - \lambda_\alpha \circ \mathbf{d}(\tau, \sigma_3)$ or $\tau \mapsto w(\tau, \sigma_3) - \lambda_{\alpha+1} \circ \mathbf{d}(\tau, \sigma_3)$ changes sign at $\tau = \tau_3$. In either case the condition (A) is satisfied and $\tau \mapsto w(\tau, \sigma_3)$ attains a local minimum. The condition (E) can be violated for $s < s_3$ only if $\tau \mapsto w(\tau, \sigma_3)$ attains a local minimum at $\tau = \tau_3$. It is therefore sufficient to consider the following cases:

(A_1) $(w(\tau_3, \sigma_3), \mathbf{d}(\tau_3, \sigma_3)) = (\lambda_\alpha \circ \mathbf{d}(\tau_3, \sigma_3), \mathbf{d}(\tau_3, \sigma_3))$ is a Chapman–Jouguet point on $\mathscr{H}_{\mathbf{e}(\sigma_3)}$;

(A_2) $(w(\tau_3, \sigma_3), \mathbf{d}(\tau_3, \sigma_3)) = (\lambda_\alpha \circ \mathbf{d}(\tau_3, \sigma_3), \mathbf{d}(\tau_3, \sigma_3))$ is not a Chapman–Jouguet point on $\mathscr{H}_{\mathbf{e}(\sigma_3)}$;

(A_3) $(w(\tau_3, \sigma_3), \mathbf{d}(\tau_3, \sigma_3)) = (\lambda_{\alpha+1} \circ \mathbf{d}(\tau_3, \sigma_3), \mathbf{d}(\tau_3, \sigma_3))$;

and the case (B).

In the cases (A_2) and (A_3) the construction stops. In the case (A_1) we can continue the curve $s \mapsto \mathbf{z}_\alpha(s; \mathbf{u}_0)$ by the α-hodograph $s \mapsto \mathbf{e}_2(s)$ such that $\mathbf{e}_2(s_3) = \mathbf{z}_\alpha(s_3; \mathbf{u}_0)$. The corresponding α-groups involve two α-c.s. waves separated by a shock with $w = \lambda_\alpha(\mathbf{u}_+) = \lambda_\alpha(\mathbf{u}_-)$.

In the case (B) the following assertions are true:

(I) $(w(\tau_3, \sigma_3), \mathbf{d}(\tau_3, \sigma_3)) = (\lambda_\alpha(\mathbf{u}_0), \mathbf{z}_\alpha(s_3; \mathbf{u}_0)) \in \mathscr{H}_{\mathbf{u}_0}$ and the α-group reduces for $s = s_3$ to a single α-shock;

(II) its propagation speed $w = \lambda_\alpha(\mathbf{u}_0) \geqslant \lambda_\alpha(\mathbf{z}_\alpha(s_3; \mathbf{u}_0)), w \leqslant \lambda_{\alpha+1}(\mathbf{z}_\alpha(s_3; \mathbf{u}_0))$;

(III) if $\mathbf{z}_\alpha(s_3; \mathbf{u}_0)$ lies on a (connected) arc $\mathscr{J}$ of $\tilde{\mathscr{H}}_{\mathbf{u}_0}$ containing $\mathbf{u}_0$, then $d\lambda_\alpha[\mathbf{r}_\alpha]$ changes sign an odd number of times on $\mathscr{J}$.

In order to reach more satisfactory conclusions in the case (B) it is necessary to make some additional assumptions. Suppose that

(I′) $(\lambda_\alpha(\mathbf{u}_0), \mathbf{z}_\alpha(s_3; \mathbf{u}_0)) \in \mathscr{H}^\alpha_{\mathbf{u}_0}$;

(III′) $d\lambda_\alpha[\mathbf{r}_\alpha]$ changes sign only once on $\mathscr{J}$, viz. at $\mathscr{J} \cap \Sigma_\alpha$.

Let $s \mapsto (w(s), \mathbf{d}(s))$ be a regular parametrization of $\mathscr{J}$. We can adjust the parametrization of $\mathscr{J}$ in such a way that $\mathbf{z}_\alpha(s_3; \mathbf{u}_0) = \mathbf{d}(s_3)$. In view of Theorems 3.1.10 and 3.1.21 $s \mapsto w(s)$ has exactly one extremum on $]s_3, 0[$, say, at s_4, and it is a maximum. In view of (III′) $w(s) - \bar{\lambda}_\alpha(s)$ changes sign at s_4. This fact, together with (II) implies that the shocks with $\mathbf{u}_+ = \mathbf{u}_0$, $\mathbf{u}_- = \mathbf{d}(s)$, $s < s_3$, satisfy the conditions (L_α) and (E). We set $\mathbf{z}_\alpha(s; \mathbf{u}_0) = \mathbf{d}(s)$ for $s < s_3$ in the case (B). It is to be noted that the condition (L_α) fails to be satisfied on the arc $\mathbf{d}(]s_3, 0[)$ since $\bar{\lambda}_\alpha(s) > w(s) > \lambda_\alpha(\mathbf{u}_0)$ (with $\mathbf{u}_- = \mathbf{u}_0$, $\mathbf{u}_+ = \mathbf{d}(s)$) for $s > s_4$ and $w(s) > \lambda_\alpha(\mathbf{u}_0), \bar{\lambda}_\alpha(s)$ for $s < s_4$.

The same construction will be applied when the arc $s \mapsto \mathbf{e}_1(s)$ defined above intersects some Σ_α, *mutatis mutandis.* The corresponding α-group consists of two α-shocks with an α-c.s.w. between them. In the case (A_1) an additional α-c.s.w. appears on the right of the α-group. In the case (B): $\mathbf{z}_\alpha(s_3; \mathbf{u}_0) = \mathbf{d}(s_1)$, the α-c.s.w. shrinks to a half-line and the two adjacent α-shocks coincide: $w(s_1) = w(\tau_3, \sigma_3)$. It follows immediately that the right state of the right shock and the left state of the left shock satisfy the Rankine–Hugoniot equations with $w = w(s_1)$. Hence $\mathbf{z}_\alpha(s_3; \mathbf{u}_0) \in \tilde{\mathscr{H}}_{\mathbf{u}_0}$ again and $w(\cdot)$ has a maximum on $]s_1, s_3[$. Assuming that $\mathbf{z}_\alpha(s_3; \mathbf{u}_0) \in \tilde{\mathscr{H}}^\alpha_{\mathbf{u}_0}$ and $d\lambda_\alpha[\mathbf{r}_\alpha] \circ \mathbf{d}$ changes sign only once on $]s_1, s_3[$ we conclude that $\mathbf{z}_\alpha(s; \mathbf{u}_0) = \mathbf{d}(s)$ for $s < s_3$ and the corresponding α-group consists of a single α-shock satisfying (L_α) and (E). The arc of $\mathscr{H}^\alpha_{\mathbf{u}_0}$ between s_1 and s_3 does not satisfy (E), since $w(s) > w(s_1) = w(s_3)$ for $s \in]s_1, s_3[$.

The cases we have discussed are sufficient to complete the construction of the one-parameter family $\mathbf{z}_\alpha(s; \mathbf{u}_0)$.

Remark

In the case of α-waves satisfying (GL_α) the curve $\mathbf{u} = \mathbf{z}_\alpha(s; \mathbf{u}_{\alpha-1})$ consists of an arc of $\tilde{\mathscr{H}}^\alpha_{\mathbf{u}}\mathbf{u}_{\alpha-1}$ and of an arc of the α-hodograph through $\mathbf{u}_{\alpha-1}$. For $|\mathbf{u}_\alpha - \mathbf{u}_{\alpha-1}|$ small enough both arcs are one-dimensional submanifolds of $\mathscr{H}$ with a regular parametrization (recall equations (3.1.48)). Hence for small

$|\mathbf{u}_\alpha - \mathbf{u}_{\alpha-1}|$ the value of s is also small.

A similar situation is encountered when the α-waves are linearly degenerate, since $\mathbf{u} = \mathbf{z}_\alpha(s; \mathbf{u}_{\alpha-1})$ obviously coincides with the α-hodograph through $\mathbf{u}_{\alpha-1}$.

Suppose that the α-waves are either linearly degenerate or genuinely non-linear, for $\alpha \in \{1, \dots, n\}$. It follows then from the above remarks and from equation (3.1.95) that forma small x $\{|\mathbf{u}_\alpha - \bar{\mathbf{u}}|\| \; \alpha = 0, 1, \dots, n\}$, $\bar{\mathbf{u}} \in \mathscr{H}$, equation (3.1.94) admits a unique solution $s_1, \dots, s_n$. Hence the Riemann problem has a unique solution with small oscillation.

3.1.8 Existence and Uniqueness in the Small for Smooth Motions of an Unbounded Medium

Suppose that the elastic body $\mathscr{B}$ fills the whole space in its reference configuration $\varkappa$, $\varkappa(\mathscr{B}) = \boldsymbol{R}^3$. Suppose also that the initial conditions $\boldsymbol{\chi}(0, \xi) = \boldsymbol{\chi}_o(\xi)$, $\boldsymbol{\chi}_{,t}(0, \xi) = \mathbf{v}_o(\xi)$, $\xi \in \boldsymbol{R}^3$, are continuous. We expect that the motion $\boldsymbol{\chi}(t, \cdot)$ remains continuous up to the occurrence of the first gradient catastrophe. In fact it is also true that the motion in a sufficiently small period of time is determined uniquely by the initial data and depends continuously on them. We shall quote a theorem of Hughes, Kato and Marsden (1976) to this effect, complementing the information of Section 3.1.2 in this respect.

For each time t in a sufficiently small interval $[-\varepsilon, \varepsilon]$ we assume that $\boldsymbol{\chi}(t, \cdot)$ belongs to an appropriate Sobolev space, and in fact

$$\boldsymbol{\chi} \in \mathscr{C}^0([-\varepsilon, \varepsilon], W) \cap \mathscr{C}^1([-\varepsilon, \varepsilon], \mathscr{X}), \tag{3.1.107}$$

where $\mathscr{Y} \subset \mathscr{X}$ continuously and densely and W is open in $\mathscr{Y}$. Equation (3.1.107) says that $t \mapsto \boldsymbol{\chi}(t, \cdot)$ is continuous from $[-\varepsilon, \varepsilon]$ to W while its time derivative $\boldsymbol{\chi}_{,t}$ is continuous from $[-\varepsilon, \varepsilon]$ to $\mathscr{X}$. Differentiation with respect to time lowers the differentiability of $(t, \xi) \mapsto \boldsymbol{\chi}(t, \xi)$ by one, hence we expect that $\mathscr{X} = H^s(\boldsymbol{R}^3)$, $\mathscr{Y} = H^{s+1}(\boldsymbol{R}^3)$, $H^j := W^{j,2}$.

For technical reasons we assume that $\boldsymbol{\chi}(t, \xi) = \boldsymbol{\chi}_0(\xi) + \mathbf{u}(t, \xi)$ and consider a slightly generalized form of the equations of motion:

$$\varrho_o u^k{}_{,tt} = A_k{}^\alpha{}_l{}^\beta(t, \xi, \mathbf{u}, \nabla\mathbf{u}) u^l{}_{,\xi^\alpha \xi^\beta} + f_k(t, \xi, \mathbf{u}, \nabla\mathbf{u}) \tag{3.1.108}$$

For generality we also assume that $\varkappa(\mathscr{B}) = \boldsymbol{R}^m$, $m = 1, 2, 3, \dots$ and that $\mathbf{u}$ takes its values in $\boldsymbol{R}^p$, $p = 1, 2, 3, \dots$ Since we do not expect the displacement gradients to grow too large we need not impose the inequality $\det\nabla\boldsymbol{\chi} > 0$. Instead we assume that $\nabla\mathbf{u}(\xi, t)$ is allowed to range over a neighbourhood $\mathscr{U}$ of $\mathbf{0} \in \boldsymbol{R}^{mp}$. We assume that $\mathscr{U}$ is contractible to 0. Let $\Omega = \boldsymbol{R}^p \times \mathscr{U}$.

Theorem 3.1.30

Suppose that $s > m/2+1$ ($s > m/2$ if the equations are linear), and, for some $T > 0$,

(i) $A_k{}^\alpha{}_l{}^\beta \in \mathrm{Lip}([-T, T], \mathscr{C}_b^{s+1}(\boldsymbol{R}^m \times \Omega)), f_k \in \mathscr{C}^0([-T, T], \mathscr{C}_b^{s+1}(\boldsymbol{R}^m \times \Omega))$,

(ii) $A_k{}^\alpha{}_l{}^\beta = A_l{}^\beta{}_k{}^\alpha, k, l \leqslant p, \alpha, \beta \leqslant m$,

(iii) $\varrho_o(\xi) \geqslant m = \text{const} > 0$,

(iv) $A_k{}^\alpha{}_l{}^\beta a^k a^l n_\alpha n_\beta \geqslant M|\mathbf{a}|^2|\mathbf{n}|^2, M > 0$,

(v) $\mathbf{u}_o \in H^{s+1}(\boldsymbol{R}^m, \boldsymbol{R}^p)$, $\mathbf{v}_o \in \mathscr{H}^s(\boldsymbol{R}^m, \boldsymbol{R}^p)$, $\nabla\mathbf{u}_o(\xi) \in \mathscr{U}$ for all $\xi \in \boldsymbol{R}^m$.

Then there is a nbhd $\mathscr{V}$ of $(\mathbf{u}_o, \mathbf{v}_o)$ in $\mathscr{H}^{s+1} \times \mathscr{H}^s$ and a positive number $\varepsilon > 0$, $\varepsilon < T$, such that for any $(\mathbf{u}_o', \mathbf{v}_o') \in \mathscr{V}$ equation (3.1.108) has a unique solution $t \mapsto \mathbf{u}(t, \cdot)$, $-\varepsilon \leqslant t \leqslant \mathbf{t}$, satisfying the initial conditions $\mathbf{u}(0, \cdot) = \mathbf{u}_o'$, $\mathbf{u}_{,t}(0, \cdot) = \mathbf{v}_o'$ and such that $\nabla\mathbf{u}(t, \xi) \in \mathscr{U}$ for $-\varepsilon \leqslant t \leqslant \varepsilon, \xi \in \boldsymbol{R}^m$, $(\mathbf{u}(t, \cdot), \mathbf{u}_{,t}(t, \cdot)) \in \mathscr{H}^{s+1} \times \mathscr{H}^s$ for $-\varepsilon \leqslant t \leqslant \varepsilon$.

Moreover

$$\mathbf{u} \in \mathscr{C}^r([-\varepsilon, \varepsilon], \mathscr{H}^{s+1-r}) \quad \text{for all } r \text{ such that } 0 \leqslant r \leqslant s,$$

and the mapping

$$\mathscr{V} \ni (\mathbf{u}_o', \mathbf{v}_o') \mapsto (\mathbf{u}(t, \cdot), \mathbf{u}_{,t}(t, \cdot)) \in \mathscr{H}^{s+1} \times \mathscr{H}^s$$

is continuous uniformly with respect to $t \in [-\varepsilon, \varepsilon]$.

Remarks

Uniform continuity of $A_k{}^\alpha{}_l{}^\beta$ (hypothesis (i)) ensures the validity of the Gårding lemma in the unbounded domain $\boldsymbol{R}^m$. For simplicity assume that $A_k{}^\alpha{}_l{}^\beta$ does not depend on t.

For the proof, equation (3.1.108) is interpreted as a dynamical system

$$x(t) := (\mathbf{u}(t, \cdot), \mathbf{u}_{,t}(t, \cdot)) \in \mathscr{H}^{s+1} \times \mathscr{H}^s,$$

$$\frac{dx}{dt} = A(x)x + f(t, x). \tag{3.1.109}$$

$A(x_o)x$ is obtained by substituting $(\xi, \mathbf{u}_o, \nabla\mathbf{u}_o)$ into the coefficients $A_k{}^\alpha{}_l{}^\beta$. $A(x_o)$ is a linear operator generating a $\mathscr{C}_o$-semigroup for each fixed x_o. Using the semigroup theory one solves

$$\frac{dx}{dt} = A(x_o)x + f(x, t) \tag{3.1.110}$$

for fixed initial data. The solution depends on x_o:

$$x = F(x_o)$$

The solution of (3.1.109) is then obtained as a fixed point of the nonlinear mapping F. The detailed proof is, however, quite complicated due to certain continuity requirements which can be met by a judicious choice of subsidiary spaces.

It follows from Theorem 1.1.45 that

$$\mathbf{u}(t, \cdot) \in \mathscr{C}_b^2, \quad \mathbf{u}_{,t}(t, \cdot) \in \mathscr{C}_b^1 \quad \text{if } s > m/2+1,$$

$$\mathbf{u}(t, \cdot) \in \mathscr{C}_b^1, \quad \mathbf{u}_{,t}(t, \cdot) \in \mathscr{C}_b^0 \quad \text{if } s > m/2,$$

i.e. the solutions are smooth.

Note that the only constitutive assumptions in Theorem 3.1.30 are the symmetry (ii) as well as SE (iv).

3.2 GLOBAL DISCONTINUOUS SOLUTIONS OF HYPERBOLIC SYSTEMS

3.2.1 Development of Discontinuities in Hyperbolic Systems

In Section 3.1 we introduced discontinuous solutions to the equations (3.1.25) in integral form without any justification. In this section we shall show the relevance of such solutions for the dynamics of continuous media governed by hyperbolic equations of motion.

For simplicity we consider the Cauchy problem for a single equation

$$u_t + c(u)u_x = 0, \tag{3.2.1}$$

$$u(x, 0) = \phi(x). \tag{3.2.2}$$

The problem (3.2.1–2) provides a convenient model of general nonlinear hyperbolic phenomena since it can be solved nearly explicitly in various classes of functions and the behavior of the singularities of the solutions can be closely examined. First we consider $\mathscr{C}^1$ solutions to the problem (3.2.1–2) with $\phi \in \mathscr{C}^1$.

For a $\mathscr{C}^1$ solution u of equation (3.2.1) satisfying the inequality $u_x \neq 0$ we consider the curves $u(t, x) = \text{const}$. Their slopes are given by the equation $\dfrac{dx}{dt} = -\dfrac{u_t}{u_x} = c(u)$ and therefore each curve $u(t, x) = u_0$ is a straight line

$$x - c(u_0)t = x_0. \tag{3.2.3}$$

Hence a $\mathscr{C}^1$ solution to the problem (3.2.1–2) satisfies the equation

$$\phi\big(x - c(u)t\big) = u. \tag{3.2.4}$$

Equation (3.2.4) can be solved in a nbhd of $t = 0$ and the solution can be continued until

$$-t\phi'\big(x - c(u)t\big)c'(u) = 1. \tag{3.2.5}$$

Substituting $u_0 = \phi(x_0)$ in equation (3.2.3) we obtain a one-parameter family of straight-line characteristics of the solution u. The envelope $\mathscr{E}$: $x = y(s)$, $t = \tau(s)$ of this family satisfies equations (3.2.4) and (3.2.5).

In order to see what happens beyond the envelope $\mathscr{E}$ we consider some instructive examples. For $c(u) \equiv u$ and unbounded initial data $\phi(x) \equiv -x$ equation (3.2.4) can be solved explicitly:

$$u = \frac{x}{t-1} \quad \text{for } t \neq 1. \tag{3.2.6}$$

All the characteristics $u = \text{const}$ intersect at a single point $t = 1$, $x = 0$, and the singularity $t = 1$ is in some sense a characteristic carrying infinite values of u from $x = \pm\infty$.

Consider the same equation with bounded initial data

$$\phi(x) \equiv \begin{cases} a & \text{for } x \geqslant a, \\ -x & \text{for } |x| \leqslant a, \\ -a & \text{for } x \leqslant -a, \end{cases}$$

$a > 0$. The solution is

$$u(t,x) \equiv \begin{cases} \dfrac{x}{t-1} & \text{for } x \in [-a(1-t), a(1-t)], t < 1, \\ -a & \text{for } x \leqslant -a|1-t|, \\ a & \text{for } x \geqslant a|1-t|. \end{cases} \tag{3.2.7}$$

Each point of the sector $\mathscr{S}$: $t > 1$, $x \in [a(1-t), -a(1-t)]$, lies on three characteristics carrying three different values of u, viz. $u = a$, $u = -a$, $u = \dfrac{x}{t-1}$. The solution u has three sheets over $\mathscr{S}$ and is continuous except at $t = 1$, $x = 0$,* as shown in Fig. 3.1.

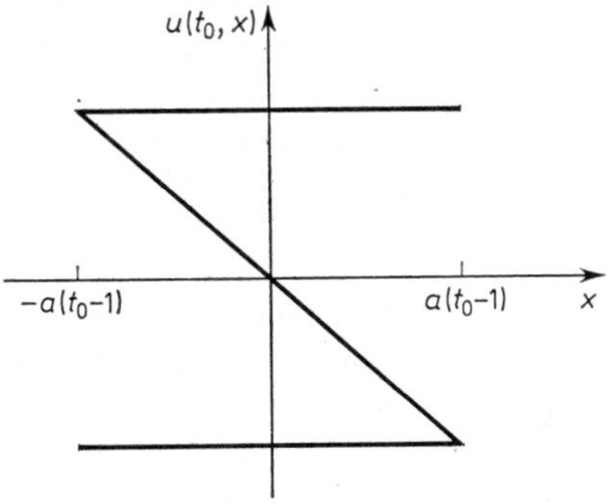

Fig. 3.1 A profile of the solution (3.2.7)

* The solution $u \in \mathscr{C}^1$ because the initial data $\phi \in \text{Lip}$ has discontinuous derivatives.

Consider finally the Cauchy problem (3.2.1–2) with $c(u) \equiv u$ and the initial data with compact support having the form of a single hump:

$$\phi(x) \equiv \begin{cases} 0 & \text{for } |x| \geqslant a, \\ a^2 - x^2 & \text{for } |x| \leqslant a, \end{cases} \tag{3.2.8}$$

$a > 0$. Equation (3.2.4) yields immediately the following explicit solution defined for $t < (2a)^{-1}$:

$$u(t, x) \equiv \begin{cases} 2t^{-2}\{2xt - 1 + \sqrt{\Delta}\} & \text{for } |x| \leqslant a, \\ 0 & \text{for } |x| \geqslant a \end{cases} \tag{3.2.9}$$

with $\Delta :\equiv (1-2xt)^2 + 4(a^2 - x^2)t^2$. The characteristics $x - (a^2 - x_0^2)t = x_0$, $x_0 \in [0, a]$, converge with increasing time and eventually form an envelope $\mathscr{E}$: $\Delta = 0$, $t \geqslant (2a)^{-1}$. The envelope $\mathscr{E}$ is the projection on the (t, x)-plane of the fold of the two-sheeted solution

$$u_\pm = 2t^{-2}\{2xt \pm \sqrt{\Delta}\} \quad \text{for } t > (2a)^{-1},\ \Delta > 0. \tag{3.2.10}$$

The lower sheet u_- attains the value 0 at $x = a$. Thus for $t > (2a)^{-1}$, $\Delta > 0$ and $x > a$ the continuous solution u assumes three different values 0, u_+, u_- (Fig. 3.2). The discontinuity of u_t, u_x on $\mathscr{E}$ is generated by the nonlinearity

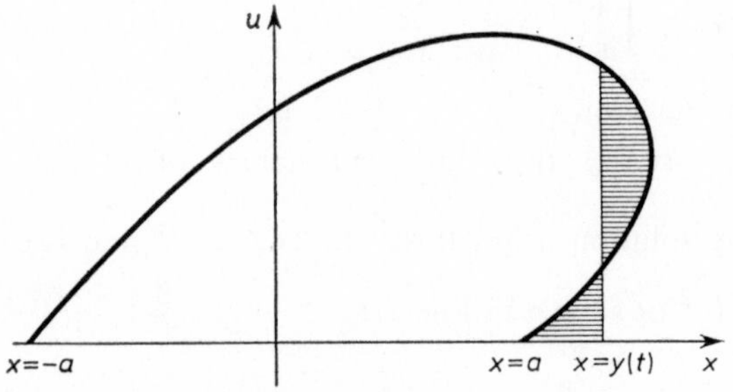

Fig. 3.2 The profile of the solution (3.2.9–10)

of equation (3.2.1), whereas the discontinuity of u_t, u_x on the half-lines $x = at$ results from the propagation of the initial discontinuity of ϕ' at $x = a$.

The profiles $x \mapsto u(t, x)$ for varying t exhibit the phenomenon of a breaking wave. Physically $u(t, x)$ can be interpreted as the velocity of a stream of free particles at the space-time point (t, x). According to the equation $u_t + uu_x = 0$ each particle moves with its initial velocity. The initially faster particles starting from $x \approx 0$ eventually overtake the slower ones starting from $x > 0$. At $t > (2a)^{-1}$ there are points (t, x), $x > a$, at which there are simultaneously faster and slower particles, represented by the non-zero sheets u_+, u_- of u. The third sheet can represent particles at rest at $x > a$.

The above phenomenon can be expected in a beam of independently moving particles. On the other hand the basic notion of a material continuum rules out any multivalued solutions $u(t, x)$ (Section 1.2). In the latter case, however nothing compels us to satisfy the differential equation (3.2.1) in the sense of the classical differential calculus. We have rather to satisfy an integral conservation law of momentum for every bounded Borel subset of the plane. In a "fluid" of free particles which does not sustain pressure the equation $u_t+uu_x = 0$ will be replaced by its integral form:*

$$\int_{x_0}^{x_1} dxu\Big|_{t_0}^{t_1} + \int_{t_0}^{t_1} dt\frac{1}{2}u^2\Big|_{x_0}^{x_1} = 0 \;\forall x_0, x_1 \in \boldsymbol{R},\; \forall t_0, t_1 \in \boldsymbol{R}_+. \tag{3.2.11}$$

We have tacitly assumed that the fluid density $\varrho = 1$.

Let us examine a piecewise $\mathscr{C}^1$ discontinuous solution of (3.2.11) defined in accordance with the recipe of Section 3.1.4. We eliminate the lower sheet u_- and connect the upper sheet u_+ of the solution (3.2.10) with $u = 0$ by a discontinuity S: $x = y(t)$, lying in the region $\Delta > 0$, $x > a$, $t > (2a)^{-1}$, above the

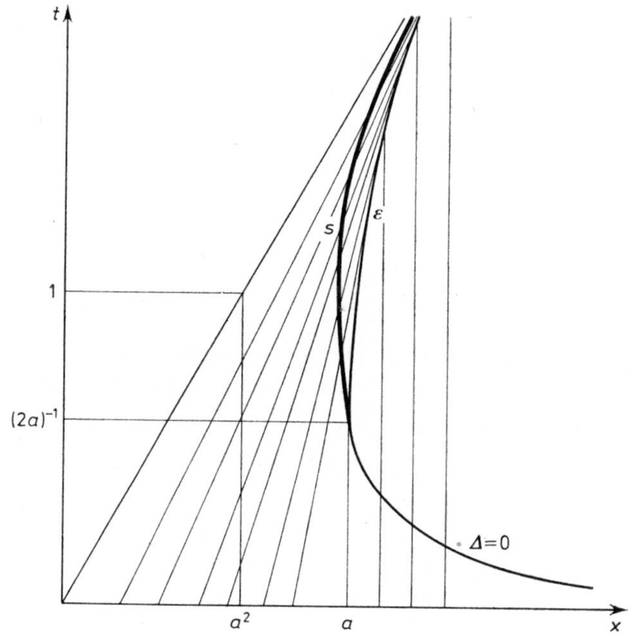

Fig. 3.3 Shock formation in the solution (3.2.9–10)

* We do not claim that equation (3.2.11) is consistent with the laws of mass and momentum conservation of a fluid.

envelope $\mathscr{E}$ (Figures 3.2 and 3.3). The discontinuous solution $\bar{u}(t, x)$ satisfies the conditions

$$\begin{aligned} &\bar{u}(t, x) = 0 \quad \text{for } x > y(t), \\ &\bar{u}(t, x) = u_+(t, x) \quad \text{for } x < y(t), \end{aligned} \tag{3.2.12}$$

the Rankine–Hugoniot equations

$$\dot{y}(t) = \frac{1}{2} u_+ \big(t, y(t)\big), \tag{3.2.13}$$

and the initial condition for the discontinuity

$$y\big((2a)^{-1}\big) = a. \tag{3.2.14}$$

At every point $(t, y(t))$ the slope of the characteristics on the left of the discontinuity is

$$\frac{dx}{dt} = u_+ \big(t, y(t)\big) > \dot{y}(t)$$

whereas on the right $\frac{dx}{dt} = 0 < \dot{y}(t)$. Hence the condition (L) is satisfied. So is (E) (see Section 3.2.3). Differentiating the equation of $\mathscr{E}$ and comparing with (3.2.13), (3.2.14) and (3.2.10) one can check that S lies above $\mathscr{E}$.

Another method of determining the function $y(\cdot)$ is suggested by the following observation. Letting $x_0 \to -\infty$, $x_1 \to \infty$ in equation (3.2.11) we obtain the following conservation law

$$\int_{-\infty}^{\infty} dx\, \bar{u}(t, x) = \int_{-\infty}^{\infty} dx\, \phi(x). \tag{3.2.15}$$

for $u(t, \cdot) \in \mathscr{L}^1$. Equation (3.2.11), (3.2.15) remain valid for multivalued solutions $u(t, x)$ in the following sense. For every fixed instant of time t we choose a parameter s on the graph of $u(t, \cdot)$ in such a way that the graph is given by the equations $x = \tilde{x}(s)$, $u = \tilde{u}(t, s)$ with a one-valued function $\tilde{u}(t, \cdot)$. We define

$$\int_{-\infty}^{\infty} dx\, u(t, x) := \int_{-\infty}^{\infty} ds\, \frac{d\tilde{x}}{ds} \tilde{u}(t, s).$$

Since equation (3.2.15) holds for the discontinuous solution $\bar{u}$ as well as for the multivalued solution u with the above definition of the left-hand side, the position of the discontinuity $x = y(t)$ is such that the shaded areas in Fig. 3.2. are equal for every $t > (2a)^{-1}$.

The above considerations carry over to the general hyperbolic systems of conservation laws. Implicit solutions of the form (3.2.4) arise in the case of simple wave solutions to quasilinear hyperbolic systems (3.1.1).

3.2.2 Non-Uniqueness of Discontinuous Solutions

In this subsection we shall continue the discussion of a single conservation law

$$u_t + f(u)_x = 0, \quad f'(u) \equiv c(u) \quad \text{for } t > 0 \tag{3.2.16}$$

in order to exhibit the role of the conditions (L) and (E). Equation (3.2.16) will be interpreted in the sense

$$\oint_{\mathscr{C}} -u\,dx + f(u)\,dt = 0 \tag{3.2.16a}$$

for every closed piecewise Lipschitz curve $\mathscr{C}$.

It was shown in Section 3.2.1 that those solutions of quasilinear hyperbolic systems which are relevant for continuum mechanics are in general discontinuous even if the initial data are smooth. We therefore seek the solutions in some class of discontinuous functions extending simultaneously the Cauchy problem to allow for discontinuous initial data.

The simplest model of a Cauchy problem with discontinuous initial data is the Riemann problem (see Section 3.1.7). It is also the starting point of some existence proofs for general Cauchy problems (see Section 3.2.8). In our case the Riemann problem is defined by

$$u(0, x) \equiv \phi(x) :\equiv \begin{cases} \alpha & \text{for } x < 0, \\ \beta & \text{for } x > 0 \end{cases} \tag{3.2.17}$$

with $\alpha, \beta \in \boldsymbol{R}$, $\alpha \neq \beta$.

To begin with we consider the simplest case where

$$f''(u) \equiv c'(u) > 0 \ \forall u \in \boldsymbol{R}. \tag{3.2.18}$$

The Riemann problem (3.2.16), (3.2.17) has an obvious solution:

$$u(t, x) \equiv \begin{cases} \alpha & \text{for } x < wt, \\ \beta & \text{for } x > wt \end{cases} \tag{3.2.19}$$

with $w = \{f(\beta) - f(\alpha)\}\{\beta - \alpha\}^{-1}$ (see equation (3.1.41)). However for $\beta > \alpha$ another solution is

$$u(t, x) \equiv \begin{cases} \alpha & \text{for } x \leqslant c(\alpha)t, \\ xt^{-1} & \text{for } c(\alpha)t < x < c(\beta)t, \\ \beta & \text{for } x \geqslant c(\beta)t. \end{cases} \tag{3.2.20}$$

Since $f''(u) > 0$ and for some $\bar{u}$, $\bar{\bar{u}}$

$$w = c(\alpha)+\frac{1}{2}f''(\bar{u})(\beta-\alpha) = c(\beta)-\frac{1}{2}f''(\bar{\bar{u}})(\beta-\alpha), \tag{3.2.21}$$

we have the inequalities

$$\begin{aligned} c(\alpha) < w < c(\beta) \quad &\text{if } \beta > \alpha \\ c(\beta) < w < c(\alpha) \quad &\text{if } \alpha > \beta. \end{aligned} \tag{3.2.22}$$

Consequently the solution (3.2.19) satisfies the Lax condition (L) iff $\alpha > \beta$. The unique solution satisfying (L) is thus given by equation (3.2.19) for $\alpha > \beta$, by equation (3.2.20) for $\alpha < \beta$.

We note that there are non-zero discontinuous solutions to the Cauchy problem with zero initial data if the condition (L) is dropped. Indeed, for $c(u) \equiv u$ a discontinuous solution is

$$u(t, x) \equiv \begin{cases} 0 & \text{for } x < -t, \\ -2 & \text{for } -t < x < 0, \\ +2 & \text{for } 0 < x < t, \\ 0 & \text{for } x > t. \end{cases}$$

The discontinuity at $x = 0$ does not satisfy (L).

The discontinuous solution (3.2.19) satisfies (E) provided $\alpha > \beta$. The condition (E) assumes the form

$$\frac{f(\beta)-f(\alpha)}{\beta-\alpha} \leqslant \frac{f(u)-f(\alpha)}{u-\alpha} \quad \text{for every } u \text{ between } \alpha \text{ and } \beta. \tag{3.2.23}$$

Let $\alpha < u < \beta$, $u = (1-\theta)\alpha+\theta\beta$, $\theta \in]0, 1[$. Since $f(u) < (1-\theta)f(\alpha)+\theta f(\beta)$, equation (3.2.23) is satisfied for $\alpha < u < \beta$. Letting $u \to \alpha$ in equation (3.2.23) we conclude that $w \leqslant c(\alpha)$. Hence, on account of (3.2.21), $w \geqslant c(\beta)$ and $\beta \leqslant \alpha$. It is easy to see that for convex f the conditions (E) and (L) are equivalent. It is not so in general.

Indeed, for $f(u) \equiv -\frac{1}{3}u^3$, $w(u_-, u_+) = -\frac{1}{3}(u_+{}^2+u_+u_-+u_-{}^2)$, $u_+ < 0$, the condition (L) is satisfied for $u_+ \leqslant u_- \leqslant -2u_+$, but the condition (E) holds only for $u_+ \leqslant u_- \leqslant -\frac{1}{2}u_+$. The point $u_- = -\frac{1}{2}u_+$ is a Chapman–Jouguet point $w = \max = c(u_-)$ on $\mathscr{H}_{u_+}$. If $u_+ = \beta < 0$, then the solution to (3.2.17) satisfying (L) and (E) has the form (3.2.20) for $\alpha < \beta$, (3.2.19) for $\beta < \alpha < -\frac{1}{2}\beta =: \gamma$ and

$$u(t, x) \equiv \begin{cases} \beta & \text{for } xt^{-1} \geqslant c(\gamma) = -\gamma^2, \\ (-xt^{-1})^{1/2} & \text{for } c(\alpha) < xt^{-1} < c(\gamma), \\ \alpha & \text{for } xt^{-1} \leqslant c(\alpha) \end{cases} \tag{3.2.24}$$

for $\alpha > \gamma$.

For a single equation (3.2.16) the condition (E) can be incorporated in the integral formulation of the problem. The latter assumes the form of a one-parameter family of integral inequalities (Vol'pert, 1967).

3.2.3 Global Formulation of the Initial-Value Problem in the Class of Discontinuous Functions. Preliminary Remarks and Definitions

Let $\mathscr{D}$ be an open domain in $\boldsymbol{R}^{m+1}$. We consider the solutions of equation (3.1.35) in a class of functions $\mathbf{u}\colon \mathscr{D} \to \mathscr{H}$ which contains all piecewise $\mathscr{C}^1$ functions. It is desirable to have some kind of Gauss–Green theorem in this class. The derivatives will be now understood in the sense of distributions, i.e.

$$\int_{\mathscr{D}} dL^{m+1}(x)\left(\mathbf{h}(x, \mathbf{u}(x))\varphi(x) + \mathbf{f}^{\mu}(x, \mathbf{u}(x))\partial_{\mu}\varphi(x)\right) = 0 \qquad (3.2.25)$$

for every test function $\varphi \in \mathscr{C}_0^{\infty}(\mathscr{D})$. We say that $\mathbf{u}$ is a **weak solution** to (3.1.35) if it satisfies equation (3.2.25) for every $\varphi \in \mathscr{C}_0^{\infty}(\mathscr{D})$. Clearly, $\mathbf{u}$ cannot be an arbitrary distribution or measurable function. A trace theorem is necessary in order to formulate the boundary and initial-value problems. Another problem arises in connection with the non-linear superpositions $\mathbf{u}(x) \mapsto \mathbf{f}^{\mu}(x, \mathbf{u}(x))$, which do not make sense for a general distribution.

A long experience in proving existence theorems for discontinuous solutions to the Riemann problem and more general Cauchy problems by means of smoothing singular perturbations of the parabolic type (the vanishing viscosity method, Section 3.4.6) or by discrete approximations (e.g. Glimm's scheme, Section 3.2.7) leads to the conclusion that weak solutions should be sought in the class of functions with bounded variation. The estimates of the approximating sequences in terms of the $\mathscr{L}^{\infty}$ norm and the total variation as well as Helly's selection theorem are crucial in both methods.

In Sections 3.2.4–3.2.6 we shall develop the theory of functions with essentially bounded variation in broad outline. In Section 3.2.6 the solutions in the class $\mathrm{BV} \cap \mathscr{L}^{\infty}$ will be defined and discussed. In Section 3.2.7 an existence proof by Glimm's scheme will be sketched. Finally, in Section 3.2.8 we shall comment on the singularities of weak solutions in the class $\mathrm{BV} \cap \mathscr{L}^{\infty}$.

In the following we need to replace some topological and differential-geometric notions by their measure-theoretic counterparts. For a more general setting consult Federer (1969).

Definition 3.2.1

For an arbitrary subset $\mathscr{E} \subset \boldsymbol{R}^n$ the symbol $|\mathscr{E}|$ denotes the **outer Lebesgue measure** of $\mathscr{E}$.

If $\mathscr{E}$ is Lebesgue measurable, then $|\mathscr{E}| = L^n(\mathscr{E})$.

Definition 3.2.2

A point $x \in \boldsymbol{R}^n$ is said to be a **density point** of $\mathscr{E} \in \boldsymbol{R}^n$ if

$$\lim_{r\to 0} \frac{|\mathscr{E}\cap\mathscr{K}(x, r)|}{|\mathscr{K}(x, r)|} = 1.$$

Let $\mathscr{E}_*$ be the set of density points of $\mathscr{E}$. Clearly int$\mathscr{E} \subset \mathscr{E}_*$. Suppose that

$$\lim_{r\to 0} \frac{|\mathscr{E}\cap\mathscr{K}(x, r)|}{|\mathscr{K}(x, r)|} = 0. \tag{3.2.26}$$

Since $|\mathscr{E}\cap\mathscr{K}(x, r)|+|\mathscr{E}^c\cap\mathscr{K}(x, r)| \geqslant |\mathscr{K}(x, r)|$, we have

$$\frac{|\mathscr{E}^c\cap\mathscr{K}(x, r)|}{|\mathscr{K}(x, r)|} \leqslant 1 \quad \text{and} \quad \liminf_{r\to 0} \frac{|\mathscr{E}^c\cap\mathscr{K}(x, r)|}{|\mathscr{K}(x, r)|} \geqslant 1,$$

hence

$$\lim_{r\to 0} \frac{|\mathscr{E}^c\cap\mathscr{K}(x, r)|}{|\mathscr{K}(x, r)|} = 1$$

and x is a density point of $\mathscr{E}^c$.

Definition 3.2.3

$\mathscr{E}^*$ is the complement of the set of density points of $\mathscr{E}^c$.

The **essential boundary** $\partial^*\mathscr{E} := \mathscr{E}^*\setminus\mathscr{E}_*$.

We have $\mathscr{E}^* = ((\mathscr{E}^c)_*)^c \subset (\text{int}\mathscr{E}^c)^c = \bar{\mathscr{E}}$ and $\partial^*\mathscr{E} \subset \partial\mathscr{E}$.

Definition 3.2.4

A unit vector $\mathbf{n} \in \boldsymbol{R}^n$ is said to be the **outer normal** of the set $\mathscr{E}$ at $x_0 \in \boldsymbol{R}^n$ **in the sense of Federer** if

$$\lim_{r\to 0} \frac{|\mathscr{K}(x_0, r)\cap\mathscr{E}\cap\mathscr{H}(\mathbf{n}, x_0)|}{|\mathscr{K}(x_0, r)|} = 0$$

$$= \lim_{r\to 0} \frac{|\mathscr{K}(x_0, r)\cap\mathscr{E}^c\cap\mathscr{H}(-\mathbf{n}, x_0)|}{|\mathscr{K}(x_0\ r)|},$$

where $\mathscr{H}(\mathbf{n}, x_0) := \{x \in \boldsymbol{R}^n|\ (x-x_0)\cdot\mathbf{n} > 0\}$.

Remarks

(1) Let

$$\mathscr{E} := \{(x, y) \in \boldsymbol{R}^2 |\ (x-1)^2+y^2 < 1\} \\ \cup \{(x, y) \in \boldsymbol{R}^2 |\ (x+1)^2+y^2 < 1\}.$$

The point $(0, 0) \in \mathscr{E}^*$ although $(0, 0) \mathrel{\overline{\in}} \operatorname{int}\mathscr{E}$.

(2) Let $\mathscr{E} := \{(x, y) \in \boldsymbol{R}^2 |\ (x^2+y^2) < 4, (x-1)^2+y^2 > 1\}$. The point $(2, 0) \mathrel{\overline{\in}} \partial^*\mathscr{E}$ although $(2, 0) \in \partial\mathscr{E}$.

Definition 3.2.5

Let $\mathbf{u}$ be any function $\mathscr{E} \to \boldsymbol{R}^m$, $\mathscr{E} \subset \boldsymbol{R}^n$ and let $x_0 \in \mathscr{E}^*$. A vector $\mathbf{p} \in \boldsymbol{R}^m$ is called an **approximate limit of u at x_0 relative to $\mathscr{E}$**, $l_{\mathscr{E}}\mathbf{u}(x_0) = \mathbf{p}$, if $\forall\ \varepsilon > 0$ $x_0 \mathrel{\overline{\in}} \{x \in \mathscr{E} |\ |\mathbf{u}(x)-\mathbf{p}| > \varepsilon\}^*$.

It is easy to check the following implications:

(1) $\mathscr{A} \subset \mathscr{B}$, $x_0 \in \mathscr{A}^*$, $l_{\mathscr{B}}\mathbf{u}(x_0)$ exists $\Rightarrow l_{\mathscr{A}}\mathbf{u}(x_0)$ exists and is equal to $l_{\mathscr{B}}\mathbf{u}(x_0)$;

(2) $x_0 \in (\mathscr{A} \cap \mathscr{B})^*$, $l_{\mathscr{A}}\mathbf{u}(x_0)$ and $l_{\mathscr{B}}\mathbf{u}(x_0)$ exist $\Rightarrow l_{\mathscr{A}}\mathbf{u}(x_0) = l_{\mathscr{B}}\mathbf{u}(x_0)$.

Definition 3.2.6

For $\mathscr{E} = \boldsymbol{R}^n$ we write $l\mathbf{u}(x_0) := l_{\mathscr{E}}\mathbf{u}(x_0)$.

For $\mathscr{E} = \mathscr{H}(\mathbf{n}, x_0)$ we write $l_{\mathbf{n}}\mathbf{u}(x_0) := l_{\mathscr{E}}\mathbf{u}(x_0)$.

Definition 3.2.7

$x_0 \in \boldsymbol{R}^n$ is said to be a **regular point for u**: $\boldsymbol{R}^n \to \boldsymbol{R}^m$ if there is a unit vector $\mathbf{n} \in \boldsymbol{R}^n$ such that $l_{\mathbf{n}}\mathbf{u}(x_0)$ and $l_{-\mathbf{n}}\mathbf{u}(x_0)$ exist.

$x_0 \in \boldsymbol{R}^n$ is said to be a **jump point of u**: $\boldsymbol{R}^n \to \boldsymbol{R}^m$ if it is a regular point for $\mathbf{u}$ and

$$l_{\mathbf{n}}\mathbf{u}(x_0) \neq l_{-\mathbf{n}}\mathbf{u}(x_0) \tag{3.2.27}$$

for some unit vector $\mathbf{n} \in \boldsymbol{R}^n$.

Lemma 3.2.8

Let x_0 be a regular point for $\mathbf{u}$ and suppose that the limits $l_{\mathbf{n}}\mathbf{u}(x_0)$, $l_{-\mathbf{n}}\mathbf{u}(x_0)$ exist for some unit vector $\mathbf{n} \in \boldsymbol{R}^n$. Then there are two possibilities:

(1) if $l_{\mathbf{n}}\mathbf{u}(x_0) = l_{-\mathbf{n}}\mathbf{u}(x_0)$, then $l_{\mathbf{v}}\mathbf{u}(x_0) = l\mathbf{u}(x_0)$ exists for every unit vector $\mathbf{v}$;

(2) if $l_{\mathbf{n}}\mathbf{u}(x_0) \neq l_{-\mathbf{n}}\mathbf{u}(x_0)$, then the unit vector $\mathbf{n}$ is defined uniquely up to sign.

Proof

Ad (1). The set $\{x|\ |\mathbf{u}(x)-\mathbf{p}| > \varepsilon,\ (x-x_0)\cdot\mathbf{n} = 0\}$ has Lebesgue measure zero, hence for $\mathbf{p} = l_{\mathbf{n}}\mathbf{u}(x_0)$

$$\begin{aligned}&\left|\{x|\ |\mathbf{u}(x)-\mathbf{p}| > \varepsilon,\ (x-x_0)\cdot\mathbf{n} > 0\}\cap\mathscr{K}(x_0, r)\right|\\&+\left|\{x|\ |\mathbf{u}(x)-\mathbf{p}| > \varepsilon,\ (x-x_0)\cdot\mathbf{n} < 0\}\cap\mathscr{K}(x_0, r)\right|\\&\geqslant \left|\{x|\ |\mathbf{u}(x)-\mathbf{p}| > \varepsilon\}\cap\mathscr{K}(x_0, r)\right|\\&\text{and}\quad x_0 \mathrel{\bar\in} \{x|\ |\mathbf{u}(x)-\mathbf{p}| > \varepsilon\}^*.\end{aligned}$$

Hence $l\mathbf{u}(x_0)$ exists and $l_{\mathbf{v}}\mathbf{u}(x_0) = l\mathbf{u}(x_0)$ for every $\mathbf{v}$.

Ad (2). Suppose for definiteness that $l_{\mathbf{n}}u^1(x_0) \neq l_{-\mathbf{n}}u^1(x_0)$ and $l_{\mathbf{v}}\mathbf{u}(x_0)$ exists for some $\mathbf{v} \neq \mathbf{n}, -\mathbf{n}$. Obviously $x_0 \in (\mathscr{H}_{\mathbf{n}}\cap\mathscr{H}_{\mathbf{v}})^*$, $x_0 \in (\mathscr{H}_{-\mathbf{n}}\cap\mathscr{H}_{\mathbf{v}})^*$, with $\mathscr{H}_{\mathbf{n}} := \mathscr{H}(\mathbf{n}, x_0)$, and therefore, by the implication (2) above $l_{\mathbf{v}}\mathbf{u}(x_0) = l_{\mathbf{n}}\mathbf{u}(x_0) = l_{-\mathbf{n}}\mathbf{u}(x_0)$, a contradiction. □

Definition 3.2.9

For any function $\mathbf{u}$ defined on (a subset of) $\boldsymbol{R}^n$ the **jump set** $\Gamma(\mathbf{u})$ is the set of all the jump points of $\mathbf{u}$. For every $x_0 \in \Gamma(\mathbf{u})$ the normal $\mathbf{v}(x_0)$ is, by definition, the vector $\mathbf{n}$ defined by equation (3.2.27) (up to sign).

Lemma 3.2.10

Let f: $\boldsymbol{R}^p \to \boldsymbol{R}$ be continuous at $\mathbf{p} \in \boldsymbol{R}^p$ and let $\mathbf{u}$ be a function defined over a nbhd of x_0 with values in $\boldsymbol{R}^p$.

If $l_{\mathscr{E}}\mathbf{u}(x_0) = \mathbf{p}$ then $l_{\mathscr{E}}f\circ\mathbf{u}(x_0)$ exists and equals $f(\mathbf{p})$. In particular, for u, v: $\boldsymbol{R}^n \to \boldsymbol{R}$, $l_{\mathscr{E}}(uv)(x_0)$ exists at every point x_0 such that $l_{\mathscr{E}}u(x_0)$ and $l_{\mathscr{E}}v(x_0)$ both exist, $l_{\mathscr{E}}(uv)(x_0) = l_{\mathscr{E}}u(x_0)\cdot l_{\mathscr{E}}v(x_0)$.

Proof

For every $\varepsilon > 0$ there is a $\delta(\varepsilon) > 0$ such that $|\mathbf{u}-\mathbf{p}| < \delta(\varepsilon)$ implies that $|f(\mathbf{u})-f(\mathbf{p})| < \varepsilon$. Hence $\mathscr{F} = \{x|\ x \mathrel{\bar\in} E \text{ or } |\mathbf{u}(x)-\mathbf{p}| < \delta(\varepsilon)\} \subset \mathscr{G} = \{x|\ x \mathrel{\bar\in} \mathscr{E}$ or $|f\circ\mathbf{u}(x)-f(\mathbf{p})| < \varepsilon\}$, $\mathscr{G}_* \subset \mathscr{F}_*$, the first part of the thesis follows from the definition of the approximate limit.

Setting $\mathbf{u} = (u, v)$, $f(\mathbf{u}) = uv$ we obtain the second part of the thesis. □

3.2.4 Functions of Essentially Bounded Variation and Sets with Bounded Perimeter

In this section we construct the class of distributions whose derivatives are locally bounded Borel measures. In the following we consider the class of sets whose characteristic functions have the above property.

For an arbitrary function $f \in \mathscr{L}^\infty(\boldsymbol{R}^m, \boldsymbol{R})$, $\lambda \in \boldsymbol{R}_+$, $x \in \boldsymbol{R}^m$ let

$$W_\lambda f(x) :\equiv (\pi\lambda)^{-m/2} \int dL^m(z) \exp(-|x-z|^2/\lambda) f(z)$$
$$\equiv (\pi\lambda)^{-m/2} \int dL^m(z) e^{-|z|^2} f(x+\sqrt{\lambda} z). \tag{3.2.28}$$

Clearly

$$|W_\lambda f(x)| \leqslant \|f\|_\infty \tag{3.2.29}$$

and $W_\lambda f \in \mathscr{C}_b^\infty$. For $f \in \mathscr{C}_b^1$ integration by parts yields

$$\frac{\partial}{\partial x^\alpha} W_\lambda f = W_\lambda \frac{\partial f}{\partial x^\alpha}. \tag{3.2.30}$$

The regularizing operators W_λ have the semigroup property

$$W_\lambda W_\mu f = W_{\lambda+\mu} f \;\; \forall \lambda, \mu > 0, \quad f \in \mathscr{L}^\infty \tag{3.2.31}$$

and

$$\lim_{\lambda \to 0} W_\lambda f(x) = f(x) \quad \text{for } L^m\text{-almost all } x \in \boldsymbol{R}^m. \tag{3.2.32}$$

The convergence in (3.2.32) is uniform on compact sets if f is continuous.

If $f \in \mathscr{L}^1(\boldsymbol{R}^m, \boldsymbol{R})$, then

$$\|W_\lambda f\|_{\mathscr{L}^1} = \int dL^m |W_\lambda f| \leqslant \int dL^m |f| = \|f\|_{\mathscr{L}^1}. \tag{3.2.33}$$

Equation (3.2.32) and the Lebesgue dominated convergence theorem imply that

$$\|W_\lambda f\|_{\mathscr{L}^1} \to \|f\|_{\mathscr{L}^1},$$
$$\|W_\lambda f - f\|_{\mathscr{L}^1} \to 0 \quad \text{as } \lambda \to 0. \tag{3.2.34}$$

Equations (3.2.30) and (3.2.31) imply that

$$\nabla W_{\lambda+\mu} f(x) = W_\lambda \nabla W_\mu f(x) \quad \text{if } \mu > 0 \tag{3.2.35}$$

and consequently

$$\|\nabla W_{\lambda+\mu} f\|_{\mathscr{L}^1} \leqslant \|\nabla W_\lambda f\|_{\mathscr{L}^1} \quad \text{for every } \lambda, \mu > 0,$$

in view of equation (3.2.33). The non-increasing positive function $\lambda \mapsto \|\nabla W_\lambda f\|_{\mathscr{L}^1}$ has a limit $I(f) \in]0, \infty]$.

Definition 3.2.10

Let $\mathbf{f}^{(n)}: \mathscr{B} \to \boldsymbol{R}^p$ be a sequence of regular vector Borel measures defined on the σ-algebra of Borel sets $\mathscr{B}$. The sequence $\{\mathbf{f}^{(n)}\}$ is said to **converge weakly** to a regular vector Borel measure $\mathbf{f}: \mathscr{B} \to \boldsymbol{R}^p$,

$\mathbf{f}^{(n)} \rightharpoonup \mathbf{f}$, if for every $g \in \mathscr{C}_0^0(\boldsymbol{R}^m, \boldsymbol{R})$

$$\lim_{n \to \infty} \int g(x)\, d\mathbf{f}^{(n)} = \int g(x)\, d\mathbf{f}.$$

Theorem 3.2.11
Suppose that a sequence $\{\mathbf{f}^{(n)}\}$ of regular vector Borel measures has uniformly bounded total variation. Then a subsequence $\{\mathbf{f}^{n_k}\}$ converges weakly to some regular Borel measure **f**.

Proof
On account of the Riesz' theorem this is a special case of Helly's selection theorem for $\mathscr{X} = \mathscr{C}^0$, see Sections 1.1.3 and 3.2.10. □

Theorem 3.2.12 (De Giorgi, 1954)
Let $f \in \mathscr{L}^\infty(\boldsymbol{R}^m, \boldsymbol{R})$, $I(f) < \infty$.

Then there exists a unique Borel vector measure $\mathbf{f}: \mathscr{B} \to \boldsymbol{R}^m$ with bounded total variation, such that for every $g \in \mathscr{C}^1$, $g(x) = O(|x|^{-(m+1)})$ for $|x| \to \infty$ we have

$$\int dL^m f\nabla g = \int g d\mathbf{f}. \tag{3.2.36}$$

Remark
Equation (3.2.36) means that $-\mathbf{f}$ can be identified with the distributional gradient of f.

Proof
Let

$$\mathbf{f}^\lambda(\mathscr{E}) := -\int \chi_{\mathscr{E}} \nabla W_\lambda f dL^m \tag{3.2.37}$$

for every bounded Borel subset $\mathscr{E} \subset \boldsymbol{R}^m$.

$$|\mathbf{f}^\lambda|(\mathscr{E}) = \int \chi_{\mathscr{E}} |\nabla W_\lambda f| dL^m \quad \text{and} \quad |\mathbf{f}^\lambda|(\boldsymbol{R}^m) \leqslant I(f) < \infty.$$

Hence a sequence $\mathbf{f}^{\lambda_n} \rightharpoonup \mathbf{f}$, with $\lambda_n \to 0$,

$$|\mathbf{f}|(\boldsymbol{R}^m) \leqslant \liminf_{n\to\infty} |\mathbf{f}^{\lambda_n}| \leqslant I(f) < \infty. \tag{3.2.38}$$

For every g satisfying the hypothesis of the theorem $\int g d\mathbf{f} = \lim_{n\to\infty} \int g d\mathbf{f}^{\lambda_n}$, equation (3.2.34) implies that $W_{\lambda_n} f \to f$ in $\mathscr{L}^1_{loc}$, and hence

$$\lim_{n\to\infty} \int dL^m\, W_{\lambda_n} f(x) \nabla g(x) = \int dL^m f(x) \nabla g(x).$$

Working out the left-hand side we find that

$$-\lim_{n\to\infty} \int g(x) \nabla W_{\lambda_n} f(x) dL^m = \lim_{n\to\infty} \int g d\mathbf{f}^{\lambda_n} = \int g d\mathbf{f}.$$

□

Theorem 3.2.13

Let $f \in \mathscr{L}^\infty(\boldsymbol{R}^m, \boldsymbol{R})$ and suppose that the regular Borel measure $\mathbf{f}$ on $\boldsymbol{R}^m$ with values in $\boldsymbol{R}^m$ has bounded total variation and satisfies equation (3.2.36). Then

$$I(f) \leqslant |\mathbf{f}|(\boldsymbol{R}^m). \tag{3.2.39}$$

Proof

For $g(z) = (\pi\lambda)^{-m/2}\exp(-|x-z|^2/\lambda)$

$$\begin{aligned}|\nabla W_\lambda f(x)| &= \left|(\pi\lambda)^{-m/2}\int dL^m(z)\nabla_z \exp(-|x-z|^2/\lambda) f(z)\right| \\ &= \left|(\pi\lambda)^{-m/2}\int \exp(-|x-z|^2/\lambda)d\mathbf{f}(z)\right| \\ &\leqslant (\pi\lambda)^{-m/2}\int \exp(-|x-z|^2/\lambda)|d\mathbf{f}|(z).\end{aligned}$$

Hence by the Fubini theorem

$$\begin{aligned}\|\nabla W_\lambda f\|_{\mathscr{L}_1} &\leqslant (\pi\lambda)^{-m/2}\int dL^m \int |d\mathbf{f}|\exp(-|x-z|^2/\lambda) \\ &= \int |d\mathbf{f}| \int dL^m (\pi\lambda)^{-m/2}\exp(-|x-z|^2/\lambda) = \int |d\mathbf{f}|\end{aligned}$$

and $I(f) \leqslant |\mathbf{f}|(\boldsymbol{R}^m) < \infty$. □

Both theorems can be applied in particular to the characteristic functions $\chi_{\mathscr{E}}$ of measurable sets $\mathscr{E}$. Let

$$P(\mathscr{E}) := I(\chi_{\mathscr{E}}). \tag{3.2.40}$$

Definition 3.2.14

The positive extended number $P(\mathscr{E})$ is called the **perimeter** of the set $\mathscr{E}$.

A measurable set $\mathscr{E}$ is said to **have bounded perimeter** if $P(\mathscr{E}) < \infty$. A measurable set $\mathscr{E}$ is said to have a **locally bounded perimeter** if for every $x \in \boldsymbol{R}^m$, $r > 0$, $P(\mathscr{E} \cap \mathscr{K}(x, r)) < \infty$.

Remark

Intuitively $|\mathbf{f}| = |-\nabla\chi_{\mathscr{E}}|$ can be identified with a Dirac delta distribution concentrated on $\partial\mathscr{E}$. On account of (3.2.36) and (3.2.39)

$$P(\mathscr{E}) = \int |\nabla\chi_{\mathscr{E}}| \tag{3.2.41}$$

measures the area of $\partial\mathscr{E}$.

Let $\mathscr{D}$ be a domain in $\boldsymbol{R}^{m+1}$ and $f \in \mathscr{L}^1_{loc}(\mathscr{D}, \boldsymbol{R}) \cap \mathscr{L}^\infty(\mathscr{D}, \boldsymbol{R})$. For $g \in \mathscr{C}^\infty_o(\mathscr{D})$ equation (3.2.36) yields the formula

$$\int_{\mathscr{D}} dL^{m+1} f\nabla g = \int dL^{m+1} f\nabla g = \int g d\mathbf{f} = \int_{\mathscr{D}} g d\mathbf{f}.$$

Definition 3.2.15

We shall say that a function $u \in \mathcal{L}^1_{loc}(\mathcal{D}, \boldsymbol{R})$ belongs to $\mathrm{BV}(\mathcal{D}, \boldsymbol{R})$ if its distributional gradient ∇u is a vector Borel regular measure.

Equivalently, $u \in \mathrm{BV}(\mathcal{D}, \boldsymbol{R})$ if

(1) $u \in \mathcal{L}^1_{loc}(\mathcal{D}, \boldsymbol{R})$,*

(2) there is a vector Borel regular measure $\mathbf{m}$ of locally bounded variation, such that for every $\varphi \in \mathscr{C}^\infty_o(\mathcal{D})$

$$\int_{\mathcal{D}} dL^{m+1}(x) u \nabla \varphi = -\int_{\mathcal{D}} \varphi d\mathbf{m}(x). \tag{3.2.42}$$

We shall write ∇u for $\mathbf{m}, d\mathbf{m}$, and $\mathrm{BV}(\mathcal{D})$ for $\mathrm{BV}(\mathcal{D}, \boldsymbol{R})$.

A function $\mathbf{u} \in \mathcal{L}^1_{loc}(\mathcal{D}, \boldsymbol{R}^p)$ is said to belong to $\mathrm{BV}(\mathcal{D}, \boldsymbol{R}^p)$ if each component function u^k, $k = 1, \ldots, p$ belongs to $\mathrm{BV}(\mathcal{D}, \mathbf{R})$. In this case $\nabla \mathbf{u}$ is a matrixvalued Borel measure.

Theorems 3.2.13 and 3.2.12 say that a function $u \in \mathcal{L}^\infty(\boldsymbol{R}^{m+1}, \boldsymbol{R})$ belongs to $\overline{\mathrm{BV}}(\boldsymbol{R}^{m+1}, \boldsymbol{R})$ iff $I(u) < \infty$.

We say that $u \in \overline{\mathrm{BV}}(\mathcal{D}, \boldsymbol{R})$ if $u \in \mathrm{BV}(\mathcal{D}, \boldsymbol{R})$ and $|\nabla u|(\mathcal{D}) < \infty$.

Theorem 3.2.16

Let $u \in \mathcal{L}^1_{loc}(\mathcal{D}, \boldsymbol{R})$.

Then $u \in \overline{\mathrm{BV}}(\mathcal{D}, \boldsymbol{R})$ iff $\forall\, \varphi \in \mathscr{C}^\infty_0$

$$\left|\int dL^{m+1} u \nabla \varphi\right| \leqslant K\|\varphi\| \quad \text{for some fixed } K > 0.$$

Proof

Necessity follows from equation (3.2.42),

$$\left|\int_{\mathcal{D}} dL^{m+1} u \nabla \varphi\right| \leqslant \sup\{|\varphi(x)| \; |x \in \mathcal{D}\} |\mathbf{m}|(\mathcal{D}).$$

For sufficiency, note that there is a continuous linear functional on the Banach space $\mathscr{C}^0$ which coincides with the left-hand side of equation (3.2.42) on the linear submanifold $\mathscr{C}^\infty_o$ of $\mathscr{C}^0$, by the Hahn–Banach theorem. By the Riesz theorem there is a regular Borel measure $\mathbf{m}$ of bounded total variation, such that equation (3.2.42) holds for all $\varphi \in \mathscr{C}^0(\mathcal{D}, \boldsymbol{R})$. □

* It can be proved however that every distribution $u \in \mathcal{D}'(\mathcal{D}, \boldsymbol{R})$, whose gradient is a regular vector Borel measure, belongs to $\mathcal{L}^1_{loc}(\mathcal{D})$ (see Miranda, 1964).

Example

For $m = 0$,

$$u(x) = H(x) = \begin{cases} 1 & \text{for } x > 0 \\ 1/2 & \text{for } x = 0, \\ 0 & \text{for } x < 0 \end{cases}$$

$$\frac{\partial u}{\partial x} = \delta(x), \quad \int_R dx \, \frac{\partial \varphi}{\partial x} u = \int_{R_+} dx \, \frac{\partial \varphi}{\partial x} = \varphi(0) = \langle \delta, \varphi \rangle = \int \varphi dH$$

(the last expression is a Lebesgue–Stieltjes integral).

3.2.5 Properties of Sets with Bounded Perimeter

We now describe some properties of sets with bounded or locally bounded perimeter. For details, consult De Giorgi (1954, 1955), Miranda (1964) and Federer (1954). Equation (3.2.36) implies the formula

$$\int_{\mathscr{E}} dL^{m+1} \nabla g = -\int g \nabla \chi_{\mathscr{E}}, \tag{3.2.43}$$

valid for every measurable set $\mathscr{E}$ with $P(\mathscr{E}) < \infty$ and every $g \in \mathscr{C}_0^\infty$. The right-hand side of equation (3.2.43) is the integral of g with respect to the measure $\nabla \chi_{\mathscr{E}}$, extended over $\boldsymbol{R}^{m+1}$. Since $\left|\int g \nabla \chi_{\mathscr{E}}\right| \leqslant \int |g||\nabla \chi_{\mathscr{E}}|$ and $\mathscr{C}_0^\infty$ is dense in $\mathscr{L}^1(|\nabla \chi_{\mathscr{E}}|)$, we have $\left|\int_{\mathscr{P}} \nabla \chi_{\mathscr{E}}\right| \leqslant \int_{\mathscr{P}} |\nabla \chi_{\mathscr{E}}|$ for every bounded Borel $\mathscr{P}$.

The Radon–Nikodym theorem implies the existence of a derivative $\boldsymbol{\nu}(x) \in \mathscr{L}^1(\boldsymbol{R}^{m+1}, \boldsymbol{R}^{n+1}; \mu) \cap \mathscr{L}^\infty(\boldsymbol{R}^{m+1}, \boldsymbol{R}^{m+1}; \mu)$ of $\nabla \chi_{\mathscr{E}}$ with respect to $\mu := |\nabla \chi_{\mathscr{E}}|$. Obviously $|\boldsymbol{\nu}(x)| \leqslant 1$ μ-a.e. in $\boldsymbol{R}^{m+1}$ and

$$-\int_{\mathscr{P}} g \nabla \chi_{\mathscr{E}} = \int_{\mathscr{P}} g \boldsymbol{\nu} d\mu \tag{3.2.44}$$

for every Borel set $\mathscr{P} \subset \boldsymbol{R}^{m+1}$ and $g \in \mathscr{C}_0^0$.

Definition 3.2.17 (De Giorgi, 1955)

Let $P(\mathscr{E}) < \infty$.

The **reduced boundary** $\partial^{**}\mathscr{E}$ of $\mathscr{E}$ is the set of points $x \in \boldsymbol{R}^{m+1}$ with the following properties:

(i) $\forall\, r > 0 \; \mu(\mathscr{K}(x, r)) > 0$;

(ii) $-\lim\limits_{r \to \infty} \dfrac{\nabla \chi_{\mathscr{E}}(\mathscr{K}(x, r))}{\mu(\mathscr{K}(x, r))} = \bar{\boldsymbol{\nu}}(x)$ exists and is finite;

(iii) $|\bar{\boldsymbol{\nu}}(x)| = 1$.

By a theorem of de la Vallée–Poussin (Remark 1.1.19(2)) the limit $\bar{\mathbf{v}}(x)$ defined by (ii) exists μ-a.e. and $\bar{\mathbf{v}}(x) = \mathbf{v}(x)$ μ-a.e. on $\boldsymbol{R}^{m+1}$. We may set $\mathbf{v}(x) = \bar{\mathbf{v}}(x)$ for all $x \in \partial^{**}\mathscr{E}$.

Theorem 3.2.18 (De Giorgi, 1955)
Let $P(\mathscr{E}) < \infty$.

Every point $x \in \partial^{**}\mathscr{E}$ has the following properties:

$$\liminf_{r\to 0} \frac{|\mathscr{E}\cap\mathscr{K}(x, r)|}{|\mathscr{K}(x, r)|} > 0, \tag{3.2.45}$$

$$\liminf_{r\to 0} \frac{|\mathscr{E}^c\cap\mathscr{K}(x, r)|}{|\mathscr{K}(x, r)|} > 0, \tag{3.2.46}$$

$$\lim_{r\to 0} \frac{|\mathscr{K}(x, r)\cap\mathscr{H}(\mathbf{v}(x), x)\cap\mathscr{E}|}{|\mathscr{K}(x, r)|} = 0, \tag{3.2.47}$$

$$\lim_{r\to 0} \frac{|\mathscr{K}(x, r)\cap\mathscr{H}(-\mathbf{v}(x), x)\cap\mathscr{E}^c|}{|\mathscr{K}(x, r)|} = 0, \tag{3.2.48}$$

$$\lim_{r\to 0} \frac{\mu(\mathscr{K}(x, r)\cap\mathscr{I}(x, \varepsilon r))}{r^m\omega_m} = 1 \quad \forall\, \varepsilon > 0, \tag{3.2.49}$$

where ω_m is the Lebesgue measure of a unit ball in $\boldsymbol{R}^m$ and

$$\mathscr{I}(x, \varepsilon r) := \{y \in \boldsymbol{R}^{m+1} \big|\ |\mathbf{v}(x)\cdot(x-y)| \leqslant \varepsilon r\}.$$

For the proof, see De Giorgi (1955). □

Equations (3.2.45) and (3.2.46) imply that $\partial^{**}\mathscr{E} \subset \partial^{*}\mathscr{E}$. Equations (3.2.47) and (3.2.48) imply that $\mathbf{v}(x)$ is the outer normal in Federer's sense. Equation (3.2.49) indicates that $\partial^{**}\mathscr{E}$ is locally plane in some measure-theoretic sense.

It is obvious that for every subset $\mathscr{A} \subset \boldsymbol{R}^{m+1}$ such that $\forall\, x \in \mathscr{A}\ \exists\, r_x > 0$ $\mu(\mathscr{K}(x, r_x)) = 0$ we have $\mu(\mathscr{A}) = 0$. Hence μ is concentrated on the set of points x satisfying (i) of Definition 3.2.17. Actually, it is true that

$$\mu((\partial^{**}\mathscr{E})^c) = 0 \tag{3.2.50}$$

(De Giorgi, 1955).

Moreover, for every Borel subset $\mathscr{P}$ of $\partial^{**}\mathscr{E}$

$$\mu(\mathscr{P}) = H^m(\mathscr{P}) \tag{3.2.51}$$

(cf. (3.2.49)). Hence, from equations. (3.2.43), (3.2.44), (3.2.50) and (3.2.51)

$$\int_{\mathscr{E}} dL^{m+1}\,\nabla g = \int_{\partial^{**}\mathscr{E}} dH^m\, g\mathbf{v} \tag{3.2.52}$$

for every Borel set $\mathscr{E}$ with $P(\mathscr{E}) < \infty$ and every $g \in \mathscr{C}_0^\infty$.

Theorem 3.2.19 (Federer, 1954)
Let $\mathscr{E}$ be a Borel subset of $\boldsymbol{R}^{m+1}$.

Then $P(\mathscr{E}) < \infty$ iff

(i) $H^m(\{x \in \boldsymbol{R}^{m+1} \mid |\tilde{\boldsymbol{\nu}}(x)| = 1\}) < \infty$,

(ii) $\forall\, g \in \mathscr{C}_o^\infty$

$$\int_{\mathscr{E}} dL^{m+1}\, \nabla g = \int dH^m\, g\tilde{\boldsymbol{\nu}}, \tag{3.2.53}$$

where

$$\tilde{\boldsymbol{\nu}}(x) := \begin{cases} \text{the outer normal in the sense of Federer for } \mathscr{E} \text{ at every} \\ \quad x \in \boldsymbol{R}^{m+1} \text{ where it exists,} \\ \boldsymbol{0} \text{ at every other point } x \in \boldsymbol{R}^{m+1}. \end{cases}$$

The right-hand side integral of equation (3.2.53) extends over $\boldsymbol{R}^{m+1}$. We note that the normal in the sense of Federer exists at $x \in \boldsymbol{R}^{m+1}$ only if $x \in \partial^*\mathscr{E}$. Indeed, on account of (3.2.47) and (3.2.48)

$$\liminf_{r\to 0} \frac{|\mathscr{K}(x,r)\cap\mathscr{E}|}{|\mathscr{K}(x,r)|} \leqslant \frac{1}{2}, \quad \liminf_{r\to 0} \frac{|\mathscr{K}(x,r)\cap\mathscr{E}^c|}{|\mathscr{K}(x,r)|} \leqslant \frac{1}{2},$$

resp., and $x \,\bar{\in}\, \mathscr{E}_*$, $x \in \mathscr{E}^*$. Hence $\operatorname{supp}\tilde{\boldsymbol{\nu}} \subset \partial^*\mathscr{E}$. From (3.2.47) and (3.2.48) we conclude that $\partial^{**}\mathscr{E} \subset \operatorname{supp}\tilde{\boldsymbol{\nu}} \subset \partial^*\mathscr{E}$.

For a bounded $\mathscr{E} \subset \mathscr{K}(x_o, R)$ we conclude from (3.2.52) and (3.2.53) that $\tilde{\boldsymbol{\nu}} = \boldsymbol{\nu}$ H^m-a.e. on $\partial^{**}\mathscr{E}$ and

$$H^m(\partial^*\mathscr{E} \setminus \partial^{**}\mathscr{E}) = 0. \tag{3.2.54}$$

For a proof of (3.2.54) for arbitrary Borel $\mathscr{E}$ with bounded perimeter, consult Vol'pert (1967).

Concerning the regularity of the reduced boundaries of sets with $P(\mathscr{E}) < \infty$, we have the following theorem (Miranda, 1964):

Theorem 3.2.20
Suppose that $\mathscr{E}$ is a measurable subset of $\boldsymbol{R}^{m+1}$ with $P(\mathscr{E}) < \infty$ and $\mathscr{A}$ is a convex open subset of $\boldsymbol{R}^{m+1}$.

If there is a positive number p such that the component $\nu_1(x)$ of the normal defined in Definition 3.2.17(ii) is $\geqslant p$ μ-a.e. in $\mathscr{A}$, then there is an open subset $\Omega \subset \boldsymbol{R}^m$ and a function $f \in \operatorname{Lip}(\Omega, \boldsymbol{R})$, such that

(i) $\partial^{**}\mathscr{E} \cap \mathscr{A} = \{(f(y), y) \mid y \in \Omega\}$;

(ii) $\operatorname{Lip} f \leqslant p^{-1}(1-p^2)^{1/2}$.

Ω is defined as the projection of $\partial^{**}\mathscr{E} \cap \mathscr{A}$ onto the hyperplane $x^1 = 0$.

If $x \mapsto \mathbf{v}(x)$ is continuous on $\partial^{**}\mathscr{E} \cap \mathscr{A}$, then there is a $\mathscr{C}^1$ function f such that $f = 0$ and $\nabla f = \mathbf{v}$ on $\partial^{**}\mathscr{E} \cap \mathscr{A}$ by a well-known Whitney extension theorem.

De Giorgi (1954) proved an important inequality valid for every Borel subset $\mathscr{E}$ of $\boldsymbol{R}^{m+1}$.

Theorem 3.2.21
Let $m \geqslant 1$. The following alternative is true: either $P(\mathscr{E})^{m+1} \geqslant L^{m+1}(\mathscr{E})^m$ or $P(\mathscr{E})^{m+1} \geqslant L^{m+1}(\mathscr{E}^c)^m$.

The proof is based on the following alternative: either $\left\{\int dL^{m+1} f\right\}^m \leqslant I(f)^{m+1}$ or $\left\{\int (1-f)\, dL^{m+1}\right\}^m \leqslant I(f)^{m+1}$, in which f is an arbitrary function $f\colon \boldsymbol{R}^{m+1} \to \{0, 1\}$. The latter inequality can be proved by regularization.

In elementary geometry the length or area of closed curves and surfaces is computed by means of approximating inscribed polyhedra. It is noteworthy that this property singles out the (boundaries of) sets with bounded perimeter.

Let Σ be the space of Lebesgue measurable subsets of $\boldsymbol{R}^{m+1}$ endowed with the metric

$$\varrho(\mathscr{E}, \mathscr{E}') := L^{m+1}\left((\mathscr{E} \setminus \mathscr{E}') \cup (\mathscr{E}' \setminus \mathscr{E})\right).$$

It can be proved that Σ is a complete metric space and the family Π of all sets with polyhedral boundaries* is dense in Σ.

Theorem 3.2.22 (De Giorgi, 1954)
If $\lim_{j\to\infty} \varrho(\mathscr{A}_j, \mathscr{E}) = 0$, then $\liminf_{j\to\infty} P(\mathscr{A}_j) \geqslant P(\mathscr{E})$. If also $P(\mathscr{A}_j) \leqslant M \ \forall j$, then $\nabla \chi_{\mathscr{A}_j} \to \nabla \chi_{\mathscr{E}}$.

Proof

$$\lim_{j\to\infty} \int dL^{m+1} |\chi_{\mathscr{E}} - \chi_{\mathscr{A}_j}| = \lim_{j\to\infty} \varrho(\mathscr{A}_j, \mathscr{E}) = 0$$

and

$$\lim_{j\to\infty} \nabla W_\lambda \chi_{\mathscr{A}_j}(x) = \nabla W_\lambda \chi_{\mathscr{E}}(x) \ \forall \lambda > 0 \ \forall x \in \boldsymbol{R}^{m+1}.$$

Indeed,

$$\begin{aligned} &|\nabla W_\lambda \chi_{\mathscr{A}_j}(x) - \nabla W_\lambda \chi_{\mathscr{E}}(x)| \\ &\leqslant (\pi\lambda)^{-(m+1)/2} \int dL^{m+1} \left|\nabla_x \exp(-|x-z|^2/\lambda)\right| \, |\chi_{\mathscr{A}_j}(z) - \chi_{\mathscr{E}}(z)| \\ &\leqslant K \|\chi_{\mathscr{A}_j} - \chi_{\mathscr{E}}\|_{\mathscr{L}^1}. \end{aligned}$$

* A set $\partial\mathscr{A}$ is said to be **polyhedral** if it is contained in a the join of a finite number of hyperplanes of codimension one.

Since $\int dL^{m+1} |\nabla W_\lambda \chi_{\mathscr{A}_j}| \leqslant P(\mathscr{A}_j)$ for every $\lambda > 0$, we also have that

$$P(\mathscr{E}) = \lim_{\lambda \to 0_+} \int dL^{m+1} |\nabla W_\lambda \chi_{\mathscr{E}}|$$

$$= \lim_{\lambda \to 0_+} \int dL^{m+1} \lim_{j \to \infty} |\nabla W_\lambda \chi_{\mathscr{A}_j}|$$

The integrals in the last expression are bounded from above by $\liminf_{j\to\infty} \int dL^{m+1} |\nabla W_\lambda \chi_{\mathscr{A}_j}| \leqslant \liminf_{j\to\infty} P(\mathscr{A}_j)$. This establishes the first part of the thesis.

Let $P(\mathscr{A}_j) \leqslant M \ \forall j, g \in \mathscr{C}_o^0$, $g_\varepsilon \in \mathscr{C}_o^1$, $\sup|g - g_\varepsilon| < \varepsilon$. Then $P(\mathscr{E}) \leqslant M$, $\lim_{j\to\infty} \int g_\varepsilon \nabla \chi_{\mathscr{A}_j} = -\lim_{j\to\infty} \int dL^{m+1} \chi_{\mathscr{A}_j} \nabla g_\varepsilon = -\int dL^{m+1} \chi_{\mathscr{E}} \nabla g_\varepsilon = \int g_\varepsilon \nabla \chi_{\mathscr{E}}$, since $\chi_{\mathscr{A}_j} \to \chi_{\mathscr{E}}$ in $\mathscr{L}^1$.

Hence

$$\limsup_{j\to\infty} \left| \int g(x) \nabla \chi_{\mathscr{A}_j} - \int g(x) \nabla \chi_{\mathscr{E}} \right| \leqslant 2\varepsilon M.$$

Since $\varepsilon > 0$ is arbitrary, it follows that $\nabla \chi_{\mathscr{A}_j} \rightharpoonup \nabla \chi_{\mathscr{E}}$. □

Theorem 3.2.23 (De Giorgi, 1954)
Let $\mathscr{E}$ be a measurable subset of $\boldsymbol{R}^{m+1}$.

Then $P(\mathscr{E}) < \infty$ iff the following conditions are simultaneously satisfied:

(i) there is a positive number M and a sequence of sets $\{\mathscr{A}_j\}$ with polyhedral boundaries, such that $\varrho(\mathscr{A}_j, \mathscr{E}) \to 0$;

(ii) $H^m(\partial \mathscr{A}_j) \leqslant M$.

In this case $\lim_{j\to\infty} P(\mathscr{A}_j) = P(\mathscr{E})$.

Since $P(\mathscr{E}) < \infty$, we know that one of the two sets $\mathscr{E}, \mathscr{E}^c$ has finite Lebesgue measure. For this set an approximating sequence of polyhedra is constructed.

The conditions (i) and (ii) correspond to the original geometric definition of sets with bounded perimeter given by Cacciopoli.

3.2.6 Properties of Functions with Essentially Bounded Variation

We shall now proceed to describe the most important properties of the BV functions.

Definition 3.2.24
Let $\mathbf{f}$ be a mapping of $\boldsymbol{R}$ into $\boldsymbol{R}^p$ and let $\mathscr{I}$ be a closed interval of $\boldsymbol{R}$. The variation of $\mathbf{f}$ over $\mathscr{I}$, $\mathrm{var}(\mathbf{f}, \mathscr{I})$, is the lowest upper bound of the sums

$\sum_{k=1}^{n-1} |\mathbf{f}(t_{k+1}) - \mathbf{f}(t_k)|$ for all the finite sequences $t_1 < t_2 < \ldots < t_n$, $t_j \in \mathscr{I}$ for $j = 1, \ldots, n$, and arbitrary integer $n > 0$. We shall supplement this definition by setting $\text{var}(\mathbf{f}, \emptyset) = 0$.

Definition 3.2.25
A measurable function $\mathbf{f}\colon \boldsymbol{R} \to \boldsymbol{R}^p$ is said to have **locally bounded variation** if $\exists\, \mathbf{g}\colon \boldsymbol{R} \to \boldsymbol{R}^p$ such that $\mathbf{f} = \mathbf{g}$ a.e. and $\text{var}(\mathbf{g}, \mathscr{I}) < \infty$ for every finite interval $\mathscr{I}$.

Theorem 3.2.26 (Krickeberg, 1957)
$\mathbf{f} \in \mathscr{L}^1_{loc}(\boldsymbol{R}, \boldsymbol{R}^p)$ has locally bounded variation if the distributional derivative $\mathbf{f}'$ is a Borel regular measure $\mathbf{m}$. Moreover $|\mathbf{m}|(\mathscr{B}) \leqslant \text{var}(\mathbf{g}, \mathscr{I})$ for every Borel set $\mathscr{B} \subset \mathscr{I}$.

It is possible to choose the function $\mathbf{g}$ in such a way that $\mathbf{g}$ is right continuous and in this case

$$|\mathbf{m}|(]a, b]) = \text{var}(\mathbf{g}, [a, b]) \ \forall a, b \in \boldsymbol{R}.$$

Definition 3.2.27
For an arbitrary function $\mathbf{v}\colon \boldsymbol{R}^{m+1} \to \boldsymbol{R}^p$, $y = (y^0, y^1, \ldots, y^{m-1}) \in \boldsymbol{R}^m$, $i \in \{0, 1, \ldots, m\}$, and for an arbitrary subset $\mathscr{E} \subset \boldsymbol{R}^{m+1}$ we set

$$\mathbf{v}(i, y)(t) :\equiv \mathbf{v}(y^0, y^1, \ldots, y^{i-1}, t, y^i, \ldots, y^{m-1}),$$
$$\mathscr{E}_{(i,y)} := \{t \in \boldsymbol{R} |\ (y^0, y^1, \ldots, y^{i-1}, t, y^i, \ldots, y^{m-1}) \in \mathscr{E}\}.$$

For $a, b \in \boldsymbol{R}^{m+1}$ let $[a, b]$ denote the closed interval $a^i \leqslant x^i \leqslant b^i, i = 0, 1, \ldots \ldots, m$.

Definition 3.2.28
A function $\mathbf{u}\colon \boldsymbol{R}^{m+1} \to \boldsymbol{R}^p$ is said to **have essentially bounded variation in the sense of Tonnelli–Cesari** if

(i) $\exists\, \mathbf{v}\colon \boldsymbol{R}^{m+1} \to \boldsymbol{R}^p$, measurable and such that $\mathbf{u} = \mathbf{v}$ L^{m+1}-a.e.;

(ii) for L^m-almost all values of $y \in \boldsymbol{R}^m$ and all $i \in \{0, 1, \ldots, m\}$ $\mathbf{v}(i, y)(\cdot)$ has locally bounded variation, is right continuous and for all a, $b \in \boldsymbol{R}^{m+1}$ the function $y \mapsto \text{var}(\mathbf{v}(i, y), [a, b]_{(i,y)})$ belongs to $\mathscr{L}^1$.

Theorem 3.2.29 (Krickeberg, 1957)
$\mathbf{u} \in \mathscr{L}^1_{loc}(\boldsymbol{R}^{m+1}, \boldsymbol{R}^p)$ has essentially bounded variation in the sense of Tonnelli–Cesari iff $\nabla \mathbf{u}$ is a vector Borel regular measure.

Remarks

(1) For $m = 1$ Theorem 3.2.29 implies that for every function $\mathbf{u} \in \text{BV}(\boldsymbol{R}^2, \boldsymbol{R}^p)$ there is a function $\mathbf{v}$ such that $\mathbf{u} = \mathbf{v}$ L^2-a.e. and

(i) for almost all t the function $\mathbf{v}(t, \cdot)$ has locally bounded variation;

(ii) for every bounded interval $[a, b]$ the variation $w(t)$ of $\mathbf{v}(t, \cdot)$ over $[a, b]$ defines a function $w \in \mathscr{L}^1_{loc}$.

(2) Let $\mathscr{G}$ be a bounded subset of $\boldsymbol{R}^m$, $\mathscr{I} = \{x^0 | a < x^0 < b\}$, $\mathbf{u} \in \mathrm{BV}(\boldsymbol{R}^{m+1})$. Then $\exists\ \mathscr{F} \subset \mathscr{G}$, $L^m(\mathscr{G} \setminus \mathscr{F}) = 0$, such that

(i) $\tilde{\mathbf{u}}(x^0, y) = \lim\limits_{h \to 0} \dfrac{1}{2h} \displaystyle\int_{x^0-h}^{x^0+h} dz\, \mathbf{u}(z, y)$ exists for all $x^0 \in \mathscr{I}$, $y \in \mathscr{F}$;

(ii) $x^0 \mapsto \tilde{\mathbf{u}}(x^0, y)$ has bounded variation for all $y \in \mathscr{F}$;

(iii) $\mathbf{I}(y) := \int \varphi(x^0, y)\, d\tilde{\mathbf{u}}(x^0, y)$ exists for all $y \in \mathrm{H}$ and for every $\varphi(x^0, y)$ summable with respect to the measure $\mathbf{u}_{x^0}$, where $L^m(\mathrm{F} \setminus \mathrm{H}) = 0$;

(iv) $\displaystyle\int_{\mathscr{I} \times \mathscr{G}} \varphi \mathbf{u}_{x^0} = \int_{\mathscr{G}} dL^m(y) \mathbf{I}(y)$.

As we shall see, all the significant properties of BV functions are expressed in terms of notions which are insensitive to any modifications on sets of measure zero. In particular derivation in the distributional sense is defined in terms of a Lebesgue integral and all the limits and traces are defined in the measure-theoretic sense. Theorem 4.2.29 and the following remarks allow a convenient choice of the representative of a BV function so that it can be handled more easily by means of the familiar topological and analytic methods and concepts.

We now investigate the set of regular points of a function $\mathbf{u} \in \mathrm{BV}(\mathscr{D}, \boldsymbol{R}^p)$, $\mathscr{D} \subset \boldsymbol{R}^{m+1}$. The fine structure of a BV function will be described in terms of sets with bounded perimeter. The key fact connecting these notions is expressed in the following theorem, which will be quoted without proof.

Theorem 3.2.30 (Fleming and Rishel, 1960)
Let $\mathscr{D}$ be a bounded subset of $\boldsymbol{R}^{m+1}$ and let $u \in \mathrm{BV}(\mathscr{D}, \boldsymbol{R})$.

For almost all the set $z \in \boldsymbol{R}$ $\mathscr{E}_z := \{x \in \mathscr{D} |\ u(x) < z\}$ has bounded perimeter and

$$|\nabla u|(\mathscr{D}) = \int_{-\infty}^{\infty} dt\, P(\mathscr{E}_t) \tag{3.2.55}$$

Let

$$Tu(x_0) := \lim_{r \to 0} \frac{1}{\omega_{m+1} r^{m+1}} \int_{\mathscr{K}(x_0, r)} dL^{m+1}\, u(x),$$

$$\bar{T}u(x_0) := \limsup_{r \to 0} \frac{1}{\omega_{m+1} r^{m+1}} \int_{\mathscr{K}(x_0, r)} dL^{m+1}\, u(x).$$

Lemma 3.2.31 (Vol'pert, 1967)
Suppose that $P(\mathscr{E}) < \infty$, $P(\mathscr{E}_j) < \infty, j = 1, \ldots, n,$

$$\mathscr{E} = \bigcup_{j=1}^{n} \mathscr{E}_j, \mathscr{E}_j \cap \mathscr{E}_k \neq \emptyset \quad \text{for } j \neq k,$$

$$\Gamma_{kl} := \partial^{**}\mathscr{E}_l \cap \partial^{**}\mathscr{E}_k.$$

Then

$$\mathscr{E}_* = \bigcup_{k=1}^{n} (\mathscr{E}_k)_* \cup \bigcup_{\substack{k,l=1 \\ k \neq l}}^{n} \Gamma_{kl} \cup \mathscr{F} \quad \text{with } H^m(\mathscr{F}) = 0,$$

$$\mathscr{F} \cap \Gamma_{kl} = \emptyset, \quad \mathscr{F} \cap \mathscr{E}_{k*} = \emptyset, \quad \mathscr{E}_{p*} \cap \Gamma_{kl} = \emptyset, \quad \mathscr{E}_{k*} \cap \mathscr{E}_{l*} = \emptyset$$

for $k \neq l$.

Moreover, if $x_0 \in \Gamma_{kl}$ and $\mathbf{v}_k$, $\mathbf{v}_l$ are the outer unit normals* for $\mathscr{E}_k$, $\mathscr{E}_l$ at x_0, resp., then $\mathbf{v}_k + \mathbf{v}_l = \mathbf{0}$.

Proof

Let $\mathscr{G} := \bigcup_{k=1}^{n} \mathscr{E}_{k*} \cup \bigcup_{\substack{k,l=1 \\ k \neq l}}^{n} \Gamma_{kl}$, $\mathscr{H} := \bigcup_{k=1}^{n} \mathscr{E}_{k*} \cup \bigcup_{\substack{k,l=1 \\ k \neq l}}^{n} (\partial^*\mathscr{E}_k \cap \partial^*\mathscr{E}_l)$. Note that $x_0 \in \mathscr{E}_{k*}$ iff $T\chi_{\mathscr{E}_k}(x_0) = 1$, $x_0 \in \Gamma_{kl}$ iff $T\chi_{\mathscr{E}_k}(x_0) = T\chi_{\mathscr{E}_l}(x_0) = \frac{1}{2}$ (equations (3.2.47) and (3.2.48)). Hence $\Gamma_{kl} \cap \mathscr{E}_{p*} = \emptyset$ for all k, l, p such that $k \neq l$. Moreover $T\chi_{\mathscr{E}}(x_0) = \sum_{k=1}^{n} T\chi_{\mathscr{E}_k}(x_0) = 1$ if $x_0 \in \mathscr{G}$. Hence $\mathscr{G} \subset \mathscr{E}_*$.

The next step is to show that $\mathscr{E}_* \subset \mathscr{H}$. Indeed, for $x_0 \in \mathscr{E}_*$ we have $T\sum_{k=1}^{n} \chi_{\mathscr{E}_k}(x_0) = 1$ and hence the following alternative is true: either (1) $\exists k \leqslant n$ such that $T\chi_{\mathscr{E}_i}(x_0) = 0$ for all $i \neq k$, or (2) $\exists i, j \leqslant n$, $i \neq j$ $\bar{T}\chi_{\mathscr{E}_i}(x_0) > 0$, $\bar{T}\chi_{\mathscr{E}_j}(x_0) > 0$. In the first case $x_0 \in \mathscr{E}_{k*} \subset \mathscr{H}$. In the second case $x_0 \in \mathscr{E}_k^* \cap \mathscr{E}_l^*$ and neither of $T\chi_{\mathscr{E}_i}(x_0)$, $T\chi_{\mathscr{E}_j}(x_0)$ can exist and be equal to one i.e. $x_0 \bar{\in} \mathscr{E}_{k*} \cup \mathscr{E}_{l*}$. Hence $x_0 \in \partial^*\mathscr{E}_k \cap \partial^*\mathscr{E}_l \subset \mathscr{H}$.

It follows immediately that $\mathscr{E}^* \setminus \mathscr{G} \subset \mathscr{H} \setminus \mathscr{G}$ has zero Hausdorff measure (equation (3.2.54)).

Let $x_0 \in \partial^{**}\mathscr{E}_k \cap \partial^{**}\mathscr{E}_l$, $\mathscr{K}_{\mathbf{n}}(r) := \mathscr{K}(x_0, r) \cap \mathscr{H}(\mathbf{n}, x_0)$.

Then

$$\frac{1}{\omega_{m+1}} |\mathscr{K}_{-\mathbf{v}_k}(1) \cup \mathscr{K}_{-\mathbf{v}_l}(1)| \geqslant \frac{1}{\omega_{m+1} r^{m+1}} \int_{\mathscr{K}_{-\mathbf{v}_k}(r) \cup \mathscr{K}_{-\mathbf{v}_l}(r)} dL^{m+1} \chi_{\mathscr{E}}$$

* In the sense of Federer.

$$\geqslant \frac{1}{\omega_{m+1} r^{m+1}} \left\{ \int\limits_{\mathscr{K}_{-\nu_k}(r)} dL^{m+1} \chi_{\mathscr{E}_k} + \int\limits_{\mathscr{K}_{-\nu_l}(r} + dL^{m+1} \chi_{\mathscr{E}_l} \right\} \to 1$$

as $r \to 0$. Hence $\boldsymbol{\nu}_k + \boldsymbol{\nu}_l = \mathbf{0}$. □

Lemma 3.2.32

Let $P(\mathscr{E}) < \infty$, $P(\mathscr{E}_k^r) < \infty$ for $r = 1, 2, 3, \ldots$; $k = 1, \ldots, n_r$ and

(i) $\mathscr{E} = \bigcup_{k=1}^{n_r} \mathscr{E}_k^r$, $\mathscr{E}_k^r \cap \mathscr{E}_l^r = \emptyset$ for $k \neq l$;

(ii) $\forall k \in \{1, \ldots, n_r\}$ $\exists l \in \{1, \ldots, n_{r+1}\}$ $\mathscr{E}_l^{r+1} \subset \mathscr{E}_k^r$.

Then $\exists \mathscr{A}, \mathscr{B}$ such that

(1) $\mathscr{A} \cap \mathscr{B} = \emptyset$, $\mathscr{A} \cup \mathscr{B} \subset \mathscr{E}_*$;

(2) $H^m(\mathscr{E}_* \setminus (\mathscr{A} \cup \mathscr{B})) = 0$;

(3) $\forall x_0 \in \mathscr{A}$ $\exists \{\mathscr{E}^r\}$ such that $\forall r \, \exists k \leqslant n_r$ $\mathscr{E}^r = \mathscr{E}_k^r$, $\mathscr{E}^{r+1} \subset \mathscr{E}^r$, $x_0 \in \mathscr{E}^r$;

(4) $\forall x_0 \in \mathscr{B}$ $\exists \mathbf{v} \in \mathscr{S}^m$ $\exists s_0 \in \boldsymbol{Z}_+$ $\exists \{\mathscr{E}_+^s\}$, $\{\mathscr{E}_-^s\}$ such that $\forall s \geqslant s_0$ $\exists k, l$ $\mathscr{E}_+^s = \mathscr{E}_k^s$, $\mathscr{E}_-^s = \mathscr{E}_l^s$, $\mathscr{E}_+^{s+1} \subset \mathscr{E}_+^s$, $\mathscr{E}_-^{s+1} \subset \mathscr{E}_-^s$ and $\mathbf{v}$ is the outer normal for $\mathscr{E}_-^s$ at x_0 and the inner normal for $\mathscr{E}_+^s$ at x_0 for every $s \geqslant s_0$.

Proof

Let

$$\mathscr{A}^s := \bigcup_{k=1}^{n_s} \mathscr{E}_{k*}^s, \quad \mathscr{B}^s := \bigcup_{\substack{k,l=1 \\ k \neq l}}^{n_s} \partial^{**}\mathscr{E}_k^s \cap \partial^{**}\mathscr{E}_l^s,$$

$$\mathscr{A} := \bigcup_{k=1}^{\infty} \bigcap_{s=k}^{\infty} \mathscr{A}^s, \qquad \mathscr{B} := \bigcap_{k=1}^{\infty} \bigcup_{s=k}^{\infty} \mathscr{B}^s.$$

Ad (1). Lemma 3.2.31 implies that $\mathscr{A}^s \cup \mathscr{B}^s \subset \mathscr{E}_*$, $\mathscr{A}^s \cap \mathscr{B}^s = \emptyset$. Hence $\mathscr{A} \cup \mathscr{B} \subset \mathscr{E}_*$, $\mathscr{A} \cap \mathscr{B} = \emptyset$.

Ad (2). Let $x \in \bigcap_{s=1}^{\infty} (\mathscr{A}^s \cup \mathscr{B}^s)$. Then for all positive integers s $x \in \mathscr{A}^s \cup \mathscr{B}^s$. If $\exists s_0$ such that $\forall s \geqslant s_0$, $x \in \mathscr{A}^s$, then $x \in \mathscr{A}$. If $\forall s_0$ such that $\exists s \geqslant s_0$, $x \in \mathscr{B}^s$, then $x \in \mathscr{B}^S$, $\bigcap_{s=1}^{\infty} (\mathscr{A}^s \cup \mathscr{B}^s) \subset \mathscr{A} \cup \mathscr{B}$ and $\mathscr{E}_* \setminus (\mathscr{A} \cup \mathscr{B}) \subset \bigcup_{s=1}^{\infty} \mathscr{E}_* \setminus (\mathscr{A}^s \cup \mathscr{B}^s)$ has zero Hausdorff measure (Lemma 3.2.31).

Ad (3). Let $x_0 \in \mathscr{A}$. There is a positive integer s_0 such that for all $s \geqslant s_0$ $x_0 \in \mathscr{A}^{s_0}$. Since $\mathscr{A}^{s_0} \subset \mathscr{A}^{s_0 - 1} \subset \ldots \subset \mathscr{A}^1$, we have that $x_0 \in \bigcap_{s=1}^{\infty} \mathscr{A}^s$. Hence $\forall s$ $\exists k \in \{1, \ldots, n_s\}$ such that $x_0 \in \mathscr{E}_k^s$. Let $\mathscr{E}^s := \mathscr{E}_k^s$ in this case. Suppose now that $\mathscr{E}^{s+1} \not\subset \mathscr{E}^s$ for some s. Obviously $\exists l, p$ such that $\mathscr{E}^{s+1} = \mathscr{E}_l^{s+1} \subset \mathscr{E}_p^s$

and $p \neq k$, hence $\mathscr{E}^{s+1} \cap \mathscr{E}^s = \emptyset$. It follows that $x_0 \bar{\in} \mathscr{E}^s_* \cap \mathscr{E}^{s+1}_*$, contradicting the preceding conclusion.

Ad (4). Let $x_0 \in \mathscr{B}$. Then $\exists s_0$ such that $x_0 \in \mathscr{B}^{s_0}$ and $\exists k, l, k \neq l$, such that $x_0 \in \partial^{**}\mathscr{E}^{s_0}_k \cap \partial^{**}\mathscr{E}^{s_0}_l$. Let $\mathbf{v}$ be the inner normal for $\mathscr{E}^{s_0}_k$ at x_0.

Let $s \geqslant s_0$. Then $\exists s_1 > s$ and $\exists k_1, l_1, k_1 \neq l_1$, such that $x_0 \in \partial^{**}\mathscr{E}^{s_1}_{k_1} \cap \cap \partial^{**}\mathscr{E}^{s_1}_{l_1}$. Let $\mathbf{v}_1$ be the inner normal for $\mathscr{E}^{s_1}_{k_1}$ at x_0. For some $i, j \leqslant n_s$ $\mathscr{E}^{s_1}_{k_1} \subset \mathscr{E}^s_i, \mathscr{E}^{s_1}_{l_1} \subset \mathscr{E}^s_j$. Let $T_{r,\mathbf{n}}\mathbf{u}(x_0) := \dfrac{2}{\omega_{m+1} r^{m+1}} \displaystyle\int_{\mathscr{K}_{\mathbf{n}}(r)} dL^{m+1}\mathbf{u}(x)$ for any $\mathbf{n} \in \mathscr{S}^m$.

Since $\chi_{\mathscr{E}^{s_1}_{k_1}} \leqslant \chi_{\mathscr{E}^s_i}$, we have $T_{r,\mathbf{v}_1}\chi_{\mathscr{E}^{s_1}_{k_1}}(x_0) \leqslant T_{r,\mathbf{v}_1}\chi_{\mathscr{E}^s_i}(x_0) \leqslant 1$ and

$$T_{r,\mathbf{v}_1}\chi_{\mathscr{E}^{s_1}_{k_1}}(x_0) \to 1 \quad \text{for } r \to 0. \tag{3.2.56}$$

Hence

$$T_{r,\mathbf{v}_1}\chi_{\mathscr{E}^s_i}(x_0) \to 1 \quad \text{for } r \to 0. \tag{3.2.57}$$

Similarly, $-\mathbf{v}_1$ is the inner normal for $\mathscr{E}^{s_1}_{l_1}$ at x_0, by Lemma 3.2.31, and

$$T_{r,-\mathbf{v}_1}\chi_{\mathscr{E}^s_j}(x_0) \to 1 \quad \text{for } r \to 0. \tag{3.2.58}$$

If $i = j$, then $T\chi_{\mathscr{E}^s_i}(x_0) = \frac{1}{2}\lim_{r\to 0}(T_{r,\mathbf{v}_1} + T_{r,-\mathbf{v}_1})\chi_{\mathscr{E}^s_i}(x_0) = 1$ and $x_0 \in \mathscr{E}^s_{i*}$, $x_0 \bar{\in} \mathscr{B}$, which contradicts the hypothesis of (4). Hence $i \neq j$ and $\chi_{\mathscr{E}^s_i}(x) + \chi_{\mathscr{E}^s_j}(x) \leqslant 1 \; \forall x$. Equation (3.2.58) implies that

$$\lim_{r\to 0} T_{r,-\mathbf{v}_1}\chi_{\mathscr{E}^s_i}(x_0) = 0. \tag{3.2.59}$$

Equations (3.2.59) and (3.2.56) show that $\mathbf{v}_1$ is the inner normal for $\mathscr{E}^s_i$ at x_0. Equations (3.2.57) and (3.2.58) imply that $-\mathbf{v}_1$ is the inner normal for $\mathscr{E}^s_j$ at x_0.

The above conclusion is true in particular for $s = s_0$. Let i, j be defined as above for $s = s_0$. Since for $s = s_0$ $T\chi_{\mathscr{E}^{s_0}_p}(x_0) = \frac{1}{2}$ for $p = i, j, k, l$, and $T\chi_{\mathscr{E}}(x_0) = 1$, we conclude that at most two of the four sets $\mathscr{E}^{s_0}_i, \mathscr{E}^{s_0}_j, \mathscr{E}^{s_0}_k, \mathscr{E}^{s_0}_l$ are distinct. Now $\mathscr{E}^{s_0}_i \neq \mathscr{E}^{s_0}_j$, $\mathscr{E}^{s_0}_k \neq \mathscr{E}^{s_0}_l$ and hence either $\mathscr{E}^{s_0}_i = \mathscr{E}^{s_0}_k$ or $\mathscr{E}^{s_0}_i = \mathscr{E}^{s_0}_l$. Hence either $\mathbf{v} = \mathbf{v}_1$ or $\mathbf{v} = -\mathbf{v}_1$. Since $s \geqslant s_0$ was arbitrary in the preceding argument, we have proved that for every $s \geqslant s_0$ $\exists \mathscr{E}^s_+, \mathscr{E}^s_-$ such that $\mathbf{v}$ is the inner normal for $\mathscr{E}^+_s$ and the outer normal for $\mathscr{E}^s_-$ at x_0, with $\mathscr{E}^s_+ = \mathscr{E}^s_i$, $\mathscr{E}^s_- = \mathscr{E}^s_j$ for some i, j.

Suppose that $\mathscr{E}^{s+1}_+ \not\subset \mathscr{E}^s_+$. Then $\mathscr{E}^s_+ \cap \mathscr{E}^{s+1}_+ = \emptyset$ and $\chi_{\mathscr{E}^s_+}(x) + \chi_{\mathscr{E}^{s+1}_+}(x) \leqslant 1$ $\forall x$. But

$$\begin{aligned}&\lim_{r\to 0} T_{r,\mathbf{v}}\left(\chi_{\mathscr{E}^s_+}(x_0) + \chi_{\mathscr{E}^{s+1}_+}(x_0)\right)\\ &= \lim_{r\to 0} T_{r,\mathbf{v}}\chi_{\mathscr{E}^s_+}(x_0) + \lim_{r\to 0} T_{r,\mathbf{v}}\chi_{\mathscr{E}^{s+1}_+}(x_0) = 2,\end{aligned}$$

a contradiction. Hence $\mathscr{E}^{s+1}_+ \subset \mathscr{E}^s_+$, $\mathscr{E}^{s+1}_- \subset \mathscr{E}^s_-$. □

Theorem 3.2.33 (Vol'pert, 1967)
Let $\mathbf{u} \in BV(\mathscr{D}, \mathbf{R}^p) \cap \mathscr{L}^\infty(\mathscr{D}, \mathbf{R}^p)$.

Let $\mathscr{R}$ be the set of regular points of $\mathbf{u}$. Then $H^m(\mathscr{D} \setminus \mathscr{R}) = 0$.

Proof
It is enough to prove the theorem for an arbitrary open ball $\mathscr{K}$, $\bar{\mathscr{K}} \subset \mathscr{D}$:

$$H^m(\mathscr{K} \setminus \mathscr{R}) = 0,$$

since $\mathscr{D}$ can be covered by a denumerable family of such balls.

Let $M = \|\mathbf{u}\|_\infty$,

(i) $-M = y_1^n < y_2^n < \dots < y_{s_n}^n = M$, $n = 1, 2, \dots$,

(ii) $\forall n$, $\forall i \leqslant s_{n+1} - 1$ $\exists k \leqslant s_n - 1$ such that $[y_i^{n+1}, y_{i+1}^{n+1}] \subset [y_k^n, y_{k+1}^n]$,

(iii) $\forall n$, $\forall i \leqslant s_n - 1$ $|y_i^n - y_{i+1}^n| \leqslant \dfrac{1}{n}$.

It follows from Theorem 3.2.30 that $\{y_i^n\}$ can be chosen in such a way that $P(\mathscr{E}_i^{k,n}) < \infty$, $\mathscr{E}_i^{k,n} := \{x \in \mathscr{K} | \ u^k(x) \in]y_i^n, y_{i+1}^n]\}$, $k \leqslant p$, $i \leqslant s_n - 1$, $n = 1, 2, \dots$

Let $\mathscr{E}_{i_1 \dots i_p}^n := \bigcap_{k=1}^{p} \mathscr{E}_{i_k}^{k,n}$, where $i_1, \dots, i_p \in \{1, \dots, s_n - 1\}$. Let $\{\mathscr{E}_i^n\}$ be the family of all the non-empty sets $\mathscr{E}_{i_1 \dots i_p}^n$. Then

(iv) $P(\mathscr{E}_i^n) < \infty$,* $i = 1, \dots, q_n$, $\mathscr{E} = \mathscr{K} = \bigcup_{i=1}^{q_n} \mathscr{E}_i^n$, $\mathscr{E}_i^n \cap \mathscr{E}_j^n = \emptyset$ for $i \neq j$, $i, j \leqslant q_n$.

(v) $\forall i \leqslant q_n$, $\forall k \leqslant p$, $\exists j \leqslant s_{n-1}$ such that

$$\mathscr{E}_n^i \subset \{x \in \mathscr{K} | \ y_j^n < u^k(x) \leqslant y_{j+1}^n\},$$

(vi) $\forall n$, $\forall i \leqslant q_{n+1}$, $\exists j \leqslant q_n$ such that $\mathscr{E}_i^{n+1} \subset \mathscr{E}_j^n$.

Let $\mathscr{A}$, $\mathscr{B}$ be defined according to Lemma 3.2.32. We shall show that $\mathscr{A} \cup \mathscr{B} \subset \mathscr{R}$.

Let $x_0 \in \mathscr{A}$. Then there is a sequence $\{\mathscr{E}^n\}$ such that, for each n, $\mathscr{E}^n = \mathscr{E}_{k_n}^n$ for some $k_n \leqslant q_n$, $x_0 \in \mathscr{E}^n{}_*$, $\mathscr{E}^{n+1} \subset \mathscr{E}^n$. Let $k \leqslant p$. Then $\forall n$, $\exists j_n \leqslant s_n - 1$ such that $\mathscr{E}^n \subset \{x \in \mathscr{E} | \ y_{j_n}^n < u^k(x) \leqslant y_{j_n+1}^n\}$.

Let $\alpha^n := y_{j_n}^n$, $\beta^n := y_{j_n+1}^n$. Since $x_0 \in \mathscr{E}^{n+1}{}_*$, it follows that $\mathscr{E}^{n+1} \neq \emptyset$. Let $x_1 \in \mathscr{E}^{n+1} \subset \mathscr{E}^n$. Then $u^k(x_1) \in]\alpha^n, \beta^n] \cap]\alpha^{n+1}, \beta^{n+1}]$ and, in view of (ii), $]\alpha^n, \beta^n[\supset]\alpha^{n+1}, \beta^{n+1}[$. From (iii) it follows that $\alpha^n \nearrow \varkappa$, $\beta^n \searrow \varkappa$ for some $\varkappa \in \mathbf{R}$.

For every $\varepsilon > 0$ $\exists n$ such that $\{x \in \mathscr{E} | \ |u^k(x) - \varkappa| > \varepsilon\} \subset (\mathscr{E}^n)^c$ (set $\varepsilon \geqslant 1/n$). Moreover $x_0 \in \mathscr{E}^n{}_*$, hence $x_0 \,\bar{\in}\, \{x \in \mathscr{E} | \ |u^k(x) - \varkappa| > \varepsilon\}^*$. Consequently $\forall k \leqslant p$ $\exists \varkappa$ $lu^k(x_0) = \varkappa$ and $l\mathbf{u}(x_0)$ exists.

* The fact that a finite intersection of sets with bounded perimeter has bounded perimeter is most easily deduced from Theorem 3.2.45 below.

Let us now suppose that $x_0 \in \mathscr{B}$, and let $\boldsymbol{\nu}$, $\{E^n_+\}$, $\{\mathscr{E}^n_-\}$ be defined according to Lemma 3.2.32. Clearly $\exists\, [\alpha^n, \beta^n]$ such that $\mathscr{E}^n_+ \subset \{x \in \mathscr{E} \mid \alpha^n < u^k(x) \leqslant \beta^n\}$. Again $]\alpha^{n+1}, \beta^{n+1}[\, \subset\,]\alpha^n, \beta^n[$, $\alpha^n \nearrow \varkappa$, $\beta^n \searrow \varkappa$ for some $\varkappa \in \boldsymbol{R}$. For any $x \in \mathscr{E}^n_+$ we have $|u^k(x) - \varkappa| < 1/n$ and for $\varepsilon \geqslant 1/n$ it follows that

$$\mathscr{G}_\varepsilon := \left\{x \in \mathscr{E} \cap \mathscr{H}(\boldsymbol{\nu}, x_0) \middle| \; |u^k(x_0) - \varkappa| > \varepsilon\right\} \subset (\mathscr{E}^n_+)^c \cap \mathscr{H}(\boldsymbol{\nu}, x_0).$$

Since $\boldsymbol{\nu}$ is the inner normal for $\mathscr{E}^n_+$ at x_0, it follows that $x_0 \in \mathscr{G}_\varepsilon{}^*$. Hence $l_\nu u^k(x_0) = \varkappa$. This proves that $\forall x_0 \in \mathscr{B}$ the limits $l\mathbf{u}_\nu(x_0)$, $l_{-\nu}\mathbf{u}(x_0)$ exist and $\mathscr{A} \cup \mathscr{B} \subset \mathscr{R}$.

Hence $H^m(\mathscr{K} \setminus \mathscr{R}) \leqslant H^m(\mathscr{K} \setminus (\mathscr{A} \cup \mathscr{B})) = 0$.

Incidentally, we have also proved that the set of jump points is contained in a countable family of reduced boundaries of sets with bounded perimeter, up to a set of zero Hausdorff measure. □

Definition 3.2.34

A Borel set $\mathscr{S}$ is said to be **of class** Γ if $\exists \{\mathscr{E}_i\}$, $P(\mathscr{E}_i) < \infty$ such that $\mathscr{S} \subset \bigcup\limits_{i=1}^{\infty} \partial^* \mathscr{E}_i$.

Theorem 3.2.35

Let $\mathbf{u} \in \mathrm{BV}\ (\mathscr{D}, \boldsymbol{R}^p) \cap \mathscr{L}^\infty(\mathscr{D}, \boldsymbol{R}^p)$ and let $\Gamma(\mathbf{u})$ be the jump set of $\mathbf{u}$. Then $\exists \mathscr{S} \in \Gamma$ such that $\Gamma(\mathbf{u}) = \mathscr{S} \cup \mathscr{M}$, $H^m(\mathscr{M}) = 0$.

Proof

It is sufficient to consider $\mathbf{u}|\mathscr{K}$, $\mathscr{K}$ being an open ball such that $\bar{\mathscr{K}} \subset \mathscr{D}$. Then $\mathscr{K} = \mathscr{A} \cup \mathscr{B} \cup \mathscr{F}$ in the notations of Lemma 3.2.32 and $\Gamma(\mathbf{u}) \subset \mathscr{B} \cup \mathscr{F}$, $H^m(\mathscr{F}) = 0$, since every point of $\mathscr{A}$ is a point of approximate continuity.* The construction of $\mathscr{B}$ in Lemma 3.2.32 shows that $\mathscr{B} \subset \mathscr{S} \cup \mathscr{G}$, $H^m(\mathscr{G}) = 0$, $\mathscr{S} \in \Gamma$. □

Definition 3.2.36

Let $\varphi \in \mathscr{L}^\infty$, $\varphi(x) \geqslant 0$ L^{m+1}-a.e., $\operatorname{supp}\varphi \subset \mathscr{K}(0, 1)$,

$$\int dL^{m+1}\, \varphi(x) = 1,$$

$$\varphi_r(x) := r^{-m-1} \varphi(xr^{-1}),\ \forall r > 0,\ x \in \boldsymbol{R}^{m+1},$$

$$T(\varphi)\mathbf{u}(x_0) := \lim_{r \to 0} \int dL^{m+1}\, \varphi_r(x - x_0)\mathbf{u}(x) \tag{3.2.60}$$

* A point of approximate continuity of **u** is a regular point of **u** which is not a jump point of **u**.

for every $x_0 \in \boldsymbol{R}^{m+1}$ such that the right-hand side exists. $T(\varphi)$ will be called an **averaging operator.**

Lemma 3.2.37

Let $\mathbf{u} \in \mathrm{BV}(\mathscr{D}, \boldsymbol{R}^p) \cap \mathscr{L}^\infty(\mathscr{D}, \boldsymbol{R}^p)$ and let $\mathscr{R}$ be the set of regular points of $\mathbf{u}$. Then for every $x_0 \in \mathscr{R}$

$$T(\varphi)\mathbf{u}(x_0) = \int_{\mathscr{H}(\mathbf{v},0)} dL^{m+1}\, \varphi(x) l_{\mathbf{v}} \mathbf{u}(x_0) + \int_{\mathscr{H}(-\mathbf{v},0)} dL^{m+1}\, \varphi(x) l_{-\mathbf{v}} \mathbf{u}(x_0) \tag{3.2.61}$$

and

$$\lim_{\lambda \to 0} W_\lambda \mathbf{u}(x_0) = \tfrac{1}{2}\{l_{\mathbf{v}}\mathbf{u}(x_0) + l_{-\mathbf{v}}\mathbf{u}(x_0)\}. \tag{3.2.62}$$

Proof

$$\int_{\mathscr{H}(\mathbf{v},x_0)} dL^{m+1}\, \varphi_r(x - x_0)\mathbf{u}(x)$$

$$= \int_{\mathscr{H}(\mathbf{v},x_0)} dL^{m+1}\, \varphi_r(x - x_0)[\mathbf{u}(x) - l_{\mathbf{v}}\mathbf{u}(x_0)] + \int_{\mathscr{H}(\mathbf{v},0)} dL^{m+1}\, \varphi(x) l_{\mathbf{v}}\mathbf{u}(x_0). \tag{3.2.63}$$

We note that $\operatorname{supp}\varphi_r \subset \mathscr{K}(0, r)$, $\|\varphi_r\|_\infty \leqslant Cr^{-m-1}$ and for any $\varepsilon > 0$

$$\lim_{r \to 0} \frac{|\mathscr{K}(x_0, r) \cap \mathscr{H}(\mathbf{v}, x_0) \cap \{x \mid |\mathbf{u}(x) - l_{\mathbf{v}}\mathbf{u}(x_0)| > \varepsilon\}|}{|\mathscr{K}(x_0, r)|} = 0, \tag{3.2.64}$$

We can choose r so small that the fraction in (3.2.64) is smaller than ε. The absolute value of the first integral in (3.2.63) is bounded by

$$\frac{\varepsilon}{2} + 2\|\mathbf{u}\|_\infty Cr^{-m-1} \varepsilon |\mathscr{K}(x_0, r)|.$$

Therefore, for $r \to 0$ the first integral tends to zero. This proves (3.2.61).

For the proof of (3.2.62) choose r in the way indicated above and then choose λ so small that

$$(\pi\lambda)^{-(m+1)/2} \int_{\mathscr{K}(0,r)^c} dL^{m+1} \exp(-|x|^2/\lambda) < \varepsilon.$$

□

Definition 3.2.38

The averaging operator $T(\varphi)$ is said to be **symmetric** if $\forall\, \mathbf{n} \in \mathscr{S}^m$

$$\int_{\mathscr{H}(\mathbf{n},0)} dL^{m+1}\, \varphi(x) = \int_{\mathscr{H}(-\mathbf{n},0)} dL^{m+1}\, \varphi(x).$$

Corollary 3.2.39

If $T(\varphi)$, $T(\psi)$ are two symmetric averaging operators then, $\forall\, x_0 \in \mathscr{R}$, $T(\varphi)\mathbf{u}(x_0) = T(\psi)\mathbf{u}(x_0)$ exist and are equal.

Definition 3.2.40
Let $T(\varphi)$ be a symmetric averaging operator. We set $\bar{\mathbf{u}}(x_0) := T(\varphi)\mathbf{u}(x_0)$ whenever the right-hand side exists.

Clearly, $\bar{\mathbf{u}}$ is defined H^m-a.e. and coincides with $\mathbf{u}$ L^{m+1}-a.e.

Definition 3.2.41
Let $\mathscr{E} \subset \boldsymbol{R}^{m+1}$, $\mathbf{u} \in \mathrm{BV}$, $x_0 \in \partial^*\mathscr{E}$. We define the **inner trace** of $\mathbf{u}$ on $\partial^*\mathscr{E}$

$$\mathbf{u}_+(x_0) := l_{\mathscr{E}}\mathbf{u}(x_0), \tag{3.2.65a}$$

and the **outer trace** of $\mathbf{u}$ on $\partial^*\mathscr{E}$

$$\mathbf{u}_-(x_0) := l_{\mathscr{E}^c}\mathbf{u}(x_0). \tag{3.2.65b}$$

Remark
It is easy to check that, for every $x_0 \in \partial^{**}\mathscr{E}$, $\mathbf{u}_+(x_0) = l_{-\boldsymbol{\nu}}\mathbf{u}(x_0)$, $\mathbf{u}_-(x_0) = l_{\boldsymbol{\nu}}\mathbf{u}(x_0)$, where $\boldsymbol{\nu}$ is the outer normal for $\mathscr{E}$ at x_0.

Theorem 3.2.42 (Vol'pert, 1967)
Let $P(\mathscr{E}) < \infty$, $\mathbf{u} \in \mathrm{BV}(\boldsymbol{R}^{m+1}, \boldsymbol{R}^n)$.
Then the traces $\mathbf{u}_+$, $\mathbf{u}_-$ exist H^m-a.e. on $\partial^*\mathscr{E}$.

Proof
Consider the function $\mathbf{v} = [u^1, \ldots, u^n, \chi_{\mathscr{E}}] \in \mathrm{BV}(\boldsymbol{R}^{m+1}, \boldsymbol{R}^{n+1})$. Let $\mathscr{R}$ be the set of regular points of $\mathbf{v}$. Let $x_0 \in \partial^*\mathscr{E} \cap \mathscr{R}$. The limits $l_{\mathbf{n}}\mathbf{v}(x_0)$, $l_{-\mathbf{n}}\mathbf{v}(x_0)$ exist for some $\mathbf{n} \in \mathscr{S}^m$. Since in particular $l_{\mathbf{n}}\chi_{\mathscr{E}}(x_0), l_{-\mathbf{n}}\chi_{\mathscr{E}}(x_0)$ exist, it follows that the outer normal $\boldsymbol{\nu}$ in the sense of Federer exists at x_0 and $\mathbf{n} = \boldsymbol{\nu}$ or $\mathbf{n} = -\boldsymbol{\nu}$. Moreover $H^m((\partial^*\mathscr{E} \setminus (\partial^*\mathscr{E} \cap \mathscr{R})) \leqslant H^m(\mathscr{R}^c) = 0$. This finishes the proof. □

Lemma 3.2.43
Let $\mathbf{u} \in \mathscr{L}^1_{loc}$, $x_0 \in \partial^*\mathscr{E}$. Suppose that (1) the outer normal for $\mathscr{E}$ at x_0 exists, and (2) the inner trace $\mathbf{u}_+(x_0)$ exists.

Then $\overline{\mathbf{u}\chi_{\mathscr{E}}}(x_0)$ exists and

$$\overline{\mathbf{u}\chi_{\mathscr{E}}}(x_0) = \tfrac{1}{2}\mathbf{u}_+(x_0). \tag{3.2.66}$$

If $\mathbf{u}_-(x_0)$ also exists then $\bar{\mathbf{u}}(x_0)$ exists and

$$\bar{\mathbf{u}}(x_0) = \tfrac{1}{2}[\mathbf{u}_+(x_0) + \mathbf{u}_-(x_0)].$$

Proof
Let $\boldsymbol{\nu}$ be the inner normal for $\mathscr{E}$ at x_0.

Let $\mathbf{v} := \chi_{\mathscr{E}}\mathbf{u}$. Obviously $l_{\boldsymbol{\nu}}\mathbf{v}(x_0) = \mathbf{u}_+(x_0)$, $l_{-\boldsymbol{\nu}}\mathbf{v}(x_0) = \mathbf{0}$, on account of the definitions of $\boldsymbol{\nu}$ and $l_{\boldsymbol{\nu}}$. Hence x_0 is a regular point for $\mathbf{v}$ and $\bar{\mathbf{v}}(x_0)$ exists,

$$\bar{\mathbf{v}}(x_0) = \tfrac{1}{2}l_{\boldsymbol{\nu}}\mathbf{v}(x_0) = \tfrac{1}{2}\mathbf{u}_+(x_0).$$ □

We now state without proof the following important fact (Vol'pert and Khudiayev, 1975):

Theorem 3.2.44
If $\mathscr{E}$ is a Borel subset of $\boldsymbol{R}^{m+1}$ of zero H^m-measure and $\mathbf{u} \in \mathrm{BV}$, then $\nabla \mathbf{u}(\mathscr{E}) = \mathbf{0}$.

Theorem 3.2.45
If $u, v \in \mathrm{BV}(\boldsymbol{R}^{m+1}, \boldsymbol{R}) \cap \mathscr{L}^\infty(\boldsymbol{R}^{m+1}, \boldsymbol{R})$, then $uv \in \mathrm{BV}(\boldsymbol{R}^{m+1}, \boldsymbol{R})$ and

$$\nabla(uv) = \bar{u}\nabla v + \bar{v}\nabla u. \tag{3.2.67}$$

Proof
Let $\varphi \in \mathscr{C}_o^\infty(\boldsymbol{R}^{m+1}, \boldsymbol{R})$. For $\lambda, \mu \to 0_+$, $\int dL^{m+1}(W_\lambda u)(W_\mu v)\nabla\varphi - \int dL^{m+1}\, uv\nabla\varphi = \int dL^{m+1}\,(W_\lambda u - u)\, W_\mu v \nabla\varphi + \int dL^{m+1}\, u(W_\mu v - v)\nabla\varphi$ tends to zero as $\lambda, \mu \to 0$, since $W_\mu v \to v$, $W_\lambda u \to u$ a.e. and $|W_\lambda u| \leqslant \|u\|_\infty$, $|W_\mu v| \leqslant \|v\|_\infty$. On the other hand $\int dL^{m+1}\,(W_\lambda u)(W_\mu v)\nabla\varphi = -\int dL^{m+1}\,(W_\lambda u)\varphi\nabla W_\mu v - \int dL^{m+1}(W_\mu v)\varphi\nabla W_\lambda u$. In the limit $\mu \to 0_+$ the first integral on the right-hand side becomes $\int (W_\lambda u)\varphi\nabla v$. For $\lambda \to 0$ $W_\lambda u \to \bar{u}$ H^m-a.e., hence, by Theorem 3.2.44, ∇v-a.e. By the Lebesgue dominated convergence theorem $\lim_{\lambda\to 0}\int (W_\lambda u)\varphi\nabla v = \int \bar{u}\varphi\nabla v$. Hence $\int dL^{m+1}\, uv\nabla\varphi = -\int \varphi\bar{u}\nabla v - \int \bar{v}\nabla u$, which implies the thesis. □

Remark
Instead of using Theorem 3.2.44, we could have proved Theorem 3.2.45 with the help of Theorem 3.2.53 below.

Theorem 3.2.46 (Vol'pert, 1967)
Assume that $P(\mathscr{E}) < \infty$, $u \in \mathrm{BV}$.
Then

$$\nabla(u\chi_{\mathscr{E}}) = u_+\nabla\chi_{\mathscr{E}} + \chi_{\mathscr{E}_*}\nabla u, \tag{3.2.68}$$

$$\nabla(u\chi_{\mathscr{E}}) = u_-\nabla\chi_{\mathscr{E}} + \chi_{\mathscr{E}^*}\nabla u, \tag{3.2.69}$$

$$\chi_{\partial^*\mathscr{E}}\nabla u + (u_- - u_+)\nabla\chi_{\mathscr{E}} = \mathbf{0}, \quad u_- - u_+ \in \mathscr{L}^1_{loc}(\mathscr{H}^m). \tag{3.2.70}$$

Proof

$$\begin{aligned}\nabla(u\chi_{\mathscr{E}}) &= \nabla(u\chi_{\mathscr{E}}^2) = \overline{u\chi_{\mathscr{E}}}\nabla\chi_{\mathscr{E}} + \overline{\chi_{\mathscr{E}}}\nabla(u\chi_{\mathscr{E}})\\ &= \overline{u\chi_{\mathscr{E}}}\nabla\chi_{\mathscr{E}} + \overline{\chi_{\mathscr{E}}}^2\nabla u + \bar{u}\overline{\chi_{\mathscr{E}}}\nabla\chi_{\mathscr{E}}\end{aligned}$$

and $\overline{\chi_{\mathscr{E}}}\nabla\chi_{\mathscr{E}} = \frac{1}{2}\nabla(\chi_{\mathscr{E}}^2) = \frac{1}{2}\nabla\chi_{\mathscr{E}}$. Hence, from (3.2.67), $\frac{1}{2}\bar{u}\nabla\chi_{\mathscr{E}} + \overline{\chi_{\mathscr{E}}}\nabla u = \overline{u\chi_{\mathscr{E}}}\nabla\chi_{\mathscr{E}} + \overline{\chi_{\mathscr{E}}}^2\nabla u$. Since ∇u, $\nabla\chi_{\mathscr{E}}$ vanish on any set of H^m-measure zero, we can replace $\overline{u\chi_{\mathscr{E}}}$ by $\frac{1}{2}u_+$, $\bar{u}$ by $\frac{1}{2}(u_+ + u_-)$ in the last formula:

$$\tfrac{1}{4}(u_- - u_+)\nabla\chi_{\mathscr{E}} + (\overline{\chi_{\mathscr{E}}} - \overline{\chi_{\mathscr{E}}}^2)\nabla u = \mathbf{0}. \tag{3.2.71}$$

Now $\overline{\chi_{\mathscr{E}}} = 1$ on $\mathscr{E}_*$, $\overline{\chi_{\mathscr{E}}} = 0$ on $(\mathscr{E}^*)^c$, $\overline{\chi_{\mathscr{E}}} = \frac{1}{2}$ H^m-a.e. on $\partial^*\mathscr{E}$. Hence $\overline{\chi_{\mathscr{E}}} - \overline{\chi_{\mathscr{E}}}^2 = 0$ on $(\partial^*\mathscr{E})^c$ and $\overline{\chi_{\mathscr{E}}} - \overline{\chi_{\mathscr{E}}}^2 = \frac{1}{4}\chi_{\partial^*\mathscr{E}}$. Substituting these formulae into (3.2.71) we obtain (3.2.70).

We also have

$$\chi_{\mathscr{E}^*} = \overline{\chi_{\mathscr{E}}} + \tfrac{1}{2}\chi_{\partial^*\mathscr{E}}, \quad \chi_{\mathscr{E}_*} = \overline{\chi_{\mathscr{E}}} - \tfrac{1}{2}\chi_{\partial^*\mathscr{E}} H^m\text{-a.e.}$$

From (3.2.70) and (3.2.67)

$$\nabla(u\chi_{\mathscr{E}}) = \pm\tfrac{1}{2}\{\chi_{\partial^*\mathscr{E}}\nabla u + (u_- - u_+)\nabla\chi_{\mathscr{E}}\}$$

$$+\bar{u}\nabla\chi_{\mathscr{E}} + \overline{\chi_{\mathscr{E}}}\nabla u = \chi_{\mathscr{E}^*}\nabla u + u_-\nabla\chi_{\mathscr{E}} = \chi_{\mathscr{E}_*}\nabla u + u_+\nabla\chi_{\mathscr{E}}.$$

Concerning the local summability of $u_- - u_+$ on $\partial^*\mathscr{E}$, we have for every Borel set $\mathscr{A}$

$$\int_{\partial^*\mathscr{E}\cap\mathscr{A}} dH^m\,|u_- - u_+| \leqslant |\nabla u|(\mathscr{A}),$$ □

on account of (3.2.70).

Lemma 3.2.47
Let $u \in \mathrm{BV}$ have compact support. Then $\nabla u(\boldsymbol{R}^{m+1}) = \mathbf{0}$.

Proof
Let $u|\mathscr{K}(0, R)^c = 0$ and $\varphi \in \mathscr{C}_0^\infty(\boldsymbol{R}^{m+1})$, $\varphi|\mathscr{K}(0, 2R) = 1$. Then ∇u is concentrated on $\mathscr{K}(0, R)$ (as a distribution and hence as a measure too), and

$$\int \nabla u = \int \varphi\nabla u = -\int dL^{m+1}\, u\nabla\varphi = \mathbf{0}.$$ □

Theorem 3.2.48 (Gauss–Green theorem)
Suppose that $P(\mathscr{E}) < \infty$, $u \in \mathrm{BV}$ and the trace u_+ of u on $\partial^*\mathscr{E}$ is H^m-summable. Then

$$\nabla u(\mathscr{E}_*) = \int_{\partial^*\mathscr{E}} dH^m\, u_+\boldsymbol{\nu}, \tag{3.2.72}$$

where $\boldsymbol{\nu}$ is the outer normal on $\partial^*\mathscr{E}$ (in the sense of Federer).

Proof
Integrate (3.2.68) over $\boldsymbol{R}^{m+1}$ and use Lemma 3.2.47 as well as De Giorgi's definition of $\boldsymbol{\nu}$ on $\partial^{**}\mathscr{E}$. Note also equation (3.2.54). □

From equation (3.2.70) we have analogously

Corollary 3.2.49

Let $\mathscr{P}$ be a H^m-measurable subset of $\partial^*\mathscr{E}$, $u \in \mathrm{BV}$, $P(\mathscr{E}) < \infty$. Then

$$\nabla u(\mathscr{P}) = \int_{\mathscr{P}} dH^m\,(u_- - u_+)\mathbf{v}. \tag{3.2.73}$$

Corollary 3.2.50

Let $P(\mathscr{E}) < \infty$.

Suppose that $\partial^*\mathscr{E} = \mathscr{S}_1 \cup \mathscr{S}_2$, $\mathscr{S}_1$ and $\mathscr{S}_2$ are H^m-measurable and $\mathscr{S}_1 \cap \mathscr{S}_2 = \emptyset$, Let $u \in \mathrm{BV}$ and suppose that the inner and outer trace u_+ and u_- of u on ∂^*E are H^m-summable on $\mathscr{S}_1$, $\mathscr{S}_2$ resp. Then

$$\nabla u(\mathscr{E}_* \cup \mathscr{S}_2) = \int_{\mathscr{S}_1} dH^m\, u_+\mathbf{v} + \int_{\mathscr{S}_2} dH^m\, u_-\mathbf{v}, \tag{3.2.74}$$

where $\mathbf{v}$ is the outer normal for $\mathscr{E}$.

Proof

$u_+ - u_-$ is H^m-summable on $\partial^*\mathscr{E}$ in view of (3.2.70). Hence u_+ is H^m-summable on $\mathscr{S}_2$ as well as on $\partial^*E = \mathscr{S}_1 \cup \mathscr{S}_2$. Substitute $\mathscr{P} = \mathscr{S}_2$ in (3.2.73) and add (3.2.72). □

Theorem 3.2.51

Let $\mathscr{S}$ be a set of class Γ, $\mathbf{u} \in \overrightarrow{\mathrm{BV}}$. Then

$$\nabla \mathbf{u}(\mathscr{S}) = \int_{\mathscr{S}} dH^m\,(l_{\mathbf{v}}\mathbf{u} - l_{-\mathbf{v}}\mathbf{u})\mathbf{v},$$

where $\mathbf{v}$ is the unique normal associated with the jump of $\mathbf{u}$ at every jump point on $\mathscr{S}$ and is arbitrary elsewhere.

Proof

$\mathscr{S} \subset \bigcup_{i=1}^{\infty} \partial^*\mathscr{E}_i$, $\quad P(\mathscr{E}_i) < \infty$. $\quad$ Let $\quad \mathscr{S}_1 := \partial^*\mathscr{E}_1$, $\quad \mathscr{S}_k := \partial^*\mathscr{E}_k \setminus \bigcup_{i=1}^{k-1} \mathscr{S}_i$,

$\mathscr{P}_k := \mathscr{S} \cap \mathscr{S}_k$. Then $\mathscr{S} = \bigcup_{k=1}^{\infty} \mathscr{P}_k$, $\mathscr{P}_i \cap \mathscr{P}_j = \emptyset$ for $i \neq j$, $\mathscr{P}_k \subset \partial^*\mathscr{E}_k$,

$$\nabla u(\mathscr{P}_k) = \int_{\mathscr{P}_k} dH^m\,(\mathbf{u}_- - \mathbf{u}_+)\mathbf{v}.$$

Let $\mathscr{C}_k := \{x \in \mathscr{P}_k |\ \mathbf{u}_+(x), \mathbf{u}_-(x), \mathbf{v}(x) \text{ exist}\}$ ($\mathbf{v}(x)$ is the outer unit normal for $\mathscr{E}_k$). Then $H^m(\mathscr{P}_k \setminus \mathscr{C}_k) = 0$. For every $x \in \mathscr{C}_k$, $\mathbf{u}_-(x) = l_{\mathbf{v}}\mathbf{u}(x)$, $\mathbf{u}_+(x)$

$= l_{-\nu}\mathbf{u}(x)$, according to the remark following Definition 3.2.41. Hence every point $x \in \mathscr{C}_k$ is a regular point for $\mathbf{u}$. Moreover

$$\nabla u(\mathscr{P}_k) = \int_{\mathscr{P}_k} dH^m \, (l_\nu \mathbf{u} - l_{-\nu}\mathbf{u})\boldsymbol{\nu}. \qquad \square$$

Corollary 3.2.52
Let $\Gamma(\mathbf{u})$ be the jump set of $\mathbf{u} \in \mathrm{BV} \cap \mathscr{L}^\infty$.

Then for every bounded Borel subset $\mathscr{E} \subset \boldsymbol{R}^{m+1}$

$$\nabla \mathbf{u}\big(\Gamma(\mathbf{u}) \cap \mathscr{E}\big) = \int_{\Gamma(\mathbf{u})\cap\mathscr{E}} dH^m \, (l_\nu \mathbf{u} - l_{-\nu}\mathbf{u})\boldsymbol{\nu}. \tag{3.2.75}$$

Proof
$\Gamma(\mathbf{u}) \cap \mathscr{E}$ is a set of class Γ up to a set of H^m-measure zero (Theorem 3.2.35). Summability of the integrand on the left-hand side of (3.2.75) follows from (3.2.70). □

3.2.7 Weak Solutions in the Class $\mathrm{BV} \cap \mathscr{L}^\infty$

It follows from the preceding section that the BV-functions have many properties which make them suitable candidates for discontinuous solutions $\mathbf{u}$ to hyperbolic systems of conservation laws. On the other hand the assumption that $\mathbf{u} \in \mathscr{L}^\infty$ is natural for a solution of a hyperbolic system (see Section 3.2.1). We must however give some meaning to the left-hand side of equation (3.1.35) or, more particularly, $\mathbf{u}_t + \mathbf{f}^k(\mathbf{u})_{x^k} = \mathbf{h}(\mathbf{u})$ for any $\mathbf{u} \in \mathrm{BV} \cap \mathscr{L}^\infty$.

Since $\mathbf{u}_t$ and $\mathbf{h}(\mathbf{u})\,dL^{m+1}$ are Borel measures, $\mathbf{f}^k(\mathbf{u})_{x^k}$ should also be a Borel measure. We shall see that it is indeed so.

Let

$$\hat{g}(\mathbf{u})(x) := \int_0^1 d\theta \, g\big(\theta l_\nu \mathbf{u}(x) + (1-\theta) l_{-\nu}\mathbf{u}(x)\big) \tag{3.2.76}$$

at every regular point x of $\mathbf{u} \in \mathrm{BV} \cap \mathscr{L}^\infty$ with the approximate limits $l_\nu \mathbf{u}(x)$, $l_{-\nu}\mathbf{u}(x)$. The above expression makes sense for every $g \in \mathscr{C}^0$. At a point x of approximate continuity $\hat{g}(\mathbf{u})(x) = g(\bar{\mathbf{u}}(x))$. Note that $\hat{g}(\mathbf{u})$ is defined H^m-a.e.

Theorem 3.2.53
Let $\mathbf{u} \in \mathrm{BV} \cap \mathscr{L}^\infty(\boldsymbol{R}^{m+1}, \boldsymbol{R}^n)$, $f \in \mathscr{C}^1(\boldsymbol{R}^n, \boldsymbol{R})$.
Then $f \circ \mathbf{u} \in \mathrm{BV}$ and

$$\nabla f \circ \mathbf{u} = \sum_{k=1}^{n} \hat{f}_k(\mathbf{u}) \nabla u^k, \tag{3.2.77}$$

where $f_k = \partial f / \partial u^k$, and $\hat{f}_k(\mathbf{u})$ is summable with respect to ∇u^k.

Proof

The proof involves the use of m-dimensional measures defined by projecting the subsets $\mathscr{E} \subset \boldsymbol{R}^{m+1}$ onto all the hyperplanes and is quite lengthy. The interested reader will find it in the paper of Vol'pert (1967). We prove it for $m = 0$ in order to indicate the role of the operator $\hat{g}(\mathbf{u})$.

If $\mathbf{u} \in \mathrm{BV}(\boldsymbol{R}, \boldsymbol{R}^n)$, then $\bar{\mathbf{u}}$ is a function of bounded variation, cf. the remarks following Theorem 3.2.29, and

$$\int \varphi \nabla \mathbf{u} = \int \varphi \mathbf{u}_x = -\int \varphi_x \mathbf{u} dx = \int \varphi d\bar{\mathbf{u}}(x)$$

for every $\varphi \in \mathscr{C}_o^\infty(\boldsymbol{R}, \boldsymbol{R})$. The last expression is a Lebesgue–Stieltjes integral. We shall prove in three steps that

$$\int \varphi d\overline{f \circ \mathbf{u}} = \sum_{k=1}^{n} \int \varphi \hat{f}_k(\mathbf{u}) d\bar{u}.$$

(1) Suppose that $\operatorname{supp}\varphi \subset [a, b]$, $a = x_0 < x_1 < \dots < x_N < x_{N+1} = b$, and $u^k|[x_j, x_{j+1}] = \text{const}$ for every $k = 1, \dots, n$; $j = 1, \dots, N$. Then

$$\begin{aligned} \int \varphi d\overline{f \circ \mathbf{u}} &= \sum_{j=0}^{N} \varphi(x_j)\{\overline{f \circ \mathbf{u}}(x_j+0) - \overline{f \circ \mathbf{u}}(x_j-0)\} \\ &= \text{the same without the overbars} \\ &= \sum_{j=0}^{N} \varphi(x_j) \sum_{k=1}^{n} \hat{f}_k(\mathbf{u})(x_j)\{\bar{u}^k(x_j+0) - \bar{u}^k(x_j-0)\} \\ &= \sum_{k=1}^{n} \int \varphi \hat{f}_k(\mathbf{u})(x) d\bar{u}^k(x). \end{aligned}$$

(2) Let $\mathbf{u} \in \mathrm{BV}$, $f_k \in \mathrm{Lip}$, $k = 1, \dots, n$, $\operatorname{supp}\varphi \subset [a, b]$. For every $k \leqslant n$ there is a sequence of piecewise constant functions $\{u_r^k | r = 1, 2, \dots\}$, such that $u_r^k \rightrightarrows \bar{u}^k$ uniformly in $[a, b]$ as $r \to \infty$, and $\operatorname{totvar} u_r^k \equiv \operatorname{var}\{u_r^k\ [a, b]\} \leqslant M$ for all r, k. Hence $\int \varphi d\overline{f \circ \mathbf{u}_r} = -\int \varphi_x \overline{f \circ \mathbf{u}_r} dx = -\int \varphi_x f(\bar{\mathbf{u}}_r) dx \to -\int \varphi_x f(\bar{\mathbf{u}}) dx = \int \varphi d\overline{f \circ \mathbf{u}}$ (we have used the fact that $\overline{f \circ \mathbf{u}_r} = f_o \bar{\mathbf{u}}_r$ at all the points of approximate continuity of $\mathbf{u}$, hence a.e. on $[a, b]$). Moreover $\hat{f}_k \in \mathrm{Lip}$ implies that $\hat{f}_k(\mathbf{u}_r)(x) \rightrightarrows \hat{f}_k(\mathbf{u})(x)$ uniformly on $[a, b]$, as $r \to \infty$. Now

$$\begin{aligned} &\int \varphi \hat{f}_k(\mathbf{u}_r) d\bar{u}_r^k - \int \varphi \hat{f}_k(\mathbf{u}) d\bar{u}^k = \int \varphi \{\hat{f}_k(\mathbf{u}_r) - \hat{f}_k(\mathbf{u})\} d\bar{u}_r^k \\ &\quad + \int (\bar{u}^k - \bar{u}_r^k) d\,(\overline{\varphi \hat{f}_k(\mathbf{u})}) + \int \{\varphi \hat{f}_k(\mathbf{u}) - \overline{\varphi \hat{f}_k(\mathbf{u})}\} d(\bar{u}_r^k - \bar{u}^k) \to 0 \end{aligned}$$

as $r \to 0$, by Helly's theorem (Theorem 3.2.60 below).

(3) The result of (2) extends to $f_k \in \mathscr{C}^0$ by regularization. □

Theorem 3.2.54

Let $f \in \mathscr{C}^0(\boldsymbol{R}^n, \boldsymbol{R})$, $\mathbf{u}: \mathscr{D} \to \boldsymbol{R}^n$. Then

(i) if $\mathbf{u}_+(x_0)$ exists at $x_0 \in \partial^*\mathscr{E}$, then $(f \circ \mathbf{u})_+(x_0)$ exists and $(f \circ \mathbf{u})_+(x_0) = f((\mathbf{u}_+(x_0))$;

(ii) $\mathscr{R}(f \circ \mathbf{u}) \subset \mathscr{R}(\mathbf{u})$, $\Gamma(f \circ \mathbf{u}) \subset \Gamma(\mathbf{u})$, where $\mathscr{R}(\mathbf{v})$ denotes the set of regular points for the function $\mathbf{v}$.

Proof

Let $x_0 \in \partial^*\mathscr{E}$ be such that $\mathbf{u}_0 := \mathbf{u}_+(x_0)$ exists.

For every $\varepsilon > 0$ there is a $\delta(\varepsilon) > 0$ such that $\{\mathbf{u} \in \mathscr{R}^n |\ |f(\mathbf{u})-f(\mathbf{u}_0)| > \varepsilon\} \subset \mathscr{K}(\mathbf{u}_0, \delta(\varepsilon))^c$, and

$$\limsup_{r\to 0} \frac{|\mathscr{K}(x_0, r)\cap\{x \in \mathscr{E} |\ |f(\mathbf{u}(x))-f(\mathbf{u}_0)| > \varepsilon\}|}{|\mathscr{K}(x_0, r)\cap\mathscr{E}|}$$

$$\leqslant \lim_{r\to 0} \frac{|\mathscr{K}(x_0, r)\cap\{x \in \mathscr{E} |\ |\mathbf{u}(x)-\mathbf{u}_0| > \delta(\varepsilon)\}|}{|\mathscr{K}(x_0, r)\cap\mathscr{E}|} = 0.$$

Hence $(f \circ \mathbf{u})_+(x_0) = l_{\mathscr{E}}(f \circ \mathbf{u})(x_0) = f(\mathbf{u}_+(x_0))$.

(ii) can be established by the same kind of argument. □

Let $\mathbf{f}^\mu \in \mathscr{C}^1$, $\mathbf{h} \in \mathscr{C}^0$. We shall consider the following problem:

$$\partial_\mu \mathbf{f}^\mu(\mathbf{u}) = \mathbf{h}(\mathbf{u}), \quad \mathbf{u} \in \mathrm{BV}(\boldsymbol{R}_+^{m+1}, \boldsymbol{R}^n)\cap\mathscr{L}^\infty(\boldsymbol{R}_+^{m+1}, \boldsymbol{R}^n), \tag{3.2.78}$$

$$\mathbf{u}_+(0, x^1, \ldots, x^m) = \mathbf{v}(x^1, \ldots, x^m) \text{ for}$$

$$\text{for } \mathrm{L}^m\text{- almost all } (x^1, \ldots, x^m) \in \boldsymbol{R}^m. \tag{3.2.79}$$

The left-hand side of equation (3.2.79) is the inner trace of $\mathbf{u}$ at $x^0 = t = 0$. The right-hand side of (3.2.78) is the measure $\mathscr{E} \mapsto \int_{\mathscr{E}} dL^{m+1}\ \mathbf{h}(\mathbf{u}(x))$. The inclusion of an additional argument x in $\mathbf{f}^\mu$, $\mathbf{h}$ would not pose any serious problem.

Remarks

(1) For any $\boldsymbol{\phi} \in \mathscr{C}_0^\infty(\boldsymbol{R}^{m+1}, \boldsymbol{R}^n)$ the definition of a weak solution given by Lax is recovered:

$$\int_{\boldsymbol{R}_+^{m+1}} dL^{m+1}\,\mathbf{h}(\mathbf{u})\cdot\boldsymbol{\phi} = -\int_{\boldsymbol{R}_+^{m+1}} dL^{m+1}\,\partial_\mu\boldsymbol{\phi}\cdot\mathbf{f}^\mu(\mathbf{u}) + \int_{\boldsymbol{R}_+^{m+1}} \partial_\mu(\boldsymbol{\phi}\cdot\mathbf{f}^\mu(\mathbf{u}))$$

$$= -\int_{\boldsymbol{R}_+^{m+1}} dL^{m+1}\,\partial_\mu\boldsymbol{\phi}\cdot\mathbf{f}^\mu(\mathbf{u}) + \int_{\boldsymbol{R}^m} dL^m\,\boldsymbol{\phi}(0, x^1, \ldots, x^m)\cdot\mathbf{f}^0(\mathbf{v}), \tag{3.2.80}$$

(Theorems 3.2.48 and 3.2.54; note that $\boldsymbol{\phi}\cdot\mathbf{f}^\mu(\mathbf{u}) \in \mathrm{BV}\cap\mathscr{L}^\infty$. On account of Theorem 3.2.53 and $\overline{\mathbf{f}^\mu(\mathbf{u})} = \mathbf{f}^\mu(\mathbf{u})\ L^{m+1}$-a.e.).

(2) Equation (3.2.78) and Theorem 3.2.54 imply the integral conservation law

$$\int_{\partial^*\mathscr{E}} dH^m\, \mathbf{f}^\mu(\mathbf{u}_+)\nu_\mu = \int_{\mathscr{E}_*} dL^{m+1}\, \mathbf{h}(\mathbf{u}) \tag{3.2.81}$$

for every bounded Borel set $\mathscr{E} \subset \boldsymbol{R}_+^{m+1}$ with $P(\mathscr{E}) < \infty$. The fact that $\mathscr{E}$ is bounded ensures the applicability of equation (3.2.72).

Conversely, if (3.2.81) is satisfied for every cube in $\boldsymbol{R}_+^{m+1}$, then by the Gauss–Green theorem we recover equation (3.2.78) in the sense of equality of measures. Equation (3.2.81) is an extension of the conservation laws of continuum mechanics to the BV class. On the other hand equation (3.2.80) is remindful of the virtual work principles.

(3) Let $\Gamma(\mathbf{u})$ be the jump set of a solution $\mathbf{u} \in \mathrm{BV}\cap\mathscr{L}^\infty$ of equation (3.2.78), defined on $\boldsymbol{R}_+^{m+1}$. Since $\Gamma(\mathbf{f}^\mu(\mathbf{u})) \subset \Gamma(\mathbf{u})$, we have for every bounded Borel subset $\mathscr{E} \subset \boldsymbol{R}_+^{m+1}$:

$$\partial_\mu(\mathbf{f}^\mu \circ \mathbf{u})(\mathscr{E}) = \int_{\mathscr{E}} dL^{m+1}\, \mathbf{h}(\mathbf{u}),$$

and for every bounded subset $\mathscr{P}$ of $\Gamma(\mathbf{u})$ of finite measure H^m

$$\int_{\mathscr{P}} dH^m\, \{l_\nu \mathbf{f}^\mu(\mathbf{u}) - l_{-\nu}\mathbf{f}^\mu(\mathbf{u})\}\nu_\mu = \mathbf{0}$$

(Corollary 3.2.52). Note that $\mathscr{P}$ has zero L^{m+1} measure on account of the definition of the Hausdorff measure on p. 33. Applying Remark (2) following Theorem 1.1.18 we recover the Rankine Hugoniot equations:

$$\{l_\nu \mathbf{f}^\mu(\mathbf{u}) - l_{-\nu}\mathbf{f}^\mu(\mathbf{u})\}\nu_\mu = \mathbf{0} \quad H^m\text{-a.e. on } \Gamma(\mathbf{u}). \tag{3.2.82}$$

We now make the problem (3.2.78) and (3.2.79) more precise by making the assumption

(Z) the conditions (L) and (E) are satisfied H^m-a.e. on the jump set $\Gamma(\mathbf{u})$ of $\mathbf{u}$.

(4) Suppose that the equation

$$\partial_\mu \mathbf{g}^\mu(\mathbf{u}) = \mathbf{k}(\mathbf{u})$$

is equivalent to (3.2.78) in the class $\mathscr{C}^1$, e.g.

$$\frac{\partial g^{\mu,r}}{\partial u^s} = \sum_{p=1}^{n} C^r{}_p(\mathbf{u}) \frac{\partial f^{\mu,p}}{\partial u^s},$$

$$k^r(\mathbf{u}) = C^r{}_p(\mathbf{u}) h^p(\mathbf{u}), \quad r, s = 1, \dots, n;\ \mu = 0, 1, \dots, m, \tag{3.2.83}$$

with a non-singular matrix $\mathbf{C}(\mathbf{u})$. In the class BV these equations are no longer equivalent, since $\widetilde{\frac{\partial g^{\mu,r}}{\partial u^s}}$ need not be related to $\widetilde{\frac{\partial f^{\mu,r}}{\partial u^s}}$ by a formula (3.2.83) with a matrix $\hat{\mathbf{C}}(\mathbf{u})$ independent of μ and defined H^m-a.e.

3.2.8 Existence of Weak Solutions: Glimm's Scheme

We shall prove existence of solutions to a class of Cauchy problems in $\boldsymbol{R}^2_+$ for strictly hyperbolic systems of conservation laws. The method we shall use is known as **Glimm's scheme** (Glimm, 1965).

Let $\mathbf{u} \in \mathrm{BV} \cap \mathscr{L}^\infty$ be defined on $\boldsymbol{R}^2_+$. Then there is another function $\hat{\mathbf{u}}$, defined in $\boldsymbol{R}^2_+$, and such that $\mathbf{u} = \hat{\mathbf{u}}$ L^2-a.e., for almost every $t > 0$ the function $\hat{\mathbf{u}}(t, \cdot)$ coincides a.e. with a function of locally bounded variation. Suppose that $\mathbf{u}(t, x)$ is a weak solution to the Cauchy problem

$$\mathbf{u}_t + \mathbf{f}(\mathbf{u})_x = \mathbf{0}, \tag{3.2.84}$$

$$\mathbf{u}_+(0, x) = \mathbf{u}_0(x). \tag{3.2.85}$$

Obviously $\hat{\mathbf{u}}$ satisfies equations (3.2.84) and (3.2.85) as well. We assume that the solution is actually represented by $\hat{\mathbf{u}}$ and with this in mind we drop the hat over $\mathbf{u}$. In fact we shall obtain a solution $\mathbf{u}$ such that for **all** $t > 0$ $\mathbf{u}(t, \cdot)$ has bounded total variation.

For the initial data $\mathbf{u}_0(\cdot)$ and some $\mathbf{u}_1 \in \mathscr{H}$ we define

$$d_0 := \|\mathbf{u}_0 - \mathbf{u}_1\|_\infty (1 + \operatorname{totvar} \mathbf{u}_0), \tag{3.2.86a}$$

$$d_1 := \|\mathbf{u}_0 - \mathbf{u}_1\|_\infty + \operatorname{totvar} \mathbf{u}_0. \tag{3.2.86b}$$

Theorem 3.2.55
Suppose that equation (3.2.84) satisfies (GL_α) for $\alpha \in \{1, \ldots, n\}$. Then for every $\mathbf{u}_1 \in \mathscr{H}$ there are numbers $\delta > 0$, $K > 0$ such that for every $\mathbf{u}_0$ satisfying $d_1 < \delta$ the Cauchy problem (3.2.84–85) has a weak solution satisfying the following conditions

$$\|\mathbf{u}(t, \cdot) - \mathbf{u}_1\|_\infty \leqslant K \|\mathbf{u}_0 - \mathbf{u}_1\|_\infty, \tag{3.2.87}$$

$$\operatorname{totvar} \mathbf{u}(t, \cdot) \leqslant K \operatorname{totvar} \mathbf{u}_0, \tag{3.2.88}$$

$$\|\mathbf{u}(t_2, \cdot) - \mathbf{u}(t_1, \cdot)\|_{\mathscr{L}^1} \leqslant K |t_2 - t_1| \cdot \operatorname{totvar} \mathbf{u}_0. \tag{3.2.89}$$

Remarks

(1) Glimm's theorem has been extended to strictly hyperbolic systems which do not satisfy the conditions (GL_α) (Kuznetsov and Tupchiyev, 1975).

(2) Suppose that the Pfaffians $\langle \mathbf{l}^\alpha(\mathbf{u}), d\mathbf{u}\rangle$ are integrable and $\langle \mathbf{l}^\alpha(\mathbf{u}), d\mathbf{u}\rangle = \mu_\alpha(\mathbf{u})\, ds^\alpha(\mathbf{u})$, $\alpha = 1, \ldots, n$. This is always true if $n = 2$. Obviously $\mu_\alpha(\mathbf{u})\, ds^\alpha[\mathbf{r}_\beta] = \delta^\alpha{}_\beta$ and for $\mathscr{C}^1$ solutions equation (3.2.84) can be rewritten in the following form:

$$s^\alpha(\mathbf{u})_t + \lambda_\alpha(\mathbf{u})\, s^\alpha(\mathbf{u})_x = 0, \quad \alpha = 1, \ldots, n. \tag{3.2.90}$$

For such systems it is possible to replace the condition $d_1 < \delta$ by a weaker one $d_0 < \delta$ (see Glimm, 1965).

Proof (in broad outline)
We shall construct a sequence of functions $\mathbf{v}$ with the property that

$$\mathbf{v}(ps, x) \equiv \mathbf{v}^p_q \quad \text{for } x \in](q-1)h, (q+1)h[, p+q \text{ odd}, \tag{3.2.91}$$

$p = 0, 1, 2, \ldots$ The positive numbers s, h are fixed for the time being. The vectors $\mathbf{v}^p_q$ will be constructed by induction with respect to p.

For $p = 0$ let $\mathbf{v}^0_q = \mathbf{u}_0(qh)$. Suppose now that the vectors $\mathbf{v}^p_q$, $q+p$ odd, have been calculated for some fixed $p \geqslant 0$. We shall solve the Cauchy problem for $\mathbf{v}(t, x)$, $ps \leqslant t \leqslant (p+1)s$, with the initial data $\mathbf{v}(ps, \cdot)$. For $t-ps$ sufficiently small there is a solution which is piecewise equal to self-similar solutions $\mathbf{w}\left(\dfrac{x-qh}{t-ps};\ \mathbf{v}^p_{q-1}, \mathbf{v}^p_{q+1}\right)$, $(q-1)h \leqslant x \leqslant (q+1)h$ to the Riemann problems $(\mathbf{v}^p_{q-1}, \mathbf{v}^p_{q+1})$ with the initial discontinuities situated at the points (ps, qh), $p+q$ even. We assume that the latter solutions satisfy the conditions (L). The difficulties associated with solving a general Riemann problem (cf. 3.1.7) do not appear here since the vectors $\mathbf{v}^p_{q-1}, \mathbf{v}^p_{q+1}$ are close on account of the hypothesis that tot var $\mathbf{u}_0$ is small (for the proof that it is indeed so, see Glimm (1965)).

Every solution $\mathbf{w}(\cdot\,; \mathbf{v}^p_{q-1}, \mathbf{v}^p_{q+1})$ is constant outside a sector with the vertex at (ps, qh). We choose $s > 0$ for each given $h > 0$ in such a way that the *Courant–Friedrichs–Léwy condition*

$$\frac{s}{h} \sup\{|\lambda_\alpha(\mathbf{u})| \mid \mathbf{u} \in \Omega,\ \alpha = 1, \ldots, n\} < 1 \tag{3.2.92}$$

is satisfied. Ω is a small set such that the values of $\mathbf{v}$ lie in it. The possibility of satisfying this condition will not be considered here (see Glimm, 1965). In the following we shall consider s to be a function of the varying h, subject to (3.2.92). In view of (L) the fans of waves produced by the neighbouring discontinuities at $t = ps$ do not overlap in the strip $ps \leqslant t \leqslant (p+1)s$.

Let $a = \{a^p_q\}$ be a sequence of numbers in $[0, 1]$ for every fixed p. We set

$$\mathbf{v}^{p+1}_q := -\mathbf{w}\left(\frac{x^{p+1}_q - qh}{s};\ \mathbf{v}^p_{q-1}, \mathbf{v}^p_{q+1}\right), \quad q+p \text{ even},$$

where $x^{p+1}_q := (q-1)h + 2ha^{p+1}_q$.

Let

$$\mathbf{v}(t, x; h, a) :\equiv \mathbf{w}\left(\frac{x-qh}{t-ps}; \mathbf{v}^p_{q-1}, \mathbf{v}^p_{q+1}\right)$$

for $ps < t \leqslant (p+1)s$, $(q-1)h \leqslant x \leqslant (q+1)h$. A detailed analysis of the relations between the adjacent fans of waves produced at (ps, qh), $(ps, (q+2)h)$ and the fan at $((p+1)s, (q+1)h)$ leads to the following estimates:

$$\begin{aligned}&\|\mathbf{v}(t, \cdot\,; h, a)-\mathbf{u}_1\|_\infty \leqslant K\|\mathbf{v}(0, \cdot\,; h, a)-\mathbf{u}_1\|_\infty\\ &\leqslant K\|\mathbf{u}_0(\cdot)-\mathbf{u}_1\|_\infty, \operatorname{totvar}\mathbf{v}(t, \cdot\,; h, a) \leqslant K\operatorname{totvar}\mathbf{v}(0, \cdot\,; h, a)\\ &\leqslant K\operatorname{totvar}\mathbf{u}_0, \|\mathbf{v}(t_2, \cdot\,; h, a)-\mathbf{v}(t_1, \cdot\,; h, a)\|_{\mathscr{L}^1}\\ &\leqslant K(4s+|t_2-t_1|)\cdot\operatorname{totvar}\mathbf{v}(0, \cdot\,; h, a)\end{aligned}$$

with $K > 0$ independent of h, a.

By Helly's selection theorem* every sequence $\{h_j\}$, $h_j \to 0$, contains a subsequence $\{\bar{h}_j\}$ such that $\mathbf{v}(t, \cdot\,; \bar{h}_j, a)$ converges in $\mathscr{L}^1_{loc}$ to a function $\mathbf{v}(t, \cdot\,; a)$ with bounded variation. In general the subsequence $\{\bar{h}_j\}$ depends on t, a.

Let t_ν, $\nu = 1, 2, \ldots$ be the set of all non-negative rational numbers and suppose that

$$\mathbf{v}(t_1, \cdot\,; h^1_j, a) \to \tilde{\mathbf{v}}(t_1, \cdot\,; a) \quad \text{in } \mathscr{L}^1_{loc}.$$

Suppose also that the subsequence h^μ_j, $j = 1, 2, \ldots$, has been chosen in such a way that

$$\mathbf{v}(t_\nu, \cdot\,; h^\mu_j, a) \to \tilde{\mathbf{v}}(t_\nu, \cdot\,; a) \quad \text{in } \mathscr{L}^1_{loc}$$

for all $\nu \leqslant \mu$. We choose a subsequence $\{h^{\mu+1}_j\}$ of $\{h^\mu_j\}$ in such a way that

$$\mathbf{v}(t_{\mu+1}, \cdot\,; h^{\mu+1}_j, a) \to \tilde{\mathbf{v}}(t_{\mu+1}, \cdot\,; a) \quad \text{in } \mathscr{L}^1_{loc}.$$

Let $\bar{h}_j = h^j_j$, $j = 1, 2, \ldots$ Then

$$\mathbf{v}(t_\nu, \cdot\,; \bar{h}_j, a) \to \tilde{\mathbf{v}}(t_\nu, \cdot\,; a) \quad \text{in } \mathscr{L}^1_{loc}$$

for all $\nu = 1, 2, \ldots$ Now, for an arbitrary $t > 0$

$$\begin{aligned}&\int_{|x|\leqslant M} dx\,|\mathbf{v}(t, x\,; \bar{h}_j, a)-\mathbf{v}(t, x\,; \bar{h}_i, a)|\\ &\leqslant \int_{|x|\leqslant M} dx\,|\mathbf{v}(t, x\,; \bar{h}_j, a)-\mathbf{v}(t_\nu, x\,; \bar{h}_j, a)|\\ &\quad+\int_{|x|\leqslant M} dx\,|\mathbf{v}(t, x\,; \bar{h}_i, a)-\mathbf{v}(t_\nu, x\,; \bar{h}_i, a)|\\ &\quad+\int_{|x|\leqslant M} dx\,|\mathbf{v}(t_\nu, x\,; \bar{h}_i, a)-\mathbf{v}(t_\nu, x; \bar{h}_j, a)|.\end{aligned}$$

* Theorem 3.2.59. Convergence in $\mathscr{L}^1_{loc}$ follows from the $\mathscr{L}^\infty$ bound by the dominated convergence theorem.

The first two integrals are bounded by $K(4s_i+4s_j+2|t-t_\nu|)\cdot\text{tot var}\,\mathbf{u}_0$. For any $t>0$ choose ν so that $2K|t-t_\nu|<\varepsilon/3$. Then choose i, j so large that $4K(s_i+s_j)<\varepsilon/3$ and the last integral is also bounded by $\varepsilon/3$. Since $\varepsilon>0$ is arbitrary, there is a subsequence $\{\hat{h}_i\}$ of $\{\overline{h_j}\}$ such that $\mathbf{v}(t,\cdot\,;\hat{h}_i,a)$ converges in $\mathscr{L}^1_{loc}$ *uniformly* with respect to $t\in[0,T]$, for an arbitrary $T>0$. It follows then that the piecewise continuous functions $\mathbf{v}(\cdot\,,\cdot\,;\hat{h}_i,a)$ converge in $\mathscr{L}^1_{loc}(\boldsymbol{R}^2_+)$ to some function $\mathbf{v}(\cdot\,,\cdot\,;a)$. Note that thevalue of a has so far been kept fixed. We shall dispose of it now.

Let $\varphi\in\mathscr{C}^\infty_o(\boldsymbol{R}^2)$ and

$$\mathbf{g}(\varphi;h,a):=\int_{R^2_+}dL^2\,\{\varphi_t\mathbf{v}+\varphi_x\mathbf{f}(\mathbf{v})\}+\int_{-\infty}^{\infty}dx\,\varphi(0,x)\mathbf{u}_0(x)$$

with $\mathbf{v}=\mathbf{v}(t,x;h,a)$. In each strip $ps\leqslant t\leqslant(p+1)s$ $\mathbf{v}(\cdot\,,\cdot\,;h,a)$ is a weak solution of equation (3.2.84). Hence

$$\int_{[ps,(p+1)s[\times R}dL^2\,\{\varphi_t\mathbf{v}+\varphi_x\mathbf{f}(\mathbf{v})\}=\int_{-\infty}^{\infty}dx\,\varphi(t,x)\mathbf{v}(t,x;h,a)\Big|_{t=ps}^{t=(p+1)s-0}$$

$$\text{and }\mathbf{g}(\varphi;h,a)=\sum_{p=0}^{\infty}\mathbf{g}_p(\varphi;h,a),$$

$$\mathbf{g}_0(\varphi;h,a):=\int_{-\infty}^{\infty}dx\,\varphi(0,x)[\mathbf{u}_0(x)-\mathbf{v}(0,x;h,a)],$$

$$\mathbf{g}_p(\varphi;h,a):=\int_{-\infty}^{\infty}dx\,\varphi(ps,x)[\mathbf{v}(ps-0;h,a)-\mathbf{v}(ps;h,a)]$$

for $p\geqslant 1$.

It is not possible to show that $\mathbf{g}(\varphi;h,a)\to 0$ for $h\to 0$ and an arbitrary fixed a. Let $\mathscr{A}$ be the Cartesian product of a countable number of copies of $[0,1]$, numbered by p, q. The last step in Glimm's proof consists in showing that

$$\int_{\mathscr{A}}dL\,(a)|\mathbf{g}(\varphi;h_j,a)|^2\to 0\quad\text{for a sequence } h_j\to 0,$$

and hence, for almost all $a\in\mathscr{A}$, $\mathbf{g}(\varphi;h_j,a)\to\mathbf{0}$ as $h_j\to 0$. Now $\mathbf{g}^2=\sum_{p=0}^{\infty}\mathbf{g}_p{}^2$ $+2\sum_{p=1}^{\infty}\sum_{q<p}\mathbf{g}_p\cdot\mathbf{g}_q$ and $\mathbf{g}_p\neq\mathbf{0}$ at most for a finite number of values of p (since φ is compactly supported). The number of values of p, for which $\mathbf{g}_p\neq\mathbf{0}$ is $O(h^{-1})$. Hence it is sufficient to show that

$$\int_{\mathscr{A}}dL\,(a)\mathbf{g}_p{}^2=O(h^2),\qquad\int_{\mathscr{A}}dL\,(a)\mathbf{g}_p\cdot\mathbf{g}_q=O(h^2).$$

In fact, Glimm proves that there is a set $\mathcal{N}$ of measure zero in $\mathcal{A}$ and a sequence $h_i \to 0$ such that $\forall a \in \mathcal{A} \setminus \mathcal{N}$, $\forall \varphi \in \mathcal{C}_0^\infty(\boldsymbol{R}^2)$ $\mathbf{g}(\varphi; h_i, a) \to \mathbf{0}$. Choosing a subsequence $\{\bar{h}_i\}$ of $\{h_i\}$ in such a way that $\mathbf{v}(\cdot, \cdot; \bar{h}_j, a) \to \mathbf{v}(\cdot, \cdot; a)$ in $\mathcal{L}^1_{loc}$ we conclude that, for $a \in \mathcal{A} \setminus \mathcal{N}$, $\mathbf{v}(\cdot, \cdot; a)$ is a weak solution of the Cauchy problem. □

Practical applicability of Glimm's scheme is limited by the lack of any indication as to how to choose the correct value of $a \in \mathcal{A}$ and the right sequence $\{\bar{h}_i\}$. Concerning the first dilemma, Liu (1977) showed that Glimm's scheme converges in particular for every $a \in \mathcal{A}$ such that $a_q^p = a_p$ does not depend on q and the number of a_p with $p \leqslant k$, $a_p \in \mathcal{I}$, divided by k, tends to half the length of the interval $\mathcal{I} \subset [0, 1]$, for every $\mathcal{I} \subset [0, 1]$.

The existence proofs by Glimm's scheme hinge on a previous solution of the Riemann problem and there is good reason to suppose that it applies whenever the Riemann problem has been successfully disposed of.

We mention finally some papers in which Glimm's scheme has been applied successfully to non-homogeneous equations with $\mathbf{h}(x, \mathbf{u})$ on the right-hand side (Liu, 1979, 1980).

3.2.9 Singularities of Weak Solutions

The class of singularities exhibited by a solution $\mathbf{u} \in \mathrm{BV} \cap \mathcal{L}^\infty$ to a hyperbolic system of conservation laws can be expected to be a proper subclass of the class of singularities of an arbitrary BV function. We shall see that it is indeed so. Unfortunately there are very few studies of the singularities and their scope is too narrow.

A nearly exhaustive analysis of singularities of a weak solution $u(t, x)$ of a single conservation law satisfying (GL) was completed by Dafermos (1978). Using an explicit solution u of a single conservation law (due to Lax) Oleinik (1955, 1956) showed that its jump set $\Gamma(u)$ is the join of an at most countable family of Lipschitz continuous curves. Using the same explicit solution Schaeffer (1973) and Golubitsky and Schaeffer (1975) showed that generically a solution with smooth initial data is piecewise smooth. It is easy however to produce a solution u whose jump set $\Gamma(u)$ is dense in $\boldsymbol{R}^2_+$ by taking a monotone function u_0 with a dense set of discontinuities emitting shock waves.

In contrast to the above papers Dafermos (1978) based his study merely on the definition of a weak solution and on the Lax condition. In view of his assumption of (GL) the Lax condition is sufficient to pick out the correct solutions. The solution u can be adjusted on a set of L^2-measure zero in such a way that, for almost all values of $t > 0$, $u(t, \cdot)$ is a function of locally bounded variation (Theorem 3.2.29). We assume that, for almost all t, $u(t, \cdot)$ has locally bounded variation in the usual sense, i.e. without adjusting it on some set of zero measure for every correct value of t.* Such a function has for almost all values of $t > 0$ and for all values of x the one-sided limits $u(t, x \pm 0)$ and $u(t, x+0) \neq u(t, x-0)$ at most on a countable subset of $x \in \boldsymbol{R}$. Dafermos' version of the condition (L) reads

$$u(t, x-0) \geqslant u(t, x+0) \tag{3.2.93}$$

with $f'' > 0$.

The basic tool use by Dafermos is the notion of a generalized characteristic. For the sake of future references we dwell on its definition for a solution $\mathbf{u}(t, x)$ in $\mathrm{BV} \cap \mathscr{L}^\infty$ of a general hyperbolic system of conservation laws.

Let $\mathscr{C}$: $x = y(t)$ be a Lipschitz curve in $\boldsymbol{R}^2_+$ and let $ds^2 = (1+\dot{y}^2)dt^2$. The subset $\mathscr{E} \subset \mathscr{C}$ of irregular points $(t, y(t))$ for $\mathbf{u}$ has H^1-measure zero. Since $|dt/ds| \leqslant 1$, it follows that for almost all $t \in \operatorname{dom} y(\cdot)$ there is a vector $\nu(t)$ such that the limits $l_{\nu(t)}\mathbf{u}$, $l_{-\nu(t)}\mathbf{u}$ exist at $(t, y(t))$.

Definition 3.2.56

The Lipschitz curve $\mathscr{C}$ is said to be a **generalized α-characteristic** of equation (3.1.35) if for almost every $t \in \operatorname{dom}(y\cdot)$ $\dot{y}(t)$ lies between $\lambda_\alpha(l_{\nu(t)}\mathbf{u})$ and $\lambda_\alpha(l_{-\nu(t)}\mathbf{u})$.

We can choose $\boldsymbol{\nu}(t)$ in such a way that $\nu_x(t) \geqslant 0$ for almost all $t \in \operatorname{dom} y(\cdot)$. In the case where $n = 1$ and u satisfies the Lax condition (L) the generalized characteristic $y(\cdot)$ satisfies the inequalities

$$\lambda(l_{\nu(t)}u) \leqslant \dot{y}(t) \leqslant \lambda(l_{-\nu(t)}u).$$

For an adjusted solution considered by Dafermos this is equivalent to

$$\lambda\big(u(t, x+0)\big) \leqslant \dot{y}(t) \leqslant \lambda\big(u(t, x-0)\big), \quad x = y(t) \tag{3.2.94}$$

for almost all $t \in \operatorname{dom} y(\cdot)$.

The α-characteristics through a point $(t_0, x_0) \in \boldsymbol{R}^2_+$ fill up a conoid $\mathscr{K}$ with the vertex at (t_0, x_0). The upper (lower) envelope of the conoid $\mathscr{K}$ is called the **maximal (minimal) α-characteristic through** (t_0, x_0).

Let us now apply this notion to the weak solution $u(t, x)$ studied by Dafermos. It is almost obvious that on every Lipschitz curve $x = y(t)$ the Rankine–Hugoniot equation is satisfied for almost all values of $t \in \operatorname{dom} y(\cdot)$:

$$f\big(u(t, y(t+0))\big) - f\big(u(t, y(t-0))\big) - \dot{y}(t)\big[u\big(t, y(t+0)\big) - u\big(t, y(t-0)\big)\big] = 0. \tag{3.2.95}$$

* The solutions obtained by Glimm's scheme certainly have this property, and for **every** $t > 0$.

For the proof one can substitute the test function

$$\varphi^\varepsilon(t, x) \equiv \psi^\varepsilon(t, x)\chi^\varepsilon(t),$$

$$\chi^\varepsilon(t) :\equiv \begin{cases} 0 & \text{for } t < \sigma, \\ \dfrac{t-\sigma}{\varepsilon} & \text{for } \sigma \leqslant t < \sigma+\varepsilon, \\ 1 & \text{for } \sigma+\varepsilon \leqslant t < \tau, \\ 1+\dfrac{\tau-t}{\varepsilon} & \text{for } \tau \leqslant t < \tau+\varepsilon, \\ 0 & \text{for } \tau+\varepsilon \leqslant t, \end{cases}$$

$$\psi^\varepsilon(t, x) :\equiv \begin{cases} 0 & \text{for } x < y(t)-\varepsilon, \\ 1+\dfrac{x-y(t)}{\varepsilon} & \text{for } y(t)-\varepsilon \leqslant x \leqslant y(t), \\ 1-\dfrac{x-y(t)}{\varepsilon} & \text{for } y(t) \leqslant x \leqslant y(t)+\varepsilon, \\ 0 & \text{for } x \geqslant y(t)+\varepsilon \end{cases}$$

into the equation

$$\int dL^2\{u\varphi^\varepsilon_t+f(u)\varphi^\varepsilon{}_x\} = 0. \tag{3.2.96}$$

The Lipschitz test function φ^ε is a limit of smooth test functions and hence the validity of equation (3.2.96) can be established. Going to the limit $\varepsilon \to 0_+$ we find that

$$\int_\sigma^\tau dt\,\{f\big(u(t, y(t-0))\big)-f\big(u(t, y(t+0))\big)-\dot{y}(t)[u(t, y(t-0)) \\ -u(t, y(t+0))]\} = 0.$$

Since σ, τ are arbitrary, and the integrand is locally summable, equation (3.2.95) follows.

Equation (3.2.94) implies immediately that at every point $(t, y(t))$ of continuity $\big(u(t, y(t+0)) = u(t, y(t-0))\big)$

$$\dot{y}(t) = \lambda\big(u(t, y(t))\big). \tag{3.2.97}$$

A generalized characteristic is said to be **genuine** if (3.2.97) holds for almost all $t \in \operatorname{dom} y(\cdot)$. It follows from the considerations of Section 3.2.1 that genuine characteristics are straight lines.

The arc of a generalized characteristic through $(t_0, x_0) \in R^2_+$ defined for $t \leqslant t_0$ $(t \geqslant t_0)$ is said to be a **backward (forward) characteristic** for that point.

Dafermos uses the condition (3.2.93) to prove the following assertions:

(1) the maximal and minimal backward characteristics through every point $(t_0, x_0) \in \boldsymbol{R}_+^2$ are genuine and defined on $[0, t_0]$. Their slopes are $\lambda(u(t_0, x_0+0))$, $\lambda(u(t_0, x_0-0))$ resp.

(2) there is a unique forward characteristic through every point (t_0, x_0), $t_0 > 0$, defined on $[t_0, \infty[$ (this need not be true for $t_0 = 0$). A genuine characteristic through (t_0, x_0) can eventually run into a shock and continue along the shock up to infinity. A genuine characteristic cannot branch out from a shock in the direction of increasing time. Hence a forward characteristic through (t_0, x_0) is either genuine or a shock, or it is genuine up to some time $t_1 > t_0$ and a shock for $t > t_1$.

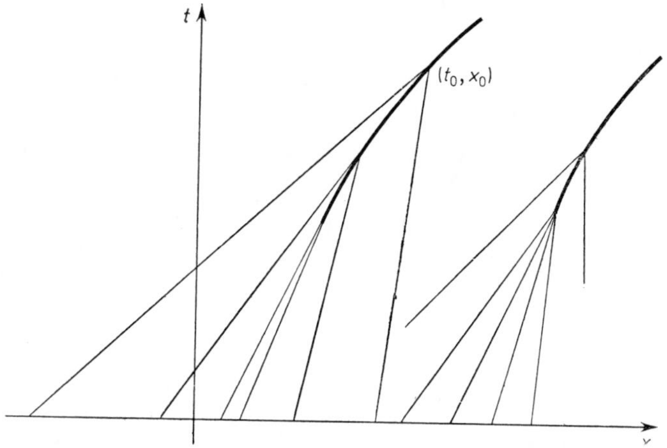

Fig. 3.4 Typical singularities of the solutions of a single convex conservation law

The singularities of $u(t, x)$ are of the following types:

(1) discontinuity curves (non-genuine generalized characteristics);

(2) shock generation points (t_0, x_0): the forward characteristic through (t_0, x_0) is a discontinuity and every backward characteristic through (t_0, x_0) is genuine;

(3) vertices of centered simple waves: there are two different backward characteristics through (t_0, x_0) and every backward characteristic contained between them is also genuine.

Clearly, the last type of singularity corresponds to an irregular point for $u(t, x)$. Another kind of irregular point for $u(t, x)$ is the intersection of two discontinuity curves.

The set of all discontinuity curves of $u(t, x)$ is at most countable. For every discontinuity $x = y(t)$ the right derivative of $y(\cdot)$ exists for all

$t \in \operatorname{dom} y(\cdot)$ and satisfies the Rankine–Hugoniot equations. The two-sided derivative $\dot{y}$ exists and is continuous except on a countable set consisting of those points which are either vertices of c.s. waves or intersection points of two discontinuity curves. At such points the strength of the discontinuity increases after the interaction.

After a suitable adjusting on a set of L^2-measure zero the solution is continuous on $\Gamma(u)^c$. If it is continuous in a nbhd of a point $(t_0, x_0) \in \boldsymbol{R}^2_+$ and $f''(u(t_0, x_0)) > 0$, then it is Lipschitz continuous in a nbhd of (t_0, x_0). The left and right states of a discontinuity $\Sigma \subset \Gamma(u)$ are also defined in the sense of topological convergence.

Let $f \in \mathscr{C}^{k+1}$, $u_0 \in \mathscr{C}^k$. If a point (t_0, x_0) of continuity of $u(t, x)$ does not lie on the envelope of a family of characteristics then $u \in \mathscr{C}^k$ in a nbhd of (t_0, x_0). In the opposite case it is a shock formation point and the discontinuity starting at (t_0, x_0) has zero strength at (t_0, x_0). The argument leading to these conclusions is based on the implicit solution discussed in Section 3.2.1. It follows then that $\Gamma(u)^c$ is open and $u|\Gamma(u)^c$ is $\mathscr{C}^k$.* Local smoothness of a discontinuity curve $x = y(t)$ as well as of the right and left states $u_+(t)$, $u_-(t)$ can be established in a similar manner for a nbhd of every point t_0 such that $(t_0, y(t_0))$ does not lie on the envelope of minimal or maximal backward characteristics. The precise conditions on the initial data which ensure that the solution is piecewise smooth are also given. For generic u_0 the solution is, however, piecewise smooth.

We now turn to the description of singularities of solutions of hyperbolic systems. DiPerna (1975) investigated the singularities of those solutions $u(t, x)$ of strictly hyperbolic systems of two conservation laws which can be obtained by Glimm's scheme. As we know, such solutions have initial data with small variation and small oscillation. DiPerna assumed that the hyperbolic system satisfied (GL_1), (GL_2) and the Glimm–Lax condition ensuring that the interaction of two α-shocks produces another α-shock and a β-c.s.w., $\beta \neq \alpha$. The restriction of the study to two conservation laws was essentially motivated by the possibility of expressing the system in the form (3.2.90), with $s^1 = J^2$, $s^2 = J^1$. The symbol J^α, $\alpha = 1,2$, denotes here the familiar Riemann–Lax invariant of the kind α (see Section 3.1.2). Thus s^2 is constant across a 1-c.s.w. and the total increase (or decrease) of s^1 across a 1-wave measures its strength. If we assume that $d\lambda_\alpha[\mathbf{r}_\alpha] > 0$, $\langle \mathbf{l}^\alpha, \mathbf{r}_\beta \rangle = \delta^\alpha{}_\beta$ and $\mu_\alpha > 0$, $\alpha, \beta = 1,2$, then $s^\alpha_- < s^\alpha_+$ for a divergent α-c.s.w. and $s^\alpha_+ < s^\alpha_-$ for a convergent α-simple wave or an α-shock. We note that the existence of s^1, s^2 is not essential for measuring the strength of the α-waves. On account of the

* More precisely, this is true for every open subset $\mathscr{U} \subset \Gamma(u)^c$, with $(t_0, x_0) \in \overline{\mathscr{U}}$.

assumption (GL_α), $\alpha = 1,2$ the strength of an α-wave could also be measured in terms of λ_α.

The generalized characteristics appear in Glimm's scheme as limits of the characteristics of the approximating functions $\mathbf{v}(\cdot, \cdot; h, a)$. Recall that the approximating functions satisfy the hyperbolic system in each strip $ps < t < (p+1)s$.

Let $x = y(t)$, $x = z(t)$ be two generalized α-characteristics through (t_0, x_0). We consider the behaviour of the absolute total variation $\mathrm{TV}s^\alpha(t)$, the increasing total variation $\mathrm{IV}s_\alpha(t)$, and the decreasing total variation $\mathrm{DV}s^\alpha(t)$ of $s^\alpha(t, \cdot)$ on $[y(t), z(t)]$ as $t \to t_0$. Let $\mathscr{D}_+^\alpha := \{(t, x) |\ t > t_0, y(t) \leqslant x \leqslant z(t)\}$. The restriction $\mathbf{u}|\mathscr{D}_+^\alpha$ of the solution $\mathbf{u}$ is said to be a **generalized divergent** α-c.s.w. if

$$\limsup_{t \to t_0} \mathrm{IV}s^\alpha(t) > 0,$$

$$\lim_{t \to t_0} \mathrm{DV}s^\alpha(t) = 0,$$

$$\lim_{t \to t_0} \mathrm{TV}s^\beta(t) = 0 \quad \text{for } \beta \neq \alpha.$$

The second and third condition imply that the contribution of α-shocks and β-shocks to the variation of $\mathbf{u}|\mathscr{D}_+^\alpha$ is negligible in the limit $t \to t_0$. According to the first condition the singularity at (t_0, x_0) is of the type of a vertex of a divergent α-c.s.w.

Similarly, let $\mathscr{D}_-^\alpha := \{(t, x) |\ t < t_0, z_\alpha(t) \leqslant x \leqslant y_\alpha(t)\}$. The restriction $\mathbf{u}|\mathscr{D}_-^\alpha$ of the solution $\mathbf{u}$ is said to be a **generalized convergent** α-c.s.w. if

$$\limsup_{t \to t_0} \mathrm{DV}s^\alpha(t) > 0,$$

$$\lim_{t \to t_0} \mathrm{IV}s^\alpha(t) = 0,$$

$$\lim_{t \to t_0} \mathrm{TV}s^\beta(t) = 0 \quad \text{for } \beta \neq \alpha.$$

According to this definition it is the convergent α-waves and α-shocks arriving at (t_0, x_0) through $\mathscr{D}_-^\alpha$ that determine the singularity.

The above two definitions make sense provided it is possible to adjust the solution $\mathbf{u}$ on a set of L^2-measure zero in such a way that the respective variations exist. It can be easily seen from the proof of Theorem 3.2.55 that every solution obtained by Glimm's scheme has this property.

Let $\mathscr{D}$ be a domain in $\boldsymbol{R}_+^2$, $(t_0, x_0) \in \bar{\mathscr{D}}$ and $\mathscr{G} \subset \boldsymbol{R}$,

$$\mathrm{var}_2(\mathscr{D}) := \sup\{\mathrm{var}(\mathbf{u}_{(2,t)}, \mathscr{I}) |\ \{t\} \times \mathscr{I} \subset \mathscr{D}\},$$

$$\mathrm{var}_1(\mathscr{D}; \mathscr{G}) := \sup\{\mathrm{var}(\mathbf{u}_{(1,x)}, \mathscr{I}) |\ \mathscr{I} \times \{x\} \subset \mathscr{D}, x \,\bar{\in}\, \mathscr{G}\}.$$

The restriction $\mathbf{u}|\mathscr{D}$ is said to **have small variation with respect to** (t_0, x_0) if there is a subset $\mathscr{G}$ of measure zero in $\boldsymbol{R}$ such that

$$\lim_{r\to 0} \operatorname{var}_2 (\mathscr{D}\cap\mathscr{K}((t_0, x_0), r)) = 0 = \lim_{r\to 0} \operatorname{var}_1 (\mathscr{D}\cap\mathscr{K}((t_0, x_0), r); \mathscr{G})$$

Theorem 3.2.57 (DiPerna, op. cit.)
Through each point $(t_0, x_0) \in \boldsymbol{R}^2_+$ there pass the generalized α-characteristics $z_\alpha(t) \geqslant y_\alpha(t)$, $\alpha = 1, 2$, defined on $[t_0-\varepsilon, t_0+\varepsilon]$, $\varepsilon > 0$, which divide a nbhd $\mathscr{U}$ of (t_0, x_0) into the sectors $\mathscr{D}^\alpha_+$, $\mathscr{D}^\alpha_-$, $\alpha = 1, 2$ (defined above) and

$$\mathscr{D}_- := \{(t, x)|\ x < y_1(t) \text{ and } t \geqslant t_0 \text{ or } x > y_2(t) \text{ and } t < t_0\},$$

$$\mathscr{D}_+ := \{(t, x)|\ x > z_2(t) \text{ and } t \geqslant t_0 \text{ or } x < z_1(t) \text{ and } t < t_0\},$$

$$\mathscr{D}_u := \{(t, x)|\ t > t_0, z_1(t) < x < y_2(t)\},$$

$$\mathscr{D}_l := \{(t, x)|\ t < t_0, z_2(t) < x < y_1(t)\}.$$

The sectors $\mathscr{D}_-$, $\mathscr{D}_+$, $\mathscr{D}_u$, $\mathscr{D}_l$ have small variation with respect to (t_0, x_0)

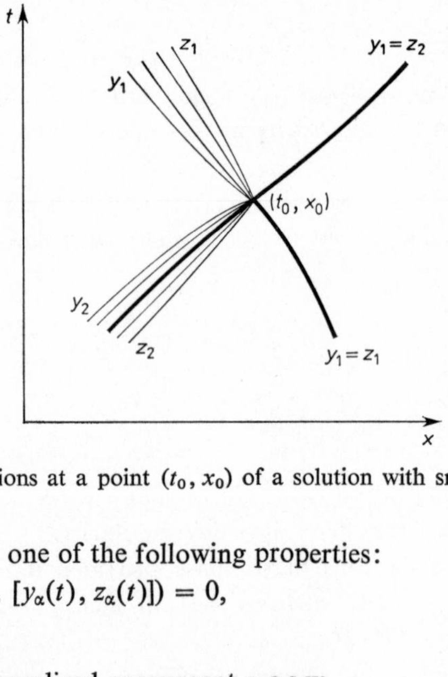

Fig. 3.5 Wave interactions at a point (t_0, x_0) of a solution with small initial oscillation

Each sector $\mathscr{D}^\alpha_-$ has one of the following properties:

(i) $\lim_{t\to t_0} \operatorname{var}(\mathbf{u}_{(2,t)}, [y_\alpha(t), z_\alpha(t)]) = 0$,

or

(ii) $\mathbf{u}|\mathscr{D}^\alpha_-$ is a generalized convergent α-c.s.w.

Each sector $\mathscr{D}^\alpha_+$ has one of the following properties:

(i) as above,

(ii′) $\mathbf{u}|\mathscr{D}^\alpha_+$ is a generalized divergent α-c.s.w.;

(iii) $\mathscr{D}^\alpha_+$ is an α-shock, $y_\alpha(t) = z_\alpha(t)$ for $t \geqslant t_0$.

Note that (i) does not exclude the possibility that $y_\alpha(t) = z_\alpha(t)$ for t close to t_0.

The domains $\mathscr{D}_-$, $\mathscr{D}_+$, $\mathscr{D}_u$, $\mathscr{D}_l$ obviously play the role of constant states in the solutions to the Riemann problems. The limits $\mathbf{u}(t_0, x_0 \pm 0)$ exist and the structure of the solution $\mathbf{u}$ in $\mathscr{U} \cap \{(t, x) | \ t > t_0\}$ is asymptotically, for $t \to t_0$, the solution of the Riemann problem with the α-group $\mathbf{u}|\mathscr{D}^\alpha_\mp$, $\alpha = 1, 2$. We also note that $\mathbf{u}(t_0, x_0 - 0) = l_{\mathscr{D}_-}\mathbf{u}(t_0, x_0)$, $\mathbf{u}(t_0, x_0 + 0) = l_{\mathscr{D}_+}\mathbf{u}(t_0, x_0)$.

We now turn to a global description of the solutions. Let $\mathscr{R}$ be the set of regular points $(t, x) \in \boldsymbol{R}^2_+$ for $\mathbf{u}$, and let $\Gamma = \Gamma(\mathbf{u})$. Then $\mathbf{u}$ is continuous (in the topological sense) on $\mathscr{R} \cap \Gamma^c$ and it is Lipschitz continuous on every open component of $\mathscr{R} \cap \Gamma^c$.

There is an at most countable set of α-shocks Γ^j_α, $j = 1, 2, \ldots$, such that

(1) $\Gamma = (\bigcup_{j=1}^{\infty} \Gamma^j_1 \cup \bigcup_{j=1}^{\infty} \Gamma^j_2) \setminus \mathscr{R}^c$;

(2) Γ^j_α, Γ^k_α, $j \neq k$ do not intersect except at their endpoints;

(3) $\mathscr{R}^c = \mathscr{C} \cup \mathscr{F} \cup \mathscr{C}_0$ is an at most countable set of points, $\mathscr{C}$ is the set of all the points $(t, x) \in \Gamma^j_\alpha \cap \Gamma^k_\beta$, $\Gamma^j_\alpha \neq \Gamma^k_\beta$ such that the jumps of $\mathbf{u}$ on Γ^j_α, Γ^k_β at (t, x) are different from zero, $\mathscr{F}$ is the set of all the endpoints $(t, x) \in \Gamma^j_\alpha$ of Γ^j_α such that the jump of $\mathbf{u}$ on Γ^j_α at (t, x) is different from zero,* $\mathscr{C}_0$ is the set of all the vertices of convergent β-c.s. waves lying on an α-shock, $\alpha, \beta = 1, 2$.

The set Γ may be dense in an open subset $\mathscr{D}$ of $\boldsymbol{R}^2_+$. It is true, however, that sum of the strengths of the discontinuities close to an α-shock within a distance $\delta > 0$ from it decreases to zero as $\delta \to 0$. More precisely, every α-shock $\Sigma \subset \Gamma$: $x = y(t)$ has the following property: if $(t_0, x_0) \in \Gamma \cap \mathscr{R}$ is not an endpoint of Σ, then

$$\forall \varepsilon > 0 \ \exists \delta > 0 \text{ such that } |t - t_0| < \delta \text{ implies that}$$

$$\operatorname{var}(\mathbf{u}_{(2,t)}, [y(t) - \delta, y(t)[) + \operatorname{var}(\mathbf{u}_{(2,t)},]y(t), y(t) + \delta]) > \varepsilon.$$

Concerning the regularity of shock discontinuities, the following is true for every $(t_0, x_0) \in \Sigma \cap \mathscr{R}$:

(a) the limits $\mathbf{u}(t_0, y(t_0) \pm 0) = l_{\pm\nu}\mathbf{u}(t_0, x_0)$ and the derivative $\dot{y}(t_0)$ exist and satisfy the Rankine–Hugoniot equations;

(b) $\mathbf{u}|\mathscr{U}_+$ and $\mathbf{u}|\mathscr{U}_-$ (see Section 3.1.4) have small variation with respect to (t_0, x_0);

(c) if Σ is an α-shock, then all the generalized α-characteristics in a nbhd of (t_0, x_0) run into Σ for $t \nearrow t_0$.

* The point $((2a)^{-1}, a)$ of the envelope $\mathscr{E}$ described in Section 3.2.1 does not belong to $\mathscr{F}$ and it is a point of approximate continuity.

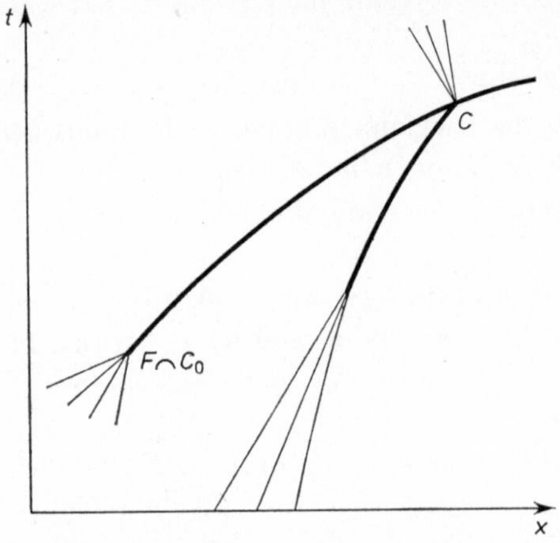

Fig. 3.6 Typical singularities of a solution

The regularity proofs in the case of $n > 1$ are considerably more complicated since we cannot use an implicit function theorem, and a detailed analysis of the approximate solutions is necessary. For the solutions of a hyperbolic system which admits the invariants s^α, $\alpha = 1, \ldots, n$, DiPerna proved interior Lipschitz regularity, as stated above.

3.2.10 Appendix: Helly's Theorems

Theorem 3.2.58 (the abstract Helly's selection theorem, Larsen (1973, p. 261))
Let $\mathscr{X}$ be a separable normed linear space and let $\mathscr{X}^*$ be its dual with the norm $\|\cdot\|$. If $\{x_n\}$ is a bounded sequence in $\mathscr{X}^*$: $\|x_n\| \leqslant M < \infty$, then a subsequence $\{x_{n_k}\}$ of $\{x_n\}$ converges to some $x \in \mathscr{X}^*$ in the weak star topology. Moreover $\|x\| \leqslant M$.

Let $\mathscr{X} = \mathscr{C}^0(\mathscr{K})$ be the Banach space of all the continuous functions $\mathscr{K} \to \boldsymbol{R}$ on a compact Hausdorff space $\mathscr{K}$, endowed with the sup norm. The celebrated Riesz' theorem says that $\mathscr{X}^*$ is the set of all Borel regular signed scalar measures on $\mathscr{K}$ endowed with the norm $\|\mu\| = v(\mu)$ (= variation of μ over $\mathscr{K}$), cf. Dunford and Schwartz (1958), Theorem 4.6.3. Combining Riesz' theorem 3.2.58 for $\mathscr{K} = \mathscr{I} \subset \boldsymbol{R}$ we get the classical second Helly's theorem (due to Bray, 1919, and Helly, 1921).

We quote both Helly's theorems in their original form (Helly, 1921).

Theorem 3.2.59 (the first theorem of Helly (Helly, 1921))
Let f_n: $\mathscr{I} \to \boldsymbol{R}$ be a sequence of functions of bounded variation defined on a (possibly unbounded) interval $\mathscr{I} \subset \boldsymbol{R}$. Suppose that

(i) $\exists\, x_o \in \mathscr{I}$ such that $|f_n(x_o)| \leqslant K \; \forall\, n \in \boldsymbol{Z}_+$;

(ii) $\operatorname{var}(f_n, \mathscr{K}) = \sup\left\{ \sum_{j=1}^{m} |f_n(x_j) - f_n(x_{j-1})| \;\middle|\; \text{all } x_0 < x_{j-1} < x_j \in \mathscr{I}' \right\}$ is bounded by a number M independent of n for every compact interval $\mathscr{I}' \subset \mathscr{I}$.

Then a subsequence $\{f_{n_k}\}$ converges pointwise on $\mathscr{I}$ to a function of bounded variation on $\mathscr{I}$.

Theorem 3.2.60 (the second theorem of Helly; Helly (1921), Bray (1919)) Let $g \in \mathscr{C}^0([a, b])$, $-\infty < a < b < \infty$, and let $\{f_n\}$ be a sequence of functions $[a, b] \to \boldsymbol{R}$ of bounded variation such that $\exists\, M > 0$ $\operatorname{var}(f_n, [a, b]) < M$, $f_n(x) \to f(x)$, $|f(x)| < \infty$, for $x \in [a, b]$. Then

$$\int_a^b g\,df_n \to \int_a^b g\,df.$$

Note that all continuous linear functionals on $\mathscr{C}^0([a, b])$ are representable by Riemann–Stieltjes integrals (Riesz and Sz.-Nagy, 1972, Chapter 3).

3.3. NON-LINEAR MODES AND GLOBAL SOLUTIONS IN DYNAMIC ELASTICITY

3.3.1 Introduction and Preliminaries

Let $\boldsymbol{\chi}(t, \cdot)$ be a one-parameter family of configurations of a simply connected hyperelastic body, $\boldsymbol{\chi} \in \mathscr{C}^2$. Setting

$$F^k{}_\alpha = \frac{\partial \chi^k}{\partial \xi^\alpha}, \quad v^k = \frac{\partial \chi^k}{\partial t}, \quad k = 1, 2, 3; \alpha = 1, 2, 3, \tag{3.3.1}$$

the equations of motion can be presented in the following form

$$\varrho_0 v^k{}_{,t} - \partial_\alpha(\varrho_0 W_{F^k{}_\alpha}) = \varrho_0 b_k, \quad k = 1, 2, 3, \tag{3.3.2a}$$

$$F^k{}_{\alpha,t} - v^k{}_{,\alpha} = 0, \quad k, \alpha = 1, 2, 3 \tag{3.3.2b}$$

for $(t, \xi) \in \mathscr{I} \times \mathscr{B}$, $\mathscr{I} = [0, T]$, $T > 0$. The initial data at $t = 0$ are assumed to satisfy the constraint

$$F^k{}_{\alpha,\beta} - F^k{}_{\beta,\alpha} = 0 \quad \text{at } t = 0; \alpha, \beta, k = 1, 2, 3. \tag{3.3.2c}$$

Equations (3.3.2b) and (3.3.2c) imply that

$$F^k{}_{\alpha,\beta} - F^k{}_{\beta,\alpha} = 0 \quad \text{in } \mathscr{I} \times \mathscr{B}, \quad \alpha, \beta, k = 1, 2, 3. \tag{3.3.2d}$$

Equations (3.3.2b) and (3.3.2d) ensure the existence of a function $\boldsymbol{\chi}: \mathscr{I} \times \times \mathscr{B} \to \boldsymbol{R}^3$, given by the formula

$$\chi^k(t, \xi) = \chi^k(t_0, \xi_0) + \int_{\mathscr{C}} F^k{}_\alpha d\xi^\alpha + v^k dt, \quad k = 1, 2, 3, \tag{3.3.3}$$

where $\mathscr{C}$ is an arbitrary piecewise $\mathscr{C}^1$ curve in $\mathscr{I} \times \mathscr{B}$, joining the fixed point $(t_0, \xi_0) \in \mathscr{I} \times \mathscr{B}$ to (t, ξ). Indeed, for any continuous functions $F^k{}_\alpha$, v^k satisfying

$$\oint_{\mathscr{C}} F^k{}_\alpha d\xi^\alpha + v^k dt = 0 \tag{3.3.4}$$

for every closed piecewise $\mathscr{C}^1$ curve $\mathscr{C}$ the right-hand side of (3.3.3) is well-defined. Moreover, the function χ^k defined by (3.3.3) satisfies equations (3.3.1). In our case $F^k{}_\alpha, v^k \in \mathscr{C}^1$ and (3.3.4) follows from (3.3.2b, d).

Equations (3.3.2a, b, d) look like a system of conservation laws. In fact equation (3.3.2a) is derivable from an integral conservation law. Equations (3.3.2b, d) in the integral form

$$\int_{\partial\mathscr{P}} dH^3(F^k{}_\alpha \nu_t - v^k \nu_\alpha) = 0,$$

$$\int_{\partial\mathscr{P}} dH^3(F^k{}_\alpha \nu_\beta - F^k{}_\beta \nu_\alpha) = 0$$

can be derived from (3.3.4) by slicing $\partial\mathscr{P}$ with $\binom{4}{2} = 6$ families of coordinate-parallel 2-dimensional planes and applying the Fubini theorem.

Let us now consider piecewise $\mathscr{C}^1$ functions $F^k{}_\alpha$, v^k with a discontinuity surface $\Sigma \subset \mathscr{I} \times \mathscr{B}$, satisfying equations (3.3.2a, b) in the weak sense. The jump equation

$$[[F^k{}_\alpha]]\nu_\beta - [[F^k{}_\beta]]\nu_\alpha = 0, \quad \alpha, \beta, k = 1, 2, 3,$$

corresponding to equation (3.3.2d), already follows from the Rankine–Hugoniot equations

$$-w[[\varrho_0 v^k]] + \nu_\alpha[[\varrho_0 W_{F^k{}_\alpha} v^k]] = 0, \quad k = 1, 2, 3, \tag{3.3.5}$$

$$-w[[F^k{}_\alpha]] - [[v^k]]\nu_\alpha = 0, \quad k, \alpha = 1, 2, 3, \tag{3.3.6}$$

corresponding to equations (3.3.2a, b). As usual, we have assumed that $\mathbf{b} \in \mathscr{L}^1_{loc}$.

Suppose that an oriented piecewise $\mathscr{C}^1$ loop $\mathscr{C}$ intersects Σ at the points $p, q \in \mathscr{I} \times \mathscr{B}$. Let $\mathscr{D}$ be any piecewise $\mathscr{C}^1$ curve* lying on Σ and joining p to q.

* Note that piecewise $\mathscr{C}^1$ curves can be replaced by piecewise Lipschitz curves in all the statements of this section.

We define the piecewise $\mathscr{C}^1$ loops $\mathscr{C}_+ = (\mathscr{C}\cap\mathscr{U}_+)\cup\mathscr{D}$, $\mathscr{C}_- = (\mathscr{C}\cap\mathscr{U}_-)\cup\mathscr{D}$, oriented in accordance with the orientation of $\mathscr{C}$. For the definition of $\mathscr{U}_+$, $\mathscr{U}_-$, see Section 3.1.4. The integral on the left-hand side of equation (3.3.4) is the sum of the corresponding integrals over $\mathscr{C}_+$, $\mathscr{C}_-$ and of the integral $\int_{\mathscr{D}} [[F^k{}_\alpha]]\, d\xi^\alpha + [[v^k]]\, dt = 0$. The first two integrals vanish by virtue of equations (3.3.2b) and the third by virtue of (3.3.6). The identity (3.3.4) has thus been established for piecewise $\mathscr{C}^1$ functions satisfying equation (3.3.2b) in the weak sense and the function $\boldsymbol{\chi}$ can be defined by equation (3.3.3). It is easy to see that equations (3.3.1) remain valid except on the discontinuity surfaces of $(\mathbf{F}, \mathbf{v})$.

We now turn to the weak solutions $(\mathbf{F}, \mathbf{v}) \in \mathrm{BV}\cap\mathscr{L}^\infty$ to equations (3.3.2a, b). For a general solution $(\mathbf{F}, \mathbf{v})$ in $\mathrm{BV}\cap\mathscr{L}^\infty$ the curvilinear integrals in (3.3.3) and (3.3.4) do not make sense unless the solution depends effectively on only two variables t, $\xi \in \boldsymbol{R}$.

Suppose, however, that we are given initial data

$$\boldsymbol{\chi}(0, \xi) = \boldsymbol{\chi}_0(\xi), \quad \boldsymbol{\chi}_0 \in W^1_{loc}, \ \xi \in \mathscr{B}, \tag{3.3.7}$$

$$\boldsymbol{\chi}_t(0, \xi) = \mathbf{v}_0(\xi), \quad \mathbf{v}_0 \in \mathscr{L}^1_{loc}, \ \xi \in \mathscr{B}. \tag{3.3.8}$$

Let $(F^k{}_\alpha, v^k) \in \mathrm{BV}\cap\mathscr{L}^\infty$ be a weak solution of the conservation laws (3.3.2a, b) satisfying the initial conditions

$$F^k{}_\alpha(0, \xi) = \chi_0{}^k{}_{,\alpha}(\xi), \tag{3.3.9}$$

$$v^k(0, \xi) = v_0{}^k(\xi), \quad \mathbf{v}_0 \in \mathscr{L}^1_{loc}. \tag{3.3.10}$$

The left-hand sides of equations (3.3.9) and (3.3.9) involve the inner traces on $t = 0$ of the respective BV functions defined on $\mathscr{I}\times\mathscr{B}$. The right-hand side of (3.3.9) involves a distributional derivative. The right-hand side of both equations belong to $\mathscr{L}^1_{loc}(\mathscr{B})$. In view of Theorem 3.2.29 we assume that they have essentially bounded variation in the sense of Tonnelli–Cesari and belong to $\mathscr{L}^\infty$.

Let

$$\chi^k(t, \xi) :\equiv \chi_0{}^k(\xi) + \int_0^t ds\, \bar{v}^k(s, \xi), \quad (t, \xi) \in \mathscr{I}\times\mathscr{B}, \tag{3.3.11}$$

where $\bar{v}^k$ is the symmetric average of v^k. The function $\bar{v}^k$ is defined H^3-a.e. in $\mathscr{I}\times\mathscr{B}$. The Fubini theorem implies that $\chi^k(t, \cdot) \in \mathscr{L}^1_{loc}(\mathscr{B})$ for every $t \in I$. We note that for almost every $\xi \in \mathscr{B}$ the function $\bar{\mathbf{v}}(\cdot, \xi)$ has locally bounded variation and

$$\lim_{\Delta\to 0} \frac{1}{2\Delta} \int_{t-\Delta}^{t+\Delta} ds\, \bar{v}^k(s, \xi) \equiv: \hat{v}^k(t, \xi)$$

exists for all $t \in \mathscr{I}$ (see the remarks following Theorem 3.2.29). Hence

$$\frac{\partial \chi^k}{\partial t}(t, \xi) = \hat{v}^k(t, \xi) \tag{3.3.12}$$

for almost all $\xi \in \mathscr{B}$ and all $t \in \mathscr{I}$, while

$$\hat{v}^k(t, \xi) = \bar{v}^k(t, \xi) \tag{3.3.13}$$

for almost all $t \in \mathscr{I}$ for almost every fixed $\xi \in \mathscr{B}$.

We now calculate the distributional derivative $\chi^k{}_{,\alpha}(t, \xi) = \partial\chi^k/\partial\xi^\alpha$. Let $\varphi \in \mathscr{C}_o^\infty(\mathscr{B})$ be an arbitrary test function vanishing in a nbhd of $\partial\mathscr{B}$. We consider the expression

$$-\int_{]0,t]\times\mathscr{B}} dL^4\, \bar{v}^k(s, \xi)\varphi_{,\alpha}(\xi). \tag{3.3.14}$$

It is well-defined since $\bar{v}^k \in \mathscr{L}^1_{loc}(\mathscr{I}\times\mathscr{B})$. By Theorems 3.2.45 and 3.2.48 as well as equation (3.3.2b) it is equal to

$$-\int_{]0,t]\times\mathscr{B}} dL^4\,(\bar{v}^k\varphi)_{,\alpha} + \int_{]0,t]\times\mathscr{B}} dL^4\, \bar{v}^k{}_{,\alpha}\varphi$$

$$= -\int_{\partial^*(]0,t]\times\mathscr{B})} dH^3\, \bar{v}^k\varphi\nu_\alpha + \int_{]0,t]\times\mathscr{B}} dL^4\, F^k{}_{\alpha,t}$$

$$= \int_{]0,t]\times\mathscr{B}} dL^4\,(F^k{}_\alpha\varphi)_{,t} = \int_{\mathscr{B}} dL^3\,[F_-{}^k{}_\alpha(t, \xi) - F_+{}^k{}_\alpha(0, \xi)], \tag{3.3.15}$$

where the subscripts $+$, $-$ denote the operations of taking the inner, outer trace, resp. On the other hand, applying the Fubini theorem to equation (3.3.14) we obtain the expression

$$-\int_{\mathscr{B}} dL^3\left\{\int_0^t ds\, \bar{v}^k(s, \xi)\right\}\varphi_{,\alpha}(\xi). \tag{3.3.16}$$

Comparing (3.3.9), (3.3.11) and (3.3.16) we find that $\chi^k{}_{,\alpha}(t, \xi) = F_-{}^k{}_\alpha(t, \xi)$. Integrating over $]0, t[\times\mathscr{B}$ we would find that $\chi^k{}_{,\alpha}(t, \xi) = F_+{}^k{}_\alpha(t, \xi)$ and hence $\chi^k{}_{,\alpha}(t, \xi) = \bar{F}^k{}_\alpha(t, \xi)$. Since $F^k{}_\alpha = \bar{F}^k{}_\alpha$ L^4-a.e., $\chi^k{}_{,\alpha}(t, \xi) = F^k{}_\alpha(t, \xi)$ in $\mathscr{I}\times\mathscr{B}$ in the sense of distributions.

Let us now calculate

$$\int_{\mathscr{I}\times\mathscr{B}} dL^4\, \varphi(\xi)\psi'(t)\boldsymbol{\chi}(t, \xi)$$

for an arbitrary $\varphi \in \mathscr{C}_0^\infty(\mathscr{B})$, $\psi \in \mathscr{C}_0^\infty(\mathscr{I})$. By the Fubini theorem

$$\int_{\mathscr{I}\times\mathscr{B}} dt\, dL^3\, \varphi(\xi)\psi'(t)\int_0^t ds\, \bar{v}^k(s,\xi)$$
$$= \int_0^T dt\, \psi'(t)\int_0^t ds \int_{\mathscr{B}} dL^3\, \bar{v}^k(s,\xi)\varphi(\xi).$$

Since $\|\mathbf{v}\|_\infty < \infty$, the inner integral is a Lipschitz continuous function of and hence we can integrate by parts in the outer integral, obtaining

$$-\int_0^T dt\, \psi(t)\int_{\mathscr{B}} dL^3\, \bar{v}^k(t,\xi)\varphi(\xi) = -\int_{\mathscr{I}\times\mathscr{B}} dt\, dL^3\, \bar{v}^k(t,\xi)\varphi(\xi)\psi(t).$$

Hence the distributional derivatives of χ^k on $\mathscr{I}\times\mathscr{B}$ are

$$\chi^k{}_{,t}(t,\xi) = v^k(t,\xi), \quad \chi^k{}_{,\alpha} = F^k{}_\alpha. \tag{3.3.17}$$

We shall henceforth look for the weak solutions $(\mathbf{F},\mathbf{v})$ of equations (3.3.2a, b) satisfying the initial conditions (3.3.9) and (3.3.10) as well as the condition

$$\det\mathbf{F} > 0 \quad L^4 \text{ a.e. in } \mathscr{I}\times\mathscr{B}$$

3.3.2 Simple and Shock Waves in Elasticity—Preliminaries

By virtue of SE (Chapter 2), for every fixed $\mathbf{n}\in\mathscr{S}^2$ the eigenvalue problem

$$\{W_{F^k{}_\alpha F^l{}_\beta}(\mathbf{F})n_\alpha n_\beta - \lambda^2\delta_{kl}\}e^l = 0, \tag{3.3.18}$$

$$\mathbf{e}^2 = \sum_{k=1}^{3}(e^k)^2 = 1$$

for the acoustic tensor has three linearly independent solutions $\mathbf{e} = \mathbf{e}_s(\mathbf{F};\mathbf{n})$, $s = 1, 2, 3$, corresponding to three positive eigenvalues $\lambda^2 = \lambda_s(\mathbf{F};\mathbf{n})^2$. We choose $\lambda_s(\mathbf{F};\mathbf{n}) > 0$.

It is possible to choose the functions λ_s, $\mathbf{e}_s$, $s = 1, 2, 3$, in such a way that the functions λ_s depend continuously on either of its arguments (Hanyga, 1981). Let $W\in\mathscr{C}^{k+2}$, $k \geqslant 0$. In a nbhd of every point $(\mathbf{F}_0,\mathbf{n}_0)\in M_+\times(R^3\setminus\{0\})$ such that $\lambda_1(\mathbf{F}_0,\mathbf{n}_0)\neq\lambda_2(\mathbf{F}_0,\mathbf{n}_0)$, $\lambda_1(\mathbf{F}_0,\mathbf{n}_0)\neq\lambda_3(\mathbf{F}_0,\mathbf{n}_0)$ the functions $\mathbf{F}\mapsto\lambda_1(\mathbf{F},\mathbf{n})$, $\mathbf{F}\mapsto\mathbf{e}_1(\mathbf{F},\mathbf{n})$ are of class $\mathscr{C}^k$, whereas the functions $\mathbf{n}\mapsto\lambda_1(\mathbf{F},\mathbf{n})$, $t\mathbf{n}\mapsto\mathbf{e}_1(\mathbf{F},\mathbf{n})$ are of class $\mathscr{C}^\infty$ and are positively homogeneous of degree one and zero, resp. (cf. Section 3.1.3).

It is convenient to consider λ_s, $\mathbf{e}_s$ as functions on $M_+\times\mathscr{S}^2$. The eigenvalue λ_1 has the same smoothness properties as above if $\lambda_1(\mathbf{F},\mathbf{n})\equiv\lambda_2(\mathbf{F},\mathbf{n})$ over an open subset of $M_+ \times \mathcal{S}^2$ containing $(\mathbf{F}_0, \mathbf{n}_0)$ and $\lambda_3(\mathbf{F}_0, \mathbf{n}_0) \neq \lambda_1(\mathbf{F}_0, \mathbf{n}_0)$. In order to see it one can apply the implicit function theorem to the square root of the determinant of the left-hand side of equation (3.3.18). In the case where λ_1 coincides with λ_2 on a subset $\mathcal{A}$ of $M_+ \times \mathcal{S}^2$ with empty interior (e.g. on a submanifold of positive codimension), the first derivatives of λ_1, λ_2 need not exist on $\mathcal{A}$. In this case one can assert the continuity of the two-dimensional space spanned by the eigenvectors $\mathbf{e}_1$, $\mathbf{e}_2$ in a nbhd of $\mathcal{A}$ (provided $\lambda_3 \neq \lambda_1$ on $\mathcal{A}$), but not the continuity of the eigenvectors $\mathbf{e}_1$, $\mathbf{e}_2$ separately (Hanyga, 1981).

Let us now consider the Rankine–Hugoniot equations (3.3.5) and (3.3.6). If $w = 0$ at some point of the discontinuity surface Σ, then

$$[[v^k]] = 0, \quad [[S_k{}^\alpha]]\nu_\alpha \equiv [[\varrho_0 W_{F^k{}_\alpha}]]\nu_\alpha = 0, \quad k, \alpha = 1, 2, 3. \tag{3.3.19}$$

A special case of discontinuities satisfying equations (3.3.19) at every point will be considered in Chapter 4.

Suppose now that $w \neq 0$ and

$$f^k := w^{-1}[[v^k]], \quad k = 1, 2, 3, \tag{3.3.20}$$

so that

$$F_+{}^k{}_\alpha = F_-{}^k{}_\alpha + f^k \nu_\alpha, \quad \alpha, k = 1, 2, 3. \tag{3.3.21}$$

Fix $\mathbf{F}_- = \mathbf{F}_0$, $\boldsymbol{\nu}$, and let $\mathbf{F} = \mathbf{F}_+$, w vary. Let

$$\overline{w}(\mathbf{f}) :\equiv W(F_0{}^k{}_\alpha + f^k \nu_\alpha). \tag{3.3.22}$$

We have

$$s_k := S_k{}^\alpha \nu_\alpha = \varrho_0 \frac{\partial \overline{w}}{\partial f^k}(\mathbf{f}) \quad \text{for } F^k{}_\alpha = F_0{}^k{}_\alpha + f^k \nu_\alpha. \tag{3.3.23}$$

At every point of a discontinuity Σ, at which $w \neq 0$, equation (3.3.21)

$$v_+^k = v_-^k - w f^k, \quad k = 1, 2, 3, \tag{3.3.24}$$

$$-[[s_k]] = w^2 \varrho_0 f^k, \quad k = 1, 2, 3, \tag{3.3.25}$$

and equation (3.3.23) are satisfied.

A **simple wave** is, by definition, a solution of equations (3.3.2) with $\mathbf{b} = \mathbf{0}$, satisfying the conditions

$$F^k{}_\alpha(t, \xi) = \hat{F}^k{}_\alpha \circ \varphi(t, y), \quad y := \nu_\alpha \xi^\alpha, \tag{3.3.26}$$

$$v^k(t, \xi) = \hat{v}^k \circ \varphi(t, y), \quad k, \alpha = 1, 2, 3, \tag{3.3.27}$$

with $\hat{v}^k$, $\hat{F}^k{}_\alpha$ differentiable and defined over some interval $\mathcal{I} = [c, d]$, $\varphi \in \mathcal{C}^1$, $\varphi(t, y) \in \mathcal{I}$, $\varphi_y(t, y) \neq 0$ for (t, y) in a domain $\mathcal{D} \subset \mathbf{R}^2_+$. We write

$$\acute{F}^k{}_\alpha = \frac{d\hat{F}^k{}_\alpha}{ds}, \quad \acute{v}^k = \frac{d\hat{v}^k}{ds}.$$

From equations (3.3.2) we get the relations

$$\varphi_t \acute{v}^k - \varphi_y W_{F^k{}_\alpha F^l{}_\beta} \acute{F}^l{}_\beta \nu_\alpha = 0, \tag{3.3.28}$$

$$\varphi_t \acute{F}^k{}_\alpha - \varphi_y \acute{v}^k \nu_\alpha = 0, \quad k, \alpha = 1, 2, 3, \tag{3.3.29}$$

whence

$$(\varphi_y{}^2 W_{F^k{}_\alpha F^l{}_\beta} \nu_\alpha \nu_\beta - \varphi_t{}^2 \delta_{kl}) \acute{v}^l = 0, \quad k = 1, 2, 3 \tag{3.3.30}$$

A solution of equation (3.3.30) is given by the formulae

$$-\varphi_t(\varphi_y)^{-1} = \iota \lambda_s(\hat{\mathbf{F}} \circ \varphi; \boldsymbol{\nu}), \tag{3.3.31}$$

$$\acute{\mathbf{v}}(\zeta) = k(\zeta) \mathbf{e}_s(\hat{\mathbf{F}}(\zeta); \boldsymbol{\nu}), \quad \zeta \in [c, d] \tag{3.3.32}$$

for some fixed $s \in \{1, 2, 3\}$ and $\iota \in \{1, -1\}$. If $\lambda_r\ (\hat{\mathbf{F}}(\zeta); \boldsymbol{\nu}) \neq \lambda_s(\hat{\mathbf{F}}(\zeta); \boldsymbol{\nu})$ for $r \neq s$, $r, s = 1, 2, 3$, $\zeta \in [c, d]$, then every solution of equation (3.3.30) is of the form (3.3.31–32). By virtue of equation (3.3.29)

$$\hat{F}^k{}_\alpha(\zeta) = \hat{F}^k{}_\alpha(c) + a^k(\zeta) \nu_\alpha, \quad \alpha, k = 1, 2, 3 \tag{3.3.33}$$

for $\zeta \in [c, d]$, and

$$\mathbf{a}(\zeta) :\equiv -\int_c^\zeta dz\, \lambda_s(\hat{\mathbf{F}}(z); \boldsymbol{\nu})^{-1} \acute{\mathbf{v}}(z).$$

Equation (3.3.33) is similar to equation (3.3.21). Both kinds of solutions must satisfy the conditions

$$\det \mathbf{F}_- > 0, \quad \det \hat{\mathbf{F}}(c) > 0, \tag{3.3.34}$$

and

$$\overset{-1}{F}_-{}^\alpha{}_k f^k \nu_\alpha > -1, \quad (\mathbf{F}(c)^{-1})^\alpha{}_k a^k \nu_\alpha > -1 \tag{3.3.35}$$

on account of the identity $\det(\mathbf{F} + \mathbf{a} \otimes \boldsymbol{\nu}) = (\det \mathbf{F})(1 + (\overset{-1}{\mathbf{F}}\mathbf{a}) \cdot \boldsymbol{\nu})$.

3.3.3 Plane Waves in Hyperelasticity

A **plane wave solution** is defined by the following equation:

$$x^k = z^k + u^k(n_l z^l, t), \quad z^k = \chi_0{}^k(\xi), \quad \mathbf{n} \in \mathcal{S}^2, \ k = 1, 2, 3. \tag{3.3.36}$$

Let $y := n_l z^l$, $f^k := u^k{}_y$. We have that

$$F^k{}_\alpha = F_0{}^l{}_\alpha(\delta^k{}_l + f^k n_l), \quad v^k = u^k{}_t, \tag{3.3.37}$$

with

$$\det \mathbf{F}_0 > 0, \quad 1 + f^k n_k > 0. \tag{3.3.38}$$

Equation (3.3.36) implies that $n_k x^k = y + n_k u^k(y, t)$. On account of (3.3.38) this equation has a unique solution

$$y = n_k z^k = \psi(n_k x^k, t). \tag{3.3.39}$$

Equation (3.3.39) and (3.3.36) imply that the solution (3.3.36) appears as a plane wave (propagating in a prestrained medium) to any observer whose Cartesian spatial coordinates are x^k, $k = 1, 2, 3$. This property of the solution is clearly invariant with respect to Galilean transformations of the observer's reference frame.

The integrands in (3.3.3), (3.3.4) are in our case equal to $dz^k + f^k dy + v^k dt$. Given two continuous vector functions $\mathbf{f}$, $\mathbf{v}$ of $(t, y) \in \mathscr{D} \subset \mathbf{R}^2$, the condition of integrability of the form $d\mathbf{u} = \mathbf{f}dy + \mathbf{v}dt$ becomes

$$\oint_{\mathscr{C}} f^k dy + v^k dt = 0, \quad k = 1, 2, 3 \tag{3.3.40}$$

for any closed $\mathscr{C}^1$ curve $\mathscr{C}$.

Equation (3.3.40) is equivalent to a conservation law and it makes sense for $f^k, v^k \in \mathrm{BV}(\mathscr{D})$ provided the values of f^k, v^k in the integrand are interpreted in the sense of inner or outer traces. If we assume its validity for every piecewise $\mathscr{C}^1$ closed curve $\mathscr{C}$, then

$$f^k{}_{,t} - v^k{}_{,y} = 0, \quad k = 1, 2, 3,$$

holds in the sense of equality of measures. For the proof, it is sufficient to apply the Gauss–Green theorem in the case where $\mathscr{C}$ is the boundary of an arbitrary square and make use of the fact that the gradients of BV functions are Borel regular measures.

Suppose now that $\mathscr{D}$ is simply connected, $f^k, v^k \in \mathrm{BV}(\mathscr{D}) \cap \mathscr{L}^\infty(\mathscr{D})$ and equation (3.3.40) is satisfied for every piecewise $\mathscr{C}^1$ closed curve $\mathscr{C}$. We define $\mathbf{u}(y, t)$ by the formula

$$u^k(y, t) = \int_{\mathscr{C}} f_+^k dy + v_+^k dt = \int_{\mathscr{C}} (-v_+^k \nu_y + f_+^k \nu_t) ds, \tag{3.3.41}$$

where $\mathscr{C}$ is an arbitrary $\mathscr{C}^1$ curve in $\mathscr{D}$ joining (t, y) to a fixed point $(t_0, y_0) \in \mathscr{D}$, (ν_t, ν_y) is a unit normal on $\mathscr{C}$ pointing into a region $\mathscr{U}_+$ on one side of $\mathscr{C}$, $ds = dH^1$ is the arc-length of $\mathscr{C}$ and f_+^k, v_+^k are the inner traces of f^k, v^k on $\mathscr{C}$ with respect to $\mathscr{U}_+$. Let f_-^k, v_-^k be the traces of f^k, v^k on $\mathscr{C}$ with respect to the opposite side $\mathscr{U}_-$ of $\mathscr{C}$. In view of (3.3.40) we have the identity

$$[[f^k]]\nu_t - [[v^k]]\nu_y = 0, \quad k = 1, 2, 3 \tag{3.3.42}$$

H^1-a.e. on $\mathscr{C}$. Hence

$$u^k(y, t) = \int_{\mathscr{C}} f_-^k dy + v_-{}^k dt$$

as well. Since the traces are locally summable on $\mathscr{C}$, it is easy to see that $du^k = \bar{f}^k dy + \bar{v}^k dt$ H^1-a.e. on $\mathscr{C}$.

Suppose that $W = \tilde{W}(F^k{}_\alpha, y)$, $F_0{}^k{}_\alpha = F_0{}^k{}_\alpha(y)$, $\varrho_0 = \varrho_0(y)$, $b_k = \bar{b}_k(\mathbf{f}, y)$. Let

$$\bar{w}(\mathbf{f}, y) :\equiv \tilde{W}((\mathbf{E}+\mathbf{f}\otimes\mathbf{n})\mathbf{F}_0(y), y), \tag{3.3.43}$$

$$\varrho_0 \bar{b}_k(\mathbf{f}, y) := \varrho_0 b_k - \frac{\partial \varrho_0}{\partial \xi^\alpha}\frac{\partial W}{\partial F^k{}_\alpha} - \varrho_0 \frac{\partial^2 W}{\partial F^k{}_\alpha \partial \xi^\alpha} - \varrho_0 \frac{\partial^2 W}{\partial F^k{}_\alpha \partial F^l{}_\beta}(\delta^l{}_p + f^l n_p)\frac{\partial F_0{}^p{}_\beta}{\partial \xi^\alpha}. \tag{3.3.44}$$

Since $\dfrac{\partial y}{\partial \xi^\alpha} = n_l F_0{}^l{}_\alpha(y)$, the right-hand side of equation (3.3.44) depends only on $\mathbf{f}$ and y. We can now define a **weak plane wave solution** by means of the formulae (3.3.36) and (3.3.41) with $f^k, v^k \in \mathrm{BV}(\mathscr{D})\cap\mathscr{L}^\infty(\mathscr{D})$ satisfying the following system of equations

$$\mathbf{f}_t - \mathbf{v}_y = \mathbf{0}, \tag{3.3.45a}$$

$$\mathbf{v}_t - \frac{\partial}{\partial y}\left(\frac{\partial \bar{w}}{\partial \mathbf{f}}\right) - \bar{\mathbf{b}}(\mathbf{f}, y) = \mathbf{0} \tag{3.3.45b}$$

in the weak sense.

Setting

$$\mathbf{U} = \begin{bmatrix} \mathbf{f} \\ \mathbf{v} \end{bmatrix} \tag{3.3.46}$$

it is easy to check that the system (3.3.45) is hyperbolic in the sense defined in Section 3.1.1. The six linearly independent right eigenvectors are now given by the formulae:

$$\mathbf{r}_{\iota s} = \begin{bmatrix} -\iota \lambda_s{}^{-1}\, \mathbf{e}_s \\ \mathbf{e}_s \end{bmatrix}, \tag{3.3.47}$$

$s = 1, 2, 3$; $\iota = 1, -1$. The left eigenvectors are given by the following expression

$$\mathbf{l}^{\iota s} = c_s[-\iota \lambda_s{}^t\mathbf{e}_s, {}^t\mathbf{e}_s], \tag{3.3.48}$$

with c_s chosen in such a way that $\langle \mathbf{l}^{\iota s}, \mathbf{r}_{\iota s}\rangle = 1$.

Remark

In the case of general three-dimensional motions of hyperelastic media the condition of hyperbolicity in the sense of Section 3.1.1 is not apparently satisfied since $\mathbf{U} = (F^k{}_\alpha, v^k)$ varies over a manifold $\mathscr{H}$ of dimension 12, while there are only six linearly independent vectors among the eigenvectors

$$\mathbf{r}_{\iota s} = \begin{bmatrix} -\iota\lambda_s{}^{-1}\nu_\alpha e_s{}^k \\ e_s{}^k \end{bmatrix}, \quad \iota = 1, -1, \; s = 1, 2, 3. \tag{3.3.49}$$

This difficulty can be traced back to the constraint (3.3.2d), which reduces the effective number of independent unknowns to six. All the important consequences of hyperbolicity remain valid, as will be seen in the following sections.

3.3.4 Discontinuity Waves in Hyperelasticity

We continue the study of discontinuity waves in hyperelastic media in somewhat more detail.

In view of equation (3.3.21) we may restrict our attention to the set of $\mathbf{F}_+ = \mathbf{F} = (\mathbf{E}+\mathbf{f}\otimes\mathbf{n})\mathbf{F}_0(\xi)$ with $\mathbf{n}\cdot\mathbf{f} > -1$, $\mathbf{F}_0(\xi) = \mathbf{F}_-(\xi)$. We use the following simplified notations:

$$\mathbf{e}_s(\mathbf{f}) = \mathbf{e}_s\big((\mathbf{E}+\mathbf{f}\otimes\mathbf{n})\mathbf{F}_0; {}^t\mathbf{F}_0\mathbf{n}\big), \tag{3.3.50}$$

$$\lambda_{\iota s}(\mathbf{f}) = \iota\lambda_s\big((\mathbf{F}+\mathbf{f}\otimes\mathbf{n})\mathbf{F}_0; {}^t\mathbf{F}_0\mathbf{n}\big), \tag{3.3.51}$$

suppressing all the arguments with fixed values: $\mathbf{F}_0, \xi, \mathbf{n}$. Let $\mathscr{F} := \{\mathbf{f}\in \boldsymbol{R}^3 |\; n_k f^k > -1\}$, $\mathscr{H} := \{\mathbf{U} = (\mathbf{f}, \mathbf{v}) \in \boldsymbol{R}^6 |\; \mathbf{f}\in\mathscr{F}\}$, $\mathscr{F}_0 := \{\mathbf{f}\in\boldsymbol{R}^3 |\; \lambda_r(\mathbf{f}) \neq \lambda_s(\mathbf{f})$ for all $s\neq r$, $s, r = 1,2,3\}$, $\pi\colon \mathscr{H}\to\mathscr{F}$; $(\mathbf{v},\mathbf{f})\mapsto\mathbf{f}$, $\mathscr{H}_0 := \pi^{-1}(\mathscr{F}_0)$. $\mathscr{F}_0$ is open in $\mathscr{F}$ $\pi\colon \boldsymbol{R}\times\mathscr{H}\to\boldsymbol{R}_+\times\mathscr{F}$; $(\lambda,\mathbf{v},\mathbf{f})\to(\lambda^2,\mathbf{f})$.

Suppose that $\mathscr{F}_0$ contains $\mathbf{0}$. Applying the results of Section 3.1.5 to the Rankine–Hugoniot equations (3.3.23–25) one readily obtains the one-dimensional submanifolds $\mathscr{H}^s_{\mathbf{U}_0} \subset \boldsymbol{R}\times\mathscr{H}_0$, $s = \pm1, \pm2, \pm3$, bifurcating from the trivial solution $\mathbf{U} = \mathbf{U}_0 = (\mathbf{0}, \mathbf{v}_0)$, $w\in\boldsymbol{R}$, at the points $w = \lambda_s(\mathbf{0})$.

One can also apply the methods of Section 3.1.5 to the equation

$$\frac{\partial\overline{w}}{\partial f^k}(\mathbf{f}) - \frac{\partial\overline{w}}{\partial f^k}(\mathbf{0}) = w^2 f^k, \quad k = 1,2,3, \tag{3.3.52}$$

(see equations (3.3.25), (3.3.23) with $\nu_\alpha = F_0{}^k{}_\alpha n_k$) on the manifold of $(w^2, \mathbf{f}) \in \overline{\boldsymbol{R}}_+\times\mathscr{F}_0$. The set $\hat{\mathscr{H}}'$ of non-trivial solutions ($\mathbf{f}\neq\mathbf{0}$) of equation (3.3.52) is non-empty and contains the submanifolds $\hat{\mathscr{H}}^s$ bifurcating from $\mathbf{f} = \mathbf{0}$ at the points $w^2 = \lambda_s(\mathbf{0})^2$, $s = 1, 2, 3$. The mapping

$$(\overline{\boldsymbol{R}}_+\times\mathscr{F}_0) \ni (w^2, \mathbf{f}) \mapsto \overline{w}_{\mathbf{f}}(\mathbf{f}) - \overline{w}_{\mathbf{f}}(\mathbf{0}) - w^2\mathbf{f} \tag{3.3.53}$$

is a submersion except at those points $(\lambda_s(\mathbf{f})^2, \mathbf{f}) \in \boldsymbol{R}_+ \times \mathscr{F}_0$, at which condition

$$\langle \mathbf{e}_s(\mathbf{f}), \mathbf{f}-\mathbf{f}_0 \rangle \neq 0 \tag{3.3.54}$$

fails to be satisfied. Note that the mapping (3.3.53) also fails to be a submersion on $\overline{\boldsymbol{R}_+} \times (\mathscr{F} \setminus \mathscr{F}_0)$. It follows from equations (3.3.21)–(3.3.25) that

(i) $\tilde{\pi}(\mathscr{H}'_{\mathbf{U}_0}) = \hat{\mathscr{H}}'$ for every $\mathbf{U}_0 = (\mathbf{0}, \mathbf{v}_0)$;

(ii) for every $\mathbf{v}_0 \in \boldsymbol{R}^3$ and every $(w^2, \mathbf{f}) \in \hat{\mathscr{H}}'$ there are two points $(w_i, \mathbf{f}, \mathbf{v}_i) \in \mathscr{H}'$, $i = 1, 2$, with $w_2 = -w_1 < 0$ and $\mathbf{v}_i$ given by the formula

$$v_i^k = v_0{}^k - w_i f^k, \quad k = 1, 2, 3, \; i = 1, 2, \tag{3.3.55}$$

(see equation (3.3.24)).

For every $\mathbf{U}_0 = (\mathbf{0}, \mathbf{v}_0)$ the mapping $\tilde{\pi}$ brings the points of $\mathscr{H}^S_{\mathbf{U}_0}$ into a one-one correspondence with the points of $\hat{\mathscr{H}}^{|s|}$, $s = \pm 1, \pm 2, \pm 3$.

Suppose that $(\mathbf{0}, \mathbf{f}_1) \in \hat{\mathscr{H}}'$ for some $\mathbf{f}_1 \in \mathscr{F}$, $\mathbf{f}_1 \neq \mathbf{0}$. Since $\overline{w}_{\mathbf{f}}(\mathbf{f}_1) = \overline{w}_{\mathbf{f}}(\mathbf{0})$, $\overline{w}_{\mathbf{ff}}(\theta \mathbf{f}_1)\mathbf{f}_1 = \mathbf{0}$ for some $\theta \in [0, 1]$, in contradiction of SE. Therefore $w^2 \neq 0$ on $\hat{\mathscr{H}}'$.

The integral curves of the vector field $\mathbf{e}_s$ on $\mathscr{F}_0$ will be referred to as the **reduced s-hodographs.** They are projections of the s-hodographs defined by the vector fields $\mathbf{r}_s, \mathbf{r}_{-s}$ on $\mathscr{H}_0$, $s = 1, 2, 3$.

Adapting the main results of Section 3.1.5 for equation (3.3.52), we arrive at the following:

Theorem 3.3.1

Suppose that $\overline{w} \in \mathscr{C}^3$ and $\mathbf{0} \in \mathscr{F}_0$.

Then the sets $\mathscr{H}'_{\mathbf{U}_0}$, $\mathbf{U}_0 = (\mathbf{0}, \mathbf{v}_0)$, and $\hat{\mathscr{H}}'$ are non-empty for every $\mathbf{v}_0 \in \boldsymbol{R}^3$, and $\hat{\mathscr{H}}' = \tilde{\pi}(\mathscr{H}'_{\mathbf{U}_0})$.

If for every point $(\lambda_s(\mathbf{f})^2, \mathbf{f}) \in \hat{\mathscr{H}}'$, $s = 1, 2$ or 3, the inequality (3.3.54) is satisfied, then the following assertions are true.

(1) the sets $\mathscr{H}_{\mathbf{U}_0} := \mathscr{H}'_{\mathbf{U}_0} \cup \{(\lambda_s(\mathbf{0}), \mathbf{0}, \mathbf{v}_0) | \; s = \pm 1, \pm 2, \pm 3\}$ and $\hat{\mathscr{H}} := \hat{\mathscr{H}}' \cup \{(\lambda_s(\mathbf{0})^2, \mathbf{0}) | \; s = 1, 2, 3\}$ are one-dimensional submanifolds of $\boldsymbol{R} \times \mathscr{H}_0$ and $\boldsymbol{R}_+ \times \mathscr{F}_0$ resp.;

(2) $\mathscr{H}_{\mathbf{U}_0} \supset \mathscr{H}^s_{\mathbf{U}_0}$, $\hat{\mathscr{H}} \supset \hat{\mathscr{H}}^{|s|}$, $\tilde{\pi}(\mathscr{H}^s_{\mathbf{U}_0}) = \hat{\mathscr{H}}^{|s|}$ for $s = \pm 1, \pm 2, \pm 3$, and $\tilde{\pi} | \mathscr{H}^s_{\mathbf{U}_0}$ is one-one;

(3) for $s = 1, 2, 3$ either of the two arcs of $\hat{\mathscr{H}}^s$ beginning at $(\lambda_s(\mathbf{0})^2, \mathbf{0})$ satisfies one of the following conditions:

(A) it escapes from $\boldsymbol{R}_+ \times \mathscr{F}_0$ or runs to infinity through $\boldsymbol{R}_+ \times \mathscr{F}_0$,

(B) it contains a point $(\lambda_r(\mathbf{0})^2, \mathbf{0})$ with $r \neq s$;

(4) if $\overline{w} \in \mathscr{C}^4$ in a nbhd of $\mathbf{0}$, then the projection $\tilde{\mathscr{H}}^s$ of $\hat{\mathscr{H}}^s$ onto $\mathscr{F}_0$ osculates the reduced s-hodograph through $\mathbf{0}$ at least to second order, and the

derivative of w with respect to an arbitrary parameter on $\hat{\mathscr{H}}^s$ $\dot{w} = (1/2)\dot{\lambda}_s$ at $(\lambda_s(0)^2, \mathbf{0})$.

All the theorems, lemmas and corollaries of Section 3.1.5 remain valid if $\mathscr{H}_{\mathbf{u}_0}$, $\mathscr{H}^{\alpha}_{\mathbf{u}_0}$, w, λ_α, $\mathbf{r}_\alpha$ are replaced by $\hat{\mathscr{H}}$, $\hat{\mathscr{H}}^s$, $w^2, \lambda_s{}^2$, $\mathbf{e}_s$ and the hypothesis (ii) of Theorem 3.1.9 is replaced by (3.3.54).

3.3.5 Discontinuity Waves in Isotropic Hyperelastic Media

We assume throughout this section that the configuration $z^k = \chi_0{}^k(\xi)$ is isotropic at the point ξ under consideration. In view of this special choice of $\mathbf{F}_0 = \nabla\boldsymbol{\chi}_0$ we assume that $\mathbf{F}_- = (\mathbf{E}+\mathbf{f}_0\otimes\mathbf{n})\mathbf{F}_0$. Equations (3.3.52) must be replaced now by

$$\overline{w}_{\mathbf{f}}(\mathbf{f})-\overline{w}_{\mathbf{f}}(\mathbf{f}_0) = w^2(\mathbf{f}-\mathbf{f}_0). \tag{3.3.52'}$$

We shall denote by $\hat{\mathscr{H}}_{f_0}$ the subset of $R_+ \times R^3$ consisting of nontrivial solutions of equation (3.3.52') and of its bifurcation points.

Let the Cartesian coordinates x^k, z^k in the space of the observer be chosen in such a way that $n_1 = 1$, $n_k = 0$ for $k = 2, 3$. In this case

$$\overline{w}(\mathbf{f}; \xi) \equiv \tilde{w}(f, g; \xi) \tag{3.3.56}$$

with

$$f := f^1, \quad g := [(f^2)^2+(f^3)^2]^{1/2}, \quad \mathbf{i} := g^{-1}\begin{bmatrix}0\\ f^2\\ f^3\end{bmatrix}. \tag{3.3.57}$$

The irrelevant argument ξ will be omitted throughout this section.

Easy calculations yield the following formulae:

$$\overline{w}_{\mathbf{f}} = \tilde{w}_f\mathbf{n}+\tilde{w}_g\mathbf{i}, \tag{3.3.58}$$

$$\overline{w}_{\mathbf{ff}} = \begin{bmatrix}\tilde{w}_{ff} & \tilde{w}_{fg}{}^t i\\ \tilde{w}_{fg}\mathbf{i} & g^{-1}\{\tilde{w}_g\mathbf{E}+(\tilde{w}_{gg}g-\tilde{w}_g)\mathbf{i}\otimes\mathbf{i}\}\end{bmatrix}. \tag{3.3.59}$$

In equation (3.3.59) $\mathbf{i} \in \mathscr{S}^1$ denotes the projection of the vector defined by the third equation of (3.3.57) onto the subspace corresponding to the coordinates x^2, x^3, and $\mathbf{E} = \mathrm{id}_{R^2}$. A few simple algebraic manipulations lead to the following equation for the eigenvalues $\lambda_s{}^2$:

$$0 = \det(\overline{w}_{\mathbf{ff}}-\lambda^2\mathbf{E})$$

$$\equiv \det\begin{bmatrix}\tilde{w}_{ff}-\lambda^2 & \tilde{w}_{fg}i_2 & 0\\ 0 & g^{-1}\tilde{w}_g-\lambda^2 & \left(\dfrac{f^2}{f^3}+\dfrac{f^3}{f^2}\right)(g^{-1}\tilde{w}_g-\lambda^2)\\ \tilde{w}_{fg}i_3 & (\tilde{w}_{gg}-g^{-1}\tilde{w}_g)i_2 i_3 & g^{-1}\tilde{w}_g-\lambda^2\end{bmatrix}. \tag{3.3.60}$$

Hence one of the solutions is

$$\lambda^2 = \lambda_C{}^2 :\equiv g^{-1}\tilde{w}_g > 0. \tag{3.3.61}$$

The inequality in (3.3.61) follows from SE.

Expanding the right-hand side of equation (3.3.60) and dividing it by $\lambda^2 - g^{-1}\tilde{w}_g$ we obtain

$$(\tilde{w}_{ff} - \lambda^2)(\tilde{w}_{gg} - \lambda^2) - \tilde{w}_{fg}{}^2 = 0, \tag{3.3.62}$$

whose roots are

$$\lambda^2_{\pm 1} = \frac{1}{2}\{(\tilde{w}_{ff} + \tilde{w}_{gg}) \pm \sqrt{(\tilde{w}_{ff} - \tilde{w}_{gg})^2 + 4\tilde{w}_{fg}{}^2}\}. \tag{3.3.63}$$

SE implies that $\lambda_{\pm 1}{}^2 > 0$ and $\tilde{w}_{ff} + \tilde{w}_{gg} > 0$.

The eigenvector $\mathbf{e}_C$ corresponding to $\lambda_C{}^2$ is proportional to the cofactors of the second row of the matrix $\overline{w}_{\mathbf{ff}} - \lambda_C{}^2\mathbf{E}$.

Hence

$$\mathbf{e}_C = \mathbf{n} \times \mathbf{i} \tag{3.3.64}$$

and

$$d\lambda_C{}^2[\mathbf{e}_C] \equiv 0. \tag{3.3.65}$$

The mode C is linearly degenerate. The reduced C-hodographs are circles in the planes orthogonal to $\mathbf{n}: f = f_0, g = g_0$. On account of equations (3.3.65) and (3.3.47)

$$\mathbf{v} - \mathbf{v}_0 = -\iota\lambda_C(f_0, g_0)(\mathbf{f} - \mathbf{f}_0) \tag{3.3.66}$$

on the C-hodographs, $\iota = 1, -1$.

It is natural to expect that λ_C is bounded in the limit $g \to 0$. This would imply that

$$\tilde{w}(f, g) = w_0(f) + g^2 w_1(f, g), \quad w_0, w_1 \in \mathscr{C}^3. \tag{3.3.67}$$

In fact $\overline{w}$ is invariant with respect to arbitrary rotations of $\mathbf{f}$ in the plane orthogonal to $\mathbf{n}$. Therefore every derivative of $\tilde{w}$ which is continuous at $g = 0$ and involves an odd number of differentiations with respect to g vanishes at $g = 0$. For $\tilde{w} \in \mathscr{C}^3$ this implies (3.3.67).

We set

$$\lambda_P := \lambda_{\iota(\mathbf{f})} > 0, \quad \lambda_S := \lambda_{-\iota(\mathbf{f})} > 0, \tag{3.3.68}$$

where $\iota(\mathbf{f}) := \operatorname{sgn}\{(w_0)_{ff} - 2w_1\}_{g=0}$. It is easy to see that

$$\lambda_C, \lambda_S \to 2w_1(f, 0) \quad \text{when } g \to 0. \tag{3.3.69}$$

It follows that strict hyperbolicity fails to hold on the one-dimensional submanifold $g = 0$.

The eigenvectors corresponding to λ_P^2, λ_S^2 are given by the expressions $\mathbf{e}_P = \mathbf{e}_{\iota(\mathbf{f})}$, $\mathbf{e}_S = \mathbf{e}_{-\iota(\mathbf{f})}$. For $g \neq 0$ the eigenvectors $\mathbf{e}_\iota$ are proportional to the cofactors of the first row of the matrix $(w_{\mathbf{ff}} - \lambda_\iota^2\mathbf{E})$. Hence

$$\mathbf{e}_\iota = [(\tilde{w}_{gg} - \lambda_\iota^2)^2 + \tilde{w}_{fg}^2]^{-1/2} \begin{bmatrix} \tilde{w}_{gg} - \lambda_\iota^2 \\ -\tilde{w}_{fg} i_2 \\ -\tilde{w}_{fg} i_3 \end{bmatrix}, \tag{3.3.70}$$

$\iota = 1, -1$. For $g \to 0$ at fixed $\mathbf{i}$, $\mathbf{e}_S \to \pm\mathbf{i}$, according to the sign of $\tilde{w}_{fg}$ at small $g \neq 0$, $\mathbf{e}_P \to -\iota(\mathbf{f})\mathbf{n}$. The reduced *S*-hodographs through a point $(f, 0, 0)$, $f \in \boldsymbol{R}$, form a paraboloid (up to a diffeomorphism) with the symmetry axis $\mathbf{n}$. The reduced *P*-hodographs through $(f, 0, 0)$ coincide with the f^1 axis, up to orientation.

Let $\mathbf{f} = f\mathbf{n} + g\mathbf{i}$, $\mathbf{f}_0 = f_0\mathbf{n} + g_0\mathbf{i}_0$, $g_0 > 0$, $\mathbf{i}\cdot\mathbf{i}_0 = \cos\phi$. By virtue of equation (3.3.58) equations (3.3.52′) assume the following form:

$$\tilde{w}_f(f, g) - \tilde{w}_f(f_0, g_0) = w^2(f - f_0), \tag{3.3.71}$$

$$\tilde{w}_g(f, g) - \tilde{w}_g(f_0, g_0)\cos\phi = w^2(g - g_0\cos\phi), \tag{3.3.72}$$

$$\tilde{w}_g(f, g)\cos\phi - \tilde{w}_g(f_0, g_0) = w^2(g\cos\phi - g_0). \tag{3.3.73}$$

Adding and subtracting equations (3.3.72), (3.3.73) one derives for $\cos\phi \neq \pm 1$ two equations,

$$\tilde{w}_g - \tilde{w}_g^0 = w^2(g - g_0)$$

and

$$\tilde{w}_g + \tilde{w}_g^0 = w^2(g + g_0).$$

The superscript 0 indicates that the function has been evaluated at (f_0, g_0). It follows immediately that $w^2 = \lambda_C(f_0, g_0)^2 = \lambda_C(f, g)^2$. The corresponding discontinuity waves are exceptional and include the degenerate simple waves $f = \text{const} = f_0$, $g = \text{const} = g_0$ (recall equation (3.3.65) and Lemma 3.1.11).

For $\phi = 0$ we obtain two equations

$$\begin{aligned} &\tilde{w}_f(f, g) - \tilde{w}_f^0 - w^2(f - f_0) = 0, \\ &\tilde{w}_g(f, g) - \tilde{w}_g^0 - w^2(g - g_0) = 0. \end{aligned} \tag{3.3.74}$$

It follows from the implicit function theorem that non-trivial solutions of quations (3.3.74) bifurcate from the trivial one $(f, g) = (f_0, g_0)$ at most at the points $w^2 = \lambda_P(f_0, g_0)^2$, $w^2 = \lambda_S(f_0, g_0)^2$.

Assuming that $\lambda_P \neq \lambda_S$ and $\tilde{w} \in \mathscr{C}^3$ we conclude from Theorem 15.2 of Prodi and Ambrosetti (1973) that there is a $\mathscr{C}^1$ curve of non-trivial solutions

$$\begin{aligned} w(s)^2 &= \lambda_P^0 + \psi(s), \\ f(s) &= (\tilde{w}_{ff}^0 - (\lambda_P^0)^2)s + (\tilde{w}_{ff}^0 - (\lambda_S^0)^2)\varphi_P(s), \\ g(s) &= -\tilde{w}_{fg}^0(s + \varphi_P(s)) \end{aligned} \tag{3.3.75}$$

with $\psi_P(0) = 0$, $\varphi_P(s) = o(s)$, bifurcating from $w^2 = (\lambda_P{}^0)^2$. The $\mathscr{C}^1$ curve bifurcating from $w^2 = (\lambda_S{}^0)^2$ is given by the same formulae with the subscript P replaced by S.

For $\phi = \pi$ we have to solve the equations

$$\begin{aligned} \tilde{w}_f(f, g) - \tilde{w}_f{}^0 - w^2(f - f_0) &= 0, \\ \tilde{w}_g(f, g) + \tilde{w}_g{}^0 - w^2(g + g_0) &= 0. \end{aligned} \tag{3.3.76}$$

At least the point $f = f_0$, $g = g_0$, $w^2 = (\lambda_C{}^0)^2$ satisfies these equations. Consider the equation

$$\begin{bmatrix} \tilde{w}_{ff} - w^2 & \tilde{w}_{fg} \\ \tilde{w}_{fg} & \tilde{w}_{gg} - w^2 \end{bmatrix} \begin{bmatrix} \dot{f} \\ \dot{g} \end{bmatrix} - \begin{bmatrix} f - f_0 \\ g - \iota g_0 \end{bmatrix} \dot{\overline{w^2}} = \begin{bmatrix} a \\ b \end{bmatrix}, \tag{3.3.77}$$

$\iota = 1, -1$ for $\phi = 0, \pi$, resp., for the unknown $\dot{f}$, $\dot{g}$, $\dot{\overline{w^2}}$ and arbitrary $a, \in b\boldsymbol{R}$. We shall assume that the condition (3.3.54) is satisfied at the points $w^2 = \lambda_s{}^2$ for $s = S, P$. In this case equation (3.3.77) has a solution for every $a, b \in \boldsymbol{R}$ and $\iota = 1, -1$, and the mappings on the left-hand side of equation (3.3.74) and (3.3.76) are submersions. It follows then that

(1) non-trivial solutions of equation (3.3.74) lie on two one-dimensional $\mathscr{C}^1$ submanifolds of $\boldsymbol{R}_+ \times \boldsymbol{R}^2$, given by equations (3.3.75) in a nbhd of (f_0, g_0);

(2) the solutions of equation (3.3.76) lie on a one-dimensional $\mathscr{C}^1$ submanifold through $((\lambda_C{}^0)^2, f_0, g_0)$.

It is noteworthy that the point $\mathbf{f}_0^- := f_0\mathbf{n} - g_0\mathbf{i}_0$ lies on the intersection of two one-dimensional submanifolds of $\hat{\mathscr{H}}_{\mathbf{f}_0}$. One of these submanifolds lies in the plane $\phi = \pi$ and the other is given by the equations $f = f_0$, $g = g_0$, $\phi \in [0, 2\pi[$. This bifurcation can be traced back to the fact that $w^2 = \lambda_C(\mathbf{f}_0^-)^2$ and $\langle \mathbf{e}_C(\mathbf{f}_0^-), \mathbf{f}_0^- - \mathbf{f}_0 \rangle = 0$, so that the condition (3.3.54) fails to be satisfied.

Suppose that a point $(w^2, f, 0)$ satisfies equations (3.3.74). In this case it also satisfies equations (3.3.76), $w^2 = (\lambda_C{}^0)^2$ and

$$\tilde{w}_f(f, 0) - \tilde{w}_f{}^0 - (\lambda_C{}^0)^2(f - f_0) = 0. \tag{3.3.78}$$

In particular $f = f_0$ satisfies equation (3.3.78). If it is assumed that

$$(\lambda_C{}^0)^2 \neq (\lambda_P{}^0)^2 \quad \text{for } g_0 \neq 0, \tag{3.3.79}$$

then the derivative of the left-hand side of equation (3.3.78) with respect to f at $(f_0, 0)$ is $\tilde{w}_{ff}(f_0, 0) - \lambda_C(f_0, 0)^2 \neq 0$. Hence for g_0 small enough equation (3.3.78) has a solution $f_1 = f_0 + O(g_0)$. Since $\lambda_C{}^0 = \lambda_S{}^0$ for $g_0 = 0$, it follows that for g_0 small enough the arc of $\hat{\mathscr{H}}_{\mathbf{f}_0}$ defined by equations (3.3.76) is connected to the arc $\hat{\mathscr{H}}^S_{\mathbf{f}_0}$ bifurcating from $((\lambda_S{}^0)^2, \mathbf{f}_0)$ at a point $((\lambda_C{}^0)^2, f_1\mathbf{n})$. Figure 3.7 shows a typical projection $\mathscr{H}_{\mathbf{f}_0}$ of $\hat{\mathscr{H}}_{\mathbf{f}_0}$ onto $\mathscr{F}$.

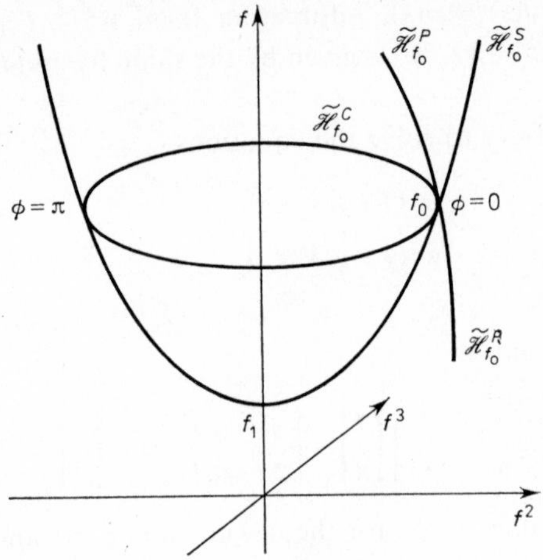

Fig. 3.7 The manifold $\hat{\mathscr{H}}_{\mathbf{f}_0}$, $g_0 \neq 0$

Let us now consider the set $\hat{\mathscr{H}}_{\mathbf{f}_0}$ for $g_0 = 0$. For $g \neq 0$ equations (3.3.52′) assume the following form:

$$\tilde{w}_f(f, g) - \tilde{w}_f(f_0, 0) = w^2(f - f_0), \tag{3.3.80}$$

$$w^2 = g^{-1}\tilde{w}_g(f, g) \tag{3.3.81}$$

and the vector $\mathbf{i}$ is now an arbitrary unit vector orthogonal to $\mathbf{n}$. A solution of equations (3.3.80–81) is given by $f = f_0$, $g = 0$, $w^2 = \lambda_C(f_0, 0)^2$. Substituting equation (3.3.81) into equation (3.3.80) we obtain an equation for (f, g):

$$\tilde{w}_f(f, g) - \tilde{w}_f(f_0, 0) - g^{-1}\tilde{w}_g(f, g)(f - f_0) = 0. \tag{3.3.82}$$

Expanding the left-hand side of (3.3.82) we have

$$[\tilde{w}_{ff}{}^0 - (\lambda_C{}^0)^2](f - f_0) + \left[\frac{1}{2}\,\tilde{w}_{fff}{}^0 - w^0_{ggf}\right](f - f_0)^2$$
$$+ \tfrac{1}{2}\, w^0_{ggf} g^2 + \ldots = 0, \tag{3.3.83}$$

where the superscript zero indicates that the function has been evaluated at $(f_0, 0)$. By assumption

$$\tilde{w}_{ff}{}^0 - (\lambda_C{}^0)^2 = (\lambda_P{}^0)^2 - (\lambda_C{}^0)^2 \neq 0. \tag{3.3.84}$$

By the implicit function theorem, in a nbhd of $(f_0, 0)$ equation (3.3.82) has a unique solution of the form

$$f - f_0 = \varphi(g) \equiv bg^2 + O(g^3) \tag{3.3.85}$$

with $b := -w^0_{1f}(\tilde{w}_{ff}{}^0 - 2w_1{}^0)^{-1}$. The set $\tilde{\mathscr{H}}^S_{\mathbf{f}_0}$ defined by equation (3.3.85) is diffeomorphic to a paraboloid and is symmetric with respect to the rotations around $\mathbf{n}$.

Remarks

(1) In the linear theory of elasticity $\lambda_P > \lambda_S = \lambda_C$, hence $b < 0$.

(2) If $(\lambda_P{}^0)^2 = (\lambda_C{}^0)^2$, but the coefficient of $(f-f_0)^2$ in equation (3.3.83) does not vanish, then it follows from the Malgrange preparation theorem (Golubitsky and Guillemin, 1973) that in a nbhd of $(f_0, 0)$ equation (3.3.82) is equivalent to an equation

$$(f-f_0)^2+\varphi(g)(f-f_0)+\psi(g) = 0 \tag{3.3.86}$$

with $\varphi(g) = dg^2+O(g^3)$, $\psi(g) = cg^2+O(g^3)$, $\operatorname{sgn} c = \operatorname{sgn}\{\frac{1}{2}\tilde{w}_{fff}{}^0 - 2w_{1f}^0\}w_{1f}^0$. For small g equation (3.3.86) has no solutions if $d^2 < c^2$ and two solutions if $d^2 > c^2$.

Let us now consider the points $(w^2, f, 0)$ of $\hat{\mathscr{H}}_{\mathbf{f}_0}$, $g_0 = 0$. In this case equation (3.3.81) drops out. Equation (3.3.80) with $g = 0$ exhibits a bifurcation point at $w^2 = \lambda_P(f_0, 0)^2$. The corresponding branch $\hat{\mathscr{H}}_{\mathbf{f}_0}^P$ of $\hat{\mathscr{H}}_{\mathbf{f}_0}$ is a one-dimensional submanifold of $\boldsymbol{R}_+ \times \mathscr{F}$ consisting of all the points $(w^2, f\mathbf{n})$ with

$$w^2 = \frac{\tilde{w}_f(f, 0)-\tilde{w}_f(f_0, 0)}{f-f_0} \quad \text{for } f \neq f_0 \tag{3.3.87}$$

and $w^2 = \lambda_P(f_0, 0)^2$ for $f = f_0$. Since $\tilde{w}_{ff} > 0$ by virtue of SE, the right-hand side of equation (3.3.87) is always positive.

The behaviour of $\hat{\mathscr{H}}_{\mathbf{f}_0}$ at large distances from the bifurcation points largely depends on the specific material. For $g_0 \neq 0$ it cannot be excluded that the submanifolds $\hat{\mathscr{H}}_{\mathbf{f}_0}^P$ and $\hat{\mathscr{H}}_{\mathbf{f}_0}^S$ join somewhere in $\boldsymbol{R}_+ \times \mathscr{F}$.

3.3.6 Shocks and Simple Waves in Isotropic Hyperelasticity. The Riemann Problem

In this section we shall be concerned with those properties of the sets $\mathscr{H}_{\mathbf{u}_0}$ which are related to the conditions (L) and (E).

The theory of discontinuity waves developed in Section 3.3.5 is sufficient for the analysis of weak shocks and of some special discontinuity waves. The properties of strong discontinuity waves depend, however, to a large extent on the particular material and consequently it is not possible to go far beyond the results obtained in Sections 3.1.5 and 3.1.6 without making some specific assumptions about the constitutive relations. The case of $g_0 = 0$ requires a special analysis.

We shall need somewhat stronger assumptions than those of the preceding section. When dealing with c.s. waves and with the Riemann problem we shall tacitly assume that $\mathbf{z} = \boldsymbol{\chi}_0(\xi)$ is an isotropic reference configuration at every point ξ. In accordance with common experience we shall assume that $\lambda_P{}^2 > \lambda_S{}^2$, $\lambda_P{}^2 > \lambda_C{}^2$, and $\lambda_S{}^2 \neq \lambda_C{}^2$ except on the manifold $g = 0$.

For $(w^2, \mathbf{f}) \in \mathscr{H}_{\mathbf{f}_0}$, $g_0 \neq 0$, $g \neq 0$ the results of Section 3.1 can be applied directly. For $g_0 \neq 0$, $g = 0$ it is advisable to reverse the roles of $\mathbf{f}_0$ and $\mathbf{f}$ and use the results we obtain below. The analysis is carried out most conveniently on the submanifolds $\hat{\mathscr{H}}_{\mathbf{f}_0} \subset \boldsymbol{R}_+ \times \mathscr{F}$. Indeed, 1°, $\lambda_s{}^2$ does not depend on $\mathbf{v}$ and $d\lambda_s{}^2[\mathbf{e}_s]$ conveniently replaces $d\lambda_\alpha[\mathbf{r}_\alpha]$ of Section 3.1; 2°, at the sonic points $\tilde{\mathscr{H}}_{\mathbf{f}_0}$ is tangent to a reduced s-hodograph. One can use equations (3.3.47), (3.3.48) and (3.3.55) to lift $\hat{\mathscr{H}}_{\mathbf{f}_0} \subset \boldsymbol{R}_+ \times \mathscr{F}_0$ to $\mathscr{H}_{\mathbf{u}_0} \subset \boldsymbol{R} \times \mathscr{H}_0$.

The case of $g_0 = 0$ is not covered by the results of Section 3.1. The points $(w^2, \mathbf{f}) \in \hat{\mathscr{H}}^P_{\mathbf{f}_0} \cup \hat{\mathscr{H}}^S_{\mathbf{f}_0}$ satisfy a system of two equations

$$\begin{aligned} &\tilde{w}_f(f, g) - \tilde{w}_f(f_0, 0) - w^2(f - f_0) = 0, \\ &\tilde{w}_g(f, g) - \tilde{w}_g(f_0, 0) - w^2(g - g_0) = 0, \end{aligned} \tag{3.3.88}$$

which has the form (3.1.43). The relevant eigenvalues are $\lambda_P{}^2$ and $\lambda_S{}^2$. By assumption $\lambda_P{}^2 \neq \lambda_S{}^2$. In the variables (w^2, f, g) the sets $\hat{\mathscr{H}}^P_{\mathbf{f}_0}$ and $\hat{\mathscr{H}}^S_{\mathbf{f}_0}$ are represented by two one-dimensional submanifolds and the theory of critical points developed in Section 3.1 can be applied to them.

Since we have assumed that $\lambda_C{}^2 \neq \lambda_S{}^2$ outside $g = 0$, the functions $w^2 - \lambda_S{}^2$ and $w^2 - \lambda_P{}^2 = \lambda_C{}^2 - \lambda_P{}^2$ do not change sign on $\hat{\mathscr{H}}^S_{\mathbf{f}_0}$. Consequently w^2 cannot have an extremum on $\hat{\mathscr{H}}^S_{\mathbf{f}_0}$ and $w^2 - (\lambda_C{}^0)^2 = w^2 - (\lambda_S{}^0)^2$ cannot vanish on $\hat{\mathscr{H}}^S_{\mathbf{f}_0}$ except at $\mathbf{f}_0$.

For small g

$$\begin{aligned} \lambda_C{}^2 - (\lambda_C{}^0)^2 &= 2(2bw_{1,f}{}^0 + w_{1,gg}{}^0)g^2 + \dots, \\ \lambda_S{}^2 - w^2 &= 4(bw_{1,f}{}^0 + w_{1,gg}{}^0)g^2 + \dots, \\ \lambda_S{}^2 - (\lambda_S{}^0)^2 &= 6(\tfrac{4}{3}bw_{1,f}{}^0 + w_{1,gg}{}^0)g^2 + \dots, \end{aligned} \tag{3.3.89}$$

so that we have the following chain of implications for small g:

$$w^2 = \lambda_C{}^2 > (\lambda_C{}^0)^2 = (\lambda_S{}^0)^2 \Rightarrow \lambda_S{}^2 > (\lambda_S{}^0)^2 \Rightarrow \lambda_S{}^2 > w^2, \tag{3.3.90}$$

and

$$\lambda_S{}^2 < w^2 \Rightarrow \lambda_S{}^2 < (\lambda_S{}^0)^2 \Rightarrow w^2 = \lambda_C{}^2 < (\lambda_C{}^0)^2 = (\lambda_S{}^0)^2 \tag{3.3.91}$$

(we have used the assumption that $\lambda_P^0 > \lambda_S^0$, i.e. $b < 0$).

From Theorem 3.1.10 it follows that for small g

$$\lambda_S{}^2 > w^2 \Rightarrow w^2 > (\lambda_S{}^0)^2, \quad w^2 < (\lambda_S{}^0)^2 \Rightarrow w^2 > \lambda_S{}^2,$$

and therefore the inequalities (3.3.90) are mutually equivalent for small g.

Similarly, the inequalities (3.3.91) are mutually equivalent for small g. The conclusion of the preceding paragraph implies that the inequality on the extreme left and on the extreme right of (3.3.90) or (3.3.91) is satisfied everywhere on $\hat{\mathscr{H}}^S_{\mathbf{f}_0}$ if it holds for small g. Hence the middle inequality of (3.3.90) or (3.3.91) is valid everywhere on $\hat{\mathscr{H}}^S_{\mathbf{f}_0}$ if it is valid for small g.

Definition 3.3.2

A discontinuity with $\mathbf{f}_+ = \mathbf{f}_0 = (f_0, \mathbf{0})$, $w > 0$, and $(w, \mathbf{f}_-) \in \hat{\mathscr{H}}^S_{\mathbf{f}_0}$ (resp. with $\mathbf{f}_- = \mathbf{f}_0 = (f_0, \mathbf{0})$, $w > 0$ and $(w, \mathbf{f}_+) \in \hat{\mathscr{H}}^S_{\mathbf{f}_0}$) is said to be a CS_+**-shock** if the inequalities (3.3.90) and $w < \lambda_P{}^0$ hold (resp., if the inequalities (3.3.91) hold*).

Definition 3.3.3

A c.s. wave satisfying the equations $\mathbf{f}' = k\mathbf{e}_S$, $yt^{-1} = \lambda_S > 0$ with $g = 0$ on one side will be called an S_0^+**-c.s.wave**.

Let us compare CS_+-shocks with S_0^+-c.s. waves.

If the inequalities (3.3.90) hold for $\mathbf{f}_0 = (f_0, \mathbf{0})$, then $\mathbf{f}_0 = (f_0, \mathbf{0})$ can appear only as the right state $\mathbf{f}_+$ of a CS_+-shock or as a left state of an S_0^+-c.s. wave. If the inequalities (3.3.91) hold for $\mathbf{f}_0$, then $\mathbf{f}_0$ can appear only as the left state $\mathbf{f}_-$ of a CS_+-shock or as the right state of an S_0^+-c.s.w.

The CS_--shocks are defined analogously, except that $w < 0$ and the subscripts "+" and "−" exchange their positions. For an S_0^--wave we assume that $yt^{-1} = -\lambda_S < 0$. The assertions for CS_--shocks and S_0^--waves follows from those of the preceding paragraph by the symmetry $y \to -y$.

If $\mathbf{f}_0 = (f_0, \mathbf{0})$ and $(w^2, \mathbf{f}) \in \hat{\mathscr{H}}^P_{\mathbf{f}_0}$, then $\mathbf{f} = (f, \mathbf{0})$. Conversely, every discontinuity with $g_+ = 0 = g_-$ satisfies the relations $(w^2, \mathbf{f}_+) \in \hat{\mathscr{H}}^P_{\mathbf{f}_-}$, $(w^2, \mathbf{f}_-) \in \hat{\mathscr{H}}^P_{\mathbf{f}_+}$. The methods of Section 3.1 carry over to such discontinuities without change.

Definition 3.3.4

A discontinuity with $g_+ = 0 = g_-$ is said to be a P_0**-shock** if it satisfies (L) and (E).

We also note that a c.s.w. satisfying the equations $\mathbf{f}' = k\mathbf{e}_P$ and $yt^{-1} = \pm\lambda_P$ with $g = 0$ on one side corresponds to P_0-shocks. It will be called a P_0**-c.s.w.**

* Note that $w = \lambda_C < \lambda_P$ automatically.

Definition 3.3.5

A P_0-**wave** is a P_0-c.s.w. or a P_0-shock.

Let us consider now the structure of a solution $\mathbf{f}(yt^{-1}), \mathbf{v}(yt^{-1})$ to a Riemann problem for an isotropic hyperelastic medium. We assume that $\overline{w}$ in the reference configuration $\mathbf{z} = \boldsymbol{\chi}_0(\xi)$ does not depend on $\mathbf{z}$ and $\overline{\mathbf{b}} = \mathbf{0}$. In this case there are simple wave solutions to equations (3.3.45).

Definition 3.3.6

A solution $\mathbf{f}(yt^{-1}), \mathbf{v}(yt^{-1})$ is said to be **admissible** if

(1) all the discontinuities with $g_+ \neq 0$, $g_- \neq 0$ and those with $g_+ = 0 = g_-$ satisfy the conditions (L) and (E);

(2) all the discontinuities with $g_+ = 0 \neq g_-$ or $g_+ \neq 0 = g_-$ are CS_+- and CS_--shocks.

An admissible solution consists of P-, S-shocks ($g_+ \neq 0 \neq g_-$), P- and S-c.s. waves, C-exceptional discontinuities, P_0-shocks, P_0-c.s. waves, CS_+- and CS_--shocks, S_0^+- and S_0^--c.s. waves.

If $g(yt^{-1}) \neq 0\ \forall yt^{-1} \in \boldsymbol{R}$, then the values of the solution lie in $\mathscr{H}_0$ and the results of Section 3.1. apply. Suppose now that, for some value $\zeta_0 > 0$ of yt^{-1}, $g(\zeta_0) = 0$. It is easy to deduce from the inequalities (3.3.90), (3.3.91) that the following situations are possible for the region $\mathscr{D}_+ = \{yt^{-1} > 0\}$. $\mathscr{D}_+$ contains

(1) a CS_+-wave with $g_+ = 0$ on the right of it as well as a group of P_0-waves on the right of it, or

(2) an S_0^+-c.s. wave with $g = 0$ on the left of it and a sonic S-wave and a group of P-waves to the right of it, or

(3) an S_0^+-wave with $g = 0$ on the right and a group of P_0-waves on the right, or

(4) a CS_+-wave with $g_- = 0$ and a group of P-waves ahead of it. The situations (1) and (2) are possible if the inequalities (3.3.90) hold at $yt^{-1} = \zeta_0$, the situations (3) and (4) if the alternative inequalities (3.3.91) hold at $yt^{-1} = \zeta_0$. In each case a CS_+-wave represents two degrees of freedom for fixed $\mathbf{f}(\zeta_0) = \mathbf{f}_0$, viz. the values of g and $\mathbf{i} = (\cos\phi, \sin\phi)$. Qualitatively, this yields the correct number of degrees of freedom for solving the Riemann problem.

Suppose that the state on the left of a CS-shock or an S_0-c.s.w. is $(\mathbf{f}_0, \mathbf{v}_0)$, with $g_0 = 0$. Then the state $\mathbf{f}_1$ on the right depends on two parameters g, ϕ:

$$\mathbf{f}_1 = \mathbf{z}(g, \phi; \mathbf{u}_0), \quad \mathbf{u}_0 = (\mathbf{f}_0, \mathbf{v}_0)$$

and the vectors $\lim_{g\to 0} \frac{\partial \mathbf{z}}{\partial g}(g, \phi; \mathbf{u}_0) = \mathbf{i}$, $\lim_{g\to 0} \frac{\partial \mathbf{z}}{\partial \phi}(g, \phi; \mathbf{u}_0)$ are orthogonal. Consequently the Riemann problem with small $|\mathbf{u}_L - \mathbf{u}_R|$ has exactly one solution with the property that all the groups of waves in it are weak.

3.3.7 Discontinuities and Plane Waves in Incompressible Isotropic Hyperelastic Materials

Plane wave motion in an incompressible hyperelastic medium is governed by the following system of equations:

$$(\varrho_0 v^k)_t - \frac{\partial}{\partial y}\left(\varrho_0 \frac{\partial \overline{w}}{\partial f^k}\right) = -\frac{\partial \overline{p}}{\partial y} n_k, \tag{3.3.92a}$$

$$f^k{}_{,t} - v^k{}_{,y} = 0, \quad k = 1, 2, 3 \tag{3.3.92b}$$

for the unknown functions f^k, v^k, $\overline{p}$ of (t, y). The solutions are subject to the constraint

$$f^k n_k = 0, \tag{3.3.93}$$

(recall equation (3.3.37)). The right-hand side of equation (3.3.92a) is the pressure term $-\partial_\alpha(p\overset{-1}{F^\alpha}{}_k \det\mathbf{F}) \equiv -(\det\mathbf{F})\frac{\partial p}{\partial x^k} \equiv -\det\mathbf{F}\frac{\partial p}{\partial y} n_k$ (the last identity follows from (3.3.93)).

Suppose that $n_3 = 1$, $n_1 = 0 = n_2$, $\varrho_0 =$ const. Equation (3.3.93) implies that $f^3 = 0$ and, by virtue of (3.3.92b), $v^3 = w(t)$. Disregarding rigid motions we assume that $w(t) \equiv 0$. The third equation (3.3.92a) (for $k = 3$) can be satisfied by a suitable choice of $p(t, y)$. Hence the system (3.3.92) reduces to

$$v^k{}_{,t} - (\hat{w}_{f^k})_y = \hat{b}_k(\mathbf{f}, y),$$

$$f^k{}_{,t} - v^k{}_{,y} = 0, \quad k = 1, 2 \tag{3.3.94}$$

with $\hat{w}(f^1, f^2) \equiv \overline{w}(f^1, f^2, 0)$, $\hat{\mathbf{b}}(\mathbf{f}, y) \equiv \overline{\mathbf{b}}(f^1, f^2, 0, y)$. The formulae (3.3.46)–(3.3.48) remain valid except that $\mathbf{f}$, $\mathbf{v}$, $\mathbf{e}_s$ are two-component vectors now.

For isotropic materials

$$\hat{w}(\mathbf{f}) = \varphi(n) = \varphi_0 + n^2\varphi_1(n), \tag{3.3.95}$$

$$n := [(f^1)^2 + (f^2)^2]^{1/2}, \quad \mathbf{i} := n^{-1}\begin{bmatrix} f^1 \\ f^2 \end{bmatrix}, \tag{3.3.96}$$

$$\hat{w}_{\mathbf{ff}} = \left(\varphi_{nn} - \frac{\varphi_n}{n}\right)\mathbf{i}\otimes\mathbf{i} + \frac{\varphi_n}{n}\mathbf{E}. \tag{3.3.97}$$

The characteristic speeds and the corresponding eigenvectors are

$$\lambda_C^2 = \frac{\varphi_n}{n} = 2\varphi_1 + n\varphi_1', \quad \lambda_C > 0, \quad \mathbf{e}_C = \mathbf{i}_3 \times \mathbf{i}, \tag{3.3.98}$$

$$\lambda_L^2 = \varphi_{nn} = 2\varphi_1 + 4n\varphi_1' + n^2\varphi_1'' > 0, \quad \lambda_L > 0, \quad \mathbf{e}_L = \mathbf{i}. \tag{3.3.99}$$

At $n = 0$ $\lambda_L^2 = \lambda_C^2$. If $3\varphi_1' + n\varphi_1'' \neq 0$, then strict hyperbolicity holds off $n = 0$. We assume that the above inequality is satisfied.

Clearly

$$d\lambda_C^2[\mathbf{e}_C] = 0 \tag{3.3.100}$$

and

$$d\lambda_L^2[\mathbf{e}_L] = \varphi_{nnn}. \tag{3.3.101}$$

We define $\iota = \operatorname{sgn}\varphi_{nnn}$ (it may depend on n). Let ι_0 be the value of ι for small $n \neq 0$. Since $\varphi_n - n\varphi_{nn} = 0$ at $n = 0$ and $(d/dn)(\varphi_n - n\varphi_{nn}) = -n\phi_{nnn}$, it follows that

$$\iota_0(\lambda_L^2 - \lambda_C^2) > 0 \tag{3.3.102}$$

for small $n \neq 0$ and hence for all $n \neq 0$. We also note that

$$\frac{d\lambda_C^2}{dn} = \frac{1}{n}(\lambda_L^2 - \lambda_C^2) \tag{3.3.103'}$$

which implies that

$$\operatorname{sgn}\frac{d\lambda_C^2}{dn} = \iota_0. \tag{3.3.103}$$

The reduced C-hodographs are circles centered at $\mathbf{f} = \mathbf{0}$. The reduced L-hodographs are half-lines through $\mathbf{f} = \mathbf{0}$. There are only two kinds of c.s. waves, viz. L_+-c s. waves $(y/t = \lambda_L)$ and L_--c.s. waves $(y/t = -\lambda_L)$. The state $n = 0$ can appear at most as an extreme right or left state in the domain of a c.s. wave. C-c.s. waves degenerate into the exceptional C-discontinuities on account of (3.3.100).

For the analysis of L-shocks we use the notation of Sections 3.3.4 and 3.3.5. Since incompressibility implies that $\det\mathbf{F}_+ = \det\mathbf{F}_-$, we assume that $\det\mathbf{F}_0 = \det\mathbf{F}_+ = \det\mathbf{F}_-$. Hence we must have $\mathbf{f}_0 \cdot \mathbf{n} = \mathbf{f} \cdot \mathbf{n} = 0$. Let $n_3 = 1$, $n_1 = 0 = n_2$, so that $f^3 = 0$ as before. Equations (3.3.52'), (3.3.55) assume the form

$$\frac{\partial \hat{w}}{\partial f^k}(\mathbf{f}) - \frac{\partial \hat{w}}{\partial f^k}(\mathbf{f}_0) = w^2(f^k - f^k{}_0), \quad k = 1, 2, \tag{3.3.104}$$

$$v^k - v^k{}_0 = -w(f^k - f^k{}_0), \quad k = 1, 2. \tag{3.3.105}$$

In the case of an isotropic medium equation (3.3.104) can be rewritten in the form:

$$\varphi_n \mathbf{i} - \varphi_n{}^0 \mathbf{i}_0 = w^2(n\mathbf{i} - n_0 \mathbf{i}_0). \tag{3.3.106}$$

Let $\mathbf{i} \cdot \mathbf{i}_0 = \cos\phi$. Contracting (3.3.106) with $\mathbf{i}$ and $\mathbf{i}_0$ in turn and combining the results we conclude that

$$(\varphi_n{}^0 - w^2 n_0)(1-\cos^2\phi) = 0 = (\varphi_n - w^2 n)(1-\cos^2\phi). \tag{3.3.107}$$

Equations (3.3.107) are satisfied if either

$$w^2 = \frac{\varphi_n{}^0}{n_0} = \frac{\varphi_n}{n} \tag{3.3.108}$$

or $\phi = 0, \pi$. The former case corresponds to the exceptional C-discontinuities which merely rotate the vector $\mathbf{f}$ through an angle ϕ. In the latter case the length of the vector $\mathbf{f}$ changes according to the equations

$$\varphi_n - \varphi_n{}^0 = w^2(n - n_0), \quad \phi = 0 \tag{3.3.109}$$

or

$$\varphi_n + \varphi_n{}^0 = w^2(n + n_0), \quad \phi = \pi. \tag{3.3.110}$$

Equations (3.3.109), (3.3.110) yield w^2 as a function of n, with $w^2 \to \varphi_{nn}(n_0)$ as $n \to n_0$ in the first case. In both cases w^2 assumes the same value at $n = 0$, viz. $w^2 = \dfrac{\varphi_n{}^0}{n_0}$. Hence the curves $\mathscr{H}_0^L$ for L-shocks are straight lines through $\mathbf{f} = \mathbf{0}$ and w^2 varies continuously at $\mathbf{f} = \mathbf{0}$.

Let $F(w^2, n) :\equiv w^2(n-n_0) - \varphi_n(n) + \varphi_n(n_0)$. Since $F'_{w^2} = n - n_0$, F is a submersion except in a neighbourhood of $n = n_0$.

We shall prove that $P := \big(\lambda_L(n_0)^2, n_0\big)$ is a bifurcation point. Indeed, $F'_n = w^2 - \varphi_{nn}$, hence $\ker F'_n(P) = \boldsymbol{R}$, $\operatorname{coker} F'_n(P) = \boldsymbol{R}$, $M = F''_{nw^2}(P) = 1$, and the hypotheses of the lemma of Prodi and Ambrosetti (see Section 3.1.5) are satisfied. By virtue of the bifurcation theorems quoted in Section 3.1.5 the solution of (3.3.109) is a one-dimensional submanifold of the half-plane $\boldsymbol{R}^2_+$, going to infinity in $\boldsymbol{R}^2_+$. It is clear that n is a good parameter on the Hugoniot defined by (3.3.109). The solution of (3.3.110) is a continuation of the solution of (3.3.109) beyond $n = 0$.

Let $\varphi \in \mathscr{C}^3$. For small $|n - n_0|$

$$w^2 = \frac{\varphi_n - \varphi_n^0}{n - n_0} = \varphi_{nn}\big(\vartheta n + (1-\vartheta)n_0\big), \quad 0 \leqslant \vartheta \leqslant 1,$$

is equal to the value of $\lambda_L{}^2$ somewhere between n and n_0. If $\varphi_{nnn}(n_0) \neq 0$ and $\iota_1 = \operatorname{sgn} \varphi_{nnn}(n_0)$, then for small $|n - n_0|$ the inequality

$$\varphi_{nn}(n) > w^2 > \varphi_{nn}(n_0)$$

holds for $\iota_1(n - n_0) > 0$.

In a nbhd of $n = 0$

$$w^2 = \frac{\varphi_n{}^0 \pm \varphi_n}{n_0 \pm n} \cong \frac{\varphi_n^0}{n_0} + \frac{n}{n_0}\left\{\pm\left(\frac{\varphi_n}{n} - \frac{\varphi_n{}^0}{n_0}\right) - \frac{\varphi_n}{n_0}\right\} + \dots \quad (3.3.111)$$

where the upper (lower) sign refers to equation (3.3.110) ((3.3.109)). The last term in the braces is $O(n^2)$ and can be neglected. Since $n < n_0$, equation (3.3.103) implies that $\iota_0\left(w^2 - \frac{\varphi_n{}^0}{n_0}\right) > 0$ holds for $\phi = 0$ and $\iota_0\left(w^2 - \frac{\varphi_n{}^0}{n_0}\right) < 0$ holds for $\phi = \pi$. Hence w^2 is monotone in a nbhd of $n = 0$.

Let us now consider the $\hat{\mathscr{H}}^L, \hat{\mathscr{H}}^C$ centered at $n_0 = 0$. Equation (3.3.106) implies that

$$w^2 = \frac{\varphi_n}{n}, \quad \phi \text{ is arbitrary.} \quad (3.3.112)$$

In fact in the cases $n = 0$ and $n_0 = 0$ the angle ϕ is meaningless, since the shock joins a state $\mathbf{f} \neq \mathbf{0}$ of arbitrary polarization to the state $\mathbf{f} = \mathbf{0}$. The shock propagates with the speed of the C-discontinuities in the state $n \neq 0$.

Concerning the Chapman–Jouguet points we note that for $n_0 \neq 0$

$$\dot{\overline{w^2}} = \dot{n}(n-n_0)^{-1}(\lambda_L{}^2 - w^2), \quad (3.3.113)$$

and $\dot{n} \neq 0$ according to our parametrization. Hence $\dot{w} = 0$ implies that $w^2 = \lambda_L(n)^2$. We also note that $w_1{}^2 = \varphi_{nn}(n_1)$, $n_1 \neq n_0$, implies that φ_{nnn} vanishes somewhere between n_0 and n_1. Indeed, let

$$f(n) :\equiv \varphi_n - \varphi_n{}^0 - \varphi_{nn}(n)(n-n_0), \quad (3.3.114)$$

$f'(n) = -\varphi_{nnn}(n)(n-n_0)$. Since $f(n_0) = 0 = f(n_1)$, it follows that $\varphi_{nnn}(n_2) = 0$ for some n_2 between n_0 and n_1. In particular it may be the case that $\varphi_{nnn}(n) = 0$ for all n between n_0 and n_1. In this case $f(n) \equiv 0$ on the interval between n_0 and n_1 and $w^2 = \lambda_L(n)^2 = \lambda_L(n_0)^2$ (the last equality follows from (3.3.113)). If $\varphi_{nnn}(n)$ does not vanish on the whole interval between n_0 and n_1, then it changes sign at some n_2 on this interval. From (3.3.113) it is also clear that $w^2 - \lambda_L{}^2$ changes sign at a Chapman–Jouguet point (w^2 = extremum), in accordance with the general theory.

For $n_0 = 0$ we note that $\dot{\overline{w^2}} \neq 0$ on account of (3.3.112), (3.3.103′). Also, for every $n > 0$ we have the inequality $w^2 > \lambda_C(0)^2 = \lambda_L(0)^2$ if $\iota_0 = 1$ and the opposite inequality holds if $\iota_0 = -1$. Let $\iota_0 = +1$. In this case we define a CL_+ shock by $\mathbf{f} = \mathbf{0}$ on the right-hand side $\mathbf{f} \neq \mathbf{0}$ on the left-hand side (in the case of a CL_- shock $\mathbf{f} = \mathbf{0}$ on the left-hand side). In the case $\iota_0 = -1$ the state $\mathbf{f} = \mathbf{0}$ appears on the left-hand side of a CL_+-shock and on the right-hand side of a CL_--shock.

A solution $\mathbf{f}(y/t)$, $\mathbf{v}(y/t)$ is admissible if it consists of C-discontinuities, c.s. waves, L_+ and L_- discontinuities with $n \neq 0$ on both sides satisfying (L) and CL_+, CL_- shocks.

Let $\iota_0 = +1$. We consider an admissible solution $\mathbf{f}(y/t)$, $\mathbf{v}(y/t)$ with $n(\zeta_0) = 0$ for some value $\zeta_0 > 0$ of y/t. Since $\iota_0 = 1$, φ_{nn} grows with n for small n and an L_+-c.s. wave can appear at most on the right-hand side of ζ_0. In this case we can take $\zeta_0 = \varphi_{nn}(0)^{1/2}$. On the other hand a CL_+ shock can appear at most on the left-hand side of ζ_0 at $y/t = w = (\varphi_n/n)^{1/2} > \varphi_{nn}(0)^{1/2}$. It follows immediately that the state $n(\zeta_0) = 0$ cannot be adjacent to an L_+-c.s. wave and a CL_+ shock simultaneously. Hence the state $n(\zeta_0) = 0$, $\zeta_0 > 0$, can appear either on the left-hand side of an L-group or on the right-hand side of a CL_+-shock. In the former case it is adjacent to an L_+-c.s. wave, possibly preceded by some other L-waves, and it extends up to $y = 0$. In the latter case it extends up to $t = 0$ and the only wave in the quadrant $y/t > 0$ of the half-plane $t \geqslant 0$ is a CL_+-shock.

Summing up, there are three possibilities for the structure of an admissible solution in the quadrant $y/t > 0$.

(1) $n(\zeta) \neq 0$ for $\zeta > 0$ and the quadrant $\zeta > 0$ contains (in general) an L_+-group and a C_+-discontinuity. Across the L_+-group the polarization $\mathbf{i} = n^{-1}\mathbf{f}$ remains constant while the length of the vector $\mathbf{f}$ varies from the value n_r on the right to n_0 on the left. At the C_+-discontinuity the polarization $\mathbf{i}$ rotates through a finite angle $\Delta\phi_+$ while $n = n_0$ remains constant.

(2) $n_r = 0$ and the quadrant $\zeta > 0$ contains only an L_+-group of waves joining $\mathbf{f}_r = \mathbf{0}$ on the right to some $\mathbf{f}_0{}^+$ in a sector $0 < \zeta < \varepsilon$.

(3) $n_r \neq 0$, the quadrant $\zeta > 0$ contains only a CL_+-wave and $\mathbf{f} = \mathbf{0}$ in a sector $0 < \zeta < \varepsilon$.

The corresponding possibilities for the quadrant $\zeta < 0$ follow by symmetry.

We also consider the variation of the velocity $\mathbf{v}$. Across an L_+-c.s. wave the variation of $\mathbf{v}$ is given by the formula

$$\mathbf{v}(\zeta_1) - \mathbf{v}(\zeta_0) = \pm \int_{\zeta_0}^{\zeta_1} d\zeta \, \lambda_L(n(\zeta)) n'(\zeta) \mathbf{i}. \tag{3.3.115}$$

Across an L-discontinuity

$$\mathbf{v}_1 - \mathbf{v}_0 = -w(n_1 - n_0)\mathbf{i} \tag{3.3.116}$$

and across a $C_\pm$-discontinuity

$$\mathbf{v}_1 - \mathbf{v}_0 = \mp \lambda_C(n) n (\mathbf{i}_1 - \mathbf{i}_0) \tag{3.3.117}$$

on account of (3.3.105).

Finally, we consider the possibility of a discontinuity at $y = 0$. From equation (3.3.104) and (3.3.105) it follows that

$$\hat{w}_{\mathbf{f}}(+0) = \hat{w}_{\mathbf{f}}(-0), \tag{3.3.118}$$

$$\mathbf{v}(+0) = \mathbf{v}(-0). \tag{3.3.119}$$

In an isotropic medium $\hat{w}_{\mathbf{f}}(\mathbf{f}) = \varphi_n(n)\mathbf{i}$, hence $\varphi_n\big(n(+0)\big) = \varphi_n\big(n(-0)\big)$. Since $\varphi_{nn} > 0$, equation (3.3.118) implies that

$$n(+0) = n(-0) =: n_0, \tag{3.3.120}$$

$$\phi(+0) = \phi(-0) =: \phi_0 \quad \text{if } n_0 > 0 \tag{3.3.121}$$

with $\mathbf{i} = [\cos\phi \;\; \sin\phi]$.

The values of $n(+0)$, $\phi(+0)$, $\mathbf{v}(+0)$ depend on $\mathbf{f}_r$, $\mathbf{v}_r$ as well as on two parameters specifying the strength and composition of the waves in $\zeta > 0$, if we consider the case (3), described above, as a degenerate limiting situation. A possible strategy of solving the Riemann problem

$$(\mathbf{f}(0, y), \mathbf{v}(0, y)) = \begin{cases} (\mathbf{f}_l, \mathbf{v}_l) & \text{for } y > 0, \\ (\mathbf{f}_r, \mathbf{v}_r) & \text{for } y < 0 \end{cases} \tag{3.3.122}$$

consists in calculating n_0, ϕ_0 from equation (3.3.119) and then determining the composition of the waves in either quadrant from the values of n_0, n_r, $\phi_0 - \phi_r$ and n_0, n_l, $\phi_0 - \phi_l$, resp. In terms of a function f, determined by equations (3.3.115) and (3.3.116) and the specific properties of the L-groups for the given function φ,

$$\begin{aligned} \mathbf{v}_0 &= \mathbf{v}_r - f(n_0, n_r)\mathbf{i}_r - n_0\,\lambda_C(n_0)(\mathbf{i}_0 - \mathbf{i}_r) \\ &= \mathbf{v}_l + f(n_0, n_l)\mathbf{i}_l + n_0\,\lambda_C(n_0)(\mathbf{i}_0 - \mathbf{i}_l). \end{aligned}$$

Hence

$$\mathbf{v}_r - \mathbf{v}_l = f(n_0, n_l)\mathbf{i}_l + f(n_0, n_r)\mathbf{i}_r + n_0\,\lambda_C(n_0)(2\mathbf{i}_0 - \mathbf{i}_l - \mathbf{i}_r)$$

and

$$\begin{aligned} (\mathbf{v}_r - \mathbf{v}_l)\cdot\mathbf{i}_r &= f(n_0, n_r) + f(n_0, n_l)\cos(\phi_r - \phi_l) \\ &\quad + n_0\,\lambda_C(n_0)\big(2\cos\Delta - \cos(\phi_r - \phi_l) - 1\big), \\ (\mathbf{v}_r - \mathbf{v}_l)\cdot\mathbf{i}_l &= f(n_0, n_l) + f(n_0, n_r)\cos(\phi_r - \phi_l) \\ &\quad + n_0\,\lambda_C(n_0)\big(2\cos(\Delta + \phi_r - \phi_l) - \cos(\phi_r - \phi_l) - 1\big), \\ \Delta &:= \phi_0 - \phi_r. \end{aligned} \tag{3.3.123}$$

The existence and uniqueness of solutions for small $|\mathbf{f}_r - \mathbf{f}_l|$, $|\mathbf{v}_r - \mathbf{v}_l|$ can be established by the method of the preceding section.

It is possible to prove (see Hanyga (1976) for thermoelasticity) that for collinear data $\mathbf{f}_r$, $\mathbf{f}_l$, $\mathbf{v}_r$, $\mathbf{v}_l$ the solution vectors $\mathbf{f}(y/t)$, $\mathbf{v}(y/t)$ are also collinear with the data. On the other hand the function $f(\cdot\,, n_r)$ is typically increasing for every $n_r \geqslant 0$. Indeed, it is easy to see that $f'_{n_0}(n_0, n_r)$ is equal either to $\lambda_L(n_0) > 0$ or to $w+n_0\, dw/dn_0$. In the latter case we deduce from equation (3.3.113) that $f'_{n_0} = [2(n_0-n_1)]^{-1}[(n_0-2n_1)w+n_0 w^{-1}\lambda_L(n_0)^2]$. n_1 denotes the value of n on the other side of the shock adjacent to n_0. If $n_0 > n_1$, then $f'_{n_0} \geqslant w > 0$ on account of the inequality $w < \lambda_L(n_0)$. The case $n_0 < n_1$ is more involved. It is easy to obtain some existence and uniqueness theorems for collinear data if f is either increasing for every n_r or decreasing for every n_r. In fact, equations (3.3.123) reduce to a single equation $v_r - v_l = f(n_0, n_r) + f(n_0, n_l)$ with exactly one solution n_0 for some range of data. The solution (n_0, v_0) lies on the intersection of the curves $v_0 = v_l + f(n_0, n_l)$, $v_0 = v_r - f(n_0, n_r)$ in the (n_0, v_0)-plane. These curves have opposite slopes. Unfortunately such solutions with constant polarization cannot be extended to non-collinear data by means of an implicit function argument. This follows from the fact that the Jacobian of equation (3.3.123) with respect to the unknown n_0, Δ vanishes at $\phi_r - \phi_l = 0$, $\Delta = 0$. Apparently any deeper existence results for the Riemann problem depend on the specific constitutive properties of the body.

3.4 CONVEX EXTENSIONS

3.4.1 Introductory Remarks

In Section 3.1 we noted that the Cauchy problem in $\boldsymbol{R}_+^{m+1}$ is well-posed for symmetric hyperbolic and strictly hyperbolic systems. In Section 3.3 we saw that the equations of motion of a hyperelastic medium are in general not strictly hyperbolic. It is now our task to show and exploit the fact that they are equivalent to a symmetric hyperbolic system.

There is good reason for the equations of physics and continuum mechanics to be symmetric hyperbolic. The point is that the equations of motion of a continuum (or field equations) imply an additional differential equation or inequality. The latter expresses the energy conservation law or the fact that dissipation is always non-negative. The additional equation or inequality enjoys a property of convexity, which is related to stability, e.g. the SE condition. The system of equations consisting of the original equations of motion and the additional inequality is then called a convex extension of the original system of equations.

The equation of energy conservation was often used in the past to prove uniqueness, existence and continuous dependence of solutions on the initial data for linear equations of mathematical physics (Rellich, Schauder and others). Convex extensions will be used here to prove that a given quasilinear system of equations is equivalent to a symmetric hyperbolic one (Section 3.4.2) and to investigate the stability and uniqueness of weak solutions (Sections 3.4.3–3.4.5). For weak shock waves the conditions (E), (L) imply that the passage of a shock wave through a portion of material gives rise to an instantaneous positive dissipation (or loss of energy).

3.4.2 Convex Extensions and Symmetrization of Hyperbolic Systems

We consider the system (3.1.1) with an invertible $\mathbf{A}^0(\mathbf{u}, x)$, satisfying the following condition of **local solvability**:

(LS) For an arbitrary analytic function $\mathbf{v}(x^1, \ldots, x^m)$ with values in $\mathscr{H}$, defined in a nbhd of an arbitrary point $(x_0^1, \ldots, x_0^m)$, the Cauchy problem $\mathbf{u}(0, x^1, \ldots, x^m) \equiv \mathbf{v}(x^1, \ldots, x^m)$ for equation (3.1.1) has a $\mathscr{C}^1$ solution $\mathbf{u}(x^0, x^1, \ldots x^m)$, defined in a nbhd of $(0, x_0^1, \ldots, x_0^m)$.

Suppose that the system (3.1.1) has another property:

(A) Every function $\mathbf{u} \in \mathscr{C}^1$ satisfying equation (3.1.1) also satisfies the inequality

$$\partial_\mu \varphi^\mu(\mathbf{u}, x) \leqslant \chi(\mathbf{u}, x), \tag{3.4.1}$$

where φ_μ, χ are some fixed $\mathscr{C}^2$ functions defined on $\mathscr{H} \times \boldsymbol{R}_+^{m+1}$.

Working out equation (3.4.1) explicitly and substituting equation (3.1.1) into it we get the inequality

$$\frac{\partial \varphi^0}{\partial u^r}\{\tilde{b}^r - \tilde{A}^{l,r}{}_s u^s{}_{,x^l}\} + \frac{\partial \varphi^k}{\partial u^r} u^r{}_{,x^k} \leqslant \bar{\chi} \tag{3.4.2}$$

with

$$\bar{\chi}(\mathbf{u}, x) :\equiv \chi(\mathbf{u}, x) - \frac{\partial \varphi^\mu}{\partial x^\mu}(\mathbf{u}, x), \quad \tilde{\mathbf{A}}^l := (\mathbf{A}^0)^{-1}\mathbf{A}^l, \quad l = 1, \ldots, m,$$

$$\tilde{\mathbf{b}} := (\mathbf{A}^0)^{-1}\mathbf{b}. \tag{3.4.3}$$

In view of the assumption (LS), for any numbers a^r, $c^r{}_k$ we can find a $\mathscr{C}^1$ function $\mathbf{u}(x)$ defined in a nbhd of an arbitrarily chosen point $x_0 = (0, x_0^1, \ldots$ $\ldots, x_0^m)$, satisfying equations (3.1.1) and $u^r(x_0) = a^r$, $u^r{}_{,x^k}(x_0) = c^r{}_k$, provided $(a^r) \in \mathscr{H}$. Hence (A) implies that the following inequalities

$$\frac{\partial \varphi^0}{\partial u^r}(\mathbf{u}, x)\tilde{b}^r(\mathbf{u}, x) \leqslant \bar{\chi}(\mathbf{u}, x), \tag{3.4.4a}$$

$$\frac{\partial \varphi^0}{\partial u^r}(\mathbf{u}, x)\tilde{A}^l(\mathbf{u}, x)^r{}_s = \frac{\partial \varphi^l}{\partial u^s}(\mathbf{u}, x), \quad l = 1, \ldots, m,$$

$$s = 1, \ldots, n \tag{3.4.4b}$$

are satisfied identically with respect to $(x, \mathbf{u}) \in \overline{R^{m+1}_{+}} \times \mathscr{H}$. (The choice of $x^0 = 0$ for the initial data was made for notational convenience only.)

Equations (3.4.4) imply that the following conservation law is satisfied by any $\mathscr{C}^1$ solution of equations (3.1.1):

$$\partial_\mu \varphi^\mu(\mathbf{u}, x) = \frac{\partial \varphi^0}{\partial u^r} \tilde{b}^r + \frac{\partial \varphi^\mu}{\partial x^\mu} \equiv : \psi(\mathbf{u}, x) \leqslant \chi(\mathbf{u}, x). \tag{3.4.5}$$

In the case of a dissipative material endowed with entropy density equation (3.4.1) can represent the Clausius–Duhem inequality and equation (3.4.5) is then the entropy balance.

In general the conditions (3.4.4a) and (3.4.4b) are non-trivial. In particular the system of equations (3.4.4b) for the functions φ^μ, $\mu = 0, \ldots, m$ is over-determined unless $m = 1$, $n \leqslant 2$. In some cases of physical interest (e.g. in hydrodynamics, Rozhdestvenskii, 1957) there is exactly one non-trivial inequality (3.4.1) satisfying (A), up to some trivial modifications.

Let

$$y_r(\mathbf{u}, x) :\equiv ((\mathbf{A}^0)^{-1})^s{}_r \frac{\partial \varphi^0}{\partial u^s} \quad \text{for } r = 1, \ldots, n, \quad y_{n+1} \equiv -1. \tag{3.4.6}$$

We assume now that the equations (3.1.1) have the particular form of conservation laws (3.1.35) with $f^{\mu,r} \in \mathscr{C}^2$, $(\mathbf{A}^\mu)^r{}_s = \dfrac{\partial f^{\mu,r}}{\partial u^s}$, $b^r = h^r - \dfrac{\partial}{\partial x^\mu} f^{\mu,r}$, $\mu = 0, \ldots, m$; $s, r = 1, \ldots, n$.

Let

$$f^{\mu,n+1} := \varphi^\mu, \quad h^{n+1} := \psi, \quad b^{n+1} := \psi - \frac{\partial \varphi^\mu}{\partial x^\mu} \equiv \frac{\partial \varphi^0}{\partial u^r} \tilde{b}^r. \tag{3.4.7}$$

We have then the identity

$$\sum_{r=1}^{n+1} y_r \frac{\partial f^{\mu,r}}{\partial u^s} \equiv 0 \quad \text{for } s = 1, \ldots, n. \tag{3.4.8}$$

and

$$\sum_{r=1}^{n+1} y_r \partial_\mu f^{\mu,r} = \sum_{r=1}^{n+1} y_r h^r \tag{3.4.9}$$

is satisfied in the sense of an identity with respect to x^μ, u^r, $u^r{}_{,\mu}$.

We now make the convexity assumption (Friedrichs, 1974):

$$\text{(C)} \quad -\sum_{s,s'=1}^{n} \sum_{r=1}^{n+1} y_r \frac{\partial^2 f^{0,r}}{\partial u^s \partial u^{s'}} \zeta^s \zeta^{s'} \tag{3.4.10}$$

is a positive definite quadratic form with respect to (ζ^s) for every $(\mathbf{u}, x) \in \mathscr{H} \times \times \overline{R^{m+1}_{+}}$.

Definition 3.4.1
A system of $n+1$ equations

$$\partial_\mu f^{\mu,r} = h^r, \quad r = 1, \ldots, n+1 \tag{3.4.11}$$

satisfying the conditions (3.4.9) and (C) for some $y_r(\mathbf{u}, x)$, $r = 1, \ldots, n+1$, is called a **convex extension** of (3.1.35).

Following Friedrichs, we define

$$B^\mu{}_{rs}(\mathbf{u}, x) :\equiv \sum_{s'=1}^{n+1} \frac{\partial y_{s'}}{\partial u^s} \frac{\partial f^{\mu,s'}}{\partial u^r}, \tag{3.4.12}$$

$$d_s(\mathbf{u}, x) :\equiv \sum_{s'=1}^{n+1} \frac{\partial y_{s'}}{\partial u^s} b^{s'}, \quad r, s = 1, \ldots, n. \tag{3.4.13}$$

It is obvious that every $\mathscr{C}^1$ solution of (3.1.35) satisfies the following equation

$$\mathbf{B}^\mu(\mathbf{u}, x)\mathbf{u}_{,x^\mu} = \mathbf{d}(\mathbf{u}, x). \tag{3.4.14}$$

Now

$$B^\mu{}_{rs} = \sum_{s'=1}^{n+1} \frac{\partial}{\partial u^s}(y_{s'} A^{\mu,s'}{}_r) - \sum_{s'=1}^{n+1} y_{s'} \frac{\partial^2 f^{\mu,s'}}{\partial u^r \partial u^s}. \tag{3.4.15}$$

The first term on the right-hand side of (3.4.15) vanishes identically, hence $B^\mu{}_{rs} = B^\mu{}_{sr}$ and $B^0{}_{rs}\zeta^r\zeta^s$ is positive definite on account of (C). Consequently the system (3.4.14) is symmetric hyperbolic.

We now prove that every $\mathscr{C}^1$ solution of equation (3.4.14) satisfies equation (3.1.35). Since

$$\sum_{r,s=1}^{n} \sum_{s'=1}^{n+1} \frac{\partial y_{s'}}{\partial u^s} \frac{\partial f^{0,s'}}{\partial u^r} \zeta^r\zeta^s = B^0{}_{rs}\zeta^r\zeta^s \tag{3.4.16}$$

is positive definite with respect to (ζ^r), the rank of the $n \times (n+1)$ matrix $\left[\frac{\partial y_{s'}}{\partial u^s}\right]$, $s' = 1, \ldots, n+1$, $s = 1, \ldots, n$, equals n. Suppose now that there are some functions $\mu^s(\mathbf{u}, x)$ such that

$$y_{s'} \equiv \sum_{s=1}^{n} \mu^s \frac{\partial y_{s'}}{\partial u^s}, \quad s' = 1, \ldots, n+1.$$

Then

$$0 = \sum_{s=1}^{n} \sum_{s'=1}^{n+1} \mu^s \frac{\partial y_{s'}}{\partial u^s} \frac{\partial f^{0,s'}}{\partial u^r} = \sum_{s=1}^{n} B^0{}_{rs}\mu^s$$

and $\mu^s \equiv 0$ for $s = 1, \ldots, n$. This proves that the $(n+1)\times(n+1)$ matrix $\left[\frac{\partial y_{s'}}{\partial u^s}, y_{s'}\right]$, $s' = 1, \ldots, n+1$, $s = 1, \ldots, n$, is invertible. Now, equations (3.1.14) and the identity (3.4.9) imply equations (3.1.35). We have thus proved

Theorem 3.4.2
If the system (3.4.11) is a convex extension of (3.1.35) then the system (3.1.35) is equivalent to the symmetric hyperbolic system (3.4.15) in so far as the $\mathscr{C}^1$ solutions are concerned.

Theorems 3.4.2 and 3.1.4 imply that the Cauchy problem in $\overline{R^{m+1}_+}$ is well-posed for smooth solutions in the small if the system (3.1.35) admits a convex extension.

Geometrically $\mathscr{H}$ is an n-dimensional manifold with some global coordinate system (u^r). The functions $\mathbf{f}^\mu(\cdot, x)$, $\mu = 0, \ldots, m$, map $\mathscr{H}$ into some affine space and we look for solutions $\mathbf{u}$ such that $\mathbf{f}^0 \circ \mathbf{u}$ belongs to BV. We shall show that the conditions (3.4.4) and (C) are invariant with respect to the coordinate transformations on $\mathscr{H}$ (which may be also local). This fact is very helpful while proving that the condition (C) is satisfied in some specific case.

The invariance of the condition (3.4.4) follows immediately from the fact that $(\tilde{\mathbf{A}}^l)^r{}_s = \frac{\partial u^r}{\partial f^{0,s'}} \frac{\partial f^{l,s'}}{\partial u^s}$. For the invariance of (C), let $\mathbf{u} = \mathbf{k}(\mathbf{v})$ be a coordinate transformation with $\mathbf{k} \in \mathscr{C}^1$

$$-\sum_{r=1}^{n+1} y_r \frac{\partial^2(f^{0,r}\circ \mathbf{k})}{\partial v^s \partial v^{s'}} = -\sum_{r=1}^{n+1} y_r \frac{\partial}{\partial u^{p'}}\left(\frac{\partial f^{0,r}}{\partial u^p}\frac{\partial k^p}{\partial v^s}\right)\frac{\partial k^{p'}}{\partial v^{s'}}$$

$$= -\sum_{r=1}^{n+1} y_r \frac{\partial^2 f^{0,r}}{\partial u^{p'}\partial u^p}\frac{\partial k^p}{\partial v^s}\frac{\partial k^{p'}}{\partial v^{s'}},$$

since $\sum_{r=1}^{n+1} y_r \frac{\partial f^{0,r}}{\partial u^p} \equiv 0$.

Suppose that the function $\mathbf{f}^0$ is globally invertible. In this case the mapping $\mathbf{f}^0 : \mathscr{H} \to R^n$ determines a global coordinate system on $\mathscr{H}$. Using this coordinate system we obtain a special form of equation (3.1.35), viz.

$$\mathbf{u}_t + \mathbf{f}^k(\mathbf{u})_{,x^k} = \mathbf{h}. \qquad (3.4.17)$$

The condition (C) is then equivalent to the convexity of φ^0:

$$\sum_{s,s'} \frac{\partial^2 \varphi^0}{\partial u^s \partial u^{s'}} \zeta^s \zeta^{s'} > 0 \quad \text{if} \sum_s (\zeta^s)^2 > 0. \tag{3.4.18}$$

Suppose that φ^0 does not depend on x, $\varphi^0 = \varphi^0(\mathbf{u})$. Let

$$v_s := \frac{\partial \varphi^0}{\partial u^s}(\mathbf{u}), \quad s = 1, \ldots, n \tag{3.4.19}$$

On account of (3.4.18) equation (3.4.19) admit a unique solution $\mathbf{u} = \mathbf{k}(\mathbf{v})$, $\mathbf{k} \in \mathscr{C}^1$, in a nbhd of every $(\mathbf{u}, \mathbf{v})$ satisfying equation (3.4.19). We shall assume for simplicity that (3.4.19) is injective. Let

$$\psi^0(\mathbf{v}) :\equiv v_s k^s(\mathbf{v}) - \varphi^0(\mathbf{k}(\mathbf{v})). \tag{3.4.20}$$

Clearly

$$k^s(\mathbf{v}) = \frac{\partial \psi^0}{\partial v_s}.$$

Let

$$g^\mu(\mathbf{v}) :\equiv v_s f^{\mu,s}(\mathbf{k}(\mathbf{v})) - \varphi^\mu(\mathbf{k}(\mathbf{v})), \quad \mu = 0, 1, \ldots, m. \tag{3.4.21}$$

On account of equations (3.4.4)

$$\frac{\partial g^\mu}{\partial v_s} = f^{\mu,s} \circ \mathbf{k}, \quad \mu = 0, 1, \ldots, m; \; s = 1, \ldots, n. \tag{3.4.22}$$

Moreover $\dfrac{\partial^2 g^0}{\partial v_r \partial v_s} = \dfrac{\partial^2 \psi^0}{\partial v_s \partial v_r}$ (cf. equations (3.4.22) and (3.4.20)) is a positive definite symmetric matrix.

The invariance of the conditions (3.4.4) and (C) with respect to the coordinate transformations on $\mathscr{H}$ implies the following:

Theorem 3.4.3

If the function $\mathbf{f}^0$ is injective and the system (3.1.35) has a convex extension such that (3.4.19) is one-to-one, then equation (3.1.35) are equivalent to a symmetric hyperbolic system of conservation laws

$$\partial_\mu \left(\frac{\partial g^\mu}{\partial v_r} \right) = \tilde{b}_r(x, \mathbf{v}), \quad r = 1, \ldots, n. \tag{3.4.23}$$

with $\dfrac{\partial^2 g^0}{\partial v_r \partial v_s}$ positive definite.

In view of (3.4.22) the equivalence of equations (3.4.23) and (3.1.35) remains valid in the class of weak solutions.

Systems of equations of the form (3.4.23) were considered by Godunov (1961). Such systems always admit a convex extension by means of the equation

$$\partial_\mu \left(\sum_{r=1}^{n} v_r \frac{\partial g^\mu}{\partial v_r} - g^\mu \right) = \sum_{r=1}^{n} v_r \tilde{b}^r - \frac{\partial g^\mu}{\partial x^\mu} . \tag{3.4.24}$$

3.4.3 Convex Extensions of a Single Hyperbolic Equation

Let us construct a convex extension of equation (3.2.1). Let $f'(u) \equiv c(u)$ and

$$g'(u) :\equiv uc(u). \tag{3.4.25}$$

Every $\mathscr{C}^1$ solution $u(t, x)$ of equation (3.2.1) satisfies another equation

$$\frac{1}{2}(u^2)_t + g(u)_x = 0 \tag{3.4.26}$$

and hence also the integral conservation law

$$\oint_{\mathscr{C}} -\frac{1}{2} u^2 dx + g(u) dt = 0 \tag{3.4.27}$$

for every closed piecewise Lipschitz curve $\mathscr{C}$ in the domain of u. Equations (3.2.1) and (3.4.26) constitute a convex extension of equation (3.2.1).

More specifically, let $u(t, x)$ be a piecewise $\mathscr{C}^1$ function with a single discontinuity $x = y(t)$, satisfying equation (3.2.16a) with $c(u) \equiv u, f(u) \equiv \frac{1}{2}u^2$. From equation (3.1.41) we conclude that

$$\dot{y}(t) = \frac{1}{2}(u_+ + u_-)(t, y(t)) \tag{3.4.28}$$

for almost all t. Suppose that the same function satisfies equation (3.4.27) with $g(u) \equiv \frac{1}{3}u^3$. It follows immediately that

$$\dot{y}(t) = \frac{2}{3} \frac{u_+{}^2 + u_+ u_- + u_-{}^2}{u_+ + u_-} . \tag{3.4.29}$$

Since u_+ and u_- take on arbitrary values in equations (3.4.28) and (3.4.29), these equations are incompatible. Hence a weak solution of (3.2.16a) does not in general satisfy equation (3.4.27).

Suppose now that $u(t, x)$ satisfies equation (3.2.16a) and a weaker condition than (3.4.27):

$$\int_{\mathscr{C}} \left(\frac{1}{2} u^2 \nu_t + g(u)\nu_x\right) ds \leqslant 0, \tag{3.4.30}$$

where $\mathscr{C}$ is an arbitrary piecewise Lipschitz curve of finite arclength, bounding a region $\mathscr{D} \subset \boldsymbol{R}^2$, (ν_t, ν_x) is the outer normal with respect to $\mathscr{D}$ and ds is the arclength element. Equations (3.1.34) and (3.1.40) imply that

$$-\dot{y}(t)\left[\left[\frac{1}{2} u^2\right]\right](t) + [[g(u)]](t) \geqslant 0 \tag{3.4.31}$$

for almost all $t \in \operatorname{dom} y(\cdot)$. Substituting (3.4.28) into the left-hand side and working out the resulting expression we find that

$$\zeta(u_+, u_-) :\equiv \frac{1}{12}(u_- - u_+)^3 \geqslant 0. \tag{3.4.32}$$

Hence $u_- \geqslant u_+$, in accordance with the conditions (E) and (L), Section 3.2.3. Note also that ζ is of third order with respect to the shock intensity $|u_- - u_+|$.

A condition like (3.4.30), derived from a convex extension, is usually called an **entropy condition** (Lax, 1971).

The entropy condition (3.4.30) for $f(u) \equiv -\frac{1}{3}u^3$, $g(u) \equiv -\frac{1}{4}u^4$, $u_+ < 0$, viz.

$$\zeta(u_+, u_-) \equiv -\frac{1}{12}(u_- + u_+)(u_- - u_+)^3 \geqslant 0,$$

is satisfied in particular for $u_+ \leqslant u_- \leqslant -u_+$ and therefore it is weaker than (L) and (E) jointly.

3.4.4 The Entropy Condition for General Strictly Hyperbolic Systems of Conservation Laws

Although the entropy condition is too weak to ensure uniqueness, it is more convenient to handle when general weak solutions in $\mathrm{BV} \cap \mathscr{L}^\infty$ are considered.

Let (3.4.5) be a convex extension of (3.1.35). The associated **entropy condition** (EN) is the inequality

$$\int_{\partial^* \mathscr{P}} dH^m \, \nu_\mu \varphi^\mu(x, \mathbf{u}_+) \leqslant \int_{\mathscr{P}} dL^{m+1} \, \chi(x, \mathbf{u}). \tag{3.4.33}$$

Here $\mathbf{u} \in BV \cap \mathscr{L}^{\infty}(\mathscr{D}, \mathscr{H})$, $\mathscr{P}$ is a subset of $\mathscr{D}$ of bounded perimeter, ν_μ is the outer normal on $\partial^*\mathscr{P}$ in the sense of Federer and $\mathbf{u}_+$ is the internal (or external) trace of $\mathbf{u}$ on $\partial^*\mathscr{P}$. We assume that $\varphi^\mu \in \mathscr{C}^1$, $\chi \in \mathscr{C}^0$.

By the Gauss–Green theorem

$$\int_{\mathscr{P}} \partial_\mu \varphi^\mu \leqslant \int_{\mathscr{P}} \chi dL^{m+1}. \tag{3.4.34}$$

The right-hand side is a regular Borel measure of $\mathscr{P}$ since $\chi \in \mathscr{L}^1_{loc}(\mathscr{D}, \boldsymbol{R})$ (Lemma 1.1.63). Equation (3.4.34) holds in particular for every cube in $\boldsymbol{R}^{m+1}$, hence for every open subset $\mathscr{P}$ of $\boldsymbol{R}^{m+1}$, and hence for every Borel subset $\mathscr{P}$ (Section 1.1.2). On account of the monotonicity of the integral

$$\int_{\mathscr{D}} (\varphi^\mu \partial_\mu \phi + \chi\phi) dL^{m+1} = \int_{\mathscr{D}} (-\partial_\mu \varphi^\mu + \chi dL^{m+1})\phi \geqslant 0 \tag{3.4.35}$$

for every non-negative test function $\phi \in \mathscr{C}_0^\infty(\mathscr{D}, \boldsymbol{R})$. The expression in the middle is the integral of ϕ with respect to the measure $\mathscr{E} \mapsto -(\partial_\mu \varphi^\mu)(\mathscr{E}) + \int_{\mathscr{E}} \chi dL^{m+1} =: \mu(\mathscr{E})$.

Conversely, suppose that

$$\int_{\mathscr{D}} (\varphi^\mu \partial_\mu \phi + \chi\phi) dL^{m+1} \geqslant 0 \tag{3.4.36}$$

holds for every non-negative $\phi \in \mathscr{C}_0^\infty(\mathscr{D}, \boldsymbol{R})$. Let $\mathscr{P}$ be any bounded measurable subset of $\mathscr{D}$. Since $\chi_{\mathscr{P}} \in \mathscr{L}^1(\mathscr{D}, \boldsymbol{R})$, $\mathscr{C}_0^\infty(\mathscr{D}, \boldsymbol{R})$ is dense in $\mathscr{L}^1(D, \boldsymbol{R})$ and the integral is a continuous functional on $\mathscr{L}^1(\mathscr{D}, \boldsymbol{R})$, equation (3.4.35) implies (3.4.34). We have thus proved

Lemma 3.4.4

The entropy condition (3.4.33) is equivalent to (3.4.36).

From equation (3.1.34) and (3.1.40) we also deduce that

$$\zeta(\mathbf{u}_+, \mathbf{u}_-) :\equiv -w[\varphi^0(\mathbf{u}_-) - \varphi^0(\mathbf{u}_+)] + n_k[\varphi^k(\mathbf{u}_-) - \varphi^k(\mathbf{u}_+)] \geqslant 0 \tag{3.4.37}$$

has to be satisfied H^m-almost everywhere on every isolated discontinuity of a piecewise $\mathscr{C}^1$ solution.

Definition

The function ζ defined by equation (3.4.37) will be called the **dissipation function.**

Let $\mathbf{u}_- = \mathbf{u}_0$ and $\mathbf{n} \in S^{m-1}$ be fixed, $(w, \mathbf{u}_+) = (w(s), \mathbf{d}(s)) \in \mathscr{H}_{\mathbf{u}_0}$, $(w(0), \mathbf{d}(0)) = (\lambda_\alpha(\mathbf{u}_0), \mathbf{u}_0)$ and

$$\zeta(s) :\equiv w(s)[\varphi^0(\mathbf{d}(s)) - \varphi_0^0] - n_k[\varphi^k(\mathbf{d}(s)) - \varphi_0^k], \tag{3.4.38}$$

where $\varphi_0^\mu := \varphi^\mu(\mathbf{u}_0)$, $\mathbf{f}_0^\mu := \mathbf{f}^\mu(\mathbf{u}_0)$. From equations (3.1.43), (3.4.6) and (3.4.8)

$$\zeta(s) = \sum_{r=1}^{n+1} y_r(\mathbf{d}(s))\{-w(s)[f^{0,r}(\mathbf{d}(s)) - f_0^{0,r}] + n_k[f^{k,r}(\mathbf{d}(s)) - f_0^{k,r}]\}. \tag{3.4.39}$$

We shall calculate $\dot{\zeta}(s) = \dfrac{d\zeta}{ds}$, $\ddot{\zeta}(s) = \dfrac{d^2\zeta}{ds^2}$,

$$\zeta^{(k)}(s) = \frac{d^k\zeta}{ds^k}, \quad k \geqslant 1.$$

On account of (3.1.43), (3.4.6) and (3.4.7)

$$\dot{\zeta}(s) = -\dot{w}(s) \sum_{r=1}^{n+1} y_r(f^{0,r} - f_0^{0,r})|_{\mathbf{u}=\mathbf{d}(s)}. \tag{3.4.40}$$

Expanding the right-hand side in Taylor series at $\mathbf{u} = \mathbf{d}(s)$ up to second order and noting (3.4.7) as well as (C) we conclude that

$$\dot{\zeta}(s) = 0 \text{ implies either } \dot{w}(s) = 0 \text{ or } \mathbf{d}(s) = \mathbf{u}_0, \tag{3.4.41}$$

and

$$\operatorname{sgn}\dot{\zeta}(s) = -\operatorname{sgn}\dot{w}(s) \quad \text{if } \mathbf{d}(s) \neq \mathbf{u}_0. \tag{3.4.42}$$

Furthermore

$$\ddot{\zeta}(s) = -\dot{w}\sum_{r=1}^{n+1} \dot{y}_r[f^{0,r} - f_0^{0,r}] - \ddot{w}\sum_{r=1}^{n+1} y_r[f^{0,r} - f_0^{0,r}], \tag{3.4.43}$$

where $\dot{y}_r = \dfrac{d}{ds} y_r \circ \mathbf{d}(s)$. Hence

$$\zeta(s) = O(s^3) \quad \text{for } s \to 0. \tag{3.4.44}$$

This result generalizes our observation following equation (3.4.32).

More generally, if $\dot{w}(s) = O(s^k)$, $k \geqslant 0$, for $s \to 0$, then $\zeta(s) = O(s^{k+3})$. Also, $w^{(k-1)}(s_0) = \ldots = \dot{w}(s_0) = 0$ implies that

$$\zeta^{(k)}(s_0) = -w^{(k)}(s_0)\sum_{r=1}^{n+1} y_r[f^{0,r} - f_0^{0,r}]_{\mathbf{u}=\mathbf{d}(s_0)}. \tag{3.4.45}$$

Hence

$$\operatorname{sgn}\zeta^{(k)}(s_0) = -\operatorname{sgn} w^{(k)}(s_0), \quad \zeta^{(l)}(s_0) = 0 \text{ for } l \leqslant k-1. \tag{3.4.46}$$

In particular we have

Theorem 3.4.5

(1) $-\zeta(\cdot)$ has the same extrema on $\mathscr{H}_{\mathbf{u}_0}$ as $w(\cdot)$;
(2) if $w(s) = \text{const}$ on $\mathscr{H}^{\alpha}_{\mathbf{u}_0}$, then $\zeta(s) \equiv 0$.

Remark

Concerning the behaviour of φ^{μ} on the α-hodographs, note that

$$\lambda_{\alpha}(\mathbf{u}, \mathbf{n}) d\varphi^0 [\mathbf{r}_{\alpha}(\mathbf{u}, \mathbf{n})] = n_k d\varphi^k [\mathbf{r}_{\alpha}(\mathbf{u}, \mathbf{n})]$$

on account of equations (3.4.4) and (3.1.3). Hence

$$\lambda_{\alpha}(\mathbf{e}(s), \mathbf{n}) \frac{d\varphi^0 \circ \mathbf{e}}{ds} = n_k \frac{d\varphi^k \circ \mathbf{e}}{ds}$$

and $\varphi^0 = \text{const}$ on each α-hodograph $\mathbf{e}(\cdot)$ if $\varphi^k \equiv 0$.

We now apply Theorm 3.4.5 to check whether the solution to the Riemann problem constructed in Section 3.1.7 satisfies the entropy condition associated with the convex extension (3.4.33).

First of all we note that, on account of equation (3.4.42), for weak α-shocks ($s \approx 0$) the condition (3.4.37) implies that $w \leqslant \lambda_{\alpha}(\mathbf{u}_-)$, and hence, by Theorems 3.1.9, 3.1.10 $w \geqslant \lambda_{\alpha}(\mathbf{u}_+)$. Referring to the notations of Section 3.1.7 the condition (EN) is satisfied on $[0, s_1]$ (in fact, also on $[0, s_1 + \varepsilon]$ for some $\varepsilon > 0$).

It is somewhat more difficult to show that the shocks connecting $\mathbf{e}(\sigma)$ on the left to $\mathbf{z}_{\alpha}(s)$ on the right also satisfy (EN). On the arc $\tau \mapsto (w(\tau, \sigma), \mathbf{d}(\tau, \sigma))$ of $\mathscr{H}_{\mathbf{e}(\sigma)}$ comprised between the points $(\hat{\lambda}_{\alpha}(\sigma), \mathbf{e}(\sigma))$ and $(\hat{\lambda}_{\alpha}(\sigma), \mathbf{z}_{\alpha}(s))$ the function $\tau \mapsto \zeta(\tau, \sigma)$ has a minimum. Moving along $\mathscr{H}_{\mathbf{e}(\sigma)}$ it is not easy to check whether the value of ζ at $(\hat{\lambda}_{\alpha}(\sigma), \mathbf{z}_{\alpha}(s))$ is greater or equal to zero. We shall prove, however, that it is true under an additional assumption.

To this end we note that $(\hat{\lambda}_{\alpha}(\sigma), \mathbf{e}(\sigma)) \in \mathscr{H}_{\mathbf{z}}$, where $\mathbf{z} := \mathbf{z}_{\alpha}(s)$. Let $\mathscr{J}$ be the arc of $\mathscr{H}_{\mathbf{z}}$ extending from $(w, \mathbf{z})$ to $(\hat{\lambda}_{\alpha}(\sigma), \mathbf{e}(\sigma))$ and given parametrically by $\tau \mapsto (\bar{w}(\tau), \bar{\mathbf{d}}(\tau))$, $\tau > 0$, $\bar{\mathbf{d}}(0) = \mathbf{z}$. Let $\tau \mapsto \bar{\zeta}(\tau)$ be the dissipation function on $\mathscr{J}$. Since $\mathbf{z}$ is a state on the right of the shock, $\mathbf{z} = \mathbf{u}_+$, the entropy condition reads now $\bar{\zeta}(\tau) \leqslant 0$. We make the following assumption:

(Z) $\mathscr{J} \subset \mathscr{H}^{\alpha}_{\mathbf{z}}$ and $d\lambda_{\alpha}[\mathbf{r}_{\alpha}]$ changes sign on $\mathscr{J}$ only once, viz. at $\bar{\mathbf{d}}(\tau_1) \in \tilde{\mathscr{J}} \cap \Sigma_{\alpha}$.

Let $(\hat{\lambda}_{\alpha}(\sigma), \mathbf{e}(\sigma)) = (\bar{w}(\tau_2), \bar{\mathbf{d}}(\tau_2))$. We are looking for a number $\tau_2 > 0$ such that

$$\bar{w}(\tau_2) = \lambda_{\alpha}(\bar{\mathbf{d}}(\tau_2)) > \lambda_{\alpha}(\mathbf{z}). \tag{3.4.47}$$

Suppose that

$$\lambda_\alpha(\bar{\mathbf{d}}(\tau)) < \bar{w}(\tau) < \lambda_\alpha(\mathbf{z}) \tag{3.4.48}$$

for small $\tau > 0$. Then $\bar{w}$ is a decreasing function for small $\tau > 0$. If $(\bar{w}(\tau_2), \bar{\mathbf{d}}(\tau_2))$ is not a Chapman–Jouguet point on $\mathscr{H}_{\mathbf{z}}$, then $\tau_2 = \tau_1$ (Corollary 3.1.19 (iv)) and $\bar{w}(\tau_2) < \lambda_\alpha(\mathbf{z})$, in contradiction of (3.4.47). If $(\bar{w}(\tau_2), \bar{\mathbf{d}}(\tau_2))$ is a Chapman–Jouguet point, then there is no Chapman–Jouguet point on $]0, \tau_2[$ and $\bar{w}(\tau_2) = \min \bar{w} < \lambda_\alpha(\mathbf{z})$, hence equations (3.4.47) fail to be satisfied again. It follows that

$$\lambda_\alpha(\mathbf{z}) < \bar{w}(\tau) < \lambda_\alpha(\bar{\mathbf{d}}(\tau)) \tag{3.4.49}$$

for small $\tau > 0$. Since there is no Chapman–Jouguet point on $]0, \tau_2[$, we have $\bar{w}'(\tau) \geqslant 0$ and $\bar{\zeta}'(\tau) \leqslant 0$ for $\tau \in]0, \tau_2[$. Hence $\bar{\zeta}(\tau_2) \leqslant 0$ and the entropy condition is satisfied, *q.e.d.*

Suppose now that we have arrived at the point $(\lambda_\alpha(\mathbf{u}_0), \mathbf{z}_\alpha(s_3; \mathbf{u}_0)) \in \mathscr{H}_{\mathbf{u}_0}$, in the notations of Section 3.1.7. For $s_3 < s < 0$ the α-shock on the right of the α-c.s.w. satisfies the inequality

$$\zeta = \hat{\lambda}_\alpha(\sigma)[\varphi^0(\mathbf{z}_\alpha(s; \mathbf{u}_0)) - \varphi^0(\mathbf{e}(\sigma))] - n_k[\varphi^k(\mathbf{z}_\alpha(s; \mathbf{u}_0)) - \varphi^k(\mathbf{e}(\sigma))] \geqslant 0. \tag{3.4.50}$$

For $s \to s_3\, \mathbf{e}(\sigma) \to \mathbf{u}_0$, $\mathbf{z}_\alpha(s; \mathbf{u}_0) \to \mathbf{z}_\alpha(s_3; \mathbf{u}_0)$ and therefore the α-shock connecting $\mathbf{u}_0$ on the left to $\mathbf{z}_\alpha(s_3; \mathbf{u}_0)$ on the right satisfies the entropy condition $\zeta \geqslant 0$. Since $\dot{w}(s_3) < 0$, we also have $\dot{\zeta}(s_3) > 0$ and the entropy condition $\zeta(s) \geqslant 0$ is satisfied for $s \leqslant s_3$.

In the case of a triple wave shrinking to a single α-shock the left shock satisfies the entropy inequality

$$\zeta_1 = w(s_1)[\varphi^0(\mathbf{d}(s_1)) - \varphi^0(\mathbf{u}_0)] - n_k[\varphi^k(\mathbf{d}(s_1)) - \varphi^k(\mathbf{u}_0)] > 0 \tag{3.4.51}$$

and in the limit $\hat{\lambda}_\alpha(\sigma) \to w(s_3) = w(s_1)$, $\mathbf{e}(\sigma) \to \mathbf{d}(s_1)$, $\mathbf{z}_\alpha(s; \mathbf{u}_0) \to \mathbf{d}(s_3)$ the dissipation on the right α-shock

$$\zeta_2 = \hat{\lambda}_\alpha(\sigma)[\varphi^0(\mathbf{z}_\alpha(s; \mathbf{u}_0)) - \varphi^0(\mathbf{e}(\sigma))] - n_k[\varphi^k(\mathbf{z}_\alpha(s; \mathbf{u}_0)) - \varphi^k(\mathbf{e}(\sigma))] > 0$$

tends to

$$w(s_1)[\varphi^0(\mathbf{z}_\alpha(s_3; \mathbf{u}_0)) - \varphi^0(\mathbf{d}(s_1))] - n_k[\varphi^k(\mathbf{z}_\alpha(s_3; \mathbf{u}_0)) - \varphi^k(\mathbf{d}(s_1))] \geqslant 0 \tag{3.4.52}$$

Adding (3.4.51) and (3.4.52) we find that

$$\zeta_3 = w(s_3)[\varphi^0(\mathbf{z}_\alpha(s_3; \mathbf{u}_0)) - \varphi^0(\mathbf{u}_0)] - n_k[\varphi^k(\mathbf{z}_\alpha(s_3; \mathbf{u}_0)) - \varphi^k(\mathbf{u}_0)] > 0, \tag{3.4.53}$$

q.e.d. Hence (EN) is satisfied under the plausible hypothesis (Z).

Whenever the assumptions of Theorems 3.1.9, 3.1.10 are satisfied it is possible to prove that for every $\mathbf{u}_1 \in \mathscr{H}$ there is a constant $\delta > 0$ such that every solution $\mathbf{u} \in \mathrm{BV} \cap \mathscr{L}^\infty$ satisfying the condition (L) and the inequality $\|\mathbf{u}-\mathbf{u}_1\|_\infty < \delta$ also satisfies the conditions (E) and (EN). We sketch the proof below.

Let $\tilde{\mathscr{H}}$, $\tilde{\mathscr{H}}_0$ be the manifolds defined in Section 3.1.5 and let $\tilde{\mathscr{H}}_0^\alpha$ be the set of points $(w, \mathbf{u}_+, \mathbf{u}_-) \in \tilde{\mathscr{H}}_0$ satisfying (L_α) with strict inequalities, $\tilde{\mathscr{H}}_0^+ := \bigcup_{\alpha=1}^{n} \tilde{\mathscr{H}}_0^\alpha$, $\tilde{\mathscr{H}}^+ := \tilde{\mathscr{H}}_0^+ \cup \partial\tilde{\mathscr{H}}$. Suppose that the hypotheses of Theorem 3.1.10 are satisfied for $\mathbf{u}_- = \mathbf{u}_0$, $\mathbf{u}_+ = \mathbf{d}(s)$, with $w^{(l)}(0) = 0$ if $l \leqslant k$, $w^{(k+1)}(0) \neq 0$, $k \geqslant 0$. Then $\tilde{\mathscr{H}}_{\mathbf{u}_0}^\alpha$ is non-empty if either k is even or k is odd and $w^{(k+1)}(0) < 0$, $\tilde{\mathscr{H}}_{\mathbf{u}_0}^\alpha := \{(w, \mathbf{u}_+, \mathbf{u}_-) \in \tilde{\mathscr{H}}_0^\alpha | \; \mathbf{u}_- = \mathbf{u}_0\}$.

Consider any point (t_0, x_0) of discontinuity of $\mathbf{u}$ with $l_\nu \mathbf{u} = \mathbf{u}_+$, $l_{-\nu}\mathbf{u} = \mathbf{u}_-$, $\nu_x \geqslant 0$, satisfying the inequalities (L_α). It follows that $(w, \mathbf{u}_+, \mathbf{u}_-) \in \tilde{\mathscr{H}}_0^+$, $|\mathbf{u}_+ - \mathbf{u}_1| < \delta$, $|\mathbf{u}_- - \mathbf{u}_1| < \delta$. Let $\varepsilon > 0$ satisfy the inequalities $\varepsilon < |\lambda_\alpha(\mathbf{v}) - \lambda_{\alpha-1}(\mathbf{v})|$, $\varepsilon < |\lambda_\alpha(\mathbf{v}) - \lambda_{\alpha+1}(\mathbf{v})|$ for every $\mathbf{v}$ in a nbhd $|\mathbf{v}-\mathbf{u}_1| < \delta'$, $\delta' > \delta$. Because of the continuity of $w(\mathbf{u}_+, \mathbf{u}_-)$, $\lambda_\alpha(\mathbf{u}_+)$, $\lambda_\alpha(\mathbf{u}_-)$ on $\tilde{\mathscr{H}}$, $|w(\mathbf{u}_+, \mathbf{u}_-) - \lambda_\alpha(\mathbf{u}_+)| < \varepsilon$, $|w(\mathbf{u}_+, \mathbf{u}_-) - \lambda_\alpha(\mathbf{u}_-)| < \varepsilon$, $(w, \mathbf{u}_+) \in \tilde{\mathscr{H}}_{\mathbf{u}_-}^\alpha$ provided δ is sufficiently small. We prove that for sufficiently small $\delta > 0$ the whole arc $s \mapsto (w(s), \mathbf{d}(s))$ of $\tilde{\mathscr{H}}_{\mathbf{u}_-}^\alpha$ joining $\mathbf{u}_-$ to $\mathbf{u}_+ = \mathbf{d}(s_0)$, $s_0 > 0$, satisfies the inequalities $|w(s) - \lambda_\alpha(\mathbf{u}_-)| < \varepsilon$, $|w(s) - \lambda_\alpha(\mathbf{d}(s))| < \varepsilon$.

On account of equations (3.1.48)–(3.1.50)

$$|\mathbf{d}(s) - \mathbf{u}_-|^2 = s^2 + \boldsymbol{\gamma}(\psi(s), s)^2 = s^2(1 + o(1)^2), \quad s \in [0, s_0],$$

$$|s_0| < \mu,$$

if we adjust the metric on $\mathscr{H}$ in such a way that the vector fields $\mathbf{r}_\beta(\mathbf{u})$ form an orthonormal basis of $T_{\mathbf{u}}\mathscr{H}$, $\mathbf{u} \in \mathscr{H}$. Let $K = \sup\{o(1)^2 | s \in [0, s_0]\}$. If we choose s_0 so small that $(1+K)s_0{}^2 < \varepsilon^2$, then equations (3.4.66) hold for all $s \in [0, s_0]$ and the arc $s \mapsto (w(s), \mathbf{d}(s))$, $s \in [0, s_0]$, does not contain any sonic points. Hence, by Theorems 3.1.17 and 3.4.5 the functions $w(\cdot)$ and $\zeta(\cdot)$ are monotone on $[0, s_0]$ and the shock $(w, \mathbf{u}_+, \mathbf{u}_-)$ satisfies (E) and (EN).

It is more difficult but possible to prove that for $\mathbf{u}_-$ in a nbhd of $\mathbf{u}_1$ the numbers μ and K can be chosen independent of $\mathbf{u}_-$. The proof requires a strengthening of the bifurcation Theorem 15.2 of Prodi and Ambrosetti (1973), which is beyond the scope of this book. Taking $\delta > 0$ small enough we have $|s_0| < \mu$, $\varepsilon(1+K)^{-1/2}$ if $|\mathbf{u}_+ - \mathbf{u}_-| < 2\delta$. Hence we have

Lemma 3.4.6

Let $\mathbf{u}_1 \in \mathscr{H}$, $(w, \mathbf{u}_+, \mathbf{u}_-) \in \tilde{\mathscr{H}}_0^+$ and $|\mathbf{u}_+ - \mathbf{u}_1| < \delta$, $|\mathbf{u}_- - \mathbf{u}_1| < \delta$. If $\delta > 0$ is sufficiently small then the shock $(w, \mathbf{u}_+, \mathbf{u}_-)$ satisfies (L_α), (E) and (EN) for some $\alpha \in \{1, \ldots, n\}$.

Remark

In the references quoted *passim* in this chapter it is usual to assume that for every $\alpha \in \{1, \ldots, n\}$ one of the following statements is true:

(GN_α) $d\lambda_\alpha[\mathbf{r}_\alpha] \neq 0$ (genuine non-linearity);

(LD_α) $d\lambda_\alpha[\mathbf{r}_\alpha] \equiv 0$ (linear degeneracy).

(LD_α) implies that $w(s) = \lambda_\alpha(\mathbf{d}(s)) = \lambda_\alpha(\mathbf{u}_-)$, $\zeta(s) \equiv 0$ on $\mathscr{H}^\alpha_{\mathbf{u}_-}$. Hence every α-shock is an exceptional discontinuity, $w = \lambda_\alpha(\mathbf{u}_+) = \lambda_\alpha(\mathbf{u}_-)$, and (EN) is satisfied with the equality sign.

(GN_α) is, however, ineffective without additional information about the global behaviour of the submanifolds $\mathscr{H}^\alpha_{\mathbf{u}_-}$. It usually works provided it is known additionally that for all $\alpha, \beta \in \{1, \ldots, n\}$.

(I) $\mathscr{H}^\alpha_{\mathbf{u}_0} \neq \mathscr{H}^\beta_{\mathbf{u}_0}$ for $\alpha \neq \beta$;

(II) $w(s) \neq \lambda_{\alpha-1}(\mathbf{u}_-)$, $\lambda_{\alpha+1}(\mathbf{d}(s))$ on that arc $\mathscr{J}$ of $\mathscr{H}^\alpha_{\mathbf{u}_-}$ extending from $(\lambda_\alpha(\mathbf{u}_-), \mathbf{u}_-)$ to infinity, on which (L_α) is satisfied for small $|s|$.

If (II) fails to hold then $w(\cdot)$ neednot decrease on $\mathscr{J}$ for arbitrarily large $|s|$. The arc $\mathscr{J}$ can exhibit a Chapman–Jouguet point at which $w(s) - \lambda_{\alpha+1}(\mathbf{d}(s))$ changes sign. Additional complications can be expected if the regularity assumptions of Theorems 3.1.10, 3.1.16 and 3.1.17 fail to be satisfied.

On the other hand, for $|\mathbf{u} - \mathbf{u}_1|$ small enough the argument presented above replaces (GN_α) and (LD_α).

3.4.5 A Rank-One Convex Extension in Non-Linear Hyperelasticity. Physical Meaning of Dissipation

Every $\mathscr{C}^1$ solution of equations (3.3.2) satisfies the equation of energy conservation

$$\varrho_0\{\tfrac{1}{2}\mathbf{v}^2 + W(\mathbf{F}, \xi)\}_{,t} - \partial_\alpha(\varrho_0 W_{F^k_{\ \alpha}} v^k) = \varrho_0 b_k v^k. \tag{3.4.54}$$

The function $\varphi^0 := \varrho_0[\frac{1}{2}\mathbf{v}^2 + W(\mathbf{F}, \xi)]$ is not a convex function of the variables $(\mathbf{v}, \mathbf{F})$ and the condition (C) fails to be satisfied.* For a fixed vector $\boldsymbol{\nu}$ however the function $\mathbf{f} \mapsto \tilde{w}(\mathbf{f}) :\equiv W((\mathbf{E} + \mathbf{f} \otimes \boldsymbol{\nu})\mathbf{F}_0, \xi)$ is convex in view of (SE). Hence all the methods developed in Sections 3.4.1–3.4.4 can be applied to plane waves in hyperelastic media. In particular the transformation (3.4.19), (3.4.21) amounts to the Legendre transformation with the variable

* This difficulty has been circumvented in the case of small strains in an isotropic medium by means of a trick due to F. John (1977).

f replaced by $\mathbf{s} = \tilde{w}_{\mathbf{f}}$ and the strain energy $\tilde{w}(\mathbf{f})$ by the complementary energy $\mathbf{s} \cdot \mathbf{f} - \tilde{w}$. This transformation is possible since $\tilde{w}_{\mathbf{f}}$ is locally (and possibly also globally) invertible.Note, however, that the transformation $\mathbf{F} \mapsto \mathbf{S} = \varrho_0 \dfrac{\partial W}{\partial \mathbf{F}}$ is not invertible.

The methods of Section 3.4.4 and of Section 3.4.7 below can however be applied to general three-dimensional problems of hyperelasticity. Indeed, it follows from Theorem 3.4.8 in Section 3.4.7 below that the dissipation associated with a solution $\mathbf{u} = (\mathbf{v}, \mathbf{F})$ of equations (3.3.2) is concentrated on the jump set $\Gamma(\mathbf{u})$ of the solution $\mathbf{u}$. On the other hand we know that equation (3.3.10) holds H^3-a.e. on $\Gamma(\mathbf{u})$. Equation (3.4.10) remains valid with

$$\begin{aligned} y_k &= v^k, \quad k = 1, 2, 3. \\ y_r &= \varrho_0 W_{F^k{}_\alpha} \quad \text{for } r = 4, \dots, 12, \\ y_{13} &= -1 \end{aligned} \tag{3.4.55}$$

and

$$-\sum_{r=1}^{13} y_r(\mathbf{v}, \mathbf{F})[f^{0,r}(\mathbf{v}, \mathbf{F}) - f^{0,r}(\mathbf{v}_0, \mathbf{F}_0)]$$

$$= -\left\{(\mathbf{v} - \mathbf{v}_0)^2 + \frac{\partial^2 W(\tilde{\mathbf{F}})}{\partial F^k{}_\alpha \partial F^l{}_\beta} f^k f^l n_\alpha n_\beta\right\} < 0$$

for $(\mathbf{v}, \mathbf{F}) \in \mathscr{H}_{(\mathbf{v}_0, \mathbf{F}_0)}$, $(\mathbf{v}, \mathbf{F}) \neq (\mathbf{v}_0, \mathbf{F}_0)$ on account of (3.3.10) and SE. Hence the function ζ, equation (3.4.38), has all the properties SE. Hence the function ζ, equation (3.4.38), has all the properties listed in Section 3.4.4.

In particular a weak α-shock, $\alpha = 1, \dots, 6$ satisfies

$$\zeta(\mathbf{u}_+, \mathbf{u}_-) = -w[[\varrho_0(\tfrac{1}{2}\mathbf{v}^2 + W(\mathbf{F}, \xi))]] - n_\alpha[[\varrho_0 W_{F^k{}_\alpha} v^k]] > 0. \tag{3.4.56}$$

Equation (3.4.36) says that the change of internal and kinetic energy of a portion of material which has been swept by a shock wave is not entirely accounted for by the energy flux $-\varrho_0 W_{F^k{}_\alpha} v^k$ and by the energy production $\varrho_0 b_k v^k$. Since $\mathbf{u}_+$ ($\mathbf{u}_-$ resp.) is the undisturbed state if $w > 0$ ($w < 0$ resp.), the shock wave is an energy sink.

Physically a shock wave is an idealization of a propagating continuous transition zone with high velocity gradients giving rise to viscous dissipation. The effect of viscosity is to change the hyperbolic equations of elastodynamics into a parabolic system. Putting it in an abstract way we have a hyperbolic system with a parabolic singular perturbation:

$$\mathbf{u}_t + \mathbf{A}^\alpha(\mathbf{u}, \xi)\mathbf{u}_{,\xi^\alpha} = \varepsilon \partial_\alpha (\mathbf{B}^{\alpha\beta}(\mathbf{u}, \xi)\, \mathbf{u}_{,\xi^\beta}), \tag{3.4.57}$$

where the symmetric part of the matrix $\mathbf{B}^{\alpha\beta} n_\alpha n_\beta$ is non-negative for every $\mathbf{n} \in \mathscr{S}^2$. The solutions of the parabolic equation (3.4.57) are smooth, but in the limit $\varepsilon \to 0_+$ they develop discontinuities.

Suppose that the matrices $\mathbf{A}^\alpha$, $\mathbf{B}^{\alpha\beta}$ do not depend on ξ. In this case equation (3.4.57) has traveling plane wave solutions $\mathbf{u}_\varepsilon(y - wt)$, $y = n_\alpha \xi^\alpha$. The function $\mathbf{u}_\varepsilon$ satisfies an ordinary differential equation:

$$-w\mathbf{u}_{\varepsilon'} + n_\alpha \mathbf{A}^\alpha(\mathbf{u}_\varepsilon)\mathbf{u}_{\varepsilon'} = \varepsilon\left(n_\alpha n_\beta \mathbf{B}^{\alpha\beta}(\mathbf{u}_\varepsilon)\mathbf{u}_\varepsilon'\right)'. \tag{3.4.58}$$

Let z be a "stretched coordinate": $\bar{\mathbf{u}}(z) := \mathbf{u}_\varepsilon(\varepsilon z)$. The function $\bar{\mathbf{u}}$ satisfies a **shock structure equation**, which does not involve ε:

$$-w\bar{\mathbf{u}}' + n_\alpha \mathbf{A}^\alpha(\bar{\mathbf{u}})\bar{\mathbf{u}}' = \left(n_\alpha n_\beta \mathbf{B}^{\alpha\beta}(\bar{\mathbf{u}})\bar{\mathbf{u}}'\right)'. \tag{3.4.59}$$

Let $\bar{\mathbf{u}}$ be a solution of equation (3.4.59) which tends to some constants: $\bar{\mathbf{u}}'(a) \to \mathbf{0}$, $\bar{\mathbf{u}}'(b) \to \mathbf{0}$, $\bar{\mathbf{u}}(a) \to \mathbf{u}_-$, $\bar{\mathbf{u}}(b) \to \mathbf{u}_+$ as $a \to -\infty$, $b \to \infty$. In general the solution $\bar{\mathbf{u}}$ as well as the possible pairs $(\mathbf{u}_+, \mathbf{u}_-)$ depend on the matrix $\mathbf{B}^{\alpha\beta}$ of dissipation coefficients.

Suppose now that $\mathbf{A}^\alpha$ does not depend on ξ and $\mathbf{A}^\alpha(\mathbf{u})\,d\mathbf{u} = d\mathbf{f}^\alpha(\mathbf{u})$. In this case we can integrate equation (3.4.59):

$$[[-w\bar{\mathbf{u}} + n_\alpha \mathbf{f}^\alpha(\bar{\mathbf{u}})]]|_a^b = n_\alpha n_\beta \mathbf{B}^{\alpha\beta}\bar{\mathbf{u}}'|_a^b. \tag{3.4.60}$$

In the limit $a \to -\infty$, $b \to \infty$ we obtain the relation $(w, \mathbf{u}_+) \in \mathscr{H}_{\mathbf{u}_-}$, which does not involve the coefficients $\mathbf{B}^{\alpha\beta}$. Moreover, for $\varepsilon \to 0_+$,

$$\mathbf{u}_\varepsilon(y - wt) = \bar{\mathbf{u}}\left(\varepsilon^{-1}(y - wt)\right) \to \begin{cases} \mathbf{u}_- & \text{if } y - wt < 0, \\ \mathbf{u}_+ & \text{if } y - wt > 0. \end{cases} \tag{3.4.61}$$

This simple but remarkable result explains why the dissipation coefficients do not influence the Rankine–Hugoniot equations although dissipation is implicitly present in every shock.

Physically it is possible to introduce some small viscosity $\varepsilon\mathbf{B}^{\alpha\beta}$ in an arbitrary non-conservative hyperbolic system (3.1.1) but the limiting discontinuous solution will remember the particular viscous dissipation matrix. The profile $\bar{\mathbf{u}}$ becomes then a significant feature of the shock wave which may be studied theoretically and experimentally.

In the case of plane waves in elastic media the formal "vanishing viscosity" (3.4.57) can be chosen to represent the familiar model of physical viscosity:

$$\mathbf{v}_t + (\tilde{w}_\mathbf{f})_y = \varepsilon\left(\mathbf{B}(\mathbf{f})\mathbf{v}_y\right)_y, \quad \mathbf{s} = \tilde{w}_\mathbf{f}, \quad \mathbf{v}\cdot\mathbf{B}(\mathbf{f})\mathbf{v} > 0 \;\; \forall\, \mathbf{v} \neq \mathbf{0},$$

$$\mathbf{f}_t - \mathbf{v}_y = \mathbf{0}.$$

Setting $\mathbf{f} = \bar{\mathbf{f}}(\varepsilon^{-1}(y - wt))$, $\mathbf{v} = \bar{\mathbf{v}}(\varepsilon^{-1}(y - wt))$, integrating the resulting ordinary differential equation and eliminating $\bar{\mathbf{v}}$ we obtain the following shock structure equations

$$w^2\bar{\mathbf{f}}+\tilde{w}_{\mathbf{f}}\bar{\mathbf{f}} = -w\mathbf{B}(\bar{\mathbf{f}})\bar{\mathbf{f}}',$$

$$\bar{\mathbf{v}} = -w\bar{\mathbf{f}}.$$

Although the Rankine–Hugoniot equations do not depend on the particular form of **B**, the conditions (E), (EN) and (L) bear some relation to viscous phenomena. This topic will be resumed in the next section.

3.4.6 Remarks About Small Viscosity

Singular perturbations of the parabolic type (3.4.57) have sometimes been used to prove existence of solutions to hyperbolic systems of conservation laws (3.1.35). The method is known as the **vanishing viscosity approximation.** The first step consists in proving that $||\mathbf{u}^\varepsilon||_\infty \leqslant M$, tot var $\mathbf{u}^\varepsilon \leqslant M$ for some $M > 0$. By Helly's theorem* a sequence $\mathbf{u}^{\varepsilon_n}$ converges a.e. boundedly to a function of bounded variation as ε_n tends to zero. On the other hand the derivatives of $\mathbf{u}^{\varepsilon_n}$ are not bounded for $\varepsilon_n \to 0$. In the case of a single function u^ε satisfying an initial-value problem for $u_t^\varepsilon+f(u^\varepsilon)x = \varepsilon u^\varepsilon{}_{xx}$ it is possible to show that $||u^\varepsilon{}_x||_\infty < C\varepsilon^{-1}$ (see Rozhdestvenskii and Yanenko, 1978, Section 4.2).

This method was successfully applied to the existence proofs for a single conservation law (Oleinik, 1957, 1959; Vol'pert, 1967; Rozhdestvenskii and Yanenko, 1978 (using potentials), Kruzhkov, 1970 (for $m > 1$)). In the case of the Burgers equation $u_t+uu_x = \varepsilon u_{xx}$ the solution of an arbitrary initial-value problem can be obtained in closed form by means of the Hopf–Cole transformation and the limits $\varepsilon \to 0$ can be investigated by elementary methods (Hopf, 1950; Whitham, 1974). Tupchiyev (1964, 1972b) and Dafermos (1973, 1974) applied the vanishing viscosity method to prove the existence of self-similar solutions to the Riemann problem. In order to preserve the symmetry of the hyperbolic problem they used a dissipation matrix of the form $\varepsilon t\mathbf{B}(\mathbf{u})$. The method proved particularly successful for $n \leqslant 2$ equations.

It was repeatedly suggested that the only (physically) admissible shock waves are those which are limits of smooth solutions with vanishing viscosity. At least some properties of the class of admissible solutions singled out on this basis do not depend on the particular choice of the dissipation coefficients $\mathbf{B}^{\alpha\beta}$. These include the familiar conditions (EN), (E), (L) as we shall see below.

* Theorem 3.2.59.

In some cases the stable and unstable manifolds of the critical points $\mathbf{u}_+$, $\mathbf{u}_-$ of the ordinary differential equations (3.4.60), modelling single shocks, do not depend on the choice of $\mathbf{B}^{\alpha\beta}$. Following this idea Kulikovskii (1962) proved that a single α-shock is a limit of smooth solutions $\mathbf{u}^\varepsilon$ provided $|w-\lambda_\alpha(\mathbf{u}_+)|$ is small enough and the condition (GL_α) is satisfied. Godunov (1961a) pointed out however that the class of shocks which are limits of functions $\mathbf{u}^\varepsilon$ satisfying equations (3.4.58) can in general depend on the dissipation coefficients $\mathbf{B}^{\alpha\beta}$ even though the limiting hyperbolic system is conservative.

Starting from the idea that two states $\mathbf{u}_-$, $\mathbf{u}_+$ with $(w, \mathbf{u}_+) \in \mathscr{H}_{\mathbf{u}_-}$ can be connected by a shock iff equation (3.4.59) has a finite number of solutions $\bar{\mathbf{u}}_k(z)$, $k = 1, \ldots, p$, with $\bar{\mathbf{u}}_1(-\infty) = \mathbf{u}_-$, $\bar{\mathbf{u}}_k(\infty) = \bar{\mathbf{u}}_{k+1}(-\infty)$, $k = 1, \ldots, p-1$, $\mathbf{u}_p(\infty) = \mathbf{u}_+$, $(w, \bar{\mathbf{u}}_k(\pm\infty)) \in \mathscr{H}_{\mathbf{u}_-}$, Tupchiyev (1964, 1966, 1973a) formulated a condition restricting the class of admissible discontinuities. Tupchiyev's admissibility condition is expressible in terms of the dimensions of the stable and unstable manifolds of each pair of the neighbouring points $\bar{\mathbf{u}}_k(-\infty)$, $\bar{\mathbf{u}}_\eta(\infty)$ on $\tilde{\mathscr{H}}^\alpha_{\mathbf{u}_-}$. For a dissipation matrix $\mathbf{B} = b(\mathbf{u})\mathbf{E}$ $b > 0$, the dimensions of the stable and unstable manifolds of equation (3.4.59) at $\bar{\mathbf{u}}_k(\infty) = \mathbf{u}_k$ are determined by the number of inequalities $w > \lambda_\beta(\mathbf{u}_k)$, $\beta \in \{1, \ldots, n\}$, provided $w \neq \lambda_\beta(\mathbf{u}_k)$ $\forall\, \beta$. For $w = \lambda_\beta(\mathbf{u}_k)$ the dimensions of the stable and unstable manifolds depend on the character of the sonic point $(w, \mathbf{u}_k) \in \mathscr{H}_{\mathbf{u}_-}$ (w = extr or $\bar{\lambda}_\beta$ = max or $\bar{\lambda}_\beta$ = min). These relations remain valid for some larger class of dissipation matrices $\mathbf{B}$ (Tupchiyev, 1973). Tupchiyev proved existence and uniqueness of solutions to the Riemann problem subject to the above admissibility condition, assuming, however, that the functions $\mathbf{f}^\mu$ are holomorphic. Kuznetsov and Tupchiyev (1975) used this result to prove existence of solutions to more general initial-value problems by means of Glimm's scheme.

Independently, several authors (Lyubimov, 1961; Kulikovskii, 1962; Germain, 1962–3, and others) studied the existence and properties of shock waves in hydrodynamics and magnetohydrodynamics on the assumption that a shock wave should be a limit of a smooth solution with the discontinuity smeared out by a dissipation matrix determined by the physical model at hand. The method applied in these papers hinges on the possibility of presenting the equations of motion in the form (3.4.23).

We now illustrate the connections between the vanishing viscosity and the conditions (E), (EN) and (L). Following Kruzhkov (1970) and Lax (1971), we show that a weak solution obtained by the vanishing viscosity approximations with $\mathbf{B} = \mathbf{E}$ satisfies the entropy condition (EN).

Consider a sequence $\mathbf{u}_n$ of functions satisfying

$$(\mathbf{u}_n)_{,t} + \mathbf{f}^k(\mathbf{u}_n)_{,x^k} = \varepsilon_n \Delta \mathbf{u}_n \tag{3.4.62}$$

with $\varepsilon_n \to 0$ and

$$\Delta \mathbf{u} := \sum_{k=1}^{m} \frac{\partial^2 \mathbf{u}}{(\partial x^k)^2}.$$

Suppose that $\mathbf{u}_n$ converges to a function $\mathbf{u}$ boundedly a.e. For an arbitrary test function $\varphi \in \mathscr{C}_o^\infty(\mathscr{K})$, $\mathscr{K}$ compact,

$$\left|\int dL^{m+1}\, \varphi \Delta \mathbf{u}_n\right| = \left|\int dL^{m+1}\, \mathbf{u}_n \Delta \varphi\right| \leqslant ||\Delta\varphi||_\infty ||\mathbf{u}_n||_{\mathscr{L}^1(\mathscr{K})} < M$$

and $\varepsilon_n \Delta \mathbf{u}_n \to 0$ in the topology of the distribution space $\mathscr{D}'$. Hence $\mathbf{u}$ satisfies equations (3.1.35) in the distributions sense (with $\mathbf{f}^0 = \mathrm{id}$). In view of equations (3.4.4) equation (3.4.62) implies that $\varphi^0(\mathbf{u}_n)_{,t} + \varphi^k(\mathbf{u}_n)_{,x^k} = \sum_{r=1}^{n} \frac{\partial \varphi^0}{\partial u^r}(\mathbf{u}_n)\Delta u_n{}^r$.

On the other hand (C) implies that

$$\Delta(\varphi^0 \circ \mathbf{u}_n) = \sum_r \frac{\partial \varphi^0}{\partial u^r} \Delta u_n{}^r + \sum_k \sum_{r,s} \frac{\partial^2 \varphi^0}{\partial u^r \partial u^s} u^r_{,x^k} u^s_{,x^k}$$

$$\geqslant \sum_r \frac{\partial \varphi^0}{\partial u^r} \Delta u_n{}^r.$$

Hence

$$\varphi^0(\mathbf{u}_n)_{,t} + \varphi^k(\mathbf{u}_n)_{,x^k} \leqslant \varepsilon_n \Delta(\varphi^0 \circ \mathbf{u}_n) \to 0 \quad \text{in } \mathscr{D}'$$

and the limit $\mathbf{u}$ satisfies the entropy condition in the weak sense. This proof carries over to the plane wave solutions in elasticity.

Let us consider now a single shock $u(x-wt)$ satisfying the conservation law (3.2.16). Suppose that $u(x-wt)$ is the limit of a sequence of solutions $u_n(x-wt) = \bar{u}(\varepsilon_n^{-1}(x-wt))$, of the equation $u_{,t} + f(u)_{,x} = \varepsilon_n b(u) u_{,xx}$, $b(u) > 0$, $\varepsilon_n \to 0_+$. We shall show that the shock $u(x-wt)$ satisfies the conditions (L) and (E).

Indeed, we have

$$F(\bar{u}(z)) :\equiv -w[\bar{u}(z) - u_-] + [f(\bar{u}(z)) - f(u_-)] = b(\bar{u}(z))\bar{u}' \qquad (3.4.63)$$

in view of (3.4.61). Suppose that the function $u \mapsto F(u)$ changes sign at u_1 between u_- and u_+. In this case u_1 is a critical point of equation (3.4.63) and either all the solutions $\bar{u}(z)$ enter this point or all the solutions leave it. Hence $F(u)$ cannot change sign at a point between u_- and u_+.

Equation (3.4.63) implies that

$$b(\bar{u}) \frac{d}{dz} (\bar{u}(z) - u_-)^2 = 2(\bar{u}(z) - u_-) F(\bar{u}(z)).$$

Since $b > 0$ and F does not change sign between u_- and u_+, it is clear that $\underline{u}(z)$ moves away from u_- all the time as $z \to \infty$ and $\frac{d}{dz}(\bar{u}(z)-u_-)^2 \geqslant 0$. Hence $(\bar{u}(z)-u_-)F(\bar{u}(z)) \geqslant 0$ and similarly $(\bar{u}-u_+)F(\bar{u}) \leqslant 0$. Dividing these inequalities by $(\bar{u}-u_-)^2$ and $(\bar{u}-u_+)^2$ resp. we find that

$$\frac{f(\bar{u})-f(u_+)}{\bar{u}-u_+} \leqslant w = \frac{f(u_-)-f(u_+)}{u_- - u_+} \leqslant \frac{f(\bar{u})-f(u_-)}{\bar{u}-u_-}$$

for every $\bar{u}$ between u_- and u_+. This is the familiar condition (E) in our case. In the limit $\bar{u} \to u_+$ or $\bar{u} \to u_-$ the conditions (L) are obtained.

In the case of two conservation laws it is possible to prove directly that a shock which is a limit of smooth solutions with viscosity $\varepsilon b(\mathbf{u})\mathbf{E}$, $b > 0$, $\varepsilon \to 0_+$, cannot satisfy the inequality $\lambda_\alpha(\mathbf{u}_-) < w < \lambda_\alpha(\mathbf{u}_+)$ for any $\alpha \in \{1, 2\}$ (Rozhdestvenskii and Yanenko, 1978).

Let us now consider a discontinuity wave Σ satisfying equations (3.4.23) with a strictly convex function g^0. Let $(-w, \mathbf{n})$ be the normal at $x_0 \in \Sigma$ pointing into $\mathscr{U}_+$. We define the **generating function** (see Kulikovskii, 1962):

$$P = wg^0 - n_k g^k + c_r u^r, \tag{3.4.64}$$

with summation over $k = 1, \ldots, m$, $r = 1, \ldots, n$. We shall choose the constants c_r in such a way that the Rankine–Hugoniot equations for the discontinuity at x_0 assume the following form:

$$\frac{\partial P}{\partial u^r}(\mathbf{u}_-) = 0 = \frac{\partial P}{\partial u^r}(\mathbf{u}_+), \quad r = 1, \ldots, n. \tag{3.4.65}$$

Let

$$Q(w, \mathbf{u}_+, \mathbf{u}_-) :\equiv P(\mathbf{u}_+) - P(\mathbf{u}_-) \equiv w\Big\{g^0(\mathbf{u}_+) - g^0(\mathbf{u}_-) -$$

$$- \frac{\partial g^0}{\partial u^r}(\mathbf{u}_-)(u_+{}^r - u_-{}^r)\Big\} - n_k\Big\{g^k(\mathbf{u}_+) - g^k(\mathbf{u}_-) - \frac{\partial g^k}{\partial u^r}(\mathbf{u}_-)(u_+{}^r - u_-{}^r)\Big\}.$$

We shall fix $\mathbf{u}_- = \mathbf{u}_0$ and consider the function $s \mapsto q(s) :\equiv Q(w(s), \mathbf{d}(s), \mathbf{u}_0)$, $(w(s), \mathbf{d}(s)) \in \mathscr{H}_{\mathbf{u}_0}$. On account of the Rankine–Hugoniot equations

$$\dot{q}(s) = \dot{w}(s)\Big\{g^0(\mathbf{d}(s)) - g^0(\mathbf{u}_0) - \frac{\partial g^0}{\partial u^r}(\mathbf{u}_0)(u^r - u_0^r)\Big\}$$

and

$$\operatorname{sgn} \dot{q}(s) = \operatorname{sgn} \dot{w}(s) \in \{1, -1, 0\}. \tag{3.4.66}$$

Moreover, if $w^{(l)}(s_0) = 0$ for $l < k$, then $q^{(l)}(s_0) = 0$ for $l < k$ and

$$\operatorname{sgn} q^{(k)}(s_0) = \operatorname{sgn} w^{(k)}(s_0). \tag{3.4.67}$$

Suppose that $\mathbf{d}(0) = \mathbf{u}_0$, $w(0) = \lambda_\alpha(\mathbf{u}_0)$ and $\dot{w}(0) \neq 0$. Then $q(s) = O(s^3)$. If $w^{(l)}(0) = 0$ for $l = 1, \ldots, k$, $k \geqslant 1$, then

$$q(s) = O(s^{k+3}). \tag{3.4.68}$$

Let us now consider a solution $\bar{\mathbf{u}}(n_k x^k - wt)$ of the corresponding equation with vanishing viscosity:

$$\frac{\partial}{\partial x^\mu}\left(\frac{\partial g^\mu}{\partial u^r}\right) = \frac{\partial}{\partial x^k}\left(B_{rs}^{kl}(\mathbf{u})\frac{\partial u^s}{\partial x^l}\right). \tag{3.4.69}$$

Substituting $\mathbf{u} = \bar{\mathbf{u}}(z)$, $z = n_k x^k - wt$ and integrating we get

$$-\frac{\partial P}{\partial u^r}(\bar{\mathbf{u}}(z)) = B_{rs}^{kl}(\bar{\mathbf{u}}(z))n_k n_l \frac{d\bar{u}^s}{dz}. \tag{3.4.70}$$

Suppose that

$$B_{rs}^{kl}(\mathbf{u})n_k n_l a^r a^s > 0 \tag{3.4.71}$$

for all $\mathbf{n} \neq \mathbf{0}$, $\mathbf{a} \neq \mathbf{0}$, and all $\mathbf{u}$. Equation (3.4.70) implies that the function $z \mapsto P(\bar{\mathbf{u}}(z))$ is decreasing:

$$-\frac{dP(\bar{\mathbf{u}}(z))}{dz} = B_{rs}^{kl}(\bar{\mathbf{u}}(z))n_k n_l \frac{d\bar{u}^r}{dz}\frac{d\bar{u}^s}{dz} > 0. \tag{3.4.72}$$

Assuming that the limits $\lim\limits_{z \to = -\infty} \bar{\mathbf{u}}(z) = \mathbf{u}_-$, $\lim\limits_{z \to \infty} \bar{\mathbf{u}}(z) = \mathbf{u}_+$ exist and are finite, we conclude that

$$\infty > P(\mathbf{u}_-) - P(\mathbf{u}_+) = \int_{-\infty}^{\infty} dz B_{rs}^{kl}(\bar{\mathbf{u}}(z))n_k n_l \frac{d\bar{u}^r}{dz}\frac{d\bar{u}^s}{dz} > 0.$$

Hence

$$P(\mathbf{u}_-) > P(\mathbf{u}_+) \tag{3.4.73}$$

and

$$\frac{d\bar{\mathbf{u}}}{dz} \to 0 \quad \text{for } z \to \pm\infty.$$

On account of (3.4.66) the condition (3.4.73), derived from the small viscosity approximation, implies the condition (L) for weak discontinuity waves (up to the first Chapman–Jouguet point on each $\mathscr{H}_{\mathbf{u}_0}$).

We now establish a relationship between the topology of the level surfaces $P(\mathbf{u}) = \text{const}$ at the critical points of equation (3.4.70) and the (L) condition. From (3.4.64) we have

$$\left[\frac{\partial^2 P}{\partial u^r \partial u^s}\right] = w\left[\frac{\partial^2 g^0}{\partial u^r \partial u^s}\right] - n_k\left[\frac{\partial^2 g^k}{\partial u^r \partial u^s}\right] = w\mathbf{A} - \mathbf{B}. \tag{3.4.74}$$

The characteristic speeds $\lambda(\mathbf{u})$ of equation (3.4.23) are given by the equation $\det(\lambda\mathbf{A}-\mathbf{B}) \equiv \det[(\lambda-w)\mathbf{E}-(\mathbf{A}^{-1/2}\mathbf{B}\mathbf{A}^{-1/2}-w\mathbf{E})]\det\mathbf{A} = 0$. Hence the eigenvalues of $\mathbf{A}^{-1/2}\mathbf{B}\mathbf{A}^{-1/2}-w\mathbf{E}$ coincide with λ_k-w and the number of positive/negative eigenvalues of $w\mathbf{A}-\mathbf{B}$ is equal to the number of negative/positive numbers λ_k-w. At each critical point of P such that the quadratic form (3.4.74) of P is non-degenerate (i.e. $w \neq \lambda_k$ for all $k = 1, \ldots, n$) we can find a local coordinate system

$$(v^1, \ldots, v^n) = \varphi(u^1, \ldots, u^n)$$

such that

$$P(\mathbf{u}) = \sum_{r=1}^{p} (v^r)^2 - \sum_{r=p+1}^{n} (v^r)^2 =: Q(\mathbf{v}). \tag{3.4.75}$$

At a critical point of P the eigenvalues of $\left[\dfrac{\partial^2 P}{\partial u^r \partial u^s}\right]$ are identical with those of $\left[\dfrac{\partial^2 Q}{\partial u^r \partial u^s}\right]$. Hence (3.4.75) is equivalent to the inequalities $\lambda_1 < \lambda_2 < \ldots$ $\ldots < \lambda_p < w < \lambda_{p+1} \ldots < \lambda_n$.

Suppose now that two non-degenerate critical points $\mathbf{u}_-$, $\mathbf{u}_+$ of P can be joined by a shock satisfying (L). In this case $\mathbf{u}_-$ is of a type $\sum_{r=1}^{p-1} (v^r)^2 - \sum_{r=p}^{n} (v^r)^2$, while $\mathbf{u}_+$ is of the type $\sum_{r=1}^{p} (v^r)^2 - \sum_{r=p+1}^{n} (v^r)^2$.

We shall now be concerned with an inverse problem. Suppose that two critical points of P satisfy (3.4.73). Is it possible to join them by a solution $\bar{\mathbf{u}}(z)$ of equation (3.4.70) with some viscosity matrix?

For any fixed unit vector $\mathbf{n}$ the tensor $B_{rs}(\mathbf{u}) :\equiv B_{rs}^{kl}(\mathbf{u})n_k n_l$ defines a Riemannian structure on $\boldsymbol{R}^n$. Equation (3.4.70) says that the trajectories $\bar{\mathbf{u}}(z)$ are everywhere orthogonal to the level surfaces $P =$ const in the sense of this Riemannian structure. Let $\mathbf{u} = \bar{\mathbf{u}}(z)$, $z \in \boldsymbol{R}$, be any $\mathscr{C}^1$ curve joining two critical points $\mathbf{u}_-$, $\mathbf{u}_+$ of P, equation (3.4.65), such that $-\dfrac{dP}{du^r}\bar{u}^{r\prime}(z) > 0$. In particular the above inequality implies that $\bar{\mathbf{u}}(z)$ is everywhere transversal to the level surfaces of P. In a nbhd of every point $\bar{\mathbf{u}}(z)$, $-\infty < z < \infty$, we can construct a coordinate system $(P, v^2, \ldots, v^n)$. Let $\mathbf{e}_1 := \bar{\mathbf{u}}'(z)$, $\mathbf{e}_k := \left.\dfrac{\partial}{\partial v^k}\right|_{\mathbf{u}=\bar{\mathbf{u}}(z)}$ for $k = 2, \ldots, n$, and let $\mathbf{Q}$ be a matrix whose columns are $\mathbf{e}_1, \ldots, \mathbf{e}_n$. On account of the transversality $\mathbf{Q}$ is invertible. Although $\mathbf{Q}$ is defined on $\bar{\mathbf{u}}(z)$ only locally, it can be pieced together by means of a partition of unity to yield a smooth matrix function $\mathbf{Q}(z)$ defined for $-\infty < z < \infty$, except near $\mathbf{u}_-$ and $\mathbf{u}_+$.

Let $\mathbf{B} := {}^t\mathbf{Q}^{-1}\mathbf{Q}^{-1}$. Since $\mathbf{e}_k = \mathbf{Q}\mathbf{s}_k$, $s_k{}^l := \delta_k{}^l$, we have $\mathbf{e}_k \cdot \mathbf{B}\mathbf{e}_l = \mathbf{s}_k \cdot \mathbf{s}_l = \delta_{kl}$. In particular $\bar{\mathbf{u}}'(z)$ is $\mathbf{B}$-orthogonal to the level surfaces of P (except in an arbitrarily small nbhd of $\mathbf{u}_+$ and $\mathbf{u}_-$). By the Whitney extension theorem (see Narasimhan, 1971) $\mathbf{B}$ can be extended to a function of $\mathbf{u}$ defined on a nbhd of $\bar{\mathbf{u}}(z)$. In fact $\mathbf{B}$ can also be defined in a nbhd of $\mathbf{u}_-$, $\mathbf{u}_+$ (Godunov, 1962). The curve $\bar{\mathbf{u}}(z)$ is a solution of (3.4.70) with a positive definite viscosity matrix.

Godunov (1962) applied a similar argument to show that

(1) the condition (L) may in general admit two different shocks $(\mathbf{u}_1, \mathbf{u}_2)$ and $(\mathbf{u}_3, \mathbf{u}_2)$ for given values of w and c_r;

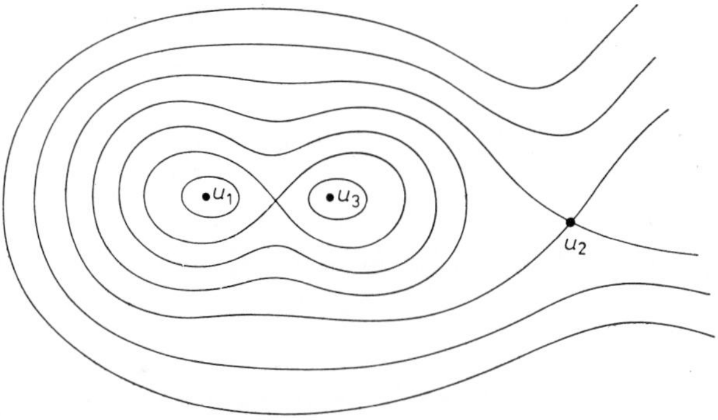

Fig. 3.8 A phase diagram for elastic waves with viscosity

(2) according to the choice of the viscosity matrix the first of these shocks can be implemented by a viscous profile and the other cannot, or vice versa. In the case considered by Godunov $n = 2$ and the equations describe a barotropic gas. The level curves of P look as shown in Fig. 3.8. The points $\mathbf{u}_1$, $\mathbf{u}_3$ correspond to local minima of P and $\mathbf{u}_2$ is a saddle, so that $w < \lambda_1^-, \lambda_2^-$ at $\mathbf{u}_1, \mathbf{u}_3$, $\lambda_1^+ < w < \lambda_2^+$ at $\mathbf{u}_2$.

In view of the simplicity of equations (3.4.70) they were widely applied in connection with dissipation in hydrodynamics and magnetohydrodynamics. The existence of a potential P greatly facilitates the analysis of the critical points of the ordinary differential equations modelling shock waves.

The same method will now be applied to the theory of plane waves in elasticity. Let $F^k{}_\alpha = f^k \nu_\alpha + F_0{}^k{}_\alpha$, with ν_α and $F_0{}^k{}_\alpha$ constant, $\tilde{w}(\mathbf{f}) := W(f^k \nu_\alpha + F_0{}^k{}_\alpha)$, $\varrho_0(\xi) = \text{const}$, $y = \nu_\alpha \xi^\alpha$. The basic equations of plane wave motion in elasticity, equation (3.3.42), may be now rewritten as

$$\left(\frac{\partial g^0}{\partial \mathbf{v}}\right)_t + \left(\frac{\partial g^1}{\partial \mathbf{v}}\right)_y = \mathbf{b}(\mathbf{f}),$$

$$\left(\frac{\partial g^0}{\partial \mathbf{s}}\right)_t + \left(\frac{\partial g^1}{\partial \mathbf{s}}\right)_y = \mathbf{0}. \tag{3.4.76}$$

with $\mathbf{s} = \tilde{w}_{\mathbf{f}}(\mathbf{f})$ equivalent to $\mathbf{f} = \tilde{\mathbf{f}}(\mathbf{s})$,

$$g^0(\mathbf{s}, \mathbf{v}) \equiv \psi^0(\mathbf{s}, \mathbf{v}) \equiv \frac{1}{2}\mathbf{v}^2 + \mathbf{s}\cdot\tilde{\mathbf{f}}(\mathbf{s}) - \tilde{w}(\tilde{\mathbf{f}}(\mathbf{s}))$$
$$g^1(\mathbf{s}, \mathbf{v}) \equiv -\mathbf{s}\cdot\mathbf{v}. \tag{3.4.77}$$

On account of (3.4.73) a state $\mathbf{u}_+ = (\mathbf{v}_+, \mathbf{s}_+)$ on the right can be connected by a shock to $\mathbf{u}_- = (\mathbf{v}_-, \mathbf{s}_-)$ on the left only if $wg^0(\mathbf{v}_-, \mathbf{s}_-) - g^1(\mathbf{v}_-, \mathbf{s}_-) > wg^0(\mathbf{v}_+, \mathbf{s}_+) - g^1(\mathbf{v}_+, \mathbf{s}_+)$ holds in addition to the Rankine–Hugoniot equations. For weak discontinuities the new condition is equivalent to (L). For strong discontinuities it follows from the assumption that each shock is an idealization of a continuous transition zone with high velocity gradient generating non-negligable viscosity.

Since $[[\mathbf{v}]] + w[[\mathbf{f}]] = \mathbf{0}$, the last inequality reduces to a simple inequality involving the Lagrangian:

$$w\left\{\frac{1}{2}\mathbf{v}_-^2 - \tilde{w}(\mathbf{f}_-)\right\} > w\left\{\frac{1}{2}\mathbf{v}_+^2 - \tilde{w}(\mathbf{f}_+)\right\}. \tag{3.4.78}$$

Remark

Although equation (3.4.71) is not valid in elasticity with the usual viscous term, we have

$$B_{kl}(\bar{\mathbf{f}})\frac{d\bar{v}^k}{dz}\frac{d\bar{v}^l}{dz} > 0 \quad \text{unless} \quad \frac{d\bar{v}^k}{dz} = 0, \quad k = 1, 2, 3.$$

Existence of finite limits of $\bar{\mathbf{f}}, \bar{\mathbf{v}}$ for $z \to \pm\infty$ implies that $\frac{d\bar{v}^k}{dz} \to 0$ for $z \to \pm\infty$.

Hence $\dfrac{d\bar{f}^k}{dz} \to 0$ as well.

3.4.7 Dissipation, Uniqueness and Stability of Weak Solutions

Let $\mathbf{u} \in \mathrm{BV} \cap \mathscr{L}^\infty$ be a weak solution of equations (3.1.35). Suppose that equations (3.1.35) satisfy condition (A) of Section 3.4.2 with $\varphi^\mu \in \mathscr{C}^1$, $\chi \in \mathscr{C}^0$ and ψ is given by equation (3.4.5). Clearly $\psi \in \mathscr{C}^0$. In view of equations (3.4.4) and (3.2.77)

$$\int_{\mathscr{P}} (\varphi^0{}_{,t} + \varphi^k{}_{,x^k}) = \int_{\mathscr{P}\setminus(\mathscr{P}\cap\Gamma)} \Big\{\widehat{d\varphi^0(\mathbf{u})}[\mathbf{u}_t] + \widehat{d\varphi^0(\mathbf{u})[\mathbf{A}^k(\mathbf{u})\mathbf{u}_{x^k}]}$$
$$+ \frac{\partial\varphi^k}{\partial x^k}\Big\} + \int_{\mathscr{P}\cap\Gamma} (\varphi^0{}_{,t} + \varphi^k{}_{,x^k}) = \int_{\mathscr{P}\setminus(\mathscr{P}\cap\Gamma)} d\varphi^0(\mathbf{u})[\mathbf{u}_t + \mathbf{A}^k(\mathbf{u})\mathbf{u}_{x^k}]$$

$$+\frac{\partial \varphi^k}{\partial x^k}\Bigg\} - \int_{\mathscr{P}\cap\Gamma} dH^m \{\Delta_\nu \varphi^0 \nu_0 + \Delta_\nu \varphi^k \nu_k\} = \int_{\mathscr{P}\setminus(\mathscr{P}\cap\Gamma)} dL^{m+1}\, \psi(x, \mathbf{u})$$

$$- \int_{\Gamma\cap\mathscr{P}} dH^m \{\Delta_\nu \varphi^0 \nu_0 + \Delta_\nu \varphi^k \nu_k\}, \tag{3.4.79}$$

where $\mathscr{P}$ is an arbitrary bounded Borel subset of $\boldsymbol{R}_+^{m+1}$ and Γ is the jump set of $\mathbf{u}$, $\Delta_\nu \varphi^\mu := l_{-\nu}\varphi^\mu - l_\nu \varphi^\mu$. Hence (3.4.5) implies that

$$\Delta_\nu \varphi^0 \nu_0 + \Delta_\nu \varphi^k \nu_k \geqslant 0 \tag{3.4.80}$$

H^m-a.e. on Γ. Equation (3.4.80) is a generalization of (3.4.37).

Definition 3.4.7

The **dissipation** of a solution $\mathbf{u}$ on a bounded Borel subset $\mathscr{P}$ of $\boldsymbol{R}_+^{m+1}$ is given by the following expression:

$$D(\mathbf{u})(\mathscr{P}) := \int_{\mathscr{P}} (\chi dL^{m+1} - \partial_\mu \varphi^\mu).$$

Theorem 3.4.8

For every bounded Borel set

$$D(\mathbf{u})(\mathscr{P}) = \int_{\mathscr{P}\setminus(\mathscr{P}\cap\Gamma)} dL^{m+1}\,(\chi - \psi) + \int_{\mathscr{P}\cap\Gamma} dH^m \{\Delta_\nu \varphi^0 \nu_0 + \Delta_\nu \varphi^k \nu_k\}. \tag{3.4.81}$$

In the following we assume for simplicity that $\chi = \psi$. It follows then from (3.4.81) that the measure $D(\mathbf{u})$ is singular with respect to the Lebesgue measure L^{m+1} and is concentrated on the jump set Γ of $\mathbf{u}$. We also assume throughout this section that φ^0, φ^k do not depend explicitly on t, x^k, $\chi = \psi = 0$.

Let

$$\alpha(\mathbf{u}, \mathbf{v}) :\equiv \varphi^0(\mathbf{u}) - \varphi^0(\mathbf{v}) - d\varphi^0(\mathbf{v})[\mathbf{u} - \mathbf{v}],$$

$$\beta^k(\mathbf{u}, \mathbf{v}) :\equiv \varphi^k(\mathbf{u}) - \varphi^k(\mathbf{v}) - d\varphi^0(\mathbf{v})[\mathbf{f}^k(\mathbf{u}) - \mathbf{f}^k(\mathbf{v})] \tag{3.4.82}$$

for $\mathbf{u}, \mathbf{v} \in \mathscr{H}$, $k = 1, \ldots, m$. For every fixed $\mathbf{v} \in \mathscr{H}$ the conservation law for $\mathbf{u}(t, x)$,

$$\alpha(\mathbf{u}, \mathbf{v})_{,t} + \beta^k(\mathbf{u}, \mathbf{v})_{,x^k} = 0, \tag{3.4.83}$$

provides a convex extension of equations (3.1.35). Indeed, the left-hand side of equation (3.4.83) equals $\varphi^0_{,t} + \varphi^k_{,x^k} - d\varphi^0(\mathbf{v})[\mathbf{u}_{,t} + \mathbf{f}^k(\mathbf{u})_{,x^k}]$, $\mathbf{f}^0 = \mathrm{id}$. Hence every $\mathscr{C}^1$ solution of equations (3.1.35) satisfies equation (3.4.83). Since

$\mathbf{u} \mapsto \alpha(\mathbf{u}, \mathbf{v}) - \varphi^0(\mathbf{u})$ is a linear function, the left-hand side of equation (3.4.83) satisfies (C). Moreover, (C) implies that

$$\alpha(\mathbf{u}, \mathbf{v}) > 0 \quad \text{unless } \mathbf{u} = \mathbf{v}. \tag{3.4.84}$$

Theorem 3.4.9

Suppose that $\hat{\mathbf{u}} \in \mathrm{BV} \cap \mathscr{L}^\infty$ is a solution of equations (3.1.35) defined in $[0, T] \times \boldsymbol{R}^m =: \mathscr{S}(T)$, satisfying the condition (EN) and $\hat{\mathbf{u}}(0, x) = \mathbf{v}$ for almost all $x \in \mathscr{K}(a, r)$, $a \in \boldsymbol{R}^m$, $r > 0$, in the sense of traces. Then there is a positive number c such that $\bar{\mathbf{u}}(t, x) = \mathbf{v}$ for $t \in [0, T[$ and for almost all x satisfying the inequality $|x-a| \leqslant r - ct$. The function $\bar{\mathbf{u}}$ denotes a symmetric average of $\hat{\mathbf{u}}$. The number c depends only on $||\hat{\mathbf{u}}||_\infty$.

Proof

Let $\varDelta := \{(t, x) \in \boldsymbol{R}_+^{m+1} |\ t \in]0, t_0], |x-a| \leqslant r - ct\}$, $t_0 < T$, $t_0 < rc^{-1}$. Since $0 \geqslant \varphi^0_{,t} + \varphi^k_{,x^k} = \tilde{\alpha}_{,t} + \tilde{\beta}^k_{,x^k}$,

$$\tilde{\alpha}(t, x) := \alpha(\hat{\mathbf{u}}(t, x), \mathbf{v}), \quad \tilde{\beta}^k(t, x) := \beta^k(\hat{\mathbf{u}}(t, x), \mathbf{v}),$$

the Gauss–Green theorem yields

$$\int_{\mathscr{K}(a, r-ct_0)} dL^m\, \tilde{\alpha}(t_0, x) - \int_{\mathscr{K}(a,r)} dL^m\, \tilde{\alpha}(0, x)$$
$$+ \int_0^{t_0} dt \int_{\partial\mathscr{K}(a, r-ct)} dH^{m-1} \left\{ c\tilde{\alpha}(t, x) + \frac{x^k - a^k}{|x-a|} \tilde{\beta}^k(t, x) \right\} \leqslant 0,$$

where $\tilde{\alpha}$, $\tilde{\beta}^k$ have been calculated by substituting the appropriate traces of $\hat{\mathbf{u}}$ into α, β^k. Now $\tilde{\alpha}(t, x) > 0$ whenever $\mathbf{u}(t, x) \neq \mathbf{v}$ and $\tilde{\beta}^k(t, x) = 0$ in the opposite case. Since α vanishes exactly to the second order and β^k vanishes at least to the second order when $\mathbf{u} \to \mathbf{v}$, and $||\hat{\mathbf{u}}||_\infty < \infty$, we may choose c so large that $|\tilde{\beta}^k \tilde{\alpha}^{-1}| \leqslant c$ for $k = 1, \ldots, m$. With this choice the last integral is non-negative and the second one vanishes. On account of (3.4.84) the outer trace of $\hat{\mathbf{u}}$ on $t = t_0$ vanishes for almost all $x \in \mathscr{K}(a, r - ct_0)$. Repeating the same argument for $\varDelta \cap (]0, t_0[\times \boldsymbol{R}^m)$ instead of $\varDelta$ we conclude that the inner trace of $\hat{\mathbf{u}}$ on $t = t_0$ vanishes for almost all $x \in \mathscr{K}(a, r - ct_0)$. This implies the thesis. □

Henceforth we restrict our attention to the case $m = 1$.

Let $\mathbf{v} = \text{const} \in \mathscr{H}$ and let $\hat{\mathbf{u}} \in \mathrm{BV} \cap \mathscr{L}^\infty$ be a weak solution of equations (3.1.35) in $\mathscr{S}(T)$ satisfying (EN). Let $\Pi_{abt_0} :=]0, t_0] \times [a, b]$, $t_0 < T$. We apply equation (3.4.79) to the convex extension $\alpha(\mathbf{u},\mathbf{v})_{,t}+\beta(\mathbf{u},\mathbf{v})_{,x}=0$ and $\mathscr{P}=\Pi_{abt_0}$:

$$\int_a^b dx\,\{\tilde{\alpha}(t_0,x)-\tilde{\alpha}(0,x)\}+\int_0^{t_0} dt\,\{\tilde{\beta}(t,b)-\tilde{\beta}(t,a)\}$$

$$=-\int_{\Gamma\cap\Pi_{abt_0}} dH^1\,\{\Delta_{\mathbf{v}}\tilde{\alpha}\nu_t+\Delta_{\mathbf{v}}\tilde{\beta}\nu_x\}\leqslant 0. \tag{3.4.85}$$

Assume that $\mathbf{u}(0,x)=\mathbf{v}=$ const outside $[a_0,b_0]$, $-\infty<a_0<b_0<\infty$, in the sense of traces. By Theorem 3.4.9 the second integral vanishes for a, b sufficiently large, $a<0$, and the integral $\int_{-\infty}^{\infty} dx\tilde{\alpha}(t_0,x)$ exists. Hence

$$\int_{-\infty}^{\infty} dx\,\tilde{\alpha}(t_0,x)\leqslant\int_{-\infty}^{\infty} dx\,\tilde{\alpha}(0,x), \tag{3.4.86}$$

$$\int_{\Gamma\cap\mathscr{S}(t_0)} dH^1\,\{\Delta_\nu\tilde{\alpha}\nu_t+\Delta_\nu\tilde{\beta}\nu_x\}\leqslant\int_{-\infty}^{\infty} dx\,\tilde{\alpha}(0,x), \tag{3.4.87}$$

where $\tilde{\alpha}(t_0,x)$ has been obtained by substituting the outer trace of $\mathbf{u}$ on $t=t_0$. The same inequalities will be obtained for the inner trace of $\mathbf{u}$ on $t=t_0$ if Π_{abt_0} is replaced by its interior.

Similar inequalities can be established for $m>1$.

Let us choose the normal $\boldsymbol{\nu}$ on Γ in such a way that $\nu_x\geqslant 0$ H^1-a.e. on Γ. Since $\dfrac{dt}{ds}=\nu_x$, $\dfrac{dx}{ds}=-\nu_t=w\dfrac{dt}{ds}$ H^1-a.e. on Γ, equations (3.4.86) and (3.4.78) can be rewritten in the following form:

$$\left|\int_{-\infty}^{\infty} dx\,\{\tilde{\varphi}^0(t_0,x)-\varphi^0(\mathbf{v})\}\right|\leqslant K, \tag{3.4.88}$$

$\tilde{\varphi}^0(t,x):=\varphi^0((\mathbf{u}(t,x))$, and

$$\sum_{i=1}^{N}\int_0^{t_0} dt\,\{-w_i[[\tilde{\alpha}]]_i+[[\tilde{\beta}]]_i\}\leqslant K,\quad N\leqslant\infty. \tag{3.4.89}$$

In equation (3.4.89) we have tacitly used the fact that the number of discontinuity lines intersecting any line $t=$ const is at most countable. This follows from the fact that the function $\mathbf{u}$ can be adjusted on a set of L^2 measure zero in such a way that for almost all t the function $x\mapsto\mathbf{u}(t,x)$ is of bounded variation. The sum extends over all the discontinuities of $\mathbf{u}(t,\cdot)$. Note that the integrand in (3.4.89) is non-negative. Equations (3.4.44), (3.4.48b) and (3.4.85) imply that a solution **u** with small oscillation satisfies an estimate

$$\sum_{i=1}^{N} \int_{0}^{t_0} dt\, |[\mathbf{u}]_i(t)|^3 \leqslant K. \tag{3.4.90}$$

The estimates (3.4.86)–(3.4.90) provide a basis for studying the decay of the solution for large t.

Following DiPerna (1979) we study the uniqueness and stability of a Lipschitz solution $\mathbf{v} \in \mathrm{Lip} \cap \mathscr{L}^\infty$ of equations (3.1.35) in the class of all the solutions $\mathbf{u} \in \mathrm{BV} \cap \mathscr{L}^\infty$. We assume that $m = 1$. Let $\gamma(\mathscr{E}) := \int_{\mathscr{E}} \tilde{\alpha}_{,t} + \tilde{\beta}_{,x}$, $\theta(\mathscr{E}) := \int_{\mathscr{E}} \tilde{\varphi}^0_{,t} + \tilde{\varphi}^1_{,x}$ for every bounded Borel set $\mathscr{E}$, with $\tilde{\alpha}(t, x) :\equiv \alpha(\mathbf{u}(t, x), \mathbf{v}(t, x))$, $\tilde{\beta}(t, x) :\equiv \beta\mathbf{u}((t, x), \mathbf{v}(t, x))$, $\tilde{\varphi}^\mu(t, x) := \varphi^\mu(\mathbf{u}(t, x), \mathbf{v}(t, x))$, $\mu = 0, 1$. We denote by $\Gamma(\mathbf{u})$, $\Gamma(\mathbf{v})$ the jump sets of **u**, **v**, resp.

Theorem 3.4.10 (DiPerna, 1979)
Suppose that $\mathbf{v} \in \mathrm{Lip}(\mathscr{S}(T)) \cap \mathscr{L}^\infty(\mathscr{S}(T))$, $\mathbf{u} \in \mathrm{BV}(\mathscr{S}(T)) \cap \mathscr{L}^\infty(\mathscr{S}(T))$ are weak solutions of (3.1.35) and **u** satisfies (EN). Then there is a constant $c > 0$, depending on $||\mathbf{u}||_\infty$, $||\mathbf{v}||_\infty$, and a constant $K > 0$, depending on T, $||\mathbf{u}||_\infty$, $||\mathbf{v}||_\infty$ and $\mathrm{Lip}\, \mathbf{v}(0, \cdot)$, such that for every a, b, $-\infty < a < b < \infty$,

$$\int_a^b dx\, |\mathbf{u}_\pm(t, x) - \mathbf{v}(t, x)|^2 \leqslant K \int_{a-ct}^{b-ct} dx\, |\mathbf{u}(0, x) - \mathbf{v}(0, x)|^2,$$

where $\mathbf{u}_+(t, \cdot)$, $\mathbf{u}_-(t, \cdot)$ are the outer and inner traces of **u** on $t =$ const, and $\mathbf{u}(0, x)$ is the inner trace of **u** on $t = 0$.

Proof
Let $\mathscr{E}$ be a Borel subset of $\Gamma(\mathbf{u}) \cap \Gamma(\mathbf{v})^c$, $\Gamma(\mathbf{v})^c := \mathscr{S}(T) \setminus \Gamma(\mathbf{v})$. On account of equations (3.4.79)

$$\gamma(\mathscr{E}) = -\int_{\mathscr{E}} \{dH^1\, \Delta_\nu \tilde{\alpha} \nu_t + \Delta_\nu \tilde{\beta} \nu_x\} = \theta(\mathscr{E}).$$

Let $\mathscr{E}$ be a Borel subset of $\Gamma(\mathbf{u})^c \cap \Gamma(\mathbf{v})^c$. By Theorem 3.2.53 $\gamma = \widehat{\alpha_\mathbf{u}}[\mathbf{u}_t] + \widehat{\alpha_\mathbf{v}}[\mathbf{v}_t] + \widehat{\beta_\mathbf{u}}[\mathbf{u}_x] + \widehat{\beta_\mathbf{v}}[\mathbf{v}_x]$, where the subscripts **u**, **v** indicate the corresponding partial derivatives. On $\Gamma(\mathbf{u})^c \cap \Gamma(\mathbf{v})^c$ the hats can be dropped and

$$\gamma = (\beta_\mathbf{v} - \alpha_\mathbf{v} \mathbf{A}(\mathbf{v}))[\mathbf{v}_x], \tag{3.4.91}$$

since $\beta_{\mathbf{u}}-\alpha_{\mathbf{u}}\mathbf{A}(\mathbf{u}) = \mathbf{0}$ (equation (3.4.4)). Now

$$\beta_{\mathbf{v}}[\mathbf{w}] = -d\varphi^1(\mathbf{v})[\mathbf{w}]+d\varphi^0(\mathbf{v})[\mathbf{A}(\mathbf{v})\mathbf{w}]+d^2\varphi^0(\mathbf{v})[\mathbf{w}, \mathbf{f}(\mathbf{v})-\mathbf{f}(\mathbf{u})],$$
$$\alpha_{\mathbf{v}}[\mathbf{w}] = d^2\varphi^0(\mathbf{v})[\mathbf{w}, \mathbf{u}-\mathbf{v}].$$

Here $d^2\varphi(\mathbf{v})[\mathbf{a}, \mathbf{b}] := \dfrac{\partial^2\varphi}{\partial v^i \partial v^j}\, a^i b^j$. In view of equation (3.4.4) and the identity $d^2\varphi^0[\mathbf{A}\mathbf{w}, \mathbf{y}] = d^2\varphi^0[\mathbf{A}\mathbf{y}, \mathbf{w}]$ (Theorem 3.4.2)

$$\gamma = -d^2\varphi^0(\mathbf{v})[Q_{\mathbf{v}}\mathbf{f}(\mathbf{u}), \mathbf{v}_x], \tag{3.4.92}$$

where

$$Q_{\mathbf{v}}\mathbf{f}(\mathbf{u}) := \mathbf{f}(\mathbf{u})-\mathbf{f}(\mathbf{v})-\mathbf{A}(\mathbf{v})(\mathbf{u}-\mathbf{v}) = 0(|\mathbf{u}-\mathbf{v}|^2). \tag{3.4.93}$$

Hence for an arbitrary Borel subset $\mathscr{E} \subset \Gamma(\mathbf{u})\cap\Gamma(\mathbf{v})^c$

$$\gamma(\mathscr{E}) = \theta(\mathscr{E}) - \int_{\mathscr{E}} dL^2\, d^2\varphi^0(\mathbf{v})[Q_{\mathbf{v}}\mathbf{f}(\mathbf{u}), \mathbf{v}_x]$$
$$\leqslant -\int_{\mathscr{E}} dL^2\, d^2\varphi^0(\mathbf{v})[Q_{\mathbf{v}}\mathbf{f}(\mathbf{u}), \mathbf{v}_x], \tag{3.4.94}$$

on account of the fact that θ is concentrated on $\Gamma(\mathbf{u})$ and $d^2\varphi^0(\mathbf{v})[Q_{\mathbf{v}}\mathbf{f}(\mathbf{u}), \mathbf{v}_x]$ is absolutely continuous with respect to $L^2(\mathbf{v}_x \in \mathscr{L}^1_{loc})$.

Let $\mathscr{E} := \{(t, x)|\ t \in [0, t_0],\ x \in [a-ct, b+ct]\}$, and let c be the constant defined in the proof of Theorem 3.4.9.

$$\gamma(\mathscr{E}) = \int_0^{t_0} dt\ \{c\tilde{\alpha}(t, a-ct)-\tilde{\beta}(t, a-ct)-c\tilde{\alpha}(t, b+ct)-\tilde{\beta}(t, b+ct)\}+$$
$$+\int_a^b dx\, \tilde{\alpha}_-(t_0, x) - \int_{a-ct_0}^{b+ct_0} dx\, \tilde{\alpha}(0, x) \geqslant \int_a^b dx\, \tilde{\alpha}_-(t_0, x) - \int_{a-ct_0}^{b+ct_0} dx\, \tilde{\alpha}(0, x), \tag{3.4.95}$$

where $\tilde{\alpha}_-(t_0, x) = \alpha(\mathbf{u}_-(t_0, x), \mathbf{v}(t_0, x))$ and $\mathbf{u}_-$ is the outer trace of $\mathbf{u}$ on $t = t_0$. The same inequality holds for the inner trace $\mathbf{u}_+$ of $\mathbf{u}$ on $t = t_0$. Since $\mathbf{u}, \mathbf{v}, \mathbf{v}_x \in \mathscr{L}^\infty$ and $\Gamma(\mathbf{v}) = \emptyset$, equation (3.4.93) implies that

$$\gamma(\mathscr{E}) \leqslant K' \int_{\mathscr{E}} dL^2\, |\mathbf{u}-\mathbf{v}|^2 \leqslant K \int_{\mathscr{E}} dL^2\, \alpha(\mathbf{u}, \mathbf{v}) \tag{3.4.96}$$

for some $K, K' > 0$ depending on $||\mathbf{u}||_\infty, ||\mathbf{v}||_\infty$, Lip$\mathbf{v}$. Combining equations (3.4.95) and (3.4.96) and applying the Fubini theorem to the right-hand side of (3.4.96) we find that

$$\int_a^b dx\, \tilde{\alpha}(t_0, x) \leqslant \int_{a-ct_0}^{b+ct_0} dx\, \tilde{\alpha}(0, x) + K\int_0^{t_0} dt \int_{a-ct}^{b+ct} dx\, \tilde{\alpha}(t, x). \tag{3.4.97}$$

By Gronwall's lemma

$$\int_a^b dx\, \tilde{\alpha}(t_0, x) \leqslant e^{Kt_0} \int_{a-ct_0}^{b+ct_0} dx\, \tilde{\alpha}(0, x), \tag{3.4.98}$$

with the right-hand side defined in terms of either trace of $\mathbf{u}$ on $t = t_0$. Hence follows the thesis. □

The above theorems can be easily generalized to $m > 1$, with the interval $[a, b]$ replaced by a compact subset $\mathscr{K}$ of $\boldsymbol{R}^m$ with bounded perimeter. The set $\mathscr{K}_t$, corresponding to $[a-ct, b+ct]$ is then obtained by shifting each point $x \in \partial\mathscr{K}$ to $(ct, x+\mathbf{v}(x)ct)$.

Theorem 3.4.10 remains true in the case of an inhomogeneous system of equations

$$\begin{aligned} \mathbf{u}_{,t} + \mathbf{f}^k(\mathbf{u})_{,x^k} &= \mathbf{b}(\mathbf{u}), \\ \mathbf{v}_{,t} + \mathbf{f}^k(\mathbf{v})_{,x^k} &= \mathbf{b}(\mathbf{v}). \end{aligned} \tag{3.4.99}$$

Indeed, in this case

$$\begin{aligned} \gamma &= (\beta^k{}_{\mathbf{v}} - \alpha_{\mathbf{v}} \mathbf{A}^k(\mathbf{v}))[\mathbf{v}_{,x^k}] + \alpha_{\mathbf{u}}[\mathbf{b}(\mathbf{u})] + \alpha_{\mathbf{v}}[\mathbf{b}(\mathbf{v})] \\ &= -d^2\varphi^0(\mathbf{v})[Q_{\mathbf{v}}\mathbf{f}^k(\mathbf{u}), \mathbf{v}_{,x^k}] + (d\varphi^0(\mathbf{u}) - d\varphi^0(\mathbf{v}))[\mathbf{b}(\mathbf{u}) - \mathbf{b}(\mathbf{v})] \\ &\quad + (d\varphi^0(\mathbf{u}) - d\varphi^0(\mathbf{v}))[\mathbf{b}(\mathbf{v})] - d^2\varphi^0(\mathbf{v})[\mathbf{b}(\mathbf{v}), \mathbf{u}-\mathbf{v}] = O(|\mathbf{u}-\mathbf{v}|^2) \end{aligned}$$

on $\Gamma(\mathbf{u})^c \cap \Gamma(\mathbf{v})^c$ and the rest of the proof remains unchanged. Note that the condition (EN) reads now $\tilde{\varphi}^0_{,t} + \tilde{\varphi}^k_{,x^k} \leqslant \psi \equiv d\varphi^0[\mathbf{b}]$.

3.4.8 Uniqueness in the Class BV for the Equations of Elastodynamics

With some additional difficulties the uniqueness Theorem 3.4.10 can be extended to the case of a bounded hyperelastic body. On account of the energy conservation (3.4.54) we shall impose the following **energy dissipation inequality** (EN)

$$\varphi^0{}_{,t} + \partial_\mu \varphi^\mu \leqslant \chi \tag{3.4.100}$$

(in the weak sense), with

$$\chi := \varrho_0 b_k v^k, \tag{3.4.101a}$$

$$\varphi^\mu := -\varrho_0 W_{F^k{}_\mu} v^k \equiv -S_k{}^\mu v^k, \quad \mu = 1, 2, 3, \tag{3.4.101b}$$

$$\varphi^0 := \varrho_0 \left[\frac{1}{2}\mathbf{v}^2 + W(\mathbf{F}, \xi)\right], \quad W \in \mathscr{C}^2 \tag{3.4.101c}$$

and

$$\partial_\mu f(\mathbf{F}, \mathbf{v}, \xi) := \frac{\partial f}{\partial \xi^\mu} + \widehat{\frac{\partial f}{\partial F^k{}_\nu}} F^k{}_{\nu,\mu} + \widehat{\frac{\partial f}{\partial v^k}} v^k{}_{,\mu}.$$

For the physical meaning of (3.4.100), see Section 3.4.5. As we know from Section 3.4.7, the difference between the right-hand side and the left-hand side of (3.4.100) is a measure supported by the jump set Γ of a weak solution $(\mathbf{F}, \mathbf{v})$ of (3.3.2) (note that $\psi = \chi$ in this case). In particular (3.4.98) is satisfied with an equality sign if $(\mathbf{F}, \mathbf{v}) \in \mathrm{Lip}$.

The method we shall apply in this section will again be based on the key identity (3.4.4). We check this identity directly. Let us rewrite equations (3.3.2) in the form

$$F^k{}_{\mu,t} - (v^k \delta^\nu{}_\mu)_{,\nu} = 0, \tag{3.4.102a}$$

$$v^k{}_{,t} - (\varrho_0{}^{-1} S_k{}^\mu)_{,\mu} = b_k - (\varrho_0{}^{-1})_{,\mu} S_k{}^\mu, \tag{3.4.102b}$$

resembling equations (3.1.35) with $\mathbf{f}^0 = \mathrm{id}$. As usual, $\mathbf{A}^\mu = D\mathbf{f}^\mu$ and D denotes the differentiation with respect to $(\mathbf{F}, \mathbf{v})$ at fixed ξ. For an arbitrary tangent vector $\mathbf{w} = (\mathbf{H}, \mathbf{u})$ at $(\mathbf{F}, \mathbf{v})$ we have

$$d\varphi^0[\mathbf{A}^\mu \mathbf{w}] = -\varrho_0 v^k W_{F^k{}_\mu F^l{}_\nu}(\mathbf{F}, \xi) H^l{}_\nu - S_k{}^\mu u^k,$$

$$d\varphi^\mu[\mathbf{w}] = -S_k{}^\mu u^k - \varrho_0 v^k W_{F^k{}_\mu F^l{}_\nu}(\mathbf{F}, \xi) H^l{}_\nu$$

on account of (1.2.87). Hence follows the identity (3.4.4).

From (3.4.8) and (3.4.101) we can now calculate α, β^μ:

$$\tilde{\alpha}(t, \xi) = \alpha(\mathbf{v}, \mathbf{F}; \tilde{\mathbf{v}}, \tilde{\mathbf{F}}; \xi) \equiv \tfrac{1}{2}\varrho_0(\mathbf{v} - \tilde{\mathbf{v}})^2 + \varrho_0 W(\mathbf{F}, \xi) - \varrho_0 W(\tilde{\mathbf{F}}, \xi) - \tilde{S}_k{}^\mu (F^k{}_\mu - \tilde{F}^k{}_\mu), \tag{3.4.103a}$$

$$\tilde{\beta}^\mu(t, \xi) = \beta^\mu(\mathbf{v}, \mathbf{F}; \tilde{\mathbf{v}}, \tilde{\mathbf{F}}; \xi) = -(S_k{}^\mu - \tilde{S}_k{}^\mu)(v^k - \tilde{v}^k), \tag{3.4.103b}$$

with $\tilde{S}_k{}^\mu = \varrho_0 W_{F^k{}_\mu}(\tilde{\mathbf{F}}, \xi)$, $S_k{}^\mu = \varrho_0 W_{F^k{}_\mu}(\mathbf{F}, \xi)$. The functions $\tilde{\alpha}(t, \xi)$, $\tilde{\beta}^\mu(t, \xi)$ are thus obtained by substituting the functions $\mathbf{v}(t, \xi)$, $\tilde{\mathbf{v}}(t, \xi)$, $\mathbf{F}(t, \xi)$, $\tilde{\mathbf{F}}(t, \xi)$ into α, β^μ.

For $(\tilde{\mathbf{v}}, \tilde{\mathbf{F}}) \in \mathrm{Lip}$ and $(\mathbf{v}, \mathbf{F}) \in \mathrm{BV} \cap \mathscr{L}^\infty$, both satisfying equations (3.4.102) with the body forces $b_k, \tilde{b}_k \in \mathscr{L}^1_{loc}$, resp., we obtain after some algebra the following formula:

$$\begin{aligned}\tilde{\alpha}_{,t} + \tilde{\beta}^\mu{}_{,\mu} = {} & \varrho_0(v^k - \tilde{v}^k)(b_k - \tilde{b}_k) \\ & + [S_k{}^\mu - \tilde{S}_k{}^\mu - \varrho_0 W_{F^k{}_\mu F^l{}_\nu}(\tilde{\mathbf{F}}, \xi)(F^l{}_\nu - \tilde{F}^l{}_\nu)]\tilde{v}^k{}_{,\mu} \\ & + [\varrho_0 \widehat{W_{F^k{}_\mu}(\mathbf{F}, \xi)} - \overline{S_k{}^\mu}] v^k{}_{,\mu}.\end{aligned} \tag{3.4.104}$$

The bar over v^k, $S_k{}^\mu$ has the same meaning as in equation (3.2.67). We have made use of the fact that $\tilde{v}^k{}_{,\mu}$, b_k cannot be concentrated on a set of Lebesgue measure zero. Note that the last term on the right-hand side of equation (3.4.104) neednot vanish on the jump set Γ of $(\mathbf{v}, \mathbf{F})$. Now, using equation (3.4.102) again, we have for $(\mathbf{v}, \mathbf{F}) \in \mathrm{BV} \cap \mathscr{L}^\infty$:

$$\varphi^0{}_{,t} + \partial_\mu \varphi^\mu = \varrho_0 b_k v^k + [\varrho_0 \widehat{W_{F^k{}_\mu}(\mathbf{F}, \xi)} - \overline{S_k{}^\mu}] v^k{}_{,\mu}.$$

Hence (3.4.100) implies that the measure

$$\widehat{[\varrho_0 W_{F^k\mu}(\mathbf{F}, \xi) - \overline{S_k^{\mu}}]v^k{}_{,\mu}} \leqslant 0, \tag{3.4.105}$$

and

$$\begin{aligned}\tilde{\alpha}_{,t} + \tilde{\beta}^{\mu}{}_{,\mu} \leqslant\; & \varrho_0(v^k - \tilde{v}^k)(b_k - \tilde{b}_k) \\ & + [S_k^{\mu} - \tilde{S}_k^{\mu} - \varrho_0 W_{F^k{}_\mu F^l{}_\nu}(\tilde{\mathbf{F}}, \xi)(F^l{}_\nu - \tilde{F}^l{}_\nu)]\tilde{v}^k{}_{,\mu}\end{aligned} \tag{3.4.106}$$

Integrating the left-hand side of (3.4.106) over $]0, t[\times \mathscr{B}_*$ and applying the Fubini theorem we get the expression

$$\int_0^t ds \int_{\partial^*\mathscr{B}} dH^2\, \tilde{\beta}^{\mu}\nu_{\mu} + \int_{\mathscr{B}} dL^3\, [\tilde{\alpha}(t, \xi) - \tilde{\alpha}(0, \xi)], \tag{3.4.107}$$

in which the integrands should be understood in the sense of inner traces. We assume their summability with respect to H^3. ν_μ is the outer normal on $[0, T]\times \partial^{**}\mathscr{B}$.

Let $t_k = S_k^{\mu}\nu_\mu$, $\tilde{t}_k = \tilde{S}_k^{\mu}\nu_\mu$ on $[0, T]\times \partial^*\mathscr{B}$, $T > 0$, in the sense of inner traces. The functions t_k, $\tilde{t}_k$ as well as the inner traces of v^k, $\tilde{v}^k$ are defined H^3-a.e. on $[0, T]\times \partial^*\mathscr{B}$ and pointwise bounded by the $\mathscr{L}^\infty$ norms of S_k^{μ}, $\tilde{S}_k^{\mu}$, v^k, $\tilde{v}^k$. According to Theorem 3.2.10 the inner trace $\tilde{\beta}^{\mu}\nu_\mu$ can be expressed in terms of the products of these traces. It is also pointwise bounded and summable on $[0, T]\times \partial^*\mathscr{B}$.

We shall assume that

$$(t_k - \tilde{t}_k)(v^k - \tilde{v}^k) \leqslant 0 \quad H^3\text{-a.e.} \quad \text{on } [0, T]\times \partial^*\mathscr{B}, \tag{3.4.108}$$

which means that the external surface loads are dissipative. In particular (3.4.108) is satisfied if $\partial^*\mathscr{B} = \Sigma_1 \cup \Sigma_2$, $H^2(\Sigma_1 \cap \Sigma_2) = 0$,

$$t_k = \tilde{t}_k \quad \text{on } [0, T]\times \Sigma_1, \tag{3.4.109a}$$

$$v^k = \tilde{v}^k \quad \text{on } [0, T]\times \Sigma_2. \tag{3.4.109b}$$

Equations (3.4.103b), (3.4.106–17) imply the following lemma.

Lemma 3.4.11

Suppose that $P(\mathscr{B}) < \infty$, $\mathscr{B}$ is bounded, $T > 0$ and $(\mathbf{F}, \mathbf{v}) \in \mathrm{BV}\cap\mathscr{L}^\infty$, $(\tilde{\mathbf{F}}, \tilde{\mathbf{v}}) \in \mathrm{Lip}$ both satisfy equations (3.3.2) on $[0, T]\times \mathscr{B}$ with the body forces b_k, $\tilde{b}_k$ resp., $b_k, \tilde{b}_k \in \mathscr{L}^1_{loc}([0, T]\times \mathscr{B})$.

Then the traces of the function $\tilde{\alpha}$ defined by equation (3.4.103) at $t \leqslant T$ satisfy the inequality

$$\begin{aligned}\int_{\mathscr{B}} dL^3\, [\tilde{\alpha}(t, \xi) - \tilde{\alpha}(0, \xi)] \leqslant & \int_0^t ds \int_{\mathscr{B}} dL^3\, (b_k - \tilde{b}_k)(\tilde{v}^k - \tilde{v}^k) \\ & + \int_{[0, t]\times\mathscr{B}} [S_k^{\mu} - \tilde{S}_k^{\mu} - \varrho_0 W_{F^k{}_\mu F^l{}_\nu}(\mathbf{F}, \xi)(F^l{}_\nu - \tilde{F}^l{}_\nu)]\tilde{v}^k{}_{,\mu}.\end{aligned} \tag{3.4.110}$$

Remark

The above inequality holds for either trace of $\tilde{\alpha}$ at t and $t = 0$. Hence it also holds for $\bar{\tilde{\alpha}}(t, \xi)$. Indeed, we can integrate over $[0, t]$ or $[0, t[$ without affecting the right-hand side.

Let $\Lambda > 0$ be the Lipschitz constant for $\tilde{\mathbf{v}}$. The measure $\tilde{v}^k{}_{,\mu}$ in (3.4.110) is absolutely continuous with respect to the Lebesgue measure L^4,

$$v^k{}_{,\mu} = f^k{}_\mu dL^4, \quad f^k{}_\mu \in \mathscr{L}^1_{loc}([0, T]\times\mathscr{B}, L^4), ||f^k{}_\mu||_\infty \leqslant \Lambda \tag{3.4.111}$$

By the Fubini theorem the second integral on the right-hand side of (3.4.110) is equal to

$$\int_0^t ds \int_{\mathscr{B}} dL^3\, f^k{}_\mu[S^\mu_K - \tilde{S}^\mu_K - \varrho_0 W_{F^k{}_\mu F^l{}_\nu}(\tilde{\mathbf{F}}, \xi)(F^l{}_\nu - \tilde{F}^l{}_\nu)]. \tag{3.4.112}$$

We can rewrite the integrand of the double integral (3.4.112) in the form

$$\tfrac{1}{2} f^k{}_\mu \varrho_0 W_{F^k{}_\mu F^l{}_\nu F^m{}_\varrho}(\tilde{\mathbf{F}} + \vartheta\boldsymbol{\Delta}, \xi)\Delta^l{}_\nu \Delta^m{}_\varrho, \qquad 0 \leqslant \vartheta(t, \xi) \leqslant 1,$$

$\Delta^l{}_\nu := F^l{}_\nu - \tilde{F}^l{}_\nu$, assuming that the third-order derivatives of W with respect to $\mathbf{F}$ are continuous functions of $\mathbf{F}$ and ξ. Let $M_1 := \max\{||\mathbf{F}||_\infty, ||\tilde{\mathbf{F}}||_\infty\}$. Since $|\vartheta\mathbf{F} + (1-\vartheta)\tilde{\mathbf{F}}| \leqslant M_1$ L^4-a.e., $\varrho_0 W_{F^k{}_\mu F^l{}_\nu F^m{}_\varrho}(\tilde{\mathbf{F}} + \vartheta\boldsymbol{\Delta}, \xi)$ is bounded L^4-a.e.

Since the integrals on the right-hand side of (3.4.110) are equivalent to four-dimensional integrals with respect to L^4, we shall replace $\boldsymbol{\Delta}$, $\mathbf{F}$, $\mathbf{v}$ by their symmetric averages $\bar{\boldsymbol{\Delta}}$, $\bar{\mathbf{F}}$, $\bar{\mathbf{v}}$ defined H^3-a.e. By the definition of a symmetric average $|\bar{\mathbf{F}}(t, \xi)| \leqslant ||\mathbf{F}||_\infty$, $|\bar{\mathbf{v}}(t, \xi)| \leqslant ||\mathbf{v}||_\infty$ for every $(t, \xi) \in [0, T)\times\mathscr{B}$ such that $\bar{\mathbf{F}}$, $\bar{\mathbf{v}}$ is defined. Since $\mathscr{B}$ is bounded, $\bar{\mathbf{F}}(t, \cdot)$, $\bar{\mathbf{v}}(t, \cdot) \in \mathscr{L}^\infty(\mathscr{B}, L^3)$ $\subset \mathscr{L}^2(\mathscr{B}, L^3)$ for all $t \in [0, T]$. Let

$$y(t) :\equiv [||\bar{\mathbf{F}}(t, \cdot) - \tilde{\mathbf{F}}(t, \cdot)||_2{}^2 + ||\bar{\mathbf{v}}(t, \cdot) - \tilde{\mathbf{v}}(t, \cdot)||_2{}^2]^{1/2}. \tag{3.4.113}$$

From the Fubini theorem it follows that $y \in \mathscr{L}^2([0, T])$. Summing the inequalities $\Delta^l{}_\nu \Delta^m{}_\varrho \leqslant \frac{1}{2}[(\Delta^l{}_\nu)^2 + (\Delta^m{}_\varrho)^2]$ we conclude that $|\sum_{l,\nu,m,\varrho} \Delta^l{}_\nu \Delta^m{}_\varrho|$ $\leqslant 9\sum_{l,\nu}(\Delta^l{}_\nu)^2$. Using this and the preceding inequalities we can estimate the integral (3.4.112) by

$$K'\Lambda \int_0^t ds\, y(s)^2,$$

for some $K' > 0$.

The first integral on the right-hand side of (3.4.110) is bounded from above by

$$\int_0^t ds\, g(s) y(s)$$

with $g(s) :\equiv ||\mathbf{b}(s, \cdot) - \tilde{\mathbf{b}}(s, \cdot)||_2$. Clearly, by the Fubini theorem $\mathbf{b}$, $\tilde{\mathbf{b}}$ are defined for almost every $s \in [0, T]$ as elements of $\mathscr{L}^1(\mathscr{B}, L^3)$ and belong to $\mathscr{L}^1([0, T]; \mathscr{L}^1(\mathscr{B}, L^3))$. For the validity of the last estimate we shall assume that

$$\mathbf{b}, \tilde{\mathbf{b}} \in \mathscr{L}^2\left([0, T]; \mathscr{L}^2(\mathscr{B}, L^3)\right). \tag{3.4.114}$$

The inner traces $\mathbf{v}(0, \cdot)$, $\mathbf{F}(0, \cdot)$ also belong to $\mathscr{L}^2(\mathscr{B}, L^3)$, e.g.

$$||\mathbf{v}(0, \cdot)||_2 \leqslant [L^3(\mathscr{B})]^{1/2} ||\mathbf{v}(0, \cdot)||_\infty \leqslant [L^3(\mathscr{B})]^{1/2} ||\mathbf{v}||_\infty .$$

Hence

$$\int_{\mathscr{B}} dL^3\, \tilde{\alpha}(0, \xi) \leqslant K_1 ||\mathbf{v}(0, \cdot) - \tilde{\mathbf{v}}(0, \cdot)||_2{}^2 + K_2 ||\mathbf{F}(0, \cdot) - \tilde{\mathbf{F}}(0, \cdot)||_2{}^2$$

$$\leqslant K y_0{}^2$$

with $K > 0$ depending on $||\mathbf{v}||_\infty$, $||\tilde{\mathbf{v}}||_\infty$, $||\mathbf{F}||_\infty$, $||\tilde{\mathbf{F}}||_\infty$ and

$$y_0 := [||\mathbf{v}(0, \cdot) - \tilde{\mathbf{v}}(0, \cdot)||_2{}^2 + ||\mathbf{F}(0, \cdot) - \tilde{\mathbf{F}}(0, \cdot)||_2{}^2]^{1/2}.$$

Summing up, we have proved

Lemma 3.4.12

Suppose that $\mathscr{B}$ is bounded, $P(\mathscr{B}) < \infty$, and the hypotheses of Lemma 3.4.11 are satisfied. Moreover, assume (3.4.112) and suppose that the third-order derivatives of W with respect to $\mathbf{F}$ are continuous functions of $(\mathbf{F}, \xi)$. Then

$$\int_{\mathscr{B}} dL^3\, \tilde{\alpha}(t, \xi) \leqslant K y_0{}^2 + \int_0^t ds\, g(s) y(s) + K' \Lambda \int_0^t ds\, y(s)^2 \tag{3.4.115}$$

for every $t \in [0, T]$, with K, K' depending on the $\mathscr{L}^\infty$ norms of the solutions and Λ defined above, $K, K', \Lambda > 0$.

The next step requires an estimate of the left-hand side of (3.4.115) from below in terms of $y(t)$. Since W is not a convex function of $\mathbf{F}$, such an estimate can only be based on SE and the Gårding lemma. This fact imposes severe restrictions on the boundary-value problem.

For the validity of the Gårding lemma we must assume that $\partial\mathscr{B}$ is regular (Section 1.1). Hence $\partial^*\mathscr{B} = \partial\mathscr{B}$. For convenience we shall assume that $\mathscr{B}$ is open. Let

$$\boldsymbol{\chi}(t, \xi) := \boldsymbol{\chi}_0(\xi)+\int_0^t ds\, \overline{\mathbf{v}}(s, \xi), \tag{3.4.116a}$$

$$\tilde{\boldsymbol{\chi}}(t, \xi) := \tilde{\boldsymbol{\chi}}_0(\xi)+\int_0^t ds\, \tilde{\mathbf{v}}(s, \xi) \tag{3.4.116b}$$

with $\boldsymbol{\chi}_0, \tilde{\boldsymbol{\chi}}_0 \in \mathscr{L}^1_{loc}(\mathscr{B})$ (see Section 3.3.1). Clearly, for every $t \in [0, T]$ $\boldsymbol{\chi}(t, \cdot)$, $\tilde{\boldsymbol{\chi}}(t,\cdot) \in \mathscr{L}^1_{loc}(\mathscr{B})$. We know that $\boldsymbol{\chi}$, $\tilde{\boldsymbol{\chi}}$ and their derivatives $\mathbf{v}$, $\mathbf{F}$, $\tilde{\mathbf{v}}$, $\tilde{\mathbf{F}}$ belong to $\mathscr{L}^2([0, T]\times\mathscr{B})$ and hence $\boldsymbol{\chi}, \tilde{\boldsymbol{\chi}} \in \mathscr{H}^1([0, T]\times\mathscr{B})$. This would imply that $\boldsymbol{\chi}(t,\cdot)$, $\tilde{\boldsymbol{\chi}}(t, \cdot) \in \mathscr{H}^{1/2}(\mathscr{B})$ by the trace theorem in $\mathscr{H}^1$. We shall prove more, viz.

Lemma 3.4.13
Let $(\mathbf{v}, \mathbf{F}) \in \mathrm{BV}\cap\mathscr{L}^\infty$ be a solution of equations (3.3.2), and let $\boldsymbol{\chi}$ be defined by equation (3.4.116a), with $F^k{}_\alpha(0, \cdot) = {\chi_0}^k{}_{,\alpha}$. Then $\boldsymbol{\chi}(t, \cdot) \in \mathscr{H}^1(\mathscr{B})$ and $\chi^k(t, \cdot)_{,\alpha} = \overline{F}^k{}_\alpha(t, \cdot)$.

Proof

Let $\hat{\boldsymbol{\chi}}(t, \xi) :\equiv \int_0^t ds\, \overline{\mathbf{v}}(s, \xi)$, $\xi \in \mathscr{B}$, $\varphi \in \mathscr{C}_o^\infty(\mathscr{B})$, $I := [0, t]$. Using the Fubini theorem, the Leibniz rule in the BV class: $(v^k\varphi)_{,\alpha} = v^k\varphi_{,\alpha}+v^k{}_{,\alpha}\varphi$, $(\varphi F^k{}_\alpha)_{,s} = \varphi F^k{}_{\alpha,s}$,* the Gauss-Green theorem, equation (3.3.2a) as well as Theorem 3.2.10 about the trace of a product of BV functions we have

$$\begin{aligned}
-\int_{\mathscr{B}} dL^3\, \hat{\chi}^k(t, \xi)\varphi_{,\alpha} &= -\int_{I\times\mathscr{B}} dL^4\, v^k(s, \xi)\varphi_{,\alpha} \\
&= -\int_{\partial^*(I\times\mathscr{B})} dH^3\, v^k_+\, \varphi\nu_\alpha + \int_{I\times\mathscr{B}} F^k{}_{\alpha,s}\varphi \\
&= \int_{I\times\mathscr{B}} F^k{}_{\alpha,s}\varphi = \int_{I\times\mathscr{B}} (F^k{}_\alpha\varphi)_{,s} \\
&= \int_{\mathscr{B}} dL^3\, [F^k{}_\alpha(t\pm 0, \varepsilon)-F^k{}_\alpha(\pm 0, \varepsilon)]\varphi.
\end{aligned}$$

* It is understood here that φ is also a function of $(s, \xi) \in [0, T]\times\mathscr{B}$, $\varphi_{,s} \equiv 0$, $\varphi_{,\alpha} \in \mathscr{L}^1_{loc}([0, T]\times\mathscr{B})$, $\varphi = \overline{\varphi}$. Note also that the trace of φ vanishes on $I\times\partial^*\mathscr{B}$, while $\nu_\alpha = 0$ on the hyperplanes $s = 0, t$.

Since the left-hand side of this sequence of identities does not depend on the alternative choices $I = [0, t]$, $I = [0, t[$, $I =]0, t]$, $I =]0, t[$, it follows that

$$\hat{\chi}^k(t, \cdot)_{,\alpha} = F^k{}_\alpha(t \pm 0, \cdot) - F^k(\pm 0, \cdot).$$

Each of the four possible combinations of signs is admissible on the right-hand side of the above formula.* Hence also

$$\hat{\chi}^k(t, \cdot)_{,\alpha} = \overline{F}^k{}_\alpha(t, \cdot) - F^k{}_\alpha(0, \cdot). \tag{3.4.117}$$

Therefore $\chi^k(t, \cdot)_{,\alpha} = \overline{F}^k{}_\alpha(t, \cdot)$ and $\chi^k(t, \cdot) \in H^1(\mathscr{B})$ for all $t \in [0, T]$. □

Let $\tilde{\nabla} := \left(\dfrac{\partial}{\partial t}, \dfrac{\partial}{\partial \xi^\alpha}\right)$. Since

$$\left| \int_{]0, T[\times \mathscr{B}} dL^4 \, \boldsymbol{\chi} \otimes \tilde{\nabla}\varphi \right| \leqslant \max\{\|F^k{}_\alpha\|_\infty, \|v^k\|\} \cdot \sup|\varphi| T L^3(\mathscr{B})$$

for every $\varphi \in \mathscr{C}_0^\infty(]0, T[\times \mathscr{B})$, it is clear that $\boldsymbol{\chi}, \boldsymbol{\chi}_0 \in \mathrm{BV}(]0, T[\times \mathscr{B})$. We set $\boldsymbol{\chi}(t, \xi) = \boldsymbol{\chi}_0(\xi) = \mathbf{v}(t, \xi) = 0$ for $\xi \,\overline{\in}\, \mathscr{B}$. After the extension $\boldsymbol{\chi}, \boldsymbol{\chi}_0, \mathbf{v} \in \mathrm{BV} \cap \mathscr{L}^\infty$, by Theorem 3.2.16. Our aim now is to prove that the Dirichlet boundary condition $\boldsymbol{\chi}(t, \cdot) - \tilde{\boldsymbol{\chi}}(t, \cdot) = \mathbf{0}$ on $\partial\mathscr{B}$ (in the sense of traces in $\mathscr{H}^1(\mathscr{B})$) implies that $\mathbf{v}$ vanishes on $]0, T[\times \partial^*\mathscr{B}$ in the sense of traces in BV.

Lemma 3.4.14

Suppose that $\partial\mathscr{B}$ is regular, $\mathbf{u} \in \mathrm{BV} \cap \mathscr{L}^\infty([0, T] \times \mathscr{B})$, $\mathbf{u}(t, \cdot) \in \mathscr{H}^1(\mathscr{B})$ for al $t \in [0, T]$. Let $\gamma\mathbf{u}(t, \cdot)$ be the trace of $\mathbf{u}(t, \cdot)$ on $\partial\mathscr{B}$ for every $t \in [0, T]$.

Then

(i) the inner trace of $\mathbf{u} \in \mathrm{BV}$ on $]0, T[\times \partial\mathscr{B}$ is equal to $\gamma\mathbf{u}(t, \cdot)$ for H^3-almost all $(t, \xi) \in]0, T[\times \partial\mathscr{B}$;

(ii) in particular, if $\mathbf{u}(t, \xi) = \int_0^t ds\,\bar{}\,(s, \xi)$ for some $\mathbf{v} \in \mathrm{BV} \cap \mathscr{L}^\infty$, $\gamma\mathbf{u}(t, \cdot) = 0$ on $\partial\mathscr{B}$ for $t \in [0, T]$, then the inner trace of $\mathbf{v}$ vanishes on $]0, T[\times \partial\mathscr{B}$.

Proof

According to Theorem 1.1.48, for almost every $\xi_1 \in \partial\mathscr{B}$, $\gamma\mathbf{u}(t, \xi_1)$ is the limit of $\mathbf{u}(t, \xi)$, $\xi \in \mathscr{B}$, the limits being taken along a family of curves transversal with respect to $\partial\mathscr{B}$. For convenience we shall rectify $\partial\mathscr{B}$ piecewise. Thus let $(\zeta^1, \zeta^2, \zeta^3) = \varphi(\xi^1, \xi^2, \xi^3)$, $\varphi \in \mathscr{C}^1$, be a local coordinate system in a nbhd $\mathscr{U}$

* The arbitrariness of the choice of traces in the above formula follows from equations (3.3.2). Intuitively, let us consider the possibility that $F^k{}_\alpha(s, \xi) = F^k_{(1)\alpha}(s, \xi) + H(s-t) \times F^k_{(2)\alpha}(s, \xi)$. Then $v^k{}_{,\alpha} = F^k{}_{\alpha,s} = F^k_{(1)\alpha,s} + H(s-t) F^k_{(2)\alpha,s} + \delta(s-t) F^k_{(2)\alpha}$, and v^k is of the form $v^k{}_{(1)} + \delta(s-t) v^k{}_{(2)}$, which contradicts the assumption that $\mathbf{v} \in \mathscr{L}^1_{loc}([0, T] \times \mathscr{B})$.

of $\xi_0 \in \partial\mathscr{B}$ such that $\varphi(\mathscr{U}\cap\partial\mathscr{B})$ is defined by the equation $\zeta^1 = 0$ and $\zeta^1 > 0$ in $\mathscr{B}\cap\mathscr{U}$. Let $\mathscr{V} := \varphi(\mathscr{U})$, $\mathscr{V}' := \{\zeta' \in \mathbf{R}^2 | (0, \zeta') \in \mathscr{V}\}$, $\bar{\mathbf{u}}(t, \xi) \equiv \mathbf{w}(t, \varphi(\xi))$. By Theorem 1.1.52 $\mathbf{w}(t, \cdot) \in H^1(\mathscr{B})$ for all $t \in [0, T]$. From Theorem 3.2.16 it follows that $\mathbf{w} \in \mathrm{BV}\cap\mathscr{L}^\infty$ as well.

Let $\boldsymbol{\nu} = (+1, 0, 0)$ be the unit inner normal on $\partial\mathscr{B}$ in the ζ coordinates. By a theorem of Vol'pert (1967, Section 11.1), in his notation,

$$\mathbf{w}_{\pm\nu}(t, 0, \zeta') := \lim_{h\to 0_+} \int_0^t d\vartheta\, \mathbf{w}(t, \pm\vartheta h, \zeta')$$

exists for almost all $(t, \zeta') \in [0, T]\times\mathscr{V}'$, $\mathbf{w}_{-\nu}(t, 0, \zeta') = 0$. By another theorem of Vol'pert (1967, Section 11.2), on account of the assumed extension of $\mathbf{v}$

$$\mathbf{w}_+(t, 0, \zeta') = 2\bar{\mathbf{w}}(t, 0, \zeta') = \lim_{h\to 0_+} \int_0^1 d\vartheta\, \mathbf{w}(t, \vartheta h, \zeta').$$

Since $\mathbf{w} = \bar{\mathbf{u}}$ is bounded on its domain of definition, we can interchange the limit and the integration, concluding that the inner trace

$$\mathbf{w}_+(t, 0, \zeta') = \gamma\mathbf{w}(t, 0, \zeta')$$

for all $t \in]0, T[$ and almost all $\zeta' \in \mathscr{V}'$. Since $\varphi \in \mathscr{C}^1$,

$$\mathbf{u}_+(t, \xi_0) = \gamma\mathbf{u}(t, \xi_0)$$

for all $t \in]0, T[$ and H^2-almost all $\xi_0 \in \partial\mathscr{B}$ (see Federer, 1969, Section 3.2 for the Hausdorff measure of a smooth or Lipschitzian surface). This proves (i).

For the proof of (ii) we shall also work in the ζ coordinates. Let $\mathbf{v}(t, \xi) \equiv \mathbf{y}(t, \zeta)$. Now $\lim_{h\to 0_+} \int_0^1 d\vartheta\, \mathbf{w}(t, \vartheta h, \zeta') = 0$ for almost all $\zeta' \in \mathscr{V}'$ and all $t \in]0, T[$.

Let $\mathscr{Q}$ be a cube in $\mathscr{V}'$. Applying the Fubini theorem and exchanging limits with integrations ($\mathbf{y}$ is bounded) we have

$$\int_{[t-\delta, t+\delta]\times\mathscr{Q}} ds\, d\zeta' \lim_{h\to 0_+} \int_0^1 d\vartheta\, \mathbf{y}(s, \vartheta h, \zeta') = \int_{\mathscr{Q}} d\zeta' \lim_{h\to 0_+} \int_0^1 d\vartheta \int_{t-\delta}^{t+\delta} ds\, \mathbf{y}(s, \vartheta h, \zeta')$$

$$= \int_{\mathscr{Q}} d\zeta' \lim_{h\to 0_+} \int_0^1 d\vartheta\, [\mathbf{w}(t+\delta, \vartheta h, \zeta') - \mathbf{w}(t-\delta, \vartheta h, \zeta'] = 0.$$

Dividing the left-hand side by $2\delta L^2(\mathscr{Q})$ and letting $\delta \to 0$, $L^2(\mathscr{Q}) \to 0$, $\delta/L^2(\mathscr{Q}) \to \lambda > 0$, we conclude that $\lim_{h\to 0_+} \int_0^1 d\vartheta\, \mathbf{y}(t, \vartheta h, \zeta') = 0$ for almost all $(t, \zeta') \in [0, T]\times\mathscr{V}'$, which implies the thesis of (ii). □

We are now ready to prove the main theorem of this section (the basic idea is due to Dafermos (1980)).

Theorem 3.4.15
Let $\mathscr{B}$ be a bounded set of finite perimeter with a regular boundary. We shall assume that $\varrho_0 \in \mathscr{C}^0$, $\varrho_0(\xi) \geqslant \varrho > 0$ for almost all $\xi > \mathscr{B}$, while $W \in \mathscr{C}^1$ and the third-order derivatives of W with respect to $\mathbf{F}$ are continuous functions of $(\mathbf{F}, \xi)$.

Suppose that $(\mathbf{F}, \mathbf{v}) \in \mathrm{BV} \cap \mathscr{L}^\infty$ and $(\tilde{\mathbf{F}}, \tilde{\mathbf{v}}) \in \mathrm{Lip}$ are defined on $[0, T] \times \mathscr{B}$ and satisfy equations (3.3.2), while $(\mathbf{F}, \mathbf{v})$ also satisfies (3.4.100) in the weak sense. Noting that $\boldsymbol{\chi}$, $\tilde{\boldsymbol{\chi}}$, defined by equations (3.4.116), have the property that $\boldsymbol{\chi}(t, \cdot)$, $\bar{\boldsymbol{\chi}}(t, \cdot) \in \mathscr{H}^1(\mathscr{B})$ for every $t \in [0, T]$ we assume that $\boldsymbol{\chi}(t, \cdot) - \tilde{\boldsymbol{\chi}}(t, \cdot) = \mathbf{0}$ on $\partial\mathscr{B}$ in the sense of traces.

Then there is a constant $\delta > 0$, depending on $\mathscr{B}$, W, $\tilde{\mathbf{F}}$, such that

$$||\mathbf{F}-\tilde{\mathbf{F}}||_\infty < \delta \tag{3.4.118}$$

implies that $y(t)$, defined by (3.4.113) satisfies the inequality (3.4.124) with positive constants M, A, Q, depending on $\mathscr{B}$, W and $\tilde{\mathbf{F}}$, $\tilde{\mathbf{v}}$.

Proof
For the proof, we have to estimate $\int_{\mathscr{B}} dL^3 \tilde{\alpha}(t, \xi)$ from below.

Noting that

$$\begin{aligned}\alpha(\mathbf{v}, \mathbf{F}; \tilde{\mathbf{v}}, \tilde{\mathbf{F}}; \xi) &= \tfrac{1}{2}\varrho_0(\mathbf{v}-\tilde{\mathbf{v}})^2 \\ &+ \tfrac{1}{2}\varrho_0 W_{F^k{}_\mu F^l{}_\nu}(\tilde{\mathbf{F}}, \xi)(F^k{}_\mu - \tilde{F}^k{}_\mu)(F^l{}_\nu - \tilde{F}^l{}_\nu) + O(|\mathbf{F}-\tilde{\mathbf{F}}|^3)\end{aligned} \tag{3.4.119}$$

and $W_{F^k{}_\mu F^l{}_\nu}(\tilde{\mathbf{F}}(t, \xi), \xi)$ is a continuous function of ξ, we conclude from SE and the preceding lemmas that the Gårding lemma can be applied to $\boldsymbol{\chi}(t, \cdot) - \bar{\boldsymbol{\chi}}(t, \cdot)$:

$$\begin{aligned}&\int_{\mathscr{B}} dL^3\, \tilde{\alpha}(t, \xi) \geqslant \tfrac{1}{2}\varrho ||\bar{\mathbf{v}}(t, \cdot) - \tilde{\mathbf{v}}(t, \cdot)||_2{}^2 \\ &+ (\lambda' - \mu\delta)||\bar{\mathbf{F}}(t, \cdot) - \tilde{\mathbf{F}}(t, \cdot)||_2{}^2 - \varkappa||\boldsymbol{\chi}(t, \cdot) - \tilde{\boldsymbol{\chi}}(t, \cdot)||_2{}^2, \\ &\varrho, \lambda', \mu, \delta > 0, \quad ||\mathbf{F}-\tilde{\mathbf{F}}||_\infty \leqslant \delta.\end{aligned}$$

The term involving δ comes from the lower bound of $O(|\mathbf{F}-\tilde{\mathbf{F}}|^3)$.

Let $\delta > 0$ be so small that $\lambda := \lambda' - \mu\delta > 0$.

For the last term of the Gårding inequality we have

$$||\boldsymbol{\chi}(t, \cdot) - \tilde{\boldsymbol{\chi}}(t, \cdot)||_2{}^2 = \int_{\mathscr{B}} dL^3 \left\{\int_0^t ds[\bar{\mathbf{v}}(s, \xi) - \tilde{\mathbf{v}}(s, \xi)] + \boldsymbol{\chi}_0(\xi) - \tilde{\boldsymbol{\chi}}_0(\xi)\right\}^2$$

$$\leqslant \int_{\mathscr{B}} dL^3 \Big\{\int_0^t ds\,[\bar{\mathbf{v}}(s,\xi)-\tilde{\mathbf{v}}(s,\xi)]\Big\}^2 + \int_{\mathscr{B}} dL^3\,[\boldsymbol{\chi}_0(\xi)-\tilde{\boldsymbol{\chi}}_0(\xi)]^2$$

$$+2\Big\|\int_0^t ds\,[\bar{\mathbf{v}}(s,\cdot)-\tilde{\mathbf{v}}(s,\cdot)]\Big\|_2\,\|\boldsymbol{\chi}_0-\tilde{\boldsymbol{\chi}}_0\|_2$$

$$\leqslant 2\int_{\mathscr{B}} dL^3 \Big\{\int_0^t ds\,[\bar{\mathbf{v}}(s,\xi)-\tilde{\mathbf{v}}(s,\xi)]\Big\}^2 + 2\int_{\mathscr{B}} dL^3 [\boldsymbol{\chi}_0(\xi)-\tilde{\boldsymbol{\chi}}_0(\xi)]^2. \quad (3.4.120)$$

In view of the Friedrichs inequality (1.1.49) the last integral in (3.4.120) can be estimated from above by $k\|\mathbf{F}_0-\tilde{\mathbf{F}}_0\|_2{}^2$, $\mathbf{F}_0 = \nabla\boldsymbol{\chi}_0$, $\tilde{\mathbf{F}}_0 = \nabla\tilde{\boldsymbol{\chi}}_0$, $k > 0$.

For the first integral on the right-hand side of (3.4.120) the Fubini theorem yields the inequality

$$\Big|\int_{\mathscr{B}} dL^3 \int_0^t ds \int_0^t ds'\,[\bar{\mathbf{v}}(s,\xi)-\tilde{\mathbf{v}}(s,\xi)]\cdot[\bar{\mathbf{v}}(s',\xi)-\tilde{\mathbf{v}}(s',\xi)]\Big|$$

$$= \Big|\int_0^t ds \int_0^t ds' \int_{\mathscr{B}} dL^3\,\bar{\mathbf{v}}\,[(s,\xi)-\tilde{\mathbf{v}}(s,\xi)]\cdot[\bar{\mathbf{v}}(s',\xi)-\tilde{\mathbf{v}}(s',\xi)]\Big|$$

$$\leqslant \int_0^t ds \int_0^t ds'\,\|\bar{\mathbf{v}}(s,\cdot)-\tilde{\mathbf{v}}(s,\cdot)\|_2\|\bar{\mathbf{v}}(s',\cdot)-\tilde{\mathbf{v}}(s',\cdot)\|_2$$

$$= \Big\{\int_0^t ds\,\|\bar{\mathbf{v}}(s,\cdot)-\tilde{\mathbf{v}}(s,\cdot)\|_2\Big\}^2.$$

The last integral can be estimated by means of the general inequality $(\int_0^t ds\,a(s))^2 = (\int_0^t ds\,a(s)\cdot 1)^2 \leqslant t\int_0^t ds\,a(s)^2$. Hence

$$\int_{\mathscr{B}} dL^3\,[\boldsymbol{\chi}(t,\xi)-\tilde{\boldsymbol{\chi}}(t,\xi)]^2 \leqslant 2t\int_0^t ds\; y(s)^2+2ky_0{}^2, \quad k>0. \quad (3.4.121)$$

Let $L := \min\{\tfrac{1}{2}\varrho, \lambda\}$, $M^2 := (2k\varkappa+K)/L$, $2P := K'\Lambda$, $2Q := \varkappa/L$, $2N := L^{-1}$. Combining all the above estimates with (3.4.110) we have

$$y(t)^2 \leqslant M^2y_0{}^2+2\int_0^t ds\,[(P+2Qt)y(s)^2+Ng(s)y(s)] \equiv: z(t)^2. \quad (3.4.122)$$

We define $z(t)$ in such a way that $z(t) \geqslant 0$ for all $t\in[0,T]$.

It remains to solve the differential inequality (3.4.122). In this regard we note that $y(t)^2 \leqslant z(t)^2$, $2P\int_0^t ds\, y(s)^2 \leqslant z(t)^2$, and

$$2\dot{z}(t)z(t) = (2P+4Qt)y(t)^2+2Ng(t)y(t)+4Q\int_0^t ds\; y(s)^2$$

$$\leqslant (2P+4Qt)z(t)^2+2Ng(t)z(t)+(2Q/P)z(t)^2.$$

For $t > 0$ $z(t) > 0$, hence it follows that

$$\dot{z}(t) \leqslant (A+2Qt)z(t)+Ng(t) \tag{3.4.123}$$

with $A := P+Q/P$.

Equation (3.4.123) implies that

$$\dot{z}(t) = -h(t)+(A+2Qt)z(t)+Ng(t) \tag{3.4.124}$$

with some $h(t) \geqslant 0$. Let $z(t) \equiv R(t)\,\exp(At+Qt^2)$. For $R(t)$ we have the initial-value problem

$$\dot{R} = [Ng(t)-h(t)]\exp(-At-Qt^2), \quad R(0) = z(0) = My_0.$$

Hence

$$R(t) = \int_0^t ds\,[Ng(s)-h(s)]\,e^{-As-Qs^2}+R(0) \leqslant R(0)+\int_0^t ds\,Ng(s)\,e^{-As-Qs^2}$$

and

$$y(t) \leqslant z(t) \leqslant My_0\,e^{At+Qt^2}+N\int_0^t ds\; g(s)\,e^{A(t-s)+Q(t^2-s^2)}. \tag{3.4.125}$$

This proves the theorem. □

Corollary 3.4.16

Let the hypotheses of Theorem 3.4.15 hold. A solution $(\tilde{\mathbf{v}}, \tilde{\mathbf{F}}) \in \mathrm{Lip}$ of equations (3.3.2) with the body forces $\tilde{b}_k \in \mathscr{L}^1_{loc}([0, T]\times\mathscr{B})\cap\mathscr{L}^2([0, T], \mathscr{L}^2(\mathscr{B}))$ is unique in the class of solutions $(\mathbf{v}, \mathbf{F}) \in \mathrm{BV}\cap\mathscr{L}^\infty$ satisfying (3.4.100), (3.4.118), the Dirichlet boundary conditions on $\partial\mathscr{B}$ and the initial conditions for $\mathbf{F}$, $\mathbf{v}$.

Remarks

(1) The use of the Gårding lemma imposes the requirement of continuity of $\varrho_0 W_{F^k{}_\mu F^l{}_\nu}(\tilde{\mathbf{F}}, \xi)$ in (3.4.119). This requirement is met provided $\tilde{\mathbf{F}} \in \mathrm{Lip}$. The estimate of the remainder of the Taylor expansion of α is based on an *a priori* bound on $||\mathbf{F}-\tilde{\mathbf{F}}||_\infty$.

(2) For $(\mathbf{F}, \mathbf{v}) \in \mathrm{Lip}$ the restriction (3.4.118) can be lifted. Indeed, using the Taylor expansion up to terms of first degree with a remainder in the integral form we have

$$\alpha = \tfrac{1}{2}(\mathbf{v}-\tilde{\mathbf{v}})^2 + \int_0^1 d\vartheta\,(1-\vartheta)\,W_{F^k{}_\mu F^l{}_\nu}(\tilde{\mathbf{F}} + \vartheta(\mathbf{F}-\tilde{\mathbf{F}}), \xi)(F^k{}_\mu - \tilde{F}^k{}_\mu)(F^l{}_\nu - \tilde{F}^l{}_\nu). \tag{3.4.126}$$

The entries of the matrix defined by the integral on the right-hand side of (3.4.126) are continuous functions of ξ for every $t \in [0, T]$ and the Gårding lemma can be applied. Hence a solution $(\tilde{\mathbf{F}}, \tilde{\mathbf{v}}) \in \mathrm{Lip}$ is unconditionally unique in the class Lip with given initial and Dirichlet boundary conditions.

3.5 BLOW-UP IN NON-LINEAR ELASTODYNAMICS

3.5.1 Introduction. Another Formulation of the Law of Energy Dissipation

In Section 3.5 we shall consider the motion of an elastic body $\mathscr{B}$ as a mapping $t \mapsto \boldsymbol{\chi}^{(t)}$, $\boldsymbol{\chi}^{(t)} = \boldsymbol{\chi}_0 + \mathbf{u}^{(t)}$, $\boldsymbol{\chi}_0 \colon \mathscr{B} \to \boldsymbol{R}^3$

$$\mathbf{u}\colon \bar{\boldsymbol{R}}_+ \to \mathscr{X};\ t \mapsto \mathbf{u}^{(t)}, \tag{3.5.1}$$

which assigns to each time a configuration $\boldsymbol{\chi}^{(t)}\colon \mathscr{B} \mapsto \boldsymbol{R}^3$ (in a fixed reference frame), specified by a function $\mathbf{u}^{(t)}$ belonging to a Banach space $\mathscr{X}$. The Banach space $\mathscr{X}$ is defined by the regularity requirements and the homogeneous Dirichlet boundary conditions on the clamped part Σ_1 of $\partial\mathscr{B}$. With this interpretation it is possible to regard the equations of motion as (a weak form of) an ordinary differential equation in $\mathscr{X}$.

Having defined the notion of a weak solution of the ODE mentioned above we shall look for a function $g\colon \mathscr{X} \to \boldsymbol{R}$ which has the following properties:

(1) for some weak solutions $t \mapsto \mathbf{u}^{(t)}$ the function

$$t \mapsto g(\mathbf{u}^{(t)}) \geqslant h(t) \tag{3.5.2}$$

and $h(t) \to \infty$ as $t \to t_0 > 0$;

(2) $g(\mathbf{u}^{(t)}) < \infty$ for every t in the domain of definition of the weak solution.

A weak solution $t \mapsto \mathbf{u}^{(t)}$ is said to **blow up** (in finite time) if it has the property (3.5.2) for some function g satisfying the condition (2) (and for some finite $t_0 > 0$). If $t_0 < \infty$ then the domain of the solution does not include

$t > t_0$. It is a well-known fact in the classical theory of ODEs that the solutions cease to exist only by escaping from the manifold in which they are allowed to take their values.

The blow-up theorems assert that the weak solutions satisfying some specified initial conditions blow up in finite or infinite time. They are hardly ever accompanied by the appropriate theorems asserting existence in the small. However some uniqueness and continuous dependence results can be established by related techniques.

In the case of non-linear elasticity the blow-up theorems are proved under some assumptions on the constitutive equations. They admit two alternative interpretations:

(I) constitutive equations with some specified properties are not admissible from a physical point of view;

(II) at the time of the blow-up some non-elastic behaviour (e.g. fracture or plastic flow) sets in.

Either of these interpretations is acceptable provided the notion of the weak solution (including the choice of $\mathscr{X}$) is large enough to allow for all physically acceptable singularities (e.g. shock waves), compatible with the basic hypotheses of elasticity.

One of the most important tools used in the proofs is the energy dissipation law in the form of the inequality

$$E(t) \leqslant E(0) \quad \text{for every } t > 0 \text{ in the domain of } t \mapsto \mathbf{u}^{(t)}. \tag{3.5.3}$$

The total energy

$$E(t) := \int_{\mathscr{B}} dL^3\, \varrho_0 \left[\frac{1}{2}\mathbf{v}^2 + W(\mathbf{F}, \xi)\right] + \Phi(t) \tag{3.5.4}$$

is the sum of the kinetic energy, the strain energy and the potential energy of the external loads. For potential loads $\varrho_0 \mathbf{b} = -dq(\mathbf{x}, \xi)/d\mathbf{x}$, $\xi \in \mathscr{B}$, $\mathbf{t} = -dr(\mathbf{x}, \xi)/d\mathbf{x}$, $\xi \in \Sigma_2$ we have

$$\Phi(t) = \int_{\mathscr{B}} dL^3\, q\left(\boldsymbol{\chi}^{(t)}(\xi), \xi\right) + \int_{\Sigma_2} dH^2\, r\left(\boldsymbol{\chi}^{(t)}(\xi), \xi\right). \tag{3.5.5}$$

The key property of Φ is

$$\dot{\Phi}(t) = -\int_{\mathscr{B}} dL^3\, \varrho_0 b_k v^k - \int_{\Sigma_2} dH^2\, t_k v^k. \tag{3.5.6}$$

Let us consider the functional

$$\mathrm{Vol}(t) := \int_{\mathscr{B}} dL^3 \det \mathbf{F}, \quad F^k{}_\alpha := \chi^{(t)k}{}_{,\alpha}. \tag{3.5.7}$$

We have

$$\frac{d}{dt}\operatorname{Vol}(t) = \int_{\mathscr{B}} dL^3 \det \mathbf{F} \overset{-1}{F^{\alpha}}{}_k \dot{F}^k{}_{\alpha} = \int_{\mathscr{B}} dL^3 \det \mathbf{F} \operatorname{div} \mathbf{v}$$

$$= \int_{\partial\chi^{(t)}(\mathscr{B})} dH^2\, v^k n_k \tag{3.5.8}$$

($\mathbf{n}$ is the unit outer normal on $\partial\chi^{(t)}(\mathscr{B})$). Hence for a body immersed in a liquid exerting a pressure p (independent of time)

$$\mathbf{Tn} = \mathbf{t} = -p\mathbf{n} \quad \text{on } \chi^{(t)}(\partial\mathscr{B}) \tag{3.5.9}$$

and

$$\Phi(t) = p\operatorname{Vol}(t). \tag{3.5.10}$$

Indeed,

$$\dot{\Phi}(t) = -\int_{\partial\chi^{(t)}(\mathscr{B})} dH^2\,(-pn_k)v^k \tag{3.5.11}$$

We now tentatively derive (3.5.3) from the condition (EN), viz. (3.4.100). Integrating (3.4.100) over $[0, T]\times\mathscr{B}$, $T > 0$, we get the inequality

$$\int_{\mathscr{B}} dL^3\,[\bar{\varphi}^0(T, \xi) - \bar{\varphi}^0(0, \xi)] \leqslant \int_{[0,T]\times\Sigma_2} dH^2\, dt S_k{}^{\alpha}\nu_{\alpha} v^k$$

$$+ \int_{[0,T]\times\mathscr{B}} dL^4\, \varrho_0 b_k v^k = -\Phi(T) + \Phi(0),$$

i.e. (3.5.3). In the middle expression the transition from BV traces to the traces in the Banach space $\mathscr{X}$ must be effected by a technique already applied in Section 3.4.6.

3.5.2 Weak Solutions in the Sense of Knops, Levine and Payne

Since the main goal is to prove non-existence it is desirable to work on a notion of a solution as weak as possible. The formulation given in this section differs somewhat from the original one worked out by Knops *et al.* (1974).*

Let $\mathbf{u}(t, \xi) \equiv \chi(t, \xi) - \chi_0(\xi)$ satisfy, in a sense to be specified below, the following equations:

$$\varrho_0 \ddot{u}^k = S_k{}^{\alpha}{}_{,\alpha} \quad \text{in } [0, T]\times\mathscr{B},\ k = 1, 2, 3,\ \varrho_0(\xi) > 0, \tag{3.5.12}$$

$$\mathbf{u}(t, \cdot)|\Sigma_1 = \mathbf{0} \quad \text{for } t \in [0, T], \tag{3.5.13}$$

$$S_k{}^{\alpha} n_{\alpha}|\Sigma_2 = 0 \quad \text{for } t \in [0, T],\quad k = 1, 2, 3, \tag{3.5.14}$$

* Their formulation does not allow for shock waves and some papers contain minor mistakes of a "technical" kind.

with Σ_1, Σ_2 sufficiently regular, $\Sigma_1 \cap \Sigma_2 = \emptyset$, $\bar{\Sigma}_1 \cup \bar{\Sigma}_2 = \partial\mathscr{B}$, $H^2(\Sigma_1) > 0$. The set $\mathscr{B}$ is assumed to be bounded and open while $\partial\mathscr{B}$ is sufficiently regular for the Gauss–Green theorem and the trace theorems to be applicable. The initial conditions are

$$\mathbf{u}(0, \xi) = \mathbf{u}_0(\xi), \quad \mathbf{u}_t(0, \xi) = \mathbf{v}_0(\xi), \quad \xi \in \mathscr{B}. \tag{3.5.15}$$

We also quote the constitutive relation

$$S_k{}^\alpha = \varrho_0 W_{F^k{}_\alpha}(\mathbf{F}). \tag{3.5.16}$$

In some cases it is convenient to rewrite equations (3.5.12), (3.5.16) in a modified form:

$$\varrho_0 \ddot{u}^k = \left(A_k{}^\alpha{}_l{}^\beta(\xi) u^l{}_{,\beta}\right)_{,\alpha} + \tilde{S}_k{}^\alpha(\nabla\mathbf{u}, \xi)_{,\alpha}, \quad k = 1, 2, 3, \tag{3.5.17}$$

$$\tilde{S}_k{}^\alpha = \varrho_0 \tilde{W}_{F^k{}_\alpha}(\nabla\mathbf{u}, \xi), \tag{3.5.18}$$

$$\varrho_0 W(\mathbf{F}, \xi) = \tfrac{1}{2} A_k{}^\alpha{}_l{}^\beta(\xi) \tilde{F}^k{}_\alpha \tilde{F}^l{}_\beta + \varrho_0 \tilde{W}(\tilde{\mathbf{F}}, \xi), \quad \mathbf{F} = \mathbf{F}_0 + \tilde{\mathbf{F}}, \tag{3.5.19}$$

where the "non-linear part" $\tilde{W}(\tilde{\mathbf{F}}, \xi)$ of the specific strain energy is singled out by an inequality

$$2(1+2K)\tilde{W}(\tilde{\mathbf{F}}, \xi) \geqslant \tilde{W}_{\tilde{F}^k{}_\alpha} \tilde{F}^k{}_\alpha, \quad K > 0. \tag{3.5.20}$$

We now extend the operator $E := -S_k{}^\alpha(\nabla\mathbf{u}+\mathbf{F}_0, \xi)_{,\alpha}$ defined on $\mathscr{D} := \{\mathbf{v} \in \mathscr{C}^2(\bar{\mathscr{B}}; \boldsymbol{R}^3) |\ \mathbf{v}|\Sigma_1 = \mathbf{0},\ S_k{}^\alpha(\nabla\mathbf{v}+\mathbf{F}_0, \xi)\nu_\alpha|\Sigma_2 = 0,\ k = 1, 2, 3\}$, $\mathbf{F}_0 = \nabla\boldsymbol{\chi}_0$, using the methods presented in Chapter 2. For easy reference we shall repeat the procedure briefly.

Let $H := \mathscr{L}^2(\mathscr{B}; \boldsymbol{R}^3)$, $\mathscr{V} := \{\mathbf{v} \in W^{1,p}(\mathscr{B}; \boldsymbol{R}^3) |\ \mathbf{v}|\Sigma_1 = \mathbf{0}\}$. The latter space will be endowed with the norm $||\mathbf{v}||_{\mathscr{V}} := ||\nabla\mathbf{v}||_p$, which is reasonable on account of the Friedrichs lemma and the assumption that $H^2(\Sigma_1) > 0$. The gradient operator $L(\supset \nabla)$ on $\mathscr{V}$ is an isometry $L: \mathscr{V} \to \mathscr{L}^p(\mathscr{B}; \boldsymbol{R}^9)$. The adjoint operator $L^\dagger$ takes $\mathscr{L}^q(\mathscr{B}; \boldsymbol{R}^9)$, $q^{-1} = 1 - p^{-1}$, to $\mathscr{V}^*$. Suppose that $\tilde{S}$ is the Nemytskii operator mapping a matrix function $\xi \mapsto \tilde{F}^k{}_\alpha(\xi)$ in $\mathscr{L}^p(\mathscr{B}; \boldsymbol{R}^9)$ to a matrix function $\xi \mapsto \tilde{S}_k{}^\alpha(\tilde{\mathbf{F}}(\xi), \xi)$ in $\mathscr{L}^q(\mathscr{B}; \boldsymbol{R}^9)$,

$$|\tilde{S}_k{}^\alpha(\tilde{\mathbf{F}}, \xi)| \leqslant c\left\{a(\xi) + \sum_{k,\alpha} |\tilde{F}^k{}_\alpha|^{p-1}\right\}, \tag{3.5.21}$$

$$|\tilde{W}(\tilde{\mathbf{F}}, \xi)| \leqslant c\left\{b(\xi) + \sum_{k,\alpha} |\tilde{F}^k{}_\alpha|^{s-1}\right\}. \tag{3.5.22}$$

Similar inequalities will be assumed for $S_k{}^\alpha$, W. The number p specifying the spaces must be chosen in such a way that the conditions (3.5.21) and $p \geqslant 2$ are satisfied, while $s \leqslant p$.

According to the Sobolev embedding theorems $\mathscr{V} \subset \mathscr{Y}$, $\mathscr{Y} := \{\mathbf{v} \in H^1(\mathscr{B}; \boldsymbol{R}^3) | \mathbf{v}|\Sigma_1 = \mathbf{0}\}$. $\tilde{W}$ maps $\mathscr{L}^s(\mathscr{B}; \boldsymbol{R}^9)$ into $\mathscr{L}^r(\mathscr{B}; \boldsymbol{R})$, $r^{-1} = 1 - s^{-1}$. Since $\mathscr{B}$ is bounded, $\mathscr{L}^r(\mathscr{B}; \boldsymbol{R}) \subset \mathscr{L}^1(\mathscr{B}; \boldsymbol{R})$ and the integral

$$\tilde{\mathscr{W}}(\tilde{\mathbf{F}}) := \int_{\mathscr{B}} dL^3\, \varrho_0\, \tilde{W}(\tilde{\mathbf{F}}(\xi), \xi), \tag{3.5.23}$$

makes sense for every $\tilde{\mathbf{F}}(\cdot) \in \mathscr{L}^p(\mathscr{B}; \boldsymbol{R}^9) \subset \mathscr{L}^s(\mathscr{B}; \boldsymbol{R}^9)$.

Let $A := L^\dagger SL: \mathscr{V} \to \mathscr{V}^*$. For every $\mathbf{h} \in \mathscr{V} \cap \mathscr{C}^1(\bar{\mathscr{B}}; \boldsymbol{R}^3)$ and $\mathbf{v} \in \mathscr{C}^2(\bar{\mathscr{B}}; \boldsymbol{R}^3)$ satisfying the condition $\mathbf{v}|\Sigma_1 = \mathbf{0}$

$$\begin{aligned}\langle A\mathbf{v}|\mathbf{h}\rangle &= \langle SL\mathbf{v}|L\mathbf{h}\rangle \\ &= -\int_{\mathscr{B}} dL^3\, S_k^{\alpha}(\mathbf{F}_0 + \nabla\mathbf{v}, \xi)_{,\alpha} h^k + \int_{\Sigma_2} dH^2\, S_k^{\alpha}(\mathbf{F}_0 + \nabla\mathbf{v}, \xi)\nu_\alpha h^k.\end{aligned}$$

The last term vanishes for $\mathbf{v} \in \mathscr{D}$ and therefore $A = E$ on $\mathscr{D}$. Moreover a function $\mathbf{v} \in \mathscr{C}^2(\bar{\mathscr{B}}; \boldsymbol{R}^3)$ satisfying $\mathbf{v}|\Sigma_1 = \mathbf{0}$ belongs to $\mathscr{D}$ iff it satisfies the boundary condition $S_k^{\alpha}(\mathbf{F}_0 + \nabla\mathbf{v}, \xi)\nu_\alpha = 0$ on Σ_2.

We assume that $\varrho_0, A_k{}^{\alpha}{}_l{}^{\beta} \in \mathscr{L}^\infty$ and define the symmetric operators P, N on $\mathscr{H}$, $\mathscr{H}^1(\mathscr{B}; \boldsymbol{R}^3)$, resp.,

$$(\mathbf{w}|N\mathbf{v}) := \int_{\mathscr{B}} dL^3 \sum_{k,l,\alpha,\beta} A_k{}^{\alpha}{}_l{}^{\beta} w^k{}_{,\alpha} v^l{}_{,\beta}, \tag{3.5.24}$$

$$(\mathbf{w}|P\mathbf{v}) := \int_{\mathscr{B}} dL^3\, \varrho_0 v^k w^k. \tag{3.5.25}$$

Note that $(\mathbf{v}|P\mathbf{v}) = 0$ implies that $\mathbf{v} = \mathbf{0}$ a.e. in $\mathscr{B}$.

We now proceed to define a weak solution in the sense of Knops, Levine and Payne.

Definition 3.5.1

$\mathbf{u}: [0, T[\to \mathscr{V}$ is said to be a **weak solution (in the sense of KLP)** if the following conditions are satisfied:

(i) $\mathbf{u}: [0, T[\to \mathscr{V}$ is weakly continuous* and $\mathbf{u}(0) = \mathbf{u}_0$;

(ii) $\dot{\mathbf{u}}: [0, T[\to \mathscr{H}$ is weakly summable, $\dot{\mathbf{u}}(0) = \mathbf{v}_0$ and for every $\boldsymbol{\phi} \in H$, $t_1, t_2 \in [0, T[$,

$$(\mathbf{u}(t_2)|P\boldsymbol{\phi}) - (\mathbf{u}(t_1)|P\boldsymbol{\phi}) = \int_{t_1}^{t_2} ds\, (\dot{\mathbf{u}}(s)|P\boldsymbol{\phi}); \tag{3.5.26}$$

* Since $\mathscr{V} \subset \mathscr{Y}$ and $\mathscr{X}^1 \subset W^{1,q}$, $q = (1 - p^{-1})^{-1} \leqslant 2$, so is $\mathbf{u}: [0, T[\to \mathscr{Y}$.

(iii) the functions $t \mapsto (\dot{\mathbf{u}}(t)|P\dot{\mathbf{u}}(t))$, $t \mapsto (\mathbf{u}(t)|N\mathbf{u}(t))$ and $t \mapsto \langle L\mathbf{u}(t)|\tilde{S}L\mathbf{u}(t)\rangle$ are uniformly bounded on compact subintervals of $[0, T[$;

(iv) for every pair of weakly continuous functions $\mathbf{v}$: $[0, T[\to \mathscr{V}$, $\dot{\mathbf{v}}$: $[0, T[\to \mathscr{H}$ satisfying equation (3.5.26) (with $\mathbf{v}$ replacing $\mathbf{u}$) and $0 \leqslant t_1 < t_2 < T$

$$(P\dot{\mathbf{u}}|\mathbf{v})|_{t_1}^{t_2} = \int_{t_1}^{t_2} ds\, \{(P\dot{\mathbf{u}}|\dot{\mathbf{v}}) - (\mathbf{v}|N\mathbf{u}) - \langle L\mathbf{v}|\tilde{S}L\mathbf{u}\rangle\}; \tag{3.5.27}$$

(v) $E(t) :\equiv \frac{1}{2}(\dot{\mathbf{u}}|P\dot{\mathbf{u}}) + \frac{1}{2}(\mathbf{u}|N\mathbf{u}) + \tilde{\mathscr{W}}(L\mathbf{u}) \leqslant E(0).$ (3.5.28)

and $E(\cdot)$ is non-increasing.

Note that $(\mathbf{v}|N\mathbf{u}) + \langle L\mathbf{v}|\tilde{S}L\mathbf{u}\rangle = \langle L\mathbf{v}|SL\mathbf{u}\rangle$, $\frac{1}{2}(\mathbf{u}|N\mathbf{u}) + \tilde{\mathscr{W}}(L\mathbf{u}) = \mathscr{W}(L\mathbf{u})$.

Finally, equation (3.5.20) can be rewritten in an abstract form:

$$2(1+2K)\tilde{\mathscr{W}}(L\mathbf{u}) \geqslant \langle L\mathbf{u}|\tilde{S}L\mathbf{u}\rangle, \quad K > 0. \tag{3.5.29}$$

The hypotheses of the blow-up theorems imply that $\tilde{\mathscr{W}}(L\mathbf{u}) < 0$. Hence $\tilde{W}(\tilde{\mathbf{F}}, \xi) < 0$ for some $\tilde{\mathbf{F}} \in \mathrm{im} L\mathbf{u}$. For an $\tilde{\mathbf{F}}$ satisfying $\tilde{W}(\tilde{\mathbf{F}}, \xi) < 0$ equation (3.5.29) means essentially that $\tilde{W}(\theta\tilde{\mathbf{F}}, \xi) = |\theta|^k \tilde{W}(\tilde{\mathbf{F}}, \xi)$, $k > 2$, i.e. the term $\tilde{W}$ of (3.5.19) is responsible for non-linear effects.

Consider the function

$$t \mapsto f(t) :\equiv (\mathbf{u}(t)|P\mathbf{u}(t)). \tag{3.5.30}$$

It is differentiable and

$$f'(t) = 2(\dot{\mathbf{u}}(t)|P\mathbf{u}(t)) = 2(P\dot{\mathbf{u}}(t)|\mathbf{u}(t)). \tag{3.5.31}$$

From equation (3.5.27)

$$f'(t) - f'(t_1) = 2\int_{t_1}^{t} ds\, \{(P\dot{\mathbf{u}}(s)|\dot{\mathbf{u}}(s)) - \langle L\mathbf{u}(s)|SL\mathbf{u}(s)\rangle\}$$

and in view of (iii) f' is uniformly Lipschitz continuous on compact subintervals of $[0, T[$. Consequently it is differentiable a.e. on $[0, T[$ with

$$f''(t) = 2(P\dot{\mathbf{u}}|\dot{\mathbf{u}}) - 2\langle L\mathbf{u}|SL\mathbf{u}\rangle. \tag{3.5.31'}$$

3.5.3 Comparison of the Weak Solutions in the Sense of KLP with the BV Solutions

Let $(\mathbf{v}, \mathbf{F})$ be a weak solution of equations (3.3.2) in the sense of Sections 3.2 and 3.3. Let $\dot{\mathbf{u}}(t) := \mathbf{v}(t+0, \cdot)$ for all $t \in [0, T]$. For bounded $\mathscr{B}$ and $p > 2$

$$\mathbf{v} \in \mathscr{L}^\infty([0, T] \times \mathscr{B}) \subset \mathscr{L}^p([0, T] \times \mathscr{B}) \subset \mathscr{L}^2([0, T] \times \mathscr{B})$$

and by the Fubini theorem $\dot{\mathbf{u}}(t) \in \mathscr{H}$, $||\dot{\mathbf{u}}(t)||_\infty \leqslant K$ for almost all t. Since

$$\int_{t_0}^{t_1} dt |(\dot{\mathbf{u}}(t)|P)| \leqslant \int_{[t_0, t_1] \times \mathscr{B}} dL^4 \left| \sum_{k=1}^{3} \varrho_0(\xi) v^k(t, \xi) w^k(\xi) \right|$$

for every $\mathbf{w} \in \mathscr{H} \subset \mathscr{L}^1$, $\dot{\mathbf{u}}(\cdot)$ is locally weakly summable.* The function $t \mapsto (\dot{\mathbf{u}}(t)|P\dot{\mathbf{u}}(t)) = \int_{\mathscr{B}} dL^3\, \varrho_0 \mathbf{v}(t, \xi)^2 \leqslant M\|\mathbf{v}\|_\infty{}^2$ is uniformly bounded.

Let $u_0{}^k{}_{,\alpha} = F^k{}_\alpha(0, \cdot)$ (the right-hand side being defined in the sense of inner traces) and let $\mathbf{u}(t)$ be defined by equation (3.5.26) with $\mathbf{u}(0) = \mathbf{u}_0$. Set $\phi = \varrho_0^1 L\psi$, $\psi \in \mathscr{C}_0^\infty(\mathscr{B})$ in equation (3.5.26). It follows that $(L\mathbf{u}(t_1)|\psi) - (L\mathbf{u}_0|\psi)$ equals

$$-\int_{]0,t_1]\times\mathscr{B}} \frac{\partial}{\partial\xi^\alpha}[v^k(t, \xi)\psi(\xi)] + \int_{]0,t_1]\times\mathscr{B}} v^k{}_{,\alpha}\psi = \int_{]0,t_1]\times\mathscr{B}} F^k{}_{\alpha,t}\psi$$

$$= \int_{\mathscr{B}} dL^3\, [F^k{}_\alpha(t_1+0, \xi) - F^k{}_\alpha(0, \xi)]\psi(\xi)$$

and $L\mathbf{u}(t_1) = [F^k{}_\alpha(t_1+0, \cdot)]$, Hence the functions $\mathscr{W}(L\mathbf{u}(t)) = \int_{\mathscr{B}} dL^3 W \times (\mathbf{F}(t+0, \xi), \xi)$, $\langle L\mathbf{u}(t)|SL\mathbf{u}(t)\rangle = \int_{\mathscr{B}} dL^3 F^k{}_\alpha(t+0, \xi) S_k{}^\alpha(t+0, \xi)$ are uniformly bounded.

3.5.4 Growth Properties and Non-Existence for Non-Monotone Operators

Theorem 3.5.2 (Knops and Straughan, 1975)
Suppose that

(i) $\langle L\mathbf{u}|SL\mathbf{u}\rangle \leqslant 0\ \forall \mathbf{u} \in \mathscr{V}$,
(ii) $(\mathbf{u}_0|P\mathbf{v}_0) > 0$.

Then

$$(\mathbf{u}|P\mathbf{u}) \geqslant (\mathbf{u}_0|P\mathbf{u}_0) + 2t(\mathbf{u}_0|P\mathbf{v}_0) + \frac{(\mathbf{u}_0|P\mathbf{v}_0)}{(\mathbf{u}_0|P\mathbf{u}_0)^{1/2}}\, t^2.$$

* $\dot{u}$ need not be weakly continuous in general, although this property is plausible for a weak solution $(\mathbf{v}, \mathbf{F})$ with the jump set Γ not too large and satisfying (L). Indeed, for $\psi = \varrho_0\phi \in \mathscr{C}_0^\infty(\mathscr{B})$,

$$(\dot{u}^k(t+0) - \dot{u}^k(t_1+0)|P\phi) = \int_{\mathscr{B}} dL^3[v^k(t+0, \xi) - v^k(t_1+0, \xi)]\psi(\xi)$$
$$= \int_{[t,t_1[\times\mathscr{B}} v^k{}_{,t}\psi$$

by the Gauss–Green theorem. Let $v^k{}_{,t}|\Gamma^c \in \mathscr{L}^1([0, T]\times\mathscr{B}, L^4)$. The last expression is equal to

$$\int_{[t,t_1[\times\mathscr{B}} dL^4\, y^k\psi + \int_{\Gamma\cap([t,t_1[\times\mathscr{B})} dH^3\, \nu_t \Delta_\nu v^k\psi.$$

In view of (L) the set Γ intersects the hyperplane $t = t_1$ transversally and hence for $t \to t_1$ the last integral tends to an integral over the set $\Gamma\cap(\{t_1\}\times\mathscr{B})$ of H^3-measure zero. In this case weak continuity of $\dot{u}$ follows. Fortunately, however, weak continuity of $\dot{\mathbf{u}}$ does not play any role in the proofs of blow-up theorems.

Proof

Equation (3.5.31) and (i) imply that $f''(t) \geqslant 2(P\mathbf{u}(t)|\mathbf{u}(t))$. By the Schwartz inequality

$$ff''-\tfrac{1}{2}(f')^2 \geqslant 2(\mathbf{u}|P\mathbf{u})(\dot{\mathbf{u}}|P\dot{\mathbf{u}})-2(\dot{\mathbf{u}}|P\mathbf{u})^2 \geqslant 0,$$

$(f^{1/2})'' = -\tfrac{1}{4}f^{-3/2}(f')^2+\tfrac{1}{2}f^{-1/2}f'f'' \geqslant 0$. Hence $f(t)^{1/2} \geqslant f(0)^{1/2}+t(f^{1/2})'(0)$ and

$$f(t) \geqslant f(0)+2f'(0)\,t+\tfrac{1}{2}t^2f'(0)f(0)^{-1/2}. \qquad \square$$

Theorem 3.5.2 implies that some weak solutions are unbounded in time if the condition (i) is satisfied for some $\mathbf{u} \in \mathscr{V}$ (Note that we needed the condition (i) only for $\mathbf{u} = \mathbf{u}(t)$, $0 \leqslant t < \infty$.) In the linear case the condition (i) can be satisfied if the elasticity matrix is not positive definite.

Theorem 3.5.3 (Straughan)

Suppose that

(i) $-\langle L\mathbf{u}|SL\mathbf{u}\rangle \geqslant k(\mathbf{u}|P\mathbf{u})^{1+\varepsilon}$, $k > 0$, $\varepsilon > 0$,

(ii) $(\mathbf{u}_0|P\mathbf{v}_0) > 0$.

Then no weak solution with the initial data $\mathbf{u}_0$, $\mathbf{v}_0$ is defined for all $t \geqslant 0$.

Proof

This technique of proving non-existence has been named the **method of convergent integrals.**

(i) implies that

$$f'' \geqslant 2kf^{1+\varepsilon} \geqslant 0 \quad \text{on } [0, T[, \tag{3.5.32}$$

while (ii) implies that $f' > 0$ for $t \in [0, \tau[$, $\tau < T$, provided the positive number τ is sufficiently small (remember that f' is Lipschitz continuous). Let $T' := \sup\{\tau \in [0, T[|\ f' > 0 \text{ on } [0, \tau[\}$ and suppose that $T' < T$. Multiplying equation (3.5.32) by f' and integrating we get the inequality

$$f'(t)^2 \geqslant \int_0^t ds\, 2kf(s)^{1+\varepsilon}f'(s)+f'(0)^2 \geqslant \alpha^2 f(t)^{2+\varepsilon}+\Phi^2, \tag{3.5.33}$$

for $t < T'$, with $\alpha^2 := 4k(1+\varepsilon)^{-1}$, $\Phi^2 := f'(0)^2-\alpha^2 f(0)^{2+\varepsilon}$. Since $f'(0) > 0$, Φ^2 can be made positive by choosing a sufficiently small k in (i). Hence $f'(T') > 0$. Since f' is continuous, this conclusion is in contradiction with the definition of T'. Thus $f' > 0$ on $[0, T[$.

Suppose that the solution is defined on $[0, \infty[$. Then $f(t) < \infty$ for all $t \geqslant 0$ and (3.5.33) implies that

$$t \leqslant \int_{f(0)}^{f(t)} dx\, [\Phi^2 + \alpha^2 x^{2+\varepsilon}]^{-1/2} \leqslant \int_{f(0)}^{\infty} dx\, [\Phi^2 + \alpha^2 x^{2+\varepsilon}]^{-1/2},$$

again a contradiction. □

Theorems 3.5.2 and 3.5.3 assert that for some initial conditions the displacements become unbounded in finite or infinite time. We have not investigated whether the condition (v) (Definition 3.5.1) is violated as yet. However, under the hypotheses of Theorem 3.5.3 it is possible to derive the following estimate for the kinetic energy:

$$4(\dot{\mathbf{u}}|P\dot{\mathbf{u}}) \geqslant \alpha^2 f^{1+\varepsilon} + \Phi^2 f^{-1}, \tag{3.5.34}$$

It implies that the kinetic energy tends to infinity in finite time. We cannot however conclude that the condition (v) is violated unless we know that all the terms in the expression (3.5.28) for $E(t)$ are nonnegative.

Theorem 3.5.4

Suppose that the weak solution $\mathbf{u}(\cdot)$ satisfies the inequality $\mathscr{W}(L\mathbf{u}(t)) \geqslant 0$ (as well as (v)). Then it satisfies the inequality

$$(\mathbf{u}(t)|P\mathbf{u}(t)) \leqslant [(\mathbf{u}_0|P\mathbf{u}_0) + tE(0)](1+t). \tag{3.5.35}$$

Proof

$(\mathbf{u}(t)|P\mathbf{v}) = (\mathbf{u}_0|P\mathbf{v}) + \int_0^t ds\, (\dot{\mathbf{u}}(s)|P\mathbf{v})$ for every $\mathbf{v} \in \mathscr{H}$, hence by the Schwartz inequality

$$(\mathbf{u}(t)|P\mathbf{v})^2 \leqslant [(\mathbf{u}_0|P\mathbf{u}_0) + \int_0^t ds\, (\dot{\mathbf{u}}(s)|P\dot{\mathbf{u}}(s))](1+t)(\mathbf{v}|P\mathbf{v})$$

$$\leqslant [(\mathbf{u}_0|P\mathbf{u}_0) + E(0)\,t](1+t)(\mathbf{v}|P\mathbf{v}).$$

For $\mathbf{v} = \mathbf{u}(t)$ this becomes (3.5.35). □

Corollary 3.5.5

Suppose that the hypothesis (i) of Theorem 3.5.2 is satisfied. If a weak solution $\mathbf{u}(\cdot)$ satisfies the hypothesis (ii) of Theorem 3.5.2 and $\mathscr{W}(L\mathbf{u}(t)) \geqslant 0$, then $t \mapsto (\mathbf{u}(t)|P\mathbf{u}(t))$ exhibits exactly quadratic growth (i.e. it is a polynomial of second degree in t).

Remark

Under the hypothesis (i) of Theorem 3.5.3 a weak solution satisfying the hypothesis (ii) as well as $\mathscr{W}(L\mathbf{u}(t)) \geqslant 0$ violates the condition (v) for t outside a bounded interval of the real axis.

3.5.5 The Role of the Non-Linear Part of the Stress. Concavity Methods

In this section we assume that the quadratic part of the strain energy has the correct sign:

$$(\mathbf{u}|N\mathbf{u}) > 0 \quad \text{for all } \mathbf{u} \in \mathscr{H} \text{ unless } \mathbf{u} = \mathbf{0} \tag{3.5.36}$$

and investigate the instability and blow-up resulting from some properties of the non-linear part $\tilde{\mathscr{W}}(L\mathbf{u})$ of the strain energy. We also assume throughout this section that $\tilde{\mathscr{W}}$ satisfies equation (3.5.29). Under the assumptions (3.5.24) and (3.5.36) it is still possible that $\mathscr{W}(L\mathbf{u}(t)) < 0$ and $E(t) < 0$ provided $\langle L\mathbf{u}(t)|\tilde{S}L\mathbf{u}(t)\rangle < 0$.

The blow-up theorems obtained in this section fall into two groups according to their hypotheses:

(1) $E(0) \leqslant 0$ (or even $E(0) < 0$), or

(2) $E(0) > 0$.

In the first case $\tilde{\mathscr{W}}(L\mathbf{u}_0) \leqslant 0$ (or even $\tilde{\mathscr{W}}(L\mathbf{u}_0) < 0$) and $|\tilde{\mathscr{W}}(L\mathbf{u}_0)|$ dominates over the quadratic part $\frac{1}{2}(\mathbf{u}_0|N\mathbf{u}_0) \geqslant 0$ of the energy. In view of equation (3.5.29) $\langle L\mathbf{u}_0|\tilde{S}L\mathbf{u}_0\rangle \leqslant 0$ (or < 0, resp.)

In the second case the inequality

$$\tfrac{1}{2}(\mathbf{v}|N\mathbf{v}) + \tilde{\mathscr{W}}(L\mathbf{v}) \geqslant 0 \quad \forall \mathbf{v} \in \mathscr{V} \tag{3.5.37}$$

implies (see equation (3.5.35)) that $f(t) \equiv (\mathbf{u}(t)|P\mathbf{u}(t))$ has at most quadratic growth. The blow-up is possible only if

$$\tilde{\mathscr{W}}(L\mathbf{v}) < -(\mathbf{v}|N\mathbf{v}) < 0 \tag{3.5.38}$$

and hence also $\langle L\mathbf{v}|\tilde{S}L\mathbf{v}\rangle < 0$ for some $\boldsymbol{v} \in \mathscr{V}$.

The key argument is a second-order differential inequality for the function

$$g(t) :\equiv f(t) + \beta(t+t_0)^2, \quad t \in [0, T] \tag{3.5.39}$$

with appropriately chosen constants β, $t_0 > 0$. Note that $g(t) \geqslant 0$ and $t_0 > 0$, $g(t) = 0$ imply that $\beta = 0$, $\mathbf{u}(t) = \mathbf{0}$. From equations (3.5.27) and (3.5.31)

$$\begin{aligned} g'(t) = {} & 2(\mathbf{u}_0|P\mathbf{v}_0) + 2\int_0^t ds\, [(\dot{\mathbf{u}}(s)|P\dot{\mathbf{u}}(s)) - (\mathbf{u}(s)|N\mathbf{u}(s)) \\ & - \langle L\mathbf{u}(s)|\tilde{S}L\mathbf{u}(s)\rangle] + 2\beta(t+t_0). \end{aligned} \tag{3.5.40}$$

For almost all $t \in [0, T]$:

$$g''(t) = 2(\dot{\mathbf{u}}(t)|P\dot{\mathbf{u}}(t)) - 2(\mathbf{u}(t)|N\mathbf{u}(t)) - 2\langle L\mathbf{u}(t)|\tilde{S}L\mathbf{u}(t)\rangle + 2\beta.$$

Hence (using equation (3.5.28))

$$\begin{aligned} g''(t) &= 4(1+K)(\dot{\mathbf{u}}|P\dot{\mathbf{u}}) + 4K(\mathbf{u}|N\mathbf{u}) + 2[2(1+2K)\tilde{\mathscr{W}}(L\mathbf{u}) \\ &\quad - \langle L\mathbf{u}|\tilde{S}L\mathbf{u}\rangle] - 4(1+2K)E(t) + 2\beta \end{aligned}$$

and

$$-K^{-1}g^{K+2}\frac{d^2g^{-K}}{dt^2} = gg'' - (1+K)(g')^2 = S(t)^2 + 4K(\mathbf{u}|N\mathbf{u})g(t)$$

$$-2(1+2K)[\beta + 2E(t)]g(t) + 2[2(1+2K)\tilde{\mathscr{W}}(L\mathbf{u}) - \langle L\mathbf{u}|\tilde{S}L\mathbf{u}\rangle]g(t),$$

with

$$\begin{aligned} S(t)^2 &:= 4(1+K)\{[(\mathbf{u}|P\mathbf{u}) + \beta(t+t_0)^2][(\dot{\mathbf{u}}|P\dot{\mathbf{u}}) + \beta] \\ &\quad - [(\mathbf{u}|P\dot{\mathbf{u}}) + \beta(t+t_0)]^2\} = 4(1+K)[(\mathbf{u}|P\mathbf{u})(\dot{\mathbf{u}}|P\dot{\mathbf{u}}) - (\mathbf{u}|P\dot{\mathbf{u}})^2 \\ &\quad + (\mathbf{u} + \beta(t+t_0)\dot{\mathbf{u}}|P[\mathbf{u} + \beta(t+t_0)\dot{\mathbf{u}}])] \geqslant 0. \end{aligned} \tag{3.5.41}$$

From the inequalities (3.5.28), (3.5.41) and (3.5.29) we conclude that

$$\begin{aligned} -K^{-1}g^{K+2}\frac{d^2g^{-K}}{dt^2} &= g''g - (1+K)(g')^2 \\ &\geqslant 4K(\mathbf{u}|N\mathbf{u})g - 2(1+2K)[\beta + 2E(0)]. \end{aligned} \tag{3.5.42}$$

Theorem 3.5.6

Every weak solution satisfying the inequality $E(0) < 0$ blows up for some $t_1 < \infty$ and the following estimate is true:

$$(\mathbf{u}(t)|P\mathbf{u}(t)) \geqslant \left[\frac{t_1}{t - t_1}\right]^{1/K} g(0) - \beta(t+t_0)^2.$$

Proof

Let $\beta = -2E(0) > 0$. Since $g(t) > 0$, we have $p(t) :\equiv -\dfrac{d^2g^{-K}}{dt^2} \geqslant 0$ for $t \in [0, T[$,

$$\begin{aligned} g(t)^{-K} &= -\int_0^t ds \int_0^s d\tau\, p(\tau) + \left[1 - Kt\frac{g'(0)}{g(0)}\right] g(0)^{-K} \\ &\leqslant \left[1 - Kt\frac{g'(0)}{g(0)}\right] g(0)^{-K} \end{aligned}$$

and

$$g(t)^K \geqslant \frac{g(0)^K}{1 - Ktg(0)^{-1}g'(0)}. \tag{3.5.43}$$

Choose t_0 so large that $g'(0) > 0$, equation (3.5.40). It follows that for $t \to t_1$, $t_1 = K^{-1}g(0)g'(0)^{-1} < \infty$, $g(t)^K \to \infty$, $K > 0$, which implies that $g(t) \to \infty$, $(\mathbf{u}|P\mathbf{u}) \to \infty$ for $t \to t_1$. □

Theorem 3.5.7

If $E(0) = 0$ and $(\mathbf{u}_0|P\mathbf{v}_0) > 0$, then the weak solution with the initial data $\mathbf{u}_0, \mathbf{v}_0$ blows up in finite time.

Proof

For $\beta = 0$ equation (3.5.43) follows by a similar argument. □

Other initial conditions leading to weak solutions which blow up in a finite time are given by Knops, Levine and Payne (1975).

Let $E(0) > 0$ and $\tilde{\mathscr{W}}(L\mathbf{v}) < -(\mathbf{v}|N\mathbf{v}) < 0$, $\langle L\mathbf{v}|\tilde{S}L\mathbf{v}\rangle < 0$ for some $\mathbf{v} \in \mathscr{V}$. Let $\beta = 0$. Since $\exists\, \lambda > 0$ such that $(\mathbf{u}|N\mathbf{u}) \geqslant \lambda(\mathbf{u}|P\mathbf{u}) = \lambda g(t)$ (in view of Friedrichs' inequality), equation (3.5.42) implies that

$$-K^{-1}g^{K+2}\frac{d^2g^{-K}}{dt^2} \geqslant 4K\lambda g(t)^2 - 4(1+2K)E(0)g(t). \tag{3.5.44}$$

Choose the initial data in such a way that

$$\frac{K\lambda g(0)}{(1+2K)E(0)} > 1, \quad g'(0) > 0, \tag{3.5.45}$$

e.g. $(\mathbf{u}_0|P\mathbf{v}_0) > 0$, $\quad K\lambda(\mathbf{u}_0|P\mathbf{u}_0) > (1+2K)\{\frac{1}{2}(\mathbf{v}_0|P\mathbf{v}_0) + \frac{1}{2}(\mathbf{u}_0|N\mathbf{u}_0) + \tilde{\mathscr{W}}(L\mathbf{u}_0)\}$. The last inequality can be satisfied provided $\tilde{\mathscr{W}}(L\mathbf{u}_0) < 0$ (for some $\mathbf{u}_0 \in \mathscr{V}$).

Theorem 3.5.8

Suppose that the weak solution satisfies the initial conditions (3.5.45) and $E(0) > 0$. Then it blows up in finite time.

Proof

Let $T' > 0$ be such that

$$\frac{K\lambda g(t)}{(1+2K)E(0)} \geqslant 1 \quad \text{for } t \in [0, T']. \tag{3.5.46}$$

Then $\dfrac{d^2g^{-K}}{dt^2} \leqslant 0$ on $[0, T']$ and

$$g'(t) \geqslant \left(\frac{g(t)}{g(0)}\right)^{1+K} g'(0) \geqslant kg(0) \quad \text{for } t \in [0, T'],$$

$$k := \left[\frac{(1+2K)E(0)}{K\lambda g(0)}\right]^{1+K}.$$

Consequently $g(t) \geqslant g(0)+ktg'(0)$ for all $t \in [0, T']$ and (3.5.46) holds for all $t \in \operatorname{dom} g(\cdot)$. Hence $\dfrac{d^2 g^{-K}}{dt^2} \leqslant -4g(t) < 0$ in $\operatorname{dom} g(\cdot)$. If $\operatorname{dom} g(\cdot) = \bar{\boldsymbol{R}}_+$, then equation (3.5.43) implies that $g(\cdot)$ blows up in finite time. □

Other cases of blow-up include the initial conditions satisfying the inequality $E(0) > 0$ and one of the following: (1) $(\mathbf{u}_0|P\mathbf{v}_0) = 0$, (2) $(\mathbf{u}_0|P\mathbf{v}_0) < 0$, but $|(\mathbf{u}_0|P\mathbf{v}_0)|^2 < h(\mathbf{u}_0)$ for some function h (Knops, Levine and Payne, 1975).

Theorem 3.5.8 indicates that a solution starting from a state u_0 with a negative strain energy $\tilde{\mathscr{W}}(L\mathbf{u}_0)$ can blow up even though $E(0) > 0$ and $\frac{1}{2}(\mathbf{v}|N\mathbf{v})$ is positive definite on $\mathscr{V}$. It follows that the stability criteria which ignore the non-linear term $\tilde{\mathscr{W}}(L\mathbf{u}_0)$ are not satisfactory.

Concerning the blow-up of kinetic energy it is possible to obtain a lower bound of the kinetic energy by a function $g(t)^\gamma$, $\gamma > 0$. Since $E(t) \leqslant E(0)$, it follows that $\tilde{\mathscr{W}}(L\mathbf{u}) \to -\infty$ and $\langle L\mathbf{u}|\tilde{S}L\mathbf{u}\rangle \to -\infty$ in finite time.

3.5.6 Some Other Blow-up Theorems

Suppose that

$$2K\mathscr{W}(L\mathbf{u})+\langle L\mathbf{u}|SL\mathbf{u}\rangle \geqslant 0, \quad K \geqslant 1. \tag{3.5.47}$$

Since the hypotheses of the theorems imply that $\mathscr{W}(L\mathbf{u}) < 0$, equation (3.5.47) means roughly speaking that $W(\theta\mathbf{F}, \xi) = |\theta|^k W(\mathbf{F}, \xi)$, $k \leqslant -2$ for some $\mathbf{F} \in \operatorname{im} L\mathbf{u}$ satisfying $W(\mathbf{F}, \xi) < 0$. The singularity at $\mathbf{F} = \mathbf{0}$ acts as an energy sink. This is obviously in contradiction with the expected behaviour under extreme compression: $W(\mathbf{F}, \xi) \to \infty$ when $\mathbf{F} \to \mathbf{0}$.

Theorem 3.5.9

Suppose that a weak solution satisfies equation (3.5.47) over its domain as well as the initial condition $(\mathbf{v}_0|P\mathbf{u}_0) > 0$. Then it cannot be defined for all $t \geqslant 0$.

Proof

Let $\varphi \in \mathscr{C}^2([0, T[, \boldsymbol{R})$, $\varphi(t) > 0$, $\mathbf{v}(t) := \varphi(t)\mathbf{u}(t)$, $\mathbf{w}(t) := \varphi(t)^2\mathbf{u}(t)$, $\dot{\mathbf{v}}(t) = \varphi'(t)\mathbf{u}(t)+\varphi(t)\dot{\mathbf{u}}(t)$, $\dot{\mathbf{w}}(t) = \varphi(t)\dot{\mathbf{v}}(t)+\varphi'(t)\boldsymbol{v}(t)$. Now

$$(P\dot{\mathbf{u}}|\mathbf{w})|_0^t = \int_0^t ds\,[(\dot{\mathbf{u}}(s)|P\dot{\mathbf{w}}(s))-\langle L\mathbf{w}(s)|SL\mathbf{u}(s)\rangle],$$

and hence

$$\varphi(t)^2 f'(t)-\varphi(0)^2 f'(0) = 2\int_0^t ds\,\big[(\dot{\mathbf{v}}(s)|P\dot{\mathbf{v}}(s))-(\varphi')^2\varphi^{-1}(\mathbf{v}|P\mathbf{v}) \\ -\varphi^2\langle L\mathbf{u}|SL\mathbf{u}\rangle\big].$$

On account of (3.5.47) and (v) the above expression is equal to

$$2\int_0^t ds\,\{(1-K)(\dot{\mathbf{w}}|P\dot{\mathbf{w}})-(1-K)\varphi'^2(\mathbf{u}|P\mathbf{u})+2K\varphi\varphi'(\mathbf{u}|P\dot{\mathbf{u}})$$

$$+\varphi^2[K(\dot{\mathbf{u}}|P\dot{\mathbf{u}})-\langle L\mathbf{u}|SL\mathbf{u}\rangle]\} \leqslant 2K\varphi\varphi' f|_0^t+4KE(0)\int_0^t ds\,\varphi(s)^2$$

$$-2\int_0^t ds\,[(\varphi'^2+K\varphi\varphi'')(\mathbf{u}|P\mathbf{u})]. \tag{3.5.48}$$

Let $\varphi(t)\equiv e^{\beta t}$, $\beta>0$, so that $(\varphi')^2+K\varphi\varphi''=\beta^2(1+K)\varphi^2>0$ and the last term on the right-hand side of equation (3.5.48) can be dropped:

$$e^{2\beta t}f'(t)\leqslant f'(0)+2K\beta[e^{2\beta t}f(t)-f(0)].$$

Hence

$$f'(t)=2K\beta f(t)+e^{-2\beta t}f'(0)-2K\beta e^{-2\beta t}f(0)-p(t), \quad p(t)\geqslant 0.$$

Let $f(t)\equiv g(t)e^{2K\beta t}$. Then

$$g(t)-f(0)=\frac{f'(0)-2K\beta f(0)}{2\beta(1+K)}\,(e^{-2K\beta t-2\beta t}-1)-\int_0^t ds\,p(s)e^{-2K\beta s}$$

and

$$f(t)\leqslant\frac{-f'(0)+2K\beta f(0)}{2\beta(1+K)}\,(e^{-2\beta t}-e^{2K\beta t})+f(0)\,e^{2K\beta t}$$

$$=H_0\,e^{2K\beta t}+L_0\,e^{-2\beta t}, \tag{3.5.49}$$

$$H_0:=\frac{2\beta f(0)-f'(0)}{2\beta(1+K)}.$$

If $H_0<0$, then for sufficiently large t equation (3.5.49) implies that $f(t)<0$, which is impossible. Hence the solution cannot be defined for all $t>0$ if $H_0<0$. On the other hand it is always possible to choose such a β that $H_0<0$ provided $f'(0)>0$. □

The case of an elastic body subject to pressure loading

$$T^{kl}n_l=-p_{(r)}n_k \quad \text{on } \boldsymbol{\chi}^{(t)}(\Sigma_r) \tag{3.5.50}$$

was considered by Ball (1978). The paper makes allowance for the possibility that the boundary $\partial\mathscr{B}$ has several components Σ_r, $r=1,2,\dots$, subject to pressures $p_{(r)}$. The pressure $p_{(r)}$ on each Σ_r is constant. Let Σ_r be $\mathscr{C}^1$-smooth, $r=1,2,\dots$ By Whitney's extension theorem (Narasimhan, 1968) we can find a $\mathscr{C}^1$ function $p(\xi)$ such that $p|\Sigma_r=p_{(r)}$, $r=1,2,\dots$

We shall construct an energetic extension of our problem (3.5.12), (3.5.50). Let

$$\mathscr{V} := \left\{\mathbf{u} \in W^{1,p}(\mathscr{B}) \,\middle|\, \int_{\mathscr{B}} dL^3\mathbf{u} = \mathbf{0}\right\}$$

be endowed with the norm $\|\mathbf{u}\|_{\mathscr{V}} := \|\nabla\mathbf{u}\|_p$. The integral condition involved in the definition of $\mathscr{V}$ eliminates the additional degrees of freedom associated with the possibility of rigid motion and serves to justify the choice of the norm on $\mathscr{V}$, by the Poincaré lemma. The gradient operator ∇, defined on $\mathscr{C}^\infty \cap \mathscr{V}$ can be extended to an isometry $L\colon \mathscr{V} \to \mathscr{Y}$, $\mathscr{Y} := \mathscr{L}^q(\mathscr{B}, \boldsymbol{R}^9)$.

Let

$$\mathscr{D} := \left\{\mathbf{u} \in \mathscr{C}^2(\overline{\mathscr{B}}) \,\middle|\, \int_{\mathscr{B}} dL^3\,\mathbf{u} = \mathbf{0},\right.$$

$$\left.\nu_\alpha[S_k^{\ \alpha}(\mathbf{F}_0+\nabla\mathbf{u}, \xi)+p(\det\mathbf{F})\overset{-1}{F^\alpha}{}_k] = 0\right\}.$$

We define the continuous linear form

$$g\colon L\mathbf{h} \mapsto \int_{\mathscr{B}} dL^3\, p_{,\alpha}(\det\mathbf{F})\overset{-1}{F^\alpha}{}_k h^k \tag{3.5.51}$$

on $\mathscr{R}(L) \subset \mathscr{Y}$. The definition is correct since

(1) $p_{,\alpha}(\det\mathbf{F})\overset{-1}{F^\alpha}{}_k \in \mathscr{L}^q$, $h^k \in \mathscr{L}^p$;

(2) the right-hand side of (3.5.51) vanishes if $L\mathbf{h} = \mathbf{0}$.

We shall take care of (1) later. For the proof of (2) note that $L\mathbf{h} = \mathbf{0}$ implies that $\mathbf{h} = \mathbf{0}$ in $\mathscr{L}^p \subset \mathscr{V}$ and

$$I := \int_{\mathscr{B}} dL^3\, p_{,\alpha}(\det\mathbf{F})\overset{-1}{F^\alpha}{}_k h^k = 0.$$

The form is continuous on $\mathscr{R}(L)$ on account of (1) and the Poincaré lemma. By the Hahn–Banach theorem it can be extended to a continuous linear functional $\tilde{g}$ on $\mathscr{Y}$, $\tilde{g} \in \mathscr{Y}^*$.

Let $\hat{S}\colon \mathscr{Y} \to \mathscr{Y}^*$ be the function

$$\tilde{\mathbf{F}}(\cdot) \mapsto [S_k^{\ \alpha}(\mathbf{F}_0(\cdot)+\tilde{\mathbf{F}}(\cdot), \cdot)+p(\cdot)\det\mathbf{F}(\cdot)\overset{-1}{F^\alpha}{}_k(\cdot)]+\tilde{g}, \tag{3.5.52}$$

$\mathbf{F} := \mathbf{F}_0+\tilde{\mathbf{F}}$. We must choose the number p in such a way that $\mathbf{F} \in \mathscr{L}^p$ entails $|\mathbf{F}|^2 \in \mathscr{L}^q$, $q^{-1}+p^{-1} = 1$. But $\mathbf{F} \in \mathscr{L}^p \subset \mathscr{L}^{2q}$ provided $p \geqslant 2q$ and, by an easy calculation, $p \geqslant 3$. We assume that $p \geqslant 3$ and

$$|S_k^{\ \alpha}(\mathbf{F}, \xi)| \leqslant a(\xi)+b|\mathbf{F}|^{p-1}. \tag{3.5.53}$$

With these assumptions $\hat{S}$ maps $\mathscr{Y}$ into $\mathscr{Y}^*$ continuously. Moreover the condition (1) above is also satisfied.

Let $A := L^\dagger SL$. For $\mathbf{h} \in \mathscr{V} \cap \mathscr{C}^1(\mathscr{B})$, $\mathbf{u} \in \mathscr{D}$,

$$(A\mathbf{u}|\mathbf{h}) = \langle \hat{S}L\mathbf{u}|L\mathbf{h}\rangle = \int_{\mathscr{B}} dL^3\, S_k{}^\alpha(\mathbf{F}_0+\nabla\mathbf{u},\, \xi)h^k{}_{,\alpha}$$

$$+\int_{\mathscr{B}} dL^3\, (ph^k)_{,\alpha}(\det\mathbf{F})\overset{-1}{F^\alpha}{}_k = \int_{\partial\mathscr{B}} dH^2\, [S_k{}^\alpha(\mathbf{F}_0+\nabla\mathbf{u},\, \xi)$$

$$+p(\det\mathbf{F})\overset{-1}{F^\alpha}{}_k]\nu_\alpha h^k - \int_{\mathscr{B}} dL^3\, S_k{}^\alpha(\mathbf{F}_0+\nabla\mathbf{u},\, \xi)_{,\alpha}h^k,$$

since $[(\det\mathbf{F})\hat{F}^\alpha{}_k]_{,\alpha} = [\in_{ijk} F^i{}_\alpha F^j{}_\beta \in^{\alpha\beta\gamma}]_{,\gamma}/2 = 0$ for $\mathbf{F} = \nabla(\boldsymbol{\chi}_0+\mathbf{u})$. A is an energetic extension for our problem in the sense of Chapter 2, since the surface integral vanishes.

A potential for the loads (3.5.50) is given by the formula

$$\Phi(\mathbf{u}) = \frac{1}{3}\int_{\partial\chi^{(t)}(\mathscr{B})} dH^2\, p\mathbf{u}\cdot\mathbf{n} = \frac{1}{3}\int_{\partial\mathscr{B}} dH^2\, p(\det\mathbf{F})\overset{-1}{F^\alpha}{}_k u^k\nu_\alpha. \tag{3.5.54}$$

We now check this assertion. Since we can add any constant to $\Phi(\cdot)$, let us consider

$$\Phi(\mathbf{u}) = \frac{1}{6}\int_{\partial\mathscr{B}} dH^2\, p \in_{ijk} \in^{\alpha\beta\gamma} F^i{}_\alpha F^j{}_\beta \chi^k\nu_\gamma$$

$$= \frac{1}{3}\int_{\partial\mathscr{B}} dH^2\, p(\det\mathbf{F})\overset{-1}{F^\gamma}{}_k \chi^k\nu_\gamma.$$

We have for $\mathbf{u} \in \mathscr{C}^2([0, T]\times\mathscr{B})$

$$D\Phi(\mathbf{u})[\mathbf{v}] = \frac{1}{6}\int_{\partial\mathscr{B}} dH^2\, p \in_{ijk} \in^{\alpha\beta\gamma} F^i{}_\alpha F^j{}_\beta v^k\nu_\gamma$$

$$+\frac{1}{3}\int_{\partial\mathscr{B}} dH^2\, p \in_{ijk} \in^{\alpha\beta\gamma} F_{i\alpha} v^j{}_{,\beta}\chi^k\nu_\gamma. \tag{3.5.55}$$

On account of Stokes' theorem

$$0 = \int_{\partial\mathscr{B}} dH^2\, (p \in_{ijk} \in^{\alpha\beta\gamma} F^i{}_\alpha \chi^j v^k)_{,\beta}\nu_\gamma = \int_{\partial\mathscr{B}} dH^2 \in_{ijk} \in^{\alpha\beta\gamma} p_{,\beta}\nu_\gamma F^i{}_\alpha \chi^j v^k$$

$$+\int_{\partial\mathscr{B}} dH^2\, p \in_{ijk} \in^{\alpha\beta\gamma} F^i{}_\alpha F^j{}_\beta v^k\nu_\gamma - \int_{\partial\mathscr{B}} dH^2\, p \in_{ijk} \in^{\alpha\beta\gamma} F^i{}_\alpha v^j{}_{,\beta}\chi^k\nu_\gamma.$$

The first integral on the right-hand side vanishes because $p_{,\beta} = \lambda \nu_\beta$ on $\partial\mathscr{B}$. The resulting identity can be substituted into equation (3.5.55):

$$D\Phi(\mathbf{u})[\mathbf{v}] = \frac{1}{2}\int_{\partial\mathscr{B}} dH^2\, p\, \epsilon_{ijk}\, \epsilon^{\alpha\beta\gamma}\, F^i{}_\alpha F^j{}_\beta v^k \nu_\gamma = \int_{\partial\chi^{(t)}(\mathscr{B})} dH^2\, p v^k n_k.\star$$

An alternative form of Φ is

$$\Phi(\mathbf{u}) = \int_{\mathscr{B}} dL^3\, (\det\mathbf{F})\,[p+\tfrac{1}{3}\, p_{,\gamma} \overset{-1}{F}_{\gamma k}\chi^k].$$

Indeed, the integrand equals $1/6\,(p\,\epsilon_{ijk}\,\epsilon^{\alpha\beta\gamma}\,F^i{}_\alpha F^j{}_\beta \chi^k)_{,\gamma}$. Note that $\det\mathbf{F} \in \mathscr{L}^1$ provided $\mathbf{F} \in \mathscr{L}^p$, $p \geqslant 3$.

It is clear that Φ is defined on $\mathscr{V}$ provided $p \geqslant 3$. Let

$$E(t) := \tfrac{1}{2}(\dot{\mathbf{u}}|P\dot{\mathbf{u}}) + \mathscr{W}(L\mathbf{u}) + \Phi(\mathbf{u}(t)). \qquad (3.5.56)$$

We shall define a weak solution of (3.3.2), (3.5.50) by (i)–(v) of Section 3.5.1 with (3.5.52), (3.5.54), (3.5.56).

Theorem 3.5.10 (Ball, 1978)

Let $3W(\mathbf{F}_0+\tilde{\mathbf{F}}, \xi) \geqslant W_{F^k{}_\alpha}\tilde{F}^k{}_\alpha$, and let $\mathbf{u}(\cdot)$ be a weak solution of (3.3.2), (3.5.50) with the initial data $\mathbf{u}_0$, $\mathbf{v}_0$ such that either $E(0) \leqslant 0$ and $(\mathbf{u}_0|P\mathbf{v}_0) > 0$ or $E(0) < 0$. Then $f(t) \geqslant \alpha(1-kt)^{-4}$, $\alpha, k > 0$.

Proof

$$\begin{aligned}\ddot{f}(t) &= 2(\dot{\mathbf{u}}|P\dot{\mathbf{u}}) - 2\langle SL\mathbf{u}|L\mathbf{u}\rangle = 2(\dot{\mathbf{u}}(t)|P\dot{\mathbf{u}}(t))\\ &\quad -2\int_{\mathscr{B}} dL^3\,(pu^k)_{,\alpha}(\det\mathbf{F})\overset{-1}{F^\alpha}{}_k - 2\int_{\mathscr{B}} dL^3\, W_{F^k{}_\alpha}(\mathbf{F}_0+\nabla\mathbf{u}, \xi)u^k{}_{,\alpha}\\ &\geqslant 5(\dot{\mathbf{u}}|P\dot{\mathbf{u}}) + 2\Big[3\mathscr{W}(L\mathbf{u}) - \int_{\mathscr{B}} dL^3\, W_{F^k{}_\alpha}(\mathbf{F}_0+\nabla\mathbf{u}, \xi)u^k{}_{,\alpha}\Big] - 6E(0)\\ &\geqslant 5(\dot{\mathbf{u}}|P\dot{\mathbf{u}}).\end{aligned}$$

For sufficiently small t, say for $t \in [0, \tau[$, $f(t) > 0$. Let $g(t) := f(t)^{-1/4}$. We have

$$\begin{aligned}\dot{g}(t) &= -\tfrac{1}{4}\dot{f}(t)f(t)^{-5/4},\\ \ddot{g}(t) &= -\tfrac{1}{4}\,[\ddot{f}(t)f(t) - \tfrac{5}{4}\dot{f}(t)^2]\, f(t)^{-9/4}\\ &\leqslant -\tfrac{5}{4}f^{-9/4}[(\dot{\mathbf{u}}|P\dot{\mathbf{u}})(\mathbf{u}|P\mathbf{u}) - (\dot{\mathbf{u}}|P\mathbf{u})^2] \leqslant 0.\end{aligned}$$

$\star$ As noted by Ball (1976) the potential $\Phi(\mathbf{u})$ is also applicable if some surfaces Σ_r are partially clamped: $\Sigma_r = \Sigma_r' \cup \Sigma_r''$, $\mathbf{u}, \mathbf{v}$ vanish on Σ_r', $p|\Sigma_r'' = p_{(r)}$. On $\Sigma_r'\nabla p$ need not be orthogonal to $\partial\mathscr{B}$ but $v_k = 0$.

In view of our hypotheses, $\dot{f}(0) = 2(\dot{\mathbf{u}}|P\mathbf{u}) > 0$ and hence $g(0) < 0$. Therefore

$$g(t) = \int_0^t ds\, \left[\dot{g}(0) + \int_0^s d\tau\, \ddot{g}(\tau)\right] + g(0) \leqslant g(0) + t\dot{g}(0) \quad \text{and} \quad f(t) \geqslant [g(0) + t\dot{g}(0)]^{-4},$$

which implies the thesis with $\alpha = (\mathbf{u}_0|P\mathbf{u}_0)$, $2\alpha k = (\mathbf{u}_0|P\mathbf{v}_0)$.

Let us now consider the case of $E(0) < 0$ without any condition on $(\mathbf{u}_0|P\mathbf{v}_0)$. Since $\ddot{f}(t) \geqslant -6E(0)$, it follows that $\dot{f}(t) \geqslant \dot{f}(0) - 6E(0)t$ and $\dot{f}(t_0) > 0$ for $t_0 > \dot{f}(0)/6E(0)$, For such $t_0 g(t_0) < 0$ and $E(t_0) \leqslant E(0) < 0$. We can now apply the previous argument with $t = t_0$ replacing $t = 0$. □

An example of W satisfying the hypothesis of the above theorem is provided by $W(\mathbf{F}) \equiv \mathrm{tr}(\mathbf{F}^t\mathbf{F}) + h(\det\mathbf{F})$, $sh'(s) \leqslant h(s)$ for all $s > 0$.

The behaviour of W for $|\mathbf{F}| \to \infty$ and $|\mathbf{F}| \to 0$ can be deduced from a simple physical argument (Ball, 1977a). Let us consider a cube of homogeneous elastic material subject to a homogeneous deformation $\theta\mathbf{F}$, $\theta > 0$. We also assume that the sides of the cube are equal to θ^{-1}. The total strain energy of the cube is $g(\theta) \equiv \varrho_0\theta^{-3}W(\theta\mathbf{F})$. The limit $\theta \to \infty$ corresponds to increasingly strained cubes having the same size and shape in the deformed configuration. Hence we expect that $g(\theta) \to \infty$ (unless the elastic constitutive laws cease to hold for sufficiently high values of θ). For increasing compression $\theta \to 0$ and $g(\theta) \to \infty$ again. Since $g'(\theta) = \varrho_0\theta^{-4}\{W_{F^k{}_\alpha}(\theta\mathbf{F})\theta F^k{}_\alpha - 3W(\theta\mathbf{F})\}$, the hypothesis of Theorem 3.5.10 should hold for sufficiently small θ, i.e. under compression.

Suppose that the hypothesis of Theorem 3.5.10 holds for $|\mathbf{F}| \leqslant f_{cr}$. The geometry of a compressive deformation sets a limit to the displacements $\mathbf{u} \in \mathscr{V}$. On the other hand in extension the body arrives eventually at a limit of validity of the hypothesis and the mechanism of the blow-up is switched off. The latter limit is defined in terms of $f = (\mathbf{u}|P\mathbf{u})$ by a condition $(\mathbf{u}|P\mathbf{u}) \leqslant Kf_{cr}{}^2$, on account of the Poincaré lemma.

APPENDIX TO CHAPTER 3

BALANCE EQUATIONS AND THE VWP IN THE CLASS BV

Suppose that $f^\mu \in \mathrm{BV}(\hat{\mathscr{B}})$, $\hat{\mathscr{B}} := \boldsymbol{\varkappa}(B) \times I$, $g \in \mathscr{L}^1_{loc}(\hat{\mathscr{B}})$, and

$$\int_{\partial\mathscr{P}} dH^3\, f_+{}^\mu\nu_\mu = \int_{\partial\mathscr{P}} dH^3\, f_-{}^\mu\nu_\mu = \int_{\mathscr{P}} dL^4\, g \tag{3A1}$$

holds for every 4-cell $\mathscr{P} \subset \hat{\mathscr{B}}$.* Applying the Gauss–Green theorem we also have

* We could have assumed that (3A1) holds for a.e. 4-cell. Equation (3A1) implies a continuity property of surface integrals like in Section 1.2 and (3A1) naturally extends to **all** 4-cells. Moreover f^μ has traces on *all* 4-cell boundaries. $\mathscr{P}$ may be open, closed or may include some of its faces.

$$\int_{\mathscr{P}_*} \partial_\mu f^\mu = \int_{\mathscr{P}^*} \partial_\mu f^\mu = \int_{\mathscr{P}} \mathrm{d}L^4 g \tag{3A2}$$

for every 4-cell $\mathscr{P}$. $\mathscr{P}_*$ coincides with the interior of $\mathscr{P}$ and $\overline{\mathscr{P}} = \mathscr{P}^*$. The right-hand side of (3A2) can be replaced by an integral over $\mathscr{P}_*$ or $\mathscr{P}^*$.

For every open $\mathscr{U} \subset \hat{\mathscr{B}}$ we can construct a monotone sequence of sets $\mathscr{P}_n$ which are joins of finite numbers of disjoint cells in such a way that $\mathscr{P}_n \nearrow \mathscr{U}$ $(\bigcup_n \mathscr{P}_n = \mathscr{U})$. Since $\partial_\mu f^\mu - dL^4 g$ is a measure we have

$$\int_{\mathscr{U}} \partial_\mu f^\mu - g dL^4 = 0. \tag{3A3}$$

Let $\mathscr{A}$ be any bounded Borel subset of $\hat{\mathscr{B}}$ and $\varepsilon > 0$. It follows easily from equation (3A3) and Lemma 1.1.63 that $\partial_\mu f^\mu$ is a regular Borel measure. So is gdL^4. Hence we can find a bounded open set $\mathscr{U} \supset \mathscr{A}$ and a closed set $\mathscr{F} \subset \mathscr{A}$ such that

$$v(\partial_\mu f^\mu - gdL^4)(\mathscr{U} \setminus \mathscr{A}) \leqslant v(\partial_\mu f^\mu - gdL^4)(\mathscr{U} \setminus \mathscr{F}) < \varepsilon.$$

Hence

$$\int_{\mathscr{A}} \partial_\mu f^\mu - gdL^4 = 0 \tag{3A4}$$

for any bounded Borel subset $\mathscr{A}$ of $\hat{\mathscr{B}}$. Consequently

$$\partial_\mu f^\mu = g \tag{3A5}$$

in the sense of equality of measures ($\partial_\mu f^\mu$ and gdL^4) or distributions. Hence (3A5) has the form of a VWP.

From (3A5) and the Gauss–Green theorem it follows that equation (3A1) holds for all sets $\mathscr{P} \subset \hat{\mathscr{B}}$ of bounded perimeter, with $\partial^*\mathscr{P}$ instead of $\partial\mathscr{P}$.

Equation (3A1) implies that

$$\int_{\partial^*\mathscr{P}} dH^3 \, (f_+{}^\mu - f_-{}^\mu)\nu_\mu = 0. \tag{3A6}$$

Suppose that $f_+{}^\mu$, $f_-{}^\mu$ exist at a point $y \in \partial^*\mathscr{P}$ and are not equal. Then $y \in \Gamma_{(f)}$ and

$$\int_{\partial^*\mathscr{P} \cap \Gamma(f)} dH^3 \, (f_+{}^\mu - f_-{}^\mu)\nu_\mu = 0, \tag{3A7}$$

which implies the jump equations on $\Gamma_{(f)}$.

We now also indicate an alternative approach for continuous f^μ. We proved in Section 1.2.7 that the balance equation

$$\int_{\partial\mathscr{P}} dH^3 \, \nu_\mu f^\mu = \int_{\mathscr{P}} dL^4 \, g \tag{3A8}$$

holds for all the polyhedra and some $f^\mu, g \in \mathscr{L}^1_{loc}(\hat{\mathscr{B}})$. Let $\mathscr{E}$ be a set of bounded perimeter and let $\{\mathscr{A}_n\}$ be a sequence of polyhedra such that $\varrho(\mathscr{A}_n, \mathscr{E}) \to 0$ and $P(\mathscr{A}_n) \leqslant R < \infty$ (Theorem 3.2.23). By Lemma 1.1.63 it follows that

$$\int_{\mathscr{A}_n} \mathrm{d}L^4\, g \to \int_{\mathscr{E}} \mathrm{d}L^4\, g.$$

On the other hand

$$\int_{\partial \mathscr{A}_n} \mathrm{d}H^3\, f^\mu \nu_\mu = \int f^\mu \partial_\mu \chi_{\mathscr{A}_n} \to \int f^\mu \partial_\mu \chi_{\mathscr{E}} = \int_{\partial^* \mathscr{E}} \mathrm{d}H^3\, f^\mu \nu_\mu$$

provided $f^\mu \in \mathscr{C}^0_0(\hat{\mathscr{B}})$. Hence follows (3A1) for sets $\mathscr{P}$ of bounded perimeter and continuous f^μ.

Chapter 4

Geometric aspects of elasticity

4.1 MATERIAL SYMMETRIES

4.1.1 Introduction

In this chapter we shall mainly be interested in two interrelated notions:

(1) internal stress,

(2) inhomogeneity of constitutive properties.

Internal stress is any stress that is maintained in the body under zero loads and pure Neumann BC. It is generated by a deformation that cannot be removed without a temporary destruction of continuity. A precise definition of internal and external stress (generated by surface loads) can be given only in linear elastostatics.* Some examples were discussed in Section 1.2 (Volterra dislocations). Although in non-linear elasticity the notion of internal stress is rather vague** the notion of inhomogeneity associated with it is unambiguous.

Inhomogeneity of constitutive properties refers to their explicit dependence of the material point $P \in \mathscr{B}$, i.e. to the dependence of the constitutive functions $\hat{T}^{kl}$, W on $\xi \in \Omega$. In some cases it can be removed by an appropriate choice of the reference configuration or at least by a suitable choice of local reference configurations:

$$W_\mu(F^k{}_\rho) = W_\mu\left(F^k{}_\alpha \frac{\partial \xi^\alpha}{\partial \xi^\rho}\right) \equiv W_\varkappa(F^k{}_\alpha, \xi),$$

$$\xi^\alpha = \varkappa^\alpha(P), \quad \zeta^\rho = \mu^\rho(P).$$

* By means of a decomposition of the Hilbert space $\mathscr{L}^2$ of strains into two orthogonal subspaces.

** One can introduce stress into a body satisfying homogeneous Neumann conditions $\mathbf{s} = 0$ on without disrupting continuity, e.g. by eversion of a thick hemispherical shel

If the reference configurations μ displaying the homogeneity of constitutive properties can be introduced *locally* then the inhomogeneities in the reference configuration $\varkappa$ can be attributed to the presence of isolated dislocations in this configuration. It may also be the case that a **continuous distribution of dislocations** in the reference $\varkappa$ is responsible for the inhomogeneity of constitutive properties. In this case the transformation to a reference configuration in which the constitutive functions W, $\hat{T}^{kl}$ do not depend on the material point is necessarily anholonomic. In either case dislocations appear in the double role of (1) internal stress exhibited by the body in the configuration $\varkappa$; (2) inhomogenity of constitutive functions if the configuration gradients $\mathbf{F}$ are referred to the configuration $\varkappa$.

In some cases, notably in crystalline solids, dislocations are also physical objects which interact with external loads and the interaction can be expressed in terms of **generalized pseudoforces**. Whenever a material point subject to a force $\mathbf{f}$ is infinitesimally shifted in physical space from $\mathbf{x}$ to $\mathbf{x}+\delta\mathbf{u}$, $\delta\mathbf{u} = \varepsilon\mathbf{v}$, the work $\mathbf{f}\cdot\delta\mathbf{u} = \varepsilon\mathbf{f}\cdot\mathbf{v}$ is performed on it. The vector $\mathbf{v}$ is the generator of a one-parameter group of translations acting on the (affine) physical space. There is an analogous relationship between virtual shifts of dislocations in the body and the pseudoforces acting on them. In order to give a precise meaning to the (generalized) pseudoforces acting on dislocations it would be necessary to replace translations by appropriate groups of automorphisms of the body $\mathscr{B}$ (e.g. rotations around the center in the case of a spherical body). The generators of one-parameter groups of automorphisms of $\mathscr{B}$ can be identified as elements of the Lie algebra $\mathfrak{g}$ of the group of automorphisms of $\mathscr{B}$. For any $\mathbf{W}\in\mathfrak{g}$ the work associated with an infinitesimal virtual displacement $\varepsilon\mathbf{W}$ of a dislocation is given by a linear functional $\mathscr{A}(\varepsilon\mathbf{W}) = \operatorname{tr}(\varepsilon\mathbf{F}\mathbf{W})\cdot\mathbf{F}$ can be identified as the generalized* pseudoforce correspond. Geometrically $\mathbf{F}$ belongs to the adjoint $\mathfrak{g}^*$ of $\mathfrak{g}$.

The work done in shifting a dislocation within the body can be defined as the difference of the energy $F = \mathscr{W}+\Phi$ corresponding to the shifted and initial position of the dislocation. The notion of a pseudoforce is however strongly suggested by the Lagrangian formulation of elasticity. We shall deduce the pseudoforces as a measure of the departure from Noether conservation laws with repect to the group of automorphisms** expressed in a suitably covariant way.

* The adjective "generalized" refers to the fact that the automorphism is not a translation in an affine space.

** For simplicity we ignore the difficulties associated with the nonaffine geometry of the body and consider translations of dislocations in the coordinates ξ^α. This simplification ignores non-local interaction of dislocations with the boundary etc.

The definition of Volterra dislocations (Section 4.2) is based on the existence of non-trivial material symmetries of the body. We shall review the theory of material symmetries in Section 4.1, mainly in view of their application to dislocation theory. More information on other applications of this theory (classification of materials, reduced constitutive equations) can be found in the books of Truesdell (1972), Wang and Truesdell (1973).

We concentrate on differential-geometric aspects of elasticity assuming for simplicity that all the configurations are smooth.

4.1.2 Isotropy Groups (Material Symmetries)

Reference configurations are commonly used for three logically independent purposes. Firstly, they allow a representation of the motion $\mathbf{h}$ and virtual configurations in terms of coordinates $\mathbf{x}$, ξ: $\mathbf{h}(P, t) = \boldsymbol{\chi}(\xi, t)$. Secondly, they allow replacing configuration gradients by deformation gradients $\mathbf{F} \in \mathscr{L}_+$, e.g. $F^k{}_\alpha = \chi^k{}_{,\xi^\alpha}$. Finally, they allow a definition of volume densities and surface densities (ϱ_0, $S_k{}^\alpha$) in terms of a volume and area measure which does not depend on the deformation.

The third use of reference configurations results from the traditional habit of representing 3- and 2-forms on $\mathscr{B}$ in terms of volume and surface densities. For the present purposes this use of reference configurations can be avoided by expressing the constitutive properties in terms of scalar* quantities such as the specific strain energy W and the Cauchy stress tensor T^{kl}:

$$W = W(\mathbf{F}, \xi), \tag{4.1.1}$$

$$T^{kl} = \hat{T}^{kl}(\mathbf{F}, \xi) = \hat{S}_k{}^\alpha(\mathbf{F}, \xi) F^l{}_\alpha (\det \mathbf{F})^{-1}.^{**} \tag{4.1.2}$$

On the other hand we shall exploit the possibility of choosing different references for the first and second purpose. If we identify the space with $\boldsymbol{R}^3$ then a configuration gradient $\mathbf{K}$ at a point $P \in \mathscr{B}$ is a mapping from the tangent space $T_P\mathscr{B}$ at P onto $\boldsymbol{R}^3 \cdot \overset{-1}{K}$ maps the standard basis $(1, 0, 0)$, $(0, 1, 0)$, $(0, 0, 1)$ of $\boldsymbol{R}^3$ onto a basis $\mathbf{e}_1, \mathbf{e}_2, \mathbf{e}_3$ of $T_P\mathscr{B}$ and therefore $\mathbf{K}$ is an element of the fiber of the principal frame bundle $\mathscr{F}\mathscr{B}$ over P. The coordinate system $(\xi^\alpha, F^k{}_\alpha)$ on $\mathscr{F}\mathscr{B}$ need not be a prolongation of the coordinate system (ξ^α) on $\mathscr{B}$. The covector basis ω^a, $a = 1, 2, 3$, on $T_P^*\mathscr{B}$ need not coincide with $d\xi^\alpha$, $\alpha = 1, 2, 3$, e.g.

$$\omega^a = H^a{}_\alpha(\xi) d\xi^\alpha. \tag{4.1.3}$$

* With respect to transformation of Lagrange coordinates.

** Note that $\hat{T}^{kl} = \hat{T}^{lk}$ identically on account of (1.2.86). Equation (1.2.102) shows that T^{kl} is the Piola–Kirchhoff stress referred to the area in the actual configuration $\boldsymbol{\chi} \circ \boldsymbol{\varkappa}$.

Expressing a configuration gradient in terms of the bases ω^a and $d\xi^\alpha$ we have $\bar{F}^k{}_a\omega^a = F^k{}_\alpha d\xi^\alpha$ and hence

$$F^k{}_\alpha = \bar{F}^k{}_a H^a{}_\alpha. \tag{4.1.4}$$

We also define the corresponding **reference configuration gradient**

$$\mathbf{K} = \mathscr{H}\left(\varkappa(P)\right) \circ \nabla\varkappa(P); \quad T_P\mathscr{B} \to \boldsymbol{R}^3. \tag{4.1.5}$$

at P. Clearly $\mathbf{F}\nabla\varkappa = \bar{\mathbf{F}}\mathbf{K}$. With the help of this notion we can achieve our goal. Using $\nabla\varkappa$ or $\mathbf{K}$ as the reference for configuration gradients at P we have the following two representations of constitutive equations:

$$T^{kl} = \hat{T}^{kl}(\mathbf{F}, \xi) = \tilde{T}^{kl}(\bar{\mathbf{F}}, \xi; \mathbf{H}) = \check{T}^{kl}(\bar{\mathbf{F}}, P; \mathbf{K}), \tag{4.1.6}$$

$$W = W(\mathbf{F}, \xi) = \tilde{W}(\bar{\mathbf{F}}, \xi; \mathbf{H}) = \check{W}(\bar{\mathbf{F}}, P; \mathbf{K}). \tag{4.1.7}$$

Obviously

$$\tilde{T}^{kl}(\bar{\mathbf{F}}, \xi; \mathbf{QH}) \equiv \tilde{T}^{kl}(\bar{\mathbf{F}}\mathbf{Q}, \xi; \mathbf{H}), \quad \check{T}^{kl}(\bar{\mathbf{F}}, P; \mathbf{QK})$$
$$\equiv \check{T}^{kl}(\bar{\mathbf{F}}\mathbf{Q}, P; \mathbf{K}) \ \forall \mathbf{Q} \in \mathscr{L}_+, \quad \forall \bar{\mathbf{F}} \in \mathscr{L}_+. \tag{4.1.7}$$

Since

$$T^{kl} = \hat{T}^{kl}(\mathbf{F}, \xi) = \frac{\varrho_0(\xi)}{\det \mathbf{F}} \frac{\partial W(\mathbf{F}, \xi)}{\partial F^k{}_\alpha} F^l{}_\alpha$$

it follows easily that

$$\check{T}^{kl}(\bar{\mathbf{F}}, P; \mathbf{K}) \equiv \frac{\varrho_{\mathbf{K}}}{\det \bar{\mathbf{F}}} \frac{\partial \check{W}(\bar{\mathbf{F}}, P; \mathbf{K})}{\partial \bar{F}^k{}_a} \bar{F}^l{}_a \tag{4.1.8}$$

with $\varrho_K := \varrho_0(\xi)/\det \mathbf{H}$.*

Definition 4.1.1
Let $P \in \mathscr{B}$, $\xi = \varkappa_i(P)$, $\varkappa_i\colon \mathscr{U}_i \to \boldsymbol{R}^3$. We say that $\mathbf{G} \in \mathscr{G}(\mathbf{K}, P)$ if

$$\check{T}^{kl}(\bar{\mathbf{F}}, P; \mathbf{K}) = \check{T}^{kl}(\bar{\mathbf{F}}\mathbf{G}, P; \mathbf{K}) \ \forall \bar{\mathbf{F}} \in \mathscr{L}_+, \tag{4.1.9}$$

$$\check{W}(\bar{\mathbf{F}}, P; \mathbf{K}) = W(\bar{\mathbf{F}}\mathbf{G}, P; \mathbf{K}) \ \forall \bar{\mathbf{F}} \in \mathscr{L}_+. \tag{4.1.10}$$

It is easy to see that the set $\mathscr{G}(\mathbf{K}, P)$ is a subgroup of SL(3). It is called the **isotropy group of** P **relative to the reference** $\mathbf{K}$. An element $\mathbf{G}$ of $\mathscr{G}(\mathbf{K}, P)$ is called a **material symmetry of** P **with respect to** $\mathbf{K}$.

Equation (4.1.9) implies the identities

$$\check{T}^{kl}(\bar{\mathbf{F}}, P; \mathbf{GK}) \equiv \check{T}^{kl}(\bar{\mathbf{F}}, P; \mathbf{K}), \tag{4.1.11}$$

$$\tilde{T}^{kl}(\bar{\mathbf{F}}, \xi; \mathbf{H}) \equiv \tilde{T}^{kl}(\bar{\mathbf{F}}, \xi; \mathbf{GH}) \equiv \tilde{T}^{kl}(\bar{\mathbf{F}}\mathbf{G}, \xi; \mathbf{H}). \tag{4.1.12}$$

* Let $\mathbf{K} = [\varkappa_1]_{\sim P}$, e.g. $\varkappa_1 = \mathbf{H} \circ \varkappa$ in a nbhd of P. The mass density in the configuration $\varkappa_1$ at P is then

$$\varrho_{\varkappa_1}(\varkappa_1(P)) = \varrho_{\mathbf{K}}.$$

From (4.1.8) it follows that (4.1.9) and (4.1.10) are equivalent provided $\mathscr{G}(\mathbf{K}, P) \subset \mathrm{SU}(3)$. We assume henceforth that this is the case.

Let $\tilde{\mathbf{K}} = \mathbf{Q}\mathbf{K}$, $\mathbf{Q} \in \mathscr{L}_+$, $\tilde{\mathbf{F}} = \bar{\mathbf{F}}\overset{-1}{\mathbf{Q}}$, so that $\tilde{\mathbf{F}}\tilde{\mathbf{H}} = \bar{\mathbf{F}}\mathbf{H}$, $\tilde{\mathbf{F}}\mathbf{G}\mathbf{Q} = \bar{\mathbf{F}}\overset{-1}{\mathbf{Q}}\mathbf{G}\mathbf{Q}$. Hence

$$\mathbf{G} \in \mathscr{G}(\tilde{\mathbf{K}}, P) \quad \text{iff} \quad \overset{-1}{\mathbf{Q}}\mathbf{G}\mathbf{Q} \in \mathscr{G}(\mathbf{K}, P),$$

so that

$$\mathscr{G}(\tilde{\mathbf{K}}, P) = \mathbf{Q}\mathscr{G}(\mathbf{K}, P)\overset{-1}{\mathbf{Q}}. \tag{4.1.13}$$

Equation (4.1.13) implies the existence of an "absolute" isotropy group of P:

$$\mathscr{G}(P) := \mathbf{K}^{-1}\mathscr{G}(\mathbf{K}, P)\mathbf{K} = \tilde{\mathbf{K}}^{-1}\mathscr{G}(\tilde{\mathbf{K}}, P)\tilde{\mathbf{K}}$$

of transformations $\mathbf{G}\colon T_P\mathscr{B} \to T_P\mathscr{B}$.

The basic intutitive idea behind the notion of a material symmetry $\mathbf{G}$ of P relative to $\mathbf{K}$ is that the reaction of stress at P to arbitrary deformation gradients $\bar{\mathbf{F}}$ is identical for the reference configurations $\mathbf{K}$ and $\mathbf{G}(\mathbf{K})$.

Definition 4.1.2

A material point $P \in \mathscr{B}$ is said to be **fluid** if $\mathscr{G}(\mathbf{K}, P) = \mathrm{SU}(3)$ for some reference $\mathbf{K}$ at P.

Remark

If $\mathscr{G}(\mathbf{K}, P) = \mathrm{SU}(3)$ for some reference $\mathbf{K}$, then $\mathscr{G}(\mathbf{K}, P) = \mathrm{SU}(3)$ for every reference $\mathbf{K}$ at P, on account of (4.1.13).

Definition 4.1.3

A material point $P \in \mathscr{B}$ is said to be **solid** if there is a reference $\mathbf{K}$ at P such that $\mathscr{G}(\mathbf{K}, P) \subset \mathrm{SO}(3)$.

In this case the reference $\mathbf{K}$ is said to be **undistorted.**

Since ${}^t(\mathbf{Q}\mathbf{G}\overset{-1}{\mathbf{Q}}) = {}^t\overset{-1}{\mathbf{Q}}\overset{-1}{\mathbf{G}}{}^t\mathbf{Q}$, $\mathscr{G}(\tilde{\mathbf{K}}, P) \subset \mathrm{SO}(3)$ provided $\mathbf{Q} \in \mathrm{SO}(3)$.

Definition 4.1.4

A material point $P \in \mathscr{B}$ is said to be **isotropic** if there exists a reference configuration $\bar{\mathbf{K}}$ at P such that $\mathscr{G}(\bar{\mathbf{K}}, P) \supset \mathrm{SO}(3)$.

A configuration $\bar{\mathbf{K}}$ with this property is said to be **undistorted.**

Definition 4.1.5

A material point P is said to be **isotropic solid** if it is both isotropic and solid. This means that there are two reference configurations $\mathbf{K}$, $\bar{\mathbf{K}}$ at P such that

$$\mathscr{G}(\bar{\mathbf{K}}, P) \supset \mathrm{SO}(3), \quad \mathscr{G}(\mathbf{K}, P) \subset \mathrm{SO}(3).$$

Theorem 4.1.6 (Truesdell 1972)
If P is an isotropic solid point of $\mathscr{B}$, then

$$\mathscr{G}(\mathbf{K}, P) = \mathscr{G}(\bar{\mathbf{K}}, P) = SO(3) \quad \text{for the undistorted references } \mathbf{K}, \bar{\mathbf{K}}.$$

Proof
By a theorem of Noll (1965) either $\mathscr{G}(\bar{\mathbf{K}}, P) = SO(3)$ or $\mathscr{G}(\bar{\mathbf{K}}, P) = SU(3)$, since SO(3) is a maximal proper subgroup of SU(3).

If $\mathscr{G}(\bar{\mathbf{K}}, P) = SU(3)$ then $\mathscr{G}(\mathbf{K}, P) = SU(3)$, in contradiction with the hypothesis. Hence $\mathscr{G}(\bar{\mathbf{K}}, P) = SO(3)$ and $\bar{\mathbf{K}}$ is an undistorted configuration in the sense of Definition 4.1.3. Now $\bar{\mathbf{K}} = \mathbf{QK}$, $\mathbf{Q} \in \mathscr{L}_+$,

$$SO(3) \supset \mathscr{G}(\mathbf{K}, P) = \overset{-1}{\mathbf{Q}}\mathscr{G}(\bar{\mathbf{K}}, P)\mathbf{Q} = \overset{-1}{\mathbf{Q}}SO(3)\mathbf{Q}. \tag{4.1.14}$$

Let $\mathbf{Q} = \mathbf{A}_o\mathbf{V}$ be the polar decomposition of $\mathbf{Q}$, $\mathbf{A}_o \in SO(3)$. Equation (4.1.14) implies that

$$\mathbf{V}^2\mathbf{A} = \mathbf{A}\mathbf{V}^2 \quad \forall\, \mathbf{A} \in SO(3).$$

Hence

$$\mathscr{G}(\mathbf{K}, P) = \mathbf{A}_o^{-1}SO(3)\mathbf{A}_o = SO(3) \qquad \square$$

Definition 4.1.7
$\mathbf{P} \in \mathscr{B}$ is said to be a **transversely isotropic solid** point if there is a reference $\mathbf{K}$ at P such that $\mathscr{G}(\mathbf{K}, P)$ is the group of rotations in $\boldsymbol{R}^3$ leaving the vector $(0, 0, 1)$ fixed.

A reference $\mathbf{K}$ satisfying the above condition is said to be **adjusted.**

4.1.3 Reduced Constitutive Equations

We shall discuss this topic rather briefly. For more details see Truesdell and Noll (1965), Truesdell (1972), Wang and Truesdell (1973).

Let $\mathbf{K}$ be an undistorted reference of a solid point $P \in \mathscr{B}$. For simplicity we omit the arguments P, $\mathbf{K}$ of $\check{W}$ and write $W(\mathbf{F})$ for $\check{W}(\mathbf{F}, P; \mathbf{K})$. Let $\mathbf{F} = \mathbf{V}\mathbf{A}_o$ be the polar decomposition of $\mathbf{F}$, $\mathbf{A}_o \in SO(3)$, $\mathbf{V} = {}^t\mathbf{V} > 0$, $\mathbf{B} = \mathbf{F}{}^t\mathbf{F} = \mathbf{V}^2$. On account of isotropy

$$W(\mathbf{F}) \equiv W(\mathbf{F}\overset{-1}{\mathbf{A}_o}) = W(\mathbf{V}) = \tilde{W}(\mathbf{B}). \tag{4.1.15}$$

Hence

$$T^{kl} = \varrho \frac{\partial W}{\partial F^k{}_\alpha} F^l{}_\alpha = 2\varrho \frac{\partial \tilde{W}}{\partial B^{ks}} B^{sl}. \tag{4.1.16}$$

On account of the invariance of W with respect to the left action of SO(3) on $\mathscr{L}_+$ (spatial rotations)

$$\tilde{W}(\mathbf{A}\mathbf{B}^t\mathbf{A}) \equiv \tilde{W}(\mathbf{B}) \quad \forall \mathbf{A} \in \mathrm{SO}(3) \tag{4.1.17}$$

and for all positive definite symmetric $\mathbf{B} \in \mathscr{L}$.

Every symmetric matrix $\mathbf{B}$ can be diagonalized by a suitable choice of $\mathbf{A} \in \mathrm{SO}(3)$,

$$\mathbf{B} = \overset{-1}{\mathbf{A}}\mathrm{diag}\{b_1, b_2, b_3\}\mathbf{A}. \tag{4.1.18}$$

Hence

$$\tilde{W}(\mathbf{B}) = \bar{W}(b_1, b_2, b_3) \equiv \bar{W}(b_2, b_1, b_3) \equiv \dots \star \tag{4.1.19}$$

By the Hamilton–Cayley theorem (Truesdell and Noll, 1965)

$$-\mathbf{B}^3+\mathrm{I}_{\mathbf{B}}\mathbf{B}^2-\mathrm{II}_{\mathbf{B}}\mathbf{B}+\mathrm{III}_{\mathbf{B}}\mathbf{E} = \mathbf{0}, \tag{4.1.20}$$

$$\begin{aligned}
\mathrm{I}_{\mathbf{B}} &:= \mathrm{tr}\,\mathbf{B} = b_1+b_2+b_3,\\
\mathrm{II}_{\mathbf{B}} &:= \tfrac{1}{2}[(\mathrm{tr}\,\mathbf{B})^2-\mathrm{tr}(\mathbf{B}^2)] = b_1b_2+b_2b_3+b_3b_1,\\
\mathrm{III}_{\mathbf{B}} &:= \det\mathbf{B} = b_1b_2b_3.^{\star\star}
\end{aligned} \tag{4.1.21}$$

Hence the eigenvalues b_1, b_2, b_3 of $\mathbf{B}$ satisfy the same equation:

$$-b^3+\mathrm{I}_{\mathbf{B}}b^2-\mathrm{II}_{\mathbf{B}}b+\mathrm{III}_{\mathbf{B}} = 0. \tag{4.1.22}$$

If $\mathbf{B}' = \mathbf{A}\mathbf{B}^t\mathbf{A}$, $\mathbf{A} \in \mathrm{SO}(3)$, then

$$\mathrm{I}_{\mathbf{B}'} = \mathrm{I}_{\mathbf{B}}, \quad \mathrm{II}_{\mathbf{B}'} = \mathrm{II}_{\mathbf{B}}, \quad \mathrm{III}_{\mathbf{B}'} = \mathrm{III}_{\mathbf{B}}. \tag{4.1.23}$$

Conversely, suppose that $\mathbf{B}, \mathbf{B}'$ are symmetric and satisfy equations (4.1.23). Suppose that the eigenvalues b_1, b_2, b_3 are all different. From (4.1.23) it follows that equation (4.1.22) has three different roots b_1, b_2, b_3 and $\{b_1', b_2', b_3'\} \subset \{b_1, b_2, b_3\}$ (b_k', $k = 1, 2, 3$, denote the eigenvalues of $\mathbf{B}'$). If two of the eigenvalues b_1', b_2', b_3' coincide then (4.1.21), (4.1.23) imply that equation (4.1.22) has two equal roots, which leads to a contradiction. Hence b_1', b_2', b_3' are a permutation of b_1, b_2, b_3. If $b_1 = b_3 \neq b_2$ then equation (4.1.22) has two equal roots by the same argument and $\{b_1', b_2', b_3'\} \subset \{b_1, b_2\}$. Two of b_1', b_2', b_3' are distinct, again by the same argument etc. Hence (4.1.23) implies that $\{b_1, b_2, b_3\} = \{b_1', b_2', b_3'\}$. On account of (4.1.19)

$\star$ For the invariance with respect to permutations take

$$\mathbf{A} = -\begin{bmatrix} 0 & 1 & 0 \\ 1 & 0 & 0 \\ 0 & 0 & 1 \end{bmatrix} \quad \text{etc.}$$

$\star\star$ A general definition of the invariants $\mathrm{I}_k(\mathbf{B})$ of an $n\times n$ matrix $\mathbf{B}$ is $\det(\lambda\mathbf{E}+\mathbf{B}) = \sum_{k=0}^{n} \lambda^{n-k} I_k(\mathbf{B})$ for all real λ. Here $I_1(\mathbf{B}) = \mathrm{I}_{\mathbf{B}}$, $\mathrm{II}_{\mathbf{B}} = \mathrm{I}_2(\mathbf{B})$, $\mathrm{III}_{\mathbf{B}} = \mathrm{I}_3(\mathbf{B})$.

$$W = \bar{\bar{W}}(\mathrm{I}_{\mathbf{B}}, \mathrm{II}_{\mathbf{B}}, \mathrm{III}_{\mathbf{B}}). \tag{4.1.24}$$

Equation (4.1.24) implies (4.1.19) immediately but the converse implication is not trivial since the determinant of the transformation (4.1.21) vanishes for some values of b_1, b_2, b_3. For an algebraic function $\tilde{W}$ equation (4.1.24) follows from standard theorems on algebraic invariants (cf. Spencer, 1971).

From equation (4.1.24) it follows that

$$T^{kl} = 2\varrho\,\frac{\partial \bar{\bar{W}}}{\partial \mathrm{I}_{\mathbf{B}}}\,B^{kl} + \frac{\partial \bar{\bar{W}}}{\partial \mathrm{II}_{\mathbf{B}}}\,[\mathrm{I}_{\mathbf{B}} B^{kl} - (\mathbf{B}^2)^{kl}] + \frac{\partial \bar{\bar{W}}}{\partial \mathrm{III}_{\mathbf{B}}}\,\det\mathbf{B}\,\delta^{kl},$$

by equation (4.1.16). More concisely,

$$\mathbf{T} = a(\mathrm{I}_{\mathbf{B}}, \mathrm{II}_{\mathbf{B}}, \mathrm{III}_{\mathbf{B}})\mathbf{E} + b(\mathrm{I}_{\mathbf{B}}, \mathrm{II}_{\mathbf{B}}, \mathrm{III}_{\mathbf{B}})\mathbf{B} + c(\mathrm{I}_{\mathbf{B}}, \mathrm{II}_{\mathbf{B}}, \mathrm{III}_{\mathbf{B}})\mathbf{B}^2. \tag{4.1.25}$$

On account of the Hamilton–Cayley equation we also have the constitutive equation

$$\mathbf{T} = \bar{a}(\mathrm{I}_{\mathbf{B}}, \mathrm{II}_{\mathbf{B}}, \mathrm{III}_{\mathbf{B}})\overset{-1}{\mathbf{B}} + \bar{b}(\mathrm{I}_{\mathbf{B}}, \mathrm{II}_{\mathbf{B}}, \mathrm{III}_{\mathbf{B}})\mathbf{E} + \bar{c}(\mathrm{I}_{\mathbf{B}}, \mathrm{II}_{\mathbf{B}}, \mathrm{III}_{\mathbf{B}})\mathbf{B}. \tag{4.1.26}$$

We also note that

$$\det\mathbf{B} = \tfrac{1}{6}(\mathrm{tr}\,\mathbf{B})^3 - \tfrac{1}{2}(\mathrm{tr}\,\mathbf{B})\mathrm{tr}(\mathbf{B}^2) + \tfrac{1}{3}\mathrm{tr}(\mathbf{B}^3). \tag{4.1.27}$$

Since both sides of (4.1.27) are invariant with respect to the transformation $\mathbf{B} \mapsto \mathbf{A}\mathbf{B}\overset{-1}{\mathbf{A}}$, it is enough to verify (4.1.27) for $\mathbf{B} = \mathrm{diag}\,\{b_1, b_2, b_3\}$.

The invariants $\mathrm{tr}\,\mathbf{B}$, $\mathrm{tr}(\mathbf{B}^2)$, $\mathrm{tr}(\mathbf{B}^3)$ are commonly used as an integral invariant basis for algebraic invariants of $\mathbf{B}$ under $\mathbf{B} \mapsto \mathbf{A}\mathbf{B}{}^t\mathbf{A}$, $\mathbf{A} \in SO(3)$ (see Spencer, 1971).

Let $P \in \mathscr{B}$ be a transversely isotropic solid point with an adjusted reference $\mathbf{K}$, $\bar{W}(\mathbf{F}) = \check{W}(\mathbf{F}, P; \mathbf{K})$. For an algebraic function $\bar{W}$ it can be proved that $\bar{W}(\mathbf{F}) \equiv \tilde{W}(\mathbf{f}, B_1^{kl})$, $B_1^{kl} := \sum_{\alpha=1}^{2} F^k{}_\alpha F^l{}_\alpha$, $f^k := F^k{}_3$ (Spencer, 1971). On account of the invariance of W with respect to spatial rotations $\tilde{W}$ is an isotropic function of $\mathbf{f}$, $\mathbf{B}_1$. Hence (Spencer, 1971)

$$\tilde{W}(\mathbf{f}, \mathbf{B}_1) = \bar{\bar{W}}(\mathrm{tr}\,\mathbf{B}_1, \mathbf{f}^2, \mathbf{f}\cdot\mathbf{B}_1\mathbf{f}, \mathbf{f}\cdot\mathbf{B}_1^2\mathbf{f})$$

provided $\tilde{W}$ is an algebraic function of $\mathbf{f}$, $\mathbf{B}_1$. Note that rank $\mathbf{B}_1 = 2$, hence $\det\mathbf{B}_1 = 0$ and

$$-\mathbf{B}_1{}^2 + \mathrm{I}_{\mathbf{B}_1}\mathbf{B}_1 - \mathrm{II}_{\mathbf{B}_1}\mathbf{E} = \mathbf{0} \quad \text{on } (\ker\mathbf{B}_1)^\perp, \tag{4.1.28}$$

so that ($\operatorname{tr}\mathbf{B}_1^2$) can be expressed in terms of $\operatorname{tr}\mathbf{B}_1$. Other invariants listed by Spencer (1971) in Section 2.3 drop out on account of the symmetry of $\mathbf{B}_1$ and equation (4.1.28).

Let us finally consider a fluid point $P \in \mathscr{B}$, $W(\mathbf{F}) :\equiv \check{W}(\mathbf{F}, P; \mathbf{K})$. Since $\det\left[\overset{-1}{\mathbf{F}}_1 \mathbf{F}_2\right] = 1$ iff $\det\mathbf{F}_1 = \det\mathbf{F}_2$, it follows that

$$W(\mathbf{F}) = \tilde{W}(\det\mathbf{F}) = \tilde{\tilde{W}}(\varrho), \qquad \varrho = \varrho_0/\det\mathbf{F}.$$

Also

$$T^{kl} = \varrho\tilde{W}'(\det\mathbf{F})\overset{-1}{F^{\alpha}}{}_{k}F^{l}{}_{\alpha} = \varrho\tilde{W}'(\det\mathbf{F})\,\delta^{k}{}_{l} = [-p(\varrho)\mathbf{E}]^{kl}. \tag{4.1.29}$$

It follows that a body consisting of fluid points only (i.e. a fluid body) sustains hydrostatic stress (pressure) only.

In this connection we note the following important facts:

(1) the strain energy $W(\mathbf{F}, \xi) = \tilde{W}(\det\mathbf{F}, \xi)$ of a fluid body is polyconvex iff $\tilde{W}$ is a convex function of the first argument;

(2) the strain energy W of a fluid body cannot satisfy SE. Indeed

$$\frac{\partial^2 W}{\partial F^k{}_\alpha \partial F^l{}_\beta} a^k a^l k_\alpha k_\beta = \left[\Delta\frac{d}{d\Delta}\left(\Delta\frac{d\tilde{W}}{d\Delta}\right) - \Delta\frac{d\tilde{W}}{d\Delta}\right]\left(\overset{-1}{F_k{}^\alpha} a^k k_\alpha\right)^2,$$

$\Delta = \det\mathbf{F}$, and $\overset{-1}{F^{\alpha}}{}_{k}a^k k_\alpha = 0$ is possible for $\mathbf{a}$, $\mathbf{k} \neq 0$.

Polyconvexity of strain energy of an isotropic solid body was discussed by Ball (1977a).

4.2 DISLOCATIONS

4.2.1 Isolated Volterra Dislocations

Let $\boldsymbol{\chi}$: $\Omega \to \boldsymbol{R}^3$ be a deformation referred to the reference configuration $\xi = \varkappa(P)$.

Let Σ be a sufficiently regular (e.g. Lipschitzian) orientable surface in $\boldsymbol{\chi}(\Omega)$. We assume that there is an open connected set $\mathscr{U} \subset \boldsymbol{\chi}(\Omega)$ containing Σ such that

(1) $\mathscr{U}\setminus\Sigma$ is not connected and consists of two components $\mathscr{U}_+$, $\mathscr{U}_-$;

(2) the field of unit normal vectors $\mathbf{n}$ (defined H^2-a.e. on Σ) points into $\mathscr{U}_+$;

(3) χ has an inverse $\overset{-1}{\chi}$ with the following properties:

(i) $\overset{-1}{\chi}|\mathscr{U}_+$, $\overset{-1}{\chi}|\mathscr{U}_-$ are both $\mathscr{C}^1$;

(ii) $\overset{-1}{\chi}$ and $\nabla\overset{-1}{\chi}$ have jump discontinuities on Σ;

(iii) the Cauchy stress field $\mathbf{T}$ is continuous on $\mathscr{U}_+$, $\mathscr{U}_-$ and has a jump discontinuity on Σ.

The equation of momentum conservation in the static case implies that $\mathbf{T}(\mathbf{x})$ satisfies the jump equations (Chapter 3):

$$[[T^{kl}(\mathbf{x})]]n_l(\mathbf{x}) = 0 \ \forall \mathbf{x} \in \Sigma. \tag{4.2.1}$$

The solutions of equlibrium equations of elasticity satisfying the conditions (1), (2) and (3) are known as **Somigliana dislocations.**

The stronger condition

$$[[T^{kl}(\mathbf{x})]] = 0 \ \forall \mathbf{x} \in \Sigma \tag{4.2.2}$$

can be satisfied by taking advantage of material symmetries of the body. With this in view we assume that

(a) the body $\mathscr{B}$ (or at least its part $\mathscr{V}$ corresponding to $\mathscr{U}$) is homogeneous in the reference configuration $\xi = \boldsymbol{\varkappa}(P)$, i.e.

$$\hat{\mathbf{T}}(F^k{}_\alpha, \xi) = \tilde{\mathbf{T}}(F^k{}_\alpha, \xi; \delta^\alpha{}_\beta) \equiv \hat{\mathbf{T}}(F^k{}_\alpha) \ \forall \mathbf{F} \in \mathscr{L}_+;$$

(b) the isotropy group of each $P \in \mathscr{V}$ relative to $\nabla\boldsymbol{\varkappa}(P)$ contains some fixed subgroup $\mathscr{G}$ of SU(3);

(c) the limits $\mathbf{F}_+$, $\mathbf{F}_-$ of $\nabla\overset{-1}{\chi}$ on Σ satisfy the relation

$$F_-{}^k{}_\alpha(\mathbf{x}) = F_+{}^k{}_\beta(\mathbf{x})G^\beta{}_\alpha(\mathbf{x}), \quad [G^\alpha{}_\beta(\mathbf{x})] \in \mathscr{G} \quad \text{for } \forall \mathbf{x} \in \Sigma.$$

The assumptions (a), (b) and (c) imply that $T^{kl}(\mathbf{x}) = \hat{T}^{kl}(F^m{}_\alpha)$ is a continuous function of $\mathbf{x} \in \mathscr{U}$ and satisfies equation (4.2.2).

If $\mathscr{G}$ is a discrete group then the continuity of $\nabla\overset{-1}{\chi}_\pm$ on Σ implies that $\mathbf{G}(\mathbf{x}) = \mathbf{G} = \text{const}$,

$$\overset{-1}{\chi}_-(\mathbf{x}) \equiv \mathbf{G}\overset{-1}{\chi}_+(\mathbf{x}) + \mathbf{b}, \quad \mathbf{G} \in \mathscr{G}, \ \mathbf{b} \in \boldsymbol{R}^3 \text{ for all } \mathbf{x} \in \Sigma. \tag{4.2.3}$$

Equation (4.2.3) can be proved by means of a convenient choice of local coordinates y^1, y^2, y^3 in a nbhd of $\mathbf{x} \in \Sigma \subset \mathscr{U}$ in such a way that Σ is given locally by $y^1 = \text{const}$, and integrating with respect to y^2, y^3. Note that equation (4.2.3) implies a discontinuity of $\overset{-1}{\chi}$ in every non-trivial case. Hence it is necessary to assume (a). Otherwise $T^{kl}(\mathbf{x}) = \hat{T}^{kl}(\mathbf{F}(\mathbf{x}), \overset{1}{\chi}(\mathbf{x}))$ would be in general a discontinuous function of $\mathbf{x}$.

We define the functions

$$\overset{-1}{\boldsymbol{\chi}}_{+}(\mathbf{x}) :\equiv \begin{cases} \overset{-1}{\boldsymbol{\chi}}(\mathbf{x}) & \text{for } \mathbf{x} \in \mathscr{U}_{+}, \\ \overset{-1}{\mathbf{G}}[\overset{-1}{\boldsymbol{\chi}}(\mathbf{x})-\mathbf{b}] & \text{for } \mathbf{x} \in \mathscr{U}_{-}, \end{cases} \tag{4.2.4}$$

$$\overset{-1}{\boldsymbol{\chi}}_{-}(\mathbf{x}) :\equiv \begin{cases} \overset{-1}{\boldsymbol{\chi}}(\mathbf{x}) & \text{for } \mathbf{x} \in \mathscr{U}_{-}, \\ \mathbf{G}(\overset{-1}{\boldsymbol{\chi}}\mathbf{x})+\mathbf{b} & \text{for } \mathbf{x} \in \mathscr{U}_{+}. \end{cases} \tag{4.2.5}$$

Either of these functions is continuous on $\mathscr{U}$.

Suppose now that $\boldsymbol{\chi}(\Omega)\setminus\Sigma$ is connected and $\overset{-1}{\boldsymbol{\chi}}$ is continuous on $\boldsymbol{\chi}(\Omega)\setminus\Sigma$. Let us follow a path from $\mathscr{U}_{+}$ to $\mathscr{U}_{-}$ which does not intersect Σ. It is obvious that $\overset{-1}{\boldsymbol{\chi}}_{+}$ goes over continuously into $\overset{-1}{\boldsymbol{\chi}}_{-}$. We can extend $\overset{-1}{\boldsymbol{\chi}}_{-}$ continuously back into $\mathscr{U}_{-}$ by means of equation (4.2.5). Iterating this procedure we shall obtain a multivalued function $\overset{-1}{\boldsymbol{\chi}}_{*}$ which is continuous on $\boldsymbol{\chi}(\Omega)\setminus\mathscr{C}$, where $\mathscr{C} := \partial\Sigma\setminus\partial\boldsymbol{\chi}(\Omega)$. We have shown in Chapter 1 how such multivalued inverse configurations can be interpreted in terms of local Lagrange coordinates. $\overset{-1}{\boldsymbol{\chi}}_{+}$, $\overset{-1}{\boldsymbol{\chi}}_{-}$ are two consecutive sheets of $\overset{-1}{\boldsymbol{\chi}}_{*}$ over $\mathscr{U}$. The relation between them is

$$\overset{-1}{\boldsymbol{\chi}}_{-}(\mathbf{x}) = \mathbf{G}\overset{-1}{\boldsymbol{\chi}}(\mathbf{x})+\mathbf{b}, \quad \mathbf{G} \in \mathscr{G},\ \mathbf{b} \in \boldsymbol{R}^3,\ \forall \mathbf{x} \in \mathscr{U}. \tag{4.2.6}$$

The deformation gradient $\nabla\boldsymbol{\chi} = (\nabla\overset{-1}{\boldsymbol{\chi}}_{*})^{-1}$ is a multivalued function of $\mathbf{x} \in \boldsymbol{\chi}(\Omega)\setminus\mathscr{C}$, but its values on different sheets over a point P differ by the left action of an element of the isotropy group. Hence the stress field $\mathbf{T}$ corresponding to $\nabla\boldsymbol{\chi}$ is univalued and continuous on $\boldsymbol{\chi}(\Omega)\setminus\mathscr{C}$.

For the functions $\mathbf{T}$, $\overset{-1}{\boldsymbol{\chi}}_{*}$ it is immaterial where we place the cut Σ, it is only the dislocation line $\mathscr{C}$ that matters. The dislocation line $\mathscr{C}$ is either closed or has both ends on $\partial\boldsymbol{\chi}(\Omega)$. The solution has in general a singularity at $\mathscr{C}$.

We shall now extend the above results to an arbitrary $\mathbf{G} \subset SO(3)$.

Let $\overset{-1}{\boldsymbol{\chi}}_{*}$, $\nabla\overset{-1}{\boldsymbol{\chi}}_{*}$ be continuous multivalued functions on $\partial\boldsymbol{\chi}(\Omega)\setminus\mathscr{C}$ with values in $\boldsymbol{R}^3$, $\mathscr{L}_{+}$ resp. We assume that any two consecutive sheets $\overset{-1}{\boldsymbol{\chi}}_{+}$, $\overset{-1}{\boldsymbol{\chi}}_{-}$ of $\overset{-1}{\boldsymbol{\chi}}_{*}$ satisfy the condition

$$\nabla\boldsymbol{\phi}(\xi) \in SO(3)\ \forall \xi \in \Omega, \tag{4.2.7}$$

$$\phi^{\alpha}(\xi) :\equiv \overset{-1}{\chi}{}_{-}^{\alpha}(\boldsymbol{\chi}_{+}(\xi)) =: \xi^{\alpha}-u^{\alpha}(\xi). \tag{4.2.8}$$

Since $\nabla\overset{-1}{\boldsymbol{\chi}}_*(\xi) \in \mathscr{L}_+$ it is clear that $\overset{-1}{\boldsymbol{\chi}}_+$ is locally invertible and $\boldsymbol{\chi}_+$ denotes its local inverse.

Since $\delta_{\alpha\beta}\phi^\alpha{}_{,\xi^\gamma}\phi^\beta{}_{,\xi^\delta} = \delta_{\gamma\delta}$ in $\mathscr{U}$, it follows that

$$u^\gamma{}_{,\xi^\beta} + u^\beta{}_{,\xi^\gamma} - u^\alpha{}_{,\xi^\beta} u^\alpha{}_{,\xi^\gamma} \equiv 0 \quad \text{in } \mathscr{U} \tag{4.2.9}$$

(summation over $\alpha = 1, 2, 3$). For simplicity we shall write $u^\alpha{}_{,\beta}$, $u^\alpha{}_{,\beta\gamma}$ for $u^\alpha{}_{,\xi^\beta}$, $u^\alpha{}_{,\xi^\beta\xi^\gamma}$ resp. Differentiate equation (4.2.9) with respect to ξ^ε:

$$u^\gamma{}_{,\beta\varepsilon} + u^\beta{}_{,\gamma\varepsilon} - u^\alpha{}_{,\beta\varepsilon} u^\alpha{}_{,\gamma} - u^\alpha{}_{,\beta} u^\alpha{}_{,\gamma\varepsilon} \equiv 0. \tag{4.2.10}$$

Exchanging γ, ε and substracting the resulting identity from (4.2.10) we have

$$u^\gamma{}_{,\beta\varepsilon} - u^\varepsilon{}_{,\gamma\beta} - u^\alpha{}_{,\beta\varepsilon} u^\alpha{}_{,\gamma} + u^\alpha{}_{,\beta\gamma} u^\alpha{}_{,\varepsilon} \equiv 0. \tag{4.2.11}$$

Exchanging β, ε in (4.2.10) and adding the resulting identity to (4.2.10) we get

$$(\delta^\alpha{}_\gamma - u^\alpha{}_{,\gamma}) u^\alpha{}_{,\beta\varepsilon} = 0.$$

Since $\mathbf{F}_+$, $\mathbf{F}_-$ are invertible, so is the matrix $[\delta^\alpha{}_\gamma - u^\alpha{}_{,\gamma}]$. Hence $u^\alpha{}_{,\beta\varepsilon} \equiv 0$ in $\mathscr{U}$ and $\boldsymbol{\phi}$ is a linear function of ξ. Hence we conclude that equation (4.2.6) holds.

Assuming (a) and (b) with $\mathscr{G} \subset \mathrm{SO}(3)$ we have constructed a multivalued solution $\overset{-1}{\boldsymbol{\chi}}_*$ with a multivalued gradient $\nabla\overset{-1}{\boldsymbol{\chi}}_*$ and a univalued stress field T^{kl}. Introducing a cut Σ we can satisfy (c) as well.

We are now ready to define a **Volterra dislocation** as a configuration $\boldsymbol{\chi} \circ \boldsymbol{\varkappa}$ of the body $\mathscr{B}$ such that

(I) $\boldsymbol{\varkappa}$ is a homogeneous configuration;

(II) $\overset{-1}{\boldsymbol{\chi}}$ is a sheet of a multivalued mapping $\overset{-1}{\boldsymbol{\chi}}_*\colon \boldsymbol{\chi} \circ \boldsymbol{\varkappa}(\mathscr{B})\setminus\mathscr{C} \to \boldsymbol{R}^3$ over $\boldsymbol{\chi} \circ \boldsymbol{\varkappa}(\mathscr{B})\setminus\Sigma$, $\mathscr{C} = \partial\Sigma\setminus\partial\boldsymbol{\chi} \circ \boldsymbol{\varkappa}(\mathscr{B})$, satisfying (4.2.6);

(III) $\mathbf{T}$ satisfies the equilibrium equation of elasticity

$$T^{kl}{}_{,x^l} + \varrho b_k = 0$$

with appropriate BC.

The condition (I) or (a) is invariant with respect to the transformation $\boldsymbol{\varkappa} \mapsto \boldsymbol{\mu} = \mathbf{L} \circ \boldsymbol{\varkappa} + \mathbf{a}$, $\mathbf{L} \in \mathrm{SL}(3)$, $\mathbf{a} \in \boldsymbol{R}^3$, $\zeta^\alpha = L^\alpha{}_\beta \xi^\beta + a^\alpha$. Under such transformations $(\mathbf{G}, \mathbf{b}) \mapsto (\mathbf{LGL}^{-1}, \mathbf{L}(\mathbf{b} - \mathbf{GL}^{-1}\mathbf{a}))$ in (4.2.6).

The property $\mathbf{G} = \mathbf{E}$ of a Volterra dislocation is invariant under such transformations. The Volterra dislocations with $\mathbf{G} = \mathbf{E}$ are called **Burgers' dislocations.**

4.2.2 Linearized Theory of Dislocations

The considerations of this subsection are only heuristic and have no bearing on the considerations of the following subsections.

The existence of solutions representing Volterra dislocations in nonlinear elastic bodies is an open problem since we have to satisfy the condition (III) of the preceding subsection. The solution of this problem depends on the particular constitutive equations.*

On the other hand explicit solutions representing Volterra dislocations are known for some linear elastic bodies (e.g. for isotropic linear elastic bodies). Since every elastic body is more or less apparently nonlinear, it is desirable to relate such solutions to the Volterra dislocations in nonlinear elastic bodies.

Let $\lambda \mapsto \hat{\mathbf{T}}(\mathbf{F}, \lambda)$, $\lambda \in [0, 1]$, be a one-parameter family of constitutive equations for a family of bodies $\mathscr{B}_\lambda$, expressed in terms of a family of reference configurations $\boldsymbol{\varkappa}_\lambda\colon \mathscr{B}_\lambda \to \Omega_\lambda$. We assume that the bodies $\mathscr{B}_\lambda$ are isotropic and the configuration gradients $\nabla\boldsymbol{\varkappa}_\lambda(P)$ are undistorted (with respect to $\hat{\mathbf{T}}(\cdot\,, \lambda)$) for every $P \in \mathscr{B}_\lambda$, $\lambda \in [0, 1]$, and also that $\hat{\mathbf{T}}(\mathbf{F}, \cdot)$ is $\mathscr{C}^1$.

Suppose that $\boldsymbol{\chi}_\lambda\colon \Omega_\lambda \to \mathscr{U}$ (onto) is a one-parameter family of configurations such that $\lambda \mapsto \overset{-1}{\boldsymbol{\chi}_\lambda}(\mathbf{x})$ is $\mathscr{C}^1$ for all $\mathbf{x} \in \mathscr{U}$,

$$\overset{-1}{\boldsymbol{\chi}_\lambda}(\mathbf{x}) = \overset{-1}{\boldsymbol{\chi}_0}(\mathbf{x}) - \lambda\mathbf{u}(\mathbf{x}) + o(\lambda), \tag{4.2.12}$$

$$\operatorname{div}\hat{\mathbf{T}}\left(\nabla\overset{-1}{\boldsymbol{\chi}_\lambda}(\mathbf{x})^{-1}, \lambda\right) + \varrho_\lambda \mathbf{b} = 0, \qquad \varrho_\lambda := \varrho_0 \det\nabla\overset{-1}{\boldsymbol{\chi}_\lambda}, \tag{4.2.13}$$

and $\boldsymbol{\chi}_\lambda$ satisfies some mixed BC (possibly depending on λ in a smooth way).

Differentiating (4.2.13) with respect to λ at $\lambda = 0$ we obtain the linear differential equation

$$\frac{\partial}{\partial x^l}\left[C^{klpq}(\mathbf{x}) \sum_{\alpha=1}^{3} \left(\overset{-1}{\chi}{}^{\alpha}_{0,x^p} u^{\alpha}_{0,x^q} + \overset{-1}{\chi}{}^{\alpha}_{0,x^q} u^{\alpha}{}_{,x^p}\right)\right]$$

$$- \varrho_0 \det\left(\nabla\overset{-1}{\boldsymbol{\chi}_0}\right) (\nabla\boldsymbol{\chi}_0)^l{}_\alpha (\nabla\mathbf{u})^\alpha{}_l b_k + \frac{\partial}{\partial x^l}\frac{\partial \hat{T}^{kl}}{\partial\lambda}\left(\nabla\overset{-1}{\boldsymbol{\chi}_0}(x)^{-1}, \lambda\right)\Big|_{\lambda=0} = 0, \tag{4.2.14}$$

$$C^{klpq} = -\left.\frac{\partial \hat{T}^{kl}}{\partial \overset{-1}{B}_{pq}}\right|_{\lambda=0}. \tag{4.2.15}$$

Suppose that $\boldsymbol{\chi}_0 = \mathrm{id}$ while $\boldsymbol{\chi}_\lambda$, $\lambda > 0$, represent Volterra dislocations with a fixed dislocation line $\mathscr{C} \subset \mathscr{U}$. Equation (4.2.14) assumes the form

$$\frac{\partial}{\partial x^l}\left[C^{klpq}(\mathbf{x})(u^p{}_{,x} + u^q{}_{,x^p})\right] + \varrho_0 b_k' = 0 \tag{4.2.16}$$

and C^{klpq} can be expressed in terms of two Lamé constants λ', μ'.

* The *universal static solutions* defined by Ericksen (See Wang and Truesdell, 1973) satisfy the equilibrium equations for all the elastic constitutive equations whose groups contain a fixed subgroup $\mathscr{G}$ of SL(3). They are, however, affine and hence cannot represent dislocations.

Since $\overset{-1}{\boldsymbol{\chi}}_\lambda \circ \boldsymbol{\chi}_\lambda = \mathrm{id}$, it follows that

$$\frac{d\overset{-1}{\boldsymbol{\chi}}_\lambda}{d\lambda} \circ \boldsymbol{\chi}_\lambda + \overset{-1}{\mathbf{F}}_\lambda \frac{d\boldsymbol{\chi}_\lambda}{d\lambda} = 0, \quad \mathbf{F}_\lambda = \nabla \boldsymbol{\chi}_\lambda .$$

For $\lambda > 0$ $\overset{-1}{\mathbf{F}}_\lambda$ is multivalued on $\mathscr{U} \setminus \mathscr{C}$. Hence $-\dfrac{d\overset{-1}{\boldsymbol{\chi}}_\lambda}{d\lambda} = \mathbf{u} + \ldots$ is multivalued. We shall assume that $\mathbf{u}$ and $\nabla \mathbf{u}$ are multivalued with the branching curve $\mathscr{C}$. The tensor function $\overset{-1}{\mathbf{B}}_\lambda = \overset{t-1-1}{\mathbf{F}_\lambda \mathbf{F}_\lambda}$ is univalued on $\mathscr{U} \setminus \mathscr{C}$, hence

$$\left.\frac{d\left(\overset{-1}{\mathbf{B}}_\lambda\right)_{pq}}{d\lambda}\right|_{\lambda=0} = -(u^p{}_{,x^q} + u^q{}_{,x^p}) \text{ is univalued on } \mathscr{U} \setminus \mathscr{C}.$$

Let $\mathbf{u}_+$, $\mathbf{u}_-$ denote two consecutive sheets of $\mathbf{u}$ over a domain $\mathscr{V} \subset \mathscr{U} \setminus \mathscr{C}$, $\bar{\mathbf{u}}(\mathbf{x}) \equiv \mathbf{u}_-(\mathbf{x}) - \mathbf{u}_+(\mathbf{x})$. Since $\bar{u}^p{}_{,x^q} + \bar{u}^q{}_{,x^p} = 0$, it follows that

$$\mathbf{u}_-(\mathbf{x}) - \mathbf{u}_+(\mathbf{x}) = \bar{\mathbf{u}}(\mathbf{x}) \equiv \mathbf{W}\mathbf{x} + \mathbf{b}, \quad \mathbf{W} = -{}^t\mathbf{W}, \ \mathbf{b} \in R^3. \tag{4.2.17}$$

Solutions of (4.2.16) satisfying (4.2.17) are known in linear elasticity as Volterra dislocations. In the case of a transversely isotropic elastic body $\mathbf{W}$ should be replaced by an element of the Lie algebra of the corresponding isotropy group.

4.2.3 Comments on Volterra Dislocations. Continuous Distributions of Dislocations

Geometrically, the possibility of Volterra dislocations is in close connection with nontrivial material symmetries of the body. Physically, it has two aspects. Firstly, a Volterra dislocation generates a stress field that is not removable by continuous deformations of the body. This property of dislocational stress is related to the fact that $\overset{-1}{\boldsymbol{\chi}}_*$, $\nabla \overset{-1}{\boldsymbol{\chi}}_*$ are multivalued. As pointed out in Chapter 1, a multivalued inverse $\xi^\alpha = \overset{-1}{\chi^\alpha}_*(\mathbf{x})$ implies that the Lagrange variables (ξ^α) are assigned bijectively to material points only locally, i.e. each material point P occupying a position x in the space, lies in an infinite number of domains $\mathbf{U}_i$ of local reference configurations $\boldsymbol{\varkappa}_i : \mathscr{U}_i \to R^3; P \mapsto (\xi^\alpha)$. If a subbody $\mathscr{U}_i$ is placed in the configuration $\boldsymbol{\varkappa}_i$ then it does not exhibit any stress. This operation presupposes that $\mathscr{U}_i$ has been cut out from the body $\mathscr{B}$. In order to release stress in the entire body it is necessary to cut it along a surface Σ described in the preceding subsection.

Secondly, the body $\mathscr{B}$ in the configuration $x^p = \chi^p \circ \boldsymbol{\varkappa}(P)$ involving a dislocation can be subjected to further deformations. Since they can be chosen in a variety of ways, we shall refer to them as **virtual**. Using $\nabla(\boldsymbol{\chi} \circ \boldsymbol{\varkappa})$ as the reference for virtual configuration gradients we can express the resulting stress by means of the constitutive equation

$$\begin{aligned} T^{kl} &= \hat{T}^{kl}_{\boldsymbol{\chi}\circ\boldsymbol{\varkappa}}({F^k}_p, \mathbf{x}) :\equiv \hat{T}^{kl}_{(\boldsymbol{\varkappa})}({F^k}_p {\chi^p}_{,\xi^\alpha}) \\ &= \check{T}^{kl}(\mathbf{F}, P; \nabla(\boldsymbol{\chi} \circ \boldsymbol{\varkappa})(P)). \end{aligned} \tag{4.2.18}$$

In the dislocated reference configuration $\boldsymbol{\chi} \circ \boldsymbol{\varkappa}$ the body exhibits inhomogeneous constitutive properties. Moreover, there is no global reference configuration diffeomorphic to the virtual ones in which the dependence of $\hat{T}^{kl}$ on material points disappears.

We base the theory of continuous distributions of dislocations on the second aspect of dislocations. Let $\xi^\alpha = \varkappa^\alpha(P)$ be the dislocated reference configuration. We again seek a reference for configuration gradients such that the constitutive functions W, $\hat{T}^{kl}$ depend on the corresponding deformation gradients only. In the case of a single Volterra dislocation (or a discrete set of Volterra dislocations) such a reference is provided by the gradients of some *locally defined* configurations $\boldsymbol{\varkappa}_i\colon \mathscr{U}_i \to \boldsymbol{R}^3$. In the case of a continuous distribution of dislocations such a reference can be constructed in terms of (local) *anholonomic* coordinates $(\xi^\alpha, {F^k}_p)$ on the principal frame bundle $\mathscr{F}\mathscr{B}$. Equivalently, there is a (locally defined) field of maps $\mathbf{K}(\xi)\colon T_P\mathscr{B} \to \boldsymbol{R}^3$, $\xi = \boldsymbol{\varkappa}(P)$, such that $W = \tilde{W}(\mathbf{F}, \xi; {\delta^\alpha}_\beta) \equiv W(\mathbf{F}\overset{-1}{\mathbf{K}}(\xi))$. The anholonomic coordinates referred to above are $\tilde{\mathbf{K}}(\in \mathscr{F}_P\mathscr{B}) \mapsto \xi = \boldsymbol{\varkappa}(P)$, $\tilde{\mathbf{F}} = \tilde{\mathbf{K}}\overset{-1}{\mathbf{K}}(\xi) \in \mathscr{L}_+$.

4.2.4 Material Atlases

Definition 4.2.1

Two material points $P, P' \in \mathscr{B}$ are said to be **isomorphic** if there is a reference configuration gradient $\mathbf{K}$, $\mathbf{K}'$ at P, P' resp., such that the response W, $\hat{T}^{kl}$ of P, P' to a deformation gradient $\mathbf{F}$ referred to $\mathbf{K}, \mathbf{K}'$ resp. is given by the same function of $\mathbf{F} \in \mathscr{L}_+$:

$$\check{T}^{kl}(\mathbf{F}, P; \mathbf{K}) = \check{T}^{kl}(\mathbf{F}, P'; \mathbf{K}') \ \forall \mathbf{F} \in \mathscr{L}_+,$$

$$\check{W}(\mathbf{F}, P; \mathbf{K}) = \check{W}(\mathbf{F}, P'; \mathbf{K}') \ \forall \mathbf{F} \in \mathscr{L}_+.$$

It follows immediately that $\mathscr{G}(\mathbf{K}, P) = \mathscr{G}(\mathbf{K}', P')$.

Definition 4.2.2

A subbody $\mathscr{V}$ of $\mathscr{B}$ is said to be **materially homogeneous** if for every $P \in \mathscr{V}$ we can find a reference $\mathbf{K}(P)$ at P that $\check{T}^{kl}(\mathbf{F}, P; \mathbf{K}(P)) \equiv T^{kl}_{\mathscr{V}}(\mathbf{F})$. In this case we shall also say that $\mathbf{K}(\cdot)$ is a **homogeneous reference** over $\mathscr{V}$.

On account of (4.1.8) the condition stated in Definition 4.2.2 is satisfied if $\check{W}(\mathbf{F}, P; \mathbf{K}(P)) = W_{\mathscr{V}}(\mathbf{F})$, $\varrho_{\mathbf{K}(P)} = \text{const}$.

Let $\{(\mathscr{U}_i, \boldsymbol{\varkappa}_i) | i \in I\}$ be a family of local reference configurations whose domains cover $\mathscr{B}$. A homogeneous reference $P \mapsto \mathbf{K}(P)$, defined on $\mathscr{V} \subset \mathscr{B}$, is said to be **$\mathscr{C}^k$-smooth** if for every $i \in I$ such that $\mathscr{V} \cap \mathscr{U}_i \neq \emptyset$ the map $\xi \mapsto \mathbf{K}(\overset{-1}{\boldsymbol{\varkappa}_i}(\xi)) \nabla \boldsymbol{\varkappa}_i(\overset{-1}{\boldsymbol{\varkappa}_i}(\xi))^{-1}$ from $\boldsymbol{\varkappa}_i$ $(\mathscr{V} \cap \mathscr{U}_i)$ into $\mathscr{L}_+$ is $\mathscr{C}^k$-smooth. Note that a $\mathscr{C}^k$-smooth homogeneous reference $\mathbf{K}$ is a $\mathscr{C}^k$ section of the tensor bundle $(R^3 \times \mathscr{V}) \otimes T_* \mathscr{V}$ (this is a tensor product of two vector bundles over $\mathscr{V}$). It is clear that $\mathscr{G}(\mathbf{K}(P), P) =: \mathscr{G}_{\mathbf{K}}$ does not depend on $P \in \mathscr{V}$.

Suppose that there is a family $\{(\mathbf{K}_j, \mathscr{V}_j) |\ j \in J\}$ of homogeneous references K_j over $\mathscr{V}_j$ whose domains $\mathscr{V}_j$ cover $\mathscr{B}$. Let

$$\hat{T}^{kl}(\mathbf{F}, P; \mathbf{K}_j(P)) =: \hat{T}^{kl}_{(j)}(\mathbf{F}) \ \forall \mathbf{F} \in \mathscr{L}_+. \tag{4.2.19}$$

Suppose that $\mathscr{V}_i \cap \mathscr{V}_j \neq \emptyset$ for some $i, j \in J$ and let $\mathbf{K}_i(P) = \mathbf{Q}(P)\mathbf{K}_j(P)$ for $P \in \mathscr{V}_i \cap \mathscr{V}_j$. Since $\hat{T}^{kl}_{(i)}(\mathbf{F}\mathbf{Q}(P)) \equiv \check{T}^{kl}(\mathbf{F}\mathbf{Q}(P), P; \mathbf{K}_i(P)) \equiv \check{T}^{kl}(\mathbf{F}, P; \mathbf{K}_j(P)) \equiv \hat{T}^{kl}_{(j)}(\mathbf{F})$ does not depend on $P \in \mathscr{V}_i \cap \mathscr{V}_j$, we have for $P, P' \in \mathscr{V}_i \cap \mathscr{V}_j$

$$\hat{T}^{kl}(\mathbf{F}\mathbf{Q}(P'); P; \mathbf{K}_i(P)) \equiv \hat{T}^{kl}(\mathbf{F}\mathbf{Q}(P), P; \mathbf{K}_i(P)),$$

hence $\check{T}^{kl}(\mathbf{F}\mathbf{Q}(P)^{-1}\mathbf{Q}(P'), P; \mathbf{K}_i(P)) = \check{T}^{kl}(\mathbf{F}, P; \mathbf{K}_i(P))$ $\forall \mathbf{F} \in \mathscr{L}_+$ and $Q(P)^{-1}\mathbf{Q}(P') \in \mathscr{G}_{\mathbf{K}_i}$. Fixing P we have $\mathbf{Q}(P') = \mathbf{L}\mathbf{G}(P'), \mathbf{L} := \mathbf{Q}(P), \mathbf{G}(P') := \overset{-1}{\mathbf{L}}\mathbf{Q}(P') \in \mathscr{G}_{\mathbf{K}_i}$. Hence

$$\begin{aligned} &\mathbf{K}_j(P) = \mathbf{L}_{ji}\mathbf{G}_{ji}(P)\mathbf{K}_i(P), \\ &\mathbf{G}_{ji}(P) \in \mathscr{G}_{\mathbf{K}_i}, \quad \mathbf{L}_{ji} \in \mathbf{L}_+ \text{ on } \mathscr{V}_i \cap \mathscr{V}_j \end{aligned} \tag{4.2.20}$$

and

$$\hat{T}^{kl}_{(j)}(\mathbf{F}) \equiv \hat{T}^{kl}_{(i)}(\mathbf{F}\mathbf{L}_{ji}\mathbf{G}_{ji}(P)) \equiv \hat{T}^{kl}_{(i)}(\mathbf{F}\mathbf{L}_{ji}). \tag{4.2.21}$$

$\hat{T}^{kl}_{(i)}$ coincides with $\hat{T}^{kl}_{(j)}$ iff $\mathbf{L}_{ji} \in \mathscr{G}_{\mathbf{K}_i}$. In this case we say that **$\mathbf{K}_i$, $\mathbf{K}_j$ are compatible.** Since

$$\hat{T}^{kl}_{(j)}(\mathbf{F}) \equiv \hat{T}^{kl}_{(i_1)}(\mathbf{F}\mathbf{L}_{ji_1}) \equiv \ldots \equiv \hat{T}^{kl}_{(j)}(\mathbf{F}\mathbf{L}_{ji_1}\mathbf{L}_{i_1 i_2}] \ldots \mathbf{L}_{i_k j}),$$

it follows that $\mathbf{L}_{ji_1}\mathbf{L}_{i_1 i_2} \ldots \mathbf{L}_{i_k j} \in \mathscr{G}_{\mathbf{K}_j}$ for any chain of domains $\mathscr{V}_j \cap \mathscr{V}_{i_1} \neq \emptyset$, $\mathscr{V}_{i_1} \cap \mathscr{V}_{i_2} \neq \emptyset, \ldots, \mathscr{V}_{i_k} \cap \mathscr{V}_j \neq \emptyset$.

It is desirable to have the same constitutive function $\hat{T}^{kl}_{(i)} = \hat{T}^{kl}$ for all $i \in J$. For every pair $i, j \in J$ we may replace $\mathbf{K}_i$, $\mathbf{K}_j$ by $\mathbf{K}'_j = \mathscr{L}^{-1}_{ji}\mathbf{K}'_i$, $\mathbf{K}'_i = \mathbf{K}_i$ so that

$$\hat{\mathbf{T}}'_{(i)}(\mathbf{F}) \equiv \check{\mathbf{T}}(\mathbf{F}, P; \mathbf{K}'_i(P)) \equiv \check{\mathbf{T}}(\mathbf{F}, P; \mathbf{K}'_j(P)) \equiv \hat{\mathbf{T}}'_{(j)}(\mathbf{F}).$$

On a global scale one might expect some topological (cohomological) obstacles to removing the $\mathbf{L}_{ji}$. We shall however make the following assumption:

(#) For every two points $P, P' \in \mathscr{B}$ there are configuration gradients $\mathbf{K}$, $\mathbf{K}'$ at P, P' such that $\forall \mathbf{F} \in \mathscr{L}_+$

$$\check{\mathbf{T}}(\mathbf{F}, P; \mathbf{K}) \equiv \check{\mathbf{T}}(\mathbf{F}, P'; \mathbf{K}'), \quad \check{W}(\mathbf{F}, P; \mathbf{K}) \equiv \check{W}(\mathbf{F}, P'; \mathbf{K}')$$

(i.e. any two points of $\mathscr{B}$ are isomorphic).

The assumption (#) amounts to the assertion that every point of $\mathscr{B}$ has essentially the same constitutive properties, provided its configuration in the reference is accounted for.

Theorem 4.2.3

If the body $\mathscr{B}$ satisfies (#) and is locally smoothly homogeneous then $\mathscr{B}$ can be equipped with a family $\{(\mathscr{V}_j, \mathbf{K}'_j) | \ j \in J\}$ of pairwise compatible smooth homogeneous references $\mathbf{K}'_j$ defined over $\mathscr{V}_j$.

Proof

Fix $P_o \in \mathscr{V}_0$ and some $P_j \in \mathscr{V}_j$ for every $j \in J$. Suppose that $\tilde{\mathbf{K}}_{oj}$, $\tilde{\mathbf{K}}_j$ are pairs of configuration gradients at P_o, P_j mentioned in (#), $\mathbf{I}_j := \tilde{\mathbf{K}}_{oj}^{-1} \circ \tilde{\mathbf{K}}_j$. Then for arbitrary $\mathbf{Q} \in \mathscr{L}_+$

$$\begin{aligned}\check{\mathbf{T}}(\mathbf{F}, P_j; \mathbf{Q}\tilde{\mathbf{K}}_{oj} \circ \mathbf{I}_j) = \check{\mathbf{T}}(\mathbf{FQ}, P_j; \tilde{\mathbf{K}}_j) &= \check{\mathbf{T}}(\mathbf{FQ}, P_o; \tilde{\mathbf{K}}_{oj}) \\ &= \check{\mathbf{T}}(\mathbf{F}, P_o; \mathbf{Q}\tilde{\mathbf{K}}_{oj}),\end{aligned}$$

i.e.

$$\check{\mathbf{T}}(\mathbf{F}, P_j; \mathbf{K} \circ \mathbf{I}_j) \equiv \check{\mathbf{T}}(\mathbf{F}, P_o; \mathbf{K})$$

for an arbitarary configuration gradient $\mathbf{K}$ at P_o.

Let $\{(\mathscr{V}_j, \mathbf{K}'_j) | \ j \in J\}$ be a family of smooth homogeneous references such that $\mathscr{V}_j$, $j \in J$, cover $\mathscr{B}$. $\mathbf{G}_{ji}$ in equation (4.2.20) are obviously smooth.

Let $\mathbf{K}_o$ be a fixed configuration gradient at P_o and

$$\mathbf{K}_j(P) := \mathbf{L}_j \circ \mathbf{K}'_j(P), \quad \mathbf{L}_j := \mathbf{K}_o \circ \mathbf{I}_j \circ \overset{-1}{\mathbf{K}'_j}(P_j).$$

Then

$$\begin{aligned}\check{\mathbf{T}}(\mathbf{F}, P; \mathbf{K}_j(P)) \equiv \check{\mathbf{T}}(\mathbf{F}, P; \mathbf{L}_j \circ \mathbf{K}'_j(P)) &\equiv \check{\mathbf{T}}(\mathbf{FL}_j, P_j; \mathbf{K}'_j(P)) \\ &\equiv \check{\mathbf{T}}(\mathbf{FL}_j, P_j; \mathbf{K}'_j(P_j)) \equiv \check{\mathbf{T}}(\mathbf{F}, P_j; \mathbf{K}_o \circ \mathbf{I}_j) \\ &\equiv \check{\mathbf{T}}(\mathbf{F}, P_o; \mathbf{K}_o)\end{aligned}$$

and the right-hand side does not depend on $j \in J$. □

Definition 4.2.4

A **material atlas** over an elastic body $\mathscr{B}$ is a maximal collection α

$= \{(\mathbf{K}_j, \mathscr{V}_j) |\ j \in J\}$ of pairwise compatible* smooth homogeneous references. It exists if the hypotheses of Theorem 4.2.3 are satisfied.

According to Theorem 4.2.3 every locally homogeneous elastic body satisfying (#) can be endowed with a material atlas $\mathfrak{a}$ and $\check{\mathbf{T}}(\mathbf{F}, P; \mathbf{K}_j(P)) \equiv \mathbf{T}_\mathfrak{a}(\mathbf{F})$ for all $j \in J$, $P \in \mathscr{V}_j$. The argument leading to (4.2.20) shows that for any two such atlases $\mathfrak{a} = \{(\mathbf{K}_j, \mathscr{V}_j) |\ j \in J\}$, $\mathfrak{a}' = \{(\mathbf{K}'_i, \mathscr{V}'_i) |\ i \in I\}$ there is a fixed $\mathscr{L} \in \mathscr{L}_+$ such that for every pair $(i, j) \in I \times J$ such that $\mathscr{V}'_i \cap \mathscr{V}_j \neq \emptyset$

$$\mathbf{K}'_i(P) = \mathbf{L} \circ \mathbf{G}_{ij}(P) \circ \mathbf{K}_j(P), \quad \mathbf{G}_{ij}(P) \in \mathscr{G}_{\mathbf{K}_j} = \mathscr{G}_\mathfrak{a}, \ \forall P \in \mathscr{V}'_i \cap \mathscr{V}_j \tag{4.2.22}$$

and

$$\mathbf{T}_{\mathfrak{a}'}(\mathbf{F}) \equiv \mathbf{T}_\mathfrak{a}(\mathbf{FL}), \quad \mathscr{G}_{\mathfrak{a}'} = \mathbf{L} \mathscr{G}_\mathfrak{a} \overset{-1}{\mathbf{L}}. \tag{4.2.23}$$

4.2.5 Linear Connections on Crystalline Bodies

According to the heuristic considerations of Section 4.2.3 the presence of a continuous distribution of dislocations (in a subbody $\mathscr{U} \subset \mathscr{B}$) manifests itself in the impossibility of finding integrable homogeneous references $\mathbf{K}_j$, $\mathscr{V}_j$ such that $\bigcup_j \mathscr{V}_j \supset \mathscr{U}$. A reference $(\mathbf{K}_j, \mathscr{V}_j)$ is said to be **integrable** if $\exists \varkappa_j : \mathscr{V}_j \to R^3$ such that $\mathbf{K}_j = \nabla \varkappa_j$. In this case a material atlas $\mathfrak{a}$ for $\mathscr{B}$ cannot contain integrable $(\mathbf{K}_j, \mathscr{V}_j)$ with $\mathscr{V}_j$ covering $\mathscr{U}$. Since any two material atlases differ by a constant linear transformation of R^3, this property holds for every material atlas if it holds for one of them.

It is our objective in Section 4.2 to derive some geometric objects which measure the departure from integrability (= the intensity of dislocations) and which represent the effect of dislocations in some equations governing the behaviour of the body.

The easiest starting point is provided by the theory of dislocations in crystalline solids.

Definition 4.2.5
A material point $P \in \mathscr{B}$ is said to be **crystalline** if it is a solid point and the isotropy group $\mathscr{G}(\mathbf{K}, P)$ relative to an undistorted reference $\mathbf{K}$ at P is discrete.

It follows immediately that the isotropy group $\mathscr{G}(\mathbf{K}, P)$ relative to an arbitrary reference $\mathbf{K}$ at P is discrete. If a point $P_o \in \mathscr{B}$ is crystalline and the body satisfies the condition (#) then every point $P \in \mathscr{B}$ is crystalline and $\mathscr{G}(\mathbf{K}, P) = \mathscr{G}(\mathbf{K}_o, P_o)$ for some $\mathbf{K}$ at P.

* For simplicity we consider any two homogeneous references with non-overlapping domains as compatible.

We assume that $\mathscr{B}$ satisfies (#), is crystalline and locally smoothly homogeneous.

Consider two local homogeneous references $(\mathbf{K}_i, \mathscr{U}_i)$, $(\mathbf{K}_j, \mathscr{U}_j)$ such that $\mathscr{U}_i \cap \mathscr{U}_j \neq \varnothing$. They neednot be compatible. Since $\mathscr{G}_j := \mathscr{G}(\mathbf{K}_j(P), P)$ is a discrete subgroup of SL(3), it follows from equation (4.2.15) that

$$\mathbf{K}_i(P) = \mathbf{L}_{ij}\mathbf{K}_j(P) \quad \text{for } P \in \mathscr{U}_i \cap \mathscr{U}_j. \tag{4.2.24}$$

Taking advantage of (4.2.24) we shall define the parallel transport of a vector $\mathbf{v} \in T_{P_0}\mathscr{B}$ along a curve $\mathscr{C}$ in $\mathscr{B}$. Without loss of generality we can assume that there is a global reference configuration $\boldsymbol{\varkappa}: \mathscr{B} \to \boldsymbol{R}^3$. We use the coordinate system (ξ^α) provided by $\boldsymbol{\varkappa}$ on $\mathscr{B}$.

Let $\mathscr{C}$ be given by the equations

$$\xi^\alpha = \phi^\alpha(s), \quad \phi^\alpha(0) = \varkappa^\alpha(P_o) = \xi_o{}^\alpha, \quad s \geqslant 0.$$

Let us cover $\mathscr{C}$ by a chain of homogeneous references with overlapping domains: $P_o \in \mathscr{V}_{i_1}$, $\mathscr{C} \cap \mathscr{V}_{i_{k+1}} \cap \mathscr{V}_{i_k} \neq \varnothing$. For any $s \geqslant 0$ such that $\boldsymbol{\phi}(s) \in \boldsymbol{\varkappa}(\mathscr{V}_{i_1})$ and arbitrary $\mathbf{v} \in T_{P_0}\mathscr{B}$ we shall define

$$\mathbf{v}(s) = \mathbf{K}_{i_1}(P(s))^{-1}\mathbf{K}_{i_1}(P_o)\mathbf{v}, \quad P(s) := \overset{-1}{\boldsymbol{\varkappa}}(\boldsymbol{\phi}(s)). \tag{4.5.25}$$

Let $P_1 = P(s_1) \in \mathscr{C} \cap \mathscr{V}_{i_1} \cap \mathscr{V}_{i_2}$, $\mathbf{v}_1 = \mathbf{v}(s_1)$. For every $P(s) \in \mathscr{V}_{i_2}$ we define

$$\mathbf{v}(s) := \mathbf{K}_{i_2}(P(s))^{-1}\mathbf{K}_{i_2}(P_1)\mathbf{v}_1. \tag{4.2.26}$$

On account of (4.2.24) the definitions (4.2.25), (4.2.26) of $\mathbf{v}(s)$ for $P(s) \in \mathscr{V}_{i_1} \cap \mathscr{V}_{i_2}$ are equivalent. If $P_2 = P(s_2) \in \mathscr{C} \cap \mathscr{V}_{i_1} \cap \mathscr{V}_{i_2}$ $\mathbf{v}_2 = \mathbf{v}(s_2)$, then

$$\mathbf{v}(s) := \mathbf{K}_{i_2}(P(s))^{-1}\mathbf{K}_{i_2}(P_1)\mathbf{v}_1 = \mathbf{K}_{i_2}(P(s))^{-1}\mathbf{K}_{i_2}(P_2)\mathbf{v}_2.$$

Hence the definition of $\mathbf{v}(s)$ on $\mathscr{V}_{i_2}$ does not depend on the choice of $P_1 \in \mathscr{C} \cap \cap \mathscr{V}_{i_1} \cap \mathscr{V}_{i_2}$. The construction can be continued recursively in an obvious way. By the same argument $\mathbf{v}(\cdot)$ does not depend on the choice of the covering $\mathscr{V}_{i_1}, \mathscr{V}_{i_2}, \ldots$ of $\mathscr{C}$.

We claim that $\mathbf{v}(s)$ defined in this way is a parallel transport of $\mathbf{v} \in T_{P_0}\mathscr{B}$ along $\mathscr{C}$. Indeed, for fixed $\mathbf{v} \mapsto \mathbf{v}(s)$ is an isomorphism of $T_{P_0}\mathscr{B}$ and $T_{P(s)}\mathscr{B}$.

We now derive a differential equation satisfied by $\mathbf{v} = v^\alpha \dfrac{\partial}{\partial \xi^\alpha}$. Let $\mathbf{K}_i(\overset{-1}{\boldsymbol{\varkappa}}(\xi)) = \mathbf{H}_i(\xi)\nabla\boldsymbol{\varkappa}(\xi)$ $\forall i \in I$. Since

$$v^\alpha(s) = \left(\mathbf{H}_{i_k}(\xi(s))^{-1}\right)^\alpha{}_a H_{i_k}(\xi_{k-1})^a{}_\beta v_{k-1}{}^\beta$$

for $P(s) \in \mathscr{U}_{i_k}$, we calculate

$$\frac{dv^\alpha(s)}{ds} := \lim_{\varepsilon \to 0} \frac{1}{\varepsilon}\Big[v^\alpha(\xi(s+\varepsilon)) - \left(\mathbf{H}_{i_k}(\xi(s+\varepsilon))^{-1}\right)^\alpha{}_a H_{i_k}(\xi(s))^a{}_\beta v^\beta(\xi(s))\Big] \tag{4.2.27}$$

for $\varepsilon \neq 0$ sufficiently small. Hence

$$\frac{dv^\alpha}{ds} = [v^\alpha{}_{,\xi^\gamma} + \Gamma^\alpha{}_{\gamma\beta}(\xi(s))v^\beta]\frac{d\xi^\gamma}{ds}, \tag{4.2.28}$$

$$\begin{aligned}\Gamma^\alpha{}_{\gamma\beta} &= -\left(\mathbf{H}_{i_k}(\xi(s))^{-1}\right)^\alpha{}_{a,\xi^\gamma} H_{i_k}(\xi(s))^a{}_\beta \\ &= \left(\mathbf{H}_{i_k}(\xi(s))^{-1}\right)^\alpha{}_a H_{i_k}(\xi(s))^a{}_{\beta,\xi^\gamma}.\end{aligned} \tag{4.2.29}$$

Note that the right-hand side of (4.2.27) is the difference between a vector at $P(s+\varepsilon)$ and a vector at $P(s)$ transported to $P(s+\varepsilon)$, both expressed in terms of the basis $\partial/\partial\xi^\alpha|_{P(s+\varepsilon)}$. Hence (4.2.28) is a covariant derivative and on account of the construction of $\mathbf{v}(s)$

$$\frac{dv^\alpha}{ds} = 0.$$

It is easy to verify by direct calculation that the right-hand side of (4.2.29) does not depend on the gauges $\mathbf{H}(\xi) \mapsto \mathbf{L} \circ \mathbf{H}(\xi)$ and has the transformation properties of connection symbols (Section 1.1).

Simple algebra also leads to the conclusion that the curvature $R^\alpha{}_{\beta\gamma\delta}$ of the connection $\Gamma^\alpha{}_{\beta\gamma}$ vanishes. This is however obvious from the definition (4.2.25), (4.2.26) of the parallel transport since these formulae extend to a neighbourhood $\mathscr{V}_{i_1} \cup \mathscr{V}_{i_2} \cup \dots \cup \mathscr{V}_{i_r}$ of $\mathscr{C}$ (the connection is locally integrable, Section 1.1).

The torsion $T^\alpha{}_{\beta\gamma}$ of the connection $\Gamma^\alpha{}_{\beta\gamma}$

$$T^\alpha{}_{\beta\gamma} = \Gamma^\alpha{}_{[\beta\gamma]} \tag{4.2.30}$$

need not vanish. We shall clear up its meaning (see also Section 1.1).

If $\mathbf{v}(s)$ results from the parallel transport of a vector along a curve $\mathscr{C} \subset \mathscr{V}_i$, then $H_i{}^a{}_\alpha(\xi(s))v^\alpha(s) = w^a = \text{const}$, $a = 1, 2, 3$. Hence $\mathbf{H}_i(\xi)$ (i) maps each $T_\xi\boldsymbol{\varkappa}(\mathscr{V}_i)$ linearly onto $\boldsymbol{R}^3$; (ii) maps parallel vectors at ξ, $\xi' \in \boldsymbol{\varkappa}(\mathscr{V}_i)$ onto the same vector $\mathbf{w} \in \boldsymbol{R}^3$. Suppose that $\mathbf{K}_i = \nabla\boldsymbol{\mu}$, $\boldsymbol{\mu}: \mathscr{V}_i \to \boldsymbol{R}^3$, or equivalently $H_i{}^a{}_\alpha(\xi) = \dfrac{\partial\phi^a}{\partial\xi^\alpha}(\xi)$, $\boldsymbol{\phi}: \boldsymbol{\varkappa}(\mathscr{V}_i) \to \boldsymbol{\mu}(\mathscr{V}_i)$. A mapping $\boldsymbol{\phi}$ with this property exists locally iff $H_i{}^a{}_{[\alpha,\xi^\beta]} = 0$, i.e. iff $T^\alpha{}_{\beta\gamma} = 0$ in $\boldsymbol{\varkappa}(\mathscr{V}_i)$. Hence $T^\alpha{}_{\beta\gamma} = 0$ is the condition of integrability of the material charts $(\mathscr{V}_i, \mathbf{K}_i)$.

4.2.6 Continuous Distributions of Dislocations and Volterra Dislocations

In order to compare isolated Volterra dislocations with continuous distributions of dislocations we shall need a visual interpretation of curvature and torsion. In view of subsequent applications we shall consider an arbitrary connection with the connection symbols (and possibly non-vanishing curvature).

Let $\mathscr{C}$ be a curve in $\mathscr{B}$, given parametrically by $\xi^\alpha = \bar{\xi}^\alpha(s)$, $s \in [0, 1]$, in terms of the coordinates (ξ^α). We map the curve $\mathscr{C}$ as well as the tangent spaces at $P \in \mathscr{C}$ onto a curve $\tilde{C} \subset \boldsymbol{R}^3$: $\zeta^a = \bar{\zeta}^a(s)$, $s \in [0, 1]$, and the tangent spaces $T_{\bar{\xi}(s)}\boldsymbol{R}^3$ in the following way:

(i) the tangent vector $\dfrac{d\bar{\xi}^\alpha}{ds}(s)$ at $\bar{\xi}^\alpha(s)$ goes over into $\dfrac{d\bar{\zeta}^a}{ds}(s)$, $s \in [0, 1]$;

(ii) if the vectors $\mathbf{v}_i$, $i = 1, 2, 3$ at $\bar{\xi}^\alpha(s)$ go over into $\tilde{\mathbf{v}}_i \in \boldsymbol{R}^3$, $i = 1, 2, 3$, resp., then $\sum_{i=1}^{3} c^i\mathbf{v}_i$ goes over into $\sum_{i=1}^{3} c^i\tilde{\mathbf{v}}_i$;

(iii) if the vectors $\mathbf{v}$ at $\bar{\xi}^\alpha(s)$ and $\mathbf{v}'$ at $\bar{\xi}^\alpha(s')$ are related by parallel transport along $\mathscr{C}$ then their images $\tilde{\mathbf{v}}$, $\tilde{\mathbf{v}}'$ in $\boldsymbol{R}^3$ coincide (note that $T_\zeta\boldsymbol{R}^3 = \boldsymbol{R}^3$ in a natural way).

For the construction of $\tilde{\mathscr{C}}$ let us choose three arbitrary linearly independent vectors $\mathbf{v}_1, \mathbf{v}_2, \mathbf{v}_3$ at $T_{\bar{\xi}(0)}\boldsymbol{\varkappa}(\mathscr{B})$. Let $\mathbf{e}_1, \mathbf{e}_2, \mathbf{e}_3$ be a basis of $\boldsymbol{R}^3$. We assign $\mathbf{e}_i$ to $\mathbf{v}_i$, $i = 1, 2, 3$. By parallel transport along $\mathscr{C}$ we produce three linearly independent vectors $\mathbf{v}_i(s)$ at $\bar{\xi}^\alpha(s)$, for every $s \in [0, 1]$. According to (iii) the image of $\mathbf{v}_i(s)$ is $\mathbf{e}_i$, for $i = 1, 2, 3$, $s \in [0, 1]$. Suppose that

$$\frac{d\bar{\xi}^\alpha}{ds}(s) = \sum_{i=1}^{3} c_i(s)v_i{}^\alpha(s).$$

The coefficients $c^i(s)$ are uniquely determined by this equation. Hence, on account of (i), (ii)

$$\frac{d\bar{\zeta}^a}{ds} = \sum_{i=1}^{3} = c^i(s)e_i{}^a, \quad a = 1, 2, 3. \tag{4.2.31}$$

Integrating these equations we obtain $\tilde{\mathscr{C}}$ up to a translation.

The curve $\tilde{\mathscr{C}}$ depends on the choice of $\mathbf{v}_i$, $\mathbf{e}_i$ and on the initial conditions for (4.2.31). Hence it is defined up to a non-singular linear transformation and a translation.

In particular for the parallel transport (4.2.26) we can take $\mathbf{e}_i = \mathbf{K}_{i_1}(P(s))\mathbf{v}_i(s)$ on $\mathscr{V}_{i_1}$, $\mathbf{e}_i = \mathbf{L}_{12}\mathbf{K}_{i_2}(P(s))\mathbf{v}_i(s)$ on $\mathscr{V}_{i_2}$ etc., with $\mathbf{L}_{12} = \mathbf{K}_{i_1}(P(s))\mathbf{K}_{i_2}(P(s))^{-1} = \text{const}$ ... In the coordinates (ξ^α), (ζ^a) on $\mathscr{B}$, $\boldsymbol{R}^3$, resp.

$$e_i{}^a = H_{i_1}(\bar{\xi}(s))^a{}_\alpha v_i{}^\alpha(s) = (L_{12})^a{}_b H_{i_2}(\bar{\xi}(s'))^b{}_\beta v_i{}^\beta(s') \ldots,$$

for $P(s) \in \mathscr{V}_{i_1}$, $P(s') \in \mathscr{V}_{i_2}$, ...

Consider a closed loop $\mathscr{C} \subset \mathscr{V}_i \subset \mathscr{B}$. Suppose that $\mathbf{H}_i = \nabla\phi$, $\phi\colon \boldsymbol{\varkappa}(\mathscr{V}_i) \to \mathscr{U}_i \subset \boldsymbol{R}^3$, $(\xi^\alpha) \mapsto (\zeta^a)$ diffeomorphically. ϕ may be considered as a coordinate transformation on $\mathscr{V}_i$. In the new coordinates (ζ^a) the vectors $\mathbf{v}_i(s)$ have the components e_i^a. The components $\dfrac{d\tilde{\zeta}^a}{ds} = \dfrac{\partial\phi^a}{\partial\xi^\alpha}\dfrac{d\xi^\alpha}{ds} = \sum_{i=1}^{3} c^i(s)e_i^a$ of the tangent vector of $\mathscr{C}$ coincide with the components of the tangent vector of $\tilde{\mathscr{C}}$. Let $\bar{\zeta}(0) = \boldsymbol{\phi}(\bar{\xi}(0))$. The curve $\tilde{\mathscr{C}}$ coincides with the image of $\mathscr{C}$ in the coordinates (ζ^a) and hence it is closed.

Suppose that the closed loop $\mathscr{C}$ lies in a simply connected domain $\mathscr{U} \subset \mathscr{V}_{i_1} \cup \mathscr{V}_{i_2} \cup \ldots \cup \mathscr{V}_{i_r}$. We decompose $\mathscr{C}$ algebraically into a number of loops $\mathscr{C}_\alpha$ such that each $\mathscr{C}_\alpha$ is contained in some $\mathscr{V}_i$ (Fig. 4.1). Since curvature vanishes parallel transport of a vector $\mathbf{v}_i$ at $\bar{\xi}(0)$ yields a unique vector field

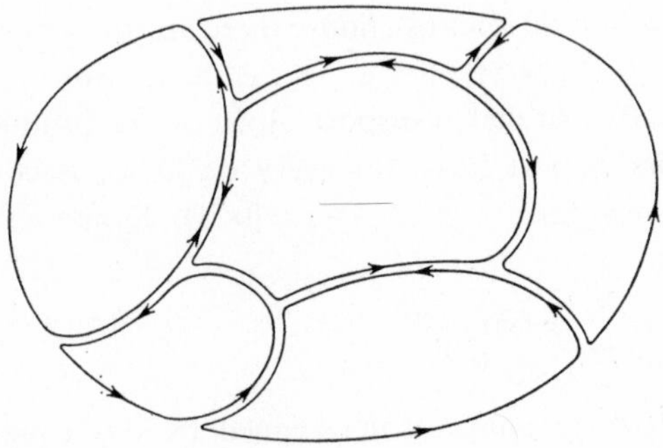

Fig. 4.1 Decomposition of a closed oriented curve

$\mathbf{v}_i(\xi)$ in $\mathscr{U}$, for $i = 1, 2, 3$. Applying the construction of the preceding paragraphs we obtain from each $\mathscr{C}_\alpha$ a closed loop $\tilde{\mathscr{C}}_\alpha$. In our case the image $\tilde{\mathscr{C}}$ of $\mathscr{C}$ is an algebraic sum of $\tilde{\mathscr{C}}_\alpha$, hence it is closed.

For a possibly non-integrable $(\mathbf{K}_i, \mathscr{U}_i)$ consider a small "rectangle" $\mathscr{C}$ in $\mathscr{U}_i$ with the vertices ξ_o^α, $\xi_o^\alpha+\varepsilon_1^\alpha$, $\xi_o^\alpha+\varepsilon_1^\alpha+\varepsilon_2^\alpha$, $\xi_o^\alpha+\varepsilon_2^\alpha$ (the ordering is important). Let ζ_o^a be the image of ξ_o^α in $\boldsymbol{R}^3$. The image of $\xi_o^\alpha+\varepsilon_1^\alpha$ is then

$$\begin{aligned}\zeta_1^a &= \zeta_o^a + \int_0^1 ds H^a{}_\alpha(\xi_o+s\varepsilon_1)\,\varepsilon_1^\alpha \\ &= \zeta_o^a + H^a{}_\alpha(\xi_o)\,\varepsilon_1^\alpha + \tfrac{1}{2}H^a{}_{\alpha,\xi^\beta}(\xi_o)\,\varepsilon_1^\alpha\varepsilon_1^\beta + O(|\varepsilon_1|^3).\end{aligned}$$

The image of the rectangle $\mathscr{C}$ is a curve with *five* vertices ζ_o^a, ζ_1^a, ζ_2^a, ζ_3^a, ζ_4^a,

$$\begin{aligned}\zeta_4^a = \zeta_o^a &+ H^a{}_\alpha(\xi_o)\,\varepsilon_1^\alpha + \tfrac{1}{2}H^a{}_{\alpha,\xi^\beta}(\xi_o)\,\varepsilon_1^\alpha\varepsilon_2^\beta + H^a{}_\alpha(\xi_o+\varepsilon_1)\,\varepsilon_2^\alpha \\ &+ \tfrac{1}{2}H^a{}_{\alpha,\xi^\beta}(\xi_o+\varepsilon_1)\,\varepsilon_2^\alpha\varepsilon_2^\beta - H^a{}_\alpha(\xi_o+\varepsilon_1+\varepsilon_2)\,\varepsilon_1^\alpha\end{aligned}$$

$$+\tfrac{1}{2}H^a{}_{\alpha,\xi^\beta}(\xi_o+\varepsilon_1+\varepsilon_2)\,\varepsilon_1{}^\alpha\varepsilon_1{}^\beta - H^a{}_\alpha(\xi_o+\varepsilon_2)\,\varepsilon_2{}^\alpha$$
$$+\tfrac{1}{2}H^a{}_{\alpha,\xi^\beta}(\xi_o+\varepsilon_2)\,\varepsilon_2{}^\alpha\varepsilon_2{}^\beta + O(|\varepsilon_1|^3+|\varepsilon_2|^3).$$

Hence

$$b^a := \zeta_4{}^a - \zeta_o{}^a = -2H^a{}_{[\alpha,\xi^\beta]}(\xi_o)\,\varepsilon_1{}^\alpha\varepsilon_2{}^\beta + O(|\varepsilon_1|^3+|\varepsilon_2|^3)$$
$$= 2H^a{}_\gamma T^\gamma{}_{\beta\alpha}\,\varepsilon_1{}^\alpha\varepsilon_2{}^\beta + O(|\varepsilon_1|^3, |\varepsilon_2|^3). \quad (4.2.32)$$

The image $\tilde{\mathscr{C}}$ of the closed rectangle $\mathscr{C}$ is not closed if $b^a \neq 0$.

The vector $(b^a) \in \boldsymbol{R}^3$ depends on the choice of the reference $\mathbf{K}_i$. Its image in $T_{\xi_o}\varkappa(\mathscr{B})$

$$b^\gamma = 2T^\gamma{}_{\beta\alpha}\,\varepsilon_1{}^\alpha\varepsilon_2{}^\beta + \ldots \quad (4.2.33)$$

is, however, uniquely defined.

Equation (4.2.33) remains valid for *all* linear connections, whether they are locally integrable or not (Cartan, 1928, 1946; Schouten, 1951). In this case the vector $(b^\alpha) \in T_{\xi_o}\varkappa(\mathscr{B})$ is the counterimage of $(b^a) \in \boldsymbol{R}^3$ under the mapping of tangent spaces on $\mathscr{C}$ onto $\boldsymbol{R}^3$ (for an alternative definition of b^α, see Rashevskii (1953)).

We shall make another important remark concerning *general* linear connections. Let us consider parallel transport $\mathbf{v}(s)$ of a vector $\mathbf{v}(0) \in T_{\xi_o}\varkappa(\mathscr{B})$ along the rectangle $\mathscr{C}$. It is fairly easy to check that

$$v^\alpha(1) - v^\alpha(0) = -R^\alpha{}_{\beta\gamma\delta}\,\varepsilon_1{}^\gamma\varepsilon_2{}^\delta v^\beta(0) + O(|\varepsilon_1|^3, |\varepsilon_2|^3) \quad (4.2.34)$$

(Schouten, 1951). If $\mathbf{K}$ is a homogeneous reference over $\mathscr{V}$, $(\mathbf{K}, \mathscr{V}) \in \mathfrak{a}$, $P = \overset{-1}{\varkappa}(\xi_o) \in \mathscr{V}$, $\mathbf{w}(s) = \mathbf{K}(P)\mathbf{v}(s)$ for $s = 0, 1$, then

$$w^a(1) - w^a(0) = -R^a{}_{(\varphi)\beta\gamma\delta}\,\varepsilon_1{}^\gamma\varepsilon_2{}^\delta w^b(0) + \ldots,$$
$$[R^a{}_{(\varphi)b\gamma\delta}\,\varepsilon_1{}^\gamma\varepsilon_2{}^\delta] \in \mathfrak{a}, \quad (4.2.35)$$

where $\mathfrak{a}$ denotes the Lie algebra of the group $\mathscr{G}$ (Section 1.1.3).

Let us now construct a counterpart of the mapping $\mathscr{C} \mapsto \tilde{\mathscr{C}}$ in the case of a single Volterra dislocation. Comparing equations (4.2.18) and (4.2.19) it becomes obvious that $\nabla\overset{-1}{\boldsymbol{\chi}}{}^*$ plays the role of $\mathbf{K}$. Let $\Gamma_1 \subset \varkappa(\mathscr{B})$ be the dislocation line (the branching curve of the multivalued mappings $\overset{-1}{\boldsymbol{\chi}}_*$, $\nabla\overset{-1}{\boldsymbol{\chi}}_*$). Let $\mathscr{C}$ be a closed oriented curve encircling Γ_1 once. In order to avoid ambiguity we fix an internal orientation on Γ_1 and adjust the orientation of $\mathscr{C}$ by the rule of a dextrorotatory screw. Since $\overset{-1}{\boldsymbol{\chi}}_*$ is a multivalued mapping, the image $\tilde{\mathscr{C}} := \overset{-1}{\boldsymbol{\chi}}_*(\mathscr{C})$ of $\mathscr{C}$ is not in general a closed curve. Its endpoints differ by a transformation (4.2.6). On the other hand the mapping $\overset{-1}{\boldsymbol{\chi}}_*: \mathscr{C} \to \tilde{\mathscr{C}}$ together with the mappings $\nabla\overset{-1}{\boldsymbol{\chi}}_*(\xi)$ of the tangent spaces satisfies the conditions (i), (ii), (iii) above. For a comparison with (4.2.34) we note that parallel transport along $\mathscr{C}$ is defined by the equation $\nabla\overset{-1}{\boldsymbol{\chi}}_*{}^a{}_\alpha(\bar{\xi}(s))v^\alpha(s) = \text{const}$. Hence in general $v^\alpha(1) \neq v^\alpha(0)$ and in fact $\mathbf{G}\nabla\overset{-1}{\boldsymbol{\chi}}_*(\bar{\xi}(0))\mathbf{v}(1) = \nabla\overset{-1}{\boldsymbol{\chi}}_*(\bar{\xi}(0))\mathbf{v}(0)$, $G \in \mathscr{G}$. This is to be confronted with (4.2.35).

Let Γ_i be either an oriented simple closed curve in $\boldsymbol{\varkappa}(\mathscr{B})$ or an oriented arc with both ends on $\partial\boldsymbol{\varkappa}(\mathscr{B})$, $i = 1, \ldots, n$. Suppose that $\Omega := \boldsymbol{\varkappa}(\mathscr{B}) \setminus \bigcup_{i=1}^{n} \Gamma_i$ is a connected set. A multiple-valued mapping $\overset{-1}{\boldsymbol{\chi}}_*: \Omega \to \boldsymbol{R}^3$ involves n Volterra dislocations if (i) for every $\xi \in \Omega$ there is a homomorphism H_ξ of the fundamental group $\pi_\xi(\Omega)$ of Ω into the semidirect product $\mathscr{G} \times \boldsymbol{R}^3$,* (ii) for the homotopy class C_i of those loops at ξ which encircle a single curve Γ_i exactly

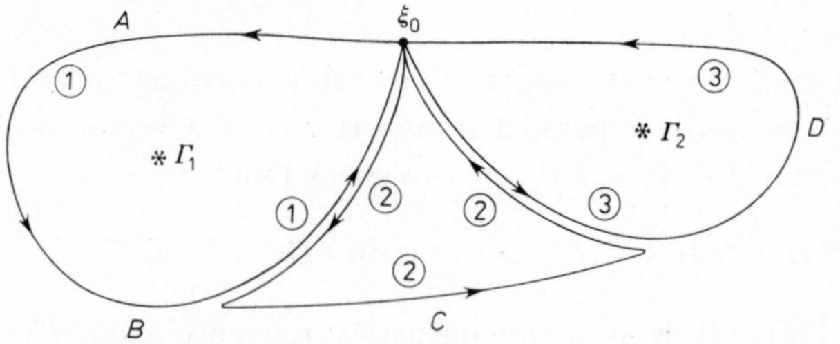

Fig 4.2 Decomposition of a closed oriented curve encircling two dislocations

once and in accordance with the rule of the dextrorotatory screw $H_\xi(C_i) = (\mathbf{A}_i, \mathbf{a}_i) \in \mathscr{G} \times \boldsymbol{R}^3$ (the right-hand side does not depend on ξ), (iii) for the homotopy class of loops at ξ which do not encircle any Γ_i, $i = 1, \ldots, n$, $H_\xi(C) = (\mathbf{E}, \mathbf{0})$.

The loop $\xi_o ABCD \xi_o$ shown in Fig. 4.2 can be decomposed into the product of loops (3), (2), (1) (in this order in the usual notation of group multiplication; in the reverse order if the direction of motion along the curve is concerned).

* We recall that $\pi_\xi(\Omega)$ denotes the group of homotopy classes of oriented simple closed curves at ξ with group multiplication defined as in Section 1.1.3.

The semi-direct product $\mathscr{G} \times \boldsymbol{R}^3$ is defined by the multiplication rule $(\mathbf{A}_1, \mathbf{a}_1)(\mathbf{A}_2, \mathbf{a}_2) = (\mathbf{A}_1\mathbf{A}_2, \mathbf{a}_1 + \mathbf{A}_1\mathbf{a}_2)$.

Note that in general neither $\pi_\xi(\Omega)$ nor $\mathscr{G} \times \boldsymbol{R}^3$ is Abelian.

The loop (2) is homotopically trivial. Hence $H_{\xi_0}(\xi_o ABCD\xi_o) = (\mathbf{A}_2\mathbf{A}_1, \mathbf{a}_2 + \mathbf{A}_2\mathbf{a}_1)$. This formula is readily generalized to the case of a loop encircling an arbitrary number of dislocation lines Γ_i.

Let us return to the connection (4.2.29). A comparison of (4.2.33) with (4.2.6) suggests an interpretation of the connection (4.2.29) in terms of a continuous distribution of Burgers' dislocations. We make this analogy more explicit by means of appropriate mathematical constructions.

Let us consider an elastic body $\mathscr{B}$ in the actual configuration $\boldsymbol{\chi} \circ \boldsymbol{\varkappa}$. Let $\xi = \boldsymbol{\varkappa}(P)$ be a reference configuration with a dense array of Burgers' dislocations, $\xi^\alpha = \chi^\alpha{}_o(y^a)$ in terms of local homogeneous coordinates (y^a). We need a physically justified measure of vector length and area in $\boldsymbol{\varkappa}(\mathscr{B})$.

Let $F^k{}_\alpha := \chi^k{}_{,\xi^\alpha}$, $C_{\alpha\beta} = F^k{}_\alpha F^k{}_\beta$, $C^{\alpha\beta} := \overset{-1}{F^\alpha{}_k}\overset{-1}{F^\beta{}_k}$ (summation over $k = 1, 2, 3$). The length of a vector $\mathbf{v} = v^\alpha \partial/\partial\xi^\alpha$ is $[C_{\alpha\beta}\, v^\alpha v^\beta]^{1/2}$. The area df of a surface element Σ spanned by two small vectors $\varepsilon_1{}^\alpha$, $\varepsilon_2{}^\alpha$, $\varepsilon_1{}^\alpha \neq \lambda\varepsilon_2{}^\alpha$, is defined as the Hausdorff measure of the surface element Σ' spanned by the vectors $F^k{}_\alpha\varepsilon_1{}^\alpha$, $F^k{}_\alpha\varepsilon_2{}^\alpha$ in the Euclidean space $(\boldsymbol{R}^3, \delta_{ij})$. By (n_k) we denote the unit normal on Σ' chosen in such a way that $(n_\alpha, \varepsilon_1{}^\alpha, \varepsilon_2{}^\alpha)$* form a right-handed system, $n_\alpha = F^k{}_\alpha n_k$. We have

$$df n_k = \epsilon_{klm} F^l{}_\beta F^m{}_\gamma\, \varepsilon_1{}^\beta \varepsilon_2{}^\gamma = (\det \mathbf{F})\overset{-1}{F^\alpha{}_k}\, \epsilon_{\alpha\beta\gamma}\varepsilon_1{}^\beta\varepsilon_2{}^\gamma, \tag{4.2.36}$$

$$C^{\alpha\beta} n_\alpha n_\beta = 1,$$

$$df n_\alpha = [\det[C_{\rho\sigma}]]^{1/2}\, \epsilon_{\alpha\beta\gamma}\, \varepsilon_1{}^\beta \varepsilon_2{}^\gamma. \tag{4.2.37}$$

In a nbhd of every $\xi \in \boldsymbol{\varkappa}(\mathscr{B})$ we classify the dislocation lines according to the unit tangent vector of the dislocation line (compatible with the orientation) and the Burgers' vector $\mathbf{b}$ (equation (4.2.6)). For definiteness we assume that $\mathbf{b} = \overset{-1}{\boldsymbol{\chi}_o}(\bar{\xi}(1)) - \overset{-1}{\boldsymbol{\chi}_o}(\bar{\xi}(0))$ for a simple closed curve $\mathscr{C}$ encircling the dislocation line and oriented according to the rule of the dextrorotary screw (the parameter s of $\mathscr{C}$: $\xi = \bar{\xi}(s)$ grows in the direction of the orientation). Let $\gamma_A(\xi)\, df$ denote the number of dislocation lines $(\mathbf{l}_A, \mathbf{b}_A)$ intersecting a surface element $df\mathbf{l}_A$ at ξ. A surface element Σ at ξ with the area df and normal n_k (in the coordinates x^k) is intersected by $\gamma_A|\mathbf{n}\cdot\mathbf{l}_A|\, df$ dislocation lines $(\mathbf{l}_A, \mathbf{b}_A)$.

Let $\mathscr{C} = \partial\Sigma$ be a rectangle spanned by the vectors ε_1, ε_2, oriented in accordance with $\mathbf{n}$. We map $\mathscr{C}$ onto $\tilde{\mathscr{C}}$ according to (i)–(iii). From the considerations concerning deformations with several Volterra dislocations we

* More correctly, $(n_k, F^k{}_\alpha\varepsilon_1{}^\alpha, F^k{}_\alpha\varepsilon_2{}^\alpha)$ form a right-handed basis.

know that the endpoints of $\tilde{\mathscr{C}}$ differ by a vector at ξ equal to the sum of the Burgers' vectors $\mathbf{b}$ of the dislocation lines encircled by $\mathscr{C}$:

$$y_4{}^a - y_0{}^a = \sum_A \gamma_A(\xi)\mathbf{n}\cdot\mathbf{l}_A b_A{}^a df \tag{4.2.38}$$

(sum over all classes of dislocation lines at ξ).

If $\mathscr{C}$ encircles a dislocation line according to the rule of the right/left-handed screw then $\mathbf{b}_A$ enters (4.2.38) with the sign plus/minus. The sign of $\mathbf{n}\cdot\mathbf{l}_A$ accounts for this fact.

Let us now turn to the connection $\Gamma^\alpha{}_{\beta\gamma}$, equation (4.2.29). Let $T^k{}_{lm}$ be the torsion of $\Gamma^\alpha{}_{\beta\gamma}$ expressed in the coordinates $x^k = \chi^k(\xi^\alpha)$. Let $\mathscr{C}$, $\tilde{\mathscr{C}}$ be defined as before. The endpoints of $\tilde{\mathscr{C}}$ diverge by a vector

$$\begin{aligned} y_4{}^a - y_0{}^a &= -2H^a{}_\alpha(\xi)\overset{-1}{F}{}^\alpha{}_k F^l{}_\beta F^m{}_\gamma T^k{}_{lm}\,\varepsilon_1{}^\beta\varepsilon_2{}^\gamma = H^a{}_\alpha(\xi)\overset{-1}{F}{}^\alpha{}_k A^{kp} n_p df \\ &= H^a{}_\alpha A^{\alpha\sigma} n_\sigma df = H^a{}_\alpha \tilde{A}^{\alpha\sigma}\,\epsilon_{\sigma\beta\gamma}\,\varepsilon_1{}^\beta\varepsilon_2{}^\gamma, \end{aligned} \tag{4.2.39}$$

where

$$A^{kp} := -\,\epsilon^{plm}T^k{}_{lm} \tag{4.2.40}$$

is known as the **Nye tensor of Burgers' dislocation density.** The tensors $A^{\alpha\sigma}$, $\tilde{A}^{\alpha\sigma}$ are obtained by the coordinate transformation $(x^k)\mapsto(\xi^\alpha)$ of A^{kp} treated as an absolute tensor or tensor density of weight $+1$ resp.

Comparing (4.2.38) and (4.2.39) we arrive at a formula which explains the Nye tensor in physical terms:

$$A^{kp} = \sum_A \tilde{\gamma}_A(\mathbf{x}) b_A{}^k l_A{}^p \tag{4.2.41}$$

with

$$\gamma_A = \tilde{\gamma}_A \circ \boldsymbol{\chi},\ b_A{}^k := F^k{}_\alpha \overset{-1}{H}{}^\alpha{}_a b_A{}^a. \tag{4.2.42}$$

4.2.7 Connections in the Case of Continuous Isotropy Groups

Suppose that the isotropy groups of an elastic body $\mathscr{B}$ have a nontrivial component of unity ($\neq \{\mathbf{E}\}$). In this case the local gauges $\mathbf{K}_i(\xi)\mathbf{K}_j(\xi)^{-1} = \mathbf{G}(\xi)$ on $\mathscr{U}_i \cap \mathscr{U}_j$ depend on ξ and the definition (4.2.25–26) of parallel transport breaks down.

Let us consider an elastic body $\mathscr{B}$ with arbitrary isotropy groups. We assume however that there is a material atlas $\mathfrak{a}'$ (Definition 4.2.4, Theorem 4.2.3) on $\mathscr{B}$. We choose a fixed atlas $\mathfrak{a}$ and define a subbundle $\mathscr{F}$ of $\mathscr{F}\mathscr{B}$ with fibers

$$\mathscr{F}_P = \{\mathbf{K}(P) \mid \text{for all } (\mathbf{K}, \mathscr{V}) \in \mathfrak{a} \text{ such that } P \in \mathscr{V}\}.$$

The reader can check easily that $\mathscr{F}$ is a principal bundle with the structure group $\mathscr{G} = \mathscr{G}_{\mathfrak{a}}$. In Section 1.1 we proved that there is at least one linear connection on $\mathscr{F}$. Every connection Γ on every subbundle $\mathscr{F}$ constructed in this way from some material atlas $\mathfrak{a}'$ will be called a **material connection.**

Even for a fixed subbundle $\mathscr{F}$ the linear connection Γ on $\mathscr{F}$ is not unique but we shall see later that all the important equations involving a material connection Γ on $\mathscr{F}$ remain true for all the other connections on $\mathscr{F}$.

For simplicity we use the abbreviated expressions **Γ-lift, Γ-parallel transport** etc. meaning a lift or parallel transport defined in terms of the connection Γ.

Let Γ be a linear connection on $\mathscr{F}$ and let $\mathbf{K}(t)$ be a Γ-lift of a curve $P(t) \in \mathscr{B}$ to $\mathscr{F}$. Obviously $\mathbf{K}(t) = \mathbf{G}_i(t)\mathbf{K}_i(P(t))$, $\mathbf{G}_i(t) \in \mathscr{G}$ for $P(t) \in \mathscr{V}_i$, $(\mathbf{K}_i, \mathscr{V}_i) \in \mathfrak{a}$. Hence

$$
\begin{aligned}
\check{\mathbf{T}}(\mathbf{F}, P(t); \mathbf{K}(t)) &= \check{\mathbf{T}}(\mathbf{F}, P(t); \mathbf{G}_i(t)\mathbf{K}_i(P(t))) \\
&= \check{\mathbf{T}}(\mathbf{F}, P(t); \mathbf{K}_i(P(t))) = \check{\mathbf{T}}_{\mathfrak{a}}(\mathbf{F}) \\
&= \check{\mathbf{T}}(\mathbf{F}, P(0); \mathbf{K}_{i_0}(P(0))) = \check{\mathbf{T}}(\mathbf{F}, \mathbf{P}(0); \mathbf{K}(0)),
\end{aligned} \tag{4.2.43}
$$

Hence for every Γ-lift $\mathbf{K}(t) \in \mathscr{F}$ the references $\mathbf{K}(0)$, $\mathbf{K}(t)$ at $P(0)$, $P(t)$ are equivalent.

In the case of a discrete group $\mathscr{G} = \mathscr{G}_{\mathfrak{a}}$ $\mathscr{F}$ is the union of a discrete collection of disjoint 3-dimensional manifolds projecting submersively onto $\mathscr{B}$. Every smooth horizontal distribution $\mathscr{H}$ on $\mathscr{F}$ is locally tangent to a section $\mathbf{K}_i(P)$, $(\mathbf{K}_i, \mathscr{V}_i) \in \mathfrak{a}$. Hence a horizontal lift $\mathbf{K}(t)$ of a curve $P(t) \in \mathscr{B}$ into $\mathscr{F}$ must locally have the form $\mathbf{K}_i(P(t))$ for some $i \in I$. This brings us back to the parallel transport (4.5.25).

Let $\mathscr{B}$ be a solid elastic body with a material atlas $\mathfrak{a}'$, $P_o \in \mathscr{V} \subset \mathscr{B}$, $(\mathbf{K}, \mathscr{V}) \in \mathfrak{a}'$. Let $\mathbf{K}_o$ be an undistorted reference at P_o, $\mathbf{K}_o = \mathbf{LK}(P_o)$, $\mathbf{L} \in \mathrm{SL}(3)$. The material atlas $\mathfrak{a} := \mathbf{L}\mathfrak{a}' = \{(\mathbf{L} \circ \mathbf{K}, \mathscr{V}) \mid (\mathbf{K}, \mathscr{V}) \in \mathfrak{a}'\}$ consists of smooth undistorted uniform references, $\mathscr{G} = \mathscr{G}_{\mathfrak{a}} \subset \mathrm{SO}(3)$, equation (4.2.23). Let $(\mathbf{K}_i, \mathscr{V}_i) \in \mathfrak{a}$. We can consider $\mathbf{K}_i(P)$ as a mapping of $T_P\mathscr{B}$ onto the space $\boldsymbol{R}^3$ endowed with the Euclidean metric δ_{ij}. Let $\langle \mathbf{a}, \mathbf{b} \rangle = \sum_{i=1}^{3} a^i b^i$, $\mathbf{a}, \mathbf{b} \in \boldsymbol{R}^3$,

$$
g(P)(\mathbf{v}, \mathbf{w}) = \langle \mathbf{K}_i(P)\mathbf{v}, \mathbf{K}_i(P)\mathbf{w} \rangle \quad \text{for } \mathbf{v}, \mathbf{w} \in T_P\mathscr{B}, P \in \mathscr{V}_i. \tag{4.2.44}
$$

If $(\mathbf{K}_j, \mathscr{V}_j) \in \mathfrak{a}$, $P \in \mathscr{V}_i \cap \mathscr{V}_j$, then $\mathbf{K}_j(P) = \mathbf{AK}_i(P)$, $\mathbf{A} \in \mathrm{SO}(3)$. Consequently the right-hand side of (4.2.44) does not depend on the choice of $(\mathbf{K}_i, \mathscr{V}_i) \in \mathfrak{a}$, provided $P \in \mathscr{V}_i$. g is a Riemannian metric on $\mathscr{B}$.

Let $\mathscr{F}$ be the subbundle of $\mathscr{F}\mathscr{B}$ associated with the atlas $\mathfrak{a}$ of undistorted uniform references. Let Γ be a connection on $\mathscr{F}$. For any curve $P(t) \in \mathscr{B}$ its Γ-lifts $\mathbf{K}(t) \in \mathscr{F}_{P(t)}$. Let $\mathbf{v}_o, \mathbf{w}_o \in T_{P(0)}\mathscr{B}$, $\bar{\mathbf{v}}_o = \mathbf{K}(0)\, \mathbf{v}_o$, $\bar{\mathbf{w}}_o = \mathbf{K}(0)\mathbf{w}_o \in \boldsymbol{R}^3$.

The Γ-parallel transports $\mathbf{v}(t)$, $\mathbf{w}(t)$ of $\mathbf{v}_o$, $\mathbf{w}_o$ along $P(t)$ satisfy the equations $\mathbf{K}(t)\mathbf{w}(t) = \bar{\mathbf{w}}_o, \mathbf{K}(t)\,\mathbf{v}(t) = \bar{\mathbf{v}}_o$. Let $P(t) \in \mathscr{V}_i$, $(\mathbf{K}_i, \mathscr{V}_i) \in \mathfrak{a}$, for some t. Since $\mathbf{K}(t) \in \mathscr{F}_{P(t)}$, $\mathbf{K}_i(P(t)) \in \mathscr{F}_{P(t)}$, we must have that $\mathbf{K}(t) = \mathbf{A}(t)\mathbf{K}_i(P(t))$, $\mathbf{A}(t) \in \mathscr{G} \subset SO(3)$ and $g(P(t))(\mathbf{v}(t), \mathbf{w}(t)) = \langle \mathbf{K}(t)\mathbf{v}(t), \mathbf{K}(t)\mathbf{w}(t)\rangle = \langle \bar{\mathbf{v}}_o, \bar{\mathbf{w}}_o\rangle$ $=$ const. Hence the metric g is invariant under Γ-parallel transport.

A connection $\Gamma^{\alpha}{}_{\beta\gamma}$ preserving a Riemannian metric $g_{\alpha\beta}$ is uniquely determined by its torsion. In fact, setting $g^{\alpha\gamma}g_{\gamma\beta} = \delta^{\alpha}{}_{\beta}$,

$$\left\{ {\alpha \atop \beta\gamma} \right\} := \tfrac{1}{2} g^{\alpha\sigma}(g_{\gamma\sigma,\xi^{\beta}} + g_{\beta\sigma,\xi^{\gamma}} - g_{\beta\gamma,\xi^{\sigma}}),$$

we have

$$\Gamma^{\alpha}{}_{\beta\gamma} = \left\{ {\alpha \atop \beta\gamma} \right\} + T^{\alpha}{}_{\beta\gamma} - T_{\beta\gamma}{}^{\alpha} + T_{\gamma}{}^{\alpha}{}_{\beta}, \tag{4.2.45}$$

with $T_{\gamma}{}^{\alpha}{}_{\beta} = g^{\alpha\sigma}g_{\gamma\tau}T^{\tau}{}_{\sigma\beta}$ etc. (Schouten, 1954). Among the connections (4.2.45), there is a unique torsionfree connection $\Gamma^{\alpha}{}_{\beta\gamma} = \left\{ {\alpha \atop \beta\gamma} \right\}$ on $\mathscr{F}\mathscr{B}$, and it is the Riemannian connection for g. It need not be a connection on $\mathscr{F}$ however.

Suppose that $\mathscr{B}$ is an isotropic solid body with a material atlas $\mathfrak{a}'$. Let $\mathbf{K}_o$ be the undistorted reference at $P_o \in \mathscr{B}$, $\mathscr{G}(\mathbf{K}_o, P_o) = SO(3)$ (Theorem 4.1.6). Proceeding as before we can construct a material atlas $\mathfrak{a}$ such that $\mathscr{G} = \mathscr{G}_{\mathfrak{a}} = SO(3)$. Let $\mathscr{F}$ be the corresponding subbundle of $\mathscr{F}\mathscr{B}$. We know that every connection on $\mathscr{F}$ preserves the Riemannian metric g defined by equation (4.2.44). Suppose now that a linear connection Γ on $T\mathscr{B}$ (and the associated connection Γ on $\mathscr{F}\mathscr{B}$) preserves g. For any Γ-lift $\mathbf{K}(t) \in \mathscr{F}\mathscr{B}$ of a curve $P(t) \in \mathscr{B}$

$$g(P(t))(\mathbf{K}(t)^{-1}\mathbf{v}_o, \mathbf{K}(t)^{-1}\mathbf{w}_o) =: B(\mathbf{v}_o, \mathbf{w}_o) \ \forall \mathbf{v}_o, \mathbf{w}_o \in \mathbf{R}^3$$

is independent of t. For $P(t) \in \mathscr{V}_i$, $(\mathbf{K}_i, \mathscr{V}_i) \in \mathfrak{a}$,

$$\langle \mathbf{K}_i(P(t))\mathbf{K}(t)^{-1}\mathbf{v}_o, \mathbf{K}_i(P(t))\mathbf{K}(t)^{-1}\mathbf{w}_o\rangle = B(\mathbf{v}_o, \mathbf{w}_o) \quad \forall \mathbf{v}_o, \mathbf{w}_o \in \mathbf{R}^3.$$

Hence, choosing $\mathbf{K}(0) = \mathbf{K}_{i_o}(P(0))$ for some i_o such that $P(0) \in \mathscr{V}_{i_o}$ and $(\mathbf{K}_{i_o}, \mathscr{V}_{i_o}) \in \mathfrak{a}$, we see that $\mathbf{K}_i(P(t))\mathbf{K}(t)^{-1} = \mathbf{A}(t) \in SO(3)$ and $\mathbf{K}(t) \in \mathscr{F}_{P(t)}$, since the action of SO(3) on fibers is transitive. The distribution $\mathscr{H}$ of the connection Γ satisfies the requirements of transversality to fibers of $\mathscr{F}$ and L_g-invariance for $g \in SO(3)$. Hence Γ is a connection on $\mathscr{F}$.

We have thus proved that every connection Γ on $\mathscr{F}\mathscr{B}$ preserving g is a connection on $\mathscr{F}$. Among the connections Γ on $\mathscr{F}\mathscr{B}$ preserving g is the Riemannian connection $\left\{ {\alpha \atop \beta\gamma} \right\}$. Hence an isotropic solid body admits exactly one torsion-free connection, which is a Riemannian connection.

Before proceeding to the analysis of transversely isotropic elastic bodies we make a few preliminary remarks. Let $\mathscr{G} = \mathscr{G}_{\mathfrak{a}} \subset SO(3)$ and let $\mathbf{S}$ be a positive definite symmetric tensor, invariant under $\mathscr{G}$: $A^{a}{}_{c}S_{ab}A^{b}{}_{d} = S_{cd}$. Since

${}^t\mathbf{A} = \overset{-1}{\mathbf{A}}$, in matrix notation this amounts to the commutation relation $\mathbf{AS} = \mathbf{SA}\ \forall \mathbf{A} \in \mathscr{G}$. If $\mathbf{S}$ has this property then the Riemannian metric

$$\bar{g}_{\alpha\beta}(P) = S_{ab} K_i{}^a{}_\alpha(P) K_i{}^b{}_\beta(P) \quad \text{for } P \in \mathscr{V}_i,\ (\mathbf{K}_i, \mathscr{V}_i) \in \mathfrak{A}$$

is independent of i and invariant under $\mathscr{G}(\nabla\varkappa(P), P)$. It is also invariant under parallel transport by any connection on $\mathscr{F}$:

$$S_{ab} K_i{}^a{}_\alpha\big(P(t)\big) K_i{}^b{}_\beta\big(P(t)\big) \overset{-1}{K}{}^\alpha{}_c(t) \overset{-1}{K}{}^\beta{}_d(t) = S_{cd} = \text{const.}$$

Following Wang and Truesdell (1973) we call every metric invariant under isotropy groups an **intrinsic metric**.

Let $\mathscr{G}$ be the group of special orthogonal transformations leaving a vector $\mathbf{k} \in \boldsymbol{R}^3$ invariant. Let $\mathbf{S}$ be a positive definite symmetric tensor commuting with $\mathscr{G}$. If $\mathbf{j}$ is an eigenvector of $\mathbf{S}$ then $\mathbf{Aj}$ is also an eigenvector of $\mathbf{S}$, corresponding to the same eigenvalue. If $\mathbf{k}+\mathbf{j}$ ($\mathbf{j} \perp \mathbf{k}, \mathbf{j} \neq \mathbf{0}$) is an eigenvector of $\mathbf{S}$ then the set of eigenvectors of $\mathbf{S}$ includes a cone $\mathbf{k}+\mathbf{Aj}$, $\mathbf{A} \in \mathscr{G}$. This cone is not a subspace of $\boldsymbol{R}^3$. Such a situation is impossible. Hence $\mathbf{Sk} = \alpha\mathbf{k}$, $\mathbf{Sj} = \beta\mathbf{j}$ $\mathbf{Aj} \perp \mathbf{k}$, $\mathbf{S} = \alpha\mathbf{k}\otimes\mathbf{k} + \beta(\mathbf{E} - \mathbf{k}\otimes\mathbf{k})$.

Let $\mathscr{B}$ be a solid body with an atlas $\mathfrak{a}$ of undistorted uniform references, $\mathscr{G} = \mathscr{G}_{\mathfrak{a}} \subset \mathrm{SO}(3)$, and let $\mathscr{F}$ be the corresponding subbundle of $\mathscr{F}\mathscr{B}$. For every connection Γ on $\mathscr{F}$ and every intrinsic metric $\bar{g}_{(r)}$ of the form described above

$$\Gamma^\alpha{}_{\beta\gamma} = \left\{{\alpha \atop \beta\gamma}\right\}_{(r)} + T_{(r)}{}^{\alpha}{}_{\beta\gamma} - T_{(r)\beta\gamma}{}^\alpha + T_{(r)\gamma}{}^\alpha{}_\beta,$$

where $\{\cdot\}_{(r)}$ denotes the Riemannian connection for $\bar{g}_{(r)}$ and $T^\alpha{}_{\beta\gamma}$ is the torsion of Γ. The subscript "r" on T refers to the fact that $\bar{g}_{(r)}$ has been used to raise or lower indices. If there is a torsion-free connection Γ on $\mathscr{F}$, then it is the Riemannian connection for every intrinsic metric $\bar{g}_{(r)}$ of the form described above.

Suppose that $\mathscr{B}$ is a transversely isotropic solid elastic body. Let us gauge the atlas $\mathfrak{a}$ on $\mathscr{B}$ in such a way that $\mathscr{G} = \mathscr{G}_{\mathfrak{a}}$ is the group or special orthogonal transformations leaving the vector $\mathbf{k}_o = (0, 0, 1) \in \boldsymbol{R}^3$ fixed. The "absolute" isotropy group $\mathscr{G}(P)$ acting in $T_P\mathscr{B}$ leaves $\mathbf{k}(P) := \mathbf{K}_i(P)^{-1}\mathbf{k}_o$ fixed for $P \in \mathscr{V}_i$, $(\mathbf{K}_i, \mathscr{V}_i) \in \mathfrak{a}$. Note that $\mathbf{K}_i(P)^{-1}\mathbf{k}_o = \mathbf{K}_j(P)^{-1}\mathbf{k}_o$ for $P \in \mathscr{V}_i \cap \mathscr{V}_j$. If Γ is a connection on $\mathscr{F}$ and $\mathbf{K}(t)$ is a Γ-lift of a curve $P(t) \in \mathscr{B}$ then

$$\begin{aligned}\mathbf{K}(t)^{-1}\mathbf{K}(0)\mathbf{k}\big(P(0)\big) &= \mathbf{K}(t)^{-1}\mathbf{K}(0)\mathbf{K}_{i_o}\big(P(0)\big)^{-1}\mathbf{k}_o = \mathbf{K}(t)^{-1}\mathbf{k}_o \\ &= \mathbf{K}\big(P(t)\big)^{-1}\mathbf{K}\big(P(t)\big)\mathbf{K}(t)^{-1}\mathbf{k}_o = \mathbf{K}\big(P(t)\big)^{-1}\mathbf{k}_o \\ &= \mathbf{k}\big(P(t)\big).\end{aligned}$$

Conversely, if Γ is a connection on $\mathscr{F}\mathscr{B}$ and Γ-parallel transport preserves an intrinsic metric $\bar{g}_{(r)}$ of the kind described above as well as the vector field $\mathbf{k}(\cdot)$, then $\mathbf{K}(t)\mathbf{K}\big(P(t)\big)^{-1} \in \mathscr{G}$ for $P(t) \in \mathscr{V}$, $(\mathbf{K}, \mathscr{V}) \in \mathfrak{a}$ and therefore Γ is a connection on $\mathscr{F}$.

Let $\mathscr{B}$ be a transversely isotropic solid elastic body and suppose that the Riemannian connections associated with all the intrinsic Riemannian metrics $\bar{g}_{(r)}$ of the kind described above coincide. We prove that the Riemannian connections are connections on $\mathscr{F}$. Since every such connection preserves a metric $\bar{g}_{(r)}$, in view of the preceding paragraph it is enough to prove that it preserves $\mathbf{k}(\cdot)$ as well. Consider two intrinsic metrics $g, \bar{g}$ of the kind described above, $\bar{g}(P)(\mathbf{k}(P), \mathbf{k}(P)) = \alpha g(P)(\mathbf{k}(P), \mathbf{k}(P)), \bar{g}(P)(\mathbf{u}, \mathbf{v}) = \beta g(P)(\mathbf{u}, \mathbf{v})$ for $\mathbf{u}, \mathbf{v}$ g-orthogonal to $\mathbf{k}(P)$, $\bar{g}(P)(\mathbf{u}, \mathbf{k}(P)) = 0$ for $\mathbf{u}$ g-orthogonal to $\mathbf{k}(P)$, $\alpha, \beta > 0$. For arbitrary $\beta > 0$ such $\bar{g}, g$ can be constructed by an appropriate choice of $\mathbf{S}$. Consider two points $P, P' \in \mathscr{B}$ connected by a curve $\mathscr{C}$ and let T be the parallel transport from $T_P\mathscr{B}$ to $T_{P'}\mathscr{B}$ along $\mathscr{C}$ defined by the Riemannian connection of g, $\bar{g}$. Let $\mathbf{T}\mathbf{k}(P) = \gamma\mathbf{k}(P') + \mathbf{u} =: \mathbf{k}'$, $g(P')(\mathbf{k}(P'), \mathbf{u}) = 0$. We have the following equations

$$g(P)(\mathbf{k}(P), \mathbf{k}(P)) = g(P')(\mathbf{k}', \mathbf{k}') = \gamma^2 g(P')(\mathbf{k}(P'), \mathbf{k}(P')) + \\ + g(P')(\mathbf{u}, \mathbf{u}),$$

$$\beta g(P)(\mathbf{k}(P), \mathbf{k}(P)) = \bar{g}(P)(\mathbf{k}(P), \mathbf{k}(P)) = \bar{g}(P')(\mathbf{k}', \mathbf{k}') \\ = \gamma^2 \beta g(P')(\mathbf{k}(P'), \mathbf{k}(P')) + g(P')(\mathbf{u}, \mathbf{u}).$$

For $\beta \neq 1$ we conclude that $g(P')(\mathbf{u}, \mathbf{u}) = 0$ and hence $\mathbf{u} = \mathbf{0}$.

We have proved that a transversely isotropic solid elastic body admits a torsion-free connection on $\mathscr{F}$ if and only if the Riemannian connections associated with all the intrinsic metrics $\bar{g}_{(r)}$ coincide. In this case they are equal to the unique torsion-free connection Γ on $\mathscr{F}$.

4.2.8 Curvature, Torsion and Dislocation Density in the General Case

We recall from Section 1.1 that (1) a curvature-free connection is locally integrable, i.e. the corresponding distribution $\mathscr{H}$ is locally tangent to local sections $\mathbf{K}(\cdot)$ of $\mathscr{F}$; (2) a curvature-free and torsion-free connection corresponds to a distribution $\mathscr{H}$ which is locally tangent to sections of $\mathscr{F} \subset \mathscr{F}\mathscr{M}$ of the form $(\xi^\alpha) \mapsto (\xi^\alpha, \partial\phi^a/\partial\xi^\alpha)$. In the latter case we can introduce local coordinates $\zeta^a = \phi^a(\xi^\alpha)$ on $\mathscr{B}$; in these coordinates $\Gamma^a{}_{bc} = 0$, i.e. the connection is flat. Indeed, (1) implies that the parallel transport $v^\alpha(t)$ of a vector $v^\alpha(0)$ satisfies the equation $\dfrac{\partial\phi^a}{\partial\xi^\alpha}(\bar{\xi}^\beta(t))v^\alpha(t) = w^a = \text{const}$, and w^a are the components of $\mathbf{v}^\alpha$ in the coordinate system (ζ^a).

(1) and (2) also imply that there are local uniform references $\mathbf{K}(\cdot)$ which are locally integrable ($\mathbf{K} = \nabla\boldsymbol{\varkappa}$); their domains cover $\mathscr{B}$. In the references $\boldsymbol{\varkappa}$

$\mathscr{B}$ is apparently homogeneous (at least locally). Physically, it has at most isolated Volterra dislocations.*

Suppose that $\mathscr{B}$ is simply connected. The existence of a torsion-free connection Γ on $\mathscr{F}$ implies that $\mathscr{F}$ has a global section $P \mapsto \mathbf{K}_o(P)$ and hence is trivial. The trivializing map is $\varphi_o: \mathbf{K} \mapsto (\sigma(\mathbf{K}), \mathbf{K}\mathbf{K}_o(P)^{-1})$. If Γ is torsion-free as well then there is a global reference $\boldsymbol{\varkappa}_o: \mathscr{B} \to \boldsymbol{R}^3$ such that $\mathbf{K}_o = \nabla\boldsymbol{\varkappa}_o$ and hence the body is homogeneous.

Suppose that $\mathscr{B}$ is locally homogeneous. Then it can be covered by the domains $\mathscr{V}_i$ of the uniform references $\nabla\boldsymbol{\varkappa}_i$, $\boldsymbol{\varkappa}_i: \mathscr{V}_i \to \boldsymbol{R}^3$. The proof of existence of connections in Section 1.1.3 and the following remark show that there is at least one torsion-free connection on $\mathscr{B}$. Suppose additionally that $\mathscr{B}$ is a solid elastic body. Let $\boldsymbol{\varkappa}$ be an arbitrary reference configuration $\mathscr{B} \to \boldsymbol{R}^3$ and let $\boldsymbol{\varkappa}_i = \boldsymbol{\phi}_i \circ \boldsymbol{\varkappa}$, $\xi^\alpha = \varkappa^\alpha(P)$. The body $\mathscr{B}$ has a locally Euclidean intrinsic metric

$$g_{\alpha\beta} = \sum_{a=1}^{3} \frac{\partial\phi^a}{\partial\xi^\alpha}\frac{\partial\phi^a}{\partial\xi^\beta}.$$

There is a connection Γ on $\mathscr{F}$ of the form (4.2.45) which is torsion-free. This connection is the Riemannian connection $\left\{{\alpha \atop \beta\gamma}\right\}$ for $g_{\alpha\beta}$. In the local coordinates $\zeta^a = \phi^a(\xi^\alpha)$ the components g_{ab} of g are constant and the connection symbols vanish. Hence for every locally homogeneous elastic *solid* there is at least one curvature-free torsion-free connection on $\mathscr{F}$ (Wang and Truesdell, 1973).

We close this section by some remarks on the interpretation of connections on $\mathscr{F}$. The reader should have noted the remarkable affinities between the definition of isolated Volterra dislocations and the definition of material atlases and associated subbundles $\mathscr{F}$. Both concepts bear the same relationship to the homogeneity of constitutive properties and to the isotropy groups. As in Section 4.2.6, we shall make these analogies more explicit.

In the case of an elastic body with isotropy groups containing a non-trivial component of identity the same inhomogeneity field represented by a material atlas $\mathfrak{a}$ or the related subbundle $\mathscr{F}$** is in some sense represented by every possible linear connection Γ on $\mathscr{F}$. Neither the curvature nor the torsion are uniquely determined. On the other hand referring to (4.2.6), (4.2.33) and (4.2.35) we would like to associate curvature and torsion with density of Vclterra dislocations. Curvature takes account of the homogeneous part

* The fact that every closed circuit $\mathscr{C}$ in $\mathscr{B}$ satisfies equation (4.2.6) follows from our assumption about the existence of a material atlas, $\mathscr{G} = \mathscr{G}_{\mathfrak{a}}$.

** The material atlas is defined only up to a constant linear transformation. This indeterminacy is relatively trivial.

of (4.2.6) (see (4.2.35)) and torsion accounts for the Burgers' vector **b** in (4.2.6).

Non-uniqueness of curvature and torsion does not contradict physical interpretation in terms of dislocation density. In classical dislocation theory it is known that a **disclination** ($\mathbf{G} \neq \mathbf{E}$ in (4.2.6)) is equivalent to some array of Burgers' dislocations ($\mathbf{G} = \mathbf{E}$) while a Burgers' dislocation can be considered as a dipole of disclinations. Hence the same inhomogeneity field (= continuous distribution of dislocations) can be decomposed in various ways into disclinations and Burgers' dislocations.*

Our theory of dislocation densities in terms of connections has some interesting implications. A crystalline solid admits continuous distributions of Burgers' dislocations and at most isolated disclinations (if it is multiply connected).

Isotropic solids admit inhomogeneity fields that can be split into disclinations and Burgers' dislocations in various ways. The inhomogeneity field in an isotropic solid can, however, be represented by a unique torsion-free connection Γ. The curvature of this connection is the Riemann curvature of an intrinsic metric g and can be identified with what is known as **strain incompatibility**. Indeed, let $C_{\alpha\beta} = F^a{}_\alpha F^a{}_\beta$ be the right Cauchy–Green tensor for the anholonomic (= non-integrable) deformation $F^a{}_\alpha$ removing the inhomogeneities. The curvature referred to above is the Riemann curvature of the metric $C_{\alpha\beta}$. If it is possible to remove inhomogeneities by local deformations $\xi^\alpha \to \zeta^a = \phi^a(\xi^\alpha)$ then $R = 0$. Thus strain incompatibility is in fact the only source of continuously distributed inhomogeneities in isotropic solids.

In the case of crystalline or transversely isotropic solids there is another obstacle to removing inhomogeneities: some preferred directions must become parallel in the homogeneous configuration. This is precisely the reason why in general there is no connection on $\mathscr{F}$ with vanishing torsion.

The splitting of incompatibility into torsion and curvature seems a bit artificial. We show (without going into details) that it can be avoided. For this purpose one can extend the principal frame bundle $\mathscr{F}\mathscr{B}$ to an affine frame bundle $\mathscr{A}\mathscr{B}$ over $\mathscr{B}$ (Lichnerowicz, 1976). The structure group of $\mathscr{A}\mathscr{B}$ is the affine group, i.e. the semidirect product $\mathrm{SL}(3) \times \boldsymbol{R}^3$. The subbundle $\mathscr{F}$ of $\mathscr{F}\mathscr{B}$ automatically extends to a subbundle $\mathscr{A} \subset \mathscr{A}\mathscr{B}$ with the structure group $\mathscr{G} \times \boldsymbol{R}^3$, $\mathscr{G} = \mathscr{G}_{\mathfrak{a}}$. Every linear connection on $\mathscr{F}$ determines a corresponding connection on $\mathscr{A}$.

* In the case of an isolated disclination ($\mathbf{G} \neq \mathbf{E}$) the Burgers' vector is not uniquely defined either.

The Lie algebra of $\mathscr{G}\times \boldsymbol{R}^3$ is the direct sum $\mathfrak{a}\oplus \boldsymbol{R}^3$. Correspondingly, the curvature has components of the form $R^a_{(\varphi)b\alpha\beta}$ and $T^a_{(\varphi)\alpha\beta}$. The latter can be identified with the torsion of the linear connection on $\mathscr{F}$.

The elements of $\mathscr{A}_P$ are reference frames with variable origins. The associated bundle can be interpreted as the bundle of vectors $\mathbf{v}\in T\mathscr{B}$ with specified **attachment points** $\mathbf{a}\in T\mathscr{B}$. If a vector $\mathbf{v}(0)\in T_{P(0)}\mathscr{B}$ attached at $\mathbf{a}(0)\in T_{P(0)}\mathscr{B}$ undergoes parallel transport along a closed curve at $P(0)$, then its attachment point shifts by (4.2.33), while the vector rotates according to (4.2.34). The displacement of the arrow end of the attached vector $(\mathbf{a},\mathbf{v})$ is the sum of (4.2.33) and (4.2.44). Such a formula was given by Kunin (1965) without a precise justification.

4.3 PSEUDOFORCES

4.3.1 The Lagrangian of Elastodynamics and the Noether Theorems

Dislocations in crystalline solids are well-defined and directly observable physical objects. What is more important, they interact with other physical objects including stress (or stress-inducing loads). This interaction is commonly expressed in terms of a "force". Physically, however, a force is associated with virtual shifts of material objects in *space-time* through the notion of *work* (or energy variation). In the case of dislocations we must consider energy variations associated with shifts of dislocations with respect to material points, i.e. on an abstract manifold $\mathscr{B}$. Due to this fact one runs into some difficulties. In order to avoid confusion we use the word "pseudoforce" in this more general situation.

Equations of motion of an elastic body (1.2.84), (1.2.85), (1.2.87) can be derived from the stationarity of

$$\int_{\varkappa(\mathscr{B})} dL^3\, L(\xi,\boldsymbol{\chi}(\xi,t),\nabla\boldsymbol{\chi}(\xi,t),\boldsymbol{\chi}_{,t}(\xi,t)),$$

with the Lagrangian L given by the formula

$$L(\xi,\boldsymbol{\chi},\nabla\boldsymbol{\chi},\boldsymbol{\chi}_{,t}) = \tfrac{1}{2}\varrho_0\boldsymbol{\chi}_{,t}{}^2-\varrho_0 W(\nabla\boldsymbol{\chi},\xi)-\varrho_0 P(\boldsymbol{\chi},\xi). \tag{4.3.1}$$

Throughout this section we assume that $\boldsymbol{\chi}\in\mathscr{C}^2$. Although such an assumption is not very realistic it is in the spirit of differential geometry and variational calculus which will be used here.

Let us consider a general Lagrangian $L(x, w(x), \nabla w(x))$ for some fields $w(x)$ which may be scalars, tensors or any geometric objects. Suppose that $I(w) = \int_\Omega dL^n L(x, w, \nabla w)$ is invariant under a one-parameter family of transformations

$$
\begin{aligned}
'x^k &= \bar{x}^k(x, w; s) = x^k + sa^k(x, w) + o(s), \quad \bar{x}^k \in \mathscr{C}^1, \\
'w^\alpha &= \bar{w}^\alpha(x, w; s) = w^\alpha + sv^\alpha(x, w) + o(s), \quad \bar{v}^\alpha \in \mathscr{C}^1.
\end{aligned} \tag{4.3.2}
$$

Since $w^\alpha(x) + sv^\alpha(x, w(x)) + \ldots = \tilde{w}^\alpha(x+sa) = \tilde{w}^\alpha(x) + \tilde{w}^\alpha{}_{,x^k} sa^k + \ldots$ the variation of the *functions* $w^\alpha(\cdot)$ is

$$\delta w^\alpha(x) := \tilde{w}^\alpha(x) - w^\alpha(x) = s[v^\alpha - w^\alpha{}_{,x^k} a^k] + \ldots$$

Note that $\Omega = \Omega(s)$. Hence, setting $I(\tilde{w}) = \tilde{I}(s)$,

$$
\begin{aligned}
I := \tilde{I}'(0)s &= \int_\Omega dL^n \left[\frac{\partial L}{\partial w^\alpha} - \frac{d}{dx^k}\frac{\partial L}{\partial w^\alpha{}_{,x^k}}\right]\delta w^\alpha \\
&\quad + \int_{\partial\Omega} dH^{n-1} n_k \left[sLa^k + \frac{\partial L}{\partial w^\alpha{}_{,x^k}}\delta w^\alpha\right],
\end{aligned}
$$

$$\frac{d}{dx^k} := \frac{\partial}{\partial x^k} + w^\alpha{}_{,x^k}\frac{\partial}{\partial w^\alpha} + w^\alpha{}_{,x^k x^l}\frac{\partial}{\partial w^\alpha{}_{,x^l}}$$

Variational problems involve appropriate BC for w^α on $\partial\Omega$. The content of a variational problem does not change if the functional I changes by a constant dependent on the BC. Since the Euler–Lagrange equations for w^α are of second order we may admit that

$$\delta I = s\int_{\partial\Omega} dH^n \phi^{-1}(x, w(x)) n_l = s\int_\Omega dL^n \frac{d\phi^l}{dx^l}(x, w(x)).$$

If ϕ^l is a fixed function of $x \in \partial\Omega$ under the given BC, then the variational problem is invariant under the transformations (4.3.2) if

$$
\begin{aligned}
&\frac{d}{dx^k}\left[La^k + \frac{\partial L}{\partial w^\alpha{}_{,x^k}}(v^\alpha - w^\alpha{}_{,x^l} a^l)\right] - \frac{d\phi^l}{dx^l} \\
&= \left[\frac{d}{dx^k}\frac{\partial L}{\partial w^\alpha{}_{,x^k}} - \frac{\partial L}{\partial w^\alpha}\right][v^\alpha - w^\alpha{}_{,x^k} a^k].
\end{aligned} \tag{4.3.3}
$$

If in addition $w^\alpha(\cdot)$ is a solution of the Euler–Lagrange equations for L, then the right-hand side of (4.3.3) vanishes and w^α satisfies a conservation law corresponding to the left-hand side of (4.3.3). This fact is known as the **Noether theorem.**

If I is invariant under (4.3.2) for arbitrary functions $w^\alpha(\cdot)$ satisfying such BCs that the surface integrals assume a fixed value, then L is a null-Lagrangian (Ball, 1976) or an **independent integrand** in the terminology of Rund (1966).

The Lagrangian (4.3.1) does not depend explicitly on time. Hence I is invariant under the transformations $t \mapsto t+a$, with $\phi^l \equiv 0$, and every solution of the equations of motion satisfies the conservation law

$$\frac{d}{dt}\left[L - \frac{\partial L}{\partial \chi^k{}_{,t}} \chi^k{}_{,t}\right] - \frac{d}{d\xi^\mu}\left[\frac{\partial L}{\partial \chi^k{}_{,\xi^\mu}} \chi^k{}_{,t}\right] = 0,$$

i.e.

$$\frac{dE}{dt} = \frac{d}{d\xi^\mu}\left[\varrho_o \frac{\partial W}{\partial \chi^k{}_{,\xi^\mu}} \chi^k{}_{,t}\right], \tag{4.3.4}$$

$$E := \varrho_o\left[\frac{1}{2}\boldsymbol{\chi}_{,t}{}^2 + W + P\right]. \tag{4.3.5}$$

Suppose now that the Lagrangian (4.3.1) does not depend explicitly on $\mathbf{x} = \boldsymbol{\chi}$. In this case

$$\frac{d}{d\xi^\mu}\frac{\partial L}{\partial \chi^k{}_{,\xi^\mu}} + \frac{d}{dt}\frac{\partial L}{\partial \chi^k{}_{,t}} = 0, \tag{4.3.6}$$

i.e. we recover the equations of motion $\varrho_o \chi^k{}_{,tt} - S_k{}^\mu{}_{,\xi^\mu} = 0$ for vanishing body loads. A deviation from the conservation law (4.3.6) is the body force per unit volume in the reference configuration:

$$\varrho_o b_k = \varrho_o \chi^k{}_{,tt} - S_k{}^\mu{}_{,\xi^\mu}. \tag{4.3.7}$$

Suppose now that the body $\mathscr{B}$ is homogeneous in the reference configuration $\boldsymbol{\varkappa}$, i.e. ϱ_o, W and P do not depend on $\xi = \boldsymbol{\varkappa}(P)$. The Noether theorem yields the conservation law

$$\frac{d}{d\xi^\nu}\left[L\delta^\nu_\mu - \frac{\partial L}{\partial \chi^k{}_{,\xi^\nu}} \chi^k{}_{,\xi^\mu}\right] - \frac{d}{dt}\left[\frac{\partial L}{\partial \chi^k{}_{,t}} \chi^k{}_{,\xi^\mu}\right] = 0 \tag{4.3.8}$$

or

$$\frac{d\pi_\mu}{dt} - \frac{d}{d\xi^\nu}\Sigma^\nu{}_\mu = 0, \tag{4.3.9}$$

$$\pi_\mu := \varrho_o \chi^k{}_{,t} \chi^k{}_{,\xi^\mu}, \tag{4.3.10}$$

$$\Sigma^\nu{}_\mu := L\delta^\nu{}_\mu + S_k{}^\nu \chi^k{}_{,\xi^\mu}. \tag{4.3.11}$$

Let $\tilde{\beta}_\mu$ be the deviation from the conservation law (4.3.9) in the general case:

$$\frac{d\pi_\mu}{dt} - \frac{d}{d\xi^\nu}\Sigma^\nu{}_\mu = \tilde{\beta}_\mu. \tag{4.3.12}$$

Working out the left-hand side of (4.3.12) and taking into account the equations of motion one arrives readily at the formula

$$\tilde{\beta}_\mu = -\frac{\partial L}{\partial \xi^\mu}, \tag{4.3.13}$$

which can be compared with the formula $\varrho_0 b_k = -\varrho_0 \dfrac{\partial P}{\partial x^k}(\boldsymbol{\chi}, \xi) = \dfrac{\partial L}{\partial \chi^k}$. The forces b_k account for the inhomogeneity of space due to the x-dependent potential field P, while the "pseudoforce" $\tilde{\beta}_\mu$ accounts for the inhomogeneity of the body associated with a ξ-dependent strain energy.

4.3.2 Covariant Expressions for Pseudoforces

We assume that $\mathscr{B}$ is a solid, $\mathfrak{a}$ is a material atlas of undistorted uniform references on $\mathscr{B}$ and $\mathscr{F}$ is the corresponding subbundle of $\mathscr{F}\mathscr{B}$. We also assume that

$$\varrho_{\mathbf{K}_i} = \varrho_0/\det \mathbf{H}_i = \text{const}. \tag{4.3.14}$$

First of all, we note that (4.3.13) is not a true measure of inhomogeneity of the body. The expression (4.3.13) vanishes only if (1) $\mathscr{B}$ is homogeneous *and* the reference configuration $\boldsymbol{\varkappa}$ is homogeneous: $\varrho_0 = \text{const}$, $W = W(\mathbf{F})$, (2) $\partial P/\partial \xi^\alpha = 0$. On the other hand equation (4.3.12) is not covariant. We shall remedy both shortcomings by rewriting (4.3.12) in terms of covariant derivatives with respect to an arbitrary connection Γ on $\mathscr{F}$. Let

$$\beta_\alpha := \frac{d\pi_\alpha}{dt} - \Sigma^\beta{}_{\alpha;\beta}. \tag{4.3.15}$$

Since $\Sigma^\beta{}_\alpha$ is a (1,1)-tensor density of weight $+1$,

$$\Sigma^\beta{}_{\alpha;\beta} = \Sigma^\beta{}_{\alpha,\beta} + 2\Gamma^\beta_{[\beta\gamma]}\Sigma^\gamma{}_\alpha - \Gamma^\gamma{}_{\beta\alpha}\Sigma^\beta{}_\gamma. \tag{4.3.16}$$

We shall prove that

$$\beta_\alpha = 2S_k{}^\gamma F^k{}_\beta[\Gamma^\beta_{[\gamma\alpha]} + \Gamma^\varrho_{[\gamma\varrho]}\delta^\beta{}_\alpha]. \tag{4.3.17}$$

Firstly, one can easily verify that the right-hand side of (4.3.15) and (4.3.17) are equal for the local connections Γ_i:

$$\Gamma^\alpha_{i\beta\gamma} = H_i{}^a{}_{\gamma,\xi^\beta} \overset{-1}{H}{}_i{}^\gamma{}_a$$

(see (4.2.29)). For this purpose it is enough to note that

$$\frac{\partial L}{\partial \xi^\alpha} = L H_i{}^a{}_{\beta,\xi^\alpha} \overset{-1}{H}{}_i{}^\beta{}_a + S_k{}^\gamma F^k{}_\beta H_i{}^a{}_{\gamma,\xi^\alpha} H_i{}^\beta{}_a$$

on account of (4.3.1), (1.2.87), (4.3.14). Since the right-hand sides of (4.3.15) and (4.3.17) are $\mathscr{F}$-linear functions of the connection symbols, they remain equal for global connections of the form

$$\Gamma^{\alpha}{}_{\beta\gamma} = \sum_i f_i \Gamma^{\alpha}_{i\beta\gamma} \tag{4.3.18}$$

(see Section 1.1.3).

Let $\tilde{\Gamma}$ be any connection on $\mathscr{F}$. We prove that the right-hand sides of (4.3.15), (4.3.17) are equal in this case too. Indeed,

$$T_{\mathfrak{a}}{}^{kl} = \varrho \frac{\partial W_{\mathfrak{a}}}{\partial F^{k}{}_{a}} F^{l}_{a} \tag{4.3.19}$$

and

$$W_{\mathfrak{a}}(\mathbf{FG}) = W_{\mathfrak{a}}(\mathbf{F}) \ \forall \mathbf{G} \in \mathscr{G}_{\mathfrak{a}} = \mathscr{G}, \quad \mathbf{F} \in \mathrm{SL}(3) \tag{4.3.20}$$

Differentiating the identity (4.3.20) with respect to the group parameters we have

$$T^{kl} \overset{-1}{F^{a}}{}_{l} F^{k}{}_{b} A^{b}{}_{a} = 0 \tag{4.3.21}$$

for every $\mathbf{A} \in \mathfrak{g}$ = the Lie algebra of $\mathscr{G}$.

For any two connections Γ, $\tilde{\Gamma}$ on $\mathscr{F}$ and a local section $\mathbf{K} = \mathbf{H}\nabla\varkappa$ of $\mathscr{F}$ we have shown in Section 1.1.2 that

$$[H^{b}{}_{\alpha}(\tilde{\Gamma}^{\alpha}{}_{\beta\gamma} - \Gamma^{\alpha}{}_{\beta\gamma}) \overset{-1}{H^{\gamma}}{}_{a}] \in \mathfrak{g}. \tag{4.3.22}$$

Consider the equation expressing the equality of the right-hand side of (4.3.15) and (4.3.17). Taking into account (4.3.16) and bringing all the terms involving the connection symbols to one side one gets two terms, $S_{k}{}^{\gamma} F^{k}{}_{\beta} \Gamma^{\beta}{}_{\alpha,\gamma}$ and $L\Gamma_{\alpha}{}^{\beta}{}_{\beta}$. Equations (4.3.19), (4.1.2) and (4.3.22) imply that the first term is equal to $S_{k}{}^{\gamma} F^{k}{}_{\beta} \tilde{\Gamma}^{\beta}{}_{\alpha\gamma}$. On account of our assumption that $\mathscr{G} \subset \mathrm{SO}(3)$ the matrices $\mathbf{A} \in \mathfrak{g}$ are skew-symmetric and $\tilde{\Gamma}^{\beta}{}_{\alpha\beta} - \Gamma^{\beta}{}_{\alpha\beta} = 0$. Since we have proved that the right-hand side of (4.3.15) and (4.3.17) are equal for the connections (4.3.18), they are equal for arbitrary connections on $\mathscr{F}$.

β_{α} is a covector field on $\mathscr{B}$ and it cannot be reduced to zero by a transformation of Lagrange variables. Although the right-hand side of (4.3.15) and (4.3.17) remain equal for arbitrary connections Γ on $\mathscr{F}$, the covector β_{α} as given by equation (4.3.17) depends on the connection. It is uniquely determined by the inhomogeneity field in $\mathscr{B}$ iff the torsion is uniquely determined by the geometry of $\mathscr{F}$. This is the case for crystalline solids. In the case of an isotropic solid we can choose Γ to be the unique torsion-less connection, so that $\beta_{\alpha} = 0$.

In the case of a crystalline solid β_{α} is a physically well-defined quantity. The first term of (4.3.17) represents the **Peach–Köhler force** familiar in the dislocation theory. In order to see it one can express (4.3.17) in terms of the Nye tensor by means of equation (4.2.40).

We suggest the following terminology: π_α is the **pseudomomentum**, $\Sigma^\beta{}_\alpha$ is the **pseudostress**; if $\mathscr{B}$ is a crystalline solid then β_α is the **pseudoforce** acting on dislocations.* The tensor density $\Sigma^\beta{}_\alpha$ was originally introducted by Eshelby (1956, 1975) and later revisited by Chadwick (1975) under the name of the **energy-momentum tensor.****

4.3.3 Some Remarks about the Applications in Dislocation Theory. The Rice Integral

In Sections 4.2.1 and 4.2.2 pseudoforces were derived from a Lagrangian setting of elasticity. In early papers of Eshelby (e.g. Eshelby, 1956) they were derived by differentiating the energy of an elastic body (under specified BC) with respect to the dislocation position. In either case pseudoforce is a static concept. For a dynamical theory of moving dislocations we need differential equations of dislocation motion in a given stress field. In turn dislocations affect the motion of material points. One might expect that the equations governing the motion of dislocations can be explained in terms of a dynamical force exerted on them by the loads and the resulting stress. The dynamical force need not coincide with the static force derived above (cf. the case of an electron in an electromagnetic field). Moreover this approach requires another quantitative characteristic of dislocations: their inertia. Dislocation theory cannot be derived from the theory of elasticity. Equations of motion of elastic bodies determine the motion of shock waves (discontinuities of $\boldsymbol{\chi}_{,t}$, $\boldsymbol{\chi}_{,\xi^\mu}$) and higher-order waves (discontinuities of higher-order derivatives of $\boldsymbol{\chi}$). They have no bearing on the behaviour of dislocations, cracks and other discontinuities of $\boldsymbol{\chi}$. However, the concepts and equations derived in this section find many applications in the theories of cracks and dislocations.

The pseudostress tensor made its appearance in connection with the analysis of stress singularities at dislocations, crack tips etc (Eshelby, 1956;

* Our analysis of pseudoforces is adequate for bodies filling the whole space. In order to determine pseudoforces acting in a bounded body it is necessary to specify the conditions in which virtual shifts of dislocations take place, e.g. what happens to the boundary of the body and to the loads. This requirement implies that the BC should be specified.

** This term is suggested by a hardly convincing analogy with relativistic theories of matter in space-time.

Rice, 1968). In static conditions $\pi_\alpha = 0$ and in a subbody $\Omega \subset \mathscr{B}$ which does not contain inhomogeneities ($\Gamma^\varrho{}_{\mu\nu} = 0$, $\beta_\nu = 0$)

$$\frac{d\Sigma^\mu{}_\nu}{d\xi^\mu} = 0 \quad \text{in } \Omega. \tag{4.3.23}$$

This fact led Rice* to consider integrals

$$J_\nu(\mathscr{A}) = \int_{\varkappa(\mathscr{A})} dH^2\, n_\mu \Sigma^\mu{}_\nu \tag{4.3.24}$$

over surfaces $\mathscr{A} \subset \Omega$. If $\mathscr{A}_1, \mathscr{A}_2 \subset \Omega$ and $\mathscr{A}_1 \cup \mathscr{A}_2 = \partial\mathscr{V}$, $\mathscr{V} \subset \Omega$, with the account taken of the orientation, then

$$J_\nu(\mathscr{A}_1) = -J_\nu(\mathscr{A}_2). \tag{4.3.25}$$

One can thus consider a sequence of surfaces $\mathscr{A}$ approaching a singularity, e.g. a crack tip, and take advantage of (4.3.25) in evaluating the singularity.

Geometrically $J_\nu(\mathscr{A})$ is not a correctly defined quantity since equation (4.3.24) involves integration of a covector density. It is however possible to replace the ***J*-integrals** (4.3.24) by the linear **Rice functionals**

$$J(\mathscr{A})[\mathbf{v}] = \int_{\varkappa(\mathscr{A})} dH^2\, n_\mu \Sigma^\mu{}_\nu v^\nu \tag{4.3.26}$$

defined on vector fields $\mathbf{v}\colon \mathscr{A} \to T\mathscr{B}$. It follows from (4.3.11) and (1.2.101′) that $J(\mathscr{A})[\mathbf{v}]$ is a scalar.

If the vector field $\mathbf{v}$ is Γ-self-parallel on Ω: $\dfrac{\partial v^\nu}{\partial \xi^\mu} + \Gamma^\nu{}_{\mu\varrho} v^\varrho = 0$ in Ω, then

$$J(\partial\mathscr{V})[\mathbf{v}] = \int_{\varkappa(\mathscr{V})} dL^3\, (\Sigma^\mu{}_\nu v^\nu)_{,\xi^\mu} = \int_{\varkappa(\mathscr{V})} dL^3\, [\Sigma^\mu{}_{\nu;\xi^\mu} v^\nu + \Sigma^\mu{}_\nu v^\nu{}_{;\xi^\mu}]$$

$$= \int_{\varkappa(\mathscr{V})} dL^3\, \Sigma^\mu{}_{\nu;\xi^\mu} v^\nu$$

for any $\mathscr{V} \subset \Omega$. In static conditions

$$-J(\partial\mathscr{V})[\mathbf{v}] = \int_{\varkappa(\mathscr{V})} dL^3\, \beta_\nu v^\nu \tag{4.3.27}$$

represents the work performed by the pseudoforces β_ν when the inhomogeneities in $\mathscr{V}$ are shifted along the vector field v^ν. If $\partial\mathscr{V} = \mathscr{A}_1 \cup \mathscr{A}_2$ and $\beta_\nu = 0$ in $\mathscr{V}$, then

$$J(\mathscr{A}_1)[\mathbf{v}] = -J(\mathscr{A}_2)[\mathbf{v}]. \tag{4.3.28}$$

* Rice's original paper (1968) is concerned with a two-dimensional case and contour integrals.

In the case of a crystalline body and $\mathscr{V} \subset \mathscr{V}_i$, $(\mathbf{K}_i, \mathscr{V}_i) \in \mathfrak{a}$, $\mathbf{K}_i = \mathbf{H}_i \nabla \boldsymbol{\varkappa}$ there is a self-parallel vector field $v^\alpha(\xi) = \overset{-1}{H_i}{}^\alpha{}_a(\xi) w^a$, $w \in \boldsymbol{R}^3$, and we arrive at the ***J*-integrals**

$$J_a(\mathscr{A}) = \int_{\varkappa(\mathscr{A})} dH^2 n_\mu \Sigma^\mu{}_\nu \overset{-1}{H_i}{}^\nu{}_a. \tag{4.3.29}$$

We now apply the Rice functionals to isolated inhomogeneities. Roughly, an isolated inhomogeneity is a subset $\mathscr{C}$ of $\bar{\mathscr{B}}$ of Lebesgue measure zero $(L^3(\boldsymbol{\varkappa}(\mathscr{C})) = 0)$ consisting of material points which are not equivalent to the neighbouring points of $\bar{\mathscr{B}} \setminus \mathscr{C}$ either in a topological sense (e.g. $\mathscr{C} \subset \partial\mathscr{B}$) or in the sense of material properties. In the former case there are vector fields $\mathbf{v}\colon \mathscr{C} \to T\mathscr{B}$ that a shift $\varepsilon\mathbf{v}$, however small, of $\mathscr{C}$ carries it outside $\mathscr{B}$. In the latter case there are vectors $\mathbf{v} \in T_P\mathscr{B}$, $P \in \mathscr{C}$, such that a shift $\varepsilon\mathbf{v}$, however small, carries P to a point P' with different material properties. This is the case of dislocations, point defects etc.

A virtual displacement $\mathbf{v}$ of an isolated inhomogeneity $\mathscr{C}$ within the body $\mathscr{B}$ entails a change of the energy of $\mathscr{B}$. To first order in $\mathbf{v}$ this change in energy is a linear functional of $\mathbf{v}$. We now show that the linear functional is the Rice functional. More precisely, let $\mathscr{V}$ be a nbhd of an isolated inhomogeneity $\mathscr{C}$ such that $\mathscr{V} \setminus \mathscr{C}$ does not contain any other inhomogeneities (isolated or continuously distributed). Enclose the inhomogeneity $\mathscr{C}$ by a surface $\mathscr{A} \subset \mathscr{V}$. We claim that, for $\mathscr{A}$ shrinking to $\mathscr{C}$, $J(\mathscr{A})[\mathbf{v}]$ approaches the work performed on the body in the course of shifting the inhomogeneity. We try to corroborate this claim in the case of a crack tip $\mathscr{C}$.

A **crack** is essentially a (concave or internal) part of the boundary $\partial\mathscr{B}$ subject to the BC, $\mathbf{s} = \mathbf{0}$. We identify the set $\mathscr{C}$ with the surface of the crack tip and consider the effects of a small displacement $\mathbf{v}$ of the crack tip from

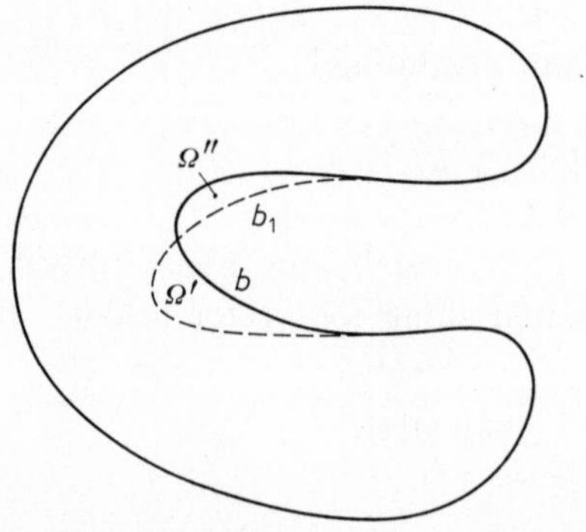

Fig. 4.3 Small displacement of a crack tip

its initial position $\mathscr{C}$ to its final position $\mathscr{C}_1$. In general the displacement can be achieved by removing a part Ω' of the body, supplying some material Ω'' and applying or releasing tractions on the crack surface. Without loss of generality we assume that $\Omega'' = \emptyset$. The displacement of $\mathscr{C}$ will be carried out in two stages: (1) Ω' is removed while the tractions on $\mathscr{C}_1$ are maintained, (2) the tractions on $\mathscr{C}_1$ are released. In the course of the second stage the body deforms from an initial configuration $\boldsymbol{\chi}_0 \circ \boldsymbol{\varkappa}$ to a final one $\boldsymbol{\chi}_1 \circ \boldsymbol{\varkappa}$ in such a way that the BC on $\partial\mathscr{B}\setminus\mathscr{C}$ are satisfied during the process. Furthermore we assume that the process is *quasistatic*, i.e. the equilibrium equations are satisfied at every moment.

Let $\mathscr{B}_1 := \mathscr{B}\setminus\Omega'$, $\partial\boldsymbol{\varkappa}(\mathscr{B}) = \bar{\Sigma}_1\cup\bar{\Sigma}_2\cup\boldsymbol{\varkappa}(\mathscr{C})$, $\partial\boldsymbol{\varkappa}(\mathscr{B}_1) = \bar{\Sigma}_1\cup\bar{\Sigma}_2\cup\boldsymbol{\varkappa}(\mathscr{C}_1)$, $\boldsymbol{\chi}|\Sigma_1 = \boldsymbol{\chi}_*$, $\mathbf{Sn} = \mathbf{s}$ on Σ_2, $\mathbf{Sn} = \mathbf{0}$ on $\boldsymbol{\varkappa}(\mathscr{C})$ before the displacement, $\mathbf{Sn} = \mathbf{0}$ on $\boldsymbol{\varkappa}(\mathscr{C}_1)$ after the displacement. Let $\boldsymbol{\chi}_\lambda$, $\lambda \in [0, 1]$, be the continuous deformation of $\boldsymbol{\varkappa}(\mathscr{B}_1)$ in the second stage. We also set $\boldsymbol{\varkappa}(\mathscr{C}_1) = \{\xi+\varepsilon\mathbf{v}(\xi)|\ \xi \in \boldsymbol{\varkappa}(\mathscr{C})\}\cup\boldsymbol{\varkappa}(\mathscr{C}')$. Suppose that the loads are potential and let

$$E = \int_{\boldsymbol{\varkappa}(\mathscr{B})} dL^3\, \varrho_0(\xi)W(\nabla\boldsymbol{\chi}(\xi), \xi) + \int_{\boldsymbol{\varkappa}(\mathscr{B})} dL^3\, P(\boldsymbol{\chi}(\xi), \xi) + \int_{\Sigma_2} dH^2\, \phi(\boldsymbol{\chi}(\xi), \xi) \tag{4.3.30}$$

be the total energy of the body and loads.

The energy increase due to the displacement is the sum of the contributions of the first and second stage:

$$E_1 - E_0 = -\int_{\boldsymbol{\varkappa}(\Omega')} dL^3\,[\varrho_0 W(\nabla\boldsymbol{\chi}_0(\xi), \xi)+P(\boldsymbol{\chi}_0(\xi), \xi)]+A, \tag{4.3.31}$$

where

$$A = \int_{\boldsymbol{\varkappa}(\mathscr{C}_1)} dH^2\, n_\alpha \int_0^1 d\lambda\, \hat{S}_k^{\ \alpha}(\nabla\boldsymbol{\chi}_\lambda, \xi)\frac{d\chi_\lambda^k}{d\lambda} \tag{4.3.32}$$

is the work performed on the body in the second stage. From Green's formula and the equilibrium equations

$$A = \int_{\boldsymbol{\varkappa}(\mathscr{B}_1)} dL^3 \int_0^1 d\lambda\,(-\varrho_0 b_k)\frac{d\chi_\lambda^k}{d\lambda} - \int_{\Sigma_2} dH^2\, s_k \frac{d\chi_\lambda^k}{d\lambda} + \int_{\boldsymbol{\varkappa}(\mathscr{B}_1)} dL^3 \int_0^1 d\lambda\, \hat{S}_k^{\ \alpha}(\mathbf{F}_\lambda, \xi)\frac{dF_\lambda{}^k{}_\alpha}{d\lambda}, \tag{4.3.33}$$

$\mathbf{F}_\lambda := \nabla\boldsymbol{\chi}_\lambda$.

Hence

$$A = \int_{\varkappa(\mathscr{B}_1)} dL^3\,[P(\chi_\lambda,\xi)+\varrho_0 W(\nabla\chi_\lambda,\xi)]\big|_{\lambda=0}^{\lambda=1}+\int_{\Sigma_2} dH^2\,\phi(\chi_\lambda,\xi)\big|_{\lambda=0}^{\lambda=1} \tag{4.3.34}$$

does not depend on the path $\lambda \mapsto \chi_\lambda$ joining the initial configuration χ_0 to the final configuration χ_1.

Let $\mathbf{S}^{(1)}(\xi) = \hat{\mathbf{S}}(\mathbf{F}_1(\xi), \xi)$. Substituting $\varrho_0 b_k = -S^{(1)}{}_k{}^\alpha{}_{,\xi^\alpha}$, $s_k = S^{(1)}{}_k{}^\alpha n_\alpha$ in equation (4.3.32) and using Green's formula again we find that

$$A = \int_{\varkappa(\mathscr{B}_1)} dL^3 \int_0^1 d\lambda\,[\hat{S}_k{}^\alpha(\mathbf{F}_\lambda(\xi),\xi)-S^{(1)}{}_k{}^\alpha(\xi)]\,\frac{dF_\lambda{}^k{}_\alpha}{d\lambda}. \tag{4.3.35}$$

In many cases equation (4.3.35) can be used for estimating A. Suppose, for example, that there is a path $\lambda \mapsto \chi_\lambda$ (satisfying the equilibrium equations and the BC) joining χ_0 to χ_1 and such that $\hat{\mathbf{S}}(\mathbf{F}_\lambda(\xi), \xi) = \lambda\mathbf{S}^{(1)}(\xi)+(1-\lambda)\mathbf{S}^{(0)}(\xi)$, $\mathbf{S}^{(0)}(\xi) := \hat{\mathbf{S}}(\mathbf{F}_0(\xi), \xi)$. Assuming the LH inequality (2.5.4), we conclude that

$$[\hat{S}_k{}^\alpha(\mathbf{F}_\lambda,\xi)-S^{(1)}{}_k{}^\alpha(\xi)]\,\frac{dF_\lambda{}^k{}_\alpha}{d\lambda} = (\lambda-1)[S^{(1)}{}_k{}^\alpha - S^{(0)}{}_k{}^\alpha]\,\frac{dF_\lambda{}^k{}_\alpha}{d\lambda}$$

$$= (\lambda-1)\frac{d\hat{S}_k{}^\alpha(\mathbf{F}_\lambda,\xi)}{d\lambda}\,\frac{dF_\lambda{}^k{}_\alpha}{d\lambda} = (\lambda-1)\frac{\partial^2(\varrho_0 W)}{\partial F^k{}_\alpha\,\partial F^l{}_\beta}\,\frac{dF_\lambda{}^k{}_\alpha}{d\lambda}\,\frac{dF_\lambda{}^l{}_\beta}{d\lambda} \leqslant 0$$

and

$$\int_0^1 d\lambda\,[\hat{S}_k{}^\alpha(\mathbf{F}_\lambda,\xi)-S^{(1)}{}_k{}^\alpha(\xi)]\,\frac{dF_\lambda{}^k{}_\alpha}{d\lambda} = \int_0^1 d\lambda\,(\lambda-1)[S^{(1)}{}_k{}^\alpha - S^{(0)}{}_k{}^\alpha]\,\frac{dF_\lambda{}^k{}_\alpha}{d\lambda}$$

$$\geqslant -[S^{(1)}{}_k{}^\alpha - S^{(0)}{}_k{}^\alpha][F_1{}^k{}_\alpha - F_0{}^k{}_\alpha].$$

Hence

$$-\int_{\varkappa(\mathscr{B}_1)} dL^3\,[S^{(1)}{}_k{}^\alpha - S^{(0)}{}_k{}^\alpha][F_1{}^k{}_\alpha - F_0{}^k{}_\alpha] < A \leqslant 0. \tag{4.3.36}$$

Equation (4.3.36) suggests that $A = O(\varepsilon^2)$. If $H^2(\boldsymbol{\varkappa}(\mathscr{C}')) = O(\varepsilon^2)$ as well then

$$E_1 - E_0 = -\int_{\varkappa(\Omega')} dL^3\,(\varrho_0 W+P)+O(\varepsilon^2) = -\varepsilon\int_{\varkappa(\mathscr{C})} dH^2\,n_\mu v^\mu(\varrho_0 W+P)$$

$$= \varepsilon\int_{\varkappa(\mathscr{C})} dH^2\,n_\mu \Sigma^\mu{}_\nu v^\nu.$$

Notation

In view of the mathematical scope of this book a rigid system of notations would be rather cumbersome. Nevertheless certain rules have been observed whenever possible.

Finite-dimensional vectors and vector-valued functions are usually—but not always—denoted by boldface lower case Roman or Greek letters. Points of an abstract manifold are denoted by Roman capitals, while points of $\boldsymbol{R}^n$ are denoted by x, ξ or (x^α), (ξ^α), etc. Matrices and matrix-valued functions are denoted by boldface Roman capitals. Sets are usually denoted by upper case script letters, occasionally by upper case German letters.

We list below some symbols appearing throughout the whole book and those which are not explained in the text.

$:=$, $:\equiv$	the symbol on the left-hand side is defined by the expression on the right-hand side
$f(x) = g(x)$	equation
$f(x) \equiv g(x)$	statement that the functions f, g are identical
$\in \ \subset \ \cup \ \cap \ \bigcup_{k=1}^{\infty}$	standard notations of set theory
$\bar{\in}$, $\not\subset$	negation of $\in$, $\subset$
$\mathscr{A}^c$	complement $\mathscr{X}\setminus\mathscr{A}$ of a subset $\mathscr{A}$ of a space $\mathscr{X}$
$\{a, b, \ldots\}$	the set of elements listed in the braces
$\{x \in \mathscr{X} \mid P(x)\}$	the set of all the elements of $\mathscr{X}$ satisfying the condition $P(x)$
$\mathscr{A}\setminus\mathscr{B}$, $\mathscr{A} \div \mathscr{B}$	$= \mathscr{A} \cap \mathscr{B}^c$, $(\mathscr{A} \cap \mathscr{B}^c) \cup (\mathscr{A}^c \cap \mathscr{B})$ resp.
$\{x_n\}$	sequence
$\bar{\mathscr{A}}$, $\partial\mathscr{A}$, $\mathring{\mathscr{A}}$	closure, boundary and interior of a set $\mathscr{A}$
$\forall$, $\exists$, $\exists!$	for every, exists, exists a unique
$\to$, $\rightharpoonup$, $\overset{*}{\rightharpoonup}$	convergence, weak convergence, weak star convergence
$\boldsymbol{R}$, $\boldsymbol{C}$	the set of all real (complex) numbers

$\boldsymbol{R}^n$	the set of real n-tuples
$\boldsymbol{Z}_+$	the set of positive integers
$\boldsymbol{R}_+, \bar{\boldsymbol{R}}_+, \boldsymbol{R}_+^{m+1}$	the set of positive, non-negative numbers, n-tuples x^k, $k = 0, \ldots, m$, $x^0 > 0$
$[a, b]$, $[a, b[$	the intervals $a^k \leqslant x^k \leqslant b^k$, $a^k \leqslant x^k < b^k$, $k = 1, \ldots, n$, etc.
$f: \mathscr{X} \to \mathscr{Y}; x \mapsto f(x)$	mapping of $\mathscr{X}$ into $\mathscr{Y}$, assigning an expression $f(x) \in \mathscr{Y}$ to $x \in \mathscr{X}$
$f \circ g(x)$	$\equiv f(g(x))$
$\overset{-1}{f}$	inverse function or mapping
$f(\mathscr{A}), f^{-1}(\mathscr{A})$	image of $\mathscr{A}$, counterimage of $\mathscr{A}$
supp$f$; im$f$	support of a function f (closure of the set where it is different from zero), image of f
$f_{,x^\alpha}$, occasionally $f_x, f_u, \partial_\alpha f$	partial derivatives $\partial f/\partial x^\alpha$, $\partial f/\partial x$, $\partial f/\partial u$, $\partial f/\partial x^\alpha$
$Df(u)[v]$	differential of the function f at u expressed as the linear functional of the tangent vector v at u
$\mathrm{id}_{\mathscr{X}}$	identity mapping $\mathscr{X} \to \mathscr{X}; x \mapsto x$
$\mathscr{S}^{n-1}; \mathscr{K}(x, r)$	the unit sphere in $\boldsymbol{R}^n$; the ball $\{y \in \boldsymbol{R}^n \mid \lvert y-x\rvert < r\}$
$\mathscr{C}_0^k(\mathscr{U}, \mathscr{V})$	the set of functions $f: \mathscr{U} \to \mathscr{V}$ with continuous derivatives up to order k ($\leqslant \infty$) (continuous functions if $k = 0$)
$\mathscr{C}_0^k(\mathscr{U}, \mathscr{V})$	the set of functions $f \in \mathscr{C}^k$ with compact support in $\mathscr{U}$
$\mathscr{C}_b^k(\mathscr{U}, \mathscr{V})$	the set of bounded functions $f: \mathscr{U} \to \mathscr{V}$ with bounded derivatives up to order k
$\mathscr{L}^p(\mathscr{U}, \mathscr{V}; \mu)$	Lebesgue space of functions $f: \mathscr{U} \to \mathscr{V}$, integrable in the power $p \geqslant 1$ with respect to the measure μ (self-evident symbols $\mathscr{U}$, $\mathscr{V}$ or μ may be omitted)
$\chi_{\mathscr{A}}$	the characteristic function of a set $\mathscr{A}$ ($\chi_{\mathscr{A}}(x) = 1$ if $\varkappa \in \mathscr{A}$, $\chi_{\mathscr{A}}(x) = 0$ if $x \bar{\in} \mathscr{A}$)
$\lVert u \rVert_{\mathscr{V}}$	norm of $u \in \mathscr{V}$ in the Banach or normed space $\mathscr{V}$
$\langle \cdot, \cdot \rangle, \langle \cdot \mid \cdot \rangle$	duality between two spaces $\mathscr{X}, \mathscr{Y}$, in particular $\mathscr{V}$ and $\mathscr{V}^*$
$\mathscr{D}(L)$	domain of an operator L
$\mathscr{R}(L)$	range of an operator L
diag$\{a, b, c, \ldots\}$	diagonal matrix with diagonal elements $a, b, c, \ldots$
$\mathbf{E}; \overset{-1}{\mathbf{A}}, \mathbf{A}^{-1}; {}^t\mathbf{A}$	unit matrix, inverse of the matrix $\mathbf{A}$, transpose of $\mathbf{A}$
det$\mathbf{A}$, tr$\mathbf{A}$	determinant, trace of a matrix, or (1,1)-tensor, $\mathbf{A}$
$M(n)$	the set of all matrices $n \times n$, $\mathscr{L} = \mathscr{M}_{3\times 3}$
L_+	the set of all the matrices with the positive determinant

L(n), SL(n)	the group of invertible $n \times n$ matrices, or matrices with positive determinants, resp.
O(3), SO(3)	the orthogonal, or special orthogonal group in n dimensions, resp.
U(n), SU(n)	the group of all the $n \times n$ matrices with $\lvert\det\mathbf{A}\rvert = 1$, or $\det\mathbf{A} = 1$, resp.

NB. In some cases vectors in $\boldsymbol{R}^n$ are treated as column or row matrices. Hence such notations as ${}^t\boldsymbol{v}$.

Abbreviations

a.a.	almost all
a.e.	almost every, almost everywhere
BC	boundary conditions
BVP	boundary value problem
o.p.s.	oriented plane set
wlsc	(sequentially) weakly lower semicontinuous
(LH)	Legendre–Hadamard
(SE)	strong ellipticity
(PC)	polyconvexity
(QC)	quasiconvexity
(GC)	growth conditions
VWP	virtual work principle

References

Adams, R. A. (1975). *Sobolev Spaces*, Academic Press, New York.

Almgren, F. J. (1968). Existence and regularity almost everywhere of solutions to elliptic variational problems among surfaces of varying topological type and singularity structure, *Ann. of Math.*, **87**, 321–391.

Antman, S. and Osborn, J. E. (1979). The principle of virtual work and integral laws of motion, *Arch. Ratl. Mech. Anal.*, **69**, 231–262.

Arthurs, A. M. (1970). *Complementary Variational Principles*, Oxford University Press, Oxford.

Babuška I. and Aziz, A. K. (1972). Survey lectures on the mathematical foundations of the finite element method, in: *The Mathematical Foundations of the FEM with Applications to Partial Differential Equations*, A. K. Aziz (ed.), Academic Press, New York.

Ball, J. M. (1977). Convexity conditions and existence theorems in non-linear elasticity, *Arch. Ratl. Mech. Anal.*, **63**, 337–403.

Ball, J. M. (1977a). Constitutive inequalities and existence theorems in non-linear elastostatics, in: *Nonlinear Analysis and Mechanics*, Vol. 1, R. J. Knops (ed.), Pitman, London.

Ball, J. M. (1978). Finite-time blow-up in non-linear problems, in: M.G. Crandall (ed.), *Nonlinear Evolution Equations*, Academic Press, New York.

Ball, J. M. (1980). Strict convexity, strong ellipticity and regularity in the calculus of variations, *Math. Proc. Cambr. Phil. Soc.*, **87**, 501–513.

Ball, J. M. (1980a). Global invertibility of Sobolev functions and the interpretration of matter,

Ball, J. M., Currie, J. C. and Olver, P. L. (1980). Null Lagrangians, weak continuity and variational problems of arbitrary order, *J. Functional Analysis*, **41**, 135–185.

Bland, D. (1969). *Non-linear Dynamic Elasticity*, Blaisdell Publ. Co., Waltham, Mass.

Boothby, W. (1975). *An Introduction to Differentiable Manifolds and Riemannian Geometry*, Academic Press, New York.

Borovikov, B. A. (1969). On the Riemann problem for a system of two quasilinear equations, *DAN SSSR* **185**, 19–21.

Bourbaki, N. (1961). *Fonctions d'une variable réelle, Éléments de mathématique*, 1ère partìe, livre IV, Hermann et Cie, Paris.

Bray, H. E. (1919). Elementary properties of the Stieltjes integral, *Ann. of Math.*, **20**, 177–186.

Busemann, H., Ewald, G. and Shephard, G. C. (1963). Convex bodies and convexity on Grassmann cones, I–IV, *Math. Ann.*, **151**, 1–41.

Caccioppoli, R. (1952). Misura ed integrazione sugli insiemi dimensionalmente orientati, *Rend. Acc. Naz. Lincei, Cl. Sci. fis. mat.., Ser. VIII*, 12, fasc. 1, 2, 3.

Cartan, É. (1928, 1946). *Leçons sur la géométrie des espaces de Riemann*, Gauthier-Villars.

Chadwick, P. (1975). Applications of an energy-momentum tensor in non-linear elastostatics, *J. Elasticity*, **5**, 249–258.

Chen, F. H. K. and Shield, R. T. (1977). Conservation laws in elasticity of the *J*-integral type, *ZAMP*, **28**, 1–22.

Cherepanov, G. P. (1977). Invariant *J*-integrals and some of their applications in mechanics, *PMM*, **41**, 399–412 (in Russian).

Dafermos, C. (1973). Solution of the Riemann problem for a class of hyperbolic systems of conservation laws by the viscosity method, *Arch. Ratl. Mech. Anal.*, **52**, 1–9.

Dafermos, C. (1974). Structure of solutions of the Riemann problem for hyperbolic systems of conservation laws, *Arch. Ratl. Mech. Anal.*, **53**, 201–217.

Dafermos, C. (1977). Generalized characteristics and the structure of solutions of hyperbolic conservation laws. *Indiana Univ. Math. J.*, **26**, 1097–1119.

Dafermos, C. (1978). Characteristics in hyperbolic conservation laws, R. J. Knops (ed.), in: *Non-linear Analysis and Mechanics*, Vol. 1, 1–58, Pitman, London.

Dafermos, C. (1979). The second law of thermodynamics and stability, *Arch. Ratl. Mech. Anal.*, **70**, 167–179.

De Giorgi, E. (1954). Su una teoria generale della misura (r−1)-dimensionale in uno spazio ad r dimensioni, *Ann. mat. pura ed appl.*, (4)**36**, 191–213.

De Giorgi, E. (1955). Nuovi teoremi relativi alle misure (r−1)-dimensionali in uno spazio ad r dimensioni, *Ricerche di matematica*, **4**, 95–113.

De Giorgi, E. (1957). Sulla differenziabilià e l'analicità delle estremali degli integrali multipli regolari, *Mem. Acad. Sci. Torino, Cl. Sci. fis. mat.*, (3) **3**, 25–43.

De Giorgi, E. (1968). Un esempio di estremali discontinue per un problema variazionale di tipo elittico, *Boll. UMI*, **1**, 135–138.

Dill, E. H. (1977). The complementary energy principle in nonlinear elasticity, *Lett. Appl. Engng Sci.*, **5**, 95–106.

DiPerna, R. J. (1975). Singularities of solutions of nonlinear hyperbolic systems of conservation laws, *Arch. Ratl. Mech. Anal.*, **60**, 75–100.

DiPerna, R. J. (1977). Decay of solutions of hyperbolic systems of conservation laws with a convex extension, *Arch. Ratl. Mech. Anal.*, **64**, 1–46.

DiPerna, R. J. (1979). Uniqueness of solutions to hyperbolic conservation laws, *Indiana Univ. Math. J.*, **28**, 137–188.

Dunford, N. and Schwartz, J. T. (1958). *Linear Operators*, Vol. I, Interscience Publ., New York.

Duvaut, G. and Lions, J.-L. (1972). *Les inéquations en mécanique et en physique*, Dunod, Paris.

D'yachenko, V. F. (1963). On the conditions of uniqueness of continuous solutions to the Riemann problem for a system of three equations. *DAN SSSR*, **153**, 1245–1248 (in Russian).

Edelen, D. G. B. (1962). The null set of the Euler-Lagrange operator, *Arch. Ratl. Mech. Anal.*, **11**, 117–121.

Ékeland, I and Témam, R. (1974). *Analyse convexe et problèmes variationnels*, Dunod, Gauthier-Villars, Paris; Engl. transl. (1976), *Convex Analysis and Variational Problems*, North-Holland, Amsterdam, American Elsevier, New York.

Eshelby, J. D. (1956). The continuum theory of lattice defects, in: F. Seitz and D. Turnbull (ed.), *Solid State Physics*, Vol. 3, Academic Press, New York, 79–144.

Eshelby, J. D. (1975). The elastic energy-momentum tensor, *J. Elasticity*, **5**, 321–335.

Federer, H. (1954). An analytic characterization of distributions whose partial derivatives are representable by measures, *Bull. AMS*, **60**, 339–341.

Federer, H. (1958). A note on the Gauss-Green theorem, *Proc. AMS*, **9**, 447–451.

Federer, H. (1969). *Geometric Measure Theory*, Springer Verlag, Berlin.

Fichera, G. (1972). Existence theorems in elasticity, C. A. in Truesdell (ed.), *Hdb. d. Phys.*, Vol. IVa/2, Springer Verlag, Berlin.

Fichera, G. (1972a). Sulla propagazione delle onde in un mezzo elastico, in: L. I. Sedov (ed.), *Continuum Mechanics and Related Problems of Analysis*, Nauka, Moscow, 556–574.

Fleming, W. H. (1960). Functions whose partial derivatives are measures, *Illinois J. Math.*, **4**, 452–478.

Fleming, W. H. and Rishel, R. (1960). An integral formula for total gradient variation, *Arch. Math.*, **11**, 218–222.

Fletcher, D. C. (1976). Conservation laws in linear elastodynamics, *Arch. Ratl. Mech. Anal.*, **60**, 329–353.

Fraeijs de Veubeke, B. M. (1972). A new variational principle for finite elastic displacements, *Int. J. Engng. Sci.*, **10**, 745–750.

Friedrichs, K. O. (1929). Ein Verfahren der Variationsrechnung das Minimum eines Integrals als das Maximum eines anderen Ausdrucks darzustellen, *Ges. Wiss. Göttingen Nachr., Math, Kl.*, 13–20.

Friedrichs K. O. (1972). Symmetric hyperbolicity and conservation laws, in: *Continuum Mechanics and Related Problems of Analysis*, Nauka, Moscow, 575–580.

Friedrichs, K. O. (1974). On the laws of relativistic electro-magneto-fluid-dynamics, *Comm. Pure Appl. Math.*, **27**, 749–808.

Funk, P. (1962). *Variationsrechnung und ihre Anwendung in Physik und Technik*, Springer Verlag, Berlin.

Gagliardo, E. (1957). Caratterizzazioni delle tracce sulla frontiera relative ad alcune classi di funzioni in *n* variabili, *Rend. Sem. Mat. Univ. Padova*, **27**, 284–305.

Gajewski, H., Gröger, K. and Zacharias, K. (1974). *Nichtlineare Operatorgleichungen und Operatordifferenzialgleichungen*, Akademie Verlag, Berlin.

Gel'fand, I. M. (1959). Problems of the theory of quasilinear equations, *UMN*, **14**, 1–87 (in Russian).

Gel'fand, I. M., Fomin, S. V. (1961). *Calculus of Variations*, Fizmatgiz, Moscow (in Russian) English transl.: 1963, Prentice Hall, Englewood Cliffs, N.J.

Germain, P. (1962–63). Théorie des ondes de choc en dynamique des gaz et en magnétodynamique des fluides, *Aérodynamique Supérieure, Notes provisoires*, **I–III**.

Giusti, E. (1969). Regolarità parziale delle soluzioni di sistemi elittici quasilineari di ordine arbitrario, *Ann. Sc. Norm. Sup. Pisa*, **23**, 115–141.

Giusti, E. (1972). Un aggiunta alla mia nota: ..., *ibid.*, **27**, 161–166.

Giusti, E. and Miranda, M. (1968). Un esempio di soluzioni discontinue per un problema di minimo relativo ad un integrale regolare del calcolo delle variazioni, *Boll. UMI*, (4), **1**, 219–226.

Glimm, J. (1965). Solutions in the large for non-linear hyperbolic equations, *Comm. Pure Appl. Math.*, **18**, 697–715.

Godunov, S. K. (1961). An interesting class of quasilinear systems, *DAN SSSR*, **139**, 521–523 (in Russian).

Godunov, S. K. (1961a). On non-unique "smearing out" of discontinuities of solutions to quasilinear systems, *DAN SSSR*, **136**, 272–273 (in Russian).

Godunov, S. K. (1962). Problems of generalized solutions of quasilinear systems and gasdynamics, *UMN*, **17**, 147–158 (in Russian).

Golubitsky, M. and Guillemin, V. (1973). *Stable Mappings and their Singularities*, Springer Verlag, New York.

Golubitsky, M. and Schaeffer, D. G. (1975), Stability of shock waves for a single conservation law, *Adv. in Math.*, **16**, 65–71.

Graves, L. M. (1939). The Weierstrass condition for multiple integral variation problems, *Duke Math. J.*, **5**, 656–658.

Green, A. E. (1973). On some general formulae in finite elastostatics, *Arch. Ratl. Mech. Anal.*, **50**, 73–80.

Guckenheimer, J. (1975). Solving a single conservation law, *Lect. Notes in Math.*, Vol. 468, Springer Verlag, New York, 108–134.

Günther, W. (1962). Über einige Randintegrael der Elastomechanik, *Abh. d. Braunschw. wiss. Ges.*, **14**, 53–72.

Gurtin, M. E. and Martins, L. C. (1976). Cauchy's theorem in classical physics, *Arch. Ratl. Mech. Anal.*, **60**, 304–324.

Hanyga, A. (1974). Evolutionary condition in the theory of detonation and deflagration waves, *Bull. Ac. Polon. Sci., sér. techn.*, **22**, 1–11.

Hanyga, A. (1975). *On the Solution to the Riemann Problem for Arbitrary Hyperbolic System of Conservation Laws*, Reports of the Institute of Fundamental Technical Research (in Polish).

Hanyga, A. (1976). On the solution to the Riemann problem for arbitrary hyperbolic systems of conservation laws, *Publs. Inst. Geophys.*, **A–1** (**98**), 1–123.

Hanyga, A. (1976a). Thermodynamics, convex extensions and the Cauchy problem for systems of conservation laws, *Bull. Acad. Polon. Sci., sér. techn.*, **24**, 25–31.

Hanyga, A. (1981). On the definition of outgoing waves and boundary-value problems of anisotropic elasticity, *Bull. Acad. Polon. Sci., sér. sci. de la terre*, **28**, 267–279.

Hanyga, A. and Seredyńska, M. (1982). The complementary energy principle in finite elasticity, *Boll. UMI, Suppl. Fisica matematica*, **2**

Hanyga, A. (1985). *Modern Thermodynamics of Continuous Media*, Polish Scientific Publishers, Warszawa (in Polish).

Hanyga, A. and Seredyńska, H. (1984). *Introduction to the Non-linear Elasticity of Three-Dimensional Bodies*, Polish Scientific Publishers, Warszawa (in Polish).

Helly, E. (1921). Über lineare Funktionaloperatoren, *Sitzungsber. d. Naturwis. Kl. Kaiserl. Ak. Wiss.*, **121**, 265–295.

Hopf, E. (1950). The partial differential equation $u_t + uu_x = \mu u_{xx}$, *Comm. Pure Appl. Math.*, **3**, 201–230.

Hughes, T. J. R., Kato, T. and Marsden, J. E. (1976). Well-posed quasilinear second-order hyperbolic system with applications to non-linear elastodynamics and general relativity, *Arch. Ratl. Mech. Anal.*, **63**, 273–295.

Hughes, T. J. R. and Marsden, J. E. (1978). Classical elastodynamics as a linear symmetric hyperbolic system, *J. Elasticity*, **8**, 97–110.

Jeffery, A. and Taniuti, T. (1964). *Non-linear Wave Propagation with Applications to Physics and Magnetohydrodynamics*, Academic Press, New York.

John, F. (1977). Finite amplitude waves in a homogeneous isotropic elastic solid, *Comm. Pure Appl. Math.*, **30**, 421–446.

Kato, T. (1975). Quasilinear equations of evolution with applications to partial differential equations, *Springer Lecture Notes*, **448**, Springer, New York, 25–70.

Kato, T. (1975a). The Cauchy problem for quasilinear hyperbolic systems, *Arch. Ratl. Mech. Anal.*, **58**, 181–205.

Knops, R. J., Levine, H. A. and Payne, L. E. (1974). Non-existence, instability and growth theorems for solutions of a class of abstract non-linear equations with applications to non-linear elastodynamics, *Arch. Ratl. Mech. Anal.*, **55**, 52–72.

Knops R. J. and Payne, L. E. (1971). Growth estimates for solutions of evolutionary equations in Hilbert spaces with applications in elastodynamics, *Arch. Ratl. Mech. Anal.*, **41**, 363–398.

Knops, R. J. and Straughan, B. (1975). Non-existence of global solutions to nonlinear Cauchy problems arising in mechanics, *Proc. Symp. Trends in Applications of Pure Mathematics to Mechanics*, Pitman, London.

Knowles, J. K. and Sternberg, E. (1972). On a class of conservation laws in linearized and finite elastostatics, *Arch. Ratl. Mech. Anal.*, **44**, 187–211.

Koshelev, A. I. (1978). Regularity of solutions of quasilinear elliptic systems, *UMN*, **33**, 3–48 (a survey) (in Russian).

Krickeberg, K. (1957). Distributionen, Funktionen beschränkter Variation und Lebesguescher Inhalt nichtparametrischer Flächen, *Ann. mat. pura ed appl.*, (4) **44**, 105–133.

Kruzhkov, K. (1970). Quasilinear equations of first order with many independent variables, *Mat. sb.* **81**, 228–225 (in Russian).

Kulikovskii, A. G. (1962). On the structure of shock waves, *PMM*, **26**, 631–641 (in Russian)

Kunin, I. A. (1965). *Dislocation Theory*. Appendix to the Russian translation of J. A. Schouten's *Tensor Analysis for Physicists*, Nauka, Moscow (in Russian).

Kuznetsov, N. N., Tupchiyev, V. A. (1975). On a generalization of Glimm's theorem, *DAN SSSR*, **221**, 287–290 (in Russian).

Larsen, R. (1973). *Functional Analysis. An Introduction*, Marcel Dekker, New York.

Lavrent'yev, M. A., Lyusternik, L. A. (1935, 1950). *A Course of Variational Calculus*, Fizmatgiz, Moscow (in Russian).

Lax, P. D. (1957). Hyperbolic systems of conservation laws, *Comm. Pure Appl. Math.*, **10**, 537–556.

Lax, P. D. (1971). Shock waves and entropy, in: E. H. Zarantonello (ed.), *Contributions to Nonlinear Functional Analysis*, Academic Press, New York, 603–634.

Lee, S. J. and Shield, R. T. (1980). Applications of variational principles in finite elasticity, *ZAMP*, **31**, 454–471.

Levine, H. A. (1974). Instability and non-existence of global solutions to non-linear wave equations of the form $P\,u_{tt} = -A\,u + F\,u$, *Trans. AMS*, **192**, 1–21.

Levine, H. A. (1975). No-existence of global weak solutions to some properly and improperly posed problems of mathematical physics: The method of unbounded Fourier coefficients, *Math. Ann.*, **214**, 205–220.

Levine, H. A. and Payne, L. E. (1975). Growth estimates and lower bounds for solutions in non-linear elastodynamics with indefinite strain energy, *J. Elasticity*, **5**, 273–285.

Lichnerowicz, A. (1976). *Global Theory of Connections and Holonomy Groups*, Noordhoff, Leyden.

Lions, J.-L. (1969). *Quelques méthodes de résolution des problèmes aux limites nonlinéaires*, Dunod, Paris.

Lions, J.-L. and Magenes, E. (1968). *Problèmes aux limites non homogènes et applications*, Vol. 1, Dunod, Paris (Engl. transl. (1972), *Non-homogeneous Boundary Value Problems and Applications*, Springer Verlag, New York).

Liu, T.-P. (1977). The deterministic version of the Glimm scheme, *Commun. math. Phys.*, **57**, 135–148.

Liu, T.-P. (1979). Quasilinear hyperbolic systems, *Commun. Math. Phys.*, **68**, 141–172.

Liu, T.-P. (1980). System of squasilinear hyperbolic partial differential equations, in: *Proc. Trends in Applications of Pure Mathematics to Mechanics, Symp. ISIMM, Heriot-Watt University 1979*, Pitman, London.

Lyubimov G. A. (1961). Structure of magnetohydrodynamical shock waves in gases with anisotropic conductivity, *PMM*, **25**, 179–186 (in Russian).

Lyusternik, L. A. (1934). On relative extrema of functionals, *Mat. sb.*, **41**, 390–401 (in Russian).

Lyusternik, L. A., Sobolev, V. I. (1965). *Elements of Functional Analysis*, Nauka, Moscow (in Russian) Engl. transl.: 1961, Ungar, New York.

Maz'ya, V. G. (1968). Examples of irregular solutions of quasilinear elliptic equations with analytic coefficients, *Functional Analysis*, 2, **3**, 53–57 (in Russian).

Meyers, N. G. (1965). Quasi-convexity and lower semicontinuity of multiple variational integrals of any order, *Trans. AMS*, **119**, 225–249.

Miranda, M. (1964). Distribuzioni aventi derivate misure ed insiemi di perimetro localmente finito, *Ann. Sc. Norm. Sup. Pisa*, **18**, 27–56.

Mizohata, S. (1969) (Engl. transl. (1973)). *The Theory of Partial Differential Equations*, Cambridge University Press, Cambridge.

Morrey, C. B. (1940). Existence and differentiability theorems for solutions of variational problems for multiple integrals, *Bull. AMS*, **46**, 434–458.

Morrey, C. B. (1952). Quasiconvexity and the lower semicontinuity of multiple integrals. *Pacific J. Math.*, **2**, 25–53.

Morrey, C. B. (1966). *Multiple Integrals in the Calculus of Variations*, Springer Verlag. Berlin.

Morrey, C, B. (1968). Partial regularity results for non-linear elliptic systems, *J. Math. Mech.*, **17**, 649–670.

Morrey, C. B. (1969). Differentiability theorems for weak solutions of non-linear elliptic differential equations, *Bull. AMS*, **75**, 684–705.

Moser, J. (1960). A new proof of de Giorgi's theorem concerning the regularity problems for elliptic differential equations, *Comm. Pure Appl. Math.*, **13**, 457–468.

Narasimhan, R. (1968). *Analysis on Real and Complex Manifolds*, North-Holland, Amsterdam.

Nash, J. (1950). Continuity of solutions of parabolic and elliptic equations, *Am. J. Math.*, **80**, 931–954.

Nečas, J. (1967). Sur la regularité des solutions variationnelles des équations elliptiques, non linéaires d'ordre $2k$ en deux dimensions, *Ann. Sc. Norm. Sup. Pisa*, **21**, 427–457.

Nečas, J. (1968). *Les méthodes directes en théorie des équations elliptiques*, Academia, Praha; Masson, Paris.

Nečas, J. (1973). Variational methods in non-linear elasticity, *Acta Polytechnica, Práce ČVUT v Práze*, IV, 129–134.

Nečas, J. (1975). On regularity of solutions to non-linear variational inequalities for second-order elliptic systems, *Rendiconti di Matematica* (2) **8** Ser. VI, 481–498.

Nečas, J. (1976). Theory of locally monotone operators modelled on the finite displacement theory for hyperelasticity, *Beitr. z. Analysis*, **8**, 103–114.

Nečas, J. (1977). Example of an irregular solution to a non-linear elliptic system with analytic coefficients and conditions for regularity, in: *Theory of Non-linear Operators, Proc. Int. Summer School Berlin, GDR, Sept. 22–26, 1975, Abh. d. Akad. Wiss. DDR*, Akademie Verlag, Berlin, 197–206.

Nečas, J., John, O. and Stará, J. (1980). Counterexample to the regularity of weak solutions of elliptic systems, *Commentationes Mathematicae Univ. Carolinae*, **21**, 145–154.

Nečas, J., Stará, J. and Švarc, R. (1978). Classical solution to a second-order non-linear elliptic system in $\boldsymbol{R}^3$, *Ann. Sc. Norm. Sup. Pisa, Cl. Scienze, Serie IV*, **5**, 605–631.

Noble, B. and Sewell, M. J. (1972). On dual extremum principles in applied mathematics, *J. Inst. Maths. Applics.*, **9**, 123–193.

Noll, W. (1965). Proof of the maximality of the orthogonal group in the unimodular group, *Arch. Ratl. Mech. Anal.*, **18**, 100–102.

Nomizu, K. (1956). *Lie Groups and Differential Geometry*, The Math. Soc. of Japan, Tokyo.

Oden, J. T. and Reddy, J. N. (1976). *Variational Methods in Theoretical Mechanics* Springer, Berlin.

Ogden, R. W. (1972). Large deformations in isotropic elasticity—On correlation of the theory and experiment for incompressible rubberlike solids, *Proc. Roy. Soc. London*, **A326**, 565–584.

Ogden, R. W. (1972a). Large deformations in isotropic elasticity—On correlation of the theory and experiment for compressible rubberlike solids, *Proc. Roy. Soc. London*, **A328**, 567–583.

Oleinik, O. A. (1954). On the Cauchy problem for non-linear equations in a class of discontinuous functions, *DAN SSSR*, **95**, 451–455 (in Russian) Engl. transl.: AMS Transl. (2) **42**, 7–12.

Oleinik, O. A. (1954a). On the Cauchy problem for non-linear equations in a class of discontinuous functions, *UMN*, **9**, 231–233 (in Russian).

Oleinik, O. A. (1955). Boundary-value problems for partial differential equations with a small parameter in the highest order coefficients and the Cauchy problem for non-linear partial differential equations in the large, *UMN*, **10**, 229–234 (in Russian).

Oleinik, O. A. (1956). Cauchy problem for non-linear partial differential equations of first order with discontinuous initial data, *Trudy Mosk. matem. obshch*, **5**, 433–454 (in Russian).

Oleinik O. A. (1956a). Discontinuous solutions of non-linear partial differential equations, *DAN SSSR*, **109**, 1098–1101 (in Russian).

Oleinik, O. A. (1957). On the uniqueness of the generalized solution of a special non-linear system of equations encountered in mechanics, *UMN*, **12**, 169–176 (in Russian).

Oleinik, O. A. (1957a). Discontinuous solutions of non-linear differential equations, *UMN* **12**, 3–73 (in Russian); Engl. transl.: *Am. Math. Soc. Transl.*, (2) **26**, 95–172.

Oleinik, O. A. (1959). On the construction of the generalized solution to the Cauchy problem for a quasilinear equation of the first order by way of introducing "vanishing viscosity" *UMN*, **14**, 160–164 (in Russian); Engl. transl.: *AMS Transl.*, **33**, 277–284.

Oleinik, O. A. (1959a). On the uniqueness and continuous dependence on the data of generalized solutions of the Cauchy problem for a single quasilinear equation, *UMN*, **14**, 165–170 (in Russian).

Peach, M. O. and Köhler, J. S. (1950). The forces exerted on dislocations and the stress field produced by them, *Phys. Rev. II*, **80**, 436–439.

Peetre, J. (1961), Another approach to elliptic boundary problems, *Comm. Pure Appl. Math.*, **14**, 711–731.

Prodi, G. and Ambrosetti, A. (1973). *Analisi nonlineare, I Quaderno*, Scuola Normale Superiore, Pisa.

Rabinowitz, P. H. (1971). A global theorem for non-linear eigenvalue problem and application, in: E. H. Zarantonello (ed.), *Contribution to Non-linear Functional Analysis*, Academic Press, New York, 11–36.

Rashevskii, P. K. (1953). *Riemannian Geometry and Tensor Analysis*, Gostekhizdat, Moscow (in Russian).

Reshetnyak, Y. G. (1968). Theorems on stability of mappings, *Siberian math. J.*, **9**, 667–684 (in Russian).

Riesz, F. and Sz. B.-Nagy, (1972). *Leçons d'analyse fonctionnelle*, 6éme éd., Akademiai Kiadó, Budapest.

Rice, J. (1968). A path independent integral and the approximate analysis of strain concentration by notches and cracks, *J. Appl. Mech.*, **35**, 379–386.

Rockafellar, R. T. (1968). Integrals which are convex functionals, I, *Pacific J. Math.*, **24**, 525–539.

Rockafellar, R. T. (1971). Integrals which are convex functionals, II, *Pacific J. Math.*, **39**, 439–469.

Rozhdestvenskii, B. L. (1957). On systems of quasilinear equations, *DAN SSSR*, **115**, 454–457 (in Russian).

Rozhdestvenskii, B. L. and Yanenko, N. N. (1978). *Systems of Quasilinear Equations and their Applications in Gasdynamics*, Nauka, Moscow (2nd ed., in Russian).

Rund, H. (1966). *The Hamilton-Jacobi Theory in the Calculus of Variations*, Van Norstrand, London.

Saks, S. (1937). *Theory of the Integral*, 2nd rev. ed., Engl. translation, Hafner Publ. Co., New York.

Schaeffer, D. G. (1973). A regularity theorem for conservation laws, *Adv. in Math.*, **11**, 368–386.

Schouten, J. A. (1951). *Tensor Analysis for Physicists*, Oxford at the Clarendon Press; the Russian translation (Nauka, Moscow, 1965) contains an appendix *Dislocation Theory* by I. A. Kunin.

Schouten, J. A. (1954). *Ricci Calculus*, Springer Verlag, Berlin.

Schwartz, L. (1957). *Théorie des distributions, I–II*, Hermann et Cie, Paris.

Sewell, M. J. (1967). On configuration-dependent loading, *Arch. Ratl. Mech. Anal.*, **23**, 327–351.

Sewell, M. J. (1969). On dual approximation principles and optimization in continuum mechanics, *Phil. Trans. Roy. Soc. London*, **A265**, 319–350.

Sod, G. A. (1978). A survey of several finite-difference methods for systems of non-linear hyperbolic conservation laws, *J. Comput. Phys.*, **27**, 1–31.

Spencer, A. J. M. (1971). Theory of Invariants, in: *Continuum Physics*, Vol. I, Part III, New York.

Sternberg, S. (1964). *Lectures on Differential Geometry*, Prentice Hall, Inc. Englewood Cliffs, N.J.

Straughan, B. (1975). Further global non-existence theorems for abstract non-linear wave equation, *Proc. AMS*, **48**, 381–390.

Synge, J. L. (1957). *The Hypercircle Method in Mathematical Physics*, Cambridge University Press, New York.

Témam, R. (1973). *Numerical Analysis*, Reidel, Dordrecht.

Truesdell, C. A. (1972). *A First Course in Rational Continuum* Mechanics, The John Hopkins University, Baltimore, Md.

Truesdell, C. A. and Noll, W. (1965). *The Non-Linear Field Theories of Mechanics, Hdb. d. Phys. III/3*, Springer Verlag, Berlin.

Tupchiyev, V. A. (1964). On the Riemann problem for a system of two quasilinear equations of first order, *Zh. vych. mat. i fiz.*, **4**, 817–825 (in Russian).

Tupchiyev, V. A. (1966). The Riemann problem for systems of quasilinear equations which do not obey convexity conditions, *Zh. vych. mat. i fiz.*, **6**, 527–547 (in Russian).

Tupchiyev, V. A. (1972). On non-isolated solutions of the Riemann problem, *Chislennye metody mekhaniki sploshnoy sredy*, **1**, 82–93 (in Russian).

Tupchiyev, V. A. (1972a). On the asymptotic properties of the solution to the Cauchy problem for the equation $\varepsilon^2 tu_{xx} = u_t + \varphi(u)_x$ degenerating in the limit $\varepsilon \to 0$ into a Riemann problem with a rarefaction wave solution, *Zh. vych. mat. i fiz.*, **12**, 770–775 (in Russian).

Tupchiyev, V. A. (1973). On the uniqueness of continuous solutions to the Riemann problem for a gradient system of equations, *Mat. zametki*, **13**, 251–258 (in Russian).

Tupchiyev, V. A. (1973a). On the vanishing viscosity method in studying the Riemann problem, *DAN SSSR*, **211**, 55–58 (in Russian).

Turner, R. E. L. (1971). Transversality in non-linear eigenvalue problems, in: E. H. Zarantonello (ed.), *Contributions to Non-linear Functional Analysis*, Academic Press, New York, 37–68.

Vainberg, M. M. (1956). *Variational Methods for the Study of Non-linear Operators*, Gostekhizdat, Moscow (in Russian); Engl. transl.: 1967, Holden-Day, San Francisco.

Vainberg, M. M. (1972). *The Variational Method and the Method of Monotone Operators in the Theory of Non-linear Operators*, Nauka, Moscow (in Russian).

Vol'pert, A. I. (1967). The space BV and quasilinear equations, *Mat. sb.*, **73**, 255–302 (in Russian); Engl. transl.: *Math. USSR-Sb.*, **2**, 225–267.

Vol'pert, A. I., Khudyayev, S. I. (1975). *Analysis in Some Classes of Discontinuous Functions and Equations of Mathematical Physics*, Nauka, Moscow (in Russian).

Wang, C.-C. and Truesdell, C. A. (1973). *Introduction to Rational Elasticity*, Noordhoff, Leyden.

Washizu, K. (1968). *Variational Methods in Elasticity and Plasticity*, Pergamon Press, New York.

Whitham, G. B. (1974). *Linear and Non-linear Waves*, J. Wiley-Interscience, New York.

Yosida, K. (1965, 1971). *Functional Analysis*, Springer Verlag, Berlin.

Index